Earth STRUCTURE

AN INTRODUCTION TO STRUCTURAL GEOLOGY AND TECTONICS

Ben A. van der Pluijm
University of Michigan

Stephen Marshak
University of Illinois at Urbana-Champaign

WCB/McGraw-Hill

A Division of The **McGraw-Hill** *Companies*

WCB/McGraw-Hill

A Division of The **McGraw·Hill** *Companies*

EARTH STRUCTURE: AN INTRODUCTION TO STRUCTURAL GEOLOGY AND TECTONICS

Paper 45# Sub New Era Recycled Matte

10 9 8 7 6 5 4 3 2 1 QPD QPD 1 2 3 4 5 6 7 8 9 10

ISBN 0-697-17234-1

Publisher: *Kevin Kane*
Sponsoring editor: *Lynne M. Meyers*
Developmental editor: *Daryl Bruflodt*
Project manager: *Cheryl R. Horch*
Designer: *Barb Hodgson*
Art editor: *Renee Grevas*
Photo editor: *Janice Hancock*
Permissions coordinator: *Gail I. Wheatley*
Cover photo: *© Doug Sherman/Geofile*
Compositor: *Shepherd, Inc.*
Typeface: *Times Roman/Helvetica*
Printer: *Quebecor Printing/Dubuque*

Library of Congress Catalog Card Number: 96-085656

http://www.mhcollege.com

Table of Contents

Preface

FOR STUDENTS

This book is concerned with deformation of rock in the Earth's lithosphere, as viewed from the atomic scale, through the grain scale, through the hand-specimen scale, through the outcrop scale, through the mountain-range scale, and through the tectonic plate scale. A deformational feature observed at one scale generally reflects processes occurring at other scales as well. We can't discuss plate collision without understanding mountains, we can't understand mountains without understanding folding and faulting, and we can't understand folding and faulting without understanding ductile and brittle deformation mechanisms at the atomic scale. This book attempts to integrate topics pertaining to all scales of rock deformation, and emphasizes the linkages between structural geology and tectonics.

The amount of material on structural geology and tectonics that has appeared over the past 150 years, and that continues to appear each month, is simply staggering. We have purposely decided to write this book with you, most likely a novice to the topic, in mind. Rather than loading the text with excessive details and peppering it with extensive referencing, we opted to present a distillation that is sufficient to develop a solid perspective on the field. The reason for this approach is to highlight the 'guts' of structural geology and tectonics, thereby providing you with a foundation for future study. Your instructor can use the text as a platform for further discussion if (s)he so desires. When reading the text, you should maintain a critical and questioning attitude toward the concepts discussed, which will not only stimulate your mind but should also aid in absorbing the material. It has been our experience that fundamentals and concepts are much better remembered when their interrelationships are recognized rather than when they are offered as a series of definitions (we were once in your place ourselves!). In some cases we may have adopted a controversial position (at least according to your instructor); maybe in future years you will prove the book's point of view right (or wrong). Those terms and definitions that we do not introduce in the main body of the text, but that are related to the topic at hand, are included in tables so that you will be able to use the book as a future reference. The index will direct you to the appropriate location in the text.

Structural geology and tectonics are a lot of fun once you have waded through the initial terminology morass, much as you'll have a better holiday experience once you have a basic knowledge of the local language (Dutch, for example). Our personal approach to teaching structural geology and tectonics is reflected in the breezy writing style. Where possible, we make use of familiar analogs (rubber bands, syrup, cars). Similarly, we have kept illustrations simple so that the point of the figure is more obvious. We hope you enjoy the text.

FOR INSTRUCTORS

Every month, perhaps a thousand pages of new ideas and observations relevant to structural geology and tectonics are published in the major scholarly journals. We, as instructors, face a massive challenge when trying to distill an introductory course out of this ever-changing and ever-growing mountain of information. We want our students to be comfortable with certain basic concepts (e.g., stress and strain theory) and yet at the same time, we want them to sense the excitement of discovery and to build their own 'big picture' of how the Earth works. And all this must be done in a few short months!

There is no one right way to teach structural geology and tectonics, or any other subject for that matter. We decided to write this book because we found that existing books in our discipline did not suit the changing needs of the courses that we teach ourselves, and we found that many other instructors shared our views. Some books try to be a lab manual and a lecture text at the same time, and others are slanted too much toward the research interests of the writer(s). Some books are organized in such a way that a reading assignment on a single topic must include splices from all over the book, and others provide much more detail than can possibly be covered in a standard semester course, so that students are, frankly, overwhelmed. We began to develop this book with the underlying view that a course textbook should be written with the *student* in mind, and that students taking a first course in structural geology and tectonics commonly do not enter the course with a lot of geologic experience under their belts.

Thus, the aim of this book is to *teach* structural geology and tectonics at an introductory level. It is not intended to discuss all the ins and outs of the subject, which are typically not taught at an introductory level anyway. With this goal in mind, we have tried to limit the amount of material covered, so that a good student will be able to digest most of the basics in a *single semester,* and still have room for a selection of specialized topics that cater to the interests of the instructor. Furthermore, we present the information in an informal and approachable tone. The book may serve as a platform on which individual instructors can develop specific topics in greater detail, or can serve as a backup to help students who are mystified by certain concepts. Wherever possible, we make use of familiar analogs (such as the interaction of cue balls on a pool table in the chapter on forces, rubber bands and syrup in the chapter on rheology, strips of rubber embedded in foam in the chapter on folding), and we relegate excessive terminology to tables. We do not clutter the narrative with references that break the flow of the text, but rather provide brief reading lists at the end of each chapter and make occasional use of footnotes. We have also deleted topics that are generally taught in laboratory sections, and refer readers to existing texts (e.g., *Basic Methods of Structural Geology,* by

S. Marshak and G. Mitra, Prentice-Hall) for such information, because these topics cannot be treated adequately within the framework of a lecture textbook.

This book is concerned with the deformation of rock in the Earth's lithosphere, as viewed from the atomic scale, through the grain scale, through the hand-specimen scale, through the outcrop scale, through the mountain-range scale, and through the tectonic plate scale. A deformational feature observed at one scale generally reflects processes occurring at other scales. Plate collision can occur, ultimately, because of atomic-scale processes in minerals. We can't understand mountains without understanding folding and faulting, and we can't understand folding and faulting without understanding deformation mechanisms, and understanding deformation mechanisms returns us to the atomic scale. This book attempts to integrate topics pertaining to all scales of rock deformation, and thus *emphasizes the linkages between structural geology and tectonics*. We conclude the book with "case studies" of selected regions, which are personal perspectives written by active researchers. Instructors may wish to choose only one or two of these to give students the flavor of tectonics synthesis.

In order to provide instructors with optimal freedom to develop a course outline of their own choosing, most chapters are self-contained modules that can be presented in a number of different sequences. Ben, for example, starts his course with a description of primary structures, faults and fractures, folds, and fabrics before introducing stress, strain, rheology, and deformation mechanisms. Steve, in contrast, teaches stress, strain, and rheology immediately after primary structures, and presents brittle deformation theory before discussing faults and fractures. We both concentrate on tectonics at the end of our courses, but tectonic implications, alternatively, could easily be interwoven with discussion of different classes of structures earlier in the course.

ABOUT THE AUTHORS

Ben van der Pluijm received his first geology training at the University of Leiden and later at Utrecht in the Netherlands, where he obtained an M.Sc. on fieldwork in northern Spain and (electron) microscopy study of phyllosilicates and calcite. He jumped the Big Water for dissertation work at the University of New Brunswick, Canada, where he worked along the rocky shores of the Newfoundland Appalachians. In 1985, he joined the faculty at the University of Michigan, where he works on field and laboratory structure/tectonics projects in low- to high-grade rocks, involving a variety of toys (eh, techniques). Like many geologists, Ben became interested in geology for its 'outdoor charm,' but especially for its unique interplay between science disciplines. He is married to Lies, and they have two boys, Wouter and Robbie.

Stephen Marshak became interested in geology while an undergraduate at Cornell University. He moved to the University of Arizona for his Master's degree and then to the Lamont-Doherty Geological Observatory of Columbia University for his Ph.D. Steve began teaching structure and tectonics at the University of Illinois in 1983, where his research interests include fold-thrust belts, Precambrian geology, and continental-interior tectonics. Over the years, he's worked on field problems in the United States, Brazil, North Africa, Italy, Australia, and Antarctica. Steve's wife, Kathy, and their two children, David and Emma, delight in the travel opportunities.

THANKS TO......

This book could not have been written without the help of the students in our classes, who, through their successes and mistakes, have shown us which explanations work and which do not. We are also grateful to our colleagues and graduate students who have provided generous dollops of advice, and from whom we have freely borrowed data and interpretations. The following persons (in alphabetical order) have commented on and/or contributed to one or more chapters: Jay Busch, Jim Cureton, Mark Fisher, Jerry Magloughlin, Klaus Mezger, Carl Richter, Mike Sandiford, John Stamatakos. Formal reviews of chapters of the text were given by David Anastasio, Stanley Cebull, Bill Dunne, Terry Engelder, Karl Karlstrom, Win Means, Jim Talbot, Adolph Yonkee, and Vincent Cronin, the complete manuscript was read by John Huntsman. Their comments were invaluable in bringing the project to completion. The editorial and production staffs at WCB/McGraw-Hill have been very helpful and accommodating. Dale Austin patiently produced the art work of chapter 19 and last-minute additions. We also wish to thank our own graduate advisors (Paul Williams and Henk Zwart, and Terry Engelder, respectively) for helping us entering this business and guiding our first (uncertain) steps. Finally, but certainly foremost, we thank our wives, Lies and Kathy, and our children, Wouter and Robbie, and David and Emma, for not grumbling too much about the absences in body and spirit that writing this book has required. To them we dedicate this book.

February 1997   Ben A. van der Pluijm, Ann Arbor, Michigan
Stephen Marshak, Urbana, Illinois

A

Fundamentals

Chapter 1

Introduction

1.1 HISTORICAL SURVEY

Did you ever take a cross-country drive? Hour after hour of tedious driving, as the highway climbed hills and then dropped into valleys. The monotonous gray rocks exposed in road cuts probably went unnoticed, right? You passed pretty scenery, but it was static and seemed to tell no story simply because you did not have a basis in your mind with which to interpret your surroundings.

It was much the same for scholars of generations past, before the establishment of modern science. The Earth was a closed book, hiding its secrets in a language that no one could translate. Certainly, ancient observers marveled at the enormity of mountains and oceans, but with the knowledge they had at hand they could do little more than dream of supernatural processes to explain the origin of these features. Gods and monsters contorted the Earth and spit flaming rock, and giant turtles and catfish shook the ground. Then, in fifteenth-century Europe, an intellectual renaissance spawned an age of discovery, during which the Earth was systematically charted. The pioneers of science cast aside a dogmatic view of the universe that had closed people's

minds for the previous millennia, and they began to systematically observe their surroundings and carry out experiments to create new knowledge. The scientific method was born.

In geology, the stirrings of discovery are evident in the ink sketches of the great artist and scientist Leonardo da Vinci (1452–1519), who carefully drew the true shapes of rock bodies in sketches to understand the natural shape of the Earth (Figure 1.1). In the seventeenth century came the first description of rock deformation. Nicholas Steno examined outcrops where the bedding of rock was not horizontal, and speculated that strata that do not presently lie in horizontal layers must have in some way been "dislocated" (the term he used for *deformed*). Perhaps Steno's establishment of the *Principle of Original Horizontality,* that rock layers are originally deposited horizontally, can be viewed as the birth of structural geology. By the beginning of the eighteenth century, the structural complexity of rocks in mountain ranges was widely recognized, as is evident in the Alpine sketches of the German artist Johannes Scheuchzer published in 1716 (Figure 1.2). It became clear that such features demanded explanation.

The pace of discovery quickened during the latter half of the eighteenth century and through the nineteenth century. Ever since the publication of James Hutton's "Theory of the Earth with Proofs and Illustrations" in 1785, the book in which the concept of *Uniformitarianism* and thenature of unconformities were proposed, there has been a group of scientists who recognized themselves as geologists. These new geologists defined the geometry of

Figure 1.1 Sketch by Leonardo da Vinci showing details of folded strata in the Mountains of Italy (ca. 1500).

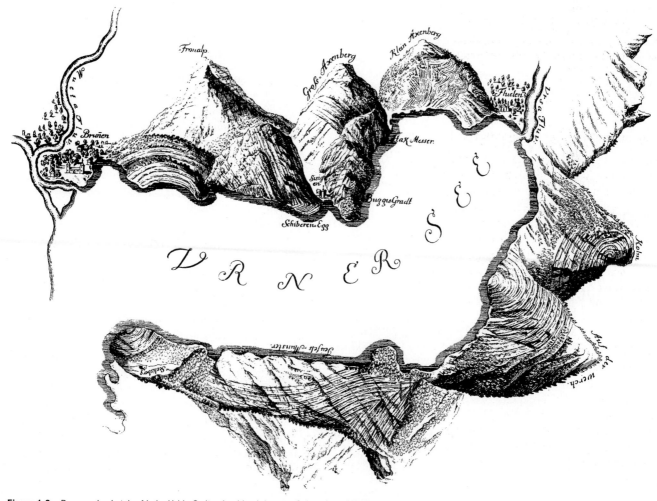

Figure 1.2 Panoramic sketch of Lake Uri in Switzerland by Johannes Scheuchzer (1716).

structures in mountain ranges, learned how to make geologic maps, discovered the processes involved in the formation of rocks, and speculated on the origins of specific structures and on mountain ranges in general.

Ideas about the origin of mountains have evolved gradually. At first, mountain ranges were thought to be a consequence of a vertical push from below, perhaps associated with intrusion of molten rock along preexisting zones of weakness; folds and faults in strata that were attributed to gravity sliding down the flanks of these uplifts (Figure 1.3). Subsequently, the significance of horizontal forces was emphasized, and geologists speculated that mountain ranges and their component structures reflected the contraction of the Earth that resulted from progressive cooling. In this model, the shrinking of the Earth led to wrinkling of the surface. One of the more notable discoveries of this period was James Hall's recognition that Paleozoic strata in the Appalachian Mountains of North America were much thicker than correlative strata in the interior of the continent. This discovery led to the development of *geosyncline theory,* a model in which deep subsiding sedimentary basins, called geosynclines, evolved into mountain ranges. Contraction theory and geosynclinal theory, or various combinations of the two, were widely accepted until the formulation of *plate tectonic theory* in the 1960s.

As the foundations of geology solidified, diverse features of rocks and mountains gained names, and the once amorphous, nondescript masses of rock exposed on our planet became history books telling the Earth's biography. Perhaps your concept of the planet has evolved rapidly as well, because of the courses in geology that you have taken thus far. Now, as you drive cross-country, you scare the daylights out of your passengers as you twist to see roadside outcrops. "Wow! Look at that fault . . . did you see that dike . . . and that fold!" The outcrops are no longer gray masses to you; now they contain patterns and shapes and fabrics. The purpose of this book is to increase your ability to interpret these features and to use them as clues to understanding the processes that have shaped and continue to change the outer layers of the Earth.

1.2 CLASSIFICATION OF GEOLOGICAL STRUCTURES

When you finished your introductory geology course, you probably had a general concept of what a *geological structure* is. The term brings to mind images of folds, faults, and cleavage. Perhaps you even had a field trip where you saw

some of these structures in the wild. These features are formed in response to pushes and pulls associated with the forces that arise from the movement of tectonic plates, or as a consequence of differential buoyancy between parts of the lithosphere. But what about bedding in a sedimentary rock and flow banding in a rhyolite flow; are these structures? And what about slump folds in a debris flow; are they structures? Well . . . yes, but the link between their formation and plate motion is not immediately clear. Maybe we need to have a more general concept of a geological structure.

The most fundamental definition of a geological structure is as follows: a *geological structure* is a geometric feature in rock whose shape, form, and distribution can be described. From this definition it is immediately obvious that there are several ways in which geologic structures can be classified or subdivided into groups. In other words, by necessity there are several different and valid classification schemes that can be used in organizing the description of geological structures. Different schemes are relevant for different purposes, so we will briefly look at the various classification schemes for geological structures that will return in subsequent chapters. At first, these various classification schemes may seem very confusing. Thus, we recommend that you start by recognizing the basic geometric classes as the foundation of your understanding. As you learn about these classes, refer back to the list below, and see how a particular geometric class fits into one or more of the classification schemes.

I.   Classification based on geometry, that is, on the shape and form of a particular structure
  • *planar (or subplanar) surface*
  • *linear feature*
  • *curviplanar surface*
  This is perhaps the most basic classification scheme. In this scheme we include the following classes of structures: joint, vein, fault, fold, shear zone, foliation, and lineation.
II. Classification based on geologic significance
  • *primary:* formed as a consequence of the formation process of the rock itself
  • *local gravity-driven:* formed due to slip down an inclined surface; slumping at any scale driven by local excess gravitational potential
  • *local density-inversion driven:* formed due to local lateral variations in rock density, causing a local buoyancy force

Figure 1.3 Model of mountain building and associated deformation as represented by G. P. Scrope (1825). The uplift is caused by intrusion of an igneous core, and the folds are generated by down-slope movement.

- *fluid-pressure driven:* injections of unconsolidated material due to sudden release of pressure
- *tectonic:* formed due to lithospheric plate interactions, due to regional interaction between the asthenosphere and the lithosphere; due to crustal-scale or lithosphere-scale gravitational potential energy and the tendency of crust to achieve isostatic compensation

The first four in this scheme can be grouped into *primary and nontectonic structures,* meaning that they are not directly related to the forces associated with moving plates. We purposely say "can" because in many circumstances these categories of structures do form in association with tectonic activity. For example, gravity sliding may be triggered by tectonically generated seismicity, and salt domes may be localized by movement of tectonic normal faults. These first four categories will be discussed in chapter 3. The fifth category of structures is very large and forms the primary focus of this book.

III. Classification based on timing of formation
- syn-formational: formed at the same time as the material that will ultimately form the rock
- penecontemporaneous: formed before full lithification, but after initial deposition
- postformational: formed after the rock has fully formed, as a consequence of phenomena not related to the immediate environment of rock formation

IV. Classification based on the process of formation (i.e., the deformation mechanism)
- *fracturing:* related to development or coalescence of cracks in rock
- *frictional sliding:* related to slip of one body of rock past another (or of grains past one another), resisted by friction
- *plasticity:* deformation by internal flow of crystals without loss of cohesion, or by nonfrictional sliding of crystals past one another
- *diffusion:* material transport in either solid-state or assisted by a fluid (dissolution)
- *combination:* combinations of deformation mechanisms contributing to the overall strain

V. Classification based on the mesoscopic cohesiveness during deformation
- *brittle:* structure forms by loss of cohesion across a mesoscopic discrete surface
- *ductile:* structure forms without loss of cohesion across a mesoscopic discrete surface
- *brittle/ductile:* deformation involves both brittle and ductile aspects

Note that the scale of observation is critical in the distinction between brittle and ductile deformation, because ductile deformation can involve microscopic scale fracturing and frictional sliding.

VI. Classification based on the strain significance, in which a reference frame must be defined (usually either the Earth's surface or the deforming layer)
- *contractional:* structure results in shortening of a region
- *extensional:* structure results in stretching of a region
- *strike-slip:* movement without either shortening or extension

Note that shortening in one direction can be, but does not have to be, accompanied by extension in a different direction, and vice versa. Also, regional deformation usually results in the vertical displacement of the Earth's surface, a component of deformation that is commonly overlooked.

VII. Classification based on the distribution of deformation in a volume of rock
- *continuous:* occurs through the rock body at all scales
- *penetrative:* occurs throughout the rock body, at the scale of observation; up close there may be spaces between the structures
- *localized:* structure is continuous or penetrative only within a definable region
- *discrete:* structure occurs as an isolated feature

Conveniently, we can consider the basic geometric classes of structures to be a manifestation of the mesoscopic cohesiveness of deformation. Joints, veins, and certain types of faults are manifestations of primarily brittle deformation, whereas cleavage, foliation, and folding are largely manifestations of ductile deformation processes. Thus, in this book we subdivide our discussions of specific structures into two parts: "Brittle Structures" (Part B) and "Ductile Structures" (Part C). As a first approximation, brittle deformation is more common in the upper part of the crust, and ductile deformation is more common in the deeper part of the crust, because ductile processes are favored under conditions of greater pressure and temperature. Also, ductile deformation is commonly a manifestation of plastic deformation, and brittle deformation is a consequence of fracturing and/or frictional sliding. However, it is important to emphasize right from the start of this book that different processes can act in the same places in the Earth. The processes that occur at any given time may reflect a variable such as *strain rate* (the rate of displacement in the rock body). For example, we will see that a sudden increase in strain rate may cause rock that is deforming in a ductile manner (by folding) to suddenly behave in a brittle manner (by fracturing).

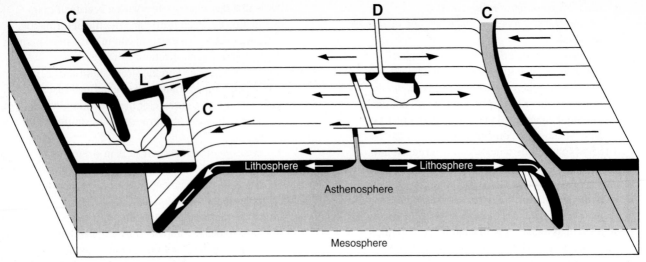

Figure 1.4 The principal features of plate tectonics. The three types of plate boundaries that arise from the relative movement (arrows) of lithospheric plates are: C—convergent boundary, D—divergent boundary, and L—lateral slip (or transform) boundary.

Ultimately, most crustal structures are a consequence of *plate tectonic activity,*[1] which is the slow (on the order of cm per year) but steady motion of the outer stiff layer of the Earth, called the *lithosphere,* over the weaker *asthenosphere.* The forces this motion generates, and especially forces created at plate-boundary interactions, cause structures to form. The three types of plate motions are *convergence, divergence,* and *lateral slip* (Figure 1.4). Without the activity arising from these plate motions (deformation, volcanism, and earthquakes), the Earth would be as dead as the Moon. There would be no distinction between continents and oceans, and there would be no mountains and no structure. In other words, plate tectonics provides the global framework to examine the significance of structures that occur on local and regional scales.

1.3   STRESS, STRAIN AND DEFORMATION

We have used the words *stress, strain* and *deformation* without definition, because these are common English words and most people have an intuitive grasp of what they mean. Stress presumably has something to do with pushing and pulling, whereas deformation has something to do with bending, breaking, stretching, or squashing. But in standard English, stress and strain are often used interchangeably; for example, advertisements for aspirin talk about "the stress and strain of everyday life," and people commonly equate strain with deformation. In structural geology these terms have more exact meanings; before proceeding further, we want to clarify their usage (and avoid headaches).

The *stress* (σ) acting on a plane is the force per unit area of the plane (σ = F/area). We will see in chapter 4 that when referring to the stress at a point in a body, a more

complicated definition is needed. *Deformation* refers to any change in shape, position, or orientation of a body resulting from the application of a differential stress (i.e., a state in which the magnitude of stress is not the same in all directions). Specifically, a general "deformation" has three components (Figure 1.5): (1) a *rotation,* which is the pivoting of a body around a fixed axis, (2) a *translation,* which is a change in the position of a body, and (3) a *strain,* which is a distortion or change in shape of a body (Chapter 4). To visualize a strain, consider the test crash of a car. In Figure 1.6a, a car is rapidly approaching a brick wall, and in Figure 1.6b, the car and the wall have attempted to occupy the same space at the same time, and have failed. Technically speaking, electron shells of the atoms composing the wall and the car have repelled each other. Since the car is made of a weaker material, the push between car and wall squashed the car, thereby resulting in a strain. In a *homogeneous strain,* the strain exhibited at one point in the body is the same as the strain at all other points. In car crashes, one would hope that the strain is *heterogeneous,* meaning that the strain is not equal throughout the body, and you are protected from the impact.

For a more geological perspective, let's examine the distorted fossil in Figure 1.7c. A living brachiopod has a symmetry axis that is perpendicular to the base, so if you find a fossil brachiopod in which the angle between the symmetry axis and the base is different from 90°, you know that the rock containing the fossil has been strained. What about translation and rotation? These components of deformation are a bit harder to recognize, but they do occur. For example, a rigid body of rock that has moved along a fault plane clearly has been translated relative to the opposing side of the fault (Figure 1.7b), and a fault block in which strata are inclined relative to horizontal strata on the opposing wall of the fault has clearly been rotated (Figure 1.7a). Such rotations occur at all scales, as emphasized by work in paleomagnetism that demonstrates that large blocks of

---

[1]*Impact structures* are one exception.

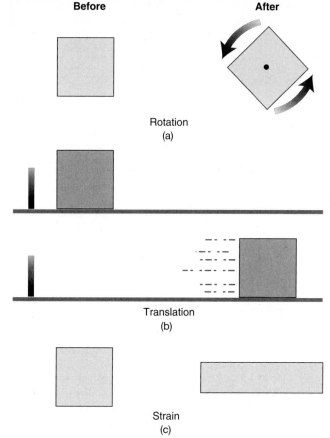

Figure 1.5 Illustrations of the three components of deformation: (a) rotation, (b) translation, and (c) strain.

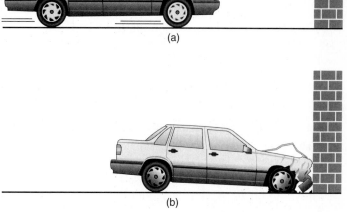

Figure 1.6 Strain in the real world. (a) A sedan approaches a brick wall in a crash test, and (b) the same sedan after impact. Note the extreme distortion of the front (i.e., inhomogeneous strain distribution).

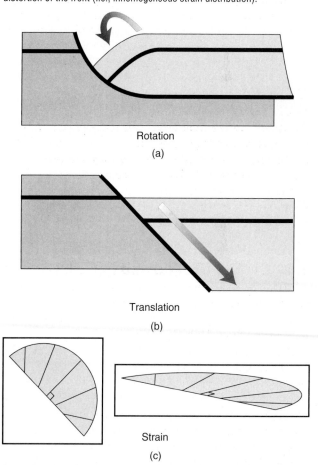

Figure 1.7 The three components of deformation shown schematically in some representative structures. (a) A rotated fault block, (b) a translated fault block, and (c) a strained fossil.

continental crust have been rotated around a vertical axis as a consequence of shear along major strike-slip faults.

Remember that in order to describe deformation, it is necessary to define a *reference frame*. The reference frame used in structural geology is loosely called the *undeformed state*. We can't know whether a rock body has been moved or distorted unless we know where it originally was and what its original shape was. Ideally, if we know both the original and final positions of an array of points in a body of rock, we can describe a deformation with mathematical precision by defining a simple coordinate transformation. For example, in Figure 1.8a, four points (labeled m, n, o, and p) define a square in a Cartesian coordinate system. If the square is sheared by stresses acting on the top and bottom surfaces, as indicated by the arrows, it changes into a parallelogram (Figure 1.8b). A strain has developed, and this strain can be described by saying that points o and p stayed fixed, and point m and n moved to point m′, and n′, respectively. In other words, coordinates of the corners of the square have been transformed. If you are mathematically adept, you will probably realize that this transformation can be described by a simple mathematical function, but we won't get into that now . . . wait until chapter 4.

In many real circumstances, we don't have a complete reference frame, so we can only partially describe a deformation. For example, at an isolated outcrop we may be able to describe strain, because of the presence of deformed fossils, but not displacement or rotation in relation to the original site of deformation. For example, a flat-lying bed of Paleozoic limestone in the Midcontinent region of the United States was at one time below sea level, and because

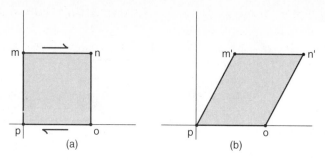

Figure 1.8 Deformation represented as a coordinate transformation. Points m and n move to new positions, m', and n'.

Table 1.1 Categories of Structural Analysis

Descriptive analysis	Involves the characterization of the shape and appearance of geological structures. It includes development of a precise vocabulary (jargon) that permits one geologist to create an image of a structure that any other geologist can understand, and development of methods for uniquely describing the orientation of a structure in three-dimensional space.
Kinematic analysis	Involves the determination of the movement paths that rocks or parts of rocks have taken during transformation from the undeformed to the deformed state. This subject includes, for example, use of features in rocks to define the direction of movement on a fault.
Strain analysis	Involves development of mathematical tools for quantifying the strain in a rock. This activity includes the search for features in rock that can be measured to define strain.
Dynamic analysis	Involves development of an understanding of how stress is related to deformation. This activity includes the invention of tools for measuring the present-day state of stress in the Earth, and of techniques for interpreting the state of stress responsible for microstructures in rocks.
Deformation-mechanism analysis	The study of processes on the grain scale to the atomic scale that allow structures to develop. This activity includes study of both fracture and flow of rock.
Tectonic analysis	The study of the relationship between structures and global tectonic processes. This activity includes the study and interpretation of regional-scale or megascopic structural features, and the study of relationships among structural geology, stratigraphy, and petrology.

---

of plate motion it was formed at a different latitude than today. But we can't immediately characterize these regional movements.[2] If, however, we see a fault offset a limestone bed by 2 meters in a quarry, we say that one side of the fault has moved 2 m relative to the other side. Thus, generally we may talk about *relative displacement and relative rotation*.

---

[2]Paleomagnetic and paleontologic (and possibly paleoclimatic) methods are used for defining plate motion in the Paleozoic. In the Mesozoic and Tertiary, ocean-floor magnetic anomalies also become available for such work, but in the Precambrian only the paleomagnetic approach remains.

1.4 STRUCTURAL ANALYSIS AND SCALES OF OBSERVATION

At this point, we know what a structure is and we know what a geologist means by deformation. We also know that there is a group of people that call themselves *structural geologists*. But what do structural geologists do? One way to gain insight into the subject of structural geology is to think about the type of work that structural geologists carry out. Perhaps not surprisingly, structural geologists do *structural analysis,* which involves many activities, as outlined in Table 1.1.

You will note by looking at Table 1.1 that in many of the definitions we refer to the *scale of observation*. For the results of a structural analysis to be interpretable, the scale of the analysis must be taken into account. For example, a bed of sandstone on the scale of a single outcrop in a mountain range may appear to be undeformed. But the outcrop may display only a small part of a huge fold that cannot be seen unless you map at the scale of the whole mountain range. Structural geologists commonly refer to three relative scales of observation, as indicated by the subjective prefixes *micro, meso,* and *macro. Micro* refers to features that are visible optically at the scale of thin sections, or may be evident only with the electron microscope. *Meso* refers to features that are visible in a single outcrop, but cannot necessarily be traced from outcrop to outcrop. *Macro* refers to features that can be traced over a region encompassing several outcrops to whole mountain ranges. In some circumstances, geologists use the prefix *mega* to refer to large deformational features, such as the movements of continents over time. Of course there are no sharp boundaries between these scales, and their usage will vary with context. In the end, a complete structural study should integrate results from as many scales of observation as possible.

Each scale of observation requires its own set of tools. For example, the optical or electron microscope is used for observations on the microscale, and satellite imaging for the macroscale. The *recognition* and *description* of rocks and their structures are of fundamental importance to field analysis, which requires a pair of eyes,[3] a hammer, a compass, and a hand lens. Field work is, in fact, pretty much a low-tech, low-budget affair unless you are working in the Higher Himalayas, Antarctica (Figure 1.9), or some similarly remote setting that requires fancy logistics (e.g., planes, helicopters, or a network of global positioning system (GPS) stations). For structural field work we record observations on lithologies and rock structures in notebooks and measure the orientation of geometric elements with a compass. The compass to a structural geologist is like the stethoscope to a doctor: it is the professional's tool (and should be worn visibly to all).

Basic geometric principles, the ways of describing geometric features and concepts related to constructions, such as structure contours and spherical projections, are extensively explained in structural geology laboratory or methods manuals. So, in this introduction we will restrict ourselves to presenting short descriptions of terms and associated concepts that are given in Table 1.2.

---

[3]Aided by corrective lenses in the case of the authors.

Table 1.2 Terminology Related to Geometry and Representation of Geological Structures

Apparent dip	Dip of a plane in an imaginary vertical plane that is not perpendicular to the strike. The apparent dip is less than or equal to the true dip.
Attitude	Orientation of a geometric element in space.
Cross section	Representation of a geometry on a plane perpendicular to the Earth's surface.
(true) dip	The slope of a surface; formally, the angle of a plane with the horizontal measured in an imaginary vertical plane that is perpendicular to the strike. (Figure 1.10a)
Dip direction	Azimuth of the horizontal line that is perpendicular to the strike. (Figure 1.10a)
Foliation	General term for a surface that occurs repeatedly in a body of rock (e.g., bedding, cleavage).
Lineation	General term for a penetrative linear element, such as the intersection between bedding and cleavage or alignment of elongate grains.
Pitch	Angle between a linear element that lies in a given plane and the strike of that plane (also *rake*). (Figure 1.10b)
Plunge	Angle of linear element with earth's surface in an imaginary vertical plane. (Figure 1.10b)
Plunge direction	Azimuth of the plunge direction. (Figure 1.10b)
Position	The geographic location of a geometric element (e.g., an outcrop).
Profile plane	Plane perpendicular to a given geometric element; for example, the plane perpendicular to the hinge line of a fold.
Rake	Angle between a linear element that lies in a given plane and the strike of that plane (also *pitch*).
Strike	Azimuth of the horizontal line in a dipping plane or the intersection between a given plane and the horizontal surface (also *trend*) (Figure 1.10a).
Trace	The line of intersection between two nonparallel surfaces.
Trend	Synonym for strike, or azimuth of any feature in a map view.

Figure 1.9 Field area in Antarctica.

1.5 SOME GUIDELINES FOR STRUCTURAL INTERPRETATION

In closing this introductory chapter, we make a few comments on structural analysis. Good scientific work carefully separates *observations* from *interpretations.* Yet, a geologic map or section without interpretation misses the unique insights of the investigator. If you spend the time collecting and digesting data, you are best suited to make the interpretations (or, at least, educated guesses). Therefore, we offer a few suggestions that may help with map interpretation, but note that they hardly do justice to the intricate process of interpretation; we merely wish to point you in the right direction.

The assumptions on which interpretations are based do not hold universally; in fact, after some field experience you may disagree with one or more of the assumptions listed here. In our experience, however, the guidelines in Table 1.3 enable a reasonable, first-order interpretation of the geometry of an area.

Each individual guideline in Table 1.3 is valid under a given set of circumstances, but remember that, except for the laws, they remain mere assumptions; no more, no less. Whenever possible, your assumptions should be tested by adding more observations, and when they continue to hold, only then may your interpretation be valid. This approach follows a proven scientific method, the *testable working hypothesis,* which eventually may lead to a boardly accepted interpretation, or a *model.* If the model is very successful it may become *law,* but this is rare in geology. Regardless, there will always be alternative interpretations or models.

Table 1.3 Some Guidelines for the Interpretation of Deformed Areas

- Strata are deposited horizontally. This is the Law of Original Horizontality, which makes bedding an internal reference frame.
- Strata follow one another in chronological, but not necessarily continuous, order.[4] This is known as the Law of Superposition.
- Separated but aligned outcrops of the same lithological sequence imply stratigraphic continuity.
- Strata occur in laterally continuous and parallel layers in a region.
- Sharp discontinuities in lithologic patterns are faults, unconformities, or intrusive contacts.
- Deformed areas can be subdivided into a number of regions that contain consistent structural attitudes (*structural domains*). For example, a folded area can be subdivided into regions with a relatively constant dip direction (and even dip), such as the limbs and hinge area of large scale folds.
- The simplest but internally consistent interpretation is most correct.[5] This is also called the *principle of least astonishment.*

[4]See the description of *facing* in Chapters 2 and 10; if younging directions are unknown, *transposition* may present complications (Chapter 12).
[5]But no simpler than that (paraphrasing Albert Einstein).

Increasingly, subsurface data (e.g., drilling, seismic profiles) are available to modern structural geologists, and they should be used to constrain your interpretation. Drilling is restricted to the upper 5–10 km of the crust, but provides actual samples of deeply buried layers that may be compared with exposed units. This is the ultimate test for the cross section you constructed. 2D and more recent 3D deep-seismic reflection imaging give us an indirect view of the deeper parts of the Earth (Figure 1.10). Seismic-reflection images are obtained by recording the travel times of sound waves that bounce off layers in the Earth. The technique is quite complicated and requires careful data processing to improve the signal-to-noise ratio and localize the reflectors. Correlation of these reflectors with features that are exposed at the surface may give information of the nature of the deep structure.

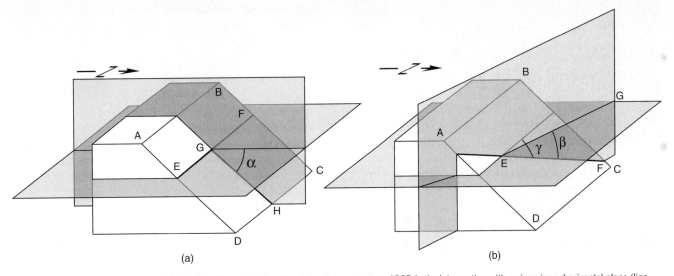

Figure 1.10 (a) Attitude of a plane: dip, dip direction and strike. The strike of dipping plane ABCD is the intersection with an imaginary horizontal plane (line EF). The dip of plane ABCD is given by the angle α, whereas line GH represents the dip direction. Note that there are two possible directions of dip for a given strike. Thus, when using dip and strike, the general direction of dip must also be included (e.g., 090°/45°N). Alternatively, the attitude of plane ABCD, using dip direction and dip, is uniquely described by 000°/45°. (b) Attitude of a line: plunge, plunge direction and pitch. Line EF lies in dipping plane ABCD. The plunge of line EF is the angle β, which is measured from the horizontal (line EG) in an imaginary vertical plane that contains both line EF and line EG; line EG is called the plunge direction. The attitude of line EF, using plunge and direction of plunge, is given by 30°/315°. The pitch (angle γ) is the angle that line EF makes with the horizontal (strike) of a plane (here plane ABCD) containing the line. Note that when the pitch of a line is recorded, the attitude of the reference plane must be given, as well as the side from which the pitch angle is measured (here: 000°/45°, 40°W).

1.6   CLOSING REMARKS

In this opening chapter of the book, we quickly traveled through the history of structural geology, the types of geological structures, the meaning of stress, strain, and deformation, and the nature of structural analysis. Of course we only scraped the surface of these topics. Our goal was to give you a first idea of what the field of structural geology and tectonics entails. Except for historical considerations, all these topics will return in subsequent chapters. The chapters of the book are broadly grouped into "Fundamentals" (Part A), describing the theory and background that are needed for the interpretation of natural structures, "Brittle Structures" and "Ductile Structures" (Parts B and C, respectively), reflecting a distinction that is based on the distribution of strain in deformed bodies, and "Tectonics and Regional Deformation" (Part D), discussing some of the fundamentals of plate tectonics and plate boundaries, and interpretations of the geology of selected regions around the world. Ultimately, we will have examined structures on all scales, ranging from the atom to the mountain belt to the whole Earth. Their relationships provide us with a relatively recent but remarkably good understanding of the tectonic evolution and inner workings of our dear planet.

ADDITIONAL READING

Lisle, R. J., 1988, *Geological structures and maps: A practical guide,* Oxford: Pergamon Press.

Loyshon, P. R., and R. J. Lisle, 1995, *Stereographic projection techniques in structural geology,* Oxford: Butterworth-Heinemann.

Marshak, S., and G. Mitra, 1988, *Basic methods of structural geology,* Englewood Cliffs, NJ: Prentice-Hall.

McClay, K., 1987, *The mapping of geological structures,* Geological Society of London Handbook, Open University Press.

Roberts, J. L., 1982, *Introduction to geological maps and structures,* Oxford: Pergamon Press.

Rowland, S. M., and Duebendorfer, E. M., 1994, *Structural analysis and synthesis,* 2nd edition, Boston: Blackwell Scientific Publications.

Chapter 2
Primary and Nontectonic Structures

2.1 INTRODUCTION

In the first chapter, we introduced a very general definition for a *geologic structure:* any definable shape or fabric in a rock body. Most of the scope of structural geology is focused on *tectonic structures,* meaning those structures that form in response to stresses generated by plate interactions (e.g., convergence, collision, rifting, strike-slip movement, plate drag, crustal-scale buoyancy; see Part D). But before we get to tectonic structures, we want to examine *primary structures,* that is, those structures that form during or shortly after the deposition of rocks, and *nontectonic struc-*

tures, that is, those structures that are not an immediate consequence of plate interactions. Primary sedimentary and igneous structures are also sometimes called nontectonic structures, because it is inferred that they form before the onset of tectonic stresses. Strictly speaking, however, this usage is misleading, because most structures are indirectly, if not directly, a consequence of tectonic activity: Examples are the creation of slopes down which sediments slide, and the occurrence of volcanic activity leading to the flow of basalt. All of these phenomena are ultimately a manifestation of movements in the Earth. Under the broad masthead of "primary and nontectonic structures" we discuss depositional, penecontemporaneous, intrusive, and gravity-slide structures for both sedimentary and igneous rocks. This heading also includes impact structures, which are discussed at the end of the chapter.

2.2 PRIMARY SEDIMENTARY STRUCTURES

When you look at an outcrop of sedimentary rock, the most obvious fabric to catch your eye is the primary layering or stratification (Figure 2.1), which is generally called *bedding.* Several terms related to stratification are listed in Table 2.1.

What defines bedding in an outcrop? In the Painted Desert of Arizona, beds are defined by spectacular variations in colors, and outcrops display garish stripes of maroon, red, green, and white. In the Grand Canyon, beds are emphasized by contrasts in resistance to erosion; sandstone and limestone beds form vertical cliff faces, whereas shale layers form shallow slopes. On the cliffs that form the east edge of the Catskill Mountains in New York State, there are abrupt contrasts in grain size between adjacent beds,

Figure 2.1 Differential erosion of bedding surfaces in the Wasatch Formation, Bryce Canyon (Utah).

Table 2.1 Some Terminology Used in Discussing Stratification

Bedding	Primary layering in a sedimentary rock, formed during deposition, manifested by changes in texture, color, and/or composition; may be emphasized in outcrop by the presence of parting.
Compaction	Squeezing unlithified sediment in response to pressure exerted by the weight of overlying layers.
Overturned beds	Beds that have been rotated past vertical in an Earth-surface frame of reference; as a consequence, facing is down.
Parting	The tendency of sedimentary layers to split or fracture along planes parallel to bedding; parting may be due to weak bonds between beds of different composition, or may be due to a bed-parallel preferred orientation of clay.
Strata	A sequence composed of layers of sedimentary rock.
Stratigraphic facing	The direction to younger strata, or, in other words, the direction to the depositional top of beds.
Younging direction	Same as stratigraphic facing.

with a coarse conglomerate juxtaposed against siltstone or shale. Strictly speaking, a bed is the smallest subdivision of a sedimentary unit.[1] It has a definable top or bottom and can be distinguished from adjacent beds by differences in grain size, composition, color, sorting, and/or by a physical parting surface. A *parting surface* is a plane of separation along which the rock splits. All of the features defining bedding, with the exception of parting, are a consequence of changes in the source of the sediment (or provenance) or the depositional environment.

In some outcrops, bedding is enhanced by the occurrence of *bedding-parallel parting*. Parting forms when beds

are unroofed (i.e., overlying strata are eroded away) and uplifted to shallower depths in the crust. As a consequence, the load pushing down on the strata decreases and the strata expand slightly. During this expansion, fractures form along weak bedding planes and define the parting. This fracturing reflects the weaker bonds between contrasting lithologies of adjacent beds, or the occurrence of a preferred orientation of sedimentary grains (e.g., mica). If a sedimentary rock has a tendency to have closely spaced partings, it is said to display *fissility*. Shale, which typically has a weak bedding-parallel fabric due to the alignment of constituent clay or mica flakes, is commonly fissile.

There are three reasons why platy grains like mica have a preferred orientation in a sedimentary rock. First, the alignment of grains can reflect settling of asymmetric bodies in the Earth's gravity field. Platy grains tend to lie down flat. To understand why, throw a deck of cards into the air: the cards fall to the floor and lie flat against the floor. It is unlikely to see one stand on its edge. Second, the

---

[1]Stratigraphers divide sequences of strata into the following units: supergroup, group, formation, member, bed. Several beds make a member, several members make a formation, and so on. Criteria for defining units are somewhat subjective. Basically, a unit is defined as a sequence of strata that can be identified and mapped at the surface or in the subsurface over a substantial region. The basis for recognizing a unit can be its age, its component sequence of lithologies, and the character of its bedding.

Table 2.2 Terms Used to Describe Types of Bedding

Thin beds	These are beds that are less then 3 cm thick.
Medium beds	These are beds that are 10–30 cm thick.
Thick beds	These are beds that are 30–100 cm thick; very thick beds are tens of m thick.
Massive beds	These are beds that are relatively thick (say, > 1 m to many meters) and show no internal layering. Massive bedding develops in sedimentary environments where large quantities of sediment are deposited very rapidly, and/or in environments where bioturbation (churning of the sediments by worms and other organisms) occurred.
Rhythmic beds	This term is used for a sequence of beds in which the contrast between adjacent beds is repeated periodically for a substantial thickness of strata.
Thinly laminated beds	These are beds that are less than 0.3 cm thick.

alignment of grains can reflect flow of the fluid in which the grains were deposited. In a moving fluid, grains are reoriented so that they are hydrodynamically stable, meaning that the traction caused by the moving fluid is not at a maximum (as would be the case if the broad face of the grain were perpendicular to the flow direction). At the same time there is a component of pressure holding the grain down, which requires that the grain be inclined slightly to the flow. Typically, grains end up being *imbricated,* meaning that they overlap one another like roof shingles. Imbrication is a useful primary sedimentary feature that can be used to define *paleocurrent direction* (i.e., the current direction at the time the sediments were deposited). Third, the alignment of grains can form as a consequence of compaction subsequent to deposition. As younger sediment is piled on top, water is squeezed out of the older sediment below and the clasts mechanically rotate into a preferred orientation, with their flat surfaces roughly perpendicular to the applied load.

Bedding in outcrops may also be highlighted by differential weathering and erosion (Figure 2.1). For example, chemical weathering (e.g., dissolution) of a sequence of strata containing alternating limestones and quartz sandstones will result in a serrated outcrop face on which the quartz-rich layers stand out in relief. Fresh limestone and dolomite are almost identical in color, but weathering of a sequence of alternating limestones and dolostones will result in a color-banded outcrop, because the dolostones tend to weather to a buff-tan color, while the limestones remain gray. Erosion of a sequence of alternating sandstones and shales may result in a stairstep outcrop face, because the relatively strong sandstone beds become vertical cliffs, whereas the relatively weak shale beds create slopes.

Geologists, characteristically, have developed a jargon for describing specific types of bedding, as defined in Table 2.2.

2.2.1 The Use of Bedding in Structural Analysis
Recognition of bedding is critical in structural analysis. Bedding provides a *reference frame* for describing defor-

mation of sedimentary rocks, because when sediments are initially deposited, they form horizontal or nearly horizontal layers—a concept referred to as the *Law of Original Horizontality.* Thus, if we look at an outcrop and see tilting or folding, what we are noticing are deviations in bedding attitude from original horizontality.

In complexly deformed metasedimentary rocks, geologists search long and hard to find subtle preserved manifestations of original bedding (e.g., variations in grain size or color). Only by finding bedding can a geologist identify the earliest-formed folds in the region. When found, bedding is labeled S_0; S_0 is pronounced "S-zero," where the "S" stands for surface. Later we will discuss other kinds of surfaces that are labeled S_1, S_2, and so on (Chapter 11).

The study of certain primary structures within beds and on bedding surfaces is useful in tectonic analysis because these structures provide important information on *depositional environment* (the setting in which the sediment was originally deposited), on *stratigraphic facing* or *younging direction* (the direction in which strata in a sequence get progressively younger), and on *current direction* (the direction in which fluid was flowing during deposition). Facing indicators allow you to determine whether a bed is right-side-up (*facing up*) or overturned (*facing down*) with respect to the Earth's surface. Recognition of facing is critical, not only for stratigraphic studies but also for structural studies. For example, the structural interpretation of a series of parallel beds in two adjacent outcrops depends on the facing—if the facing is the same in each outcrop, then the strata are probably *homoclinal,* meaning that they have a uniform dip. But if the facing is opposite, then the two outcrops must be on different limbs of a fold (see Chapter 10), whose hinge area was eroded away.

2.2.2 Graded Beds and Cross Beds
Geologists like finding graded beds and cross beds when doing field work, because these types of beds contain extra information about stratigraphic facing and possibly current directions, which are often critical for tectonic interpretations. *Graded beds* display progressive fining of clast grain

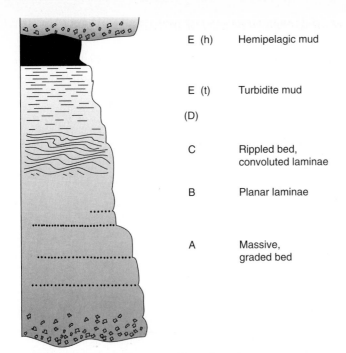

E (h)　　　Hemipelagic mud

E (t)　　　Turbidite mud

(D)

C　　　　Rippled bed,
　　　　　convoluted laminae

B　　　　Planar laminae

A　　　　Massive,
　　　　　graded bed

Figure 2.2　Graded bedding in a turbidite (or Bouma) sequence.

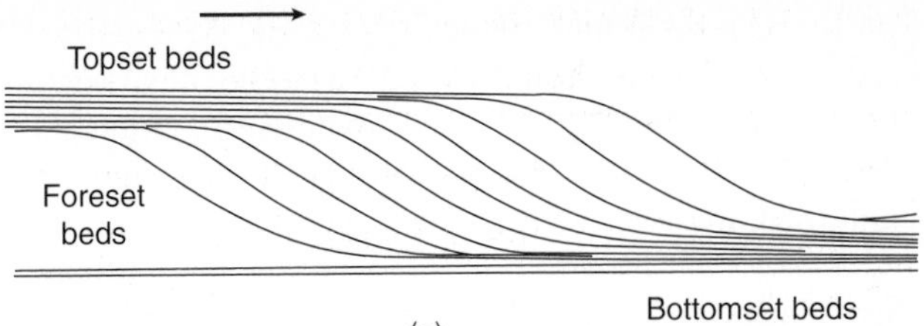

Topset beds

Foreset beds

Bottomset beds

(a)

(b)

Figure 2.3　(a) Terminology of cross bedding, and (b) Cross beds in the Coconino Sandstone, Oak Creek Canyon (Arizona).

size from the base to the top (Figure 2.2), and are a consequence of deposition from turbidity flows. A *turbidity flow* is a cloud of sediment that moves down a slope under water, because the density of the sediment-water mixture is greater than that of clean water, and denser liquids sink through less dense liquids. Turbidite flows are triggered by storms or earthquakes (because of their association with seismicity, the occurrence of turbidites may indicate that the sediment source region was tectonically active), and move down very gentle slopes at considerable speed. Typically, a flow is confined to a submarine channel or canyon; when a broadening of the channel or a decrease in slope slows the speed of a turbidity current, the sediment cloud settles. During settling, the largest grains fall first, and the finest grains last. Each turbidity flow produces a separate graded sequence or a *turbidite*. Ultimately, *pelagic sediment,* meaning deep-marine sediments like clay and plankton shells, cap the sequence. Typically, turbidites display an internal order, called a *Bouma sequence*[2] (Figure 2.2), which reflects changing hydrodynamic conditions as the turbidity current slows down.

In pre-plate tectonics geological literature (i.e., pre-Beatles), thick sequences of turbidites were referred to as *flysch,* a term originating from Alpine geology (see Chapter 19). Flysch was thought to be an orogenic deposit, meaning that it was a sequence of strata that was deposited just prior to and during the formation of a mountain range. Exactly why such strata were deposited, however, was not understood. Modern geologists now realize that the classical flysch sequences are actually turbidites laid down in a deep trench marking a convergent plate boundary (i.e., a subduction zone). Turbidite flows are common in trenches,

because the margins of trenches are slopes, and convergent margins are seismically active. After deposition, the trench turbidites are scraped up and deformed by continued convergence between plates, and may eventually be caught in a continental collision zone.

Cross beds are surfaces within a thicker, master bed that are oblique to the overall bounding surfaces of the master bed (Figure 2.3). Cross beds, which are defined by subtle partings or concentrations of grains, form when sediment moves from the windward or upstream side of a dune, ripple, or delta to a slip face on the leeward or downstream side where the current velocity is lower and the sediment settles out. Thin beds parallel to the upper bounding surface are called *topset beds,* the inclined layers deposited parallel to the slip face are called *foreset beds,* and the thin beds parallel to the lower bounding surface are called *bottomset beds.* The foreset beds, which typically are curved (concave up) and merge with the topset and bottomset beds, are the cross beds. If the topset beds and the upper part of the foreset beds are removed by local erosion, the bottomset beds of the next higher layer of sediment are juxtaposed against the foreset beds of the layer below. Thus, cross beds tend to be truncated at the upper master bedding plane, whereas they are asymptotic to the lower bedding plane. This geometry provides a clear stratigraphic facing indicator, as well as a paleocurrent direction. The current direction in a cross-bedded layer is taken to be approximately perpendicular to the strike of the foresets.

[2]Named after the sedimentologist Arnold Bouma.

2.2.3 Bed-Surface Markings

Environmental phenomena such as rain, desiccation (drying), current traction, and movement of organisms affect the surface of a bed of sediment. If the sediment is unlithified, these phenomena leave an imprint known as a *surface marking*. Table 2.3 lists some of the more common surface markings.

2.2.4 Load Casts and Disrupted Bedding

Load casts, which are also called ball-and-pillow structures, are bulbous protrusions extending downward from a sand layer into an underlying mud layer. They form prior to lithification as a consequence of a gravitational instability. In a situation where denser sand lies on top of less dense mud, a

Table 2.3 Common Bed-Surface Markings

Animal tracks	These form when critters like trilobites, worms, and lizards tromp over the surface and indent the surface (the characteristic trails of these organisms are a type of *trace fossil*).
Clast imbrication	This is the shinglelike overlapping arrangement of tabular clasts on the surface of a bed in response to a current. Imbrication develops because tabular clasts tend to become oriented so that the pressure exerted on them by the moving fluid is minimized.
Flute casts	Asymmetric troughs formed by vortices (mini-tornadoes) within the fluid that dig into the unconsolidated substrate. The troughs are deeper at the upstream end, where the vortex was stronger. They get shallower and wider at the downstream end, because the vortex dies out. Flute casts can be used as facing indicators.
Mudcracks	Desiccation of mud causes the mud to crack into an array of polygons and intervening mudcracks. Each polygon curls upwards along its margins, so that the mudcracks taper downwards and the polygons resemble shallow bowls. Mudcracks can be used as facing indicators, because an individual crack tends to taper downwards (Figure 2.4).
Raindrop impressions	Circular indentations on the bed-surface of mudstone, formed by raindrops striking the surface while it was still soft.
Ripple marks	Ridges and valleys on the surface of a bed formed as a consequence of fluid flow. If the current flows back and forth, as along a beach, the ripples are *symmetric,* but if they form in a uniformly flowing current, they are *asymmetric* (Figure 2.5). The crests of symmetric ripples tend to be pointed, whereas the troughs tend to be smooth curves. Thus, symmetric ripples are good facing indicators. Asymmetric ripples are not good facing indicators, but do provide current directions.
Traction lineation	Subtle lines on the surface of a bed formed either by trails of sediment that collect in the lee of larger grains, or by alignment of inequant grains in the direction of the current to diminish hydraulic drag.
Worm burrows	The traces of worms or other burrowing organisms that live in unconsolidated sediment. They stand out because of slight textural and color contrasts with the unburrowed rock.

Figure 2.4 Mudcracks, separating a mud layer into plates. The circular indentations are raindrop imprints.

Figure 2.5 Asymmetric ripple marks; the arrow indicates current direction during deposition.

Figure 2.6 Load cast, or ball-and-pillow structure; stakes for scale (Eifel, Germany).

Figure 2.7 Disrupted bedding in turbidites (Cantabria, Spain).

disturbance by a storm or an earthquake causes blobs of sand to sink into the underlying mud. Preserved load casts are useful stratigraphic facing indicators (Figure 2.6).

Where sand and mud layers are progressively buried, it is typical for the mud layers to compact and consolidate before the sand layers do. As a consequence, the water in the sand layer is under pressure. If an earthquake, storm, or slump suddenly cracks the permeability barrier surrounding the sand, water is released. When this happens near the Earth's surface, little mounds of sand called *sand volcanoes* erupt at the ground surface. At depth, partially consolidated

beds of mud break into pieces and settle into the underlying sand. The resulting chaotic layering is known, simply, as *disrupted bedding* (Figure 2.7).

Studies of disrupted bedding and sand volcanoes in recent lake and marsh deposits provide a basis for determining the recurrence interval of large earthquakes. In these studies, investigators dig a trench across the deposit and then look for disrupted intervals within the sequence. Radiocarbon dates on organic matter in the disrupted layer define the absolute age of disruption.

Figure 2.8 A clastic dike (Sudbury, Ontario). Note that the dike cuts across bedding.

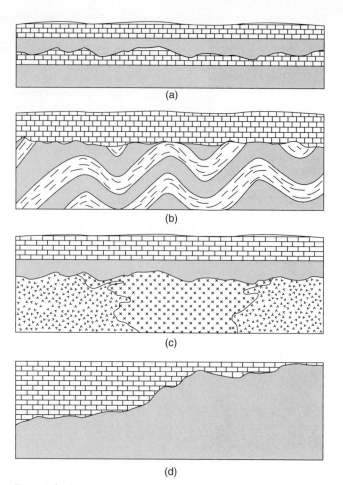

Figure 2.9 Sketches of the principal types of unconformities: (a) disconformity, (b) angular unconformity, (c) nonconformity, and (d) buttress unconformity.

2.2.5 Sedimentary Dikes

In many sedimentary environments, sand layers alternate with shale layers. During rapid compaction, as might occur during the high deposition rates characteristic of turbidite successions, water is squeezed out of the mud, compacting it and making the resulting shale relatively impermeable before the sand layers lithify. Therefore, water is trapped in the sand layers. If the mud beds become hard enough to fracture while the sand layers are still uncemented and contain water, the sand forces into cracks and intrudes into the shale. Sand effectively is squeezed into adjacent beds by the weight of overlying strata. The resulting wall-like intrusions of sand (or in some localities, even conglomerate) are called *clastic dikes* (Figure 2.8).

2.2.6 Conformable and Unconformable Depositional Contacts

Remember that a contact is the surface between two geologic units, and that there are three basic types of contacts: (1) *depositional contacts,* where a sediment layer is deposited over preexisting rock, (2) *fault contacts,* where two units are juxtaposed by a fracture on which sliding has occurred, and (3) *intrusive contacts,* where one rock body cuts across another rock body. In this section, we consider the nature of depositional contacts.

Relatively continuous sedimentation in a region leads to the deposition of a sequence of parallel sedimentary beds in which the contacts between adjacent beds do not represent substantial gaps in time (e.g., gaps that can be identified based on gaps in the fossil succession). The boundary between adjacent beds or units in such a sequence is called a *conformable contact.* For example, we say, "In eastern New York, the Becraft Limestone was deposited conformably over the New Scotland Formation." The New Scotland Formation is an argillaceous limestone representing marine deposition below wave base, whereas the Becraft Limestone is a pure coarse-grained limestone representing deposition in a shallow-marine beach environment. Bedding in the two units is parallel, and the contact between these two units is gradational.

If there is an interruption in sedimentation, such that there is a measurable gap in time between the base of the sedimentary unit and whatever lies beneath it, then we say that the contact is *unconformable.* For example, we say, "In eastern New York, the Upper Silurian Rondout Formation is deposited unconformably on the Middle Ordovician Austin Glen Formation," because Upper Ordovician and Lower Silurian strata are absent. Unconformable contacts are generally referred to simply as *unconformities,* and the gap in time represented by the unconformity (i.e., the difference in age between the base of the strata above the unconformity and the top of the unit below the unconformity) is called a *hiatus.* In order to convey a meaningful description of a specific unconformity, geologists distinguish among four types of unconformities (Figure 2.9), which are defined in Table 2.4.

Unconformities represent gaps in the rock record that can range in duration from a few years to billions of years. Examples of minor unconformities representing a short hiatus occur in high-energy depositional environments, such as a coastal region where a storm may erode a layer of

Table 2.4   Types of Unconformities

Angular unconformity	At an angular unconformity, strata below the unconformity have a different attitude than strata above the unconformity. Beds below the unconformity are truncated at the unconformity, while beds above the unconformity roughly parallel the unconformity surface. Therefore, if the unconformity is tilted, the overlying strata are tilted by the same amount. Because of the angular discordance at angular unconformities, they are quite easy to recognize in the field. Their occurrence means that the sub-unconformity strata were deformed (tilted or folded) and then were truncated by erosion prior to deposition of the rocks above the unconformity. Therefore, angular unconformities are indicative of a period of active tectonism. If the beds below the unconformity are folded, then the angle of discordance between the super- and sub-unconformity strata will change with location, and there may be outcrops at which the two sequences are coincidentally parallel.
Buttress unconformity	A buttress unconformity (also called *onlap unconformity*) occurs where beds of the younger sequence were deposited in a region of significant predepositional topography. Imagine a shallow sea in which there are islands composed of older bedrock. When sedimentation occurs in this sea, the new horizontal layers of strata terminate at the margins of the island. Eventually, as the sea rises, the islands are buried by sediment. But along the margins of the island, the sedimentary layers appear to be truncated by the unconformity. Rocks below the unconformity may or may not parallel the unconformity, depending on the pre-unconformity structure. Note that a buttress unconformity differs from an angular unconformity in that the younger layers are truncated at the unconformity surface.
Disconformity	At a disconformity, beds of the rock sequence above and below the unconformity are parallel to one another, but there is a measurable age difference between the two sequences. The disconformity surface represents a period of nondeposition and/or erosion.
Nonconformity	Nonconformity is used for unconformities at which strata were deposited on a basement of older crystalline rocks. The crystalline rocks may be either plutonic or metamorphic. For example, the unconformity between Cambrian strata and Precambrian basement in the Grand Canyon is a nonconformity.

Figure 2.10 Unconformable contacts between mid-Proterozoic Grenville gneiss and Cambrian sandstone and Pleistocene soils in southern Ontario (Canada).

Figure 2.11 Angular unconformity at Siccar Point (Scotland). The hammerhead is on the unconformity. Note that the unconformity is tilted.

sediment prior to deposition of the next younger layer of sediment, or a flood plain where several seasons pass before a spring flood is severe enough to overtop the natural levees of the river channel and deposit new sediment. Examples of great unconformities, representing millions or even billions of years, occur in the Canadian shield, where Pleistocene till buries Proterozoic and Archean gneisses (Figure 2.10).

It is an exciting experience to put your finger on major unconformities and to think about how much of Earth's history is missing at the contact. Imagine how James Hutton felt when, in the late eighteenth century, he stood at Siccar Point along the coast of Scotland (Figure 2.11), stared at the unconformity between shallowly dipping Devonian Red Sandstone and vertically dipping Silurian strata and, as the

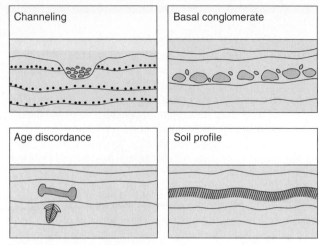

Figure 2.12 Some features used to identify unconformities (scour channels, paleosol, basal conglomerate, etc.).

Figure 2.13 Suture-like stylolites. Pocketknife for scale.

present-day waves lapped on and off the outcrop and deposited new sand, suddenly realized what the contact meant. His discovery is one of the most fundamental in geology.

How do you recognize an unconformity in the field (Figure 2.12)? Well, if it is an angular unconformity or a buttress unconformity, there is an angular discordance between bedding above and below the unconformity. A nonconformity is obvious, because crystalline rocks occur below the contact. Disconformities, however, can be a challenge to recognize. If strata in the sequence are fossiliferous, and you can recognize the fossil species and know their age, then you can recognize a gap in the fossil succession. Commonly, the unconformity may be marked by a surface of erosion, as indicated by scour features, or by a weathered zone or a *paleosol* (a soil horizon formed prior to deposition of the overlying sequence). Some unconformities are marked by the occurrence of a *basal conglomerate,* which contains clasts of the rocks under the unconformity. Recognition of such a basal conglomerate is also helpful in determining whether the contact between strata and a plutonic rock is intrusive or represents a nonconformity.

2.2.7 Compaction and Diagenetic Structures

When a clastic sediment initially settles, it is a mixture primarily of grains and water. The proportion of solid to fluid varies depending on the type of sediment. Gooey mud (i.e., clay plus water) obviously contains more water than well-packed sand. Progressive burial of sediment squeezes the water out, and the sediment compacts. Compaction results in a decrease in porosity (>50% in shale and >20% in sand) and, therefore, an increase in the density of the sediment.

Lateral variation in the amount of compaction within a given layer, or contrasts in the amount of compaction in a vertical section, is a phenomenon called *differential compaction.* Differential compaction within a layer can lead to

lateral variations in thickness, which are called *pinch-and-swell structure.* Pinch-and-swell structure can also form as a consequence of tectonic stretching, so again, you must be careful when you see the structure to determine whether it is a depositional structure or a tectonic structure.

The compaction of mud leads to development of a preferred orientation of clay in the resulting shale. Clay occurs in tiny flakes shaped like little playing cards. In a wet sediment the flakes are not all parallel to one another, as in a standing house of cards, but after compaction the flakes are essentially parallel to one another, as in a collapsed house of cards. The preferred orientation of clay flakes, as we have seen, leads to bedding plane fissility. Compaction of sand composed of equant grains causes the grains to pack together more tightly.

Compaction can also cause *pressure solution,* a process by which soluble grains preferentially dissolve along the faces at which stress is the greatest.[3] In pure limestones or sandstones, this process causes grains to suture together, meaning that the grain surfaces interlock with one another like pieces of a jigsaw puzzle. In conglomerates, the squeezing together of pebbles results in the formation of pits (indentations) on the pebble surfaces, creating pitted pebbles. In limestone and sandstone that contain some clay, the clay enhances the pressure solution process. Specifically, pressure solution occurs faster where the initial clay concentration is higher. As a result, distinct seams of clay residue develop in the rock, and these seams are called *stylolites* (Figure 2.13). In rocks with little clay (<10%), stylolites tend to look like the sutures in your skull (i.e., their trace in cross section is jagged and toothlike). The teeth are caused by the distribution of grains of different solubility along the stylolite. In rocks with more clay, the stylolites are wavy, and the teeth are less pronounced, because the clay seams become thicker than tooth amplitude.

---

[3]We will be talking more about pressure solution when we consider the process of cleavage formation (Chapter 11).

Figure 2.14 Penecontemporaneous folds in the Maranosa Arenaci (Italian Apennines).

Some of the dissolved ions removed at pressure-solved surfaces precipitate locally in the rock in veins or as cement in pore spaces, and some get transported out of the rock by moving groundwater. The proportion of reprecipitated to transported ions is highly variable. In some outcrops as much as 40% of the rock may have dissolved and been removed during formation of stylolites.

Some sedimentary rocks exhibit color banding that cuts across bedding. This color banding, which is called *Liesegang banding,* is the result of diffusion in groundwater of impurities, or of reactions leading to alternating bands of oxidized and reduced iron. It can be mistaken for bedding or cross bedding. To avoid mistaken identity, search the outcrop to determine whether sets of bands cross each other (possible for Liesegang bands, but impossible for bedding), and whether the bands are disrupted at fractures or true bedding planes, because these are places where the diffusion rate changes.

2.2.8 Penecontemporaneous Folds and Faults

If sediment layers have an initial dip, meaning a gentle slope caused by deposition on a preexisting slope, gravity acts to pull the layers down the slope. The ease with which sediments move down a slope is increased by fluid pressure in the layers, which effectively keeps the layers apart. Movement is resisted by weak electrostatic adhesion between grains, but this resistance can be readily overcome by the energy of an earthquake or storm, and the sediment will move down the slope. If the sediment completely mixes with water and becomes a turbid suspension flowing into deeper water, then all of the preexisting primary structure is lost and the grains are resedimented as a new graded bed (turbidite) farther down the slope. If the flowing mixture of sediment and water is dominantly sediment, it churns into a slurry containing chunks and clasts that are suspended in a matrix. Such slurries are called *debris flows,* and where

preserved in a stratigraphic sequence, they become matrix-supported conglomerates containing a range of clast sizes and shapes (i.e., the conglomerate is poorly sorted).

If the beds were compacted sufficiently prior to movement so that they maintain cohesion, then the movement is called *slumping.* During slumping the sedimentary layers tend to be folded and pulled apart, and are thrust over one another. The folds and faults formed during this slumping are *penecontemporaneous structures,* because they formed almost (Greek prefix *pene*) at the same time as the original deposition of the layers. Penecontemporaneous folds and faults are characteristically chaotic. These folds display little symmetry, and folds in one layer are of a different size and orientation than the folds in adjacent layers. Penecontemporaneous faults are not associated with pronounced zones of brittle fracturing (we turn to the characteristics of brittle fracturing later in the book, Chapters 6–8). Perhaps the key to recognizing slump folding in a sedimentary sequence is that the deformed interval is *intraformational,* meaning that it is bounded both above and below by relatively undeformed strata (Figure 2.14). Additionally, intervals of penecontemporaneous folds occur in a sequence that also includes debris flows and turbidites, all indicative of an unstable depositional environment. In some cases, slump folding can be mistaken for local folding adjacent to a fault. However, the opposite may also occur: tectonic folds may be mistakenly interpreted as slump structures. Not a simple matter to distinguish between the two!

We tend to think of debris flows and landslides as being relatively small structures, capable of disrupting a hillslope, and unfortunately anything built on it or downslope, but generally not much more. However, the geologic record indicates that, on occasion, truly catastrophic landslides of enormous dimension have occurred. In northern Wyoming, for example, a giant Eocene slide in association with volcanic eruption displaced dozens of mountain-sized blocks and hundreds of smaller blocks. In fact, one such large block, Hart Mountain, moved intact for several tens of km, apparently riding on a cushion of compressed air above a nearly planar subhorizontal (detachment) fault.

2.3 SALT STRUCTURES AND PASSIVE-MARGIN BASINS

Salt is a sedimentary rock that forms by the precipitation of halide and sulfate minerals (halite, gypsum, and anhydrite) from saline water. Salt deposits accumulate in any *sedimentary basin* (a low region that is the site of deposition) where saline water, such as seawater, evaporates sufficiently for solid salt to precipitate. Particularly thick salt deposits lie at the base of *passive-margin basins,* so named because they occur along tectonically inactive edges of continents. To see how these basins form, imagine a supercontinent, like Early Mesozoic Pangaea, that is being pulled apart. This process, called *rifting,* involves brittle and ductile faulting, the net result of which is to thin the continental lithosphere until it breaks and an oceanic ridge is formed. During the early stages of rifting, the rift basin is dry or contains freshwater lakes. Eventually, the floor of the rift drops below sea level and a shallow sea forms. During this stage, evaporation rates may be very high, so various salts (halite, gypsum, anhydrite) precipitate out of the seawater and are deposited on the floor of the rift. When the rift finally evolves into an open ocean, the continental margins become passive margins and gradually subside. With continued subsidence, the layer of evaporite (salt) is buried by clastic sediments and carbonates typical of continental-shelf environments. We'll discuss this tectonic environment in more detail later (Chapter 15), but for now you have a picture of a thick pile of sediment with a layer of salt near its base, which is the starting condition for the formation of *salt intrusions.*

Salt differs from other sedimentary rocks in that it is much weaker and, as a consequence, is able to flow like a viscous fluid under conditions in which other sedimentary rocks behave in a brittle fashion.[4] In some cases, deformation of salt is due to tectonic faulting or folding, but because salt is so weak, it may also deform solely in response to gravity, and thereby cause deformation of surrounding sedimentary rock. If gravity is the only reason for salt movement, the deformation resulting from the movement is called *halokinesis* and the resulting body of salt is a *salt structure.*

2.3.1 Why Halokinesis Occurs

Halokinesis begins in response to three factors: (1) development of a density inversion, (2) differential loading, and (3) existence of a slope at the base of a salt layer. All three of these factors occur in a passive-margin basin setting. Salt is a nonporous and essentially incompressible material. Thus, when it gets buried deeply in a sedimentary pile, it does not become denser. In fact, salt actually gets less dense with depth, because at greater depths it becomes warmer and expands. Other sedimentary rocks (e.g., sandstone and shale), in contrast, form from sediments that originally had high porosity and thus become denser with depth because the pressure caused by overburden makes them compact. This contrast in behavior, in which the density of other sedimentary rocks exceeds the density of salt at depths greater than about 6 km, results in a *density inversion,* meaning a situation where denser rock lies over less dense rock. Specifically, salt density is about $2200 \text{ kg} \cdot \text{m}^{-3}$, whereas the density of the sedimentary rocks directly overlying the salt is about $2500 \text{ kg} \cdot \text{m}^{-3}$. A density inversion is an unstable condition, because where it exists, the salt has positive buoyancy. *Positive buoyancy* means that in a gravity field buoyancy force causes lower density materials to try to rise above higher density materials, thereby decreasing the overall gravitational potential energy of the system.[5] A familiar example of positive buoyancy forces is the push that your hand feels when you try to hold an air-filled balloon under water. When the positive buoyancy force exceeds the strength of the salt and is sufficient to upwarp strata that lie over the salt structure, then it will contribute to the formation of the salt structure.

Differential loading of a salt layer takes place when the downward force on the salt layer caused by the weight of overlying strata varies laterally. Differential loading may result where there are primary variations in the thickness or composition of the overlying strata, primary variations in the original surface topography of the salt layer, or changes in the thickness of the overlying strata due to faulting. Regardless of its cause, differential loading creates a situation in which some parts of the salt layer are subjected to a greater vertical load than other parts, and the salt is squeezed from areas of higher pressure to areas of lower pressure. For example, imagine a layer of salt whose upper surface initially bulges upward to form a small 'dome.' The weight of a column of rock and water from sea level down to a horizontal surface in the salt layer on either side of the dome is less than the weight of the column penetrating the top of the dome, because salt is less dense than other sedimentary rocks. As a consequence, salt is squeezed into the dome, making it grow upwards. A salt layer that has provided salt for the production of a salt structure, and thus has itself been changed by halokinesis, is the *source layer.*

---

[4]In fact, dry salt moves plastically; movement of damp salt involves pressure solution. We discuss these deformation mechanisms in Chapter 11.

[5]Negative buoyancy, vice versa, is a force that causes a denser material to sink through a less dense material.

Figure 2.15 A salt glacier that originates from a salt dome (western Iran).

The combination of differential loading and buoyancy force drives salt upward through the overlying strata until it reaches the *level of neutral buoyancy,* meaning the depth at which it is no longer buoyant, that is, where it has the same density as surrounding strata. Clastic strata are sufficiently compacted so that their density equals that of salt at depths of around 450–1500 m below the surface of the basin, depending on composition. At the level of neutral buoyancy, salt may begin to flow laterally, much like a thick pile of maple syrup flows laterally over the surface of your pancakes. This process, which is driven by gravity (above the level of neutral buoyancy, the salt is subjected to a negative buoyancy force), is known as *gravity spreading.* If the salt is extruded at the land surface, it becomes a salt glacier (Figure 2.15); beneath the sea, the salt also spreads like a salt glacier, except that while this movement occurs the salt continues to be buried by new sediment. If a slope is present, the salt flows down the slope.

2.3.2 Geometry of Salt Structures Formed by Halokinesis

In response to positive buoyancy force and to differential loading, salt tends to flow upwards relative to the source bed, and therefore the source bed thins. If the source bed thins so much that it disappears and the strata above the source bed are juxtaposed against the strata below the source bed, we say that the contact between the two beds is a *primary weld.* In general, a *weld* is any contact between strata that were once separated by salt. At any given time, a region may contain salt structures at many stages of development. Geologists working with salt structures have assigned different names to these structures based on their geometry (Figure 2.16). The name assigned to a specific structure depends on its shape as observed today, but in the context of geologic time, this shape is only temporary.

Here is some salt structure terminology (Figure 2.16). A salt structure that is simply an upward bulge relative to the source layer is a *salt pillow* or *salt dome* if it is roughly symmetrical in plan, or a *salt anticline* if it is elongate in plan view. Strata overlying these structures conformably warp around the structure. If the structure continues to grow, the salt body eventually pierces the bedding in the overlying strata, at which time it becomes a *salt diapir.* Elongate diapirs in plan view are *salt walls,* whereas equant diapirs in plan are *salt stocks.* Mature salt diapirs are generally narrower at depth and broader at the top; in the case of salt stocks, the lower part is the *stem* and the upper part is the *bulb.*

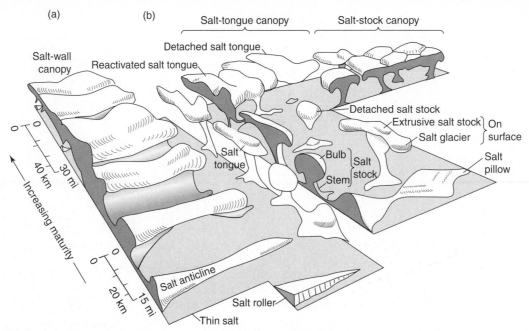

Figure 2.16 Diagram showing the stages in the formation of salt structures and associated terminology. Structural maturity and size increase toward the structures in the rear. Sequence in (a) shows structures rising from line sources, while structures in (b) originate from point sources.

In regions where the source bed was quite thick so that many diapirs form in the same region, salt walls or salt stocks may spread out and merge at a higher stratigraphic level, forming a *salt canopy*. Salt canopies may flow dominantly in one direction in response to gravity; if so, they are better termed *salt-tongue canopies*. Any salt layer that has spread laterally in a stratigraphic horizon that is higher than the original salt bed is considered to be allochthonous. Eventually, stems or walls connecting salt stocks or salt canopies to the source bed may pinch out, so that salt structures originally separated by the stem or wall are juxtaposed. The now isolated salt structure is a *detached bulb* or *detached canopy,* respectively. Interestingly, allochthonous salt canopies may serve as the source for still higher diapirs and canopies, as salt migrates ever higher in the section. If the salt diapir pierces the ground surface so that salt flows out over the land, the resulting structure, is a *salt glacier* (Figure 2.15).

Bedding within salt structures is contorted into very complex folds. Flow of the salt into bulbs results in folds that wrap around themselves and thus have circular profiles. Folding of layers within stems resembles the pleats in curtains.

2.3.3 Deformation and Sedimentation Adjacent to Salt Structures

Because salt both rises up into preexisting strata and rises during the time of deposition of overlying strata, geologists distinguish between two types of salt structure growth. If the salt rises after the overlying strata have already been deposited, then the rising salt will warp and eventually break through the overlying strata. This process is called

upbuilding. If, however, the rise of the salt relative to the source layer occurs during further deposition, the distance between the source layer and the surface of the basin also increases. This process is called *downbuilding.* As salt moves, it deforms adjacent strata and creates complex folds and local faults. When salt diapirs approach the surface, the overlying strata are arched up and therefore are locally stretched, resulting in the development of normal faults in a complex array over the crest of the salt structure (Figure 2.17).

The structural geometry of passive-margin basins is complicated because during the time that salt movement is occurring, sedimentation continues. Sedimentary layers thicken and thin as a consequence of highs and lows caused by the salt movement, and the resulting differential compaction causes further salt movement. The sedimentation pattern may change in a locality when the salt filling a salt structure drains out and flows into a structure at another locality. Thus, in regions where halokinesis occurs, it is common to find places where an arch later evolves into a basin, and vice versa.

2.3.4 Gravity-Driven Faulting and Folding in Passive-Margin Basins

Salt-structure formation is a dynamic process that is intimately linked to faulting in the overlying strata. Salt is so weak that it makes a good glide horizon on which detachment and displacement of overlying strata occurs. In fact, in most passive-margin basins, the entire package of sedimentary rocks filling the basin tends to detach and slump seaward, gliding on a detachment fault in the layer of salt at

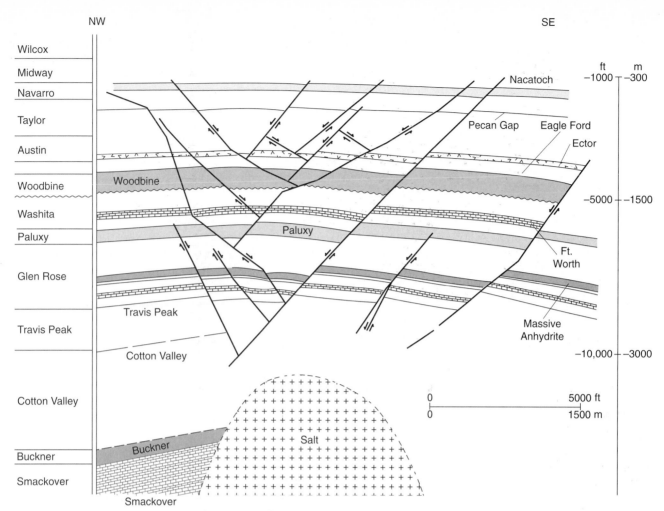

Figure 2.17 Cross section illustrating a normal fault array over the top of a salt dome in Texas.

its base. This movement resembles the slumping of sediment of a hillslope, though the scale of movement is quite different. As slumping occurs, the landward portion of the basin is stretched and is therefore broken by a series of normal faults whose dip tends to decrease with depth. This change in dip with depth makes the faults concave up; faults with this geometry are called *listric faults*. As movement occurs on a listric fault, the strata above the fault arch into rollover folds (Figure 2.18). Many listric normal faults intersect the ground surface in southern Texas, because this region is part of the passive-margin basin along the Gulf Coast. Because the faults dip south and they transport rock toward the Gulf of Mexico, they are called *down-to-the-Gulf faults*. Slip on these faults thins the stratigraphic section above the primary salt layer and thus results in differential loading of the salt layer. As a result, the salt rises beneath the fault. Initially, the resulting salt dome is symmetric, resembling an ocean wave, and is called a *salt*

roller. A salt roller may evolve into a diapir, eventually cutting the overlying fault. At the toe of the passive margin wedge, a series of thrust faults develop to accommodate displacement of the seaward-moving wedge, just as thrusts develop at the toe of a hillside slump.

2.3.5 Practical Importance of Salt Structures

Why spend so much time dealing with salt structures and passive margins? Simply because these regions are of great economic and societal importance. Passive-margin basins are major oil reservoirs, and much of the oil in these reservoirs is trapped adjacent to salt bodies. Oil rises in the upturned layers along the margins of the salt body and is trapped against the margin of the impermeable salt. More recently, salt bodies in passive-margin basins are being used as giant storage tanks for gas or oil, and are being considered as potential sites for the placement of nuclear waste.

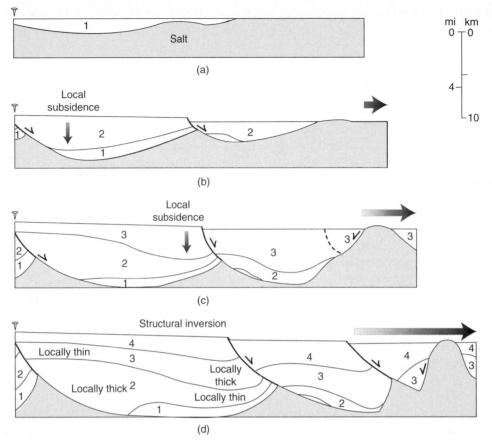

Figure 2.18 Cross sections showing "down-to-Gulf" type movement of a passive-margin salt wedge; the sections also show listric normal faults and salt rollers. The sequence a through d shows successive stages in evolution of the margin, and accompanying extension.

2.4 PRIMARY IGNEOUS STRUCTURES

You will recall from earlier courses in geology that there are two principal classes of igneous rocks, and that these classes are distinguished from each other based on the environment in which the melt cools. *Extrusive rocks* are formed either from lava that flowed over the surface of the Earth and cooled under air or water, or from ash that exploded out of a volcanic vent. *Intrusive rocks* cooled beneath the surface of the Earth. During the process of intruding, flowing, settling, and/or cooling, igneous rocks develop primary structures. In this context, we use the term "primary structure" to refer to a geometry of fabric that is a consequence of igneous processes.

Where do magmas come from? If you have had a course in igneous petrology, you know the answer to this question. But for those of you who have not, we'll quickly outline the nature of magmatic activity. Magma forms where conditions of heat and pressure cause existing rock (either in the crust or in the mantle) to melt. Commonly, only certain minerals (the ones that melt at a lower temperature) of the solid rock melt, in which case we say that the rock has undergone *partial melting*. A magma formed by partial melting has a composition that differs from that of the rock from which it was extracted. For example, 1–6% partial melt of ultramafic rock (peridotite) in the mantle yields a mafic magma that, when solidified, forms the gabbro and basalt that characterizes oceanic crust. Melting of an intermediate-composition crustal rock (diorite) yields a silicic magma that, when solidified, forms granite or rhyolite. Once formed, magma is less dense than the surrounding rock, and buoyancy forces cause it to rise. The density decrease is a consequence of the expansion that accompanies melting, the formation of gas bubbles within the magma, and the difference in composition between magma and surrounding rock.

Magma probably moves by oozing up through a network of cracks and creeping along grain surfaces. Its movement is inhibited by viscous drag, but magma pressure (the difference between the pressure within the magma and the lithostatic pressure in the surrounding rock) is so substantial that, as the magma enters the brittle crust, it can force open new cracks. Magma continues to rise until it reaches a *level of neutral buoyancy,* defined as the depth where pressure in the magma equals lithostatic pressure in the surrounding rock (buoyancy force is zero). At the level of neutral buoyancy, the magma may spread laterally to

Table 2.5   Terminology of Igneous Intrusions

Batholith	A huge bloblike intrusion; usually a composite of many plutons.
Dike	A sheet intrusion that crosscuts stratification in a stratified sequence, or is roughly vertical in an unstratified sequence.
Hypabyssal	An intrusion formed in the upper few kilometers of the Earth's crust; hypabyssal intrusions cool relatively quickly, and thus are generally fine grained.
Laccolith	A hypabyssal intrusion that is concordant with strata at its base, but bows up overlying strata into a dome or arch.
Pluton	A moderate sized bloblike intrusion (several km in diameter). Sometimes the term is used in a general sense to refer to any intrusion regardless of shape or size.
Sill	A sheet intrusion that parallels preexisting stratification in a stratified sequence, or is roughly subhorizontal in an unstratified sequence.
Stock	A small, bloblike intrusion (a few km in diameter).

form a *sheet intrusion* (dike or sill; Table 2.5), or may pool in a large *magma chamber,* which solidifies into a bloblike intrusion generally called a *pluton.* During formation of large sheet intrusions, magma may flow horizontally for tens to hundreds of kilometers. If the magma pressure is sufficiently high, the magma rises all the way to the surface of the Earth like water in an artesian well, and is extruded at a volcano.

2.4.1   Structures Associated with Sheet Intrusions

One of the key aspects of sheet intrusions that is of interest to structural geologists is their relationship to stress. In Chapter 3, we will introduce the concept of stress in some detail, but for now, we need to point out that the stress acting on a plane is defined as the force per unit area of the plane. Intuitively, therefore, you can picture that the rock of the crust is held together least tightly in the direction of the smallest stress (technically called the direction of least principal stress). Sheet intrusions, in general, form perpendicular to the direction of the least principal stress.[6] For example, in regions where the greatest stress is caused by the weight of the overlying rocks, and is therefore vertical, the least principal stress is horizontal, so vertical dikes form. Regional *dike swarms,* which are arrays of subparallel dikes occurring over broad regions of the crust, probably represent intrusion at depth in association with horizontal extension, which causes the horizontal stress to be tensile.

Not all dikes occur in parallel arrays. In the immediate vicinity of volcanoes, the ballooning of magma chambers and/or the collapse of a magma chamber locally modifies the stress field and causes a complex pattern of fractures. As a result, the pattern of dikes around a volcano (Figure 2.19) includes *ring dikes,* which are dikes

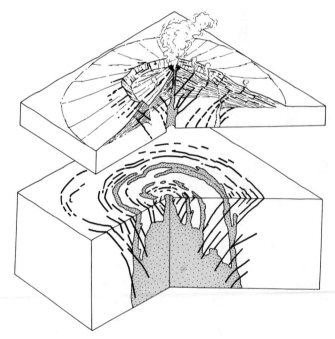

Figure 2.19   Types of sheet intrusions around a volcano.

that dip steeply away from the center of the volcano and have a circular trace in map view, *cone sheets,* which dip steeply toward the center of the volcano and have a circular trace in map view, and *radial dikes,* which are vertically dipping and run outward from the center of the volcano like spokes of a wheel. At a distance from the volcano, where the local effects of the volcano on the stress field become less important, radial dikes may change trend to become perpendicular to the regional least principal stress.

---

[6]This rule is violated if the rock contains preexisting planes of weakness (mechanical anisotropies), such as faults.

Sills, or subhorizontal sheetlike intrusions, form where local stress conditions cause the least principal stress to be vertical, and/or where there are particularly weak horizontal partings in a stratified sequence. Sill intrusion can result in the development of faults. If the width of the intrusion changes along strike, there must be differential movement of strata above the intrusion. *Laccoliths* resemble sills in that they are concordant with strata at their base; but unlike sills, laccoliths bow up overlying strata to create a dome. The well-known laccoliths of the Henry Mountains in Utah are 2–4 km in diameter.

Sheet intrusions occur in all sizes, from thin seams measured in centimeters, to the Great Dike of Zimbabwe, which is nearly 500 km long and several km wide. Considering the dimensions of large intrusions (tens to hundreds of kilometers long), it is important to keep in mind that a large volume of magma may flow past a given point in the body. Such flow may be recorded as primary structures in the rock. For example, careful examination of dike-related structures shows the presence of drag folds, scour marks, imbricated phenocrysts, and flow foliation, particularly along the walls of the intrusion.

2.4.2 Structures Associated with Plutons

The nature of primary structures found in plutonic rocks depends on the depth in the Earth at which the intrusion solidified, because these structures reflect the temperature contrast between the intrusion and the country rock[7] around it. Remember that the Earth gets warmer with depth: at the surface, the average temperature is 20°C, whereas at the center it may be as much as 4000°C. The change in temperature with depth is called the *geothermal gradient.* In the shallow crust, the geothermal gradient is in the range of 20° to 40°C/km. At greater depths, however, the gradient must be less (<10°C/km), because temperatures at the continental Moho (at about 40 km depth) are in the range of about 650°C, and temperatures at the base of the lithosphere (at about 150 km) are in the range of about 1280°C. The increased geothermal gradient near the surface is due to the concentration of radioactive elements in the minerals of silicic rocks. Granitic magma begins to solidify at temperatures between 550° to 800°C. Therefore, the temperature contrast between the magma and the surrounding country rock decreases with depth. In the case of shallow-level plutons, which intrude at depths of less than about 5 km, the contrast between magma and country rock is several hundred degrees. Contacts at the margins of such shallow intrusions are sharp, meaning that you can put your finger on the contact. Angular blocks of country rock float in the magma near the contact, and the country rock adjacent to the contact may be altered by fluids expelled by the magma or may be baked into a rock type called *hornfels.* With greater depth, the temperature contrast between the magma and

country rock decreases, and contacts are more gradational, until the country rock itself is likely to be undergoing partial melting. Minerals that melt at lower temperature (e.g., quartz and feldspar) turn to liquid, while refractory minerals (i.e., minerals that melt at higher temperatures, such as amphibole and pyroxene) are still solid, though quite soft. Movement of the melt causes the soft solid layers to be contorted into wild folds. When this mess eventually cools, the resulting rock, which is composed of a marble-cake-like mixture of light and dark contorted bands, is called a *migmatite* (Figure 2.20). Folding in migmatites is typically so complicated and chaotic that it cannot be used to define a structural history.

Many plutons exhibit an *intrusion foliation* that is particularly well developed near the margin of the pluton and is subparallel to pluton-host rock contact (Figure 2.21). This foliation is defined by alignment of inequant crystals and by elongation of chunks of country rock or early phases of pluton (called *xenoliths,* which means "foreign rocks"; Figure 2.21) that were incorporated in the magma. Such fabric is a consequence of shear of the magma against the walls of the magma chamber and of flattening of partially solidified magma along the walls of the chamber in response to pressure exerted as new magma pushes into the interior of the chamber. Because intrusion foliation forms during the formation of the rock and is not a consequence of tectonic movements, it is a primary igneous structure. Similarly, intrusion foliation can develop along the margins of dikes.

Intrusion foliations may be difficult to distinguish from schistosity resulting from tectonic deformation. Plutons tend to act as strong ('competent') blocks, so that regional deformation is deflected and concentrated along the margins of the pluton. Interpretation of a particular foliation therefore depends on regional analysis and study of the deformation microstructures (Chapter 9) in thin section. For example, if the fabric attitude stays parallel with the boundary of the intrusion even when the boundary changes, and individual grains show no evidence for solid-state deformation, the foliation is likely primary igneous in origin.[8]

2.4.3 Structures Associated with Extrusion

As basaltic lava flows across the surface of the Earth under air, the surface of the flow may wrinkle into primary folds that resemble coils of rope, or may break into a jumble of jagged blocks that resembles a breccia. Lava flows with the rope-coil surface are called *Pahoehoe flows,* and lavas with the broken-block surface are called *Aa flows.* The wrinkles in a Pahoehoe should not be mistaken for tectonic folds, and the jumble of blocks caused by *autobrecciation* (i.e., breaking up during flow) of Aa should not be mistaken for a tectonic breccia related to faulting.

---

[7]To a geologist, "country rock" is not only a type of music but a casual term for the rock that was in a locality before the intrusion came in.

[8]This is a topic of active study. The distinction between tectonic and primary foliations in plutons has proven to be quite difficult and more often than not ambiguous.

Figure 2.20 A migmatite from the North Cascades (Washington).

Figure 2.21 Increasingly deformed xenoliths at the margin of a shear zone in granite (Lepontine Alps, Switzerland).

If basaltic lava is extruded beneath seawater, the surface of the flow cools extremely quickly, and a glassy skin coats the surface of the flow. Eventually, the pressure in the glass-encased flow becomes so great that the skin punctures, and a squirt of lava pushes through the hole and then quickly freezes. The process repeats frequently, resulting in a flow composed of blobs (0.5–2 m diameter) of lava. Each blob, which is called a pillow, is coated by a rind of fine-grained to glassy material. As the pillows build out into a large pile, called a *pillow basalt*, successive pillows flow over earlier pillows, and while still soft conform to the shape of the earlier flow surface

Figure 2.22 Pillow basalt from the Point Sal ophiolite, California. The asymmetric shape of the pillows and location of the "points" indicate top and bottom.

(Figure 2.22). As a result, pillows commonly have a rounded top and a pointed bottom (the 'apex') in cross section, and this shape can be used as a stratigraphic facing criterion.

In 1902, Mt. Pelee on the Caribbean island of Martinique erupted. It was a strange eruption, for instead of lava flows, a spine of rhyolite rose day by day from the peak of the volcano. This spine, as it turned out, was like the cork of a champagne bottle slowly being worked out. When the cork finally pops out of the champagne bottle, a froth of gas and wine flows down the side of the bottle. Likewise, when, on the morning of May 2, the plug exploded off the top of Mt. Pelee, a froth of hot (800°C) volcanic gas and ash floated on a cushion of air and rushed down the side of the mountain at speeds of up to 100 km/h. This ash flow engulfed the town of St. Pierre, and in an instant almost 30,000 people were dead. When the ash stopped moving, it settled into a hot layer that welded together. Such a layer of welded tuff is called an *ignimbrite*. Ignimbrites commonly have an internal layering that forms in a unique way. Volcanic ash is composed of tiny glass shards with jagged spinelike forms. The glassy texture is a consequence of very rapid cooling. When the ash settles, the glass shards are still hot and soft, so the compaction pressure exerted by the weight of overlying ash causes the shards to flatten, thereby creating a primary foliation in ignimbrite that is comparable to bedding.

Rhyolitic lavas commonly display subtle color banding. This banding, called *flow foliation*, has been attributed to flow of the lava before complete solidification, and thus is a primary layering. The banding forms because lavas are not perfectly homogeneous materials. Since the temperature is not perfectly uniform, there may be zones in which crystals have formed, while adjacent regions are still molten. The shear resulting from movement of the lava smears out these initial inhomogeneities into subparallel bands. To visualize this, think of a bowl of vanilla cake batter into which you have dripped spoonfuls of chocolate batter. If you slowly stir the mixture, the blobs of chocolate smear out into sheets. Two chocolate blobs that were initially nearby would smear into parallel sheets with an intervening band of vanilla. In the lava, movement of the lava smears out blobs of contrasting texture into layers, which, when the rock finally freezes, have a slightly different texture than adjacent bands and thus are visible markers in outcrop. Commonly, continued movement causes previously formed layers to fold, so flow-banded outcrops typically display complex primary folds.

2.4.4 Cooling Fractures

As shallow intrusions and extrusive flows cool, they contract. Because of their fine grain size, the rocks in such bodies are susceptible to forming joints (natural cracks) in response to the thermal stress associated with cooling. Such joints are typically arranged in approximate hexagonal arrays that isolate columns of rock (Figure 2.23). Because of this geometry, this fracturing is called *columnar jointing*. Popular tourist destinations like Devils Tower in Wyoming

Figure 2.23 Columnar jointing, Devil's Postpile (California).

or Giant's Causeway in Ireland are spectacular examples. If you look closely at unweathered columnar joints, the surfaces of the joints are ribbed. We will learn later that this feature is a consequence of the way in which cracks propagate through rock. The long axes of columns are perpendicular to isotherms (surfaces of constant temperature) and thus they are typically perpendicular to the boundaries of the shallow intrusion or flow.

2.5 IMPACT STRUCTURES

Glancing at the moon through a telescope, the most obvious landforms that you see are craters. The moon has been struck countless times by meteors, and each impact has left a scar that, because the moon is tectonically inactive and has no atmosphere or water, has remained largely unchanged through succeeding eons. The Earth has undoubtedly been pummeled at least as frequently as the moon, but smaller meteors disintegrate in the atmosphere before reaching the surface, and the scars of those that did strike

the surface have been erased by erosion and tectonics. It is not surprising that the relics of impact, *impact structures,* are relatively rare on Earth. The vast majority of impacts on the moon occurred prior than about 3.5 Ga, when the solar system contained a multitude of fragments that were not yet incorporated into planets. Seventy percent of the Earth's surface is underlain by oceanic lithosphere, most of which is less than 100 million years old, and all but a relatively small portion of the continental crust has either been covered by younger strata or has been involved in orogenesis since 3.5 Ga.

Even though they are rare, and in some cases difficult to recognize, impact structures do exist on Earth. For purpose of discussion we can distinguish three categories, based on the most obvious characteristic of the impact: (1) relatively recent surficial impacts that are defined by a visible crater, (2) impacts whose record at the present Earth surface is the disruption of sedimentary strata, and (3) impacts whose record is a distinctive map-view circular structure in basement.

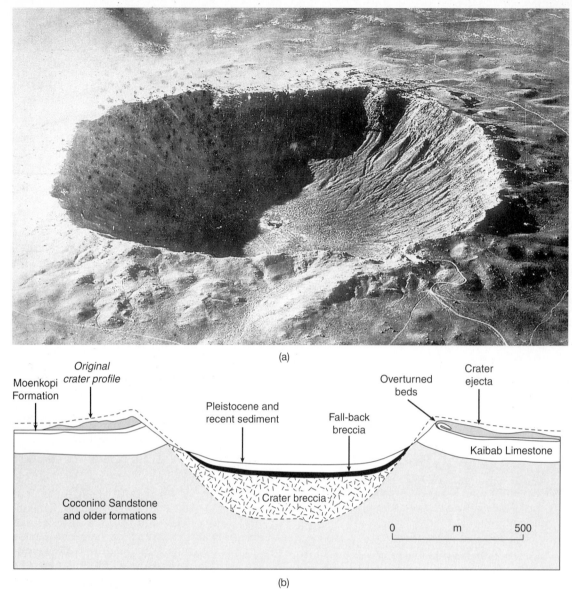

Figure 2.24 Aerial photograph (a) and cross section (b) of Meteor Crater (Arizona).

There are only about two dozen obvious surficial impact structures on the planet. The largest and perhaps the most famous of these is Meteor Crater in Arizona, which is 1.2 km in diameter, 180 m deep, and is surrounded by a raised rim about 50 m high (Figure 2.24a). As shown in the cross section (Figure 2.24b), the impact created a breccia that is about 200 m thick beneath the floor of the crater. The raised rims of the crater are not only composed of shattered rock ejected from the crater, but are also sites where sedimentary bedrock has been upturned. Ancient impact sites that are no longer associated with a surficial crater dot the Midcontinent region of the United States. These sites are defined by relatively small (less than a few kilometers across) semicircular disruption zones, in which the generally flat-lying Paleozoic strata of the region are severely fractured, faulted, and tilted. They were originally called *crypto-volcanic structures* (*crypto* is "hidden"), because it was assumed that they were the result of underlying explosive volcanism. Typically, steeply dipping normal faults, whose map traces are roughly circular, define the outer limit of these structures. These faults are crosscut by other steep faults that radiate from the center of the structure like spokes of a wagon wheel. Near the center of the structure, bedding is steeply dipping, and faulting has juxtaposed units of many different ages. Locally, the strata broke into huge blocks that jumbled together to create an *impact breccia.* Throughout this region, rock breaks into distinctive *shatter cones,* which are conelike arrays of penetrative fractures similar to the fractures that form next to a blast hole in rock (Figure 2.25). The apex of the cone points to the direction from which the impacting object came; in impact structures, shatter cones point up, confirming that they were caused by impact from above, as would be the case if the structure was due to an incoming meteor.

Figure 2.25 Shatter cones (Sudbury, Ontario).

Why do impact structures have the geometry that they do? To see why, think of what happens when you drop a pebble into water. Initially, the pebble pushes down the surface of the water and creates a depression, but an instant later, the water rushes in to fill the depression, and the place that had been the center of the depression rises into a dome. In the case of meteor impact against rock, the same process takes place. The initial impact gouges out a huge crater and elastically compresses the rock around the crater. But an instant later, the rock rebounds. At the margins of the affected zone it pulls away from the walls, creating normal faults, and in the center of the zone, it flows upward, creating the steeply tilted beds. Because rock in the near surface is brittle while this occurs, this movement is accompanied by faulting and brecciation.

The incredibly high pressures that develop during an impact create distinctive changes in the rocks of the impact site. The shock wave that passes through the rock momentarily subjects rocks to very high pressures, a condition that causes *shock metamorphism.* Shock metamorphism of quartz yields unusual high-density polymorphs, like stishovite, and characteristic deformation microstructures. In addition, the kinetic energy of impact is suddenly transformed into heat, so that rocks of the impact site are momentarily heated to temperatures as high as 1700°C. At such temperatures, the whole rock melts, only to freeze quickly into glass of the same composition as the original rock. In some cases, melt mixes with impact breccia, and injects into cracks between larger breccia fragments.

Impact structures affecting now exposed basement crystalline rocks characteristically cause distinctive circular patterns of erosion in the basement; these patterns stand out in satellite imagery. One of the best known basement impact structures is the Sudbury complex in southern Ontario, Canada. Not only are the characteristic features of impact (e.g., shatter cones) visible in the field, but the Sudbury impact was large enough to affect the whole crust and cause melting of the underlying mantle. As a result, mantle-derived magmas injected the crust in a distinctive ring at Sudbury. In places, the resulting igneous rocks contain valuable ores.

2.6 CLOSING REMARKS

This chapter has briefly described various types of structures whose formation is not an immediate consequence of the interaction of plates, of isostatic consequences, or of thickening or thinning of the crust. Because of their mode of formation, we've called them primary and nontectonic structures. A discussion of primary and nontectonic structures is a good way to start a structural geology course, because it gets you thinking about geometries and shapes. We were able to introduce these structures, at least at an elementary level, without dealing too much with the concepts of stress and strain. But before going further into tectonic structures, we must introduce stress and strain in some detail. Without understanding the meaning of these concepts, it is impossible to understand how and why tectonic structures form, even at an elementary level. Thus, stress, strain, and their relationship (rheology) are the subjects of our next chapters.

ADDITIONAL READING

Collinson, J. D., and D. B. Thompson, 1989, *Sedimentary structures* (second edition): London, Unwin Hyman, 207 p.

Jackson, M. P. A., and C. J. Talbot, 1994, Advances in salt tectonics, *in* Hancock, P. L., ed., *Continental deformation,* 159–179, Oxford, Pergamon Press.

Paterson, S. R., R. H. Vernon, and O. T. Tobisch, 1989, A review of criteria for the identification of magmatic and tectonic foliations in granitoids: *Journal of Structural Geology,* v. 11, p. 349–363.

Selley, R. C., 1988, *Applied sedimentology:* New York, Academic Press, 446 p.

Shrock, R. R., 1948, *Sequence in layered rocks,* New York, McGraw-Hill, 507 p.

Worrall, D. M., and S. Snelson, 1989, Evolution of the northern Gulf of Mexico, with emphasis on Cenozoic growth faulting and the role of salt, in *The geology of North America—An overview,* v. A: Boulder, CO, Geological Society of America, p. 97–138.

Chapter 3

Force and Stress

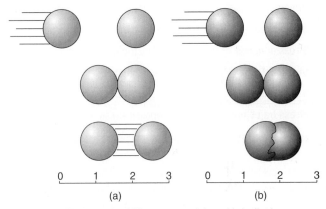

Figure 3.1   The interaction of forces on nondeformable bodies is described by classical or Newtonian mechanics (a) and that between deformable bodies by continuum mechanics (b). Imagine the difference between playing pool with regular balls and balls made of jelly.

3.1   INTRODUCTION

We are all too familiar with *stress*—it is a term that we use on many occasions. Stress from yet another homework assignment, a test, or maybe an argument with a roommate or spouse. We are also aware that we are subject to the *forces* of nature. The force of gravity tries to keep us on the Earth's surface and the force of impact destroys our car. Just like us, rocks experience the pull of gravity throughout their history, beginning with deposition and followed by deformation and erosion. Forces arising from plate interactions result in a broad range of geological structures that include fabrics and textures on the mi-

croscale, faults and folds on the mesoscale, and mountain ranges and oceans on the macroscale. In geology, the terms *force* and *stress* have very specific meaning. In this chapter we will discuss these concepts, followed by a more detailed look at the components of stress that ultimately are responsible for the formation of geological structures.

To understand the processes associated with the structural geology of rocks and regions, we must be acquainted with some fundamental principles of mechanics. *Mechanics* is concerned with the action of forces on bodies and their effect; you can say that it is the science of motion. *Classical* or *Newtonian*[1] *mechanics* describes the action of forces on rigid bodies. The equations of Newtonian mechanics explain everything from the entertaining interaction between colliding balls in a game of pool (Figure 3.1a) to the galactic dance of the planets in our solar system. However, in

---

[1]After the British physicist Sir Isaac Newton (1642–1727).

Table 3.1 Some Terminology of Force and Stress

Force	Mass times acceleration (m · a; Newton's second law); F.
Stress	Force per unit area (F/A); σ.
Anisotropic stress	At least one principal stress has a magnitude unequal to the other principal stresses (describes an ellipsoid).
Deviatoric stress	Component of the stress that remains after the mean stress is removed; this component of the stress contains the six shear stresses; σ_{dev}.
Differential stress	The difference between two principal stresses ($\sigma_1 - \sigma_3$), which by definition is ≥ 0; σ_d.
Homogeneous stress	Stress at each point in a body has the same magnitude and orientation.
Hydrostatic stress/pressure	Isotropic component of the stress; strictly: pressure at the base of a water column.
Inhomogeneous stress	Stress at each point in a body with different magnitude and/or orientation.
Isotropic stress	All three principal stresses have equal magnitude (describes a sphere).
Lithostatic stress/pressure	Isotropic pressure at depth in the Earth arising from the overlying rock column ($\rho \cdot g \cdot h$); P_l.
Mean stress	$(\sigma_1 + \sigma_2 + \sigma_3)/3$; σ_m.
Normal stress	Stress component oriented perpendicular to a given plane; σ_n.
Principal plane	Plane of zero shear stress; a principal plane contains two principal stresses and is perpendicular to the third, so three principal planes exist.
Principal stress	The normal stress on a plane with zero shear stress; three principal stresses exist, with $\sigma_1 \geq \sigma_2 \geq \sigma_3$.
Shear stress	Stress parallel to a given plane; σ_s (sometimes the symbol τ is used).
Stress ellipsoid	Geometric representation of stress; the axes of the stress ellipsoid are the principal stresses.
Stress field	The orientation and magnitudes of stresses in a body.
Stress tensor	Mathematical description of stress (stress is a second-rank tensor).
Stress trajectory	Principal stress directions in a body.

rocks we often deal with processes that result both in movement and *distortions,* that is, displacements between as well as within bodies (Figure 3.1b). The previously mentioned encounter between a wall and a car comes to mind as an unfortunate example of displacement and distortion (Figure 1.6). The theory associated with this type of material behavior is the topic of *continuum mechanics.* In continuum mechanics, a material is treated as a continuous medium; that is, there are no discontinuities that affect its behavior. Immediately this may seem inappropriate for rocks, because we know that they consist of many grains whose boundaries are material discontinuities. At the scale of these discontinuities (the grain scale), one might argue that continuum mechanics does not apply. Yet, on the scale of the rock body we can consider the system statistically homogeneous, and the predictions from continuum mechanics theory give us adequate first-order descriptions of the changes in rocks. The primary advantage of a continuum mechanics approach is that it allows for the mathematical description of deformation in relatively simple terms. However, if the behavior of rocks is dominated by small-scale discontinuities, such as is the case in fracture mechanics (chapter 6), continuum mechanics theory no longer applies. For now, however, we will consider it an adequate approach.

By the time you have reached the end of this chapter, a considerable number of terms and concepts will have appeared. Some of the more common ones are listed in Table 3.1 for reference. Before turning to the details of force and stress, we briefly look at the notion of quantities and units.

3.2 FUNDAMENTAL QUANTITIES AND UNITS

When you measure something you must first select a unit for the quantity that is to be measured. The physical properties of a material can be expressed in terms of four fundamental quantities: *mass, length, time,* and *charge.* For our current purposes we may ignore the quantity charge, which describes the electromagnetic interaction of particles. It plays a role, however, when we wish to understand the detailed behavior of materials at the atomic scale. The units of mass, length, and time are the kilogram (kg), the meter (m), and the second (s), respectively. These units follow the Système Internationale (French), and are better known as

Table 3.2 Units of Stress and Their Conversions

	Bar	Dynes cm^{-2}	Atmosphere	kg · cm^{-2}	Pascal	Pounds · in^{-2} (psi)
Bar		10^6	0.987	1.0197	10^5	14.503
Dynes cm^{-2}	10^{-6}		0.987×10^{-6}	1.919×19^{-6}	0.1	14.503×10^{-6}
Atmosphere	1.013	1.013×10^6		1.033	1.013×10^5	14.695
kg · cm^{-2}	0.981	0.981×10^6	0.968		0.981×10^5	14.223
Pascal (Pa)	10^{-5}	10	0.987×10^{-5}	1.0197×10^{-5}		14.503×10^{-5}
Pounds · in^{-2}	6.895×10^{-2}	6.895×10^4	6.81×10^{-2}	7.03×10^{-2}	6.895×10^3	

To use this table, start in the left-hand column and read along the row to the column for which a conversion is required. For example, 1 bar = 10^5 Pa or 1 Pa = 14.5×10^{-5} psi.

SI units. Throughout the text we will use SI units, but other conventions remain popular in geology (such as the kilobar). Where appropriate we will add these units in parentheses. In Table 3.2 the SI units and some common conversions are listed.

It is convenient to use symbols to describe fundamental quantities. The symbol for mass is [m], for length [l], and for time [t]. For example, velocity [v] combines the fundamental quantities of length and time; it has the units of length divided by time. In conventional symbols this is written as

$$[v] = [lt^{-1}]$$

in which the colon means "has the quantity of." We begin by looking at force.

3.3 FORCE

A force changes the velocity of an object; kick a ball to test this. Newton's first law of motion, the *Law of Inertia,* states that in the absence of a force a body moves at constant velocity or is at rest. Stated more formally: a free body moves without acceleration. The change in velocity is called acceleration [a], which is defined as velocity divided by time:

$$[a] = [vt^{-1}] = [lt^{-2}]$$

The unit of acceleration, therefore, is ms^{-2}.

Force [F], according to Newton's *Second Law of Motion,* is mass multiplied by acceleration:

$$[F] = [ma] = [mlt^{-2}]$$

The unit of force is kgms^{-2}, known as *newton* (N) in SI units. Kick a tennis ball and a basketball, and notice the different forces required to move them.

Force, like velocity, is a *vector quantity,* meaning that it has both magnitude and direction. So it is a line with an arrow at one end. Manipulation of forces therefore conforms to the rules of vector algebra. For example, a force at an angle to a given plane can be resolved into two components, one parallel and one perpendicular to that plane.

In nature we recognize four basic forces: (1) the gravitational force, (2) the electromagnetic force, (3) the nuclear or strong force, and (4) the weak force. Gravity is a special force that acts over large distances and is always attractive; for example, the ocean tides reflect the gravitational interaction between moon and Earth. The other three forces act only over short ranges (atomic scale) and can be attractive or repulsive. The electromagnetic force describes the interaction between charged particles, such as the electrons around the atomic nucleus; the strong force holds the nucleus of an atom together, and the weak force is responsible for radioactivity. It is quite possible that really only one fundamental force exists. Since Albert Einstein's[2] first efforts, great progress has been made in unifying all four forces into the *Grand Unified Theory,* but this has so far excluded gravity.

Forces that result from action of a field at every point within the body are called *body forces.* Jumping out of an airplane gives you a very vivid sensation of the presence of body forces, even if you use a parachute; gravity is an example of a body force. The magnitude of body forces is proportional to the mass of the body. Forces that act on a specific surface area in a body are called *surface forces.* They reflect the pull or push of the atoms on one side of a surface against the atoms on the other side. The magnitude of surface forces is proportional to the area of the surface. Some common examples of surface forces are the force of a cuestick that hits a pool ball, the force of expanding gases on an engine piston, the force of the jaws of a vice, and our unfortunate car incident mentioned earlier. Surface forces are of great importance in structural geology, but remember that body forces (gravity!) are always present on Earth.

Forces on a body may change the velocity of (i.e., accelerate) the body as a whole, and/or may result in a shape change of the body (i.e., accelerate one part of the body with respect to another). Clearly, force is an important

---

[2]German-born theoretical physicist (1879–1955).

concept, but it cannot be used to distinguish its effect on bodies of equal mass but with different shapes. A force applied by a sharp object has a different effect than a similar force applied by a dull object. Two practical examples may help to clarify this concept. A human body is safely supported by a water bed, but when we place a nail between the person and the water bed the effect is much more dramatic. Or, think about hitting a rock with a pointed or flat hammer using roughly equal force. The rock cracks more easily with the pointed hammer than with the flat-headed hammer; in fact, we apply this principle when we use a chisel rather than a sledge hammer to collect rock samples. These considerations of force lead us into the topic of *stress*.

3.4 STRESS

Stress (σ) is the force per unit area (A), or $\sigma = F/A$. You may, therefore, consider stress as the *intensity of force*, or a measure of how concentrated the force is. A given force acting on a small area (the pointed hammer) will have a greater intensity than that same force acting on a larger area (the flat-headed hammer), because the stress associated with the small area is greater. Those of you who remember turntables and records (ask your parents) may be familiar with this effect. The weight of the arm holding the needle is only a few grams, but the stress of the needle on the record is many orders of magnitude greater, because the contact area between the needle and the record is very small. Eventually the high stresses at the contact give rise to scratches and ticks in the records, so it is little wonder that we have moved to CD technology.

Right at the start we alert you that stress is a complex parameter, with properties that depend on the reference system. Stress on a plane is a vector quantity, but stress at a point is something more complicated (a tensor). We will therefore gradually develop the concept of stress.

Because stress is force per area, it has the fundamental quantity:

$$[\sigma] = [mlt^{-2} \cdot l^{-2}] \text{ or } [ml^{-1} t^{-2}]$$

The corresponding unit of stress is $kg \cdot m^{-1}s^{-2}$ (or $N \cdot m^{-2}$), which is called the *pascal*[3] (Pa). Instead of this SI unit, many geologists continue to use the unit *bar*, which is approximately 1 atmosphere. They are related as:

$$1 \text{ bar} = 10^5 \text{ Pa} \approx 1 \text{ atmosphere} \qquad \text{(Table 3.2)}$$

In geology you will generally encounter its larger equivalents, the kilobar and the megapascal (MPa):

$$1 \text{ kbar} = 1000 \text{ bar} = 10^8 \text{ Pa} = 100 \text{ MPa} \qquad \text{(Table 3.2)}$$

The unit gigapascal (1 GPa = 1000 MPa = 10 kbar) is generally used to describe the very high pressures that occur deep in the Earth; for example, the pressure at the core-mantle boundary (at approximately 2900 km depth) is 136 GPa, and at the center of the Earth (6370 km depth) is 364 GPa.[4] Later we will see how these values can be calculated (section 3.9).

3.5 NORMAL STRESS AND SHEAR STRESS

Stress acting on a plane is a vector quantity, meaning that it has both magnitude and direction, and is sometimes called *traction*. Stress on an arbitrarily oriented plane, however, is not necessarily perpendicular to that plane, and, like any vector, it can be resolved into components normal to the plane and parallel to the plane. The vector component normal to the plane is called the *normal stress*, for which we use the symbol σ_n (sometimes just the symbol σ is used); the vector component along the plane is the *shear stress*, and has the symbol σ_s (sometimes the symbol τ is used). In contrast to the resolution of forces, the resolution of stress into its components is not straightforward, because the area changes as a function of the orientation of the plane with respect to the stress vector. Let us first examine the resolution of stress on a plane in some detail, because as we will see this has important implications.

In Figure 3.2, stress σ has a magnitude F/AB and makes an angle θ with the top and bottom of our square. The forces perpendicular (F_n) and parallel (F_s) to the plane EF are:

$$F_n = F \cos\theta = \sigma \text{ AB} \cos\theta$$
$$= \sigma \text{ EF} \cos^2\theta \text{ (AB = EF} \cos\theta) \qquad \text{Eq. 3.1}$$
$$F_s = F \sin\theta = \sigma \text{ AB} \sin\theta = \sigma \text{ EF} \sin\theta \cos\theta$$
$$= \sigma \text{ EF} \tfrac{1}{2}(\sin 2\theta) \qquad \text{Eq. 3.2}$$

Thus the corresponding stresses are:

$$\sigma_n = F_n/EF = \sigma \cos^2\theta \qquad \text{Eq. 3.3}$$
$$\sigma_s = F_s/EF = \sigma \tfrac{1}{2}(\sin 2\theta) \qquad \text{Eq. 3.4}$$

You will notice that the equation for the normal stress and the normal force are different, as are the equations for F_s and σ_s. To graphically illustrate this difference between forces and stresses on an arbitrary plane, we plot their normalized values as a function of the angle θ in Figure 3.2c and d, respectively. In particular the observed relationship between F_s and σ_s is instructive. Both the shear force and the shear stress increase with increasing angle θ, but at 45° the shear stress reaches a maximum and then decreases while F_s continues to increase.

Thus, the stress vector acting on a plane can be resolved into vector components perpendicular and parallel to

---

[3]After the French philosopher and mathematician Blaise Pascal (1623–1662).

[4]Based on (P)reliminary (R)eference (E)arth (M)odel (PREM; Anderson, 1989).

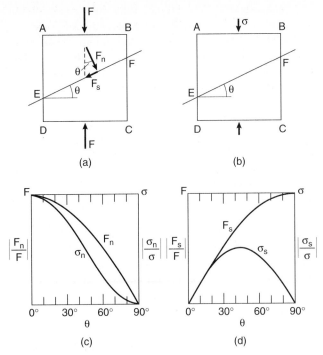

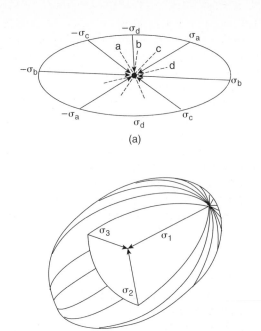

Figure 3.2 The relationship between force (F) and stress (σ). Section through a cube showing face ABCD with ribs of length AB on which a force F is applied. This force is resolved into orientations parallel (F_s) and perpendicular (F_n) to a plane that makes an angle θ with the top and bottom surface (EF is the trace of this plane). The magnitudes of vectors F_s and F_n are a function of the angle θ: $F_n = F \cdot \cos\theta$, $F_s = F \cdot \sin\theta$. The magnitude of the normal (σ_n) and shear stress (σ_s) is a function of the angle θ *and* the area: $\sigma_n = \sigma \cos^2\theta$, $\sigma_s = \sigma \frac{1}{2} (\sin 2\theta)$. (a) Force F on plane; (b) stress σ on plane; (c) normalized values of F_n and σ_n on plane with angle θ; (d) normalized values of F_s and σ_s on a plane with angle θ.

Figure 3.3 (a) A point represents the intersection of an infinite number of planes. The stresses on these planes describe an ellipse in the two-dimensional case. In three dimensions this generates the stress ellipsoid (b), defined by three mutually perpendicular principal stress axes ($\sigma_1 \geq \sigma_2 \geq \sigma_3$).

3.6.1 Stress at a Point

We shrink our three-dimensional body down to the size of a point for the analysis of the full stress state of a body. The need for this will not be immediately obvious unless you realize that a point defines the intersection of an infinite number of planes, each with a different orientation. Remember that two nonparallel planes have a line in common, so three or more nonparallel planes have only a point in common. The stress state at a point, therefore, describes the stresses acting on all these planes. In other words, when we determine the corresponding stresses for each of these planes, we will obtain the stress state at a point.

In Figure 3.3a the stresses acting on several planes are drawn. For clarity we limit our illustrations to two dimensions (i.e., a section through the body), but you will see later that this easily expands into the three-dimensional case. Because of Newton's Third Law of Motion, the stress on each plane must be balanced by one of opposite sign. You recall that the stress varies as a function of orientation, so the magnitude (vector length) of the stress for each plane is different. If we draw an envelope around these stress vectors for several of these planes (heavy line in Figure 3.3a), we obtain an ellipse. This says that the magnitude of the stress for each possible plane is represented by a point on this ellipse. Now, the same can be done in three dimensions, in which case we obtain an envelope that has the shape of an ellipsoid. This ellipsoid,

that plane, but their magnitudes vary as a function of the orientation of the plane. Let us further examine this not immediately obvious property of stress by determining the stress state for a three-dimensional body.

3.6 STRESS IN THREE DIMENSIONS, PRINCIPAL PLANES, AND PRINCIPAL STRESSES

So far we have discussed only stress acting on a single plane (i.e., the two-dimensional case), which resulted in the recognition of two vector components, the normal stress and the shear stress, with some unexpected properties. However, to describe stress on a randomly oriented plane we must consider the three-dimensional case. To minimize complications, we include the condition that the body containing our plane is at rest. So, a force is balanced by an opposing force of equal magnitude but opposite sign; this condition is known as Newton's *Third Law of Motion*. Our final Newtonian sports analogy: try to kick a ball (in this case any type will do) that rests against a wall and notice how the ball (the wall, in fact) pushes back.

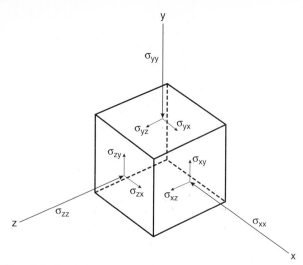

Figure 3.4 Resolution of stress into components perpendicular (three normal stresses, σ_n) and components parallel (six shear stresses, σ_s) to the three faces of an infinitesimally small cube, relative to the reference system x, y, and z.

called the *stress ellipsoid,* fully describes the stress state at a point (Figure 3.3b) and enables us to determine the stress for any plane. Like all ellipsoids, the stress ellipsoid is defined by three axes, which are called the *principal stresses.* These principal stresses have two properties: (1) they are orthogonal to each other, and (2) they are perpendicular to three planes that do not contain shear stresses; these planes are called the *principal planes of stress.* Note also that the principal stresses are vectors. So, we can describe the stress state of a body simply by specifying the orientation and magnitude of three principal stresses.

3.6.2 The Components of Stress

The orientation and magnitude of the stress state of a body can be defined in terms of its components projected in a specified Cartesian reference frame, which is defined by three mutually perpendicular coordinate axes, x, y, and z. To see this, instead of representing an infinite number of planes on which our stress acts, we consider our point to be an *infinitely small cube* whose sides are perpendicular to each of the coordinate axes, x, y, and z. We can now resolve the stress acting on each face of a cube into three components (Figure 3.4). For a face normal to the x-axis the components are σ_{xx}, which is the component *normal* to that face, and σ_{xy} and σ_{xz}, which are the two components *parallel* to that face. These last two stresses are shear stress components, acting along one of the other coordinate axes y and z, respectively. Applying this same procedure for the faces normal to y and z, we obtain a total of nine stress components (Figure 3.4):

	In the direction of		
	x:	y:	z:
stress on the face normal to x:	σ_{xx}	σ_{xy}	σ_{xz}
stress on the face normal to y:	σ_{yx}	σ_{yy}	σ_{yz}
stress on the face normal to z:	σ_{zx}	σ_{zy}	σ_{zz}

The columns, from left to right, each represent the components in the x, y, and z directions, respectively. σ_{xx}, σ_{yy}, and σ_{zz} are normal stress components; the other six are shear stress components. Of the six shear stress components, three must be equivalent (σ_{xy} to σ_{yx}, σ_{yz} to σ_{zy}, and σ_{xz} to σ_{zx}), because we specified that the body itself is at rest (if these components are unequal, the body would move, which violates this condition). So we are left with only *six independent stress components* acting on any arbitrary infinitesimal element in a stressed body:

	In the direction of		
	x:	y:	z:
stress on the face normal to x:	σ_{xx}	σ_{xy}	σ_{xz}
stress on the face normal to y:	σ_{xy}	σ_{yy}	σ_{yz}
stress on the face normal to z:	σ_{zx}	σ_{yz}	σ_{zz}

The only ingredient left is a *sign convention.* In physics and engineering, tensile stress is considered positive, and compressive stress negative. In geology, however, it is customary to make compression positive and tension negative, because compression is most common in the Earth. We will use the geologic sign convention throughout the text, but do not confuse this with the engineering sign convention used in some other geology textbooks.[5]

For any given state of stress, there is at least one set of three mutually perpendicular planes on which the shear stresses are zero. In other words, you can rotate our infinitesimal cube such that the shear stresses on its faces are zero. These three faces are the *principal planes of stress* (the same ones that we described in our stress ellipsoid; Section 3.6.1), and they intersect in three mutually perpendicular axes that are the *principal axes of stress* (which are the same as the axes of the stress ellipsoid). The stresses acting along them are called the *principal stresses* for a given point or homogeneous domain within a body.

3.6.3 Stress States

If the three principal stresses are equal in magnitude, the stress is *isotropic.* This stress state is represented by a sphere because all radii are equal. For example, the stress state of a body submerged in water is isotropic. If the principal stresses are unequal in magnitude, the stress is called *anisotropic.* By convention, the maximum principal stress is given the symbol σ_1; the intermediate and minimum principal stresses acting along the other two axes are given the symbols σ_2 and σ_3, respectively. Thus:

$$\sigma_1 \geq \sigma_2 \geq \sigma_3$$

The following relationships between the principal stresses define common *stress states:*

General triaxial stress:	$\sigma_1 > \sigma_2 > \sigma_3 \neq 0$
Biaxial (plane) stress: one axis = 0	(e.g., $\sigma_1 > 0 > \sigma_3$)
Uniaxial tension:	$\sigma_1 = \sigma_2 = 0; \sigma_3 < 0$
Uniaxial compression:	$\sigma_2 = \sigma_3 = 0; \sigma_1 > 0$
Hydrostatic stress (= pressure):	$\sigma_1 = \sigma_2 = \sigma_3$

[5]Elastic constants are typically listed using the engineering convention, so their sign needs to be reversed.

(a)

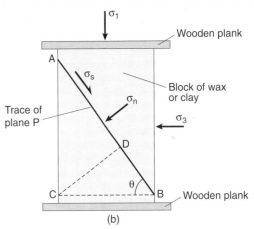

(b)

Figure 3.5 The normal and shear stresses on a plane in a stressed body as a function of the principal stresses. An illustration from the late nineteenth-century fracture experiments of Daubrée using wax is shown in (a). For our classroom experiment, a block of clay is squeezed between two planks of wood (b). AB is the trace of imaginary plane P in our body that makes an angle θ with σ_3. The two-dimensional case shown is sufficient to describe the experiment, because σ_2 equals σ_3 (atmospheric pressure).

So, we have learned that the stress ellipsoid is defined by nine components. Mathematically this is described by a 3×3 matrix (a *second-rank tensor*), but geologists like to use an ellipsoid because it is easy to visualize. However, for any mathematical operation we are better off using tensor operations, to which we will return later in this chapter (section 3.10).

3.7 RELATIONSHIP BETWEEN NORMAL STRESSES, SHEAR STRESSES, AND PRINCIPAL STRESSES

Now that we are able to express the stress state of a body by its principal stresses, we can turn to practical applications. Say we carry out a simple classroom experiment in which we compress a block of clay between two planks. What are the normal and the shear stresses on an arbitrarily oriented plane in this body? To answer this question we use essentially the same approach as before (Equations 3.1 to 3.4), but now we express these stresses in terms of the principal stress axes.

The principal stresses acting on our block of clay are σ_1 (maximum stress), σ_2 (intermediate stress), and σ_3 (minimum stress). Since we carry out our experiment under atmospheric conditions, the values of σ_2 and σ_3 will be equal, and we may simplify our analysis by neglecting σ_2 and consider only the $\sigma_1 - \sigma_3$ plane (Figure 3.5). We now take an arbitrary plane within our block that makes an angle θ with σ_3. This plane makes the trace AB in Figure 3.5. We also assume that the plane represented by AB has unit area (e.g., 1 cm^2) and length AB has unit length (e.g., 1 cm). We

can resolve AB along AC (parallel to σ_1) and along BC (parallel to σ_3). Then, by simple trigonometry, we see that the area represented by AC $= 1 \cdot \sin\theta$, and the area represented by BC $= 1 \cdot \cos\theta$.

Next we consider the forces acting on each of the surface elements represented by AB, BC, and AC. Since force equals stress times the area over which it acts, we obtain

$$\text{force on side BC} = \sigma_1 \cdot \cos\theta$$
$$\text{force on side AC} = \sigma_3 \cdot \sin\theta$$

The force on side AB consists of a normal force ($= \sigma_n \cdot 1$) and a shear force ($= \sigma_s \cdot 1$).

For equilibrium, the forces acting in the direction of AB must balance, and so must the forces acting perpendicular to AB (is parallel to CD). Hence, resolving along CD:

force $\perp$ AB = force $\perp$ BC resolved on CD + force $\perp$ AC resolved on CD, or

$$1 \cdot \sigma_n = \sigma_1 \cos\theta \cdot \cos\theta + \sigma_3 \sin\theta \cdot \sin\theta \qquad \text{Eq. 3.5}$$
$$\sigma_n = \sigma_1 \cos^2\theta + \sigma_3 \sin^2\theta \qquad \text{Eq. 3.6}$$

Substituting these trigonometric relationships in Equation 3.6:

$$\cos^2\theta = \tfrac{1}{2}(1 + \cos2\theta)$$
$$\sin^2\theta = \tfrac{1}{2}(1 - \cos2\theta)$$

and simplifying, gives:

$$\sigma_n = \tfrac{1}{2}(\sigma_1 + \sigma_3) + \tfrac{1}{2}(\sigma_1 - \sigma_3)\cos2\theta \qquad \text{Eq. 3.7}$$

and,

force parallel AB = force $\perp$ BC resolved on AB + force $\perp$ AC resolved on AB, or

$$1 \cdot \sigma_s = \sigma_1 \cos\theta \cdot \sin\theta - \sigma_3 \sin\theta \cdot \cos\theta \qquad \text{Eq. 3.8}$$

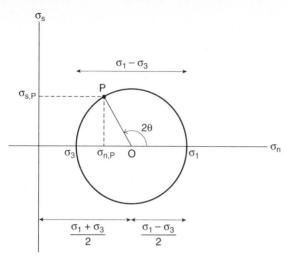

Figure 3.6 The Mohr diagram for stress. Point P represents the plane in our clay experiment in $\sigma_n - \sigma_s$ space (compare with Figure 3.5).

Note that the force perpendicular to AC resolved on AB acts in opposite direction from the force perpendicular to BC resolved on AB; hence a negative sign is needed in Equation 3.8, which further simplifies to:

$$\sigma_s = (\sigma_1 - \sigma_3)\sin\theta \cdot \cos\theta \qquad \text{Eq. 3.9}$$

Substituting this trigonometric relationship in Equation 3.9:

$$\sin\theta \cdot \cos\theta = \tfrac{1}{2}\sin 2\theta$$

gives:

$$\sigma_s = \tfrac{1}{2}(\sigma_1 - \sigma_3)\sin 2\theta \qquad \text{Eq. 3.10}$$

From Equations 3.7 and 3.10 we see that the planes of maximum normal stress are at an angle θ of 0° with σ_3 because $\cos 2\theta$ reaches its maximum value ($\cos 2\theta = 1$) at that angle. We also see that the planes of maximum shear stress lie at an angle θ of 45° with σ_3 because $\sin 2\theta$ reaches its maximum value ($\sin 2\theta = 1$) at $2\theta = 90°$ (compare these results with Figure 3.2c and d).

3.8 MOHR DIAGRAM FOR STRESS

The equations we derived for σ_n and σ_s do not allow you to get a rapid sense of their values as a function of orientation of a plane in our block of clay. Of course, a programmable calculator or simple computer program will do the job, but a convenient graphical method to solve Equations 3.7 and 3.10 was introduced over a century ago, and is known as the *Mohr diagram*[6] (Figure 3.6). A Mohr diagram is a plot, in a Cartesian reference frame defined by the σ_n and σ_s axes, of the equations defining normal stress and shear stress acting on a plane expressed in terms of the principal stresses (σ_1, σ_2, and σ_3). Before discussing the construction of the Mohr diagram, we examine the proof of this method.

---

[6]Named after the German engineer Otto Mohr (1835–1918).

If we rearrange Equations 3.7 and 3.10 and square them, we get:

$$\{\sigma_n - \tfrac{1}{2}(\sigma_1 + \sigma_3)\}^2 = \{\tfrac{1}{2}(\sigma_1 - \sigma_3)\}^2 \cos^2 2\theta \qquad \text{Eq. 3.11}$$

$$\sigma_s{}^2 = \{\tfrac{1}{2}(\sigma_1 - \sigma_3)\}^2 \sin^2\theta \qquad \text{Eq. 3.12}$$

Adding Equations 3.11 and 3.12 gives:

$$\{\sigma_n - \tfrac{1}{2}(\sigma_1 + \sigma_3)\}^2 + \sigma_s{}^2$$
$$= \{\tfrac{1}{2}(\sigma_1 - \sigma_3)\}^2 (\cos^2 2\theta + \sin^2 2\theta) \qquad \text{Eq. 3.13}$$

Using this trigonometric relationship in Equation 3.13:

$$(\cos^2 2\theta + \sin^2 2\theta) = 1$$

gives:

$$\{\sigma_n - \tfrac{1}{2}(\sigma_1 + \sigma_3)\}^2 + \sigma_s{}^2 = \{\tfrac{1}{2}(\sigma_1 - \sigma_3)\}^2 \qquad \text{Eq. 3.14}$$

Equation 3.14 has the form $(x - a)^2 + y^2 = r^2$, which is the equation for a circle with radius r and centered on the x-axis at distance a from the origin. Thus the Mohr circle has a radius $\tfrac{1}{2}(\sigma_1 - \sigma_3)$ centered on the σ axis at a distance $\tfrac{1}{2}(\sigma_1 + \sigma_3)$ from the origin, as shown in Figure 3.6; the center of the circle is called the mean stress (σ_m). You also notice from this figure that $\tfrac{1}{2}(\sigma_1 - \sigma_3)$ is the maximum shear stress. The *stress difference* ($\sigma_1 - \sigma_3$), which is called the *differential stress,* is indicated by the symbol σ_d.

3.8.1 Constructing the Mohr Diagram

To construct a Mohr diagram we define two mutually perpendicular axes: σ_n is the abscissa and σ_s is the ordinate. In our experiment the maximum principal stress (σ_1) and the minimum principal stress (σ_3) act on plane P that makes an angle θ with the σ_3 direction (Figure 3.5); in the Mohr diagram we then plot σ_1 and σ_3 on the σ-axis. The principal stress values are plotted on the σ_n axes because they are also normal stresses, but with the special condition that the shear stress is zero ($\sigma_s = 0$). We then construct a circle through points σ_1 and σ_3, with O, the midpoint, at $\tfrac{1}{2}(\sigma_1 + \sigma_3)$ as center; the radius of this circle is $\tfrac{1}{2}(\sigma_1 - \sigma_3)$. Next, we draw a radius OP such that angle POσ_1 is equal to 2θ. This step often gives rise to confusion and errors. Remember that we plot twice the angle θ, which is the clockwise angle between the plane and σ_3, in a counterclockwise sense from the σ_1-side of the σ_n-axis.[7] When this is all done, the Mohr diagram is complete, and we can read off the value of $\sigma_{n,P}$ along the σ-axis, and the value of $\sigma_{s,P}$ along the τ-axis for our plane P, as shown in Figure 3.6. We see that:

$$\sigma_{n,P} = \tfrac{1}{2}(\sigma_1 + \sigma_3) + \tfrac{1}{2}(\sigma_1 - \sigma_3)\cos 2\theta, \text{ and}$$

$$\sigma_{s,P} = \tfrac{1}{2}(\sigma_1 - \sigma_3)\sin 2\theta$$

A couple of interesting observations can immediately be made from the Mohr diagram (Figure 3.7). There are two planes oriented at angle θ and its complement (90 − θ),

---

[7]Alternative conventions for this construction are also in use, so be careful that they are not mixed.

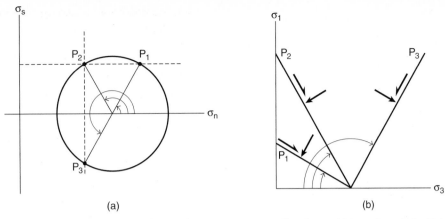

Figure 3.7 For each value of the shear stress and the normal stress there are two corresponding planes, which are shown in the Mohr diagram (a). The corresponding planes in $\sigma_1 - \sigma_3$ space are shown in (b).

which have equal shear stresses but different normal stresses. Also, there are two planes with equal normal stresses but shear stresses of opposite sign (i.e., acting in different directions within these planes).

In general, for each different orientation of a plane defined by the angle θ, there is a corresponding point on the circle. The coordinates of that point represent the normal and shear stresses on that plane. For example, when $\theta = 0°$ (plane parallel to σ_3), P coincides with σ_1, which gives $\sigma = \sigma_1$ and $\sigma_s = 0$. In other words, for any value of σ_1 and σ_3 (= σ_2 in our compression experiment), we can determine σ_n and σ_s graphically for planes that lie at an angle θ with σ_3. If we decide to change our experiment by gluing the planks to the clay block and then moving the planks apart (a tension experiment), we must use a negative sign for the least principal stress (in this case, $\sigma_1 = \sigma_2$ and σ_3 is negative). So the center O of the Mohr circle may lie on either side of the origin in Figure 3.6, but always on the σ-axis.

The Mohr diagram also nicely illustrates the attitude of planes along which the shear stress is greatest for a given state of stress. The point on the circle for which σ_s is maximum corresponds to a value of $\theta = 45$. For the same point, the magnitude of σ_s is equal to the radius of the circle, that is, $\frac{1}{2}(\sigma_1 - \sigma_3)$. Thus the $(\sigma_1 - \sigma_3)$ stress difference is twice the magnitude of the shear stress:

$$\sigma_d = 2\sigma_s$$

When there is a change in the principal stress magnitudes without a change in the differential stress, the Mohr circle moves along the σ_n-axis without changing the magnitude of σ_s. In our experiment this would be achieved by increasing the air pressure in the classroom or carrying out the experiment under water;[8] this 'surrounding' pressure is called the *confining pressure* (P_f) of the experiment. In Chapter 6 we return to the Mohr diagram for stress and the role of the confining pressure for fracturing of rocks.

3.8.2   Some Common Stress States

Now that you are somewhat familiar with the Mohr construction, we can look at its representation for the various stress states that were identified earlier (Section 3.6.3). The three-dimensional Mohr diagrams in Figure 3.8 may at first look a lot more complex than those in our earlier example, because they represent three-dimensional stress states rather than two-dimensional conditions. However, they simply are diagrams combining three individual Mohr circles for $(\sigma_1 - \sigma_2)$, $(\sigma_1 - \sigma_3)$, and $(\sigma_2 - \sigma_3)$. Each of these three Mohr circles adheres to the procedures that we used in our earlier clay experiment. Figure 3.8a shows the case for general triaxial stress in which all three principal stresses have non-zero values ($\sigma_1 > \sigma_2 > \sigma_3 \neq 0$). Biaxial (plane) stress, in which one of the principal stresses is zero (e.g., $\sigma_3 = 0$) is shown in Figure 3.8b. Uniaxial compression ($\sigma_2 = \sigma_3 = 0$; $\sigma_1 > 0$) is shown in Figure 3.8c; uniaxial tension ($\sigma_1 = \sigma_2 = 0$; $\sigma_3 < 0$) would place the Mohr circle on the other side of the σ_s-axis. Finally, hydrostatic stress is represented by a single point on the σ_n-axis of the Mohr diagram (positive for compression, negative for tension), because all three principal stresses are equal in magnitude ($\sigma_1 = \sigma_2 = \sigma_3$; Figure 3.8d).

3.9   MEAN STRESS AND DEVIATORIC STRESS

Stresses that act on a body may result in its deformation, which we will discuss in detail in Chapters 4 and 5. We can subdivide the total stress into two convenient components, the mean stress and the deviatoric stress, which are responsible for different types of deformation (Figure 3.9). The *mean stress* is defined as $(\sigma_1 + \sigma_2 + \sigma_3)/3$, for which we use the symbol σ_m, and its difference with the total stress is the *deviatoric stress* (σ_{dev}):

$$\sigma_{total} = \sigma_m + \sigma_{dev}$$

The mean stress is often called the hydrostatic component of stress, or the *hydrostatic pressure,* because a fluid is stressed equally in all directions. The magnitude of the

---

[8]Both conditions can be uncomfortable.

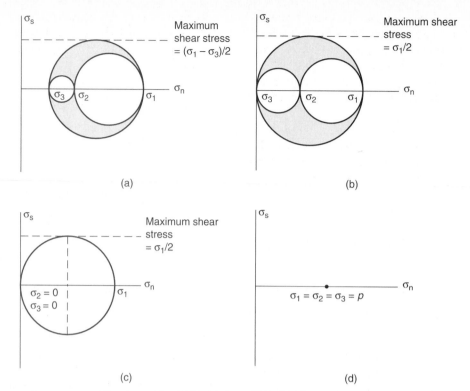

Figure 3.8 Mohr diagrams of some representative stress states: (a) triaxial stress, (b) biaxial (plane) stress, (c) uniaxial compression, and (d) hydrostatic pressure (compression).

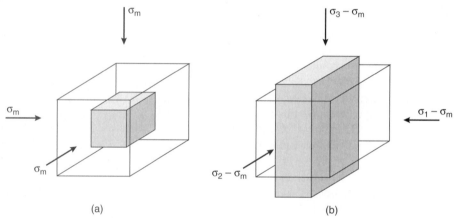

Figure 3.9 The mean (hydrostatic) and deviatoric components of the stress. (a) Mean stress causes volume change and (b) deviatoric stress causes shape change.

hydrostatic stress is equal in all directions and therefore it is an *isotropic stress* component. When we consider rocks at depth in the Earth, we generally refer to the term *lithostatic pressure,*[9] P_l, rather than the hydrostatic component of stress. The lithostatic stress component is best explained by a simple calculation. Imagine a point at a depth of 3 km in the Earth's crust without any other sources of stress present. The lithostatic pressure at this point is simply a function of the weight of the overlying rock column:

$$P_l = \rho \cdot g \cdot h \qquad \text{Eq. 3.15}$$

If ρ (density) equals 2700 kg · m^{-3}, g (gravity) is 9.8 ms^{-2}, and h (depth) is 3000 m, we get:

$$P_l = 2700 \cdot 9.8 \cdot 3000 = 79.4 \cdot 10^6 \text{ Pa}$$
$$\approx 80 \text{ MPa (800 bars).}$$

In other words, for every kilometer in the crust, the lithostatic pressure increases by approximately 27 MPa. With depth, the density of rocks increases. So one cannot use the value of 2700 kgm^{-3} for depths greater than approximately 15 km; the average density of the crust is 2900 kgm^{-3}. Deeper into the Earth the density increases considerably (as high as 13,000 kgm^{-3} in the solid inner core)[10] resulting in very high lithostatic pressures. In fact, you can estimate the density of core and mantle rocks using the values of the Earth's pressure at depth that were given in Section 3.6 and using Equation 3.15.

Because the lithostatic stress is equal in all directions, it follows that $\sigma_1 = \sigma_2 = \sigma_3$. The actual state of stress on a body at depth in the Earth is more complex than just the

---

[9]Also called overburden pressure.

[10]Based on PREM.

weight of the overlying rock column. Anisotropic stresses that arise from tectonic processes, such as the collision of continental plates or the drag of the plate on the underlying material, contribute to the total stress state at depth. The associated differential stresses, however, are typically many orders of magnitude less than the lithostatic stress, except near the surface (see Section 3.12).

So why do we divide the total stress into an isotropic (lithostatic/hydrostatic) and an anisotropic (deviatoric) component? For an explanation we return to the deformation of a stressed body. The isotropic stress component results in a *volume change* of the body (Figure 3.9a). You need only think about the consequences of increasing the water depth on a human body or the takeoff and landing of a plane (remember those painfully popping ears from the changes in pressure?). Isotropic stress is also responsible for maintaining the spherical shape of a balloon filled with air. Deviatoric stress, on the other hand, changes the *shape* of a body (Figure 3.9b). As we will see in Chapter 4, the shape change of a body is commonly measured in structural geology, and is represented by the parameter *strain*. Volume change, on the other hand, is a parameter that is considerably more difficult to detect. As with determining shape change, knowledge about the original volume of a body is a prerequisite to determining any volume change. However, such information is rare in rocks, so we need to resort to indirect approaches such as chemical variations between deformed and undeformed samples (see Chapter 10). Thus, the separation of an isotropic and anisotropic component of the stress provides a connection with the components of deformation, and strain in particular.

3.10 THE STRESS TENSOR

The stress ellipsoid is a convenient way to visualize the state of stress, but it is cumbersome for calculations. For example, it is difficult to determine the stress on a randomly chosen plane using the stress ellipsoid, or when we want the reference system to change (say, a rotation). In contrast, the stress tensor, which describes the stress state in terms of three orthogonal stress vectors, makes such determinations relatively easy. The stress tensor is the mathematical description of the stress state at a point for which we previously used an ellipsoid. So let us take a look at the stress tensor in a little more detail.

A vector is a physical quantity that has magnitude and direction, and it is visualized as an arrow with length and orientation at a point in space. A vector is represented by three coordinates in a Cartesian reference frame that we describe by a matrix consisting of three components. Figure 3.4 showed that stress at a point is a physical quantity that is defined by nine components, which is called a *second-rank tensor*. This is represented by an ellipsoid with orientation, size, and shape at a point in space. The *rank* of a tensor reflects the number of matrix components by raising three to the power of a tensor's rank; for the stress tensor this means $3^2 = 9$ components. It follows that

a vector is a first-rank tensor ($3^1 = 3$ components) and a scalar is a zero-rank tensor ($3^0 = 1$ component).

In matrix notation, the nine components of stress tensor are:

$$\begin{bmatrix} \sigma_{11} & \sigma_{12} & \sigma_{13} \\ \sigma_{21} & \sigma_{22} & \sigma_{23} \\ \sigma_{31} & \sigma_{23} & \sigma_{33} \end{bmatrix}$$

with σ_{11} oriented parallel to the 1-axis acting on a plane perpendicular to the 1-axis, σ_{12} oriented parallel to the 1-axis acting on a plane perpendicular to the 2-axis, and so on. The systematics of these nine components make for an unnecessarily long notation, so in shorthand we may write:

$$[\sigma_{ij}]$$

where i refers to the row (component parallel to the i-axis) and j refers to the column (component acting on the plane perpendicular to the j-axis).

You will notice the similarity of the stress tensor with our earlier approach to the description of stress at a point (Section 3.6.2), consisting of one normal stress (i = j) and two shear stresses (i ≠ j) for each of three orthogonal planes. The stress tensor simply is the mathematical representation of this condition. Now let us use this notation for decomposing the total stress into the *mean stress* and *deviatoric stress*:

$$[\sigma_{ij}] = \begin{bmatrix} \sigma_m & 0 & 0 \\ 0 & \sigma_m & 0 \\ 0 & 0 & \sigma_m \end{bmatrix} + \begin{bmatrix} \sigma_{11}-\sigma_m & \sigma_{12} & \sigma_{13} \\ \sigma_{21} & \sigma_{22}-\sigma_m & \sigma_{23} \\ \sigma_{31} & \sigma_{32} & \sigma_{33}-\sigma_m \end{bmatrix}$$

with $\sigma_m = (\sigma_{11} + \sigma_{22} + \sigma_{33})/3$

Decomposing the stress state in this manner emphasizes the fundamental property that the shear stresses (i ≠ j) are restricted to the deviatoric component of the stress. Because $\sigma_{ij} = \sigma_{ji}$, both the mean stress and the deviatoric stress are *symmetrical tensors*.

Once you have determined the stress tensor, it is relatively easy to change reference system. In this context you are reminded that the values of the nine stress components are a function of the reference frame. Thus, changing the reference frame (say a rotation) changes the components of the stress tensor. These transformations are greatly simplified by using tensors for stress analysis. We would need another few pages explaining vectors and matrix transformations to show some applications, but this takes us beyond the scope of the book. If you would like to see a more in-depth treatment of this topic, several useful references are given in the reading list.

3.11 A BRIEF SUMMARY OF STRESS

Let's summarize where we are in our understanding of stress. You have seen that there are two ways to talk about stress. First, you can refer to stress on a plane (or *traction*), which can be represented by a vector (a quantity with magnitude and direction) that can be subdivided into a component normal to the plane (σ_n, the *normal stress*) and a component parallel to the plane (σ_s, the *shear stress*). If the shear stress is zero, then the stress vector is perpendicular to the plane, but this is a special case—that is, a stress vector does not have to be perpendicular to the plane on which it acts. It is therefore meaningless to talk about stress without specifying the plane on which it is acting. For example, it is wrong to say "the stress at 1 km depth in the Earth is 00°/070°," but it is reasonable to say "the stress vector acting on a vertical, north-south striking joint surface is oriented 00°/070°." Note that in this example there is a shear stress acting on the joint; if the magnitude of this shear stress exceeds the frictional resistance to sliding along the joint, then there might be movement.

The stress state at a point cannot be described by a single vector. Why? Because a point represents the intersection of an infinite number of planes, and without knowing which plane you are talking about, you cannot define the stress vector. If you want to describe the stress state at a point, you must have a tool that will allow you to calculate the stress vector associated with any of the infinite number of planes. We introduced three tools for you: (1) the stress ellipsoid, (2) the three principal stress axes, and (3) the stress tensor. The stress ellipsoid is a surface containing the tails or tips (for compression and tension, respectively) of the infinite number of stress vectors associated with the infinite number of planes passing through the point, with each of the specified vectors and its opposite associated with one plane. On all but three of the planes the vectors have shear components. As a rule, there will be three mutually perpendicular planes on which the shear component is zero; the stress vector acting on each of these planes is perpendicular to the plane. These three planes are called the *principal planes of stress,* and the associated stress vectors are the *principal axes of stress,* or *principal stresses* ($\sigma_1 \geq \sigma_2 \geq \sigma_3$).

Like any ellipsoid, the stress ellipsoid has three axes, and the principal stresses lie parallel to these axes. Given the three principal stresses, you have uniquely defined the stress ellipsoid, and given the stress ellipsoid, you can calculate the stress acting on any random plane that passes through the center of the ellipsoid (which is the point for which we defined the stress state). So, the stress ellipsoid and the principal stresses give a complete description of the stress at a point. Structural geologists find these tools convenient to work with because they are easy to visualize. Thus, we often represent the stress state at a point by picturing the stress ellipsoid, or we talk about the values of the principal stresses at a location. For example, we would say that "the orientation of the maximum principal stress at the New York/Pennsylvania border trends about 070°."

For calculations, these tools are a bit awkward. A more general description of stress at a point is commonly used, a tool called the *stress tensor.* The stress tensor consists of the components of three stress vectors, each associated with a face of an imaginary cube centered in a specified Cartesian frame of reference. Each face of the cube contains two of the Cartesian axes. If it so happens that the stress vectors acting on the faces of the cube have no shear components, then by definition they are the principal stresses, and the axes in your Cartesian reference frame are parallel to the principal stresses. But if you keep the stress state constant and rotate the reference frame, then the three stress vectors will have shear components. The components of the three stress vectors projected onto the axes of your reference frame (giving one normal stress and two shear stresses) are written as components in a 3×3 matrix (a second-rank tensor). If the axes of the reference frame happen to be parallel to the principal stresses, then the diagonal terms of the matrix are the principal stresses and the off-diagonal terms are zero (the shear stresses are zero). If the axes are in any other orientation, then the diagonal terms are not the principal stresses and some or all of the off-diagonal terms are unequal to zero. Note that by using the three principal stresses or the stress ellipsoid, you are specifying a special case of the stress tensor at a point.

3.12 STRESS TRAJECTORIES AND STRESS FIELDS

By connecting the orientation of a particular stress vector (e.g., the maximum principal stress) at several points in a body, you obtain lines that show the variation in orientation of that vector within the body. These lines are called *stress trajectories.* Thus, a change in trend of stress trajectories means a change in orientation of the principal stresses. Collectively, principal stress trajectories represent the orientation of the *stress field* in a body. In some cases the magnitude of a particular stress vector is represented by varying the spacing between the trajectories. An example of the stress field in a block that is pushed on one side is shown in Figure 3.10. If the stress at each point in the field is the same in magnitude and orientation, the stress field is *homogeneous;* otherwise it is *heterogeneous.* The stress field shown in Figure 3.10 is an example of heterogeneous

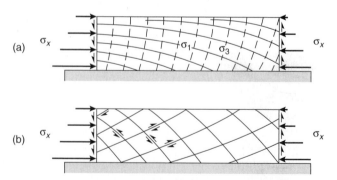

Figure 3.10 Theoretical stress trajectories of σ_1 (full lines) and σ_3 (dashed lines) in a block that is pushed from the left resisted by frictional forces at its base (a). Using the predicted angle between maximum principal stress (σ_1) and fault surface of around 30° (Coulomb failure criterion; Chapter 6), we can draw the orientation of faults, as shown in (b).

Table 3.3 Common Stress Measurement Techniques

Bore-hole breakouts	After drilling, the shape of a bore hole changes in response to stresses in the host rock. Specifically, the hole becomes elliptical, with the long axis of the ellipse parallel to minimum horizontal principal stress.
Hydrofracture	If water is pumped under sufficient pressure into a well that is sealed off, the host rock will fracture. These fractures will be parallel to the maximum principal stress, because the water pressure necessary to open the fractures is equal to the minimum principal stress.
Strain release	A strain gauge, consisting of tiny electrical resistors in a thin plastic sheet, is glued to the bottom of a bore hole. The hole is drilled deeper with a hollow drill bit (called *overcoring*), thereby separating the core to which the strain gauge is connected from the wall of the hole. The inner core expands (elastic relaxation), which is measured by the strain gauge. The direction of maximum elongation is parallel to the direction of maximum compressive stress, and its magnitude is proportional to stress via Hooke's law (see Chapter 5).
Fault-plane solutions	From records of the first motion on seismographs around the world, we can divide the world into two sectors of compression and two sectors of tension. These zones are separated by the orientation of two perpendicular planes. One of these planes is the fault plane on which the earthquake occurred. From the distribution of compressive and tensile sectors, the sense of slip on the fault can also be determined. Seismologists assume that the bisector of the two planes in the tensile sector represents the minimum principal stress, and that in the compressive field the bisector is taken to be parallel to the maximum compressive stress.

stress. Homogeneity and heterogeneity of the stress field should not be confused with isotropic and anisotropic stress. Isotropic means that the principal stresses are equal (describing a sphere); homogeneous stress implies that the orientation and shape of the stress ellipsoids are equal throughout the body. Thus, in a homogeneous stress field, all principal stresses have the same orientation and magnitude. In nature, the orientation of stress trajectories commonly varies, which arises from sources such as fractures in a rock body, contrasts in viscosity, or the complex interplay of more than one stress field (e.g., gravity).

3.13 METHODS OF STRESS MEASUREMENT

To this point, our discussion of stress has been pretty theoretical, except maybe our classroom experiment with clay and kicking a ball around. Before you forget that stress is an actual quantity rather than an abstract concept, we close this chapter with a few notes on stress measurements and one impressive application. The details of present-day stress measurement methods are explained in most engineering texts on rock mechanics. Some are briefly outlined in Table 3.3.

3.13.1 Present-Day Stress

As it turns out, it is quite difficult to obtain a reliable measure of present-day stresses in the Earth. In particular, determining the absolute magnitude of stress poses problems. Generally, stress measurements give the stress differences (the differential stress) and the orientation of the principal stresses. Methods include the analysis of earthquake focal

mechanisms, well-bore enlargements (or 'breakouts') and other *in-situ* stress measurements (Table 3.3), and the analysis of faults and fractures. Earthquake focal mechanisms define a set of two possible fault planes and slip vectors, which are assumed to be parallel to the maximum resolved shear stress on these planes. Several focal mechanisms and slip vectors on faults of different orientation are used to determine the (best-fit) principal stress axes. To analyze the orientation of exposed faults and their observed slip, we use a similar inversion approach. The elliptical distortion of vertical wells that were drilled for petroleum and gas exploration is a direct gauge of the local stress field that is widely available; the long axis of the ellipse is parallel to the horizontal, minimum principal stress. Other *in-situ* stress measurements, such as hydraulic fracturing, in which a hole is capped and pressurized by a fluid until fracturing releases the fluid pressure, all reflect the ambient stress field.

We can get an intuitive sense of differential stress magnitudes that exist in nature from a simple consideration in mountainous regions. We have all looked in awe at vertical walls of rock or steep mountains, especially when they are scaled by climbers. In the western Himalayas, vertical cliffs rise up to 2000 m above the valleys. Thus, using Equation 3.15, the vertical stress at the base of such cliffs is ~50 MPa,[11] and the horizontal stress (i.e., the atmospheric pressure) at that point is about 0.1 MPa. Not surprisingly, the differential stress is enough to flatten your finger were you to place it at that point under the wall. Present-day stress measurements using the methods in Table 3.3 typically give differential stress magnitudes that range from 50 to 150 MPa. But it is important to realize

---

[11] $2700 \times 9.8 \times 2000 = 53$ MPa

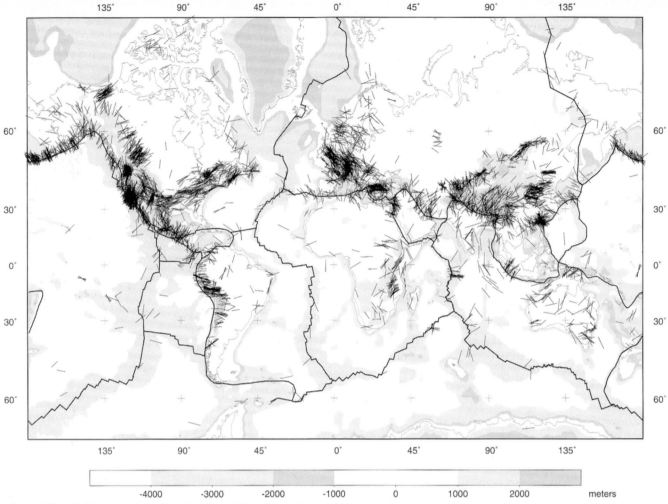

-4000   -3000   -2000   -1000   0   1000   2000   meters

Figure 3.11 World stress map showing orientations of the maximum horizontal stress superimposed on topography. On the next page, the generalized pattern shows stress trajectories for individual plates. An inward pointing arrow set reflects reverse faulting; an outward pointing arrow set reflects normal faulting; double sets indicate strike-slip faulting.

that these methods only sample stress magnitudes in the outermost part (upper crust) of the Earth. The magnitude of differential stresses deeper in the Earth can only be understood once we know something about the Earth's thermal structure and mechanisms of rock deformation (see Section 3.13.3).

3.13.2 Paleostress

If we wish to determine the ancient stress field from rocks, most of the approaches listed above are obviously not suitable. For the analysis of *paleostress* we are essentially limited to the analysis of fault and fracture data, and to microstructural approaches such as grain-size and twinning analysis. Fault-slip analysis requires some understanding of fault mechanics (Chapter 6), but, in short, it uses fault orientation and the sense of slip on that fault under the assumption that the slip direction parallels the maximum shear stress in that plane. Numerical analysis of sufficiently large sets of these fault data will provide a differential stress ratio, $(\sigma_2 - \sigma_3)/(\sigma_1 - \sigma_2)$. Microstructural methods require an understanding of crystal plastic processes,

to which we will return later (Chapter 9). These methods give complementary data on differential stress, because fault data are restricted to their occurrence in the upper crust, whereas microstructures are formed at deeper levels in the Earth (such as olivine grain-size piezometry for upper mantle rocks). Although the methods have to remain largely unexplained at this point, some of the information that is obtained from them is incorporated in the next section on stress in the Earth.

3.13.3 Stress in the Earth

Comparison between large data sets of present-day stress measurements has shown that the results are generally in good agreement as regards the orientation of the principal stresses, and that they compare reasonably well in magnitude. A broadly based application of these techniques and the information that they may provide about regional stress patterns and plate dynamics is shown in Figure 3.11. This global synthesis of stress data, part of the World Stress Map project, reflects an international effort to determine global, present-day stress patterns.

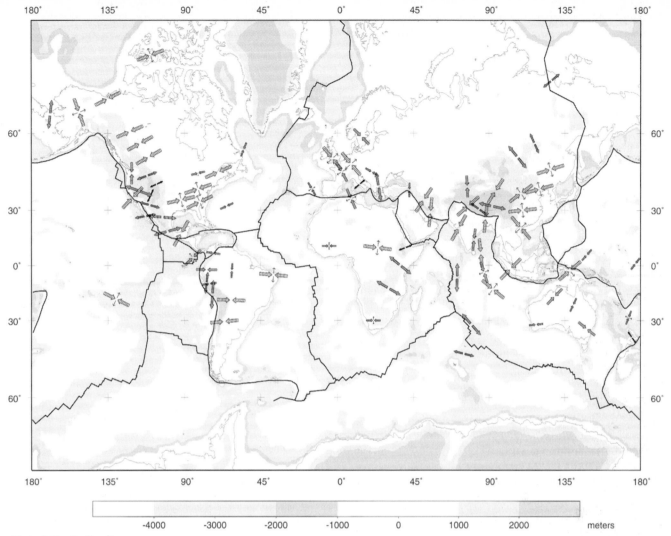

180°　135°　90°　45°　0°　45°　90°　135°　180°

-4000　　-3000　　-2000　　-1000　　0　　1000　　2000　　meters

Figure 3.11　Continued

The global stress summary map (Figure 3.11b) shows regionally systematic stress fields in the upper crust, despite the intrinsic geologic complexity one predicts for the crust. The orientation and magnitudes of horizontal principal stresses are uniform over areas hundreds to thousands of km in extent. These data also show that the upper crust is generally under compression, meaning that maximum compressive stresses are horizontal. For example, the maximum stress in the eastern half of North America is oriented approximately Northeast to Southwest, with differential stresses on the order of a hundred MPa. Areas where horizontal tensile stresses dominate are regions of active extension, such as the East-African Rift zone, the Basin and Range of western North America, and high plateaus in Tibet and western South America. The geology of these areas is discussed in Chapter 19. Using this compilation, we can divide the global stress field into *stress provinces,* which generally correspond to active geologic provinces. From this a pattern emerges that is remarkably consistent with the broad predictions from the main driving forces of plate tectonics (such as the pull of the downgoing slab in subduction zones and push at ocean ridges) and the effects of plate interactions (such as continent-continent collision). Be aware, however, that these patterns reflect only the present-day stress field. Many of the world's geologic provinces present ancient tectonic activity, with configurations and corresponding processes that are no longer active today. Therefore, the present-day global stress pattern is unrelated to this past activity. For example, today's compressive stresses in eastern North America are oriented at a high angle to the maximum paleostress predicted for the formation of the Appalachian Mountains.

What happens with depth? Differential stress cannot increase without bounds, because the rocks that compose the Earth do not have infinite strength. *Strength* is the ability of a material to support differential stress; in other words, it is the maximum stress before rocks fail by fracturing (brittle behavior) or flow (plastic behavior). A combination of present-day stress data and paleostress data, complemented by experimental data on flow of rocks and minerals (Chapter 5), provides us with generalized strength curves for the Earth. *Strength curves* show the differential stress magnitude with depth, given assumptions on the composition and temperature of rocks. At this point we include representative strength curves merely to give you an idea of the magnitudes of stress with depth. As mentioned earlier, much of how we actually obtain this information will have to wait until we examine rheology and the mechanism by which rocks flow.

You will remember from your introductory geology class that the outermost layer of the Earth, comprising the crust and part of the upper mantle, is called the *lithosphere* (see Chapter 14). The lithosphere overlies the *asthenosphere*. The strength curve across the lithosphere in Figure 3.12a, based on a quartzo-feldspathic crust and an olivine-rich upper mantle, assumes a low geothermal gradient (about 10°C/km) and a crustal thickness of about 40 km. You will notice that a sharp decrease in strength occurs around 25 km, which reflects the change from brittle to plastic flow (the *brittle-plastic transition*) in quartzo-feldspathic rocks. The properties of an olivine-rich mantle are quite different from a quartzo-feldspathic crust, and this produces a sharp increase in strength and, therefore, a return to brittle behavior at the crust-mantle boundary (the Moho). As with the crustal profile, strength decreases as plastic behavior replaces the brittle regime of mantle rocks. This behavior of rock is strongly dependent on temperature, as demonstrated by using a higher geothermal gradient (Figure 3.12b), which promotes plastic flow and reduces the strength by an order of magnitude. With a geothermal gradient of 25°C/km, the brittle-plastic transition occurs at a depth of approximately 10 km. In all cases the deeper mantle is weak, because it is characterized by high temperatures and supports differential stresses only on the order of a few MPa. As you examine Figure 3.12, make sure you remember the distinction between differential stress and lithostatic stress. The lithostatic stress (or pressure) always becomes greater with depth in the Earth, and is many orders of magnitude greater than the differential stress. For example, the lithostatic stress at 100 km depth in the Earth is several thousand MPa (use Equation 3.15), but the differential stress is only on the order of 1–10 MPa!

3.14 CLOSING REMARKS

In this chapter you have learned the fundamental aspects of force and stress. In this book we do not teach you how to work with these concepts to any great depth—this is the subject of more advanced texts (see reading list). Rather, we want you to have an intuitive sense of the difference between forces, stress on a plane, and stress at a point, and we want you to know what we are talking about when we refer to principal stresses. This is necessary because throughout the book we will need to refer to the relationship (or lack

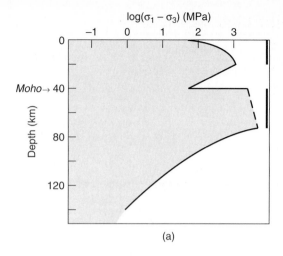

(a)

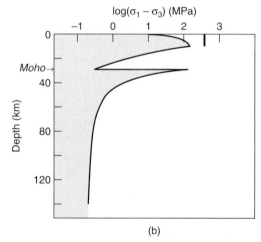

(b)

Figure 3.12 Strength curves showing the variation in differential stress magnitude with depth in the Earth for (a) a region characterized by a low geothermal gradient (e.g., Precambrian shield areas) and (b) a region with a high geothermal gradient (e.g., areas of continental extension). Differential stresses are largely based on experimental data for brittle failure and ductile flow, which change as a function of composition and temperature. In these diagrams the only compositional change occurs at the crust-mantle boundary (the Moho); in the case of additional compositional stratification, more drops and rises will be present in the strength curve. The bar at the right side of each diagram indicates where seismic activity may occur.

thereof) between the geometry and formation of a specific structure, and the orientation of the principal stresses. Let's hope we have achieved this.

Dynamic analysis, the study of stresses in a body, is a topic that goes well beyond geology. Building collapse and mass wasting are some examples that come to mind, whose disastrous effects can be minimized by adequate knowledge of stress. Such applied topics are beyond the scope of this chapter, but you have obtained the basic tools to mount the next step in our journey: the analysis of deformation and strain.

ADDITIONAL READING

Anderson, D. L., 1989, *Theory of the earth,* Oxford: Blackwell Scientific, 366 p.

Angelier, J., 1994, Fault slip analysis and palaeostress reconstruction, *in* Hancock, P. L., ed., *Continental deformation:* Oxford, Pergamon, p. 53–100.

Engelder, T., 1993, *Stress regimes in the lithosphere,* Princeton, Princeton University Press, 457 p.

Jaeger, J. C., and N. G. W. Cook, 1976, *Fundamentals of rock mechanics,* London, Chapman and Hall, 585 p.

Means, W. D., 1976, *Stress and strain—Basic concepts of continuum mechanics for geologists:* New York, Springer-Verlag, 339 p.

Nye, J. F., 1985, *Physical properties of crystals, their representation by tensors and matrices,* (second edition). Oxford, Oxford University Press, 329 p.

Turcotte, D. L., and G. Schubert, 1982, *Geodynamics— Applications of continuum physics to geological problems,* New York, J. Wiley & Sons, 450 p.

Zoback, M. L., 1992, First and second order patterns of stress in the lithosphere: The World Stress Map project: *Journal of Geophysical Research* 97, p. 11703–11728.

Chapter 4
Deformation and Strain

4.1   INTRODUCTION

The slab of rock in Figure 4.1 contains trilobites with shapes differing from the original shape of these fossils; that is, they were distorted. The fold in Figure 4.2 is another manifestation of distortion, in which originally horizontal beds are now folded. Both these fossils and the fold, and we will see many other examples, illustrate the permanent shape changes that occur in many natural rocks. The study of these distortions, which occur in response to forces acting on bodies, is the subject of deformation analysis.

Recall the force of gravity and its many expressions. It is easy pouring syrup on pancakes because of the presence of gravity, but in the space shuttle it is quite difficult to keep the syrup in place. The fact that you are able to read this text sitting down is another convenient effect of gravity; in the space shuttle you would be floating around (probably covered by syrup). As another example, forces exerted by your body enable you to move around, but they do not provide enough acceleration to leave the Earth's surface for any length of time (because gravity pulls you back). Now let us consider a more controlled experiment to analyze the response of materials to an applied force. We can easily change the shape of a cube of clay or plasticine, by the action of, say, your hands. Where forces affect the spatial geometry of a body (syrup, you, plasticine, or rocks), we enter the realm of deformation. Simply stated: deformation of a body occurs in response to forces.[1] From the examples given here, you see that this response may have many faces. In one case the body is merely displaced or rotated (e.g., getting up from the chair and moving around the room). In

---

[1]Later we will see that deformation affects stress, so there is not a simple stress-strain relationship.

Figure 4.1 Deformed trilobites (*Angelina sedgwicki*) in a Cambrian slate from Wales. Knowledge about their original shape enables quantification of the strain.

Figure 4.2 A folded turbidite on New World Island (Newfoundland). An axial plane cleavage (Chapter 11) is present in the mica-rich layers.

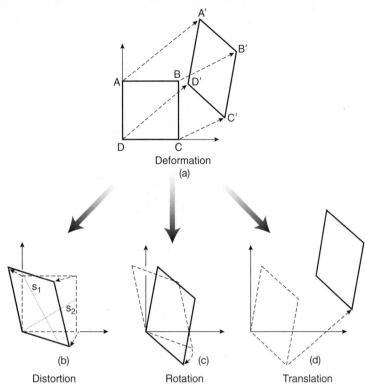

Figure 4.3 The components of deformation. The deformation of a square is subdivided into three independent components: a distortion, a rotation, and a translation. The displacement of each material point in the square, represented by the four corners of the initial square, describes the displacement field. The distortion occurs along the axes s_1 and s_2, which are the principal strain axes.

other cases the body becomes distorted (e.g., plasticine cube and syrup). In this chapter we will examine such responses both qualitatively and quantitatively.

4.2 DEFORMATION AND STRAIN

At the onset we need to distinguish deformation from a closely related term, *strain*. Deformation and strain are often used as synonyms, but they are different concepts. *Deformation* describes the collective displacements of points in a body; in other words, it describes the complete transformation from the initial to the final geometry of a body. This change can include a *translation* (movement from one place to the other), a *rotation* (spin around an axis), and a *distortion* (change in shape). Already in Chapter 1 we mentioned this. *Strain* describes the changes of points in a body relative to each other; so, it describes the distortion of a body. This distinction between deformation and strain may not be immediately obvious from these descriptions. So here's an example. In Figure 4.3 we change the shape and position of a square, say a thin slice of our plasticine cube, within an arbitrarily chosen reference frame (with axes parallel to the margins of the page). The displacement of points within the body, represented by the

four corner points of the square, is indicated by vectors. These vectors describe the displacement field of the body from the initial to the final shape. The displacement field can be subdivided into three components:

1. a distortion (Figure 4.3b)
2. a rotation (Figure 4.3c)
3. a translation (Figure 4.3d)

Each component in turn can be described by a vector field (shown for one corner only), and their sum gives the total *displacement field*. However, a change in the order of addition of the three vector components affects the final result, so deformation, similar to stress, is not a vector quantity. We will return to this later.

When the rotation and distortion components in our example are zero, we have only a translation. This translation is formally called *rigid-body translation* (RBT), because the body undergoes no shape change while it moves. For convenience, we will refer to this component simply as *translation*. When the translation and distortion components are zero, we have only a rotation of the body. Analogous to translation, we call this component *rigid-body rotation* (RBR), or simply *rotation,* and the corresponding deformation is rotational. If you recall the pool table example of the previous chapter (Figure 3.1), the deformation of the pool balls in response to the applied stress is fully described by a translation and a spin component. When translation and spin are both zero, the body undergoes distortion; this component is described by *strain.* So we see that strain is a component of deformation and not a synonym for deformation. In essence we define deformation and strain relative to a *frame of reference.* Deformation describes the complete displacement field of points in a body relative to an *external reference frame,* such as the edges of the paper on which Figure 4.3 is drawn. Strain, on the other hand, describes the displacement field of points relative to one another. So this requires only a reference within the body, that is, an *internal reference frame.* Place yourself in the square and you would be unaware of any translation in the absence of an external frame. Flying in an airplane or riding a train is an example; only by looking out of the window do you become aware of motion.[2]

One final component is missing. In Figure 4.3 we have constrained the shape change of the square by maintaining constant area. You recall that shape change results from the deviatoric component of the total stress, meaning where the principal stresses are unequal in magnitude (see Figure 3.9). The hydrostatic component of the total stress, however, also contributes to deformation by changing the area (or volume in three dimensions) of a body. Area or volume change is called *dilation*[3] and may be positive or negative, depending whether the volume increases or

---

[2]We assume constant velocity; your stomach would notice acceleration.
[3]Or *dilatation.*

decreases. Similar to strain, dilation results in changes of line lengths, but in contrast to strain the relative lengths of lines remain the same. Thus, it is convenient to include volume change as the fourth component of deformation. To summarize, deformation is described by:

1. Rigid-body translation (or translation)
2. Rigid-body rotation (or rotation)
3. Strain
4. Volume change (or dilation)

In practice, it may be difficult to determine the translational, rotational, and dilational components of deformation. Only in cases where we are certain about the original position of a body can rigid-body translation and rigid-body rotation be determined, and only when we know the original volume of a body can dilation be obtained. More commonly we know the original shape of a body, so we can quantify strain.

4.3 HOMOGENEOUS STRAIN AND THE STRAIN ELLIPSOID

Previously we learned that strain describes the distortion of a body in response to an applied stress. Strain is *homogeneous* when any two portions of the body that were similar in form and orientation before the strain are still similar in form and orientation after the strain. This is illustrated with a deck of cards, in which homogeneous strain changes a square into a parallelogram and a circle into an ellipse (Figure 4.4). We define homogeneous strain by its geometric consequences:

1. originally straight lines remain straight
2. originally parallel lines remain parallel
3. circles become ellipses (Figure 4.4b); in three dimensions: spheres become ellipsoids

When any one or more of these three restrictions do not apply, we call the strain *heterogeneous* (Figure 4.4c). Because points (1) and (2) are maintained during the deformation components of translation and rotation, deformation by definition is homogeneous if the strain is homogeneous. Conversely, heterogeneous strain implies heterogeneous deformation, which is geologically more common. Homogeneous versus heterogeneous deformation should not be confused with rotational versus nonrotational deformation, which reflects the presence of a rotational component.

You will guess that heterogeneous strain is more complex to describe than homogeneous strain, so commonly we analyze a heterogeneously strained body or region by separating it into homogeneous portions. In other words, homogeneity of deformation is a matter of scale. For example, a heterogeneously deformation feature such as a fold can be approximated by three essentially homogeneous sections, the two limbs and the hinge (see Chapter 10 for fold terminology). Given the scale dependence of homogeneity and not to complicate matters unnecessarily, we will limit our discussion of deformation in this chapter to homogeneous behavior.

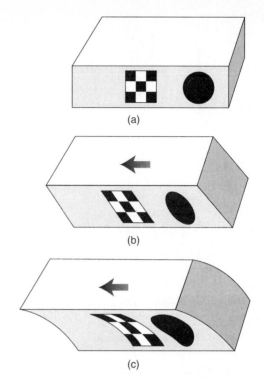

(a)

(b)

(c)

Figure 4.4 Homogeneous and heterogeneous strain. A square and a circle drawn on a stack of cards (a) transform into a parallelogram and an ellipse by sliding each card the same amount, which represents homogeneous strain (b). Heterogeneous strain (c) is produced by variable slip on the cards, for example by increasing the slip on individual cards from bottom to top.

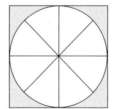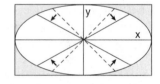

Figure 4.5 The transformation of a square to a rectangle, or a circle to an ellipse describes homogeneous strain. Two material lines remain perpendicular before and after strain; they are the principal axis of the strain ellipse (labeled x and y). The dashed lines are material lines that do not remain perpendicular after strain; they rotate toward the long axis of the ellipse.

In any homogeneously strained, three-dimensional body, there will be at least three lines of particles, or *material lines,* that do not rotate relative to each other. What is a material line? It connects features, such as an array of grains, that are recognizable throughout the body's strain history. The behavior of four material lines is illustrated in Figure 4.5 for the two-dimensional case, in which a circle changes into an ellipse. In homogeneous strain, only two orientations of material lines remain perpendicular before and after strain. These two material lines are the axes of an ellipse that is called the *strain ellipse.* Note that the lengths of these two material lines have changed from the initial to the final stage; otherwise we would not change the shape of our initial sphere. Analogously, in three dimensions we

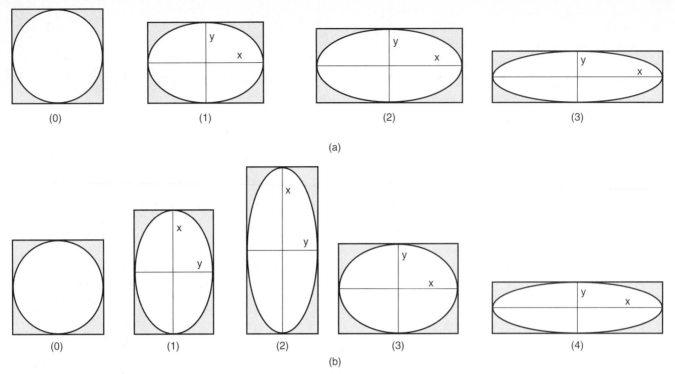

Figure 4.6 The finite strains in (a) and (b) are the same, but the strain paths are different. This illustrates the importance of understanding the incremental strain history of rocks and regions.

have three material lines that remain perpendicular after strain; they define the axes of an ellipsoid, the *strain ellipsoid*. The lines that are perpendicular before and after strain are called the *principal strain axes*. Their lengths after deformation define the strain magnitude. We will use the symbols X, Y, and Z for them, with the convention:

$$X \geq Y \geq Z.$$

As a more intuitive explanation, you may consider the strain ellipsoid as the modified shape of an initial sphere embedded in a body after the application of a homogeneous strain. Much like the stress ellipsoid introduced in the previous chapter, the strain ellipsoid is the visual representation of a second-rank tensor. However, keep in mind that the stress and strain ellipsoids are *not* the same. We will dedicate an entire chapter to the relationship between stress and strain (Chapter 5), where you will see the complexity of their relationship both in orientation and magnitude.

4.4 STRAIN PATH

The measure of strain that compares the initial and final configuration is called the *finite strain*. It is independent of the intermediate steps. Intermediate strain steps may be determined separately and are called *incremental strains*. Thus, the net effect of all incremental strains (i.e., their product) gives the finite strain. We will see later in this chapter that there are many ways to measure finite strain in

a rock, but measurement of the individual increments is more difficult. Yet, incremental strain may be more important for unraveling the deformation history of a rock or region than finite strains. Let us see why with an example, again using the two-dimensional case (Figure 4.6).

Finite strains for the distortion of a square in Figure 4.6a and b are the same, because the initial and final configurations are identical. The steps or strain increments by which these final shapes were reached, however, are very different. We say that the *strain path* of each example is different, but that the finite strains are the same. The path presented in Figure 4.6a has incremental strains that reflect the orientation of the finite strain ellipse throughout its history. The original circle merely becomes increasingly elliptical; in other words, the ratio of the long over the short axis (X/Y) becomes larger. The path in Figure 4.6b, on the other hand, shows that before reaching its final shape, the orientation of the X and Y ellipse axes during part of the history is perpendicular to the orientation of the finite strain ellipse. If we now consider this example in a geological context, the two paths represent very different histories for the deformation of the region, yet their finite strains are identical. Obviously, an important piece of information is lost without knowledge of the path, that is, the strain increments. It is therefore critical for any interpretation to distinguish finite from incremental strain. Because incremental strains are hard to determine, most structural geologists imply finite strain when they loosely discuss the "strain" of a body or a region.

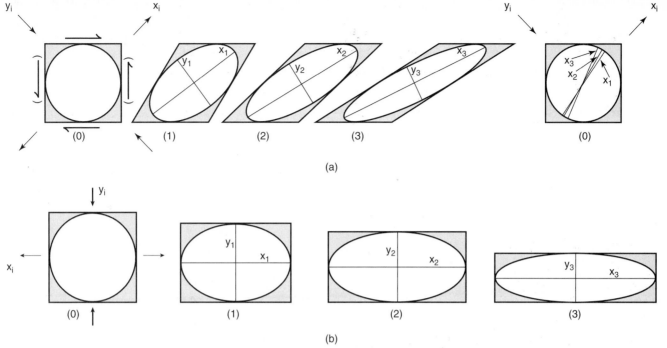

Figure 4.7 (a) Noncoaxial and (b) coaxial strain. The incremental strain axes parallel different material lines in each noncoaxial strain increment. In coaxial strain the incremental strain axes are parallel to the same material lines. Note that the magnitude of the strain axes changes with each step.

4.5 COAXIAL AND NONCOAXIAL STRAIN ACCUMULATION

In Figure 4.5 we saw that strain may involve rotation of material lines. Recall that a material line is simply made up of a series of points in a body: for example, a line of Ca atoms in a calcite crystal or an array of grains in a pure quartzite. There is no property contrast between the material line and the body as a whole; material lines are therefore *passive markers*. All material lines in the body, except those that remain perpendicular before and after a strain increment (the principal strain axes), rotate relative to each other. In the general case, the material lines that are perpendicular before and after each strain increment, however, are not necessarily the same throughout the strain history. In other words, these material lines rotate relative to the finite strain axes; this condition is called *noncoaxial strain accumulation*. The other case, in which the same material lines remain the principal strain axes at each increment, is called *coaxial strain accumulation*. These important concepts are not obvious, so we will use a classroom experiment for clarification.

Take a deck of playing or computer cards and draw a circle on the face perpendicular to the individual cards. Slide these cards past one another and notice how the initial circle changes into an ellipse (Figure 4.7). Draw the ellipse axes (= strain axes X_1 and Y_1) on the cards. Then continue to slide the cards, which will produce a more elliptical shape. Again mark the axes of this second step on the cards, but use another color or use different labels (e.g., X_2 and Y_2). Repeat this a third time so that in the end you have

three steps (increments) and three X-Y pairs. Note that the last ellipse represents the finite strain. Now, as you return the cards to the starting configuration, so restoring the original circle, you will notice that the pairs of strain axes of the three increments do not coincide. For each step a different set of material lines was perpendicular, and thus the incremental strain axes do not coincide with the finite strain axes. You will also notice that with each step the long axis of the ellipse progressively rotates toward the plane over which the cards slide. You can imagine that a very large amount of sliding will orient the long axis of the strain ellipse essentially parallel to (meaning a few degrees off) the shear plane. This experiment represents noncoaxial strain accumulation.

For the case of coaxial strain accumulation, we return to our earlier experiment with plasticine or clay (Figure 4.7b). Place a cube of this material with a circle drawn on its front surface on a table and press down on the top. When you draw the incremental strain axes at various steps, you will notice that they coincide with one another, even though the ellipticity (X/Y ratio) increases. So, with coaxial strain accumulation there is no rotation of the incremental strain axes with respect to the finite strain axes.[4]

The component describing the rotation of material lines with respect to the principal strain axes is called the *internal vorticity,* which is a measure of the degree of

---

[4]In theory we use the *infinitesimally* small incremental strain axes or the *instantaneous* strain axes rather than the incremental strain axes.

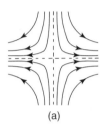
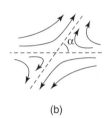
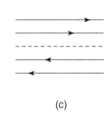

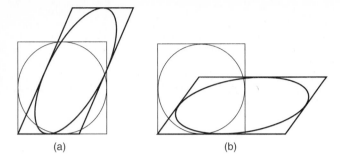

(a)   (b)

Figure 4.8 A combination of simple shear (a special case of noncoaxial strain) and pure shear (coaxial strain) is called general shear or general noncoaxial strain. Two types of general shear are called transtension (a) and transpression (b), are shown.

(a)   (b)   (c)   (d)

Figure 4.9 Particle paths or flow lines during progressive strain accumulation. These flow lines represent pure shear (a), general shear (b), simple shear (c), and rigid-body rotation (d). The cosine of the angle α is the kinematic vorticity number, W_k, for these strain histories: $W_k = 0$, $0 > W_k > 1$, $W_k = 1$, and $W_k = \infty$, respectively.

noncoaxiality. If there is zero internal vorticity, the strain history is coaxial (as in Figure 4.7b), which is otherwise known as *pure shear*. The noncoaxial strain history in Figure 4.7a describes the special case in which the distance perpendicular to the shear plane (or the thickness of our stack of cards) remains constant; this is also known as *simple shear*. In many cases a combination of simple shear and pure shear exists, which we call *general shear* or *general noncoaxial strain accumulation* (Figure 4.8). Internal vorticity is conveniently quantified by the *kinematic vorticity number,* W_k, which relates the angular velocity and stretching rate of material lines. A graphical way to understand this parameter is shown in Figure 4.9. When we track the movement of individual points within deforming bodies

relative to a reference plane, we obtain a displacement field (or *flow lines*) that enables us to quantify the internal vorticity. The angular relationship between the asymptote and the reference line defines W_k;

$$W_k = \cos \alpha$$

Thus, $W_k = 0$ for pure shear (Figure 4.9a), $0 > W_k > 1$ for general shear (Figure 4.9b), and $W_k = 1$ for simple shear (Figure 4.9c). Note that given its definition, rigid-body rotation can also be represented by the kinematic vorticity number ($W_k = \infty$, Figure 4.9d), but remember that this is a rotational component distinct from the internal vorticity. Using Figure 4.7 as an example, the deformation history shown in Figure 4.7a represents noncoaxial strain and zero rigid-body

---

Table 4.1 'Types' of Strain

Coaxial strain	The same material lines are parallel to the finite strain axes during progressive strain.
Heterogeneous strain	Any two portions of a body similar in form and orientation before strain undergo relative change in form and orientation (also: inhomogeneous or nonhomogeneous strain).
Homogeneous strain	Any two portions of a body similar in form and orientation before strain remain similar in form and orientation after strain.
Incremental strain	Strain state of one step during a progressive strain history.
Infinitesimal strain	Incremental strain of vanishingly small magnitude (a mathematical descriptor).
Finite strain	Strain that compares only the initial and final configuration; sometimes called *total strain*.
Noncoaxial strain	Material lines rotate relative to the finite strain axes during progressive strain.

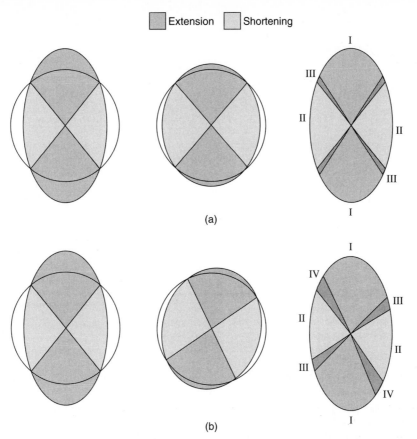

Figure 4.10 Superimposed strain. The strain ellipse contains regions in which lines are extended (screen 1) and regions where lines are shortened (screen 2) that are separated by lines of zero finite elongation. Coaxial superimposition of a strain increment (a) produces regions in which lines continue to be extended (I), regions where lines continue to be shortened (II), separated by regions where initial shortening is followed by extension (III). Noncoaxial superimposition (b) in addition produces a region where initial extension is followed by shortening (IV). Practically this means that a variety of sometimes contradictory structures can be formed, especially in noncoaxial regimes such as shear zones.

rotation. The orientation of the shear surface does not change for each step, but the incremental strain axes rotate. Thus, internal vorticity is an independent rotational component of deformation. The history in Figure 4.7b represents coaxial, nonrotational deformation.

Already, several 'types' of strain have passed before your eyes, so we summarize them in Table 4.1 before continuing with strain superimposition.

4.6 SUPERIMPOSED STRAIN

A strain path describes the superimposition of a series of strain increments (Figure 4.6). For each of these increments the strain ellipsoid is divided into regions containing material lines that extend and shorten, which are separated by planes containing lines with zero length change. Again we consider only the two-dimensional case to illustrate this (Figure 4.10). In a strain ellipse we recognize two regions of extension separating two regions of shortening along two lines of zero length change. When we coaxially superimpose a strain increment of different magnitude on the first ellipse, we get three regions (Figure 4.10a): (I) a region of continued extension, (II) a region of continued shortening,

and (III) a region of initial shortening that is now in extension. The geometry is a little more complex when the strain history is noncoaxial. In this case superimposing the same increment noncoaxially on the first strain state results in four regions (Figure 4.10b): (I) a region of continued extension, (II) a region of continued shortening, (III) a region of initial shortening that is now in extension, and (IV) a region of extension that is now in shortening. So, superimposition of strain increments may produce a complex deformation history. For example, extensional structures formed during one part of the history may become shortened during a later part of the history. As a result, progressive strain histories are able to produce outcrop patterns that at first glance may seem contradictory (e.g., extension vs. shortening).

4.7 STRAIN QUANTITIES

Now that we have discussed the fundamentals of strain, we turn to the measures of strain quantification that are typically used in structural geology. Three basic measures are used: length change or longitudinal strain, volume change or volumetric strain, and angular change or angular strain.

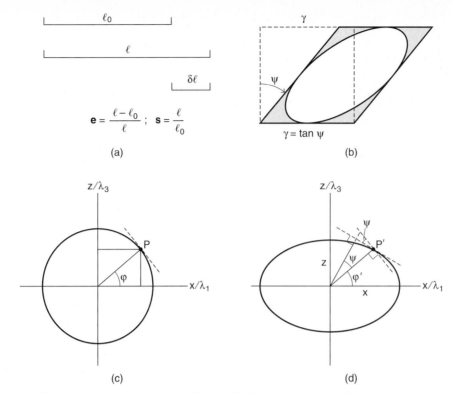

Figure 4.11 Strain quantities. The elongation, **e**, and stretch, **s**, are shown in (a); the angular shear, ψ, and the shear strain, γ, are shown in (b). In (c) the relationship between quadratic elongation, stretch, and angular shear are shown for line OP that transforms into OP′ (d) using the strain ellipse.

4.7.1 Longitudinal Strain

Longitudinal strain is defined as a change in length divided by the original length. Because we divide values of the same dimension, longitudinal strain is a *dimensionless quantity*. Longitudinal strain is expressed by the *elongation,* **e:**

$$\mathbf{e} = (1 - l_o)/l_o \text{ or } \delta l/l_o \qquad \text{Eq. 4.1}$$

with 1 the final length, l_o the original length, and δl the length change (Figure 4.11a). For example, a longitudinal strain in a stretched rod of 0.33 means 33% extension (**e** × 100%). We use the convention that negative values of **e** reflect shortening, while positive values of **e** represent extension; and we label the largest, intermediate, and shortest principal strains $\mathbf{e}_1$, $\mathbf{e}_2$, and $\mathbf{e}_3$, respectively; that is, $\mathbf{e}_1 \geq \mathbf{e}_2 \geq \mathbf{e}_3$.

4.7.2 Volumetric Strain

An equation similar to that for length changes can be given for three-dimensional (volume) change. For *volumetric strain,* Δ, this relationship is:

$$\Delta = V - V_o/V_o \text{ or } \delta V/V_o \qquad \text{Eq. 4.2}$$

with V the final volume, V_o original volume, and δV the volume change. Similar to longitudinal strain, volumetric strain is a ratio of values of the same dimension, so it is also a dimensionless quantity.

4.7.3 Angular Strain

Both longitudinal and volumetric strain are relatively straightforward parameters. Angular strains are slightly more difficult and are measured by the change in angle between two lines that were initially perpendicular. The change in angle is the *angular shear,* ψ, but more commonly its tangent is used, called the *shear strain,* γ (Figure 4.11b):

$$\gamma = \tan\psi \qquad \text{Eq. 4.3}$$

Similar to elongation and volume strain, the angular shear and the shear strain are dimensionless parameters.

4.7.4 Other Strain Quantities

In calculations, such as those associated with the Mohr circle for strain (see below), we make use of a quantity called the *quadratic elongation,* λ, which is defined as:

$$\lambda = (l/l_o)^2 = (1 + \mathbf{e})^2 \qquad \text{Eq. 4.4}$$

with 1 the final length and l_o the original length, and **e** is the elongation.

The root of the quadratic elongation is called the *stretch,* **s:**

$$\mathbf{s} = \lambda^{1/2} = l/l_o = 1 + \mathbf{e} \qquad \text{Eq. 4.5}$$

From these definitions, it is obvious that the quadratic elongation and the stretch are directly related:

$$X^2 = \lambda_1, Y^2 = \lambda_2, Z^2 = \lambda_3; \text{ and} \qquad \text{Eq. 4.6}$$

$$X = s_1, Y = s_2, Z = s_3 \qquad \text{Eq. 4.7}$$

The quadratic elongation and especially the stretch are convenient measures because they describe the lengths of the principal axes of the strain ellipsoid. This relationship between the quadratic elongation, stretches, and the strain ellipse is illustrated in Figure 4.11c. A circle with unit radius (r = 1) becomes distorted into an ellipse that is defined by the length of its axes $\sqrt{\lambda_1}$ (= X) and $\sqrt{\lambda_3}$ (= Z). As a consequence of this distortion, a line OP at an initial angle of φ with the X-axis becomes elongated (OP') with an angle φ' to the λ_1/X-axis.[5] From Figure 4.11c you can determine the relationship between φ and φ', which is described by:

$$\tan\varphi' = Z/X \cdot \tan\varphi = (\lambda_3/\lambda_1)^{\frac{1}{2}} \cdot \tan\varphi \qquad \text{Eq. 4.8}$$

or, rearranging this equation:

$$\tan\varphi = X/Z \cdot \tan\varphi' = (\lambda_1/\lambda_3)^{\frac{1}{2}} \cdot \tan\varphi' \qquad \text{Eq. 4.9}$$

In section 4.2 we introduced the concept of dilation, or volume change, as a special component of deformation in which length changes occur proportionally. Let's see how this affects Equations 4.8 and 4.9. In the two-dimensional case, the area of an ellipse is $\pi(X \cdot Z)$; thus an ellipse derived from a unit circle with area π (r = 1, so $\pi r^2 = \pi$) implies that $X \cdot Z = 1$. So, if we therefore assume zero dilation during the change from the initial circle to the ellipse, we can simplify Equations 4.8 and 4.9 because Z = 1/X. Thus:

$$\tan\varphi' = X^{-2} \cdot \tan\varphi; \text{ or} \qquad \text{Eq. 4.10}$$

$$\tan\varphi = X^2 \cdot \tan\varphi' \qquad \text{Eq. 4.11}$$

To see whether you adequately understand this, show why we may also use this exact same equation in a three-dimensional case, provided Y = 1.[6]

A final measure of strain that is most suitable for the analysis of incremental strain histories is the *natural strain.* This strain measure is no more or less 'natural' than any of the others, but it involves the natural logarithm; hence the name.[7] In contrast to the previous measures, natural strain does not compare the initial and final strain states, but rather it is the summation of individual strain increments. Recall that the elongation is defined as $\delta l/l_o$ (Equation 4.1). This also holds for incremental strains, in which l_o represents the length at the beginning of each increment. For a vanishingly small (or *infinitesimal*) increment, the elongation is defined as:

$$e_i = dl/l_o \qquad \text{Eq. 4.12}$$

The natural strain, ε, is the summation of these increments:

$$\varepsilon = \sum_{l=l_o}^{l=l} dl/l_o = \int_{l_o}^{l} dl/l_o \qquad \text{Eq. 4.13}$$

When we integrate this equation, we get:

$$\varepsilon = \ln l/l_o = \ln s \text{ (or } \varepsilon = \ln (1 + e) = \tfrac{1}{2}\ln \lambda) \qquad \text{Eq. 4.14}$$

4.8 THE MOHR CIRCLE FOR STRAIN

You already saw that the strain state is described by an ellipsoid, and strain is therefore a *second-rank tensor.* So, we can use the same mathematics for strain that we used for stress in Chapter 3, but do remember that the stress and strain ellipsoids are not the same. Another consequence of the tensor property of strain is that a Mohr circle construction can be used to represent the relationship between longitudinal and angular strain in a way similar to that for σ_n and σ_s in the Mohr diagram for stress. Particularly, the quadratic elongation, λ, and the shear strain, γ, are used, but we need to rewrite some relationships and introduce a few substitutions.

If you consider Figure 4.11c again and apply some trigonometry, then:

$$\lambda = \lambda_1 \cos^2\varphi + \lambda_3 \sin^2\varphi \qquad \text{Eq. 4.15}$$

$$\lambda = \tfrac{1}{2}(\lambda_1 + \lambda_3) + \tfrac{1}{2}(\lambda_1 - \lambda_3) \cos2\varphi; \text{ and} \qquad \text{Eq. 4.16}$$

$$\gamma = ((\lambda_1/\lambda_3) - \lambda_3/\lambda_1 - 2)^{\frac{1}{2}} \cos\varphi\sin\varphi \qquad \text{Eq. 4.17}$$

$$\gamma = -\tfrac{1}{2}(\lambda_1 - \lambda_3) \sin2\varphi \qquad \text{Eq. 4.18}$$

But this is the wrong way around! These relationships express the strain in terms of the undeformed state. We actually observe the body after strain occurred, so it is more logical to express strain in terms of the *deformed state.* We therefore need to express the equations in terms of the angle φ', which we can measure, rather than the original angle φ, which is generally unknown. To this end we introduce the parameters $\lambda' = 1/\lambda$ and $\gamma' = \gamma/\lambda$ and use the equations for double angles. We then get:

$$\lambda' = \tfrac{1}{2}(\lambda_1' + \lambda_3') - \tfrac{1}{2}(\lambda_3' - \lambda_1') \cos2\varphi' \qquad \text{Eq. 4.19}$$

$$\gamma' = \tfrac{1}{2}(\lambda_3' - \lambda_1') \sin2\varphi' \qquad \text{Eq. 4.20}$$

If you compare these equations to Equations 3.7 and 3.10 for the normal (σ_n) and shear stress (σ_s) in Chapter 3 and follow their manipulation in section 3.8, you will find that Equations 4.19 and 4.20 also describe a circle, with a radius $\tfrac{1}{2}(\lambda_3' - \lambda_1')$ that is located around a center at $\tfrac{1}{2}(\lambda_1' + \lambda_3')$ in a reference frame with γ' on the vertical and λ' on the horizontal axis. This is the *Mohr circle for strain.*

At first these manipulations may appear unnecessarily confusing, and as such they tend to obscure the application of the construction. So let's look at an example (Figure 4.12). Assume that a unit square is shortened by 50% and extended by 100% (Figure 4.12a). Thus, $e_1 = 1$ and $e_3 = -.5$, respectively; consequently, $\lambda_1 = 4$ and $\lambda_3 = .25$. Note that the area remains constant because

---

[5]It is customary to use the ' (apostrophe) to mark the deformed state.

[6]Consider that zero elongation occurs in the direction perpendicular to our ellipse axes X and Z; the case Y = 1 is, in fact, a common assumption in strain analysis.

[7]The base of the natural logarithm (ln) is e (2.72); the base 10 (log) may also be used.

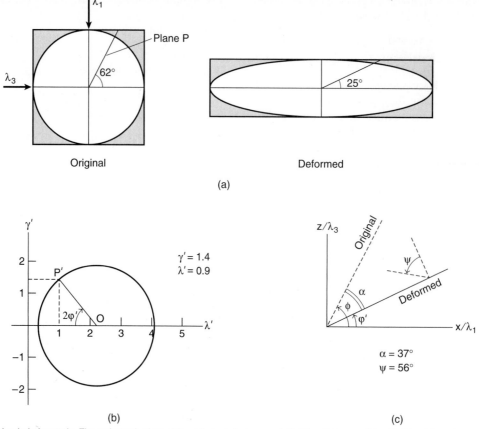

Figure 4.12 The Mohr circle for strain. The reciprocal values of the principal strains are plotted in $\gamma'\lambda'$ space (a), with $\lambda' = 1/\lambda$ and $\gamma' = \gamma/\lambda$. The corresponding rotation of line OP in XZ (or $\lambda_1\lambda_3$) space is shown in (b).

$\lambda_1^{\frac{1}{2}} \cdot \lambda_3^{\frac{1}{2}} = 1$. Using the parameter λ', we get $\lambda_1' = .25$ and $\lambda_3' = 4$. Plotting these values on the Mohr diagram results in a circle with radius $r = \frac{1}{2}(\lambda_3' - \lambda_1') = 1.9$, centered around $\frac{1}{2}(\lambda_1' + \lambda_3') = 2.1$ on the λ' axis. It is now quite simple to obtain a measure of the longitudinal strain and the angular strain for any line oriented at an angle φ' to the strain axes. Say, for a line in the $\lambda_1\lambda_3$ – plane (= XZ plane) of the strain ellipsoid at an angle of 25° to the maximum strain axis, we plot the angle $2\varphi'$ (50°) from the λ_1' end of the circle and draw line OP' (Figure 4.12b). The corresponding strain values are:

$\lambda' = .9$ and $\gamma' = 1.4$; thus

$\lambda = 1.1$ and $\gamma = 1.5$

This means that if line OP represented the long axis of a fossil (say a belemnite), it will have extended and also rotated from this original configuration. Using Equation 4.9, we can also calculate the original angle φ that our belemnite made with our reference frame:

$\varphi = \arctan((\lambda_1/\lambda_3)^{\frac{1}{2}} \cdot \tan \varphi') = 62°$

This latter calculation highlights the easily misunderstood relationship between the angular shear and the rotation angle of a particular element in a deforming body. The rotation of line OP to OP' in the deformed state occurred over an angle of 37° (62° – 25°). However, this angle is unequal to the angular shear, ψ, of that element, which is derived from Equation 4.3, giving $\psi = 56°$. We plotted these various angles in $\lambda_1\lambda_3 (= XZ)$ – space in Figure 4.12c.

You may have noticed that we use coaxial strain in our example of the Mohr circle for strain construction; that is, the incremental strain axes are parallel to the finite strain axes. The construction for noncoaxial strain adds a component of rotation to the deformation (Section 4.4).[8] In Mohr space, this rotational component moves the center of the Mohr circle off the λ' (reciprocal longitudinal strain) axis. In fact, the rotational component can be quantified from the off-axis position of the Mohr circle, but an explanation of this construction in Mohr strain space takes us well beyond the scope of our chapter. We direct you to the reading list for these details.

---

[8]This rotation is not to be confused with the line rotation in our previous example; all lines, except material lines that parallel the principal strain axes, rotate in coaxial strain.

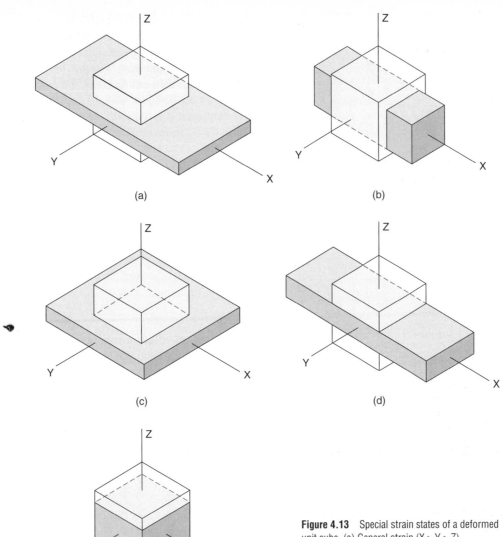

Figure 4.13 Special strain states of a deformed unit cube. (a) General strain (X > Y > Z), (b) axially symmetric extension (X > Y = Z), (c) axially symmetric shortening (X = Y > Z), (d) plane strain (X > 1 > Z), and (e) simple shortening (1 > Z).

4.9 SPECIAL STRAIN STATES

Now that we have defined the various strain parameters as well as their mathematical descriptions, it becomes instructive to examine some strain states that arise from various relationships between the principal strain axes, X, Y, and Z. These strain states are illustrated in Figure 4.13.

- *General strain*[9] is defined by the relationship X > Y > Z (Figure 4.13a). Thus all three axes have different lengths, which is therefore also called *triaxial strain.* This strain state does not imply anything about volume change.
- *Axially symmetric elongation* is a type of axial strain that is defined by X > Y = Z, which is called *axially symmetric extension* (Figure 4.13b), or X = Y > Z, called *axially symmetric shortening.* Axially symmetric extension results in a prolate strain

ellipsoid, with extension occurring only in the X direction accompanied by equal amounts of shortening in the Y and Z directions (Y/Z = 1). This geometry is sometimes referred to as a cigar-shaped ellipsoid. Axially symmetric shortening requires equal amounts of extension (X/Y = 1) in the plane perpendicular to the shortening direction, Z. The strain ellipsoid assumes an oblate or hamburger shape.

- *Plane strain* is a condition in which one of the strain axes (commonly Y) is of the same length before and after strain: X > Y(= 1) > Z (Figure 4.13d). Thus plane strain is a type of triaxial strain, but it can be conveniently described by a two-dimensional strain ellipse with axes X and Z, because no change occurs in the third dimension (Y). In many studies this particular strain state is assumed to hold.

[9]Not to be confused with *general shear* (p. 58).

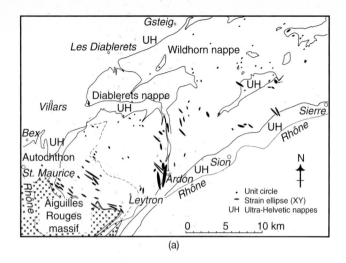

(a)

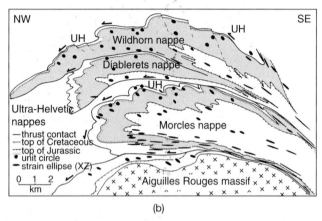

(b)

Figure 4.14 In (a) shapes of the XY sectional ellipse of the finite strain ellipsoids are shown on a map of the Helvetic Alps in Switzerland. In (b) sectional XZ ellipses are shown in a profile constructed from this map (note that the profile plane is not perpendicular to the map surface). The map and profile show a series of NW-directed, low-angle reverse faults (or thrusts) in which the long axis of the strain ellipsoid lies in the direction of thrust transport and the strain ratio generally increases with depth.

Each of these three strain states may represent constant volume conditions, which implies that $X \cdot Y \cdot Z = 1$, meaning that there is no dilation ($\Delta = 0$). This contrasts with simple elongation.

- In *simple elongation,* all points move parallel to a straight line, defined by $X > Y = Z = 1$ or $X = Y(= 1) > Z$. In these two cases, a sphere becomes a prolate ellipsoid in extension and an oblate ellipsoid in shortening (Figure 4.13e), respectively. Because two strain axes remain of equal length before and after deformation, simple elongation must involve a change in volume ($\Delta \neq 0$), that is, a volume decrease in the case of simple shortening and a volume increase in the case of simple extension.

4.10 REPRESENTATION OF STRAIN

When measuring strain in rocks, we generally want to compare these values to those obtained on other parts of the same area or in the same sample, or even compare strain data between different regions. Of course the numerical values for X, Y, and Z and their orientation will allow direct comparison, but a much more informative way, particularly for large data sets, is to use a graphical approach. Remember that strain is typically heterogeneous at the scale of whole structure, such as a fold, and certainly is so at the scale of mountains and orogens. When we determine strain at a locality, we assume that the volume of rock is homogeneous; but in reality the strain that we measure is only representative of the particular element that we use in our analysis. Nonetheless, given sufficient spatial distribution, we can make some general predictions on the state of strain in our structure or region, facilitated by a variety of ways to represent strain quantities.

4.10.1 Orientation

Perhaps the most informative way to illustrate the orientation of the strain ellipsoid is to simply plot the orientation of one or more strain axes at their location in the area under consideration. This 'area' may of course represent many scales, ranging from a thin section to an entire orogenic belt. Moreover, you are not restricted to the orientation of the ellipsoid, because the magnitude of strain can also be included, for example by using the strain ratio X/Z or the absolute strain magnitudes as a scaling tool.

A consequence of this two-dimensional mode of representation, for example strain variation in a rock slab, is that we only show sections through the strain ellipsoid, something we call *sectional strain ellipses.* Sometimes these sectional ellipses do not coincide with a plane containing one or more of the principal strain axes; that is, the axes of the sectional ellipse are not parallel to the principal strain axes. In such cases one is left with two reasonable options: (1) draw the shape of the sectional ellipse on the surface, or (2) show the orientation of the strain axes using plunge and direction of plunge. In the latter case, only two of the three axes are necessary to describe the orientation of the strain ellipsoid, because they are perpendicular to the third axes.

Figure 4.14 shows an example from an area in the Swiss Alps of Europe that illustrates the informative approach of strain data superimposed on a map and on a section. This representation method allows you to examine the regional variation in the data. In this case you see that the degree of strain generally increases with depth in a stack of thrust sheets (e.g., the ratio of X/Z increases with depth; for thrusting see Chapter 15). The orientation of the strain ellipsoid also varies through the stack, but the long axis (X) generally points in the direction of thrust transport and is parallel to the boundaries of regions of high strain ('shear zones'; Chapter 11) that separate the individual thrust sheets.

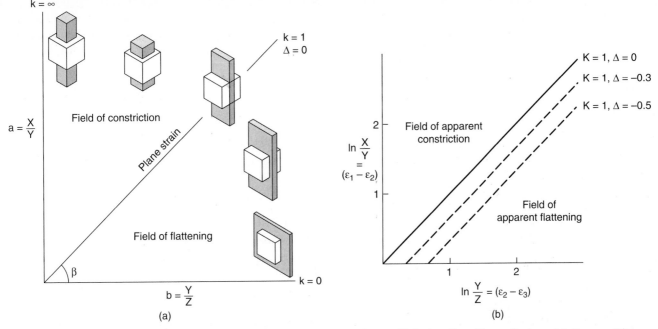

Figure 4.15 The Flinn diagram plots the strain ratios X/Y vs. Y/Z (a). In the Ramsay diagram (b), the logarithm of these ratios is used; in the case of the natural logarithm (ln), it plots $(\varepsilon_1 - \varepsilon_2)$ vs. $(\varepsilon_2 - \varepsilon_3)$. The parameters k and K represent the shape of the strain ellipsoid (see text). Volume change, Δ, produces a parallel shift of the line K = 1.

4.10.2 Shape and Intensity

The most generally used representation of strain data in structural geology is the *Flinn diagram*,[10] which allows the representation of inherently three-dimensional data in a two-dimensional plot by using ratios of the principal strain axes. Later (Section 4.11) you will see that strain analysis typically gives strain ratios rather than absolute magnitudes for the strain axes, so this axial ratio plot is very useful for the representation of strain data. In the Flinn diagram for strain (Figure 4.15a), we plot the ratio of the maximum stretch over the intermediate stretch on the vertical (a) axis, while we plot the ratio of the intermediate stretch over the minimum stretch on the horizontal (b) axis:

$$a = X/Y = (1 + \mathbf{e}_1)/(1 + \mathbf{e}_2) \qquad \text{Eq. 4.21}$$
$$b = Y/Z = (1 + \mathbf{e}_2)/(1 + \mathbf{e}_3) \qquad \text{Eq. 4.22}$$

The shape of the strain ellipsoid is represented by the parameter, k:

$$k = (a - 1)/(b - 1) \qquad \text{Eq. 4.23}$$

This parameter describes the slope of a line that passes through the origin (angle β in Figure 4.15a). A sphere would lie at the origin of this plot (coordinates 1,1), representing a = b = 1. Ellipsoid shapes are increasingly oblate for values of k approaching 0 and increasingly prolate for

values of k approaching ∞. If k = 0, the strain is uniaxially oblate (a = X/Y = 1), and if k = ∞, the strain is uniaxially prolate (b = Y/Z = 1). The value k = 1 is the special case for which a = b, which represents plane strain (Y = 1). We say that the line k = 1 separates the field of constriction ($\infty > k > 1$) from the field of flattening ($1 > k > 0$) in the Flinn diagram.[11]

A useful modification of the Flinn diagram, called the Ramsay diagram,[12] uses the natural logarithm of the ratios a and b (Figure 4.15b):

$$\ln a = \ln (X/Y) = \ln ((1 + \mathbf{e}_1)/(1 + \mathbf{e}_2)) \qquad \text{Eq. 4.24}$$
$$\ln b = \ln (Y/Z) = \ln ((1 + \mathbf{e}_2)/(1 + \mathbf{e}_3)) \qquad \text{Eq. 4.25}$$

using:

$$\ln x/y = \ln x - \ln y, \text{ and} \qquad \text{Eq. 4.26}$$
$$\text{natural strain, } \varepsilon = \ln (1 + \mathbf{e}) \qquad \text{Eq. 4.27}$$

we get:

$$\ln a = \varepsilon_1 - \varepsilon_2 \qquad \text{Eq. 4.28}$$
$$\ln b = \varepsilon_2 - \varepsilon_3 \qquad \text{Eq. 4.29}$$

The parameter k of the Flinn diagram becomes K in the Ramsay diagram:

$$K = \ln a/\ln b = (\varepsilon_1 - \varepsilon_2)/(\varepsilon_2 - \varepsilon_3) \qquad \text{Eq. 4.30}$$

---

[10]Named after the British structural geologist Derek Flinn.

[11]Later we will modify this by adding "apparent" to flattening and constriction, in view of the uncertainty related to dilation.
[12]Named after the British-born structural geologist John Ramsay.

Both logarithmic plots with base e (natural logarithm, ln) and base 10 (log) are used for the Ramsay plot. Similar to the Flinn diagram, the line K = 1 separates the fields of constriction ($\infty > K > 1$) and flattening ($1 > K > 0$) and the unit sphere lies at the origin (ln a = ln b = 0). Note that the origin in the Ramsay diagram has coordinates (0,0). There are a few advantages to the Ramsay diagram. First, small strains that plot near the origin and large strains that plot away from the origin are more evenly distributed in the Ramsay plot. Second, the Ramsay diagram allows a graphical evaluation of the incremental strain history, because equal increments of progressive strain (the strain path) plot along straight lines, while unequal increments follow curved trajectories. In the Flinn diagram, both equal and unequal increments plot along curved trajectories. Third, volume change during deformation is more readily examined using the Ramsay diagram. This last point needs some more explanation.

The foregoing description assumed that volume change (Δ) did not occur during the strain history. To consider the effect of volume change, we have to recall that if $\Delta = 0$ then $X \cdot Y \cdot Z = 1$ for a deformed unit sphere. Thus, rearranging the equation:

$$\Delta = V - V_0/V_0 \qquad \text{Eq. 4.31}$$

and substituting $V = X \cdot Y \cdot Z$ and $V_0 = 1$, we get:

$$\Delta + 1 = X \cdot Y \cdot Z = (1 + e_1) \cdot (1 + e_2) \cdot (1 + e_3) \quad \text{Eq. 4.32}$$

expressed in terms of natural strains:

$$\ln (\Delta + 1) = \varepsilon_1 + \varepsilon_2 + \varepsilon_3 \qquad \text{Eq. 4.33}$$

Further rearrangement of this expression in a form that uses the axes of the Ramsay diagram gives:

$$(\varepsilon_1 - \varepsilon_2) = (\varepsilon_2 - \varepsilon_3) - 3\varepsilon_2 + \ln (\Delta + 1) \qquad \text{Eq. 4.34}$$

Prolate and oblate ellipsoids are separated by those for plane strain conditions ($\varepsilon_2 = 0$) by:

$$(\varepsilon_1 - \varepsilon_2) = (\varepsilon_2 - \varepsilon_3) + \ln (\Delta + 1) \qquad \text{Eq. 4.35}$$

which represents a straight line ($y = mx + n$) with unit slope (m = 1, or slope angle is 45°). If $\Delta > 0$, the line intersects the ($\varepsilon_1 - \varepsilon_2$) axis (volume gain), and if $\Delta < 0$, it intersects the ($\varepsilon_2 - \varepsilon_3$) axis (volume loss). In all cases the slope of the line remains 45° and $\varepsilon_2 = 0$. In Figure 4.15b the case for various percentages of volume loss is illustrated. Consequently, we use the terms *field of apparent flattening* and *field of apparent constriction* for the areas of the diagram separated by the line representing plane strain conditions (K = 1). The term *apparent* is added because volume change has to be known before the actual strain state of any body can be determined. For example, the location of the strain ellipsoid in the 'flattening field' may represent true flattening ($\Delta = 0$) but also plane strain or even true constriction. The latter two simply reflect degrees of volume loss, with the line K = 1 coinciding with the measured

strain ellipsoid and K = 1 located to the right of the point. The Ramsay plot offers a distinction between strain increments with constant versus varying ratios of volume change over strain. In the former case the strain path is straight, while in the latter case the path is curved.

The parameters k and K describe the shape of the ellipsoid; however, the position of a point in the Flinn and Ramsay diagrams not only describes the shape of the strain ellipsoid but also reflects the *intensity* (or degree) of strain. This second aspect may not be immediately appreciated. The farther a point in the Flinn diagram is located from the origin, the greater the strain ellipsoid deviates from a sphere. But the same deviation from a sphere, and so the degree of strain, may occur for different shapes of the ellipsoid, that is, different values of k (or K). Similarly, the same shape of the strain ellipsoid may occur for different degrees of strain. The parameter that describes the intensity of strain is defined as:

$$i = ((X/Y) - 1)^2 + ((Y/Z) - 1)^2)^{1/2} \qquad \text{Eq. 4.36}$$

or in the case of natural strains:

$$I = (\varepsilon_1 - \varepsilon_2)^2 + (\varepsilon_2 - \varepsilon_3)^2 \qquad \text{Eq. 4.37}$$

The values of k (or K) and i (or I) numerically describe the complete strain state of our analysis, and offer an alternative to plotting its position in the graphical Flinn (or Ramsay) diagram.

In summary, the ability of the Flinn diagram to represent strain states and volume change, and the added ability of the Ramsay diagram to evaluate incremental strain histories, allows sufficient flexibility to graphically present most strain data. Listing the corresponding shape (k or K) and intensity (i or I) parameters allows for numerical comparisons between strain analyses in the same structure and/or those made over large regions.[13] Although several other graphical strain representations have been proposed, they generally offer little or no improvement over the Flinn or Ramsay diagrams. We do not include them here, except for a method to combine both orientation and magnitude data.

4.10.3 *Magnitude-Orientation Plot*

Instead of separating orientation and intensity data in two diagrams, it is useful to combine strain data into a single plot, the *magnitude-orientation plot*. The three principal axes of the strain ellipsoid are plotted in lower hemisphere, spherical projection, whereas their relative magnitudes are indicated by contours. The number of contours separating the axes is proportional to their difference in magnitude. Figure 4.16 shows a number of these plots using the same orientation for the three principal strain axes, but with ellipsoids of different shapes (defined by k;

---

[13]There are no simple mathematical relationships between k and K, and i and I.

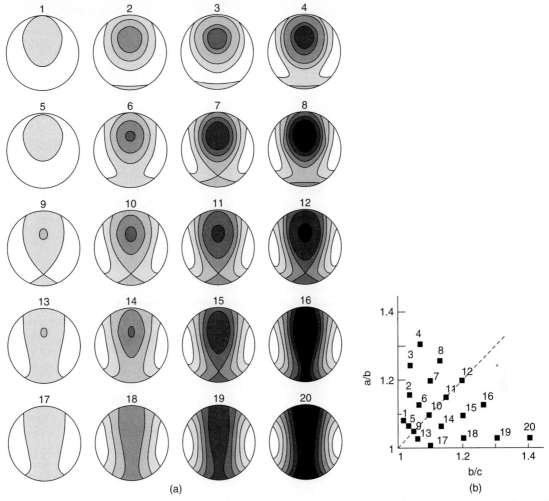

(a)   (b)

Figure 4.16 Magnitude-orientation plots (a), showing oblate, plane strain, and prolate ellipsoids of varying intensity. The maximum, intermediate, and minimum axes in all diagrams are at 60°/000°, 30°/180°, and 00°/090°, respectively. The corresponding position of each ellipsoid in a Flinn diagram is shown in (b).

see Equation 4.23) and different intensities (P = a/b, using a and b from Equations 4.21 and 4.22, respectively). As the shape changes from prolate (ellipsoids 1–8) to oblate (ellipsoids 13–20), the distribution of the maximum changes from a point maximum to a girdle distribution. As the intensity changes, that is, P increases, the number of contours increases (from left column to right column).

4.11 FINITE HOMOGENEOUS STRAIN MEASUREMENT

Finally we have reached the practical aspects of strain analysis. The measurement of strain is a common task undertaken by structural geologists to unravel the geological history of an area or to examine the variation in strain state across an individual structure. Over the years many methods have been proposed, and some have proven to be more useful and rigorous than others. Therefore, you will not find in this section an exhaustive treatment of all strain analysis methods. Rather we limit ourselves to the principles, advantages, and disadvantages of the most widely used methods for the measurement of homogeneous strain.

Advanced structural geology textbooks and several laboratory manuals discuss these and related methods in great detail and often present a step-by-step explanation of the procedures. There is little doubt that the best way to learn these methods (and appreciate their assumptions) is to apply them to natural data sets. You are therefore strongly encouraged to have a look at some of the exercises in this literature.

The rise of personal computers and peripherals (digitizers, scanners, printers, image analyzers) over the last decade adds a new and powerful dimension to strain analysis. Increasingly sophisticated programs are becoming available that will carry out many of the more time-consuming tasks. It is easy nowadays to obtain lots of strain data without knowing much about the methods. Such a "black box" approach is dangerous because you may not appreciate the inherent assumptions and limitations of the method you apply. A basic understanding of the methods is absolutely required to interpret the data. Therefore, in the next few sections we will look at some of the fundamentals of the art of strain analysis.

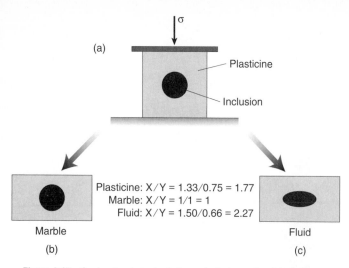

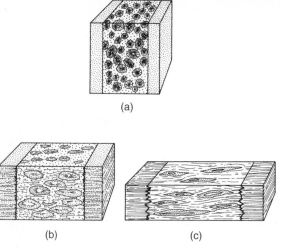

Plasticine: X/Y = 1.33/0.75 = 1.77
Marble: X/Y = 1/1 = 1
Fluid: X/Y = 1.50/0.66 = 2.27

Figure 4.17 Contrasting behavior between inclusion and matrix. In two simple experiments we deform a cube of plasticine with a marble inclusion and a fluid inclusion. If we require that the elongations in the plasticine are the same for both runs, the resulting elongations for the marble and fluid bubble are quite different. This illustrates the different response to stress of materials with mechanical contrast (or heterogeneous systems). A practical example is the determination of strain using (strong) conglomerate clasts in a (weak) clay matrix.

Figure 4.18 Ooids after (b) 25% (X/Z = 1.8) and (c) 50% (X/Z = 4.0) shortening.

4.11.1 What Are We Really Measuring in Strain Analysis?

Strain analysis attempts to quantify the magnitude and/or the orientation of the strain ellipsoid(s) in rocks and regions. In its most complete form, each strain analysis gives the lengths and orientations of the three principal strain axes. Most commonly, however, we obtain strain ratios (X/Y, Y/Z, and X/Z) from strain analysis, because we do not know the absolute dimensions of the original state. Powerful as strain quantification may appear, you should ask yourself two important questions. First, the strain analyzed may represent only a part of the total strain history that the rock or region has undergone. Rephrasing this as a question: *how complete is our strain measurement?* Second, the results of a particular strain analysis pertain only to the objects that were used for the analysis; a difference in behavior between the objects and other parts of the rock, say conglomerate clasts and their matrix, may result in a strain value for the whole rock that differs from that for the conglomerate clasts only. So our second question is: *how representative is the strain analysis* for the rock or region as a whole? We further explore this second consideration by a simple experiment that is shown in Figure 4.17.

We place a marble in a cube of plasticine. As you press down on the top, the plasticine deforms into a rectangular box. But you notice that the marble remains undistorted! So, if you use the marble for strain analysis, you will conclude zero percent strain (or X/Y = 1); yet looking at the plasticine you clearly see that strain has accumulated. In fact, the strain ratio X/Y for the plasticine is 1.8. Now let us carry out a second experiment, in which we replace the space occupied by the marble with a fluid. The result is quite different from that of the previous run. In this case both materials show finite strain, but the strain measured from the bubble, which now has the shape of an ellipsoid, is higher than that measured from the shape of the plasticine cube. These contrasting results do not present a paradox, because all answers are in their own way correct: there is zero strain for the marble, and finite strain for the fluid bubble is greater than that for the plasticine. These different results simply reflect the response to stress of materials with different strength. The plasticine is weaker than the marble, but the fluid bubble is even weaker than the plasticine. We therefore identify strain markers of two types: passive and active markers. *Passive strain markers* are elements in the body that have no mechanical significance; they deform in a manner indistinguishable to that of the whole body. For example, a circle drawn on our plasticine cube with a marker pen constitutes a passive marker. Finding pencil markings is rare in nature, where inclusions with the same properties as the matrix are the closest we will get to passive markers (e.g., quartz grains in a quartzite or oolites in a carbonate). In the case of passive markers, we may say that our body behaves as a homogeneous system for strain. *Active strain markers* have mechanical contrast with their matrix and may behave quite differently. The marble and fluid inclusions in the above experiments are examples of active markers. Conglomerate clasts in a shale matrix or garnets in mica schists are natural examples of active strain markers, representing a heterogeneous system for strain. Don't forget these simple experiments when you get to the interpretation of strain data.

The considerations of the relative role of markers need not pose crippling problems as long as the results for any strain analysis are interpreted keeping its limitations in mind. One useful clarification is to include the type of strain quantification method; for example, R_f/ϕ strain or March strain (see below). Of course, strain analysis may also lead to misinterpretation because of an uncritical confidence in 'hard numbers', about which we warn you prior to diving into the practices of strain determination.

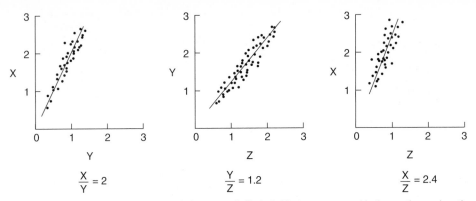

Figure 4.19 Strain from initially spherical objects. The long and short axes of elliptical objects are measured in three orthogonal sections. For convenience we assume that these sectional ellipses are parallel to the principal planes. The slope of regression lines through these points (which should intersect the origin) is the strain ratio in that section. In this example: X/Y = 2, Y/Z = 1.2, and X/Z = 2.4, which gives X/Y/Z = 2.4/1.2/1.

4.11.2 Initially Spherical Objects

Recall that homogeneous strain is defined as the change in shape from a sphere to an ellipsoid. Thus, any geologic feature with known spherical shape would be perfect for strain analysis. Indeed, some of the first analyses were carried out using spherical objects. The classic example involves ooids in (oolitic) limestones (Figure 4.18). Ooids are particles that have grown radially by accretion around a nucleus; most commonly they are calcareous in composition. These spheres become subsequently deformed into ellipsoids (Figure 4.18b and c). Some other objects that are approximately spherical at their initial formation are vesicles and amygdules in basalt flows and reduction spots (areas where chemical change occurred due to the presence of impurities) in shales. But keep in mind the possibility of depositional conditions affecting some of these strain markers; flow of lava may stretch amygdules, and compaction may affect the shape of reduction spots.

Once we are convinced that our markers are initially spherical, the principle of strain analysis of deformed spherical objects is relatively straightforward. If we observe their deformed shapes in three dimensions, we have a direct representation of the shape of the strain ellipsoid. Practically, we measure the shape of the objects in three mutually perpendicular sections. In fact, these sections do not have to be perpendicular, but this certainly simplifies the procedure. In each section we measure the long and the short axis of the sectional ellipse for several objects, which gives us ratios. We use the term *sectional ellipse* to emphasize the fact that you do not measure the true lengths of the axes of the objects, but only their lengths in section. If we combine these ratios from each of the three sections, we fully determine the shape of the strain ellipsoid. We greatly simplify our procedure by choosing the sections such that they coincide with the principal planes of strain (i.e., those planes containing two of the three principal strain axes). In this case, we directly measure the strain ratios (X/Y, Y/Z, and X/Z) in our three sections. Otherwise we need to use trigonometric routines to determine the ellipsoid.

An example is given in Figure 4.19. You measure a ratio of X/Y = 2 in one section; in the second section,

parallel to the YZ plane of the strain ellipsoid, you measure a ratio of Y/Z = 1.2. In three dimensions this gives an X/Y/Z ratio of 2.4/1.2/1. You can check this with the third section, which should give a ratio of X/Z = 2.4. Assuming that the volume remained constant (i.e., $\Delta = 0$), you can fully specify the strain state of your sample:

if, X/Y/Z = 2.4/1.2/1, and $\Delta + 1 = X \cdot Y \cdot Z = 1$ (Eq. 4.32),

then

X = 2Y, Z = Y/1.2 gives

$2Y \cdot Y \cdot (Y/1.2) = 1.7 \; Y^3 = 1$; thus Y = 0.8, and

X/Y/Z = 1.7/0.8/0.7

As an exercise, why don't you try to determine what happens to our strain ratio X/Y/Z if you allow for approximately 50% volume loss (i.e., $\Delta = -0.5$), which is a volume change that has been suggested in many slate studies. You will see that, irrespective of the amount of volume change, the strain ratios remain the same, but that the magnitude of the individual axes becomes less with volume loss (and, vice versa, greater with volume gain).[14] Consequently, the position of the strain state in the Flinn (or Ramsay) diagrams remains the same with or without volume change, and thus values of k (or K) and i (or I) also remain the same. So we get no indication of volume change from our analysis of changes in the shape of markers unless we know their original volume. To solve this problem of volume change, we generally use geochemical approaches, to which we will return in Chapter 10.

4.11.3 Initially Nonspherical Objects

Particles of all shapes that become strained have several properties in common with the finite strain ellipsoid. Material lines (arrays of particles) that are not parallel to any one of the principal strain axes will tend to rotate toward the maximum strain axis (X) and away from the minimum strain axis (Z). Secondly, material lines that are parallel to the Z-axis will undergo the greatest amount of shortening,

---

[14]X/Y/Z = 1.3/0.7/0.5

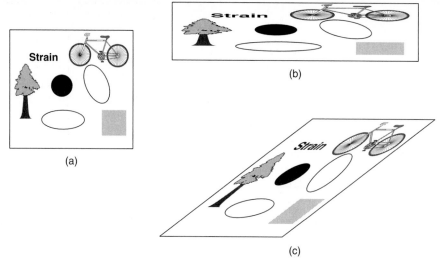

(a)

(b)

(c)

Figure 4.20 Changes in markers of various shapes (a) under homogeneous strain. (b) Coaxial, constant volume strain and (c) noncoaxial, constant volume strain. Note especially the change from the circle to an ellipse and from an initial ellipse to an ellipse of different ratio, and the relative extension and shortening of the markers.

whereas material lines that are parallel to the X-axis will have the greatest amount of extension. There are several other similarities, but for an intuitive understanding of strain analysis from nonspherical objects these two are most important. Figure 4.20 shows what happens to markers of several shapes: They change in somewhat predictable manner, but quantifying the strain from these changes is quite another matter. Quantification is greatly simplified if the initial shape of our marker is an ellipsoid (or ellipse in the two-dimensional case of Figure 4.20), because the superimposition of two ellipsoids by definition describes a third ellipsoid. Is it true that adding two tensors of the same rank gives a third tensor of that rank? Let's first do this for low-rank tensors: adding two zero-rank tensors (scalars) gives a third scalar, for example, 3 + 5 = 8, and adding two first-rank tensors (vectors) gives another vector. So this indeed is a property of tensors, and therefore adding two second-rank tensors will give another second-rank tensor.[15] In other words, superimposing the finite strain ellipsoid on an ellipsoidal body gives an ellipsoid that contains the properties of both the initial ellipsoid (initial marker shape) and the superimposed ellipsoid (finite strain ellipsoid). We already showed this graphically when discussing superimposed strain (Figure 4.10). Many objects in nature are not perfect spheres, but their shapes may be reasonably well approximated by ellipsoids; an example of such ellipsoidal objects are the clasts in a conglomerate (Figure 4.21). If you followed the material so far, you will realize that we cannot simply measure shapes in the deformed state of the ellipsoids to determine the strain without knowing the initial shape of the objects. Two techniques are commonly used to solve this problem: the center-to-center method and the R_f/Φ method.

Figure 4.21 Deformed clasts in a conglomerate (Narragansett, Rhode Island).

4.11.3.1 Center-to-Center Method

The underlying principle of the center-to-center method of strain analysis is that the distances between centers of objects are systematically related to the orientation of the superimposed strain ellipsoid.[16] For example, in the X-Y plane of the strain ellipsoid, grain centers that lie along the direction of the shortening axis (Y) tend to become closer during deformation than grain centers aligned with the extension axis, X. Measuring the distances of centers as a function of an arbitrary reference orientation produces maximum and minimum values that correspond to the orientation and the strain ratio of X and Y (Figure 4.22). Computers have greatly eased the application of this principle by using a digitizer to determine grain centers and a method to graphically analyze the results. Dividing the center-to-center distance between two objects by the sum

---

[15]Mathematically it is the product of two tensors, but to show this requires an understanding of matrix multiplication.

[16]Requiring an anticlustered initial distribution of centers.

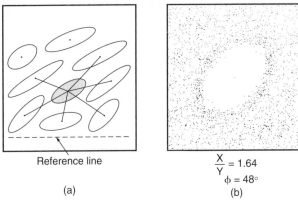

Figure 4.22 Strain from initially spherical or nonspherical objects: Center-to-center method. The distance between object centers is measured as a function of the angle with a reference line (a). A graphical representation of this method using 250 deformed ooids in a natural sample is shown in (b). The empty area defines an ellipse with the ratio X/Y; the angle of the long axis with the reference line is also given.

Reference line

(a)

$\dfrac{X}{Y} = 1.64$

$\phi = 48°$

(b)

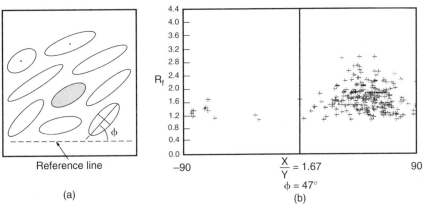

Reference line

(a)

$\dfrac{X}{Y} = 1.67$

$\phi = 47°$

(b)

Figure 4.23 Strain from initially spherical or nonspherical objects: R_f/ϕ method. The ratios of elliptical objects are plotted as a function of their orientation relative to a reference line (a). Application of this method to 250 ooids (same sample as in Figure 4.22) shows the cloud of data points that characterizes this method. From the associated equations we determine the strain ratio and the angle of its long axis with the reference line. The values obtained from using the center-to-center (Figure 4.22) and R_f/ϕ methods are indistinguishable within the resolution of these methods.

of their mean radii is a normalization procedure that better defines the shape of the ellipsoid.[17] Figure 4.22b shows the result of such a *normalized center-to-center analysis* using a computer routine.

4.11.3.2 R_f/ϕ Method

The R_f/ϕ method utilizes the systematic shape changes that occur in deformed ellipsoidal objects. In a given section we measure the long and short axes of the sectional ellipses and the orientation of the long axis with respect to a reference line (Figure 4.23a). For convenience we again assume that the section is parallel to the X-Y plane; however, this is not a prerequisite of the method. Plotting the ratio of these axes (R_f) versus the orientation (ϕ) yields a cloud of data points, reflecting the addition of the initial ellipsoid (the original shape of the body) and the finite strain ellipsoid. The maximum and minimum values of R_f are related to the initial ratio (R_o) of the objects and the strain ratio (R_s; in this example: X/Y), by:

$$R_{f\,max} = R_s \cdot R_o \qquad \text{Eq. 4.37}$$

$$R_{f\,min} = R_o/R_s \qquad \text{Eq. 4.38}$$

which gives the strain ratio, R_s, by dividing Equations 4.36 and 4.37:

$$R_s = (R_{f\,max}/R_{f\,min})^{\frac{1}{2}} \qquad \text{Eq. 4.39}$$

The maximum initial ratio is determined by multiplying Equations 4.37 and 4.38:

$$R_o = (R_{f\,max} \cdot R_{f\,min})^{\frac{1}{2}} \qquad \text{Eq. 4.40}$$

These relationships hold when the original ratio, R_o, is greater than the strain ratio, R_s ($R_o > R_s$). On the other hand, if $R_o < R_s$, then

$$R_{f\,max} = R_s \cdot R_o \text{ (same as above), but} \qquad \text{Eq. 4.41}$$

$$R_{f\,min} = R_s/R_o \qquad \text{Eq. 4.42}$$

and the strain ratio and the maximum original ratio become:

$$R_s = (R_{f\,max} \cdot R_{f\,min})^{\frac{1}{2}} \qquad \text{Eq. 4.43}$$

$$R_o = (R_{f\,max} \cdot R_{f\,min})^{\frac{1}{2}} \qquad \text{Eq. 4.44}$$

The orientation of the X-axis relative to the reference line in our example is determined by the position of $R_{f\,max}$. Of course, the Y-axis lies perpendicular to the X-axis in the section surface. Evaluation of the scatter of points in R_f/ϕ plots is greatly eased by reference curves that are used as overlays on the data. An example of the application of this method is shown in Figure 4.23b.

---

[17]Known as the Fry and Norm(alized)-Fry methods, respectively; after the British geologist Norm Fry.

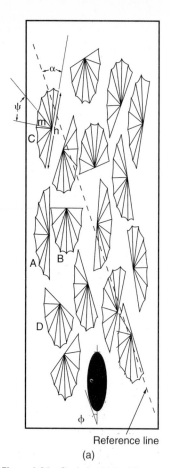

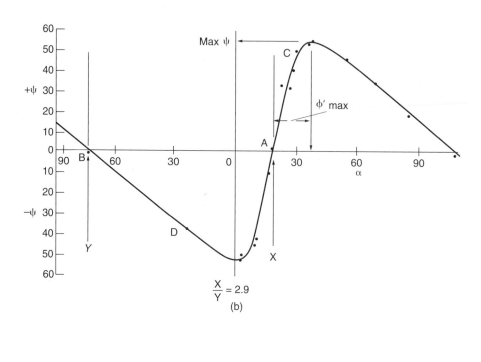

$$\frac{X}{Y} = 2.9$$

(b)

(a)

Figure 4.24   Strain is obtained from a collection of deformed bilaterally symmetric fossils (e.g., brachiopods; (a)) by constructing a graph that correlates the angular shear with the orientation (b). From this (Breddin) graph the orientation of the principal axes and the strain ratio can be determined.

A complete three-dimensional determination of the strain from ellipsoidal objects requires that the center-to-center and/or R_f/ϕ methods are applied to at least two perpendicular surfaces, or three nonperpendicular sections. If nothing else, you will now realize that strain analysis may take a considerable amount of time and effort.

4.11.4   Objects with Known Angular Relationships or Dimensions

The transformation of a sphere to an ellipsoid involves changes in line lengths as well as angular changes (Figure 4.3). Thus, methods that allow us to determine either or both of these parameters, are also suitable for strain analysis. Several approaches can be used, but we limit this section to the principles behind these approaches.

4.11.4.1   Angular Changes

Recall Equation 4.9, which describes the relationship between the original angle φ of a line, the angle after deformation (φ'), and the strain ratio (X/Y):

$$X/Y = \tan\varphi/\tan\varphi' \qquad \text{Eq. 4.9}[18]$$

The original angular relationships are well known in some geologic objects, because they have natural symmetry. One of the most suitable groups of objects is fossils, such as trilobites (Figure 4.1) and brachiopods, because they contain easily recognizable elements that are originally perpendicular. When we have several deformed objects available that contain such initial geometric relationships, we can reliably obtain both the orientation of the strain axes and the strain ratios from their distortion. One approach is schematically shown in Figure 4.24, which requires several deformed brachiopods in different orientations. The method is independent of the size of the individual fossils (big ones mixed with small ones) or even the presence of different species (as long as the symmetry element remains present), because only the angular changes in these objects are used for the analysis. In our plane of view, which often is the bedding plane, we measure the angular shear, ψ, for each deformed object. We plot the value of this angle ψ versus a reference orientation (angle α), making sure that we carefully keep track of positive and negative angles. The resulting relationship is a curve that intersects the α axis at two points (Figure 4.24b).[19] At these

---

[18]We consider only the two-dimensional case here, so X/Z becomes X/Y.

[19]This is known as a Breddin curve, after Hans Breddin.

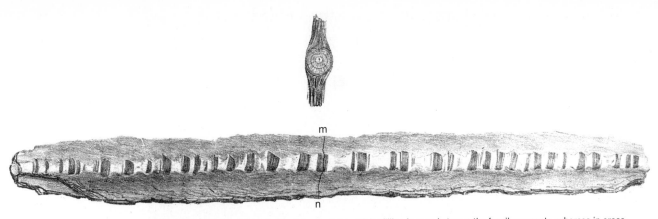

Figure 4.25 Stretched belemnite from the Swiss Alps. In longitudinal view bottom, calcite filling is seen between the fossil segments, whereas in cross-sectional view mn; top of the marker remained circular.

intersections there is zero angular shear, which coincides with the orientation of the principal strain axes in this surface, because the principal strain axes are defined as the material lines that are perpendicular before and after strain (Section 4.3). The strain ratio is determined using Equation 4.9 at conditions for maximum angular shear, which occurs at an angle $\varphi = 45°$. Substituting $\tan\phi = 1$ in Equation 4.9, gives:

$$Y/X = \tan\varphi'_{max} \qquad \text{Eq. 4.45}$$

with ϕ'_{max} the angle between the position of the principal axis and maximum angular shear on the α-axis (Figure 4.24b). In cases where too few distorted fossils are available to derive a reliable curve, the strain ratio can be determined by comparison with a set of curves that are pre-calculated for various strain ratios. Figure 4.24a also illustrates that a population of fossils with different orientations in the plane of view allow direct determination of the orientation of the principal strain axes. By definition, fossils that do not show any angular shear must have symmetry elements that coincide with the orientation of the principal strain axes (fossils A and B in Figure 4.24).

4.11.4.2 Length Changes

Change in line lengths is the final way by which we will determine the strain ellipsoid from shape changes. This is possibly the easiest method to understand, but its practical application is limited to special circumstances; so it is appropriately left to the last. A classic example on which to apply longitudinal strain analysis is shown in Figure 4.25, which illustrates two sections through a rock containing deformed belemnites. If you recall the change from a circle to an ellipse, you will remember that length changes occur in a systematic manner (Figure 4.5). Given several length changes, we can therefore determine a best-fit strain ellipsoid, or ellipse in two dimensions. For this type of analysis it is most convenient to use ratios of the longitudinal strain quantities, stretch ($s = l/l_o$) or its square, quadratic elongation ($\lambda = (l/l_o)^2$), because for orientations parallel to the principal strain axes, these quantities correspond directly to the lengths of the ellipsoid axes. Individual stretches are readily measured if the original length of an object is

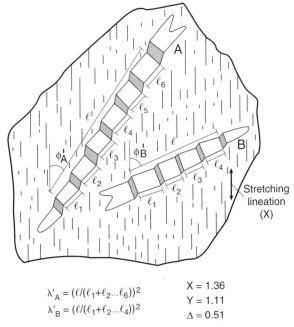

$$\lambda'_A = (\ell/(\ell_1 + \ell_2 ... \ell_6))^2$$
$$\lambda'_B = (\ell/(\ell_1 + \ell_2 ... \ell_4))^2$$

$X = 1.36$
$Y = 1.11$
$\Delta = 0.51$

Figure 4.26 Changes in line length from broken and displaced segments of once continuous fossils (a; belemnites), assuming that the extension direction (and thus φ') is known. Simultaneously solving equation $\lambda' = \lambda_1' \cos^2\varphi' + \lambda_2' \sin^2\varphi'$ (the reciprocal of Equation 4.14) for each element gives the principal strains X and Y. An advantage of this method is that volume change (Δ) can be directly determined from the relationship $X \cdot Y = 1 + \Delta$.

known. To determine the principal strain ratios we need at least two different stretches and some indication of the orientation of one of the principal strain axes within that plane (Figure 4.26), or otherwise at least three longitudinal strain measures. Similar analysis in a second, perpendicular plane will allow us to obtain the three-dimensional strain, but finding this in natural rocks is rare. Strain determinations from length (and angular) changes give two-dimensional strains, but given appropriate assumptions (e.g., Y = 1) the results can nevertheless be quite useful. For example, the belemnites in Figure 4.26 give a X/Y strain ratio of 1.23, but also indicate a volume gain within this plane of $\Delta = 0.51$. In contrast, assuming constant volume strain for the rock as a whole implies that $Z = (X \cdot Y)^{-1} = 0.66$.

Figure 4.27 An X-ray pole-figure device, consisting of an X-ray diffractometer (in this case a single-crystal device) with an attachment to rotate a sample that is located within the holder (middle) through the X-ray beam. The X-ray source is to the right, and detector on the left.

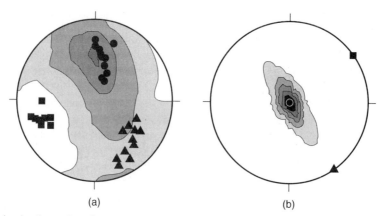

(a)   (b)

Figure 4.28 X-ray pole figures showing the c-axis preferred orientation of (a) chlorite and (b) mica in two slates; the contours are multiples of random distribution. Superimposed on these X-ray fabrics are the magnetic susceptibility fabrics from each slate sample, using a square/triangle/circle for the maximum/intermediate/minimum susceptibility axis; the minimum susceptibility axis coincides with the orientation of the c-axis.

4.11.5 *Rock Textures and Other Strain Gauges*

So far we have mainly considered strain from objects that change shape. But what about obtaining strain from rigid objects, that is, objects that react to strain without changing shape? Imagine that our plasticine cube in Figure 4.17 contains a few dozen tiny steel needles that are randomly oriented. After deformation these needles will have changed orientation, but they will not have changed shape. You already learned that reorientation is a function of the amount of strain, and that all material lines not parallel to the principal strain axes change angle in an orderly fashion. We should be able, therefore, given a few assumptions, to quantify the strain by measuring the preferred orientation of grains. Such measurements can be obtained in a number of ways, including methods that utilize the relationship between grain shape

and grain crystallography. Two powerful methods are X-ray pole-figure analysis (Figure 4.27) and magnetic anisotropy. We really do not want to discuss the technical aspects of these methods here; but in short, the pole-figure device uses the diffraction of X-rays to measure the crystallographic-preferred orientation of grains, and magnetic anisotropy measures magnetic properties of minerals as a function of their shape or crystallography. Examples of the application of these methods using a chlorite-dominated and a mica-dominated slate are shown in Figure 4.28. Typically, the preferred orientation of the minerals is determined relative to a uniformly distributed sample. In one numerical model, the *March model,* no mechanical contrast between markers and matrix is assumed, meaning that the markers respond to strain in the same way as the rock as a whole (such objects

Figure 4.29 Calcite twinning strains in eastern North America. Regional tectonic provinces are labeled, except for the Paleozoic cover sequence inland from the Appalachian-Ouachita thrust front (bold, teethed line). Strains are presented by orientation (short lines) and magnitude of e_1 in percent (negative is shortening); typically, e_1 is horizontal and perpendicular to the Appalachian-Ouachita thrust front. Twinning data from other tectonic provinces show patterns that are unrelated to Paleozoic deformation of the eastern midcontinent.

are called passive markers).[20] From theoretical considerations that we will not give, corresponding March strains are defined by:

$$e_i = \rho_i^{-1/3} - 1 \qquad \text{Eq. 4.46}$$

in which e is a 'March elongation' and ρ is obtained by the normalization of principal pole density to the average pole density for each of three perpendicular directions (i = 1–3), which defines the March strain ellipsoid. Note our use of single quotation marks around the term 'March elongation' to indicate that only under certain conditions does e agree with **e** (the elongation of Equation 4.1); regardless of this correlation, e gives a quantitative measure of the intensity of the grain fabric.

Other strain methods require some knowledge of material that will not be presented until later chapters in this book. So unless you have already studied that material,

they are mentioned here for completion only. The orientation of a foliation may be used as a strain gauge in shear zones, and provides an estimate of the shear strain given some knowledge of the width of the zone, degree of coaxiality, and origin of the foliation (Chapter 12). Fibrous overgrowths on rigid grains or fibrous vein-filling may track part of the *incremental strain* history, and thus they are incremental strain gauges (Chapter 7). This contrasts with all of the earlier methods, which are *finite strain* gauges. Finally, crystal plastic processes (Chapter 9) can give measures of finite strain. In particular, calcite twinning is a sensitive strain gauge, which records strains down to less than one percent. Figure 4.29 shows an example of strain analysis on the scale of half a continent using calcite twinning[21] in Paleozoic carbonates covering cratonic eastern North America. Note how the strain magnitudes decrease from a few percent shortening strain away from the front of the Appalachian-Ouachita

---

[20]Note that there is a strong mechanical contrast between the steel needles and plasticine in our experiment. In many natural rocks this mechanical contrast is much less, and at low strains there is little difference between the March passive-marker model and numerical models involving active markers (e.g., Jeffery model).

[21]Following Rick Groshong's calcite strain-gauge technique.

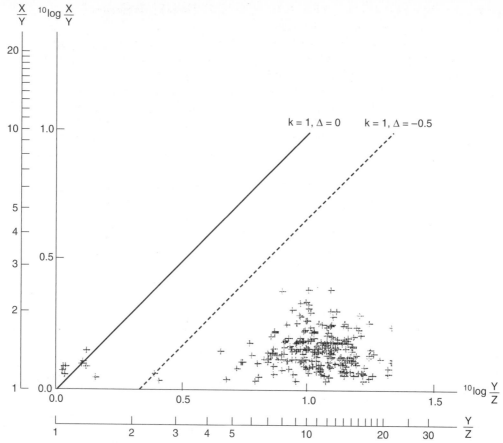

Figure 4.30 Compilation of finite strain values from natural markers in a Ramsay diagram that plots ^{10}log X/Y vs. ^{10}log Y/Z; the corresponding X/Y and Y/Z ratios are also shown. The field inside the dashed line contains approximately 1000 analyses from slates that have been omitted for clarification. Plane strain conditions at constant volume (k = 1, Δ = 0) separate the field of apparent constriction from the field of apparent flattening; the latter contains most of the strain values. Plane strain conditions accompanied by 50% volume loss are indicated by the line k = 1, Δ = −0.5.

Orogen down to less than a percent, which indicates that strain (and stress) from collisional processes at the Paleozoic margin of eastern North America was recorded well over 1000 km inland.

Several other strain gauges have been proposed, and further modifications and extensions of those presented in this chapter exist. Generally, you will have an opportunity to determine finite strain in most every area you work. But what does it all mean? We close this section on strain by trying to answer this question in a general sense.

4.11.6 What Do We Learn from Strain Analysis?

Structural geologists have produced a vast amount of finite and incremental strain data on structures ranging from the microscale (thin sections) to the macroscale (mountain belts and entire continents). The aim of most of these studies is to measure strain magnitude and its variation in a region, outcrop, or hand specimen in order to quantify the structural history. For example, a sharp increase in strain magnitude in parts of an area enable you to recognize regions of high strain, such as shear zones (Chapter 11). Strain analysis also provides constraints on the strain required to form specific geological structures, such as folds (Chapter 9), and foliations and lineations (Chapter 10),

which can then be incorporated into the larger regional history. We could give numerous examples of individual studies, but instead we will draw a few general conclusions from the large strain data base to see what such data may tell us.

Figure 4.30 is a compilation of hundreds of finite strain values that are obtained using a variety of techniques, which are plotted in a Ramsay diagram with the axes ^{10}log X/Y versus ^{10}log Y/Z. This large data base gives us a good idea of the magnitudes of strain that are typical in natural rocks. Axial ratios range from 1 to 20, and stretches are in the range 1 < X < 3, and 0.13 < Z < 1. You will also notice that most of the data points lie in the field of apparent flattening (compare with Figure 4.15), which may be interpreted as true flattening, or strain with a component of volume loss.[22] If we assume that plane strain conditions (k = 1) equally divide the number of measurements, then a volume loss on the order of 50% (Δ = −0.5) is required. You wonder, of course, where that volume (which is actual rock material) went; indeed, the role of

---

[22]Theoretically, it may also be explained by the superimposition of strain histories, such as vertical compaction followed by horizontal shortening.

mass transfer continues to be a topic of lively debate in the literature (see also Chapter 9). The problem is further aggravated when you remember that strain values reflect the strain of the marker that was used for their determination and not necessarily that of the host rock (see Section 4.11.1); in many practical examples this means that the estimates may underestimate the actual strain. Moreover, most of these data represent regional strains and exclude strains measured in shear zones. The axial ratio increases exponentially, with increasing shear strain, and reported values for γ as much as 40 (i.e., ϕ is up to 88°; Equation 4.3) represent extremely large axial strain ratios that lie well outside the plot. A wide range in magnitudes indeed.

A final aspect that is important for deformation studies is the duration interval over which a certain amount of strain accumulates, which is called the *strain rate*. In simple terms, strain rate is elongation divided by time (or $\dot{e} = \mathbf{e}/t$). However, we will wait until the chapter on rheology (Chapter 5) to discuss this important concept and its significance for understanding the deformation of rocks.

4.12 CLOSING REMARKS

The quantification of strain is a popular approach in structural geology to understand the geologic history of rocks and regions. You should be cautious, however, with the interpretation of these data, and for that matter numerical values in general. Numbers tend to provide a false sense of reliability; adequate interpretation requires a solid understanding of the underlying assumptions as well as the inherent errors. Always remember that the numbers generated are no better than the input (assumptions, measurements); stated otherwise: "garbage in, garbage out." With this (crude) qualification we proceed to the final chapter of this block on fundamentals of structural geology and tectonics. The topic, rheology, examines the relationship between stress and strain, and offers an opportunity to integrate all that we have learned so far.

ADDITIONAL READING

Elliott, D., 1972, Deformation paths in structural geology, *Geological Society of America Bulletin*, v. 83, p. 2621–2638.

Erslev, E. A., 1988, Normalized center-to-center strain analysis of packed aggregates, *Journal of Structural Geology*, v. 10, p. 201–209.

Fry, N., 1979, Random point distributions and strain measurement in rocks, *Tectonophysics*, v. 60, p. 89–105.

Groshong, R. H., Jr., 1972, Strain calculated from twining in calcite, *Geological Society of America Bulletin*, v. 83, p. 2025–2038.

Lisle, R. J., 1985, *Geological strain analysis, A manual for the R_f/ϕ method*, Oxford, Pergamon Press.

Lister, G. S., and P. F. Williams, 1983, The partitioning of deformation in flowing rock masses, *Tectonophysics*, v. 92, p. 1–33.

Means, W. D., 1976, *Stress and strain. Basic concepts of continuum mechanics for geologists*, New York, Springer-Verlag.

Means, W. D., 1990, Kinematics, stress, deformation and material behavior, *Journal of Structural Geology*, v. 12, p. 953–971.

Means, W. D., 1992, How to do anything with Mohr circles (except fry an egg): A short course about tensors for structural geologists, *Geological Society of America Short-Course Notes* (2 vols). Boulder, CO.

Oertel, G., 1983, The relationship of strain and preferred orientation of phyllosilicate grains in rocks—A review. *Tectonophysics*, v. 100, p. 413–447.

Pfiffner, O. A., and J. G. Ramsay, 1982, Constraints on geologic strain rates: Arguments from finite strain states of naturally deformed rocks, *Journal of Geophysical Research*, v. 87, p. 311–321.

Ramsay, J. G., and D. S. Wood, 1973, The geometric effects of volume change during deformation processes, *Tectonophysics*, v. 16, p. 263–277.

Ramsay, J. G., and M. I. Huber, 1983, *The techniques of modern structural geology*, v. 1, Strain analysis, London, Academic Press, 307 p.

Simpson, C., 1988, Analysis of two-dimensional finite strain, *in* Marshak, S., and Mitra, G., eds., *Basic methods of structural geology*, 333–359. Englewood Cliffs, NJ, Prentice-Hall.

Chapter 5
Rheology

5.1 INTRODUCTION

Strain is the shape change that a body undergoes in the presence of a stress. But what do we really know about the corresponding stress by measuring strain? In this final chapter of the block on "Fundamentals," we turn to the final and perhaps most challenging aspect: the relationship between stress and strain. While it is evident that there is no strain without stress, their actual relationship is not easy to define on a physical basis. In other words, stating that stress and strain in rocks are related is quite a different matter from physically determining to what extent and their actual relationship(s). In materials science

and geology to describe the ability of materials under stress to deform or to flow, using fundamental parameters such as strain rate (strain per time; Section 5.1.1) and viscosity (Section 5.3.2). These and several other concepts will be discussed in this chapter. We will look especially at their significance for an understanding of rock deformation. Up front we give a few brief, but incomplete descriptions of the most important concepts that will appear throughout this chapter (Table 5.1); they should help you to navigate through some of the initial material, until more complete descriptions can be given.

Rheology[1] is the study of flow of matter. Flow is an everyday phenomenon; in the previous chapter (Section 4.1) we used syrup on pancakes and human motion as examples of deformation. Rocks don't seem to change much by comparison, but remember that geologic processes take place over hundreds of thousands to millions of years. For example, considerable deformation has occurred in rocks along the San Andreas Fault during the last 700,000 years, and lateral displacements on the order of tens of kilometers may have occurred in the Paleozoic Appalachian fold-and-thrusts belt over a time period of a few million years (m.y.). Geologically speaking, time is available in large supply, and given sufficient amounts of it, rocks are able to flow somewhat like syrup. One everyday example of flow in what we consider solid material can be found in an

---

[1]*Rheos* (Greek) means "stream" or "flow."

Table 5.1 Brief Descriptions of Fundamental Concepts and Terms Related to Rheology

Elasticity	Recoverable (nonpermanent), instantaneous strain.
Fracturing	Deformation mechanism by which a rock body or mineral loses coherency (simultaneously breaking of many atomic bonds).
Nonlinear viscosity	Permanent strain accumulation where the stress is exponentially related to the strain rate.
Plasticity	Deformation mechanism that involves breaking of atomic bonds without the material losing coherency.
Strain rate	Rate of strain accumulation (here, elongation over time).
Viscosity	Nonrecoverable (permanent) strain that accumulates with time; the strain rate-stress relationship is linear.

unexpected place. When you look carefully at the windows in an old house, you may find that the glass distorts your view. The reason is that, with time, the glass has sagged under its own weight (driven by gravity), giving rise to a wavy image. If the window glass is very old, you find that the top part of the glass is actually thinner than the bottom part. Eventually, the glass should leave the window frame and form a glass ornament on the window sill. There is no immediate reason for concern, however, because this process will take many thousands of years, as you will learn in the next section. Let us first look at the relationship between time and strain, before we turn to examining the flow behavior of materials.

5.1.1 Strain Rate

Strain rate is the time interval it takes to accumulate a certain amount of strain; it is defined as the elongation per time,[2] and identified by the symbol $\dot{e}$:

$$\dot{e} = e/t = \delta l/(l_o t) \hspace{2cm} \text{Eq. 5.1}$$

You recall that elongation, length change divided by the original length (Equation 4.1), is a dimensionless quantity; thus the dimension of strain rate is $[t]^{-1}$; the unit is second^{-1}. This is perhaps a strange unit at first glance, so let's give an example. If 30% finite longitudinal strain ($|e| = 0.3$) is achieved in an experiment that lasts one hour (3600 seconds), the corresponding strain rate is $0.3/3600 = 8.3 \times 10^{-5}/s$. Now let's see what happens to the strain rate when we change the time interval, but maintain the same amount of finite strain:

Time interval for 30% strain	$\dot{e}$
1 day (86.4 × 10³ s)	3.5 × 10⁻⁶/s
1 year (3.15 × 10⁷ s)	9.5 × 10⁻⁹/s
1 m.y. (3.15 × 10¹³ s)	9.5 × 10⁻¹⁵/s

Thus, the value of the strain rate changes as a function of the time period over which strain accumulates. Note that the percentage of strain did not change over any of the time intervals. So what is the strain rate for a thrust that moves 50 km in 1 m.y.? Actually, it is not possible to answer this question unless the displacement is expressed relative to another dimension of the body, that is, as a strain. We try again: what is the strain rate of a 200 km long thrust sheet moving 50 km in 1 m.y.? Now we calculate a strain rate of $0.8 \times 10^{-14}/s$. In some cases, structural geologists prefer to use the *shear strain rate* rather than strain rate. The relationship between shear strain rate and (longitudinal) strain rate is:

$$\dot{\gamma} = 2\dot{e} \hspace{2cm} \text{Eq. 5.2}$$

A variety of approaches can be used to determine characteristic strain rates for geologic processes. A widely used estimate is based on the Quaternary displacement along the San Andreas Fault of California, which gives a strain rate on the order of $10^{-14}/s$. Other observations (such as isostatic uplift, earthquakes,[3] and orogenic activity) support this estimate, and typical geologic strain rates therefore lie in the range of $10^{-12} - 10^{-15}/s$. Now consider a small tectonic plate with a long dimension of 500 km at a divergent plate boundary. Using a geologic strain rate of $10^{-14}/s$, we obtain the yearly spreading rate by multiplying this dimension of the plate with $3.15 \times 10^{-7}/year$, giving 16 cm/year, which agrees well with present-day observation of plate velocities. And remember the old window glass? If glass flows at a moderate geologic strain rate of $10^{-14}/s$, a 1.5 m high window will make a 10cm high window-sill ornament in a little under 3 m.y. Finally, say your 1 cm long fingernail grows 1 cm per year, meaning a strain (growth) rate of 1/year, or $3 \times 10^{-8}/s$. We can give many more geologic examples, but at this point we want you to

---

[2]We consider only *longitudinal* strain rate here.

[3]Remember that earthquakes typically lasting only a few seconds reflect discrete displacements (on the order of 10–100 cm), but in between (i.e., most of the time) there is no seismic activity and thus little or no slip.

remember the general concept of strain rate and the typical range of 10^{-12} to 10^{-15}/s for geological processes. Note that exceptions to this typical range are such special events as meteorite impacts and explosive volcanism, which are on the order of 10^2/s to 10^{-4}/s.

5.2 GENERAL BEHAVIOR: THE CREEP CURVE

Compression tests on rock samples illustrate that the behavior of rocks to which a load is applied is far from simple. Figure 5.1a shows what is called a *creep curve*, which plots strain as a function of time. In this experiment the differential stress is held constant. Three creep regimes are observed: (1) *primary* or *transient creep*, during which strain decreases with time following very rapid initial accumulation, (2) *secondary* or *steady-state creep*, during which strain accumulation is approximately linear with time, and (3) *tertiary* or *accelerated creep*, during which strain increases with time; eventually, continued loading will lead to failure. Restating these three regimes in terms of strain rate, we have regimes of (1) decreasing strain rate, (2) constant strain rate, and (3) increasing strain rate. The strain rate in each regime is the slope along the creep curve.

Instead of continuing our creep experiment until the material fractures, we decide to remove the stress sometime during the interval of steady-state creep. The corresponding creep curve of this second experiment is shown in Figure 5.1b. A rapid drop in strain occurs when the stress is removed, after which the material *relaxes* a little more with time. Eventually there is no more change with time, but permanent strain remains. To examine this behavior of natural rocks, we will turn to simple analogies and rheological models.

5.3 RHEOLOGIC RELATIONSHIPS

In describing the various rheologic relationships, we first divide the behavior of materials into two types, *elastic behavior* and *viscous behavior*. In some cases, the flow of natural rocks may be approximated by combinations of these *linear rheologies,* in which the ratio of stress over strain or stress over strain rate is a constant. This holds largely true for mantle rocks, but correspondence between stress and strain rate for crustal rocks is best achieved by considering *nonlinear rheologies* (such as *elastic-plastic behavior*), which we discuss after linear rheologies. For each rheologic model, Figure 5.2 shows a physical analog, a creep (= strain-time) curve and a stress-strain or stress-strain rate relationship, which will assist you with the descriptions below. Such equations that describe the linear and nonlinear relationships among stress, strain, and strain rate are called *constitutive equations*.

5.3.1 Elastic Behavior

What is elastic behavior and is it relevant for rocks? Let's first look at the relevance. In the field of seismology, the

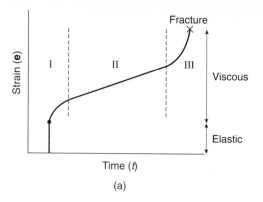

(a)

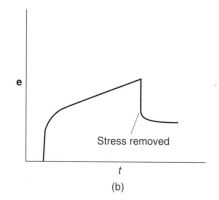

(b)

Figure 5.1 Generalized strain-time or creep curve, which shows primary (I), secondary (II), and tertiary (III) creep. Under continued stress the material will fail (a); if we remove the stress the material relaxes, but permanent strain remains (b).

study of earthquakes, elastic properties are very important. As you know, seismic waves from an earthquake pass through the Earth to seismologic stations around the world. As they travel, these seismic waves briefly deform the rocks, but after they have passed, the rocks return to their undeformed state. To understand how rocks are able to do so, we turn to a common analog: a rubber band. When you pull (apply a stress) a rubber band, it extends; and when you remove the stress, the band returns to its original shape. The greater the stress, the farther you extend the band. Of course, beyond a certain point the rubber band and the experiment painfully fails. The ability of rubber to extend lies entirely in its molecular structure. The bond lengths between atoms and their mutual angles in a crystal structure represent a state of lowest potential energy for a crystal. These bonds are able to elongate and change their relative angles to some extent, without producing permanent changes in the crystal structure. Rubber extends particularly well because it allows large changes in the angular relationships between bonds; however, this causes a considerable increase in the potential energy that is recovered when we let go of the band or when it snaps. So, once the stress is released, the molecular structure returns to its energetically most stable configuration, that is, the lowest potential energy. Similar to a rubber band, the ability of

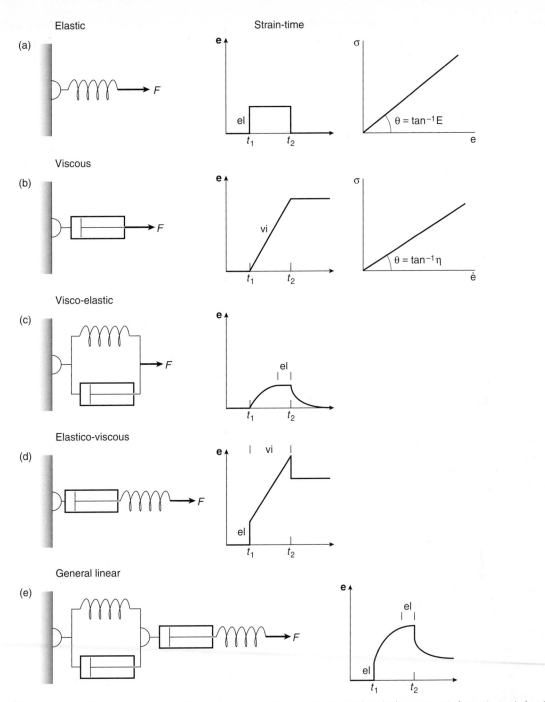

Elastic

Strain-time

(a)

Viscous

(b)

Visco-elastic

(c)

Elastico-viscous

(d)

General linear

(e)

Figure 5.2 Models of linear rheologies. Physical models consisting of strings and dash pots and associated strain-time, stress-strain, or stress-strain rate curves are given for (a) elastic, (b) viscous, (c) visco-elastic, (d) elastico-viscous, and (e) general linear behavior. A useful way to examine these models is to draw your own strain-time curves by considering the behavior of the spring and the dash pot individually, and their interaction.

rocks to deform elastically also resides in nonpermanent distortions of their crystal lattice; but unlike with rubber, the magnitude of this behavior is relatively small in rocks.

Expressing elastic behavior in terms of stress and strain, we get:

$$\sigma = E \cdot e \qquad \text{Eq. 5.3}$$

where E is a constant of proportionality called *Young's modulus* that describes the slope of the line in the σ-**e** diagram (tangent of angle θ; Figure 5.2a). The unit of this

elastic constant is Pascal (Pa = $kgm^{-1}s^{-2}$), which is the same as that of stress (recall that strain is a dimensionless quantity). Typical values of E for crustal rocks are on the order of -10^{11}Pa.[4] Linear Equation 5.3 is also known as *Hooke's law,* describing Hookean behavior.[5] Typically we use a spring as the physical model (Figure 5.2a).

---

[4]We require a negative sign here to produce a negative elongation (shortening) from applying a compressive (positive) stress.

[5]After the English physicist Robert Hooke (1635–1703).

Table 5.2 Some Representative Bulk Moduli (K) and Shear Moduli (or rigidity, G) at Atmospheric Pressure and Room Temperature in 10^5 MPa

Crystal	K	G
Iron (Fe)	1.7	0.8
Copper (Cu)	1.33	0.5
Silicon (Si)	0.98	0.7
Halite (NaCl)	0.14	0.26
Calcite ($CaCO_3$)	0.69	0.37
Quartz (SiO_2)	0.3	0.47
Olivine (Mg_2SiO_4)	1.29	0.81
Ice (H_2O)	0.073	0.025

Source: Poirier, (1985).

Table 5.3 Some Representative Poisson's Ratios (at 200 MPa confining pressure)

Basalt	0.25
Gabbro	0.33
Gneiss	0.27
Granite	0.25
Limestone	0.32
Peridotite	0.27
Quartzite	0.10
Sandstone	0.26
Schist	0.31
Shale	0.26
Slate	0.30

Source: Hatcher, (1995).

We can also write elastic behavior in terms of the shear stress:

$$\sigma_s = G \cdot \gamma \qquad \text{Eq. 5.4}$$

where G is another constant of proportionality, called the *shear modulus* or the *rigidity,* and γ is the shear strain.

The corresponding constant of proportionality in volume change (dilation) is called the *bulk modulus,* K:

$$\sigma = K \cdot ((V - V_o)/V_o) \qquad \text{Eq. 5.5}$$

A more intuitive understanding of the bulk modulus comes from its inverse, 1/K, which is the *compressibility* of a material. Representative values for the bulk and shear moduli are listed in Table 5.2.

It is quite common to use an alternative to the bulk modulus that expresses the relationship between volume change and stress, called *Poisson's ratio, ν.*[6] This elastic constant is defined as the ratio of the elongation perpendicular to the compressive stress and the elongation parallel to the compressive stress:

$$\nu = e_{\text{perpendicular}}/e_{\text{parallel}} \qquad \text{Eq. 5.6}$$

Poisson's ratio of a material describes the ability to shorten parallel to the compression direction without corresponding stretching in a perpendicular direction. Therefore the ratio ranges from 0 to 0.5 for fully compressible to fully incompressible materials, respectively. Incompressible materials maintain constant volume irrespective of the stress. Intuitively you will realize that a sponge will have a very low Poisson's ratio, while a lead cylinder will have a relatively high value. Low Poisson's ratios also imply that a lot of potential energy is stored when the material is compressed; if we remove the stress from the sponge, it will jump right back to its original shape. Values for Poisson's ratio in natural rocks typically lie in the range 0.25–0.35 (Table 5.3).

An important characteristic of elastic behavior is its *reversibility:* once you remove the stress, the material returns to its original state. Reversibility implies that the energy introduced remains available for returning the system to its original state. This energy, which is a form of potential energy, is generally called *internal strain energy.* Because the material is undistorted after the stress is removed, we call the strain *recoverable.* Thus, elastic behavior is characterized by recoverable strain. A second characteristic of elastic behavior is the instantaneous response to stress: finite strain is achieved immediately. Releasing the stress will result in an instantaneous return to a state of no strain (Figure 5.2a). Both these elastic properties, recoverable and instantaneous strain, are visible in our rubber band experiment. A summary of the elastic constants is given in Table 5.4.

Now we return to our original question about the importance of elastic behavior for rocks. With regard to finite strain accumulation, elastic behavior is relatively unimportant in naturally deformed rocks. Typically, elastic strains are less than a few percent of the total strain. So the answer to our question on the importance of elasticity depends on your point of view; a seismologist will say that elastic behavior *is* important for rocks, but a structural geologist will say that it *is not* very important.

5.3.2 Viscous Behavior

The flow of water in a river is an everyday example of *viscous behavior.* With time, the water travels farther downstream. With this second type of rheologic behavior, strain accumulates as a function of time. We describe a relationship between stress and strain rate as:

$$\sigma = \eta \cdot \dot{e} \qquad \text{Eq. 5.7}$$

where η is a constant of proportionality called *viscosity* (tan θ, Figure 5.2b) and $\dot{e}$ is the strain rate. This ideal type of viscous behavior is commonly referred to as *Newtonian*[7] or

---

[6]Named after the French mathematician Simeon-Denis Poisson (1781–1840).

[7]Named after the British physicist Isaac Newton (1642–1727).

Table 5.4 Elastic Constants

Bulk modulus (K)	Ratio of pressure and volume change.
Compressibility (1/K)	The inverse of the bulk modulus.
Elasticity (E)	Young's modulus.
Poisson's ratio (ν)	A measure of compressibility of a material. It is defined as the ratio between **e** normal to compressive stress and **e** parallel to compressive stress.
Rigidity (G)	Shear modulus.
Shear modulus (G)	Ratio of the shear stress and the shear strain.
Young's modulus (E)	Ratio of compressive stress and longitudinal strain.

Table 5.5 Some Representative Viscosities (in Pa · s)

Air	10^{-5}
Water	10^{-3}
Olive oil	10^{-1}
Honey	4
Glycerin	83
Lava	$10-10^4$
Asphalt	10^5
Pitch	10^9
Ice	10^{12}
Rock salt	10^{17}
Sandstone slab	10^{18}
Asthenosphere (upper mantle)	10^{20}
Lower mantle	10^{21}

Sources: Several sources, including Turcotte and Schubert (1982).

linear viscous behavior. Do not confuse the use of 'linear' in linear viscous behavior with that in the linear stress-strain relationship in the section on elasticity. Linear is used here to emphasize a distinction with nonlinear viscous (or non-Newtonian) behavior that we introduce later (Section 5.3.6).

To obtain the dimensional expression for viscosity, you must remember that strain rate has the dimension of $[t^{-1}]$ and stress has the dimension $[ml^{-1}t^{-2}]$. Therefore η has the dimension $[ml^{-1}t^{-1}]$. In other words, the SI unit of viscosity is the unit of stress multiplied by time, the Pa · s $(kgm^{-1}s^{-1})$. In the literature we may still find the unit Poise[8] used, with 1 Poise = 0.1 Pa · s.

The example of flowing water brings out an important characteristic of viscous behavior. Viscous flow is irreversible and produces *permanent* or *nonrecoverable strain*. The physical model for this type of behavior is the dash pot (Figure 5.2b), which really is a leaky piston that moves inside a fluid-filled cylinder.[9] The resistance encountered by the moving piston reflects the viscosity of the fluid. In the classroom you can model viscous behavior by using a tight syringe with one end open to the air. To give you a sense of the large variation, we list the viscosities of some materials in Table 5.5.

Table 5.5 shows that the viscosity of water is on the order of 10^{-3} Pa · s. How does this compare to rocks? Calculations that treat the mantle as a viscous medium produce viscosities on the order of $10^{20}-10^{22}$ Pa · s. Obviously the mantle is much more viscous than water (>20 orders of magnitude!). You can demonstrate this graphically when

calculating the slope of the lines for water and mantle material in the stress-strain rate diagram; they are 0.06° and nearly 90°, respectively. Thus there is an enormous difference between materials that flow in our daily experience, such as water and syrup, and the 'solids' that make up the Earth. Nevertheless, we may approximate the behavior of parts of the Earth as a viscous medium given the large amount of time available to geologic processes (we will return to this with modified viscous behavior in Section 5.3.4). Note that a mantle viscosity of 10^{21} Pa · s and a differential stress of 50 MPa, gives a strain rate of 5×10^{-14}/s.

5.3.3 Visco-Elastic Behavior

Consider the situation in which the deformation process is reversible, but in which strain accumulation as well as strain recovery are delayed; this behavior is called *visco-elastic behavior*.[10] A simple analog is a water-soaked sponge that is loaded on the top. The load on the soaked sponge is distributed between the water (viscous) and the sponge material (elastic). The water will flow out of the sponge in response to the load, and eventually the sponge will support the load elastically. For a physical model we place a spring (elastic behavior) and a dash pot (viscous behavior) in parallel (Figure 5.2c). When a stress is applied, both the spring and the dash pot move simultaneously. However, the dash pot retards the extension of the spring. When the stress is released, the spring will return to its original configuration, but again this movement is delayed by the dash pot.

The constitutive equation for visco-elastic behavior reflects the addition of elastic and viscous components:

$$\sigma = E \cdot \mathbf{e} + \eta \cdot \dot{\mathbf{e}} \qquad \text{Eq. 5.8}$$

---

[8]Named after the French physician Jean-Louis Poiseuille (1799–1869).
[9]Just like us you probably never heard of a dash pot before; so, a leaky old V-8 engine that uses equal amounts of oil and gas is a better known analog.

[10]Also known as firmo-viscous or Kelvinian behavior, after the Irish-born physicist William Kelvin (1824–1907).

5.3.4 Elastico-Viscous Behavior

Elastico-viscous[11] materials behave elastically at the first application of stress, but then continue to behave in a viscous manner. When the stress is removed, the elastic portion of the strain is recovered, but the viscous component remains. This behavior is modeled by placing our spring and dash pot in series (Figure 5.2d). The spring will deform instantaneously as soon as a stress is applied, after which the stress is transmitted to the dash pot. The dash pot will move at a constant rate for as long as the stress remains. When the stress is removed, the spring returns to its original state, but the dash pot remains where it stopped. The constitutive equation for this behavior, which is not derived here, is:

$$\dot{e} = \dot{\sigma}/E + \sigma/\eta \qquad \text{Eq. 5.9}$$

where $\dot{\sigma}$ is the stress per time unit (i.e., stress rate).

When the spring is extended, it stores energy that slowly relaxes as the dash pot moves, until the spring has returned to its original state. The time taken for the stress to reach $1/e$ times its original value is known as the *Maxwell relaxation time,* where e is the base of natural logarithm (e = 2.718). Stress relaxation in this situation decays exponentially. The Maxwell relaxation time is obtained by dividing the viscosity by the shear modulus (or rigidity):

$$\text{Maxwell relaxation time} = \eta/G \qquad \text{Eq. 5.10}$$

Maxwell proposed this model to describe materials that initially show elastic behavior, but given sufficient time display viscous behavior. When you recall that seismic waves are elastic phenomena (short time intervals) and that the mantle is capable of flowing in a viscous manner over geologic time (long time intervals), elastico-viscous behavior fits the properties of the Earth's mantle rather well. Thus, if we take a mantle viscosity of 10^{21} Pa · s and a rigidity of 10^{11} Pa, you get a Maxwell relaxation time for the mantle of 10^{10} seconds, or on the order of 1000 years.

5.3.5 General Linear Behavior

So far we have examined two fundamental and two combined models. With some further fine tuning we can arrive at a physical model that fairly closely approaches reality while maintaining linear rheologies. *General linear behavior* is modeled by placing the elastico-viscous and visco-elastic models in series (Figure 5.2e). Elastic strain accumulates at the first application of stress (the elastic segment of the elastico-viscous model). Subsequent behavior displays the interaction between the elastico-viscous and visco-elastic models. When the stress is removed, the elastic strain is first recovered, followed by the visco-elastic component. However, some amount of strain (permanent strain) will remain even after long time intervals (the viscous component of the elastico-viscous model). The creep

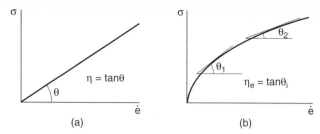

Figure 5.3 (a) Linear and (b) nonlinear viscous rheologies in a stress-strain rate plot. The viscosity is defined by the slope of the linear viscous line in (a) and the effective viscosity by the slope of the tangent to the curve in (b).

(**e**-t) curve for this type of behavior, which is shown in Figure 5.2e, closely mimics the creep curve for natural rocks (Figure 5.1b). So, general linear behavior is a reasonably close approximation of creep in natural materials using linear rheologies. We will not present here the lengthy equation describing general linear behavior, but you realize that it is a combination of visco-elastic and elastico-viscous behavior.

5.3.6 Nonlinear Behavior

The fundamental characteristic that is common to all previous rheological models is a linear relationship between strain rate and stress (Figure 5.3a): $\dot{e} \propto \sigma$. Experiments on geologic materials (say, a silicate) under elevated temperature show, however, that the relationship between strain rate and stress is often *nonlinear* (Figure 5.3b): $\dot{e} \propto \sigma^n$; with n > 1. In other words, the proportionality of strain rate and stress is nonlinear. Rather, strain rate changes as a function of stress, and vice versa. In order to understand the physical basis for nonlinear behavior, we need to understand the processes that occur at the atomic scale during the deformation of materials. This requires an extensive treatment and the introduction of many new concepts that we will not give here. We return to this aspect in detail in Chapter 9, where we examine the role of crystal defects and their mobility under conditions of stress.

A physical model representative for rocks using nonlinear behavior is shown in Figure 5.4, and is known as *elastic-plastic behavior.* In this configuration, a block and a spring are placed in series. The spring extends when a stress is applied, but only elastic (recoverable) strain accumulates until a critical stress is reached (the *yield stress*), above which the block moves and permanent strain occurs. The yield stress has to overcome the resistance of the block to move (friction), but once it moves the stress remains constant while the strain accumulates. In fact, you experience elastic-plastic behavior when towing your car on a nylon rope that allows some stretch. Removing the elastic component (a sliding block on a nonelastic rope) is called *ideal plastic behavior,*[12] but this has less relevance to rocks.

---

[11]Also called Maxwellian behavior, after the Scottish physicist James C. Maxwell (1831–1879).

[12]Or *Saint-Venant behavior,* after the French physicist A. J. C. Barre de Saint-Venant (1797–1886).

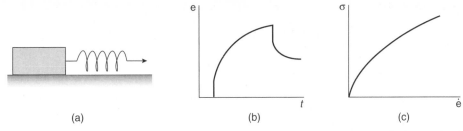

(a) (b) (c)

Figure 5.4 Nonlinear rheologies: elastic-plastic behavior. (a) The physical model consisting of a block and a spring, (b) the associated strain-time curve, (c) stress-strain rate curves.

An important consequence of nonlinear rheologies is that we can no longer talk about (Newtonian) viscosity, because as the slope of the stress-strain rate curve varies, the viscosity also varies. Nevertheless, as it is convenient for modeling purposes to define viscosity, at individual points along the curve, we define the *effective viscosity, η_e*:

$$\eta_e = \sigma/\dot{e} \qquad \text{Eq. 5.11}$$

This relationship is the same as that for viscous or Newtonian behavior (Equation 5.7), but in this case you have to remember that the value for η_e changes as the stress and/or the strain rate changes. From Figure 5.4 you see that the effective viscosity (the tangent of the slope) decreases with increasing stress and strain rate, which means that flow proceeds faster at these conditions. Thus, effective viscosity is not a material property like Newtonian viscosity, but simply a convenient description of behavior under known conditions of stress or strain rate. For this reason, η_e is also called *stress-dependent* or *strain rate-dependent viscosity*.

The constitutive equation describing the relationship between strain rate and stress for nonlinear behavior is:

$$\dot{e} = A \cdot \sigma^n \exp(-E^*/RT) \qquad \text{Eq. 5.12}$$

Equation 5.12 introduces several new parameters. A is an empirically derived constant typically in the range 100–500 kJ · mol^{-1}, and n is the stress exponent that lies in the range $1 > n > 5$ for most natural rocks. The crystal processes that enable creep are temperature dependent and need a minimum energy before they are activated (see Chapter 9); these parameters are included in the exponential part of the function[13] as the activation energy (E*) and the temperature (T in degrees K); R is the gas constant. Table 5.6 lists experimentally derived values for A, n, and E* for some common rock types.

From Table 5.6 we can draw some first-order conclusions about the relative strength of the rock types, that is, their *flow stresses*. If we assume that the strain rate remains constant at 10^{-14}/s, we can solve the constitutive equation for various temperatures. Let us first rewrite Equation 5.11 as a function of stress:

$$\sigma = (\dot{e} \cdot A)^{1/n} \exp(E^*/RT) \qquad \text{Eq. 5.13}$$

Table 5.6 Experimentally Derived Creep Parameters for Some Common Rock Types

Rock type	^{10}log A (MPa^{-n}s^{-1})	n	E* (kJ · mol^{-1})
Albite rock	18	3.9	234
Anorthosite	16	3.2	238
Clinopyroxenite	17	2.6	335
Clinopyroxenite (wet)	5.17	3.3	490
Diabase	17	3.4	260
Granite	6.4	3.4	139
Granite (wet)	7.7	1.9	137
Marble	33.2	4.2	427
Olivine rock	4.5	3.6	535
Olivine rock (wet)	4.0	3.4	444
Quartz diorite	11.5	2.4	219
Quartzite	10.4	2.8	184
Quartzite (wet)	10.8	2.6	134
Rock salt	−1.59	5.0	82

Source: Kirby and Kronenberg (1987).

If we substitute the values for A, n, and E* at constant T, we find that the differential stress value for rock salt is much less than that for any of the other rock types. This fits the observation that rock salt flows readily, as we learn from, for example, the surface occurrence of salt diapirs (Chapter 2). Limestones and marbles are also relatively weak and therefore deformation is commonly localized in these horizons. Quartz-bearing rocks, such as quartzites and granites, in turn are weaker than plagioclase-bearing rocks. Olivine-bearing rocks are among the strongest of rock types, meaning that they require large differential stresses to flow. But if this is true, how can mantle flow at differential stresses less than 10 MPa (Section 3.13.3)? The answer lies in the fact that Equation 5.13 can also be solved for various temperatures. In Figure 5.5 this is done using a "cold" geothermal gradient of 10°K/km for some of the rock types in Table 5.6.[14] This graph supports our first-order conclusion on the relative strength of various rock

---

[13]exp(a) means e^a, with e = 2.72.

[14]A "hot" geotherm is 30°K/km.

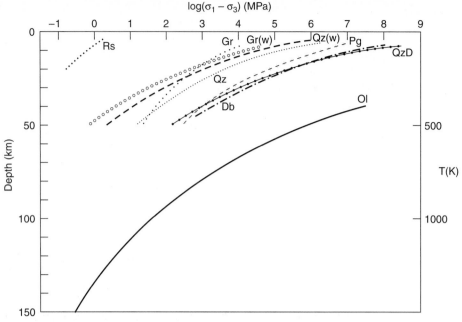

Figure 5.5 Creep strength with depth for several rock types using a geothermal gradient of 10 K/km. Rs = rock salt, Gr = granite, Gr(w) = wet granite, Qz = quartzite, Qz (w) = wet quartzite, Pg = plagioclase-rich rock, QzD = quartz diorite, O = olivine-rich rock.

types, but it also shows that with depth the strength of *all* rock types decreases significantly. The latter is an important observation for understanding the nature of deformation processes in the deeper levels of the Earth.

5.4 ADVENTURES WITH NATURAL ROCKS

Our discussion of rheology so far has been mostly abstract. We examined relationships between stress, strain, and strain rate by using simple analogs such as rubber bands and fluids, or physical models such as a spring and a dash pot. In the end we are really only interested in the behavior of natural rocks. The results from experiments on natural materials will help us to get a better appreciation of the flow of rock. The reason for doing experiments on natural rocks is twofold: (1) we observe the actual behavior of natural rocks rather than syrup or elastic bands, and (2) we can vary several parameters in our experiments, such as pressure, temperature, time, and fluid pressure, to examine their role in rock deformation. An incredible amount of experimental data is available to us and many of the principles have therefore been known for several decades. Here we will limit our discussion by looking only at experiments that each highlight one particular parameter. Combining the various responses, we can begin to understand the rheology of natural rocks.

An alternative approach to examining flow of rocks is to study the behavior in scaled experiments. Scaling brings fundamental quantities such as length [l], mass [m], and time [t] to the human scale. For example, we can use clay as a model material to study faulting, or stitching wax to examine time-dependent behavior. Each analog that is used in scaled experiments has advantages and disadvantages, and the experimentalist has to make trade-offs between geologic relevance and experimental conditions. We will not use scaled experiments in the subsequent sections of this chapter, so for a treatment on scaling you are referred to other sources.

5.4.1 The Deformation Apparatus

A deformation experiment on a rock or a mineral can be carried out quite easily by placing a small piece in a vise, but when you try this experiment you have to be careful with randomly flying chips as the material fails. If you ever cracked a hard nut in a nutcracker, you know what we mean. In rock deformation experiments we attempt to control the experiment a little better for the sake of the experimentalist, as well as for analysis and interpretation of the results. A typical deformation apparatus is schematically shown in Figure 5.6. In this rig a cylindrical rock specimen is placed in a pressure chamber, which is surrounded by a pressurized fluid that provides the *confining pressure*, P_c, to the specimen through an impermeable jacket. This experimental setup is known as a *triaxial testing apparatus*, named for the triaxial state of the applied stress in which all three principal stresses are unequal to zero. For practical reasons, two of the principal stresses are equal. In addition to the fluid that provides the confining pressure, a second fluid may be present in the specimen to provide *pore-fluid pressure*, P_f. The difference $(P_c - P_f)$ is called the *effective pressure*. Pistons at the ends of the test cylinder result in either a maximum or minimum stress along

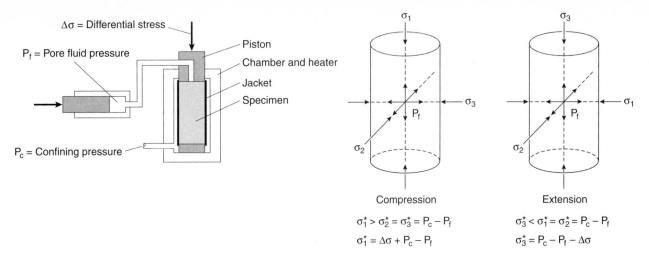

Figure 5.6 Schematic diagram of a triaxial compression apparatus and states of stress in cylindrical specimens in compression and extension tests. The values of P_c, P_f, and σ can be varied during the experiments.

the cylinder axis, depending on the magnitudes of the fluid pressure and the axial stress. The remaining two principal stresses are equal to the effective pressure. By varying any or all of the axial stress, the confining pressure, or the pore-fluid pressure, we obtain a range of stress conditions to carry out our deformation experiments. In addition, we may heat the sample during the experiment so that we can also examine the effect of temperature.

A triaxial apparatus enables us to vary stress, strain, and strain rate in rock specimens under controlled parameters of confining pressure, temperature, pore-fluid pressure, and time (duration of the experiment). What happens when we vary these parameters, and what does this tell us about the behavior of rocks under natural conditions? Before we examine this aspect, it is useful to briefly recall how these properties relate to the Earth. Both confining pressure and temperature increase with depth in the Earth (Figure 5.7). The confining pressure is obtained from relationship:

$$P_c = \rho \cdot g \cdot h \qquad \text{Eq. 5.14}$$

with ρ the density, g is gravity, and h is depth. This is the pressure from the weight of the overlying rock column, which we call *lithostatic pressure*. The temperature structure of the Earth is slightly more complex than the constant gradient of 10°C/km used in Figure 5.5. At first, temperature increases at an approximately constant rate (10–30°C/km), but then the thermal gradient becomes less (Figure 5.7). Additional complexity is introduced by the heat generated from compression at high pressures, which is reflected in the *adiabatic gradient* (dashed line in Figure 5.7). But if we limit our considerations to the crust and upper mantle, a linear geothermal gradient is a reasonable approach. In Section 5.1 we learned that geologic processes typically occur at strain rates on the order of 10^{-14}/s, with the exception of meteoric im-

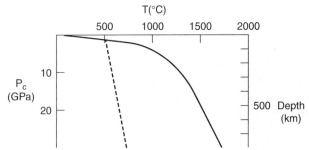

Figure 5.7 Change of temperature (T) and pressure (P_c) with depth. The dashed line is the adiabatic gradient, which is the increase of temperature resulting from increasing pressure and the associated compressibility of silicates with depth.

pacts, seismic events, and explosive volcanism. These relatively fast phenomena have important environmental and human consequences, but they play a subordinate role for the creep of natural rocks. In contrast to geological strain rates, experimental work is typically limited by the patience and life expectancy of the experimentalist. Some of the slowest experiments are carried out at strain rates of 10^{-8}/s (e.g., 30% shortening in a year), which still leaves 4 to 7 orders of magnitude difference with natural rates. Having said all this, now let's look at the effects of varying confining pressure, temperature, strain rate, and pore-fluid pressure during deformation experiments.

5.4.2 Confining Pressure

You recall from Chapter 3 (Section 3.9) that confining pressure acts equally in all directions. When we change the confining pressure during our experiments we observe a very important characteristic: with increasing confining pressure greater amounts of strain accumulate before failure occurs

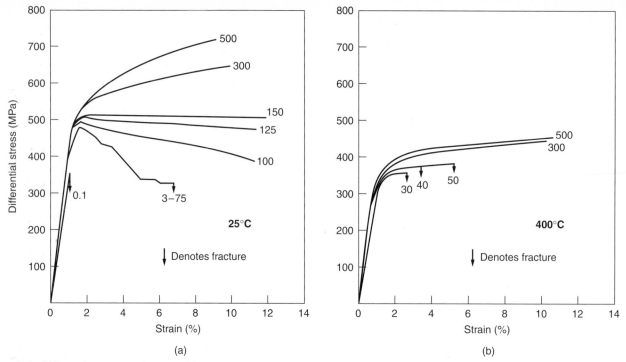

Figure 5.8 Compression stress-strain curves of Solnhofen limestone at various confining pressures (indicated in MPa) at (a) 25°C and (b) 400°C.

(Figure 5.8). In other words, increasing confining pressure increases the viscous component and the rock's ability to flow. What is the explanation for this? If you have already read Chapter 6, the Mohr circle for stress and failure criteria give an explanation, but we will assume that this material is still to come. So, let's take another approach. Clapping your hands together as part of a workout exercise is quite easy. Now do the same exercise under water, and you will notice that it is a lot harder to clap your hands. Water 'pushes' back, and in doing so it opposes the opening and closing motion of your hands.[15] Similarly, higher confining pressures increasingly resist the opening of fractures, and any shape change that occurs in the presence of a differential stress is therefore viscous (ignoring the small elastic component).

The effect of confining pressure is particularly evident at elevated temperatures (Figure 5.8b). Not all rocks respond equally to changes in confining pressure; in Figure 5.9 we compare several common rock types, which show that the change is much more pronounced in sandstones and shales than in quartzites and slates. Thus, based on experiments with varying confining pressure, it appears that larger strains can be achieved before failure with increasing depths in the Earth where we have higher lithostatic pressures.

---

[15]The analogy only emphasizes the resistance to opening of fractures; there are no differential stresses under water, so only volumetric strain can occur.

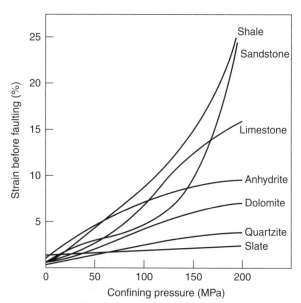

Figure 5.9 The effect of changing the confining pressure on various rock types. The amount of strain before failure (ductility) differs significantly among these common rock types.

5.4.3 Temperature

A change in temperature conditions also produces a marked change in response (Figure 5.10). Again using limestone, the block fails readily at low temperatures. At these conditions most of the strain prior to failure is recoverable (elastic).

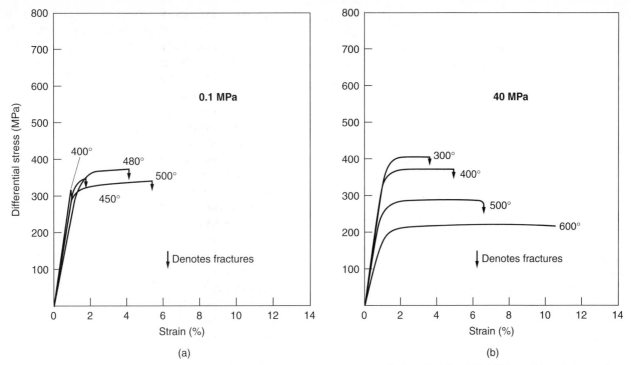

Figure 5.10 Compression stress-strain curves of Solnhofen limestone at various temperatures (indicated in °C) at (a) 0.1 MPa confining pressure and at (b) 40 MPa confining pressure.

When we increase the temperature, the elastic portion of the strain decreases while the ductility increases, which is most noticeable at elevated confining pressures (Figure 5.10b). You personally experience this temperature-dependence of flow when you try to pour syrup on pancakes in a tent in the Arctic or do the same in the Sahara: the ability of syrup to flow increases with temperature. Furthermore, the maximum stress that a rock can support until it flows (the *yield strength* of a material) decreases with increasing temperature. The behavior of various rock types and minerals under conditions of increasing temperature is given in Figure 5.11, showing that calcite-bearing rocks are much more affected than, say, quartz-bearing rocks. These experiments all demonstrate that rocks have lower strength and become more ductile with depth in the Earth where we find higher temperatures.

5.4.4 Strain Rate

It is simply impossible to carry out rock deformation experiments at geologic strain rates because they are so slow (Section 5.4.1); therefore, it is particularly important for the interpretation of experimental results to understand the effect of changes in strain rate. This is most markedly seen in experiments at elevated temperatures, such as those on marble shown in Figure 5.12. Decreasing the strain rate results in decreased rock strength and increased ductility. We again turn to an analog. If you slowly press on a small ball of Silly Putty®,[16] it spreads under the applied stress

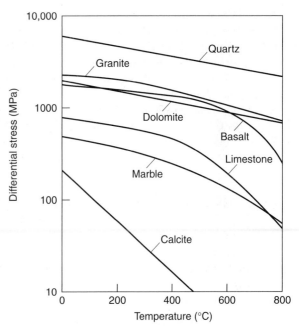

Figure 5.11 The effect of temperature changes on the compressive strength of some rocks and minerals.

(ductile flow). If, on the other hand, you deform the same ball by the blow of a hammer, the Silly Putty will shatter into many pieces (brittle failure). The ambient deformation conditions are the same, except that the strain rate differs;[17] the response is dramatically different.

---

[16]A silicone-based material that offers hours of entertainment if you are of kindergarten age or a professional structural geologist.

[17]Assuming that the stress from slow push and a rapid blow are equal.

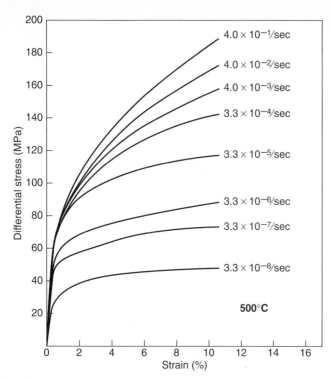

Figure 5.12 Stress vs. strain curves for extension experiments in weakly foliated Yule marble for constant strain rates at 500°C.

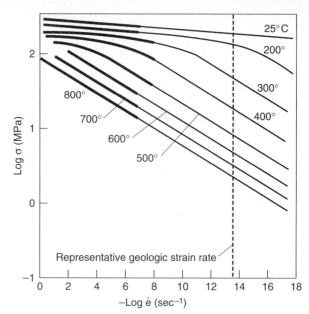

Figure 5.13 Log stress vs. −log strain-rate curves for various temperatures based on extension experiments in Yule marble. The heavy lines mark the range of experimental data; the thin lines are extrapolations to lower strain rates.

Rocks show similar effects of strain-rate variation, and they highlight a great uncertainty in experimental rock deformation. We must extrapolate experimental results over many orders of magnitude to examine strain-rate variation effects on natural materials, and this has significant consequences (Figure 5.13). For example, at a strain rate of 10^{-14}/s and a temperature of 400°C, ductile flow occurs at a differential stress of 20 MPa. At the same temperature, but at an experimental strain rate of 10^{-6}/s, the flow stresses are nearly an order of magnitude higher (160 MPa). You may have noticed that temperature changes produce similar effects as strain-rate variations in rock experiments (higher T $\propto$ slower $\dot{e}$). Thus, elevated temperatures are used to substitute for geologic strain rates. In spite of these uncertainties, the volume of experimental work and our understanding of the mechanisms of ductile flow (Chapter 9) allow reasonable extrapolations of rock deformation at geological strain rates.

5.4.5 Pore-Fluid Pressure

Natural rocks often contain a fluid phase that may originate from the depositional history or be secondary in origin (e.g., fluids released from prograde metamorphic reactions). In particular, low-grade rocks, such as sandstones and shales, contain a significant fluid component that affects their behavior under stress. To examine this fourth parameter, the deformation rig shown in Figure 5.6 con-

tains an impermeable jacket around the sample. Experiments show that increasing the pore-fluid pressure produces a drop in the strength and reduces the ductility of the sample (Figure 5.14a). In other words, rocks are weaker when the pore-fluid pressure is high. We again return to this in Chapter 6, but let us briefly try to understand this effect here. Pore-fluid pressure acts equally in all directions and thus counteracts the confining pressure, resulting in an *effective pressure* ($P_e = P_c - P_f$) that is less than the confining pressure originally imposed. Thus, we hypothesize that increasing the pore-fluid pressure will have the same effect as decreasing the confining pressure of the experiment. We put this to the test by comparing the result of two experiments on limestone in Figure 5.14a and 5.14b that vary pore-fluid pressure and confining pressure, respectively. Clearly, there is remarkable agreement between the two experiments, supporting our hypothesis.

The role of fluid content and ductility is a little more complex than immediately apparent from these experiments because of fluid-chemical effects. While ductility decreases with increasing pore-fluid pressure, the corresponding decreased strength of the material will actually promote flow. The same material with low fluid content ('dry' conditions) would resist deformation, but at high fluid content ('wet' conditions) flow may occur readily. This is nicely illustrated by looking at the deformation of minerals, such as quartz, with different H_2O content (Figure 5.15). The behavior of

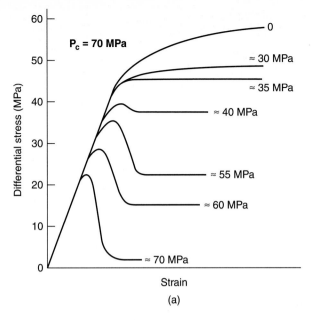

P_c = 70 MPa

(a)

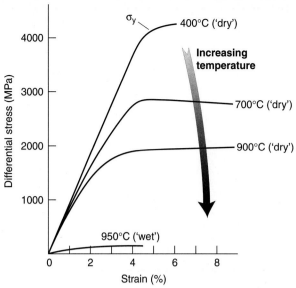

Figure 5.15 The effect of water content on the behavior of natural quartz. 'Dry and wet' refer to low and high water content, respectively. The curves also show the effect of temperature for single crystal deformation, which is similar to that for rocks (Figure 5.10).

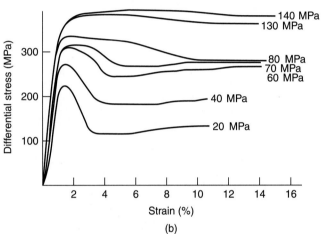

(b)

Figure 5.14 Comparing the effect of pore-fluid pressure on the behavior of limestone (a), with that of varying the confining pressure (b).

quartz is similar to that in the previous rock experiment: the strength of 'wet' quartz is only about one tenth that of 'dry' quartz at the same temperature. The reason for this weakening is chemical, and lies in the substitution of OH groups for O in the silicate crystal lattice, which strains and weakens the Si-O atomic bonds (see Chapter 6).[18] In practice, H_2O-content explains why many minerals and rocks deform relatively easily even under moderate stress conditions.

5.4.6 Work Hardening–Work Softening
Experiments on rocks bring an interesting property to light that we first noticed in the general creep curve of

[18]This is called *hydrolitic weakening* (see Chapter 6).

Figure 5.1a: the relationship between strain and time may vary in a single experiment. The strain rate may decrease, increase, or remain constant under constant stress. In fact, when we carry out experiments while keeping the strain rate constant, we typically find that the stresses necessary to continue the deformation experiment may increase or decrease, phenomena that engineers call *work hardening* (greater stress needed) and *work softening* (lower stress needed), respectively. In a way you can think of this as rock becoming stronger or weaker with increasing strain; therefore we also call this effect *strain hardening* and *strain softening*, respectively. A practical application of work hardening is the repeated rolling of metal, especially when it is heated, which gives it greater strength. While you may not realize this when you pay the repair bill after an unfortunate encounter with a moving tree, car makers use this to strengthen metal parts such as bumpers. This effect was already known in ancient times, when Japanese samurai sword makers were known to produce some of the hardest blades available using repeated heating, rolling and hammering of metal. Work softening is the opposite effect, in which the stress required to continue the experiment is less; in constant stress experiments this results in a strain rate increase. These processes may occur in the same materials as shown in, for example, Figure 5.8a. At low confining pressure a stress drop is observed after approximately 2% strain, representing work softening. At high confining pressures, the limestone displays work hardening, because increasingly higher stresses are necessary to continue the experiment. Schematically this is shown in

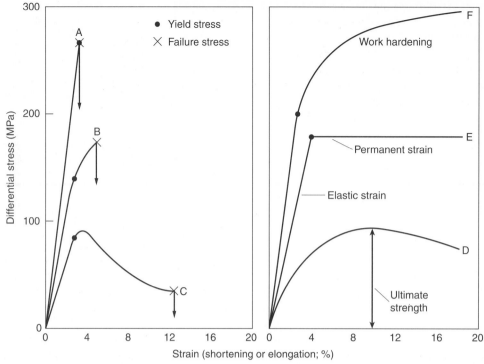

Figure 5.16 Representative stress-strain curves of brittle (A and B), brittle-ductile (C), and ductile behavior (D–F). A is elastic behavior followed immediately by failure, which represents brittle behavior. In B, a small viscous component (permanent strain) is present before brittle failure. In C, a considerable permanent strain accumulates before the material fails, which represents the transition between brittle and ductile behavior. D displays no elastic component and is work softening. E is ideal elastic-plastic behavior in which permanent strain accumulates at constant stress above the yield stress. F is the typical behavior seen in many experiments, displaying a component of elastic strain followed by permanent strain that requires increasingly higher stresses to accumulate (work hardening). The yield stress marks the stress at the change from elastic (recoverable or nonpermanent strain) to viscous (nonrecoverable or permanent strain) behavior; failure stress is the stress at fracturing.

Figure 5.16 (curve D displays work softening and curve F displays work hardening).

Work hardening and work softening are fully explained by the atomic-scale processes that enable rocks to flow. However, unless you already have some background in crystal plastic processes, any brief explanation here would have to be incomplete. We therefore defer this explanation of the physical basis of these processes until the time we discuss crystal defects and their movement (Chapter 8).

5.4.7 Significance of Experiments to Natural Conditions

We close this section on experiments with natural rocks with a table summarizing the results of varying confining pressure, temperature, fluid pressure, and strain rate (Table 5.7) during experiments on rocks and examining their significance for geological conditions.

From Table 5.7 we learn that increasing the confining pressure (P_c) and fluid pressure (P_f) have opposing effects,

Table 5.7 Effect of Experimental Parameters on Rheologic Behavior

	Effect	Explanation
High P_c	Suppresses fracturing; increases ductility; increases strength; increases work hardening	Prohibits fracturing and frictional sliding; higher stress necessary for fracturing exceeds that for ductile flow
High T	Decreases elastic component; suppresses fracturing; increases ductility; reduces strength; decreases work hardening	Promotes crystal plastic processes
Low $\dot{e}$	Decreases elastic component; increases ductility; reduces strength; decreases work hardening	Promotes crystal plastic processes
High P_f	Decreases elastic component; promotes fracturing; reduces ductility; reduces strength *or* promotes flow	Decreases P_c ($P_e = P_c - P_f$) and weakens Si-O atomic bonds

while increasing temperature (T) and lowering strain rate (ė) have the same effect. Confining pressure and temperature, which both increase with depth in the Earth, result in rocks that increasingly resist failure, while at the same time they allow larger strain accumulation, that is, they increase the ability for rocks to flow. High fluid content is more complex and may promote fracturing if P_f is high or promote flow in the case of intracrystalline fluids. From these observations we would predict that brittle behavior (fracturing) is largely restricted to the upper crust, while ductile behavior (flow) dominates at greater depth. A natural test supporting this hypothesis is the realization that faulting and earthquakes generally occur at shallow crustal levels (<15 km depth),[19] while large-scale ductile flow dominates the deeper crust and mantle (e.g., mantle convection).

5.5 CONFUSED BY THE TERMINOLOGY?

By now a baffling array of terms and concepts have passed before your eyes. If you think that you are confused, you should look at the scientific literature on this topic. Let us therefore try to get some additional order in this terminology. Table 5.8 lists brief descriptions of terms that are commonly used in the context of rheology, and contrasts the mechanical behavior of rock deformation with operative deformation mechanisms. A schematic diagram of some representative stress-strain curves summarizes the most important elements (Figure 5.16).

Perhaps the two most commonly used terms in the context of rheology are *brittle* and *ductile*. In fact, we use them to subdivide rock structures in parts B and C of this text, so let us first turn to their meaning. *Brittle behavior* describes deformation that is localized on the mesoscopic scale and involves formation of fractures. For example a fracture in a tea cup is brittle behavior. In natural rocks, brittle fracturing occurs at strains of 5% or less. In contrast, *ductile* behavior describes the ability of rocks to accumulate significant permanent strain with deformation that is distributed on the mesoscopic scale. The shape change from pressing our plasticine cube is ductile behavior. A fractured rock and a folded rock are examples of brittle and ductile behavior, respectively. These two modes of behavior do not define the mechanism by which the deformation occurs. This distinction between behavior and mechanism is an important one that we explain with a simple example.

Consider a cube that is filled with small undeformable spheres (e.g., marbles) and a second cube that is filled with spheres consisting of plasticine (Figure 5.17). If we deform these cubes into rectangular blocks, the mechanism by which this shape change is achieved is very different. In Figure 5.17a, the rigid spheres slide past one another to accommodate the shape change without distortion of the indi-

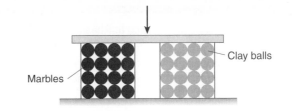

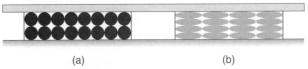

Figure 5.17 Deformation experiment of two cubes containing marbles (a) and balls of plasticine (b), showing ductile strain accumulation by different deformation mechanisms. The finite strain is equal in both cases, and the mode of deformation on the scale of the block is distributed (ductile behavior). However, the mechanism by which the deformation occurs is quite different: in (a) frictional sliding of marbles occurs, while in (b) individual plasticine balls plastically distort into ellipsoids.

vidual marbles. In Figure 5.17b, the shape change is achieved by changes in the shape of individual plasticine spheres to ellipsoids. In both cases the deformation is not localized, but distributed throughout the block (at the sphere boundaries (Figure 5.17a) and within the spheres (Figure 5.17b). Therefore, both experiments are expressions of ductile behavior, although the mechanisms by which deformation occurs are quite different.[20]

Commonly you will encounter the term *brittle-ductile transition* in the literature, but it is not always clear what is meant. For example, seismologists use this term to describe the depth below which nonsubduction zone earthquakes no longer occur. If they mean that deformation occurs by a mechanism other than faulting, this usage is incorrect because ductility is not a mechanism. Ductile behavior can occur by faulting if thousands of small cracks take up the strain. Alternatively, ductile behavior may represent deformation in which crystallographic processes are important. We have seen that ductility simply reflects the ability of a material to accumulate significant permanent strain. Clearly, we have to separate behavior (brittle vs. ductile) from mechanism (e.g., fracturing, frictional sliding, crystal plasticity, diffusion, all of which are discussed in later chapters), and in many instances terms such as brittle-ductile transition lead to unnecessary confusion. Instead, a useful solution is to contrast the mechanisms: for example, *brittle-plastic transition*, or localized versus nonlocalized deformation.

In other instances we appear to be faced with a *brittle-ductile paradox*. It is quite common in the field to find folded beds that appear to be closely associated with faults

---

[19]Excluding deep earthquakes in subduction zones, which represent special conditions (Chapter 16).

[20]These mechanisms are called grain-boundary sliding and crystal plasticity, respectively.

Table 5.8   Terminology Related to Rheology, with Emphasis on Material Behavior and Mechanisms

Brittle-ductile transition	Depth in the Earth below which brittle behavior is replaced by ductile processes (see under "On Material Behavior" below).
Brittle-plastic transition	Depth in the Earth where the dominant deformation mechanism changes from fracturing to crystal plastic processes (see "On Deformation Mechanisms" below).
Competency	Relative term comparing the resistance of rocks to flow.
Failure stress	Stress at which rocks break.
Fracturing	Deformation mechanism by which a rock body or mineral loses coherency.
Crystal plasticity	Deformation mechanism that involves breaking of atomic bonds without the material losing coherency.
Strength	Stress that a material can support before failure.
Ultimate strength	Maximum stress that a material undergoing work softening can support before failure.
Work hardening	Stress necessary to continue deformation experiment increases as strain increases.
Work softening	Stress necessary to continue deformation experiment decreases as strain increases.
Yield stress	Stress at which permanent strain replaces elastic strain.

On Material Behavior

Brittle behavior	Response of a solid material to stress during which the rock loses continuity (cohesion); brittle behavior reflects the occurrence of brittle deformation mechanisms. Brittle behavior occurs only when stresses exceed a critical value, and thus only occurs after the body has already undergone some elastic and/or plastic behavior. The stress necessary to induce brittle behavior is affected strongly by pressure, under a given temperature; brittle behavior generally does not occur at high temperatures.
Ductile behavior	A general term for the response of a solid material to stress such that the rock appears to flow mesoscopically like a viscous fluid. In a material that has deformed ductilely, strain is distributed; that is, strain develops without formation of mesoscopic discontinuities in the material. Ductile behavior can involve brittle (cataclastic flow) or plastic deformation mechanisms.
Elastic behavior	Response of a solid material to stress such that the material develops an instantaneous, recoverable strain that is linearly proportional to the applied stress; elastic behavior reflects the occurrence of elastic deformation mechanisms. Rocks can only undergo less than a few percent elastic strain before they fail by brittle or plastic mechanisms, with conditions of failure dependent on pressure and temperature during deformation.
Plastic behavior	Response of a solid material to stress such that when stresses exceed the yield strength of the material, it develops a strain without loss of continuity (i.e., without formation of fractures). Plastic behavior reflects the occurrence of plastic deformation mechanisms, and is affected strongly by temperature and requires time to accumulate.
Viscous behavior	Response of a liquid material to a stress. As soon as the differential stress becomes greater than zero, a viscous material begins to flow, and the flow rate is proportional to the magnitude of the stress. Viscous deformation takes time to develop.

On Deformation Mechanisms

Brittle deformation mechanisms	Mechanisms by which brittle deformation occurs, namely fracture growth and frictional sliding; fracture growth includes both joint formation and shear rupture formation, and sliding implies faulting. If fracture formation and frictional sliding occur at a grain scale, the resulting deformation is called cataclasis; if cataclasis results in the rock "flowing" like a viscous fluid, then the process is called cataclastic flow (see Chapter 6).
Elastic deformation mechanisms	Mechanisms by which elastic behavior occurs, namely the stretching, without breaking, of chemical bonds holding atoms or molecules together.
Plastic deformation mechanisms	The mechanisms by which plastic deformation occur, namely dislocation glide, dislocation creep (glide and climb; including recovery, recrystallization), diffusive mass transfer (grain-boundary diffusion or Coble creep, and volume diffusion or Herring-Nabarro creep); grain-boundary sliding/superplasticity (see Chapter 9).

(such as fault-propagation and fault-bend folds, see chapters 10 and 17). If we assume that faulting and folding occurred simultaneously, then we are left with a situation where both brittle faulting and ductile folding took place at essentially the same level in the Earth. How can we explain this situation and why does it only appear to be a paradox? There is no reason to expect that P_c, T, and P_f are sufficiently different over the relatively small volume of rock to account for the simultaneous occurrence of these two behavioral modes of deformation. Strain rate and therefore *strain rate gradients,* on the other hand, may vary considerably in any given body of rock. Recall the Silly Putty experiment (Section 5.4.4) in which fracturing occurred at high strain rates and ductile flow at lower rates. Similarly, faulting in nature may occur at regions of high strain rates, but some distance away the strain rate is sufficiently less to give rise to ductile folding. So strain rate gradients are an explanation for the simultaneous occurrence of brittle and ductile behavior, which resolves the apparent brittle-ductile paradox.

Competency and strength are two related terms that describe the relationship of rocks to stress. *Strength* is the stress that a material can support before failure.[21] *Competency* is a relative term that compares the resistance of rocks to flow (for example, Figures 5.5 and 5.11). Experiments and general field observations have given us a qualitative *competency scale* for rocks. Increasingly competent sedimentary rocks are:

rock salt - shale - limestone - graywacke - sandstone - dolomite.

For metamorphic/igneous rocks, the corresponding competency scale is:

schist - marble - quartzite - gneiss - granite - basalt

Note that competency is not the same as the amount of strain that accumulates in a body. Therefore, ductility contrast between materials should not be used as a synonym for competency contrast.

5.6 CLOSING REMARKS

With this chapter on rheology we conclude the block on fundamentals of rock deformation, which should give us the necessary background to examine the significance of natural deformation structures on all scales, from mountain belts to thin sections. In fact, we can predict broad rheological properties of the whole Earth, given assumptions on mineralogy and temperature structure. Figure 5.18 uses experimental data, such as those shown in this chapter, to calculate the rheologic stratification of the crust and upper mantle. Such models provide reasonable, first-order predictions on the rheology of the lithosphere, which can then be used in numerical models of whole earth dynamics.

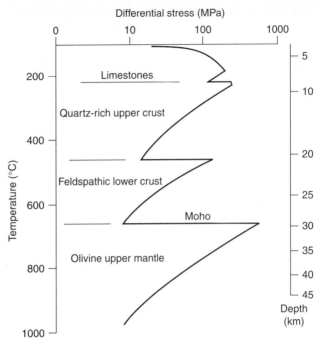

Figure 5.18 Rheological stratification of the lithosphere based on the mechanical properties of characteristic minerals. Computed lithospheric strength (i.e., the differential stress) changes not only as a function of composition, but also as a function of depth (i.e., temperature).

Much of the material on rheology took a basic approach. We mainly examined monomineralic rocks, such as marbles, as opposed to polymineralic rocks, such as granites. Intuitively you will realize that polymineralic rocks require an understanding of the behavior of each of the various constituents, but it is more complicated than that. Consider glass needles, a material that breaks quite easily, that are embedded in flexible resin (epoxy). The *composite material* is unexpectedly strong and resistant to breaking. Fiberglass, consisting of glass and resin, combines the strength of glass with the flexibility of resin. Thus, the behavior of composite materials, and therefore polymineralic rocks, is not simply a matter of knowing the behavior of its constituents.

We have chosen to subdivide tectonic structures into *brittle* and *ductile* structures. In spite of the sometimes confusing use of these terms, they do allow a convenient distinction for the behavior of natural rocks. Figure 5.19 schematically highlights the observational aspect of this

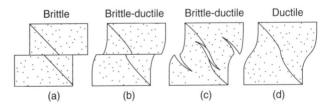

Figure 5.19 Brittle (a) to brittle-ductile (b, c) to ductile (d) deformation, reflecting the general subdivision that is used in the subsequent chapters.

---

[21]In the case of work softening, we use the term *ultimate strength* (Figure 5.16).

subdivision, without at this point inferring the mechanisms by which the features form; the subsequent chapters will take care of that. This figure also illustrates the broad separation into three types of common structures: faults, folds, and shear zones.

Aside from all the rheologic information that deformation structures provide about the Earth (and that is discussed at length in the subsequent chapters), you should not forget to simply enjoy the sheer beauty and enormity of deformation structures such as faults and folds. This is what attracted many of us to this field in the first place.

ADDITIONAL READING

Carmichael, R. S., ed., 1989, *Practical handbook of physical properties of rocks and minerals.* Boca Raton: CRC Press, 741 p.

Jaeger, J. C., and Cook, N. G. W., 1976. *Fundamentals of rock mechanics.* London: Chapman and Hall, 585 p.

Kirby, S. H., and Kronenberg, A. K., 1987, Rheology of the lithosphere: Selected topics. *Reviews of Geophysics,* v. 25, p. 1219–1244.

Nicolas, A., 1987, Principles of rock deformation, Dordrecht, D., Reidel Publishing Company.

Poirier, J.-P., 1985, *Creep of crystals.* High-temperature deformation processes in metals, ceramics and minerals. Cambridge: Cambridge University Press, 260 p.

Ranalli, G., 1987, *Rheology of the Earth.* Boston: Allen and Unwin, 366 p.

Rutter, E., 1986, On the nomenclature of mode of failure transitions in rocks: *Tectonophysics,* v. 122, p. 381–387.

B

Brittle Structures

Chapter 6

Brittle Deformation Processes

6.1 INTRODUCTION

Drop a glass on a hard floor and watch as it smashes into dozens of pieces; you have just witnessed an example of brittle deformation! Because you've probably broken a glass (or a plate or a vase) or have seen cracked sidewalks and roads, you already have an intuitive feel for what brittle deformation is all about. In the upper crust of the Earth, rocks undergo brittle deformation, creating a myriad of geologic structures that resemble those in shattered glass or concrete. To understand why these structures exist and how they form in rocks of the crust, we must first learn why and how brittle deformation takes place in materials in general. Our purpose in this chapter is to introduce the basic terminology used to describe brittle deformation, to explain the processes by which brittle deformation takes place, and to describe the physical conditions that lead to brittle deformation. This chapter, therefore, provides a basis for the discussion of specific brittle structures in Chapters 7 and 8.

6.2 BASIC VOCABULARY OF BRITTLE DEFORMATION

Recent research has changed the way geologists think about brittle deformation, and consequently, the vocabulary of brittle deformation has evolved. To avoid misunderstanding, therefore, we begin this chapter by explaining our use of terminology. Table 6.1 summarizes definitions of terms that we use in discussing brittle deformation, and also includes additional terms that you may come across when reading articles. Brittle deformation vocabulary remains controversial, and you will find that not all geologists agree on the definitions that we provide.

Brittle deformation is simply the permanent change that occurs in a solid material due to the growth of fractures and/or due to sliding on fractures once they have formed. In this definition, a *fracture* is any surface of discontinuity, meaning a surface across which the material is no longer bonded. If a fracture fills with minerals precipitated out of a hydrous solution, it is a *vein;* if it fills with rock (igneous or sedimentary) intruded from elsewhere, it is a *dike*. A *joint* is a natural fracture in rock across which there is no measurable shear displacement (Figure 6.1a). Because of the lack of shear involved in joint formation, joints can also be called *cracks* or *tensile fractures. Shear*

Table 6.1   Terminology of Brittle Deformation

Brittle deformation	The permanent change that occurs in a solid material due to the growth of fractures and/or due to sliding on fractures. Brittle deformation occurs only when stresses exceed a critical value, and thus only after a rock has already undergone some elastic and/or plastic behavior.
Brittle fault	A single surface on which movement occurs specifically by brittle deformation mechanisms.
Brittle fault zone	A band of finite width in which slip is distributed among many smaller discrete brittle faults, and/or in which the fault surface is bordered by pervasively fractured rock.
Cataclasis	A deformation process that involves distributed fracturing, crushing, and frictional sliding of grains or of rock fragments.
Crack	*Verb:* to break or snap apart; *noun:* a fracture whose displacement does not involve shear displacement (i.e., a joint or microjoint).
Fault	*Broad sense:* a surface or zone across which there has been measurable sliding parallel to the surface. *Narrow sense:* a brittle fault. The narrow definition emphasizes the distinctions among faults, fault zones, and shear zones.
Fracture	A general term for a surface in a material across which there has been loss of continuity and, therefore, strength. Fractures range in size from grain scale to continent scale.
Fracture zone	A band in which there are many parallel or subparallel fractures. If the fractures are wavy, they may anastomose with one another. (*Note:* the term has a somewhat different meaning in the context of ocean-floor tectonics.)
Healed microcrack	A microcrack that has cemented back together. Under a microscope, it is defined by a plane containing many fluid inclusions. (Fluid inclusions are tiny bubbles of gas or fluid embedded in a solid.)
Joint	A natural fracture that forms by tensile loading; that is, the walls of the fracture move apart very slightly as the joint develops. (*Note:* a minority of geologists argue that joints can form due to shear loading.)
Microfracture	A very small fracture of any type. Microfractures range in size from the dimensions of a single grain to the dimensions of a thin section.
Microjoint	A microscopic joint; microjoints range in size from the dimensions of a single grain to the dimensions of a hand specimen. Synonymous with *microcrack*.
Shear fracture	A macroscopic fracture that grows in association with a component of shear parallel to the fracture. Shear fracturing involves coalescence of microcracks.
Shear joint	A surface that originated as a joint but later became a surface of sliding. (*Note:* a minority of geologists consider a shear joint to be a joint that initially formed in response to shear loading.)
Shear rupture	A shear fracture.
Shear zone	A band of finite width in which shear strain is significantly greater than in the surrounding rock. Movement in shear zones is a consequence of ductile deformation mechanisms (cataclasis, crystal plasticity, diffusion).
Vein	A fracture filled with minerals precipitated from a water solution.

fractures, in contrast, are fractures across which there has been displacement. Commonly, geologists use the term "shear fracture" instead of "fault" when they wish to imply that the amount of shear displacement on the fracture is relatively small, and that the shear displacement accompanied the formation of the fracture in once intact rock.

In a broad sense, a *fault* is a surface or zone on which there has been measurable displacement (Figure 6.1b). In a narrower sense, geologists restrict the definition of "fault" to refer strictly to a fracture surface (i.e., a *brittle* structure) on which there has been sliding. When using the narrow definition for fault, we apply the term *fault*

(a)

(b)

Figure 6.1 (a) A geologist measuring fractures in an outcrop, near Tuross Point, on the southeastern coast of Australia. The more intensely fractured rock is a fine-grained mafic intrusive, whereas the less fractured rock is a coarse-grained silicic intrusive. (b) Brecciated limestone from the Transvaal Supergroup of South Africa. The white material is calcite vein fill that precipitated between the breccia blocks.

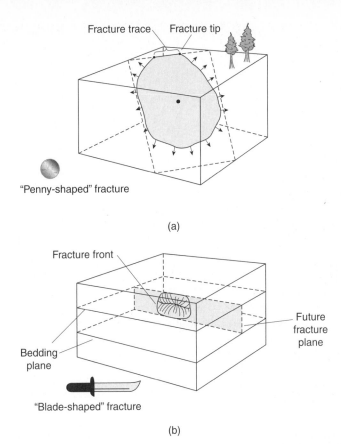

Figure 6.2 (a) A block diagram illustrating that a fracture surface terminates within the limits of a rock body. The top surface of the block is the ground surface; erosion exposes the fracture trace. Note that the trace of the fracture on the ground surface is a line of finite length (with a fracture tip at each end). The arrows indicate that this particular fracture grew radially outward from an origin. The origin is the dot in the center of the fracture plane. (b) A blade fracture that has propagated in a sedimentary layer and terminates at the bedding plane.

zone to refer either to a band of finite width across which the displacement is partitioned among many smaller faults, or to the zone of rock bordering the fault that has fractured during faulting (Figure 6.1c). We apply the term *shear zone* to a band of finite width in which shear strain is significantly greater than in the surrounding rock. Movement in shear zones can be the consequence of *cataclasis* (distributed fracturing, crushing, and frictional sliding of grains of rock or rock fragments), crystal plastic deformation mechanisms (dislocation glide, dislocation climb), and diffusion. We'll describe cataclastic shear zones in this chapter, and other types of shear zones in Chapter 12, after we have introduced plastic deformation mechanisms (Chapter 9).

Regardless of type, a fracture does not extend infinitely in all directions (Figure 6.2). Some fractures intersect the surface of a body of rock, whereas others terminate within the body. The line representing the intersection of the fracture with the surface of a rock body is the *fracture trace,* and the line separating the region of the rock that has fractured from nonfractured regions is the *fracture front.* The point at which the fracture trace terminates on the surface of the rock is the *fracture tip.* In three dimensions, some fractures are irregular surfaces, whereas others are planes that roughly resemble coins or blades (Figure 6.2).

6.3 CATEGORIES OF BRITTLE DEFORMATION

Recall that a solid is composed of atoms or ions that are bonded to one another by chemical bonds, which can be visualized as tiny springs. Each chemical bond has an equilibrium length, and the angle between any two chemical bonds connected to the same atom has an equilibrium value (Figure 6.3a). During elastic strain, the bonds holding the atoms together within the solid stretch, shorten, and/or bend, but they do not break (Figure 6.3b)! When the stress is removed, the bonds return to their equilibrium conditions and the elastic strain disappears. In other words, elastic strain is *recoverable.*

Rocks cannot develop large elastic strains; you certainly cannot stretch a rock to twice its original length and expect it to spring back to its original shape! At most, a rock can develop only a few percent strain by elastic distortion. If the stress applied to a rock is greater than the stress that the rock can accommodate by straining elastically, then one of two changes can happen: the rock deforms ductilely (strains without breaking), or the rock deforms brittlely.

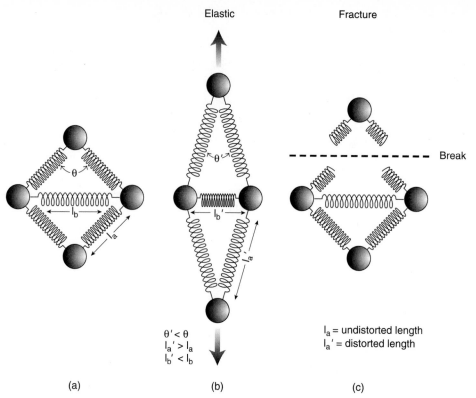

Elastic   Fracture

Break

$\theta' < \theta$
$l_a' > l_a$
$l_b' < l_b$

l_a = undistorted length
l_a' = distorted length

(a)   (b)   (c)

Figure 6.3 A sketch illustrating what is meant by stretching and breaking chemical bonds. The bonds are represented by springs, and the atoms by spheres. (a) Four atoms arranged in a lattice at equilibrium. (b) Consequences of stretching of the lattice. Some bonds stretch and some shorten. Also, the angle between pairs of bonds changes. (c) If the bonds are stretched too far, they break.

What actually happens during brittle deformation? Simply put, if the stress becomes large enough to stretch or bend chemical bonds so much that the atoms are too far apart to attract one another, then the bonds break, resulting in either formation of a fracture or slip on a preexisting fracture (Figure 6.3c). Brittle deformation is *nonrecoverable,* meaning that when the stress is removed, the deformation remains. The pattern of breakage during brittle deformation depends on stress conditions and on material properties of the rock. Brittle deformation does not involve just one process. For purposes of our discussion, we divide brittle deformation processes into four categories that are listed in Table 6.2 and illustrated in Figure 6.4.

Table 6.2   Categories of Brittle Deformation Processes

Cataclastic flow   This type of brittle deformation refers to macroscopic ductile flow as a result of grain-scale fracturing and frictional sliding distributed over a band of finite width.

Frictional sliding   This process refers to the occurrence of sliding on a preexisting fracture surface, without the significant involvement of plastic deformation mechanisms.

Shear rupture   This type of brittle deformation results in the initiation of a macroscopic shear fracture at an acute angle to the maximum principal stress when a rock is subjected to a triaxial compressive stress. We will see that shear rupturing involves growth and linkage of microcracks.

Tensile cracking   This type of brittle deformation involves propagation of cracks into previously unfractured material when a rock is subjected to a tensile stress. If the stress field is homogeneous, tensile cracks propagate in their own plane and are perpendicular to the least principal stress in the rock.

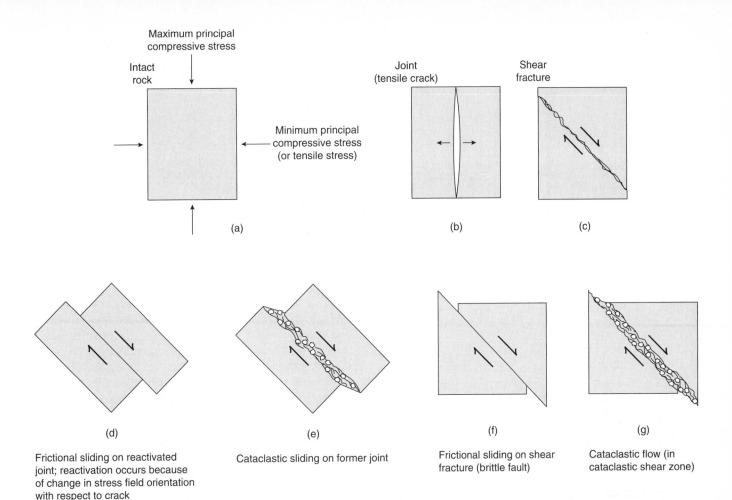

(a)

Joint (tensile crack)   Shear fracture

(b)   (c)

(d)   (e)   (f)   (g)

Frictional sliding on reactivated joint; reactivation occurs because of change in stress field orientation with respect to crack

Cataclastic sliding on former joint

Frictional sliding on shear fracture (brittle fault)

Cataclastic flow (in cataclastic shear zone)

Figure 6.4 Types of brittle deformation. (a) Orientation of the remote principal stress directions with respect to an intact rock body. (b) A tensile crack, forming parallel to σ_1 and perpendicular to σ_3 (which may be tensile). (c) A shear fracture, forming at an angle of about 30° to the σ_1 direction. (d) A tensile crack that has been reoriented with respect to the remote stresses and becomes a fault by undergoing frictional sliding. (e) A tensile crack that has been reactivated as a cataclastic shear zone. (f) A shear fracture that has evolved into a fault. (g) A shear fracture that has evolved into a cataclastic shear zone.

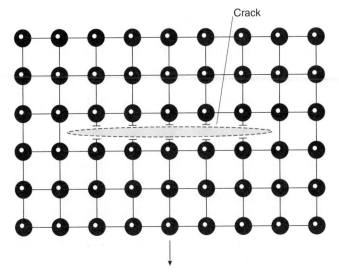

Figure 6.5 A cross-sectional sketch of a crystal lattice (balls are atoms and sticks are bonds) in which there is a crack. The crack is a plane of finite extent across which all bonds are broken.

6.4 TENSILE CRACKING

6.4.1 Stress Concentration and Griffith Cracks

Figure 6.5 illustrates a crack in rock on the atomic scale. One way for such a crack to develop would be for all the chemical bonds across the crack surface to break at once. In this case, the tensile stress necessary is equal to the strength of each chemical bond multiplied by all the bonds that had once crossed the area of the crack. If you know the strength of a single chemical bond, then you can calculate the stress necessary to break all the bonds simply by multiplying. The resulting number, known as the *theoretical strength* of rock, is about 500 to 5000 MPa—a very large stress indeed! Keeping the concept of theoretical strength in mind, we face a paradox. Measurement of rock strength in the Earth's crust shows that tensile cracking occurs at crack-normal tensile stresses of less than about 10 MPa, when the confining pressure is very low,[1] a value that is substantially

---

[1]The failure strength of rock under tension is much less than the failure strength of rock under compression; failure strength under compression depends strongly on confining pressure.

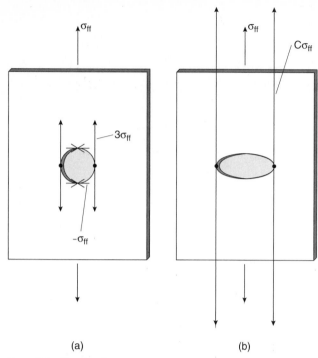

(a)   (b)

Figure 6.6 Sketches illustrating stress concentration adjacent to a hole in an elastic sheet. If the sheet is subjected to a far-field tensile stress at its ends, then stress magnitudes at the sides of the holes are equal to $C\sigma_{ff}$, where C is the stress concentration factor. (a) For a circular hole, C = 3. (b) For an elliptical hole, C > 3. The value of C depends on the eccentricity of the ellipse.

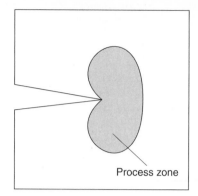

Figure 6.7 Geometry of the "process zone" at the tip of a crack. In the shaded area, stresses exceed the plastic yield strength of the material, and the rock deforms plastically.

less than the theoretical strength of rock. How can real rocks fracture at stresses that are so much lower than their theoretical strength?

The first step to resolving the *strength paradox* came when engineers studying the theory of elasticity realized that the *remote stress* or *far-field stress* (stress due to a load applied at a distance from an area of interest) gets concentrated at the sides of flaws (e.g., holes) inside an elastic material. For example, in the case of a circular hole in a vertical elastic sheet subjected to tensile stress at its ends (Figure 6.6a), the *local stress* (= stress at the point of interest) at the sides of the hole is three times the remote stress magnitude. The magnitude of the local tangential stress at the top and bottom of the hole equals the magnitude of remote stress, but is opposite in sign (i.e., is compressive). If the hole were the shape of an ellipse instead of a circle (Figure 6.6b), the amount of stress concentration would be equal to (2a + 1)/b, where 'a' and 'b' are the long and short axes of the ellipse, respectively. Thus, values for stress concentration at the ends of an elliptical hole depend on the axial ratio of the hole: the larger the axial ratio, the greater the stress concentration. For example, at the ends of an elliptical hole with an axial ratio of 8:1, stress is concentrated by a factor of 17, and at the ends of an elliptical hole with an axial ratio of 32:1, stress is magnified by a factor of 65.

With this understanding in mind, A. W. Griffith, in the 1920s, took the next step toward resolving the strength paradox when he applied the concept of stress concentration at the ends of elliptical holes to fracture development. Griffith suggested that all materials contain preexisting microcracks or flaws at which stress concentrations naturally develop, and that because of the stress concentrations that develop at the tips of these cracks, the cracks propagate and become larger even when the host rock is subjected to relatively low remote stresses. He pointed out that in a material with cracks of different axial ratios, the crack with the largest axial ratio will most likely propagate first. In other words, stress at the tips of preexisting cracks can become sufficiently large to rupture the chemical bonds holding the minerals together at the tip and cause the crack to grow, even if the remote stress is relatively small. Preexisting microcracks and flaws in a rock, which include grain-scale fractures, pores, and grain boundaries, are now called *Griffith cracks* in his honor. Thus, rocks in the Earth's crust are weak because they contain Griffith cracks.

Griffith's concept provided useful insight into the nature of cracking, but his theory did not adequately show how factors such as crack shape, crack length, and crack orientation affect the cracking process. In the 1930s, engineers developed a new approach to study the problem. In this approach, called *linear elastic fracture mechanics,* it is assumed that all cracks in a material have nearly infinite axial ratio (defined as the ratio of the long axis over the short axis); that is, all cracks are very sharp. The reason that very sharp cracks do not propagate under extremely small stresses is that the tips of all real cracks are blunted by a crack-tip *process zone,* in which the material is deformed plastically (Figure 6.7). Linear elastic fracture mechanics theory predicts that if all other factors (e.g., shape and orientation) are equal, then a longer crack will propagate before a shorter crack. We'll see why later in this chapter, when we discuss failure criteria.

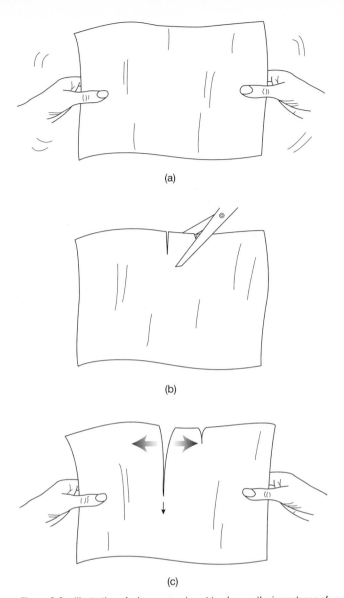

(a)

(b)

(c)

Figure 6.8 Illustration of a home experiment to observe the importance of preexisting cracks in creating stress concentrations. (a) An intact piece of paper is difficult to pull apart. (b) Two cuts, a large one and a small one, are made in the paper. (c) The larger preexisting cut propagates.

To drive home the fact that preexisting cracks affect the magnitude of stress necessary for a tensile cracking to occur, let's do a simple experiment (Figure 6.8). Take a sheet of paper and pull at both ends. You have to pull quite hard in order for the paper to tear. Now make two tears, one that is 0.5 cm long and one that is 2 cm long, in the edge of the sheet near its center, and pull again. The pull that you apply gets concentrated at the tip of the preexisting tear, and thus at this tip the strength of the paper is exceeded. You should find that it takes much less force to tear the paper, and that the paper usually tears apart by growth of the longer preexisting tear.

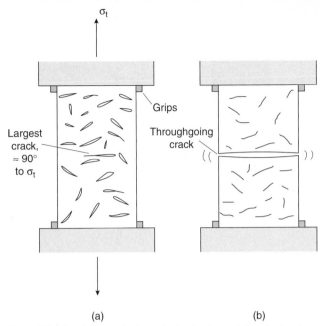

Figure 6.9 Development of a throughgoing tensile crack in a block under tension. (a) When tensile stress (σ_t) is applied, Griffith cracks open up. (b) The largest, properly oriented Griffith crack propagates to form a throughgoing crack.

Note that it is implicit in our description of crack propagation that the total area of a crack does not form instantaneously; rather, a crack initiates at a small flaw and then grows outward. If you have ever walked out on thin ice covering a pond, you are well aware of this fact. As you move away from shore, you suddenly hear a sound like the echo of a gunshot; it is the sound of a fracture forming in the ice due to the stress applied by your boot. If you have the presence of mind under such precarious circumstances to watch how the crack formed, you might notice that the crack initiated under your boot, and propagated outwards into intact ice at a finite velocity. This observation emphasizes that, at any instant, only chemical bonds at the crack tip are breaking. In other words, not all the bonds cut by the fracture broke at once, and thus the basis we used for calculating theoretical strength in the first place does not represent reality.

6.4.2 Examples of Tensile Crack Development

Tensile cracking may be induced in rock samples in a number of different ways, which we outline below.

Let's consider what happens during a laboratory experiment in which we stretch a rock cylinder along its axis under a relatively low confining pressure (Figure 6.9a), a process called *axial stretching*. As soon as the remote tensile stress is applied, preexisting microcracks in the sample open slightly, and the remote stress is magnified to create a larger local stress at the crack tips. Eventually, the stress at the tip of a crack exceeds the theoretical strength of the

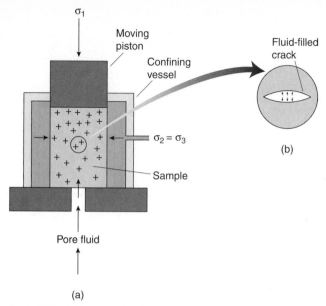

Figure 6.10 (a) Cross-sectional sketch illustrating a rock cylinder in a triaxial load rig experiment. Fluid has access to the cylinder and is able to fill cracks. (b) A fluid-filled crack that is being pushed apart from within by pore pressure.

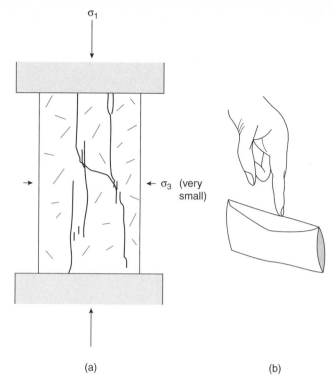

Figure 6.11 (a) A cross section showing a rock cylinder with mesoscopic cracks forming by the process of longitudinal splitting. (b) An "envelope" model of longitudinal splitting. If you push down on the top of an envelope (whose ends have been cut off), the sides of the envelope will move apart.

rock and the crack begins to grow. If the remote tensile stress stays the same after the crack begins to propagate, then the crack continues to grow, and may eventually reach the sample's margins. When the crack reaches the sample's margins, the sample *fails,* meaning it separates into two pieces that are no longer connected (Figure 6.9b).

We can induce tensile fracturing in a rock cylinder, even if the remote principal stresses are compressive, simply by increasing the fluid pressure in pores and cracks of the sample (i.e., the *pore pressure;* Figure 6.10). The outward push of a fluid in a microcrack has the effect of locally creating a tensile stress at the crack tip, and thus can cause a crack to propagate. We call this process *hydraulic fracturing.* As soon as the crack begins to grow, the volume of the crack (i.e., the amount of fluid that fills the space between the two crack walls) increases; so, if no additional fluid enters the crack, then the fluid pressure decreases. Crack propagation ceases when pore pressure drops below a value necessary to create a sufficiently large tensile stress at the crack tip, and does not begin again until the pore pressure builds up sufficiently. Therefore, tensile cracking driven by an increase in pore pressure typically occurs in pulses. We'll discuss the process of hydraulic fracturing more later in this chapter and again in chapter 7.

We can generate tensile cracks in a third way, by subjecting a rock cylinder to axial compression under conditions of very low confining pressure. Under such stress conditions, mesoscopic tensile fractures develop parallel to the cylinder axis (Figure 6.11a), a process known as *longitudinal splitting.* Longitudinal splitting is similar to tensile cracking, except that because the sample is subjected to uniaxial compression, the cracks that are not parallel or subparallel to the σ_1 direction are closed, whereas cracks that are parallel to subparallel to the σ_2 direction can open

up. To picture this, imagine an envelope standing on its edge. If you push down on the top edge of the envelope, the sides of the envelope pull apart, even if they were not subjected to a remote tensile stress (Figure 6.11b). As σ_1 increases, the tensile stress at the tips of cracks parallel to the σ_1 direction exceeds the theoretical strength of the rock, and the crack propagates.

6.4.3 Modes of Crack-Surface Displacement

Before leaving the subject of Griffith cracks, we need to address one more issue, namely, the direction in which an individual crack grows when it is loaded. So far we have limited ourselves to looking only at cracks that are perpendicular to a remote tensile stress. But what about cracks in other orientations with respect to stress, and how do they propagate? Materials scientists identify three configurations of crack loading. These configurations result in three different *modes of crack-surface displacement* (Figure 6.12). Note that the 'displacement' that we are referring to when describing crack propagation is only the *infinitesimal* movement that initiates propagation of the crack tip.

During *Mode I displacement,* a crack opens very slightly in the direction perpendicular to the crack surface; so Mode I cracks are *tensile cracks.* They form parallel to the principal plane of stress that is perpendicular to the σ_3 direction, and can grow in their plane, without bending or changing orientation. During *Mode II displacement,* the sliding mode, rock on one side of the crack surface moves

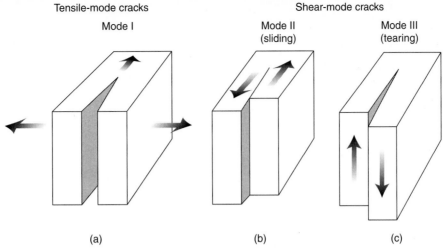

Figure 6.12 Block diagrams illustrating the three modes of crack surface displacement. (a) Mode I (tensile); (b) Mode II; (c) Mode III. Mode II and Mode III are shear modes.

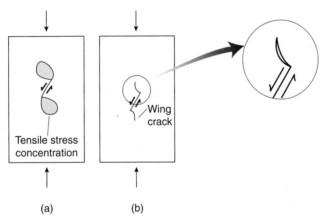

Figure 6.13 The formation of wing cracks. (a) A tensile stress concentration occurs at the ends of a crack that is being loaded by a shear stress. (b) Tensile wing cracks form in the zones of tensile-crack concentration. Inset shows an enlargement of a wing crack.

very slightly in the direction parallel to the fracture surface and perpendicular to the fracture front. During *Mode III displacement,* the tearing mode, rock on one side of the crack slides very slightly parallel to the crack but in a direction parallel to the fracture front. Mode II and Mode III cracks (together known as shear-mode cracks) do *not* grow in their plane. Rather, as they start growing, they immediately either curve and become Mode I cracks, or spawn new tensile cracks called *wing cracks;* Figure 6.13). Because Mode II and Mode III cracks cannot propagate in their own plane, faults and shear ruptures are *not* simply large Mode II or Mode III cracks, as we see below.

6.5 FORMATION OF SHEAR FRACTURES

Shear fractures differ markedly from tensile cracks. A *shear fracture* (or a *shear rupture*) is a surface across

which a rock loses continuity when the shear stress parallel to the surface is sufficiently large. Shear fractures initiate in laboratory rock cylinders at an acute angle to the remote σ_1, under conditions of confined compression ($\sigma_1 > \sigma_2 = \sigma_3$). Because there is a component of normal stress across the fracture as it forms, friction resists sliding on the fracture even during its formation. However, if the shear stress acting on the fracture continues to exceed frictional resistance to sliding, the fracture evolves into a fault on which there may be significant displacement. Shear fractures are *not* simply large shear-mode cracks, because, as we have seen, shear-mode cracks cannot grow in their own plane. This difference is, conceptually, very important.

How do shear fractures form? We can gain insight into the process of shear-fracture formation by generating shear ruptures during a laboratory triaxial loading experiment, using a rock cylinder under confined compression.

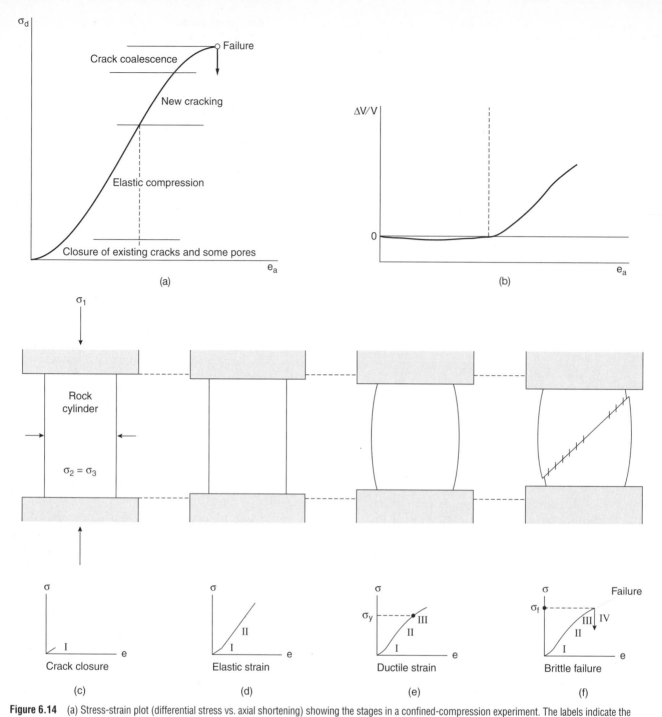

Figure 6.14 (a) Stress-strain plot (differential stress vs. axial shortening) showing the stages in a confined-compression experiment. The labels indicate the process that accounts for the slope of the curve. (b) Shows the changes in volume accompanying the axial shortening to illustrate the phenomenon of dilatancy. Note that to the left of the dashed line the sample volume decreases, whereas to the right of the dashed line the sample volume increases. (c–f) Schematic cross sections showing the mesoscopic behavior of rock cylinders during the successive stages of a confined compression experiment.

(a) Modified from *Introduction to Rock Mechanics*, by Richard E. Goodman, p. 70. Copyright 1989 John Wiley & Sons, New York. Reprinted by permission of John Wiley & Sons, Inc.

In a confined-compression triaxial-loading experiment, we take a cylinder of rock, jacket it in copper or rubber, surround it with a confining fluid in a pressure chamber, and squeeze it between two hydraulic pistons. In the experiment shown in Figure 6.14, the rock itself stays dry. During an experiment, we apply a confining pressure (σ_3) to the sides of the cylinder by increasing the pressure in the surrounding fluid, and apply an axial load (σ_1) to the ends of the cylinder by moving the pistons together at a constant rate. By keeping the value of σ_3 constant while σ_1 gradually increases, we can increase the differential stress ($\sigma_d = \sigma_1 - \sigma_3$). In an experiment, we measure the magnitude of σ_d, the change in length of the cylinder (which is the axial strain, e_a), and the change in volume of the cylinder (Δ).

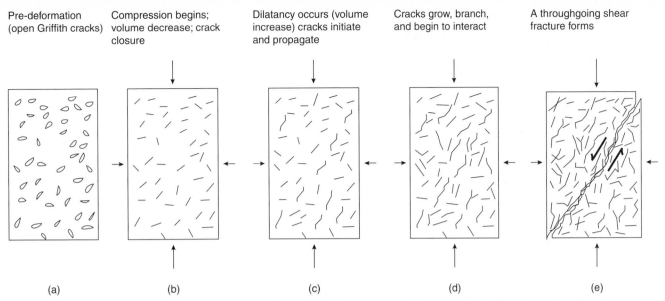

| Pre-deformation (open Griffith cracks) | Compression begins; volume decrease; crack closure | Dilatancy occurs (volume increase) cracks initiate and propagate | Cracks grow, branch, and begin to interact | A throughgoing shear fracture forms |

(a) (b) (c) (d) (e)

Figure 6.15 Schematic cross sections of a rock cylinder during successive stages in the development of a shear fracture, emphasizing behavior of Griffith cracks (cracks shown are not real dimensions). (a) Pre-deformation state. (b) Crack closure and volume decrease. (c) Crack propagation and dilatancy (volume increase). (d) Merging of cracks along the future throughgoing shear fracture. (e) Formation of the throughgoing shear fracture, with the consequent loss of cohesion of the sample.

A graph of σ_d versus e_a (Figure 6.14) shows that the experiment has four stages. In Stage I, we find that as σ_d increases, e_a also increases and that the relationship between these two quantities is a concave-up curve. In Stage II of the experiment, the relationship between σ_d and e_a is a straight line with a positive slope. During Stage I and most of Stage II, the volume of the sample decreases slightly. In Stage III of the experiment, the slope of the line showing the relation between σ_d and e_a decreases. The stress at which the curve changes slope is called the *yield strength.* During the latter part of Stage II and all of Stage III, we observe a slight increase in volume, a phenomenon known as *dilatancy.* If we had a very sensitive microphone attached to the sample, we would hear lots of popping sounds that reflect the formation and growth of microcracks. Suddenly, when $\sigma_d = \sigma_f$, a shear rupture surface develops at an angle of about 30° to the cylinder axis, and there is a stress drop. A *stress drop* in this context means that the axial stress supported by the specimen suddenly decreases, and large strain develops at a lower stress. To picture a stress drop, imagine that you're pushing a car that is stuck in a ditch. You have to push hard until the tires come out of the ditch, at which time you have to stop pushing so hard or you will fall down as the car rolls away. The value of σ_d at the instant that the shear rupture forms and the stress drops is called the *failure strength for shear rupture.* Once failure has occurred, the sample is no longer intact, and frictional resistance to sliding on the surface determines its ability to resist further deformation.

What physically happened during this experiment? During Stage I, preexisting open microcracks underwent closure. During Stage II, the sample underwent elastic shortening parallel to the axis and, because of the Poisson effect,[2] expanded slightly in the direction perpendicular to the axis (Figure 6.15). During the start of State III, tensile microcracks began to grow throughout the sample and wing cracks grew at the tips of shear-mode cracks. The initiation and growth of these cracks caused the observed slight increase in volume, and accounted for the popping noises. During Stage III, the tensile cracking intensified along a narrow band that cut across the sample at an angle of about 30° to the axial stress.[3] Failure occurred when the cracks coalesced to form a throughgoing surface, oriented at about 30° to the axial stress across which the sample loses continuity, so that the rock on one side could frictionally slide relative to the rock on the other side (Figure 6.15d and e). As a consequence, the cylinders moved together more easily and stress abruptly dropped (Stage IV).

Note that *failure strength for shear fracture* is not a definition of the stress state at which a single crack propagates, but rather is the stress state at which a multitude of cracks coalesce to form a throughgoing rupture. Also note that in some experiments, two shear ruptures form, each at 30° to the axial stress. The angle between these *conjugate fractures* is about 60°, and the acute bisectrix between the fractures is parallel to the far-field σ_1. With

---

[2]The *Poisson effect* refers to the phenomenon in which a rock that is undergoing elastic shortening in one direction extends in the direction at right angles to the shortening direction (see chapter 5). The ratio between the amount of shortening and the amount of extension is called *Poisson's ratio, ν*. Since Poisson's ratio is given by length/length, it is a dimensionless number. A typical value of ν for rock is 0.25.
[3]Under conditions of low confining pressure, the angle between the fractures and the σ_1 direction is less.

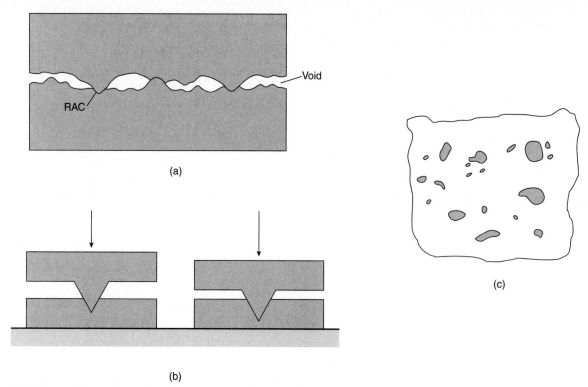

Figure 6.16 Concept of asperities and the real area of contact RAC. (a) Schematic cross-sectional close-up showing the irregularity of a fracture surface and the presence of voids and asperities along the surface. (b) Idealized asperity showing the consequence of normal stress across the fracture on the real area of contact. (c) Map of a fracture surface; the shaded areas are the real area of contact.

continued displacement, however, it is geometrically impossible for both fractures in the conjugate system to be active, because displacement on one fracture will offset the other fracture. Thus, typically only one fracture evolves into a throughgoing fault.

6.6 PROCESSES OF BRITTLE FAULTING

A *brittle fault* is surface on which measurable slip developed without much contribution by plastic deformation mechanisms. Brittle faulting happens in response to the application of a differential stress, because slip occurs in response to a shear stress parallel to the fault plane. In other words, for faulting to take place, σ_1 cannot equal σ_3, and the fault surface cannot parallel a principal plane of stress. Faulting causes a change in shape of the overall rock body that contains the fault. Hence, faulting contributes to development of regional strain. But because a brittle fault is, by definition, a discontinuity in a rock body, the occurrence of faulting does not require development of measurable ductile strain in the surrounding rock.

There are two basic ways to create a brittle fault (Figure 6.4). The first is by shear rupturing a previously intact body of rock. The second is by shear reactivation of a previously formed weak surface (e.g., a joint, a bedding surface, or a preexisting fault) in a body of rock. A preexisting weak surface may slip before the differential stress magnitude reaches the failure strength for shear rupture of intact rock (see Chapter 8). Once formed, movement on brittle faults takes place either by frictional sliding, by the growth of fault-parallel veins, or by cataclastic flow. If movement on the fault involves development of a zone in which cataclastic flow occurs, the fault may also be referred to as a *cataclastic shear zone*.

6.6.1 Frictional Sliding

Friction is the resistance to sliding on a surface. *Frictional sliding* refers to the movement on a surface that takes place when shear stress parallel to the surface exceeds the frictional resistance to sliding. Friction exists because no real surface in nature, no matter how finely polished, is perfectly smooth. The bumps and irregularities that protrude from a rough surface are called *asperities* (Figure 6.16a). When two surfaces are in contact, they touch only at the asperities, and the asperities of one surface may indent or sink into the face of the opposing surface (Figure 6.16b). The cumulative area of the asperities that contact the opposing face is the *real area of contact* (Figure 6.16c).

Figure 6.17 Calcite slip fibers on a fault surface.

In essence, asperities act like an anchor holding a ship in place. In order for the ship to drift, either the anchor chain must break, or the anchor must drag along the seafloor. Similarly, in order for one rock surface to begin sliding past another, asperities must either break off or plow a furrow or groove into the opposing surface. The stress necessary to break off an asperity or to cause it to plow depends on the real area of contact; so, as the real area of contact increases, the frictional resistance to sliding (i.e., the stress necessary to cause sliding) increases. Again, considering our ship analogy, it takes less wind to cause a ship with a small anchor to drift than it does to cause the same sized ship with a large anchor to drift. The frictional resistance to sliding is proportional to the normal stress component across the surface (a relation known as Amonton's law) because of the relation between real area of contact and friction. An increase in the normal stress component pushes asperities into the opposing wall more deeply, causing an increase in the real area of contact.

6.6.2   Slip by Growth of Fault-Parallel Veins

Not all faults that undergo displacement in the brittle field move by frictional sliding. On some faults, the opposing surfaces are separated from each other by mineral crystals (such as quartz or calcite) precipitated out of fluids present along the fault during movement (Figure 6.17). Such *fault-surface veins* may be composed of mineral fibers (needlelike crystals), blocky crystals, or both.

The process by which fault-surface veins form is not well understood. In some cases, they may form when high fluid pressures cause a crack to develop along a weak fault surface. Immediately after cracking, one side of the fault moves slightly with respect to the other and the crack then seals by precipitation of vein material. In other cases, vein formation may reflect gradual dissolution of asperities or steps on the fault surface and then transfer of ions through fluid films to sites of lower stress where mineral precipitation takes place. This second process may occur without the formation of an actual discontinuity across which the rock loses cohesion. Whether syn-slip veining or frictional sliding takes place on a fault surface probably depends on strain rate and on the presence of water. Fault-surface veining is probably common when water is present along the fault, and when movement occurs very slowly.

6.6.3   Cataclasis and Cataclastic Flow

Cataclasis refers to movement on a fault by a combination of microcracking, frictional sliding of fragments past one another, and rotation and transport of grains (Figure 6.18). To picture the process of cataclasis, imagine what happens to corn passing between two old-fashioned mill- stones. The millstones slide past each other, but in the process transform the corn into cornmeal. Cataclasis, if affecting a relatively broad band of rock, results in mesoscopic ductile strain, in which case it may also be called *cataclastic flow* (Chapter 9), because the rock over the width of the band effectively flows. To picture cataclastic flow, think of how

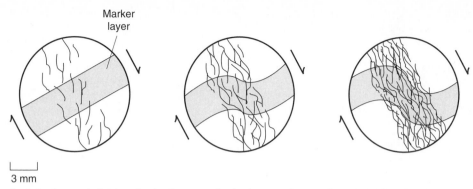

Figure 6.18 Schematic photomicrograph sketches showing the successive development of a zone of cataclasis. The lines are fracture surfaces. With progressive deformation, more fractures propagate, and fractures anastamose, outlining fracture-bounded fragments that can then undergo frictional sliding.

the cornmeal that we just produced behaves much like a fluid when poured from one container to another, even though the individual grains are undeformable.

6.7 PREDICTING INITIATION OF BRITTLE DEFORMATION

We have seen that brittle structures develop when rock is subjected to stress, but so far we have been rather vague about defining stress states at which brittle deformation occurs. Clearly, an understanding of the stress state at which brittle deformation begins is valuable not only to geologists who want to know when, where, and why brittle structures (joints, faults, veins, and dikes) develop in the Earth, but also to engineers who must be able to estimate the magnitude of stress that a building or bridge can sustain before it collapses. When we discuss brittle deformation, we are really talking about three phenomena: tensile crack growth, shear fracture development, and frictional sliding. In this section, we examine the stress conditions necessary for each of these phenomena to occur.

6.7.1 Tensile-Cracking Criteria

A *tensile-cracking criterion* is a mathematical statement that predicts the stress state at which a crack begins to propagate. All tensile-cracking criteria are based on the assumption that mesoscopic cracks grow from preexisting flaws (Griffith cracks) in the rock, because preexisting flaws cause stress concentrations. Griffith was one of the first researchers to propose a tensile-cracking criterion. He did so by looking at how energy was utilized during cracking. Griffith envisioned that a material in which a crack forms can be modeled as a thermodynamic system consisting of an elastic plate containing a preexisting elliptical crack. If a load is applied to the ends of the sheet so that it stretches and the crack propagates, the total energy of this system can be defined. Using this concept, along with several theorems from elasticity theory, Griffith devised a tensile-cracking criterion, which we present without derivation:

$$\sigma_t = [2E\gamma/(\pi(1-\nu^2)c)]^{\frac{1}{2}} \qquad \text{Eq. 6.1}$$

where σ_t = critical remote tensile stress (tensile stress at which the weakest Griffith crack begins to grow), E = Young's modulus, γ = energy used to create new crack surface, ν = Poisson's ratio, and c = half-length of the pre-existing crack. Reading this equation, we see that the critical remote tensile stress for a rock is proportional to material properties of the sample and the length of the crack.

Subsequently, researchers have utilized concepts from the engineering study of linear elastic fracture mechanics to develop tensile-cracking criteria. According to this work, the following equation defines conditions at which Mode I cracks propagate:

$$K_I = \sigma_t Y(\pi c)^{\frac{1}{2}} \qquad \text{Eq. 6.2}$$

where K_I (pronounced "K one," where the 'one' represents a Mode I crack) is the *stress intensity factor*, σ_t is the remote tensile stress, Y is a dimensionless number that takes into account the geometry of the crack (e.g., whether it is penny shaped, blade shaped, or tunnel shaped), and c is half of the crack's length. In this analysis, all cracks in the body are assumed to have very large ellipticity (i.e., cracks are assumed to be very sharp at their tips). Cracks with smaller ellipticity require higher stresses to propagate than Equation 6.1 predicts.

Equation 6.1 says that the value of K_I increases when σ increases. A crack in the sample begins to grow when K_I attains a value of K_{Ic}, which is the *critical stress intensity factor* or the *fracture toughness* (i.e., a measure of tensile strength). The fracture toughness is constant for a given material. When K_I reaches K_{Ic}, the value of σ reaches σ_t, where σ_t is the *critical tensile stress* at the instant the crack starts to grow. We can rewrite Equation 6.2 to create an equation that more directly defines the value of σ_t at the instant the crack grows:

$$\sigma_t = K_{Ic}/(Y(\pi c)^{\frac{1}{2}}) \qquad \text{Eq. 6.3}$$

Notice that, in this equation, the remote stress necessary for cracking depends on the fracture toughness, the crack shape, and the length of the crack. If other factors are equal, a longer crack generally propagates before a shorter crack. Similarly, all factors being equal, crack shape determines which crack propagates first. Note that c increases as the crack starts to grow, so crack propagation typically leads to sample failure.

Equations like Equations 6.2 and 6.3 can also be written for Mode II and Mode III cracks. By comparing equations for the three different modes of cracking, you will find that, other factors being equal, a Mode I crack (i.e., a crack perpendicular to σ_3) propagates before a Mode II or Mode III crack. However, since other factors like crack shape and length come into play, Mode II or Mode III cracks sometimes propagate before Mode I cracks in a real material. Remember that the instant they propagate, they either bend and become Mode I cracks, or they develop wing cracks at their tips; shear cracks cannot propagate significantly in their own plane.

In summary, we see that the stress necessary to initiate the propagation of a crack depends on the ellipticity, the length, the shape, and the orientation of a preexisting crack. Study of crack-propagation criteria is a very active research area, and further details concerning this complex subject are beyond the scope of this book.

6.7.2 Shear-Fracture Criteria and Failure Envelopes

A *shear-fracture criterion* is an expression that describes the stress state at which a shear rupture forms and separates a sample into two pieces. Because shear-fracture initiation in a laboratory sample inevitably leads to failure of the sample, meaning that after rupture the sample can no longer support a load that exceeds the frictional resistance to sliding on the fracture surface, shear-rupture criteria are also commonly known as *shear failure criteria*.

Charles Coulomb, an eighteenth-century French naturalist, was one of the first to propose a shear-fracture criterion. Coulomb suggested that if all the principal stresses are compressive, as is the case in a confined compression experiment, a material fails by the formation of a shear fracture, and that the shear stress parallel to the fracture surface, at the instant of failure, relates to the normal stress by the equation:

$$\sigma_s = C + \mu\sigma_n \qquad \text{Eq. 6.4}$$

where σ_s is the shear stress parallel to the fracture surface at failure, C is the *cohesion* of the rock (a constant that specifies the shear stress necessary to cause failure if the normal stress across the potential fracture plane equals zero; note that this 'C' is not the same as the 'c' in Equations 6.2 and 6.3), σ_n is the normal stress across the shear fracture at the instant of failure, and μ is a constant traditionally known as the *coefficient of internal friction*. The name for μ originally came from studies of friction between grains in unconsolidated sand and the control that such friction has on slope angles of sand piles. So, the name is essentially meaningless in the context of shear failure of a solid rock; μ should be viewed simply as a constant of proportionality. Equation 6.4, also known as

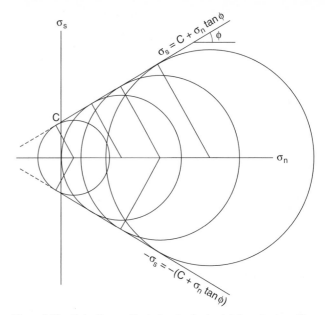

Figure 6.19 Mohr diagram illustrating the Coulomb failure envelope. The envelope is represented by two straight lines. The circles represent stress states at the instant of shear failure.

Coulomb's criterion, basically states that the shear stress necessary to initiate a shear fracture is proportional to the normal stress across the fracture surface.

The Coulomb criterion plots as a straight line on a Mohr diagram (Figure 6.19).[4] To see this, let's plot the results of four triaxial loading experiments in which we increase the axial load on a confined granite cylinder until it ruptures. In the first experiment, we set the confining pressure ($\sigma_2 = \sigma_3$) at a relatively low value, increase the axial load (σ_1) until the sample fails, and then plot the Mohr circle representing this *critical stress state,* meaning the stress state at the instant of failure on the Mohr diagram. When we repeat the experiment, using a new cylinder and starting at a higher confining pressure, we find that as σ_3 increases, the differential stress ($\sigma_1 - \sigma_3$) at the instant of failure also increases. Thus, the Mohr circle representing the second experiment has a larger diameter and lies to the right of the first circle. When we have repeated the experiment two more times and plot the four circles on the diagram, we find that they are all tangent to a straight line with a slope of μ (i.e., tan ϕ) and a y-intercept of C; this straight line is the Coulomb failure criterion. Note that we can also draw a straight line representing the criterion in the region of the Mohr diagram below the σ_n-axis.

---

[4]Recall that on the diagram, normal stresses (σ_n) plot on the X-axis and shear stresses (σ_s) plot on the Y-axis. A Mohr circle represents the stress state by indicating the values of σ_n and σ_s acting on a plane oriented at $\theta°$ to the σ_1 direction (using Convention 1, described in Chapter 3). The circle intersects the x-axis at σ_1 and σ_3 (both of which are normal stresses, because they are principal stresses), and the angle between the x-axis and a radius from the center of the circle to a point on the circle defines the angle 2θ.

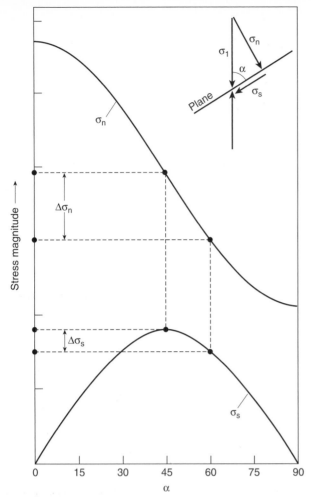

Figure 6.20 Graph illustrating the change in magnitudes of the normal and shear components of stress acting on a plane as a function of the angle α between the plane and the σ_1 direction. At $\alpha = 45°$, shear stress is a maximum, but the normal stress across the plane is quite large. At $\alpha = 30°$, the shear stress is still quite high, but the normal stress is much lower.

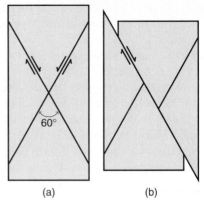

Figure 6.21 Cross-sectional sketch showing how only one of a pair of conjugate shear fractures (a) can evolve into a fault with measurable displacement (b).

A line drawn from the center of a Mohr circle to the point of its tangency with the Coulomb criterion defines 2θ, where θ is the angle between the σ_1 direction and the plane of shear fracture (typically about 30°). Because the Coulomb criterion is a straight line, this angle is constant for the range of confining pressures for which the criterion is valid. The reason for the 30° angle is evident in a graph plotting normal stress magnitude and shear stress magnitude as a function of the angle between the plane and the σ_1 direction (Figure 6.20). Notice that the minimum normal stress does not occur on the same plane as the maximum shear stress. Shear stress is at its highest on a potential failure plane oriented at 45° to σ_1, but the normal stress across this potential plane is still too large to permit shear fracturing on planes of this orientation. The shear stress is a bit lower across a plane oriented at 30° to σ_1, but

is still fairly high. However, the normal stress across the 30° plane is substantially lower, favoring shear-fracture formation.

Coulomb's criterion is an *empirical relation,* meaning that it is based on experimental observation alone, rather than on theoretical principles or knowledge of atomic-scale or crystal-scale mechanisms. This failure criterion does not relate the critical stress state to physical parameters, as does the Griffith criterion, nor does it define the state of stress at which the microcracks, which eventually coalesce to form the shear rupture, begin to propagate. The Coulomb criterion does not predict whether the fractures that form dip to the right or to the left with respect to the axis of a rock cylinder in a triaxial-loading experiment. In fact, *conjugate shear fractures,* one with a right-lateral shear sense and one with a left-lateral shear sense, may develop (Figure 6.21). The two fractures, typically separated by an angle of ~60°, correspond to the tangency points of the circle representing the stress state at failure with the Coulomb failure envelope.

The German engineer Otto Mohr conducted further studies of shear-fracture criteria and found that Coulomb's straight-line relationship works for only a limited range of confining pressures. He noted that at lower confining pressure, the line representing the stress state at failure curved to a steeper slope, and that at higher confining pressure, the line curved to a shallower slope (Figure 6.22). Mohr concluded that over a greater range of confining pressures, the failure criterion for shear rupture resembles a portion of a parabola lying on its side; this curve represents the *Mohr-Coulomb criterion* for shear fracturing. Note that this criterion is also empirical. Unlike Coulomb's straight-line relation, the change in slope of the Mohr-Coulomb failure envelope indicates that the angle between the shear fracture plane and σ_1 actually depends on the stress state. At lower confining pressures, the angle is smaller, and at high confining pressures, the angle is steeper.

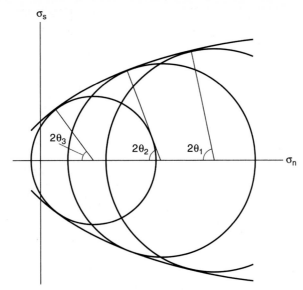

Figure 6.22 Mohr failure envelope. Note that the slope of the envelope steepens toward the σ_s-axis. Therefore, the value of 2θ is not constant (compare $2\theta_1$, $2\theta_2$, and $2\theta_3$).

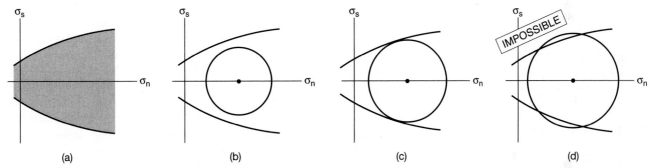

Figure 6.23 (a) A brittle failure envelope as depicted on a Mohr diagram. Within the envelope (shaded area), stress states are stable, but outside the envelope, stress states are unstable. (b) A stress state that is stable, because the Mohr circle, which passes through values for σ_1 and σ_3 and therefore defines the stress state, falls entirely inside the envelope. (c) A stress state at the instant of failure. The Mohr circle touches the envelope. The failure plane is at an angle of θ with respect to the σ_1 direction. (d) A stress state that is impossible.

The plot of the Coulomb or Mohr-Coulomb criterion (both for positive and negative values of σ_s) defines a failure envelope on the Mohr diagram. A *failure envelope* separates the field on the diagram in which stress states are 'stable' from the field in which stress states are 'unstable' (Figure 6.23). In this definition, a *stable stress state* is one that a sample can withstand without undergoing brittle failure. An *unstable stress state* is an impossible condition to achieve, for the sample will have failed by fracturing before such stress states are reached (Figure 6.23). In other words, a stress state represented by a Mohr circle that lies entirely within the envelope is stable, and will not cause the sample to develop a shear rupture. A circle that is tangent to the envelope specifies the stress state at which brittle failure occurs. Stress states defined by circles that extend beyond the envelope are unstable, and are therefore impossible within the particular rock being studied.

Can we define a failure envelope representing the critical stress at failure for very high confining pressures, very low confining pressures, or for conditions where one of the principal stresses is tensile? The answer to this question is controversial. We'll look at each of these conditions separately.

At high confining pressures, samples may begin to deform plastically. Under such conditions, we are no longer really talking about brittle deformation, so the concept of a 'failure' envelope no longer really applies. However, we can approximately represent the 'yield' envelope, meaning the stress state at which the sample begins

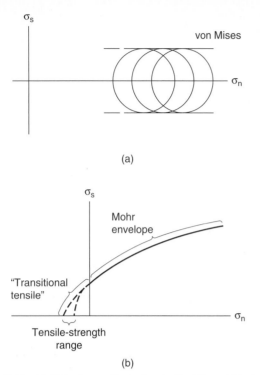

(a)

(b)

Figure 6.24 (a) Mohr diagram illustrating the Von Mises yield criteria. Note that the criteria are represented by two lines that parallel the σ_n-axis. (b) The extrapolation of a Mohr envelope to its intercept with the σ_n-axis, illustrating the "transitional-tensile" regime, and the tensile strength. Note that the tensile strength is a range of values, because the tensile strength depends on the dimensions of preexisting flaws in the deforming sample.

to yield plastically, on a Mohr diagram by a pair of lines that parallel the σ_n-axis (Figure 6.24a). This yield criterion, known as *Von Mises criterion*, indicates that plastic yielding is effectively independent of the differential stress, once the yield stress has been achieved.

If the tensile stress is large enough, the sample fails by developing a throughgoing tensile crack. The tensile stress necessary to induce tensile failure may be represented by a point, the *tensile strength* t_0, along the σ_n-axis to the left of the σ_s-axis (Figure 6.24b). As we have seen, however, the position of this point depends on the size of the flaws in the sample. Thus, even for the same rock type, experiments show that the tensile strength is very variable, and that tensile strength is best represented by a range of points along the σ_n-axis.

There are competing views as to the nature of failure for rocks subjected to tensile stresses that are less than the tensile strength. Some geologists have suggested that failure occurs under such conditions by formation of fractures that are a hybrid between tensile cracks and shear ruptures, which are *transitional-tensile fractures* or *hybrid shear fractures*. The failure envelope representing the

conditions for initiating transitional-tensile fractures is the steeply sloping portion of the parabolic failure envelope (Figure 6.24b). Most fracture specialists, however, claim that transitional-tensile fractures do not occur in nature, and point out that no experiments have yet clearly produced transitional-tensile fractures in the lab. We'll explore this issue further in Chapter 7.

Taking all of the above empirical criteria into account, we can construct a *composite failure envelope* that represents the boundary between stable and unstable stress states for a wide range of confining pressures and for conditions in which one of the principal stresses is tensile (Figure 6.25). The envelope roughly resembles a cross section of a cup lying on its side. The various parts of the curve are labeled. Starting at the right side of the diagram, we have Von Mises criteria, represented by horizontal lines; remember that the Von Mises portion of the envelope is really a plastic yield criterion, not a brittle failure criterion. The portion of the curve where the lines begin to slope effectively represents the brittle-plastic transition. To the left of the brittle-plastic transition, the envelope consists of two straight sloping lines, representing Coulomb's criterion for shear rupturing. For failure associated with the Coulomb criterion, remember that the angle between the shear rupture and the σ_1 direction is independent of the confining pressure. Closer to the σ_s-axis, the slope of the envelope steepens, and the envelope resembles a portion of a parabola. This parabolic part of the curve represents Mohr's criterion; for failure in this region, the angle between the fracture and the σ_1 direction decreases depending on how far to the left the Mohr circle touches the curve. The part of the parabolic envelope with steep slopes specifies failure criteria for supposed transitional-tensile fractures formed at a very small angle to σ_1, but as we discussed, the existence of such fractures remains controversial. The point where the envelope crosses the σ_n-axis represents the failure criterion for tensile cracking, but as we have discussed, this criterion really shouldn't be specified by a point, for the tensile strength of a material depends on the dimension of the flaws it contains. Note that for a circle tangent to the composite envelope at t_0, $2\theta = 0$, so the fracture that forms is parallel to σ_1! Also, note that there is no unique value of differential stress needed to cause tensile failure, as long as the magnitude of the differential stress (the diameter of the Mohr circle) is less than about $4t_0$, for this is the circle whose curvature is the same as the apex of the parabola.

6.7.3 Frictional Sliding Criteria

Because of friction, a certain critical shear stress must be achieved in a rock before frictional sliding initiates on a preexisting fracture. A relation defining this critical stress is the *failure criterion for frictional sliding*. Experimental work shows that failure criteria for frictional sliding, just

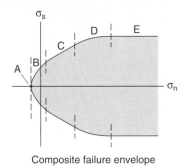

Composite failure envelope

A: Tensile failure criterion
B: Mohr (parabolic) failure criterion
C: Coulomb (straight-line) failure criterion
D: Brittle-plastic transition
E: von Mises plastic yield criterion

(a)

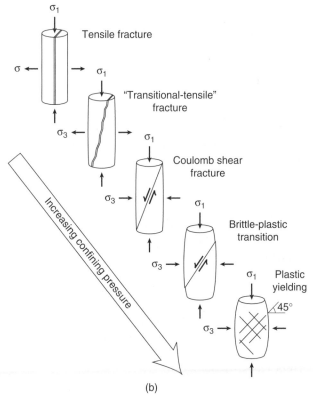

(b)

Figure 6.25 (a) A representative composite failure envelope on a Mohr diagram. The different parts of the envelope are labeled, and are discussed in the text. (b) Sketches of the fracture geometry that forms during failure. Note that the geometry depends on the part of the failure envelope that represents failure conditions, because the slope of the envelope is not constant.

like the Coulomb failure criterion for intact rock, plot as sloping straight lines on a Mohr diagram. Furthermore, a compilation of friction data from a large number of experiments using a great variety of rocks (Figure 6.26) shows

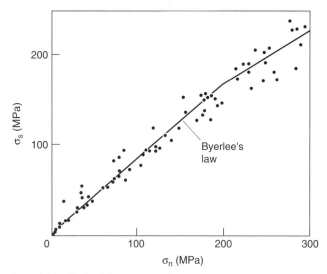

Figure 6.26 Graph of shear stress and normal stress measurements at the initiation of sliding on preexisting fractures. The best-fit line defines Byerlee's law.

that the failure criterion for frictional sliding is basically independent of rock type:

$$\sigma_s/\sigma_n = \text{constant} \qquad \text{Eq. 6.5}$$

The empirical equation known as *Byerlee's law*[5] that best fits observations depends on the value of σ_n. For $\sigma_n < 200$ MPa, the best-fitting criterion is a line described by the equation $\sigma_s = 0.85\sigma_n$, whereas for 200 MPa $< \sigma_n <$ 2000 MPa, the best fitting criterion is a line described by the equation:

$$\sigma_s = 50 \text{ MPa} + 0.6\sigma_n$$

6.7.4 Will New Fractures Form or Will Existing Fractures Slide?

Failure envelopes allow us to quickly determine whether it is more likely for an existing shear rupture to slip, or for a new shear rupture to form (Figure 6.27). For example, Figure 6.27b shows both Byerlee's frictional sliding envelope and the Coulomb shear fracture envelope for Blair Dolomite. Note that the slope and intercept of the two envelopes are different, so that for a specific range of preexisting fracture orientations, the Mohr circle representing the stress state at failure touches the frictional envelope before it touches the fracture envelope, meaning that the preexisting fracture slides before a new fracture forms.

However, preexisting fractures do not always slide before new fractures initiate. Confined compression experiments indicate that if the preexisting fracture is oriented at

---

[5]After the geophysicist J. Byerlee, who first proposed the equations in 1978.

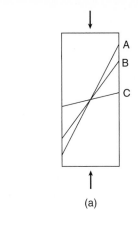

(a)

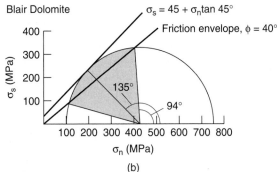

Blair Dolomite

$\sigma_s = 45 + \sigma_n \tan 45°$

Friction envelope, $\phi = 40°$

σ_s (MPa)

135°

94°

σ_n (MPa)

(b)

Figure 6.27 (a) Cross-sectional sketch showing three preexisting fractures in a rock cylinder subjected to confined compression. Surface B is the Coulomb shear fracture that would form in an intact rock. Surface A is a precut surface that would slip prior to B, because the normal stress across it is relatively low. Surface C would not slip prior to B, because it is at such a high angle to σ_1 that it is easier to initiate a new fracture than to cause surface C to slip. (b) Mohr diagram for Blair Dolomite showing how a Mohr circle would contact the frictional sliding envelope before it would contact the Coulomb envelope. The shaded wedge shows the range of values for 2θ for which the preexisting fracture would slide first; ϕ is the slope of the envelope.

a high angle (generally >75°; plane C in Figure 6.27) to the σ_1 direction, the normal stress component across the joint is so high that friction resists sliding, and it is actually easier to initiate a new shear fracture at a smaller angle to σ_1. Sliding then occurs on the new fracture. If a preexisting fracture is at very small angle to σ_1 (generally <15°; plane A in Figure 6.27), the shear stress on the surface is relatively low; so again it ends up being easier to initiate a new shear fracture than to cause sliding on a preexisting weak surface. Thus, preexisting planes at angles between 15° and 75° to σ_1 probably will be reactivated before new fractures form.

6.8 EFFECT OF ENVIRONMENTAL FACTORS ON FAILURE

The occurrence and character of brittle deformation at a given location in the Earth depends on environmental conditions (confining pressure, temperature, and fluid pressure) at that location, and on the strain rate (see Section 6.4). Conditions conducive to the occurrence of brittle deforma-

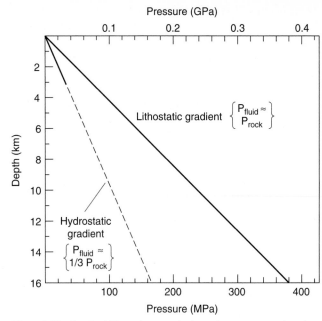

Figure 6.28 Graph of lithostatic versus hydrostatic pressure as a function of depth in the Earth's crust.

tion are more common in the upper 10–15 km of the Earth's crust. However, at slow strain rates or in particularly weak rocks, ductile deformation mechanisms can also occur in this region, as evident by the development of folds at shallow depths in the crust. Below 10–15 km, plastic deformation mechanisms dominate, but at particularly high fluid pressures or at very rapid strain rates, brittle deformation can still occur at these depths.

In this chapter, we have described brittle deformation without considering how it is affected by environmental factors. Not surprisingly, temperature, fluid pressure, strain rate, and rock anisotropy play significant roles in the stress state at failure and/or in the orientation of the fractures that form when failure occurs. Most of these factors are discussed in Section 6.4; therefore, we close this chapter on brittle deformation processes by looking mainly at the effects of fluids.

6.8.1 Effect of Fluids on Tensile Crack Growth

All rocks contain pores and cracks, and we have already seen how important these are in the process of brittle failure. In the upper crust of the Earth below the water table, these spaces, which constitute the porosity of rock, are filled with fluid. This fluid is most commonly water, though in some places it is oil or gas.

If there is a high degree of *permeability* in the rock, meaning that the fluid can flow relatively easily from pore to pore and/or in and out of the rock layer, then the pressure in a volume of pore water at a location in the crust is roughly *hydrostatic,* meaning that the pressure reflects the weight of the overlying water column (Figure 6.28). Hydrostatic (fluid) pressure is defined by the relationship $P_f = \rho g h$, where ρ is the density of water (1000 kg/cm^3), g is the gravitational constant (9.8 m/sec^2), and h is the depth. *Pore pressure,* which is the fluid pressure exerted by fluid within the pores of a rock, may exceed hydrostatic pressure if permeability is restricted. For example, the fluid trapped in a sandstone lens surrounded by impermeable shale cannot escape, so the pore pressure in the sandstone can approach *lithostatic pressure* (P_l), meaning that the pressure approaches the weight of the overlying column of rock (i.e., $P_f = P_l = \rho gh$, where $\rho = 2000$ to 3000 kg/m^3). When the fluid pressure in pore water exceeds hydrostatic pressure, we say that the fluid is *overpressured.*

How does pore pressure affect the tensile failure strength of rock? The pore pressure is an outward push that opposes inward compression from the rock, so the fluid supports part of the applied load. If pore pressure exceeds the least compressive stress (σ_3) in the rock, tensile stresses at the tips of cracks oriented perpendicularly to the σ_3 direction become sufficient for the crack to propagate. In other words, pore pressure in a rock can cause tensile cracks to propagate even if none of the remote stresses are tensile, because pore pressure can induce a crack-tip tensile stress that exceeds the magnitude of σ_3. This process is *hydraulic fracturing.* On a Mohr diagram, it can be represented by movement of Mohr's circle to the left (Figure 6.29). Note that rocks do not have to be overpressured in order for natural hydraulic fracturing to occur, but P_f must equal or exceed the magnitude of σ_3.

Another effect of fluids comes from the chemical reaction of the fluids with the minerals composing a rock. Reaction with fluids may lower the tensile stress needed to cause a crack to propagate, even if the pore fluid pressure is low. Water, for example, reacts with quartz, resulting in the substitution of OH molecules for O atoms in quartz lattice at a crack tip (Figure 6.30). Since the bond between adjacent OH's is not as strong as the bond between oxygen atoms, it breaks more easily, so it takes less remote tensile stress to cause the crack to propagate. This phenomenon is called *subcritical crack growth,* because crack propagation occurs at stresses less than the critical stress necessary to cause a crack to propagate in 'dry' rock.

6.8.2 Effect of Dimensions on Tensile Strength

Rock tensile strength is *not* independent of scale, in that larger rock samples are inherently weaker than smaller rock samples. Why? Because larger samples are more likely to contain appropriately oriented and larger Griffith cracks that will begin to propagate when a stress is applied, thereby nucleating the throughgoing cracks that result in failure of the whole sample. Equation 6.3 emphasizes this point, because the tensile stress at failure is inversely proportional to the crack half-length, c. You can imagine that if a sample is so small that it consists only of a piece of perfect crystal lattice, it will be very strong indeed. In fact, the reason that turbine blades in modern jet engines are so strong is that they are grown as relatively flawless single crystals.

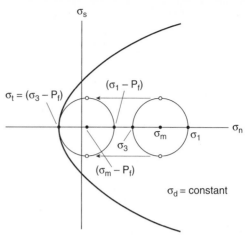

Figure 6.29 A Mohr diagram, with a Mohr failure envelope, showing how an increase in pore pressure moves the Mohr circle to the left. Note that the increase in pore pressure decreases the mean stress, but does not change the differential stress. In other words, the diameter of the circle remains constant, but its center moves to the left.

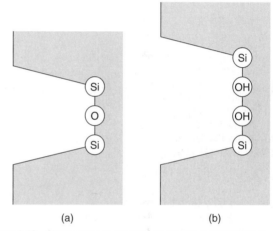

Figure 6.30 Sketch of the atoms at the tip of a crack, to illustrate the phenomenon of hydrolytic weakening. (a) At the tip of a dry crack, Si is bonded to O. (b) At the tip of a wet crack, Si is bonded to OH. The Si-O bond is stronger than the Si-OH bond.

6.8.3 Effect of Pore Pressure on Shear Failure and Frictional Sliding

We can observe the effects of pore pressure on shear fracturing by running a confined compression experiment in which we pump fluid into the sample through a hole in one of the pistons, thereby creating a fluid pressure, P_f, in pores of the sample (Figure 6.10). The fluid creating the confining pressure acting on the sample is different from and is not connected to the fluid inside the sample. The value of P_f subtracts both from the value of confining pressure (σ_3) and from the value of σ_1. So if the pore pressure increases in the sample, the mean stress decreases but the differential stress remains the same. This effect can be represented by the Coulomb failure criterion equation; P_f subtracts from the value of σ_n on the right side of the equation:

$$\sigma_s = c + \mu(\sigma_n - P_f) \qquad \text{Eq. 6.6}$$

The term $(\sigma_n - P_f)$ is commonly labeled $\sigma_n{}^*$, and is called *effective stress*.

We can easily see the effect of increasing P_f in the experiment on a Mohr diagram. When P_f is increased, the whole Mohr circle moves to the left but its diameter remains unchanged (Figure 6.29); and when the circle touches the failure envelope, shear failure occurs, even if the relative values of σ_1 and σ_3 are unchanged. In other words, a differential stress that is insufficient to break a dry rock may break a wet rock, if the fluid in the wet rock is under sufficient pressure. Thus, an increase in pore pressure effectively weakens a rock. In the case of forming a shear fracture in intact rock, pore pressure plays a role by pushing open microcracks, which coalesce to form the rupture at smaller remote stresses.

Similarly, an increase in pore pressure decreases the shear stress necessary to initiate frictional sliding on a preexisting fracture, for the pore pressure effectively decreases the normal stress across the fracture surface. Thus, as we discuss further in Chapter 8, fluids play an important role in controlling the conditions under which faulting occurs.

6.8.4 Effect of Intermediate Principal Stress on Shear Rupture

Fractures form parallel to σ_2, and so the value of σ_2 does not affect values of normal stress (σ_n) and shear stress (σ_s) across potential shear rupture planes. Thus, in this chapter we have assumed that the value of σ_2 does not have a major effect on the shear failure strength of rock, and we considered failure criteria only in terms of σ_1 and σ_3. In reality, however, σ_2 does have a relatively small effect on rock strength. Specifically, rock is stronger in confined compression when the magnitude of σ_2 is closer to the magnitude of σ_1 than when the magnitude of σ_2 is closer to the magnitude of σ_3. An increase in σ_2 has the same effect as an increase in confining pressure.

6.9 CLOSING REMARKS

Why study brittle deformation? For starters, since rocks deform brittlely under the range of pressures and temperatures usually found at or near the Earth's surface (<10–15 km), fractures pervade rocks of the upper crust. In fact, every rock outcrop that you will ever see contains fractures at some scale. Because fractures are so widespread, they play a major role in determining the permeability and strength of rock, and the resistance of rock to erosion. Therefore, fractures affect the velocity and direction of toxic waste transport, the location of an ore deposit, the durability of a foundation, the stability of a slope, the suitability of a reservoir, the safety of a mine shaft, and the form of a landscape. Moreover, fracture formation is the underlying cause of earthquakes, and contributes to the evolution of regional tectonic features. This chapter has provided an introduction to the complex and rapidly evolving subject of brittle deformation and fracture mechanics. We have tried to describe what fractures are, how they form, and under what conditions. In the next two chapters, we apply this information to developing an understanding of the major types of brittle structures: joints, veins, and faults.

ADDITIONAL READING

Atkinson, B. K., ed., 1987, *Fracture mechanics of rock,* London: Academic Press, 534 p.

Brace, W. F., and Bombolakis, E. G., 1963, A note on brittle crack growth in compression, *Journal of Geophysical Research,* v. 68, p. 3709–3713.

Engelder, T., 1993, *Stress regimes in the lithosphere,* Princeton Univ. Press, 455 p.

Jaeger, J. C., and Cook, N. G. W., 1979, *Fundamentals of rock mechanics* (third edition). Chapman and Hall, London, 593 p.

Lawn, B. R., 1993, *Fracture of brittle solids,* Cambridge: University Press, Cambridge, 378 p.

Paterson, M. S., 1978, *Experimental rock deformation—the brittle field,* Springer-Verlag, New York, 254 p.

Pollard, D. D., and Aydin, A., 1988, Progress in understanding jointing over the past century, *Geological Society of America Bulletin,* v. 100, p. 1181–1204.

Price, N. J., 1966, *Fault and joint development in brittle and semi-brittle rock,* Pergamon Press, London, 176 p.

Scholz, C. H., 1990, *Mechanics of earthquakes and faulting,* Cambridge University Press, Cambridge, 439 p.

Secor, D. T., 1965, Role of fluid pressure in jointing, *American Journal of Science,* v. 263, p. 633–646.

Chapter 7
Joints and Veins

7.1 INTRODUCTION

Visitors from around the world trek to Arches National Park in southeastern Utah to marvel at its graceful natural arches. These arches appear to have been carved through high, but relatively thin, freestanding sandstone walls.

From the air, you can see that the park contains a multitude of such walls, making its landscape resemble a sliced-up loaf of bread (Figure 7.1a). The surfaces of rock walls in Arches Park initiated as *joints,* which are natural fractures in rock across which there has been no shear displacement (see Table 7.1 for a more formal definition). Erosive processes have preferentially attacked the walls of the joints, so that today you can walk in the space between the walls. Though joints are not always as dramatic as those in Arches National Park, nearly all outcrops contain joints (Figure 7.1b). At first glance, joints may seem to be simple and featureless geologic structures, but in fact they are well worth studying, not only because of their importance in controlling landscape morphology, but also because they profoundly affect rock strength and permeability, and because they can provide a detailed, though subtle history of stress and strain in a region.

Although the basic definition of the term *joint* is non-genetic, most contemporary geologists who study joints believe that they form during Mode I loading (see Chapter 6); that is, they are tensile fractures that form perpendicular to the σ_3 trajectory and parallel to the principal plane of stress that contains the σ_1 and σ_2 directions. Not all geologists share this viewpoint, and some researchers use the term 'joint' when referring to shear fractures as well. This second usage is ambiguous, because structures that are technically faults might also be referred to as joints. Therefore, we do not use the term 'joint' in reference to a shear fracture.

(a)

(b)

(c)

Figure 7.1 Examples of joints and veins on different scales. (a) Air photo of regional joints in sandstone near Arches National Park, Utah. Note the Colorado River for scale. (b) Large joint face in Entrada sandstone near Moab, Utah. Note that thin bedded shale unit below the sandstone has much more closely spaced joints. (c) Veining in limestone exposed in a road cut near Catskill, New York.

Table 7.1 Joint Terminology

Arrest line	An arcuate ridge on a joint surface, located at a distance from the origin, where the joint front stopped or paused during propagation of the joint; also called *rib marks*.
Columnar joints	Joints that break rock into generally hexagonal columns; they form during cooling and contraction in hypabyssal intrusions or lava flows.
Conjugate system	Two sets of joints oriented such that the dihedral angle between the sets is approximately 60°.
Continuous joints	Throughgoing joints that can be traced across an outcrop, and perhaps across the countryside.
Cross joints	Discontinuous joints that cut across the rock between two systematic joints, and are oriented at a high angle to the systematic joints.
Cross-strike joints	Joints that cut across the general trend of fold hinges in a region of folded rocks (i.e., the joints cut across regional bedding strike).
Desiccation cracks	Joints formed in a layer of mud when it dries and shrinks; desiccation cracks (or *mud cracks*) break the layer into roughly hexagonal plates.
Discontinuous joints	Short joints that terminate within an outcrop, generally at the intersection with another joint.
En echelon	An arrangement of parallel planes in a zone of fairly constant width; the planes are inclined to the borders of the zone and terminate at the borders of the zone. In an *en echelon* array, the component planes are of roughly equal length.
Hackle zone	The main part of a plumose structure, where the fracture surface is relatively rough due to microscopic irregularities in the joint surface formed when the crack surfaces get deflected in the neighborhood of grain-scale inclusions in the rock, or due to off-plane cracking (formation of small cracks adjacent to the main joint surface) as the fracture propagates.
Hooking	The curving of one joint near its intersection with an earlier formed joint.
Inclusion	A general term for any solid inhomogeneity (e.g., fossil, pebble, burrow, xenolith, amygdule, coarse grain, etc.) in a rock; inclusion may cause local stress concentrations.
Joint	A natural, unfilled, planar or curviplanar fracture that forms by tensile loading (i.e., the walls of a joint move apart very slightly as the joint develops). Joint formation does not involve shear displacement.
Joint array	Any group of joints (systematic or nonsystematic).
Joint density	The surface area of joints per unit volume of rock (also referred to as *joint intensity*).
Joint origin	The point on the joint (usually a flaw or inclusion) at which the fracture began to propagate; it is commonly marked by a dimple.
Joint set	A group of systematic joints.
Joint stress shadow	The region around a joint surface where joint-normal tensile stress is insufficient to cause new joints to form.
Joint system	Two or more geometrically related sets of joints in a region.
Mirror region	Portion of a joint surface adjacent to the joint origin where the surface is very smooth; mirrors do not occur if the rock contains many small-scale heterogeneities.
Mist region	A portion of a joint surface surrounding the mirror where the fracture surface begins to roughen.
Nonsystematic joints	Joints that are not necessarily planar, and are not parallel to nearby joints.
Orthogonal system	Two sets of joints that are at right angles to each other.
Plume axis	The axis of the plume in a plumose structure.
Plumose structure	A subtle roughness on the surface of some joints (particularly those in fine-grained rocks) that macroscopically resembles the imprint of a feather.
Sheeting joints	Joints formed near the ground surface that are roughly parallel to the ground surface; sheeting joints on domelike mountains make the mountains resemble delaminating onions.
Strike-parallel joints	Joints that parallel the general trend of fold hinges in a region of folded strata (i.e., the joints parallel regional bedding strikes).
Systematic joints	Roughly planar joints that occur as part of a set in which the joints parallel one another, and are relatively evenly spaced from one another.
Twist hackle	One of a set of small *en echelon* joints formed along the edge of a larger joint; a twist hackle is not parallel to the larger joint, and forms when the fracture surface twists into a different orientation and then breaks up into segments.

(a)

(b)

Figure 7.2 Photographs of plumose structure. (a) Wavy plumose structure on a joint in siltstone. (b) Plumose structure in thin bedded siltstone; the pencil points to the joint origin.

In this chapter, we begin by describing the morphology of individual joints and the geometric characteristics of groups of joints. Then, we discuss how to study joints in the field, and how to interpret them. We conclude by describing *veins,* which are fractures filled with minerals that precipitated from a fluid (Figure 7.1c). But before we begin, we offer a note of caution. The interpretation of joints and veins remains quite controversial, and it is common for field trips that focus on these structures to end in heated debate. As you read this chapter, you'll discover why.

7.2 SURFACE MORPHOLOGY OF JOINTS

7.2.1 Plumose Structure

If you look at an exposed joint surface, you'll discover that the surface is not perfectly smooth. Rather, joint surfaces display a subtle roughness that resembles the imprint of a feather. We call this pattern *plumose structure* (Figure 7.2). Plumose structures form at a range of scales, depending on the grain size of the host rock. In very fine-grained coal, for example, components of plumose structure tend to be much smaller than in relatively coarser siltstone. Really clear examples of plumose structure form in fine-grained rocks like shale, siltstone, and basalt, but you might not see obvious plumose structure on joints in very coarse-grained rocks like granite.

Let's look at plumose structure a little more closely (Figure 7.3a). A plumose structure spreads outward from the *joint origin,* which, as the name suggests, represents the point at which the joint started to grow. Joint origins typically look like small dimples in the fracture plane (Figure 7.3b). Several distinct morphological zones surround the joint origin. In the *mirror zone,* which lies closest to the origin, the joint surface is very smooth. Further from the origin, the mirror zone merges with the *mist zone,* in which the joint surface slightly roughens. Mirror and mist zones, while well developed in

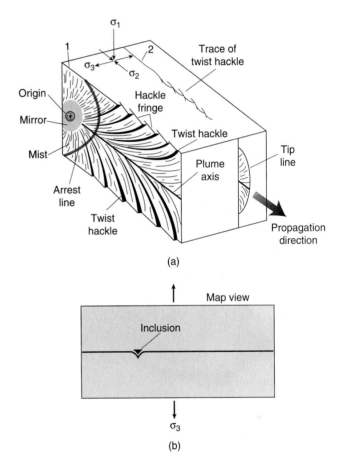

(a)

(b)

Figure 7.3 (a) Block diagram showing the various components of an ideal plumose structure on a joint. The face of joint 1 is exposed. Joint 2 is within the rock. (b) Simple cross-sectional sketch showing the dimple of a joint origin, controlled by an inclusion.

joints formed in glassy rocks, are difficult to recognize in coarser rocks. Continuing outward, the mist zone merges with the *hackle zone,* in which the joint surface is even rougher. It is the hackle zone that forms most of the plumose structure. Roughness in the hackle zone defines vague lineations, or

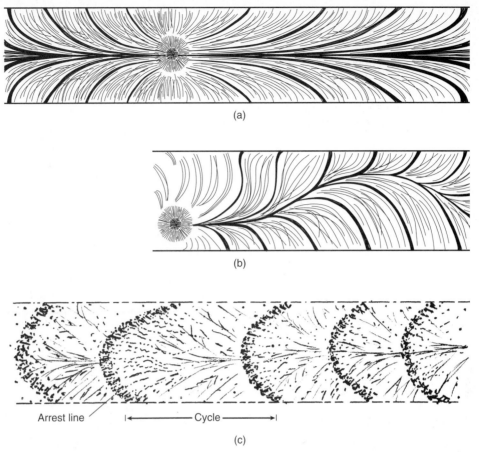

(a)

(b)

Arrest line |◄——— Cycle ———►|

(c)

Figure 7.4 Types of plumose structure. (a) Straight plume. (b) Curvy plume. (c) Plume with many arrest lines, suggesting that it opened rhythmically.

barbs, that curve away from a *plume axis,* which together with barbs compose the featherlike plume. The acute angle between the barbs and the axis points back toward the joint origin, so the plume defines the local direction of joint propagation. The median line may be fairly straight and distinct, or it may be wavy and diffuse (Figure 7.4a and b). On some joint surfaces, concentric ridges known as *arrest lines* (Figure 7.4c) form on the joint surface at a distance from the origin.

7.2.2 Why Does Plumose Structure Form?

Mode I loading of a perfectly isotropic and homogeneous material should yield a perfectly smooth, planar fracture oriented perpendicular to the remote σ_3. Real joints are not perfectly smooth for two reasons. First, real rocks are not perfectly isotropic and homogeneous, meaning that the material properties of a rock change from point to point in the rock. Inhomogeneities in rock exist because not all grains in a rock have the same composition and because not all grains are in perfect contact with one another. The presence of inhomogeneities distorts the local stress field at the tip of a growing joint, so that the principal stresses at the tip are not necessarily parallel to the remote σ_3. As a consequence, the joint-propagation path twists and tilts slightly as the joint grows.

Second, the stress field at the tip of a crack changes as the crack tip propagates. Recall from Equation 6.3 that the stress intensity at the crack tip is proportional to the length of the crack, and that the magnitude of the local tensile stress at the tip of the crack is, in turn, proportional to the stress intensity. Thus, as the crack grows, the stress intensity at the crack tip grows, up to a limiting value. Experimental work demonstrates that the velocity of crack-tip propagation is also proportional to the stress intensity. Stress magnitude and tip-propagation velocity are relatively small near the joint origin, because the crack is very short. These parameters increase away from the origin and eventually reach a maximum at a distance from the origin. If the stress magnitude at the tip exceeds a critical value, the energy available for cracking rock exceeds the energy needed to create a single surface. The excess energy goes into breaking bonds off the plane of the main joint surface, resulting in the formation of microscopic off-plane cracks that splay off the main joint. If the energy becomes excessive, the crack may actually split into two separate, parallel surfaces.

With these two reasons in mind, we can now explain why plumose structure has distinct morphological features. The dimple at the origin forms because the flaw[1] at which

---

[1]Flaws at which joints initiate include open pores, preexisting microcracks, irregularities on a bedding plane, *inclusions* (a solid inhomogeneity in a rock, like a pebble, fossil, amygdule, or concretion), or primary sedimentary structures (a sole mark or ripple).

the joint nucleated either was not perpendicular to the remote σ_3, or caused a local change in the orientation of stress trajectories, and thus, the portion of the joint that formed in the immediate vicinity of the origin was not perpendicular to the remote σ_3. As soon as the crack propagated away from the flaw, it curved into parallelism with the σ_1-σ_2 principal plane (Figure 7.3). In the mirror zone, the joint is still short, and thus the stress intensity, tensile stress magnitude, and tip-propagation velocity are all relatively small. As a consequence of the low stress, only bonds in the plane exactly perpendicular to the local σ_3 can break, so the joint surface that forms is very smooth. In the mist zone, however, the joint moves faster, stress is higher, and stress at the joint tip is sufficiently large to break off-plane bonds, thereby forming microscopic off-plane cracks that make the surface rougher than in the mirror zone. In the hackle zone, the joint tip is moving at its terminal velocity. Stresses at the crack tip are so large that larger off-plane cracks propagate, and the crack locally bifurcates at its tip to form microscopic splays that penetrate the joint walls. The roughness of the hackle zone also reflects the formation of tiny splays and warps of the joint surface where the joint tip twists or tilts as it passes an inclusion and breaks into microscopic steps. Arrest lines on a joint surface represent places where the fracture tip pauses between successive increments of propagation. The visible ridge of the arrest line, in part, represents the contrast between the rough surface of the hackle and the relatively smooth surface of the mirror/mist zone formed as the fracture begins to propagate. In part, the ridge may be analogous to the dimple formed at a crack origin. Thus, it is the twisting, tilting, and splitting occurring at the tip because of variations in local stress magnitude and orientation that cause plumose structures to form.

7.2.3 Twist Hackle

Features such as bedding planes and preexisting fractures, like 'free surfaces',[2] locally modify the orientation of principal stresses. If a growing joint enters a region where it no longer parallels a principal plane of stress (e.g., as occurs when the crack tip of a joint in a sedimentary bed approaches the bedding plane), the crack tip pivots to a new orientation. As a consequence, the joint splits into a series of small *en echelon* joints, because a joint surface cannot twist and still remain a single continuous surface. The resulting array of fractures is called *twist hackle,* and the edge of the fracture plane where twist hackle occurs is called the *hackle fringe* (Figure 7.3a). Note that if the hackle fringe intersects an outcrop face, the trace of a large planar joint within the outcrop may look like a series of small joints in an *en echelon* arrangement.

---

[2]A 'free surface' is a surface across which there is no cohesion, so it cannot transmit shear stresses. By definition, a free surface is a principal plane of stress, but if the free surface is not parallel to a principal plane of the remote stress, then the remote stress trajectories change orientation so that they are either parallel or perpendicular to the free surface. Weak bedding planes do not transmit shear stress completely, so they behave somewhat like free surfaces, and therefore affect the local stress field.

7.3 CHARACTERISTICS OF JOINT ARRAYS

7.3.1 Systematic versus Nonsystematic Joints

Systematic joints are a group of joints that are parallel or subparallel to one another, and maintain roughly the same average spacing over the region of observation (Figure 7.5a). There are no restrictions on the minimum spacing of joints for them to be systematic, but as a rule of thumb, the joints must be close enough that several are visible at the scale of observation. Systematic joints may cut through many layers of strata, or be confined to a single layer of strata. *Nonsystematic joints* have an irregular spatial distribution, they do

(a)

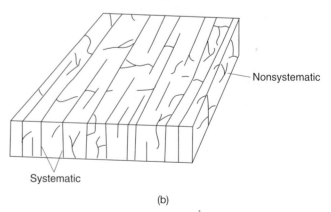

(b)

Figure 7.5 (a) Three sets of systematic joints controlling erosion in Cambrian sandstone (Kangaroo Island, Australia). (b) Block diagram showing occurrence of both systematic and nonsystematic joints in a body of rock.

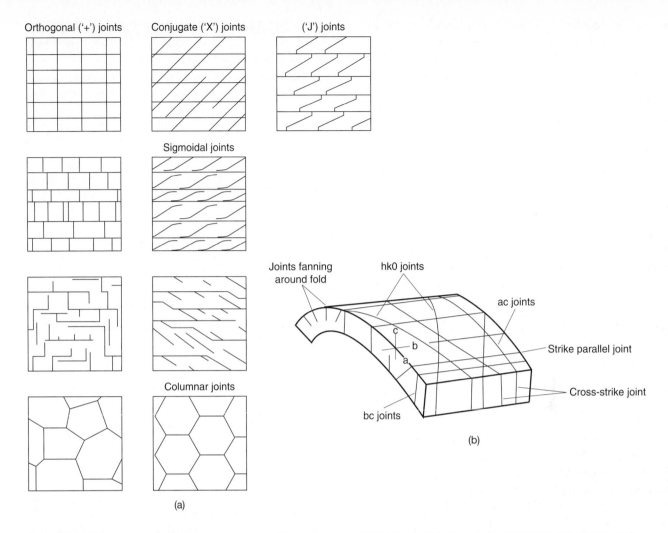

Figure 7.6 (a) Traces of various types of joint arrays on a bedding surface. (b) Idealized fold, showing arrangement of joint arrays with respect to fold symmetry axes. The "hko" label for joints that cut diagonally across the fold hinge is based on the Miller Indices from mineralogy; they refer to the intersections of the joints with the symmetry axes of the fold.

not parallel neighboring joints, and they tend to be nonplanar (Figure 7.5b). Nonsystematic joints may terminate at other joints. Both systematic and nonsystematic joints can occur in the same outcrop.

7.3.2 Joint Sets and Joint Systems

Describing groups of joints efficiently requires a fair bit of jargon. Matters are made even worse because not all authors use joint terminology in the same way, so it's good practice to define your terminology in context. We'll save the explanations of why various different groups of joints form until later in the chapter.

A *joint set* is a group of systematic joints. Two or more joint sets that intersect at fairly constant angles compose a *joint system*. The angle between two joint sets in a joint system is the *dihedral angle*. If the two sets in a system are mutually perpendicular (i.e., the dihedral angle is ~90°), we call the pair an *orthogonal system* (Figure 7.6a), and if the two sets intersect with a dihedral angle of significantly less than 90° (e.g., a dihedral angle of about 30–60°), we call the pair a *conjugate system* (Figure 7.6a). Many geologists use the terms 'orthogonal' or 'conjugate' to imply that the pair of joint sets formed at the same time. However, as you will see later in this chapter, nonparallel joint sets typically form at different times. So, we use the terms merely to denote a geometry, not a mode or timing of origin.

As shown in Figure 7.6a, many different configurations of joint systems occur. They are distinguished from one another by the nature of the intersections between sets and by the relative lengths of the joints in the different sets. In joint systems where one set consists of relatively long joints that cut across the outcrop and the other set consists of relatively short joints that terminate at the long joints, the throughgoing joints are *master joints,* and the short joints that occur between the continuous joints are *cross joints* (Table 7.1).

In the flat-lying sedimentary rocks that occur in continental interior basins and platforms (e.g., the Midwest of the eastern United States), joint sets are perpendicular to the ground surface and, therefore, to bedding, and orthogonal systems are common (e.g., Figure 7.5). In gently folded sedimentary rocks, such as along the foreland margin of a mountain range (e.g., the western side of the Appalachians), strata contain both vertical joint sets that cut across the folded layers, as well as joints that are at a high angle to bedding and fan around the folds (Figure 7.6b). Both orthogonal and conjugate systems occur in such gently folded strata. The joint sets of an orthogonal system in folded sedimentary rocks commonly have an approximate spatial relationship to folds of the region. So we can distinguish between *strike-parallel joints,* which parallel the general strike of bedding (i.e., parallel to regional fold hinges), and *cross-strike joints,* which trend at high angles (~60° to 90°) to the regional bedding strike (Figure 7.6b).[3] Conjugate systems in gently folded rocks consist of two cross-strike sets with their acute bisectrix at a high angle to the fold hinge. Keep in mind that both sets of joint systems need not form at the same time, so a conjugate geometry does not require that a system of joints conjugate shear ruptures.

In the internal portions of mountain belts, where rocks have been intensely deformed and metamorphosed, outcrops may contain so many nonsystematic joints that joint systems may be difficult to recognize or simply do not exist. In such regions, joints formed prior to deformation and metamorphism have been erased. New joints then form at different times during deformation, during uplift subsequent to deformation, or even in response to recent stress fields. Rocks in such regions are so heterogeneous that the stress field varies locally, and thus joints occur in a wide range of orientations. Nevertheless, in some cases, younger joints, meaning those formed during uplift or due to recent stress fields, may stand out as distinct sets.

Intrusive and metamorphic rocks without a strong schistosity (e.g., granite, migmatitic gneiss) commonly contain a set of joints that roughly parallels ground surface topography, and whose spacing decreases progressively toward the surface. Such joints are *sheeting joints,* or *exfolia-*

Figure 7.7 Sheeting joints in granite of the Sierra Nevada.

tion joints (Figure 7.7). If the ground surface is not horizontal, as is the case on the sloping side of a mountain, sheeting joints curve and follow the face of the mountain, thus giving the mountain the appearance of a partially delaminated onion. Rock sheets peel off the mountain along these joints, thereby creating smooth dome-shaped structures known as *exfoliation domes.* "Half Dome," the world-class challenge that draws mountain climbers to Yosemite National Park in the Sierra Nevada Mountains of California, is an exfoliation dome one-half of which was cut away by glacial erosion.

Hypabyssal intrusive igneous rock bodies (dikes and sills) and lava flows in many localities display *columnar jointing,* meaning that they have been broken into joint-bounded columns that, when viewed end on, have roughly hexagonal cross sections (Figure 7.6a). In the case of sheet intrusions, the long axis of the columns tends to be perpendicular to the boundaries of the sheet (i.e., are horizontal in dikes and vertical in sills); however, in some bodies the columns curve. The visual impression of columnar jointing catches people's imaginations, so columns tend to be dubbed with unusual names like Giant's Causeway (Ireland), Devils Postpile (California; Figure 2.23), and Devils Tower (Wyoming).

7.3.3 Cross-Cutting Relations Between Joints

The way in which nonparallel joints intersect one another provides useful information concerning the relative ages of joints. For example, if Joint A terminates at its intersection with Joint B, then Joint A is younger, because a propagating fracture cannot cross a free surface, and an open preexisting joint behaves like a free surface.[4]

A younger joint's orientation also may change where it approaches an older joint, if the older joint behaved like a free surface while the younger joint formed. Why? Remember that at or near a free surface, a Mode I fracture must be

---

[3]Note that cross-strike joints are not necessarily the same as cross joints (Table 7.1).

[4]Joints that have been cemented together, or whose faces are being tightly held together, can transmit at least some shear stress, and thus do not behave like perfect free surfaces.

either parallel or perpendicular to the surface so as to maintain perpendicularity to σ_3. Thus, near a free surface, the local stress field may differ from the remote stress field, if the free surface does not parallel a principal plane of the remote stress field. If an older joint (B) acts as a free surface, then the younger joint (A) curves in the vicinity of joint B to become parallel to the local principal plane of stress adjacent to B, unless it already happens to parallel a principal plane of stress. The way in which the younger joint curves depends on the stress field. If the local σ_3 adjacent to the older joint is parallel to the walls of the older joint, then the younger joint curves so that it is orthogonal to the first joint at their point of intersection, a relationship called *hooking,* and the junction is called a 'J' junction (Figure 7.6a). However, if the local σ_3 is perpendicular to the walls of the older joint, then the younger joint curves into parallelism with joints of the first set, and has a sigmoidal appearance (Figure 7.6a).

In some joint systems, two nonparallel joints appear to cross each other without any apparent interaction, or, in other words, are *mutually cross cutting.* Such intersections are sometimes referred to as '+' intersections, if the joints are orthogonal, or '×' intersections if they are not orthogonal (Figure 7.6c). These relationships may represent situations where: (1) the earlier joint did not act as a free surface; (2) the intersection of two younger joints at the same point on an older joint is simply coincidental; or, (3) the mutual cross-cutting relationship is an illusion—within the body of the outcrop, the older joint terminated, and the younger joint simply grew around it.

7.3.4 Joint Spacing in Sedimentary Rocks

When looking at jointing in a sequence of stratified sedimentary rock, you might notice that within a bed joints are somewhat evenly spaced. Where this occurs, we can define *joint spacing* as the average distance between adjacent members of a joint set, as measured perpendicular to the surface of the joint. Informally, geologists refer to joints as being "closely spaced" or "widely spaced" in a relative sense, but to be precise, you should describe joint spacing in units of length (e.g., 10 cm).

To understand why joints are evenly spaced, we must first have an image of how an array of joints develops in a bed. Consider the bed of sandstone shown in Figure 7.8. Ultimately (time 7), the bed contains five joints. These joints could have formed all at once, or in sequence. Experimental work suggests that joints form in sequence, that is, first joint 1, then joint 2, then joint 3, and so on. Note that when a new joint forms, its plane is at some distance greater than a minimum distance (d_m) from a preexisting joint. Formation of a joint relieves tensile stress for a critical distance d_m on either side (Figure 7.9a). The zone on

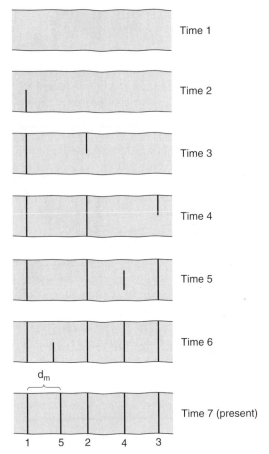

Figure 7.8 A model of the sequence of development of joints. Time 1 refers to the time before the first joint forms, and time 7 is the present day. This model suggests that joints form in a random sequence.

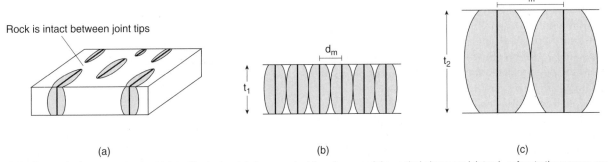

(a) (b) (c)

Figure 7.9 Stress shadow concept around joints. The horizontal planes are bedding planes, and the vertical planes are joints; d_m refers to the average spacing between joints. (a) Block diagram illustrating stress shadow (shaded area) around each joint. Note how stresses can be transmitted into the interior. Layer-parallel stress is transmitted across regions that are unfractured in the third dimension. Stresses are also exerted by tractions at bedding contacts. (b) Thin bedded sequence, containing joints with narrow stress shadows, so that the joints are closely spaced. (c) Thick bedded sequence, containing joints with wide stress shadows, so that the joints are widely spaced.

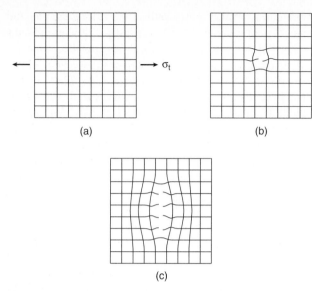

(a)

(b)

(c)

Figure 7.10 Illustration of why joint stress shadows exist. (a) A grid of springs. (b) Cutting one spring causes only a few springs to relax around the cut. (c) Cutting many springs in a row causes a wider band of springs to relax.

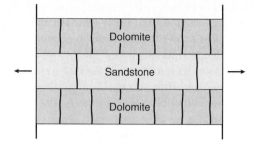

Figure 7.11 Cross-sectional sketch illustrating a multilayer that is composed of rocks with different Young's modulus. The stiffer layers (dolomite) develop more closely spaced joints.

either side of a joint in which there has been a decrease in tensile stress due to formation of the joint is called the *joint stress shadow.* Stresses sufficient to create the next joint are achieved only outside of this shadow, and are created by traction between the bed and other beds above and below it, as well as by stress transmitted within the bed beyond the fracture front of the preexisting joint. The spacing between joints is determined by the width of the joint stress shadow; so, because the shadow is about the same width for all joints in the bed, the spacing ends up being fairly constant. Joint spacing depends on four parameters: bed thickness, strain, stiffness, and tensile strength. We'll examine each of these parameters in turn.

Relation between joint spacing and bed thickness. All other parameters being equal, joints are more closely spaced in thinner beds, and are more widely spaced in thicker beds. The relationship is a reflection of joint stress shadow width, because the greater the cross-sectional length of the joint (= length of the joint trace in a plane perpendicular to bedding and the joint), the wider the stress shadow (Figure 7.9b and c). To picture why this is so, imagine a net composed of springs (Figure 7.10a). If you reach into the net and cut one spring, only a few of the neighboring springs relax (Figure 7.10b); however, if you cut many of the springs in a row, a much wider zone of neighboring springs relaxes (Figure 7.10c). In thicker beds, joint stress shadows are wider, so joints tend to be more widely spaced.

Relation between joint spacing and lithology. Recall that the *stiffness* (namely, the value of E, Young's modulus, in Hooke's law, $\sigma = E \cdot \mathbf{e}$; see Chapter 5) of a rock layer depends on lithology. Imagine a block of rock composed of three layers: dolomite, sandstone, and dolomite (Figure 7.11). Dolomite is stiffer (E is ~600 MPa) than sandstone (E is ~200 MPa). We stretch the block under brittle condi-

tions by a uniform amount so that all layers undergo exactly the same elastic elongation (**e).** The stress that develops in each bed is defined by Hooke's law, but since the stretch is the same for each bed, the magnitude of σ depends on E. Thus, beds composed of rock with a larger E develop a greater stress and fracture first. In the model of Figure 7.11, the stiffer dolomite bed probably fractured a few times before the sandstone bed fractured for the first time, so more joints develop in the dolomite bed than in the sandstone bed. In sum, for a given strain, stress is larger in a stiffer bed than in a less stiff bed; so other factors being equal, stiffer beds have smaller joint spacings.

Relation between joint spacing and tensile strength. Prediction of fracture spacing by considering E alone is not foolproof, because in some circumstances a rock with a smaller E may actually have a lower tensile strength, and thus will crack at a lower strain than a rock with a larger E, if the rock with the larger E also has a larger tensile strength. Other factors being equal, rock with the smaller tensile strength develops more closely spaced joints.

Relation between joint spacing and the magnitude of extensional strain. A bed that has been stretched more contains more joints than a bed that has been stretched less, as you might expect.

If you ever have the chance to hike down the Grand Canyon, don't forget to look at the jointing in different units as you descend. Because bedding planes tend to be weak and do not transmit shear stress efficiently, joints typically terminate at bedding planes. Because joint spacing depends on bed thickness and lithology, joint spacing varies from bed to bed. Weak, thinly bedded shales contain such closely spaced joints that they break into tiny fragments. As a consequence, they tend to form slopes. In contrast, thick sandstone beds develop only widely spaced fractures. These joints control erosion, so thick sandstone beds hold up high cliffs.

7.4 JOINT STUDIES IN THE FIELD

Before explaining how you might go about studying joints in the field, it is worth discussing *why* you might want to study joints in the field. Perhaps the most common reason that people study joints is for engineering or hydrologic applications. As we noted before, fractures affect the strength of foundations, quarrying operations, excavations, groundwater and toxic waste flow, and slope stability. For example, if you find that a region contains a systematic joint set that is oriented north-south, you could expect groundwater to flow faster in the north-south direction than in the east-west direction, or that quarrying might be easier if the quarry walls strike north-south than if they strike east-west. But study of jointing has applications to academic geologic issues as well. Geologists who are interested in tectonics might study joints to see if they provide information about paleostress fields, and geomorphologists may wish to study joints to find out if they control the drainage patterns or the orientation of escarpments. With these goals in mind, what specifically do we look for in a joint study? In most cases the questions that we ask include:

(1) Is the jointing in the outcrop systematic or nonsystematic? In other words, can we define distinct sets of planar joints and/or regularly oriented cross joints in an outcrop, or does the outcrop contain irregular and randomly oriented joints with relatively short traces (i.e., nonsystematic joints)? If nonsystematic jointing is present, is it localized or pervasive? Hypotheses on joint formation in a region depend on whether the joints are systematic or not. Systematic joints likely reflect regional tectonic stress trajectories at the time of fracturing, whereas nonsystematic joints reflect very local heterogeneities of the stress field. While nonsystematic joints may be important for determining rock strength and permeability, they provide no information on regional paleostress orientation.

(2) If joint sets are present, what are the orientations of the joint sets? If there is more than one set, is there a consistent angular relationship between the sets, such that we can describe a joint system? Information on the orientation and distribution of joint sets and systems is critical for engineering and hydrologic analyses. For example, joint sets that run parallel to a proposed road cut would create a greater rock-fall hazard than joints that are perpendicular to the cut.

(3) What is the nature of cross-cutting relationships between joints of different sets, and what is the geometry of joint intersections? Do joints cross without appearing to interact, do they curve to create J-intersections, or do they curve into parallelism with one another? Knowledge of cross-cutting relations allows one to determine whether one set of joints is older or younger than another set of joints, a determination that is critical to tectonic interpretations using joints.

(4) What is the surface morphology of the joints? Is plumose structure visible on joint surfaces, and if so, what types of plumes (wavy or straight) are visible? Are numerous arrest lines clearly evident on the joint surface? The presence of plumose structure is taken as proof that a joint propagated as a Mode I fracture, and the geometry of the plume provides clues to the way in which the joint propagated (e.g., in a single pulse, or in many distinct pulses;

pulsating growth). For example, if a joint surface contains numerous origins, the joint probably initiated at different times along its length. Joints whose surfaces contain many arrest lines probably propagated in increments. We will see that start and stop growth of a fracture may indicate that fracture growth was controlled by fluctuating fluid pressure in a rock. Are there other structural features superimposed on joints (e.g., stylolitic pits or slip lineations)? If joints display surface features other than plumose structure, the features indicate post-joint formation strain. Stylolitic pitting on a joint indicates compression and resulting pressure solution across the joint, and slip lineations suggest that the joint was reactivated as a fault later in its history.

(5) What are the dimensions of joints? In other words, are the trace lengths of joints measured in centimeters or hundreds of meters? The effect that jointing has on rock strength and rock permeability over a region is significantly affected by the dimensions of the joints. For example, large throughgoing joints that parallel an escarpment contribute more to the hazard of escarpment collapse than will short nonsystematic joints. Tectonic geologists commonly focus attention on interpretation of large joints, on the assumption that these reflect regional tectonic stress conditions, which are often of interest.

(6) What is the spacing and joint density in outcrop? By joint spacing, we mean the average distance between regularly spaced joints. Information on joint spacing provides insight into mechanical properties of rock layers, and on the fracture permeability of rock layers. By *joint density,* we mean, in two dimensions, what is the trace length of joints per unit area of outcrop, or in three dimensions, what is the area of joints per unit volume of outcrop? Joint density depends both on the length of the joints and on their spacing. Information on joint density helps define the fracture-related porosity and permeability of a rock body.

(7) How is the distribution of joints affected by lithology? In sedimentary rocks, do individual joints cut across a single bed, or do they cut across many beds, or even through the entire outcrop and beyond? In what way is joint spacing affected by bed composition? In the igneous rocks and their contact zones, is the joint spacing or style (both within an igneous body and in the country rock that was intruded) controlled by the proximity of the joint to contact? Information about the relationship between jointing and lithology can be related to physical characteristics such as Young's modulus (E) of rock, and also can help determine variations in fracture permeability as a function of position in a stratigraphic sequence. Information on the relationship between jointing and lithology may also give insight into the cause of joint formation.

(8) Are joints connected to one another or are they isolated? The connectivity of joints is critical to determining whether the joints could provide a permeable network through which fluids (e.g., contaminated groundwater, or petroleum) could flow.

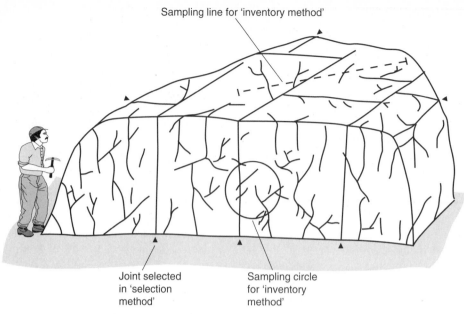

Sampling line for 'inventory method'

Joint selected in 'selection method'

Sampling circle for 'inventory method'

Figure 7.12 Joint study using the inventory and selection methods.

(9) How are joints related to other structures and fabrics? Are joints parallel to tectonic foliations? Are joints geometrically related to folds? Are joints reoriented by folding, or do they cut across the folds? Is there a relationship between joint orientation and measured contemporary stresses? Is the spacing or style of joints related to the proximity of faults? Information on the relation of joints to other structures provides insight into the tectonic conditions in which joints form and the timing of joint formation with respect to other structures in a region.

7.4.1 Dealing with Field Data about Joints

There are basically two ways to carry out a field study of joint orientation, spacing, and intensity. In the *inventory method,* you define a representative region and measure all joints that occur within the region. For example, you could draw a circle or square on the outcrop and measure all joints that occur within the circle, or you could draw a line across the outcrop and measure all joints that cross the line (Figure 7.12). The inventory method is necessary if you need to determine fracture density in a body of rock, or to provide statistics on joint data. Mathematical procedures for determining joint density in three dimensions from measurements on two-dimensional surfaces are available, but are beyond the scope of this book. You can use the inventory method for either systematic or nonsystematic joints.

The inventory method allows you to determine dominant joint orientations using statistical methods. But the problem with the inventory method is that a large number of nonsystematic fractures in an outcrop may obscure the existence of sets of systematic joints, especially if the systematic joints are widely spaced. For this reason, if you are trying to define systematic sets of fractures, the *selection method* is a more appropriate approach. In this method, you visually scan the outcrop and subjectively decide which are

the dominant sets (Figure 7.12). Then, you measure a few representative joints of each set and specify the spacing between joints in the set. Effectively, you are filtering your measurements in the field. While this technique won't permit determination of fracture density or provide statistics on joint orientation, it will allow you to define fracture systems in a region. The hazard with this procedure is that careless observers may record what they want to see, not what is really in the outcrop. Unfortunately, when observers start scanning a new outcrop, they may subconsciously look for the same joint sets that they saw in the last outcrop studied, and therefore may miss different, but important, joint sets that occur in the second outcrop.

Joint data can be recorded in a number of ways. One way is to plot the strike and dip of joints on a geologic map. If the joints are vertical, a measurement of the trend of the joint may be sufficient. You can create a powerful visual impression of joint attitudes in a region by drawing representative *joint trajectories* as trend lines on a map (Figure 7.13a). These trajectories are merely lines that represent the trends of joints; they are not necessarily the map traces of individual joints.

Statistical diagrams that show attitudes of many different joints within a given region can help you identify dominant joint orientations in a region. What at first sight may be a meaningless jumble may resolve into significant groupings. If joints in a particular region are not vertical, it is most appropriate to plot their attitudes on a contoured equal-area net; but if the joints are basically vertical, a common occurrence in flat-lying sedimentary strata, their strikes can be shown on histograms. A *histogram,* in the case of joints, indicates the number of joints whose strike falls within a particular range. On a bar histogram (Figure 7.13b), the abscissa represents bearings from 0° to 180°, and the ordinate is proportional to the number of fracture-strike measurements. On a polar

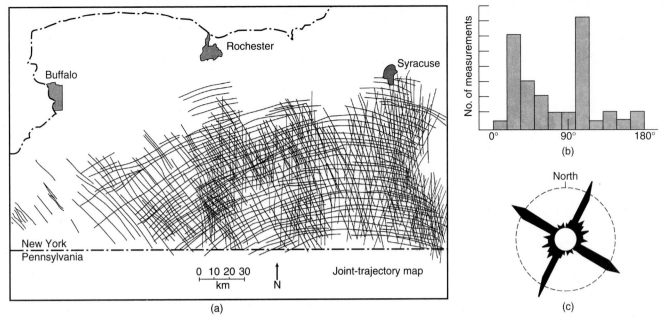

Figure 7.13 Ways of representing joint arrays. (a) Joint trajectory map. (b) Frequency diagram. (c) Rose diagram. The three examples do not portray the same data sets.

histogram, which is also called a *rose diagram,* you can show the bearings directly on the diagram (Figure 7.13c). The number of joints whose strike falls within a given range is shown by a pie-slice segment whose radius is proportional to the number or to the percentage of joints with that orientation. Rose diagrams work better than bar histographs to give you an intuitive feel for the distribution of joint attitudes.

The orientation of joints is not the only important information to record during a joint study, as you can see from the list of questions in the previous section. In modern joint studies, you should also record joint spacing, joint trace length, cross-cutting relations between joints, the relation of joints to lithology, joint surface morphology, and the relation of joints to other structures. To clarify relations among joints, it helps to make outcrop sketches in your notes.

7.5 ORIGIN AND TECTONIC INTERPRETATION OF JOINTS

Why do joints form? In Chapter 6, we learned that joints develop when stress exceeds the tensile fracture strength of a rock, and Griffith cracks begin to propagate. But under what conditions in the Earth's brittle crust are stresses sufficient to crack rocks? In this section, we describe several ideas to explain how stress states leading to joint formation develop in a rock body. But before you read further, a note of caution. When using these ideas as a basis for field interpretation of jointing, keep in mind that different joints in the same outcrop may have formed at different times and for different reasons, because once formed, a joint doesn't heal and disappear unless the rock gets metamorphosed or pervasively sheared. Further, local variations in the stress field, which are a natural feature of inhomogeneous rock, may cause joints that formed at the same time to have different orientations at different locations. Because of these factors, joint interpretation continues to challenge geologists.

7.5.1 Joints Related to Uplift and Unroofing

Lithostatic pressure, due to the weight of overlying rock, compresses rock at depth in the crust. Also, because of the Earth's geothermal gradient, rock at depth is warmer than rock closer to the Earth's surface. Regional uplift leads to erosion of the overburden and the unroofing of buried rock (Figure 7.14), which causes a change in the stress state for three reasons: cooling, Poisson effect, and membrane effect.

As the burial depth of rock decreases, it *cools and contracts.* The rock can shrink in a vertical direction without difficulty, because the Earth's surface is a free surface. But, because the rock is embedded in the Earth, it is not free to shrink elastically in the horizontal direction as much as it would if it were unconfined, so horizontal tensile stress develops in the rock. Furthermore, as the overburden diminishes, rock expands (very slightly) in the vertical direction. Therefore, because of the *Poisson effect* (see Chapter 6), it contracts in the horizontal direction. Again, because the rock is embedded in the earth, it cannot shorten in the horizontal direction as much as it would if it were unconfined, so a horizontal tensional stress develops. Uplift and unroofing effectively cause rock layers to move away from the center of the earth. The layer stretches like a membrane as its radius of curvature increases, thereby creating tensile stress in the layer, called the *membrane effect.*

If the horizontal tensional stress created by any or all of these reasons overcomes the compressive stresses due to burial and exceeds the tensile strength of the rock, it will cause

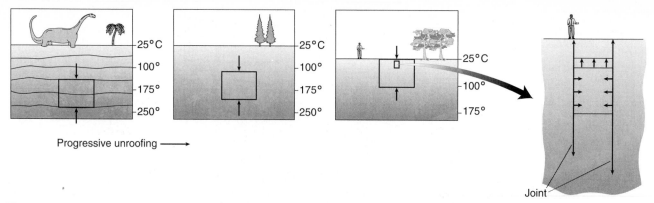

Figure 7.14 Joint formation during unroofing. As the block of rock approaches the ground surface, subsequent to the erosional removal of overburden, it expands in the vertical direction and contracts in the horizontal direction. It also cools (vertical axis shows generalized isotherms).

the rock to crack and form joints. Joints formed for the above reasons tend to be vertical because they generate a horizontal σ_3. Recall that the Earth's surface is a free surface, so it must be a principal plane of stress. Therefore, the other two principal planes of stress must be vertical. Uplift and unroofing are probably important causes of joint formation in sedimentary basins of continental interiors, which are subjected to epeirogenic movements, or in orogens that are uplifted long after collisional or convergent tectonism has ceased.

7.5.2 Formation of Sheeting Joints

Uplift and exhumation of rocks may lead to the development of sheeting joints within a few hundred meters of the Earth's surface. As we have seen, sheeting joints are commonly subhorizontal or parallel to topographic surfaces. They develop most prominently in rocks that do not contain bedding or schistosity, particularly granitic rocks.

The origin of sheeting joints is a bit problematic. At first glance, you might not expect joints to form parallel to the ground surface, because they are tensile fractures, and near the ground surface there is a compressive load perpendicular to the ground surface due to the dead weight of the overlying rock and fluid pressure is low. It appears that sheeting joints form where horizontal stress is significantly greater than the vertical load (Figure 7.15a), for in such a stress field, joints propagate parallel to the ground surface. In this regard, formation of sheeting joints may resemble cracks formed by longitudinal splitting in laboratory specimens.

The stresses causing sheeting joints may, in part, be tectonic in origin, but they may also be residual stresses. A *residual stress* is a stress that exists in a rock even if the rock is not loaded externally (e.g., in an unconfined block of rock sitting on a table). Residual stresses develop in a number of ways.[5] In the case of plutons, residual stresses de-

velop because the thermal properties (e.g., coefficient of thermal expansion) of the pluton differ from those of the surrounding wall rock, and because during cooling, the pluton cools by a greater amount than the wall rock. The pluton and the wall rock tend to undergo different elastic strains as a result of thermal changes during cooling and later unroofing (Figure 7.15b and c). Because the pluton is welded to the surrounding country rock, the differential strain creates an elastic stress in the rock. For example, if the pluton shrinks more than the wall rock, tensile stresses develop perpendicular to the wall. To picture this phenomenon, imagine a spherical balloon filling a spherical hole in a block of wood. The surface of the balloon is attached to the wall of the hole by tiny springs. If you let the balloon shrink, simulating contraction, the tiny springs stretch, simulating tensile stress. At depth, compressive stress due to the overburden counters these tensile stresses, but near the surface, residual tensile stress perpendicular to the walls of the pluton may exceed the weight of the overburden and cause sheeting joints parallel to the wall of the pluton to form.

We noted above that sheeting joints tend to parallel topography. This relationship either reflects topographic control on the geometry of joints (perhaps because the vertical load is perpendicular to the ground surface), or joint control on the shape of the land surface (because rocks spall off the mountainside at the joint surface). Geologists are not sure which phenomenon is more important.

7.5.3 Natural Hydraulic Fracturing

As we saw in Chapter 3, the three principal stresses at depth in most of the continental lithosphere are compressive. Yet joints form in these regions, and these joints may be decorated with plumose structure, indicating that they were driven by tensile stress. How can joints form if all three principal stresses are compressive? As we described in Chapter 6, the solution to this paradox came from considering the effect of pore pressure on fracturing. Simplistically, the increase in pore pressure in a preexisting crack pushes outward and causes a tensile stress to develop at the crack tip that eventually exceeds the magnitude of the least principal compressive stress. If the pore pressure is sufficiently large, a tensile stress that exceeds the magnitude of

---

[5]To picture development of residual stress in sandstone, imagine a layer of dry sand that gets deeply buried. Because of the weight of the overburden, the sand grains squeeze together and strain elastically. If, at a later time, groundwater fills the pores between the strained grains, unstrained cement may precipitate and lock the grains together. As a consequence, the elastic strain in the grains gets locked into the resulting sandstone. If unroofing later exposes the sandstone, the grains and the cement attempt to expand, but by different amounts, and as a consequence stress develops in the sandstone.

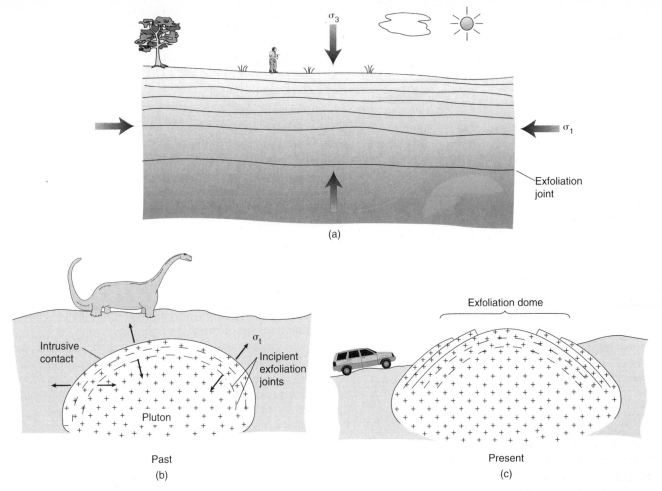

Figure 7.15 (a) Exfoliation joints forming in a location where σ_1 is horizontal and while σ_3 is vertical, near the ground surface. Note that the joints become more closely spaced closer to the ground surface. (b) Possible situation where a pluton cools and contracts more than country rock, so σ_t (tensile stress) is oriented perpendicular to the intrusive contact. (c) Later, when the pluton is exhumed, joints form parallel to the intrusive contact and create an exfoliation dome.

σ_3 develops at the tip of the crack, even if the remote principal stresses are all compressive (Figure 7.16a), and the crack propagates. This process is called *hydraulic fracturing*. Oil well engineers commonly use hydraulic fracturing to create fractures, and enhance permeability, in the rock surrounding an oil well. They create hydraulic fractures by first sealing off a portion of the well and then increasing the fluid pressure in the sealed segment until the wall rock breaks. But hydraulic fracturing also occurs in nature, due to the fluid pressure of water, oil, and gas in rock. It is this *natural hydraulic fracturing* that causes some joints to form.

If you think hard about the simplistic explanation of hydraulic fracturing that we just provided, you may be wondering whether the process implies that the pore pressure in the crack becomes greater than pore pressure in the pores of the surrounding rock. It doesn't! Pore pressure in the crack can be the same as in the pores of the surrounding rock during natural hydraulic fracturing. Thus, we need to look a little more closely at the problem to understand why pore pressure can cause joint propagation.

Imagine that a cemented sandstone contains fluid-filled pores and fluid-filled cracks (Figure 7.16b). Let's focus our attention on the crack and its walls. Because the pores and the crack are connected, the fluid pressure in the pores and the crack are the same. Fluid pressure within the crack is pushing outwards, creating an opening stress, but at the same time, the fluid pressure in the pores as well as the stress in the rock is pushing inwards, creating a closing stress. As long as the closing stress exceeds the opening stress, the crack does not propagate. If the fluid pressure increases, the opening stress increases at the same rate as the increase in fluid pressure, but the closing stress increases at a slower rate. Eventually, the opening stress exceeds the closing stress, so that the crack propagates; effectively, the outward push of the fluid in the crack creates a tensile stress at the crack tip. Why does the closing stress increase at a slower rate than the fluid pressure and the opening stress? Because grains in the rock are cemented to one another, the increase in fluid pressure in the pores cannot move the grains freely. The elasticity of the grains themselves, therefore, takes up some of the push caused by the

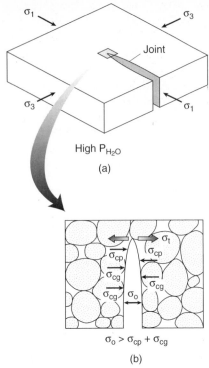

Figure 7.16 (a) Block diagram showing the stresses in the vicinity of a crack in which there is fluid pressure that exceeds the magnitude of σ_3. As a result, there is a tensile stress along the crack σ_t. (b) Enlargement of the crack tip, illustrating the poroelastic effect. The opening stress σ_0 due to fluid pressure in the crack exceeds the closing stress σ_c, which is the sum of σ_{cp}, the closing stress where a pore is in contact with the crack, and σ_{cg}, the closing stress where a grain is in contact with the crack.

fluid pressure. Thus, the closing stress acting on the fluid in the crack where it is in contact with a grain is less than where the fluid in the crack is in contact with a pore, but the outward push of the fluid in the crack is the same everywhere.[6] As a result, the net outward push exceeds the net inward push, and tensile stress develops.

Once the crack propagates, the volume of open space between the walls of the crack increases, so the fluid pressure in the crack decreases. As a consequence, the crack stops growing until an increase in fluid pressure once again allows the stress intensity at the crack tip to drive the tip into unfractured rock. Thus, the surfaces of joints formed by natural hydraulic fracturing tend to have many arrest lines.

7.5.4 Joints Related to Tectonic Deformation

During a convergent or collisional orogenic event (see Part D of this book), compressive tectonic stress affects rocks over a broad region, even into the continental interior. Joints form within the foreland of orogens during tectonism for a number of reasons.

Joints from *natural hydrofracturing* appear to form on the foreland margins of orogens during orogeny. The conclusion that the joints are syntectonic is based on two observations. First, the joints parallel the σ_1 direction

---

[6]This is known as the *poroelastic effect*.

associated with development of tectonic structures, such as folds. Second, the joints locally contain mineral fill that formed at temperatures and fluid pressures found at a depth of several kilometers in the Earth; thus, they are not a consequence of recent cracking of rocks in the near surface. The origin of such joints may reflect increases in fluid pressure within confined rock layers due to the increase in overburden resulting from thrust-sheet emplacement or from the deposition of sediment eroded from the interior of the orogen.

During an orogenic event, the maximum horizontal stress is approximately perpendicular to the trend of the orogen. As a consequence, the joints that form by syntectonic natural hydraulic fracturing are roughly perpendicular to the trend of the orogen. Because the stress state may change with time in an orogen, later formed joints may have a different strike than earlier formed joints, and the joints formed during a given event might not be exactly perpendicular to fold trends where they form. Such joint patterns are typical of orogenic foreland regions, but may also occur in continental interiors.

Joints are commonly related to faulting, and such joints fall into three basic classes. The first class is composed of regional joints that develop in the country rock due to the stress field that is also responsible for generation and/or movement on the fault itself. Since faults are usually inclined to the remote σ_1 direction, the joints that form in the stress field that causes a fault to move will not be parallel to

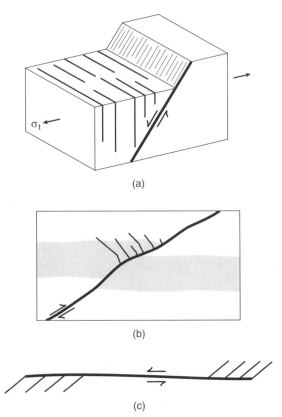

Figure 7.17 (a) Formation of joints in the hanging-wall block of a region in which normal faulting is taking place. (b) Formation of joints above an irregularity in a fault surface. (c) Pinnate joints along a fault.

the fault (Figure 7.17). The second class includes joints that develop due to the distortion of a moving fault block (Figure 7.17b). For example, the hanging wall of a normal fault block may undergo some extension, resulting in the development of joints. Or, the hanging block of a thrust fault may be warped as it moves over the fault if the underlying fault surface is not planar, and thus may locally develop tensile stresses sufficient to crack the rock. The third class includes joints that form immediately adjacent to a fault in response to tensile stresses created in the wall rock while the fault moves. Specifically, during the development of a shear rupture (i.e., a fault), an *en echelon* array of short joints forms in the rock adjacent to the rupture. These joints, which merge with the fault and are inclined at an angle of around 30°–45° to the fault surface, are called *pinnate joints* (Figure 7.17c). The acute angle between a pinnate joint and the fault indicates the sense of shear on the fault, as we will see in Chapter 8.

When the stress acting on a region of crust is released, the crust elastically relaxes to attain a different shape. This change in shape may create tensile stresses within the region that are sufficient to create *release joints*. Joints also form in relation to folding. Folded rocks may be cut by syntectonic natural hydrofractures, manifested by joints oriented at a high angle to the fold hinge, as we described above. In addition, during the development of folds in nonmetamorphic conditions, joints may also develop locally because of local tensile stresses associated with bending of the layers (Figure 7.18). Joints resulting from this process of outer-arc extension have a strike that is parallel to the trend of the fold hinge, and may converge toward the core of the fold. If development of folds results in stretching of the rock layer parallel to the hinge of the fold, then cross-strike joints may develop.

Finally, joints may develop in a region of crust that has been subjected to broad regional warping. Similar to folding, the joint formation reflects tensile stresses that develop when the radius of curvature of a rock layer changes in the brittle field.

7.5.5 Orthogonal Joint Systems

In orogenic forelands and in continental interiors, you will commonly find two systematic joint sets that are mutually perpendicular. In some cases, the joints define a *ladder pattern* (Figure 7.19a), in which the joints of one set are relatively long and the joints of the other set are relatively short cross joints that terminate at the long joints. In other cases, the joints define a *grid pattern* (Figure 7.19b), in which the two sets appear to be mutually crosscutting. The existence of such orthogonal systems has perplexed geologists for decades, because at first glance it seems impossible for two sets of tensile fractures to form at 90° to each other in the same regional stress field. Recent field and laboratory studies suggest a number of possible ways in which orthogonal systems develop, though the applicability of a given explanation to a specific region remains controversial.

In orogenic forelands, an orthogonal joint system typically consists of a strike-parallel and a cross-strike set, typically defining a ladder pattern. The two sets may have quite different origins: cross-strike joints parallel the regional maximum horizontal stress trajectory associated with folding, and thus possibly formed as syntectonic natural hydrofractures, whereas strike-parallel joints could reflect outer-arc extension of folded layers. Alternatively, the strike-parallel joints could be release joints formed when orogenic stresses relax.

Orthogonal joint systems may develop in regions of the crust that had been subjected to a regional tensile stress that was later relaxed. During the initial stretching of the region, a set of joints develops perpendicular to the regional tensile stress. When the stress is released, the region rebounds elastically, and expands slightly in the direction perpendicular to the original stretching. A new set of joints therefore develops perpendicular to the first.

Orthogonal joint systems may also develop during uplift. Imagine that a rock layer is unloaded when the overburden above it erodes away. As a result, a joint set perpendicular to the regional σ_3 develops. With continued uplift and expansion, tensile stress that develops in the layer can be relieved easily in the direction perpendicular to the existing joints; that is, they just open up. But tensile stresses cannot be relieved in the direction parallel to the existing joints; so new cross joints form, creating a ladder pattern.

Grid patterns suggest that the two joint sets initiated at roughly the same time, or that cracking episodes alternated between forming members of one set and then the other. If we assume that both joint sets form in the principal plane that is perpendicular to σ_3, we can interpret such occurrences to be related to the back-and-forth interchange of σ_2

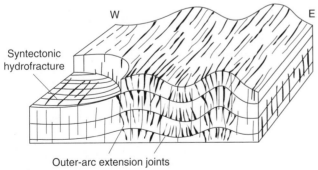

Figure 7.18 Block diagram showing outer-arc extension joints whose strike is roughly parallel to the axis of a fold.

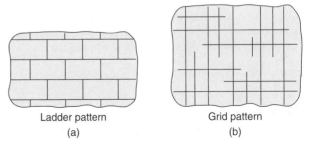

Figure 7.19 Two patterns of orthogonal joint systems. (a) Traces of joints defining a ladder pattern. (b) Traces of joints defining a grid pattern.

and σ_3 during uplift, if as a starting condition σ_2 and σ_3 are similar in magnitude. To see what we mean, imagine a region where σ_1 is vertical and σ_3 is initially north-south. When σ_3 is north-south, east-west trending joints develop. But if σ_3 switches with σ_2 and becomes east-west, then north-south trending joints develop.

7.5.6 Conjugate Joint Systems

At some localities in orogenic forelands we find that joint sets define a conjugate system in which the bisector of the dihedral angle is perpendicular to the axis of the folds. The origin of such fracture systems remains one of the most controversial aspects of joint interpretation. Based on their geometry, it was traditionally assumed that joints of conjugate systems are either shear fractures, formed at about 30° to σ_1 (representing failure when the Mohr circle touches the Coulomb failure envelope; Figure 6.19), or so-called transitional-tensile fractures, which are thought to form at angles less than 30° to σ_1 (representing failure when the Mohr circle touches the steep part of the failure envelope; Figure 6.24). Yet, if you examine the surfaces of the joints in these conjugate systems, you find in many cases that they display plumose structure, confirming that they formed as Mode I (extension) fractures. Furthermore, as we noted earlier, transitional-tensile fractures have never been created in the lab, so their very existence remains suspect. The only type of crack that is known to propagate for long distances in its own plane is a Mode I crack; shear fractures form by linkage of microcracks, not by propagation of a single shear surface in its own plane. But if the members of conjugate joint systems are not shear fractures, how do they form?

Many joint researchers now believe that both of the two nonparallel sets in the conjugate system are cross-strike joints that initially formed perpendicular to σ_3. Thus, to explain the contrast in orientation between the two sets, they suggest that the two sets formed at different times in response to different stress fields. For example, the slightly folded Devonian strata of the Appalachian Plateau in south-central New York State contain two joint sets separated by an angle of less than 60° (Figure 7.13a). The two sets are attributed to different, distinct generations of the Late Paleozoic Alleghanian orogenic phase; the maximum horizontal stress during the first generation of the orogeny was not parallel to that during the second generation of the orogeny. Geologists who favor this model for conjugate joint system formation conclude that the occurrence of slip lineations on joints of conjugate systems does not imply that the joints originated as shear fractures, but rather that they reactivated as mesoscopic faults subsequent to their formation.

7.5.7 Joint Formation in Igneous Rocks

Joints form in association with igneous rocks for several reasons. They form in the country rock adjacent to plutonic intrusions if the emplacement of the intrusion stretches the overlying rock, thereby creating tensile stress. The pattern of joints that results reflects the shape of the intrusion. For example, radial joints develop around a pluton that is roughly circular in plan view. At a distance from the pluton, such joints may curve and bend into parallelism with the regional maximum horizontal stress. Magma from the pluton may intrude during or subsequent to the joint formation, creating igneous dikes. In fact, the pressure of the magma may drive dike formation. Joints may also form immediately adjacent to an intrusion because of thermal cracking of the cooler country rock and because fluids being released from the pluton cause local hydrofracturing.

Finally, joints may form during cooling and contraction of an igneous intrusion or a lava flow. When magma either intrudes at a shallow crustal level or extrudes over the Earth's surface, it cools fairly quickly and solidifies. As the resulting rock continues to cool, it contracts. Eventually, it cools to temperatures below the brittle/plastic transition, at which time its contraction can no longer be accommodated by plastic flow, and an elastic strain develops in the rock that in turn causes tensile stress (recall Hooke's law!). When the tensile stress exceeds the strength of the rock, it fractures. Since the rock shrinks equally in all directions, the stress cannot be relieved by development of one joint set. Rather, joints form in many directions, and these intersect to outline roughly hexagonal columns, thereby creating columnar jointing. The long axis of the columns forms perpendicular to the isotherms (surfaces of equal temperature) in the cooling body (Figure 7.7). Joints bounding the columns grow incrementally along their length by the addition of small rectangular joint segments whose long dimensions are perpendicular to the overall column axis, thereby giving each face of a column a striped appearance. As we mentioned earlier, cooling and contraction of massive plutons may contribute to formation of sheet fractures.

7.5.8 Joint Trends as Paleostress Trajectories

Orientation data on jointing holds valuable information about the orientation of stress fields at the time of failure. Joints propagate normal to σ_3, so their planes define the trajectories of σ_3 in a region. In the case of vertical joints, the strike of the joint defines the trajectory of maximum horizontal stress (S_H). We don't know if S_H represents σ_1 or σ_2. The maximum principal stress could be either parallel or perpendicular to the Earth's surface, depending on the depth at which the joint formed.

7.6 LIMITS ON THE GROWTH OF A JOINT

We've discussed various ways in which joints initiate. To complete our discussion of joints we also need to address the issue of why joints stop growing. Recall from Chapter 6 that the stress intensity at the tip of a crack depends on the length of the crack. Thus, the intensity increases as the crack grows, and as long as the stress driving joint growth remains unchanged, the joint will keep growing. For this reason, joints that grow in large bodies of homogeneous rock (such as the massive beds of sandstone at Arches National Park) can become huge surfaces. But joints clearly do not propagate from one side of a continent to the other. They stop growing due to one or more of the following factors.

The joint tip *intersects a (nearly) free surface.* Joints obviously stop growing when they reach the Earth's surface. They also stop growing in the subsurface where they intersect a preexisting open fracture (joint or fault) or a weak bedding plane. Two joints that are growing towards each other but are not coplanar stop growing when they enter each other's stress shadow. In some cases, the *interaction of joint tips* causes curvature and the two joints link (Figure 7.20b). If, however, the preexisting joint is squeezed together so tightly that friction allows shear stress to be transmitted across it, or if it has been sealed by vein material, then a younger joint can cut across it.

Formation of the joint itself may cause a local *drop in fluid pressure,* because creation of the joint forms more space for fluid to occupy. The increase in space temporarily causes a drop in fluid pressure, so that the stress intensity at the joint tip becomes insufficient to drive the tip into unfractured rock. When fluid flow into the joint increases the fluid pressure to a large enough value, the joint continues to grow. Thus, as we mentioned earlier, it is characteristic for joints being driven by high fluid pressures to grow in a start-stop manner, so their surfaces have many arrest lines.

Finally, if the joint grows into a region where energy at the crack tip can be dissipated by *plastic yielding,* the joint stops growing. Similarly, propagation of a joint into a rock with a *different tensile strength* may cause it to stop growing. Also, if the joint tip enters a region where the stress intensity at the crack tip becomes too small to drive the cracking process, then the joint stops growing. The de-

crease in stress intensity may be due to a decrease in the tensile stress magnitude in the rock, or due to an increase in compressive stress that holds the joint together.

7.7 VEINS AND VEIN ARRAYS

In the vast desert ranges of western Arizona, there are few permanent residents, save for the snakes and scorpions. But at some time during the past century, almost every square meter of the rugged terrain has been trod upon by a dusty prospector in search of valuable ore deposits of gold, silver, or copper. Modern geologists mapping in the region frequently come upon traces of the prospectors from years past. Here is a pile of rusted cans, there a sardine tin containing yellowed flakes of paper that once defined a mining claim. When you poke into many of these excavations, you find that the focus of the effort, the days of agonizing labor with pick and shovel, is nothing more than a vein of milky white quartz. With every blow, hammer against rock, the prospector thought, "Quartz veins may have gold! Why not here?" Countless prospectors have asked that question and have dug and hammered (often in vain) to find an answer.

What are veins? Simply speaking, a *vein* is a fracture filled with material that precipitated from a fluid (Figure 7.1c; Table 7.2). Quartz or calcite form the most common vein fill, but other minerals do occur in veins, including

Table 7.2   Vein Terminology

Vein	A fracture that has filled with minerals precipitated from water solutions that passed through the fracture.
Vein array	A group of veins in a body of rock.
Planar systematic arrays	In a planar systematic array, the component veins are planar, are mutually parallel, and are regularly spaced.
Nonsystematic arrays	The veins in nonsystematic arrays tend to be nonplanar, and individual veins may vary in width.
***En echelon* arrays**	An *en echelon* array consists of short parallel veins that lie between two parallel enveloping surfaces and are inclined at an angle to the surfaces. Typically, the veins in an *en echelon* taper toward their terminations. The veins may be sigmoidal in cross section.
Stockwork veins	Stockwork veins are a cluster of irregularly shaped veins that occur in a pervasively fractured rock body. The veins are nonplanar arrays and occur in a range of orientations. In rock bodies with stockwork veins, as much as 40% or 50% of the outcrop is composed of vein material, and vein material may completely surround blocks of the host rock.

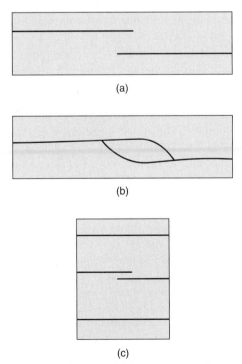

(a)

(b)

(c)

Figure 7.20   Joint terminations. (a) Joints terminating without curving when they approach each other. (b) Joints curving into each other and linking. (c) Map-view sketch illustrating how joint spacing is fairly constant because joints that grow too close together cannot pass each other.

numerous ore minerals, zeolites, and chlorite. Some veins initiated as joints, whereas others initiated as shear ruptures or as cracks formed adjacent to shear ruptures. Veins come in all dimensions; some are narrower and shorter than a strand of hair, whereas others compose massive tabular accumulations that are a couple of meters across and many meters long. Groups of veins are called *vein arrays* and can have a variety of forms, as described in Table 7.2.

7.7.1 Formation of Vein Arrays

Planar systematic arrays (Figure 7.21a) represent either mineralization of preexisting systematic joint sets, or mineralization during formation of a systematic joint set. *Stockwork vein arrays* (Figure 7.21b) form where rock has been shattered, either by the existence of locally very high fluid pressure, or as a result of local pervasive fracturing in association with folding and faulting.

En echelon vein arrays form in a couple of different ways. They may form by filling *en echelon joints* in the twist hackle fringe of a larger joint. As we saw already, the twist hackle fringe represents the breakup of a joint into short segments when it enters a region of the rock where the stress field is different. *En echelon* vein arrays also develop as a consequence of shear within a rock body that is accommodated by displacement across a fault zone (Figure 7.22). The fractures composing an *en echelon* array initiate parallel to σ_1, typically at an angle of about 45° to the borders of the shear. Fractures in an *en echelon* array tend to open as displacement across the shear zone develops, and therefore they fill with vein material. Once formed, the veins are material objects within the rock, so continued shear tends to rotate the veins and the acute angle increases. If, however, a new increment of vein growth occurs, it again initiates at approximately 45° to the shear surfaces. Therefore, the veins become sigmoidal in shape. Locally, a second generation of veins may initiate at the center of the original veins; this second set cuts obliquely across the first generation of veins. Because of the geometric relationship between *en echelon* veins and the shear sense, the orientation of shear-related *en echelon* veins can be used to determine shear sense (see Chapters 8 and 12).

7.7.2 Nature of Vein Fill (Blocky vs. Fibrous Veins)

Vein fill, the mineral crystals within a vein, is either blocky (or sparry) or fibrous. In *blocky veins,* the crystals of vein fill are roughly equant, and may exhibit crystal faces (Figure 7.23a). The occurrence of blocky veins means either that the vein was an open cavity when the mineral precipitated (this is possible only in veins formed near the Earth's surface, where rock strength is sufficient to permit a cavity to stay open or fluid pressure is great enough to hold the fracture open), that an earlier formed fibrous vein fill later recrystallized to form blocky crystals, or that there were few nucleation sites for crystals to grow from during vein formation.

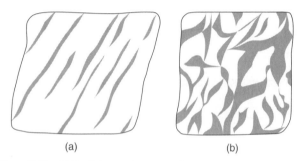

Figure 7.21 (a) Parallel array of veins. (b) Stockwork array of veins. Vein fill is dark.

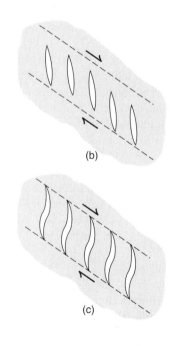

Figure 7.22 (a) *En echelon* veins in the Lachlan Orogen (southeastern Australia). (b) Formation of a simple *en echelon* array. (c) Formation of sigmoidal *en echelon* veins, due to rotation of the older central part of the veins, and the growth of new vein material at ~45° to the shear surface.

In *fibrous veins*, the crystals are very long relative to their width, so that the vein has the appearance of being spanned by a bunch of hairs (Figure 7.23b). Geologists don't fully agree on the origin of fibrous veins, but some fibrous veins may form by a *crack-seal mechanism*. The starting condition for this process is an intact rock containing pore fluid that in turn contains dissolved minerals. If the fluid pressure becomes great enough, the vein 'cracks' and a very slight opening (only microns wide) develops. This crack immediately fills with fluid; but since the fluid pressure within the open crack is less than that in the pores of the surrounding rock, the solubility of the dissolved

mineral decreases and the mineral precipitates, thereby 'sealing' the crack. The process repeats tens to hundreds of times, and each time the vein width grows slightly. During each increment of growth, existing grains in the vein act as nuclei on which the new vein material grows, and thus continuous crystals grow. Alternatively, formation of some fibrous veins may occur by a diffusion process, whereby ions migrate through fluid films on grain boundaries and precipitate at the tips of fibers while the vein walls gradually move apart. During this process, an open crack never actually develops along the vein walls or in the vein.

7.7.3 Interpretation of Fibrous Veins

Fibrous veins, in particular those consisting of calcite and quartz, are quite useful because they record information about the progressive strain history in an outcrop. There are two endmember types of fibrous veins (Figure 7.24). *Syntaxial veins* form in rocks where the vein fill is the same composition as the wall rock (e.g., quartz veins in a quartz sandstone). The vein fibers nucleate on the surface of grains in the wall rock and grow inwards to meet at a median line. Each successive increment of cracking occurs at the median line, because at this locality separate fibers

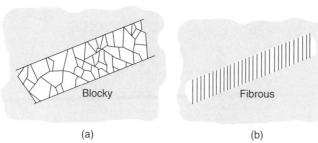

Figure 7.23 Basic vein types. (a) Blocky vein fill. (b) Fibrous vein fill.

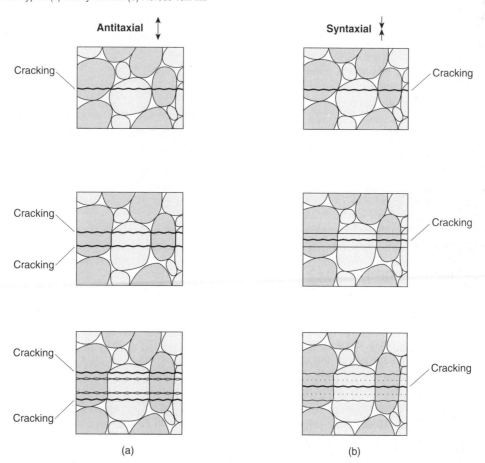

Figure 7.24 Cross-sectional sketches, at the scale of individual grains, showing the contrast in the stages of crack-seal leading to antitaxial veins and syntaxial veins. (a) Formation of antitaxial veins. The increments of cracking form along the margins of the vein, and the vein composition differs from the wall rock (i.e., the fiber crystals are not in optical continuity with the grains of the wall. During increments of cracking, tiny slices of the wall rock spall off. The slices bound the growth increments in a fiber. (b) Syntaxial vein formation. During each increment, the cracking is in the center of the vein. The composition of the fibers is the same as the grains of the wall rock (i.e., the fibers are in optical continuity with the grains of the wall rock). Optical continuity between fiber and grain means that the crystal lattice of the grain has the same orientation as the crystal lattice of the fiber. Optically continuous fibers and grains go extinct at the same time, when viewed with a petrographic microscope.

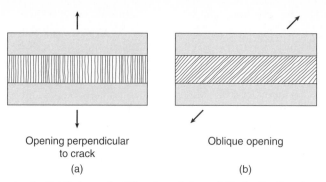

Opening perpendicular
to crack

(a)

Oblique opening

(b)

Figure 7.25 (a) Cross-sectional sketches showing that the long axis of fibers in a vein is parallel to the direction of stretching. If the stretching direction is perpendicular to the vein wall, then the fibers are perpendicular to the wall. If the stretching direction is oblique to the vein wall, then the fibers are oblique to the vein wall. Note that if the stretching direction is effectively parallel to the vein surface (i.e., the vein is a fault surface), then the fibers are almost parallel to the vein wall, and on exposed surface would be called slip lineations.

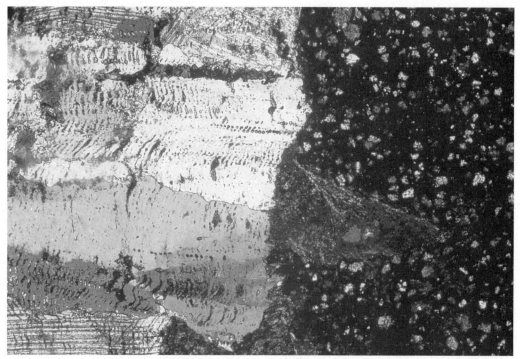

Figure 7.26 Photomicrograph of a fibrous antitaxial vein in limestone from the Aarmassif (Switzerland). The dark bands in the elongate quartz and calcite crystals parallel the vein wall. Field of view is approximately 1 mm.

meet; whereas at the walls of the grains, vein fibers and the grains of the wall rock form single continuous crystals. Each growth increment of a fiber is bounded by trails of fluid inclusions. *Antitaxial veins,* in contrast, form in rocks where the vein fill is different from the composition of the wall rock (e.g., a calcite vein in a quartz sandstone). In antitaxial veins, the increments of cracking occur at the boundaries between the fibers and the vein wall, possibly because that is where the bonds are weakest, and thus the veins grow outward from the center. Increments of growth are sometimes bounded by thin dislodged flakes of the wall rock.

In some cases, the long axis of a fiber in a fibrous vein tracks the direction of maximum extension (stretching) at the time of growth, that is, the long axis of the incremental strain ellipsoid. When the fibers are perpendicular to the walls of the vein, the vein opened in a direction perpendicular to the vein walls (Figure 7.25a). However, vein fibers oblique to the vein walls indicate that the vein opened obliquely and that there was a component of shear displacement during vein formation (Figure 7.25b). When vein fibers are sigmoidal in shape (Figure 7.26), the extension direction rotated relative to the vein wall orientation. Note that for fibers of the same shape, the order of the movement stages depends on whether the vein is antitaxial or syntaxial (Figure 7.27). For example, if the fibers in a syntaxial vein are perpendicular to the walls in the center and oblique to the walls along the margins, the early stage in vein formation had an oblique opening component while the later stage did not (remember that the fibers grow toward the center). In contrast, fibers with exactly the same shape in an antitaxial vein would indicate the opposite strain history, because in antitaxial veins the fibers grow toward the walls. The presence of a median line helps you recognize this important kinematic difference.

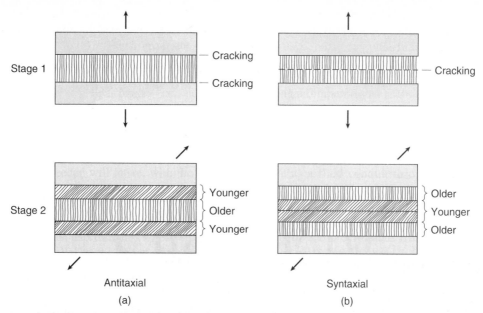

Figure 7.27 Cross-sectional sketches showing how a change in stretching direction leads to the formation of sigmoidal fibers. In this example, the opening is first perpendicular to the plane of the vein, and then is oblique to the vein. Note that because of the locus of vein fiber precipitation, (a) antitaxial and (b) syntaxial veins have different shapes.

Figure 7.28 Aerial photograph of the Duncan Lake area (Northwest Territories, Canada) showing lineaments and structural control of topography. Scale is 1:14,500.

7.8 LINEAMENTS

A geologic *lineament* is a linear feature recognized on aerial photos, satellite imagery, or topographic maps. Lineaments generally are defined only at the regional scale; that is, they are not mesoscopic or microscopic features. Structural lineaments, meaning ones that are a consequence of the localization of known geologic features, are defined by structurally controlled alignments of topographic features like ridges, depressions, or escarpments (Figure 7.28).

Structural lineaments may even be manifested by changes in vegetation, which are, in turn, structurally controlled. Most lineaments are the geomorphological manifestation of joint arrays, faults, folds, dikes, or contacts, but some remain a mystery and do not appear to be associated with obvious structures.[7] You should maintain a healthy skepticism when reading articles about lineaments that do not confirm imagery interpretations with ground truth. Some 'lineaments' that have been described in the literature turn out to be merely artifacts of how sunlight interacts with the ground surface, or are of human origin, and thus do not have geologic significance. However, study of true structural lineaments may provide insight into the distribution of structural features, ore deposits, and seismicity.

7.9 CLOSING REMARKS

In this introduction to the rapidly evolving science of joint analysis, we have tried to convey not only descriptive information about the structures, but also a sense of the controversy surrounding their interpretation. One of the common questions that students ask when beginning a field mapping project, is, "Should I pay attention to the joints and veins?" An astute advisor might answer the question philosophically, with the words, "That depends."

If the purpose of the map is to define variations in permeability, or the location of faults, or the distribution of ore deposits, or the meaning of satellite-imagery lineaments, or the composition of fluids passing through the rock during deformation, then by all means the joints and veins in the area should be studied. Perhaps you will find an interpretable variation in joint intensity within your map area, even if there is no systematics to the jointing. Joint study is particularly important in studies of rock permeability, because the rate of fluid flow through joints may exceed the rate of fluid flow through solid rock by orders of magnitude. Joints may make otherwise impermeable granite into a fluid reservoir, and may provide cross-formational permeability that permits oil to leak through an otherwise impermeable seal, or permits toxic waste in groundwater to leak across an aquitard. If the purpose of your mapping is to develop an understanding of paleostress in the region, then it may be worthwhile to study joints only if you can identify systematic sets. If the purpose of your study is to develop an understanding of stratigraphy in the map area, or interpret the history of deformation in metamorphic rocks, then joint analysis probably won't help you very much and you shouldn't give it too much time. Of course this advice might change in the coming decade, as sophistication in our understanding of the meaning of joints continues to grow.

Joints, by definition, are fractures on which there has been no shear. In the next chapter, we shift our focus to fractures on which there clearly has been shear, that is,

faults. As you study chapter 8, keep in mind the features of joints that we just described, so that you can compare them with the descriptive features of faults.

ADDITIONAL READING

Ashby, M. F., and Hallam, S. D., 1986, The failure of brittle solids containing small cracks under compressive stress states, *Acta Metallurgica,* v. 34, p. 498–510.

Atkinson, B. K., 1987, Fracture Mechanics of Rock, London, Academic Press, 534 p.

Bahat, D., and Engelder, T., 1984, Surface Morphology on Cross-Fold Joints of the Appalachian Plateau, New York and Pennsylvania, *Tectonophysics,* v. 104, p. 299–313.

Beach, A., 1975, The Geometry of en-echelon vein arrays, *Tectonophysics,* v. 28, p. 245–263.

Hancock, P. L., 1985, Brittle microtectonics: Principles and practice: *Journal of Structural Geology,* v. 7, p. 438–457.

Hodgson, R. A., Classification of structures on joint surfaces, *American Journal of Science,* v. 259, p. 493–502.

Kulander, B. R., Barton, C. C., and Dean, S. L., 1979, The application of fractography to core and outcrop fracture investigations, Report to U.S. Department of Energy, Morgantown Energy Technology Center, METC/SP-79/3, 174 p.

Narr, W., and Suppe, J., 1991, Joint spacing in sedimentary rocks, *Journal of Structural Geology,* v. 13, p. 1038–1048.

Nickelsen, R. P., and Hough, V. D., 1967, Jointing in the Appalachian Plateau of Pennsylvania, *Geological Society of America Bulletin,* v. 78, p. 609–630.

Pollard, D. D., and Aydin, A., 1988, Progress in understanding jointing over the past century, *Geological Society of America Bulletin,* v. 100, p. 1181–1204.

Price, N. J., 1966, Fault and Joint Development in Brittle and Semi-brittle Rock, London, Pergamon Press, 176 p.

Ramsay, J. G., 1980, The crack-seal mechanism of rock deformation, *Nature,* v. 284, p. 135–139.

Reches, Z., 1976, Analysis of joints in two monoclines in Israel, *Geological Society of America Bulletin,* v. 87, p. 1654–1662.

Rives, T., Rawnsley, K. D., and Petit, J.-P., 1994, Analogue simulation of natural orthogonal joint set formation in brittle varnish, *Journal of Structural Geology,* v. 16, p. 419–429.

Segall, P., 1984, Formation and growth of extensional fracture sets, *Geological Society of America Bulletin,* v. 94, p. 563–575.

---

[7]Except perhaps ancient landing strips for extraterrestrials.

Chapter 8
Faults and Faulting

8.1 INTRODUCTION

Imagine a miner in a cramped tunnel crunching eastward through a thick seam of coal. Suddenly, his pick hits hard rock. The miner chips away a bit more only to find that the seam he has been following for the past three weeks abruptly terminates against a wall of sandstone. He curses, "My flippin' seam's cut off—there's a fault with it!" From previous experience, the miner knew that because of the fault he would have to waste precious time digging a shaft up or down to intersect the coal seam again.

Geologists adopted the term *fault,* but you'll find that they use the term in different ways, in different contexts. In a general sense, a fault is any surface or zone in the Earth across which measurable *slip* (shear displacement) develops. In a more restricted sense, faults are fractures on which slip develops primarily by brittle deformation processes (Figure 8.1a). This second definition serves to distinguish 'faults' (*sensu stricto*) from 'fault zones' and 'shear zones.' We use the term *fault zone* for brittle structures in which loss of cohesion and slip occurs on several faults within a band of definable width (Figure 8.1b). Displacement in fault zones can involve formation and slip on many small, subparallel brittle faults, or slip on a principal fault off which many smaller faults (*fault splays*), diverge or slip on an anastamosing[1] array of faults (Figure 8.1c,d). *Shear zones* are ductile structures, across which a rock body does not lose mesoscopic cohesion, so that strain is distributed across a band of definable width. In ductile shear zones,

---

[1]Anastamosing refers to the geometry of a group of wavy, subparallel surfaces that merge and diverge; resembling a braid of hair.

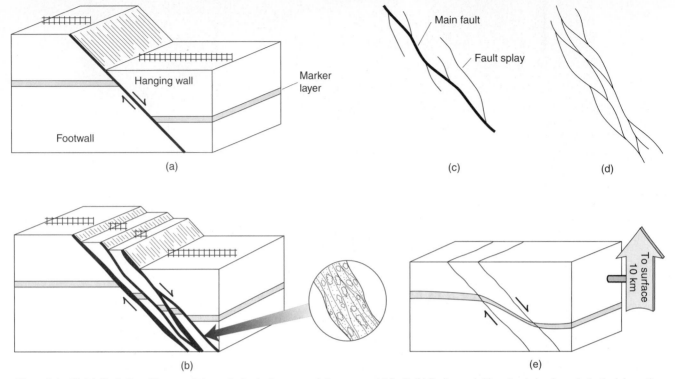

Figure 8.1 Sketch illustrating differences between faults, fault zones, and shear zones. (a) Fault. (b) Fault zone (with an inset showing cataclastic deformation adjacent to a fault surface). (c) Sketch illustrating the relation between a principal fault and fault splays. (d) Anastamosing faults in a fault zone. (e) Shear zone. The fault and the fault zone are shown to intersect the ground surface when displacement occurred; the shear zone is at depth in the crust.

rocks deform by *cataclasis,* a process involving fracturing, crushing, and frictional sliding of grains or rock fragments, or by primarily *crystal plastic* deformation mechanisms (Figure 8.1e). We describe cataclastic shear zones in this chapter, but delay discussion of other deformation mechanisms in ductile shear zones until Chapter 12.

Faults occur at all scales in the lithosphere (Figure 8.2), and geologists study them for many reasons. Faults control the spatial arrangement of rock units, so their presence creates puzzles that challenge geologic mappers. Faults may affect topography and modify the landscape. Faults affect the distribution of economic resources (e.g., oil fields and ore bodies), and control permeability of rocks and sediments, properties that, in turn, control fluid migration. Finally, faulting creates deformation (strain ± rotation ± translation) in the lithosphere during plate interactions and intraplate movements, which may cause devastating earthquakes.

Thus, fault analysis plays a role in diverse aspects of both academic and applied geology. In order to provide a basis for you to work with faults, this chapter introduces the terminology used to describe fault geometry and displacement, discusses how to represent faults on maps and cross sections, and shows you how to recognize and inter-

pret faults at the surface and in the subsurface. We conclude the chapter by introducing fault-system tectonics, the relation between faulting and resources, and the relation between faulting and earthquakes. We treat the tectonics of fault systems more fully in Part D, where the basic concepts of plate tectonics are discussed.

8.2 FAULT GEOMETRY AND DISPLACEMENT

8.2.1 Basic Vocabulary

In order to discuss faults, we first need to introduce basic fault vocabulary. To simplify the discussion, we treat a fault as a geometric surface in a body of rock. Rock adjacent to a fault surface is the *wall* of the fault, and the body of rock that moved as a consequence of slip on the fault is a *fault block.* If the fault is not vertical, you can distinguish between the *hanging-wall block,* which is the rock body above the fault plane, and the *footwall block,* which is the rock body below the fault plane (Figure 8.1a). You cannot distinguish between a hanging wall and a footwall for a vertical fault.

(a)

(b)

Figure 8.2 Photographs of faults at
different scales. (a) Microscopic faults,
showing fractured feldspar grain.
(b) Mesoscopic faults cutting thin layers in
an outcrop. (c) The trace of a fault cutting
across the countryside.

(c)

Table 8.1 Description of Fault Dip

Horizontal faults	Faults with dips of about 0°; if the fault dip is between about 10° and 0°, it is called *subhorizontal.*
Listric faults	Faults that have a steep dip close to the Earth's surface and have a shallow dip at depth; because of the progressive decrease in dip with depth, listric faults have a curved profile that is concave up.
Moderately dipping faults	Faults with dips between about 30° and 60°.
Shallowly dipping faults	Faults with dips between about 10° and 30°; these faults are also called *low-angle faults.*
Steeply dipping faults	Faults with dips between about 60° and 80°; these faults are also called high-angle faults.
Vertical faults	Faults that have a dip of about 90°; if the fault dip is between about 80° and 90°, the fault can be called subvertical.

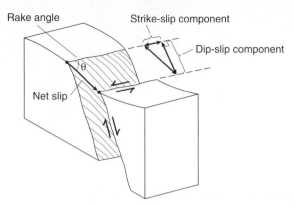

Figure 8.3 Block diagram sketch showing concept of the net-slip vector and its strike-slip and dip-slip components, as well as the rake.

To describe the attitude of a fault precisely, you need to measure the strike and dip (or dip and dip direction) of the fault. Commonly geologists use general adjectives (e.g., steep, shallow, vertical etc.) to convey an approximate image of fault dip (Table 8.1). Keep in mind that a fault is not necessarily a perfectly planar surface. Rather, it may curve and change attitude along strike and up and down dip. Faults whose dip decreases progressively with depth have been given the special name *listric faults.* Where such changes occur, a single strike and dip measurement is not sufficient to describe the attitude of the whole fault, and you should provide separate measurements for distinct segments of the fault.

When fault movement occurs, one fault block slides relative to the other, which is described by the *net slip.* You can completely describe displacement by specifying the *net-slip vector,* which connects two formerly adjacent points now on opposite walls of the fault (Figure 8.3). To describe a net-slip vector, you must specify its magnitude and orientation (plunge and bearing, or rake on a plane), and the *sense of slip* (or *shear sense*). Shear sense defines the relative displacement of one wall of the fault with respect to the other wall, that is, whether one wall went up or down, and/or to the left or right of the other wall.

Like any vector, the net-slip vector can be divided into components. Generally you will use the strike and dip of the fault as a reference frame for defining these components. Specifically, you measure the *dip-slip component* of

net slip in the direction parallel to the dip direction, and the *strike-slip component* of net slip in the direction parallel to the strike. If the net-slip vector parallels the dip direction of the fault (within ~10°), the fault is a *dip-slip fault;* if the vector roughly parallels the strike of the fault, the fault is a *strike-slip fault.* If the vector is not parallel to either dip direction of the strike, then you call the fault an *oblique-slip fault.* Oblique-slip faults have both a strike-slip and a dip-slip component of movement (Figure 8.3).

You describe the shear sense on a dip-slip fault with reference to a horizontal line on the fault, by saying that the movement is *hanging-wall up* or *hanging-wall down* relative to the footwall. Hanging-wall down faults are called *normal faults,* and hanging-wall up faults are called *reverse faults* (Figure 8.4a,b). To define sense of slip on a strike-slip fault, imagine that you are standing on one side of the fault and are looking across the fault to the other side. If the opposite wall of the fault moves to your right, the fault is *right-lateral* (or *dextral*), and if the opposite wall of the fault moves to your left, the fault is *left-lateral* (or *sinistral;* Figure 8.4c,d). Note that this determination of shear sense does not depend on which side of the strike-slip fault you are standing. Finally, you define shear sense on an oblique-slip fault by specifying whether the dip-slip component of movement is hanging wall up or down and whether the strike-slip component is right-lateral or left-lateral (Figure 8.4e–g). Commonly, further distinction between fault types is made by adding reference to the dip angle of the fault surface; we recognize high-angle (>60° dip), intermediate-angle (30–60° dip), and low-angle faults (<30° dip). We provide definitions of the basic fault types in Table 8.2, along with descriptions of other commonly used names (e.g., thrust and detachment).

The list in Table 8.2 is long, and we haven't even included the many terms within each broad grouping. You may be wondering where the terms "normal" and "reverse" come from. Perhaps normal faults were thought to be "normal" because the hanging-wall block appeared to have slipped down the fault plane, just like a person slips down a slide. Alternatively, the name "normal" came into use because most faults described by nineteenth century geologists in Europe resulted in hanging-wall down movement, and thus this type of

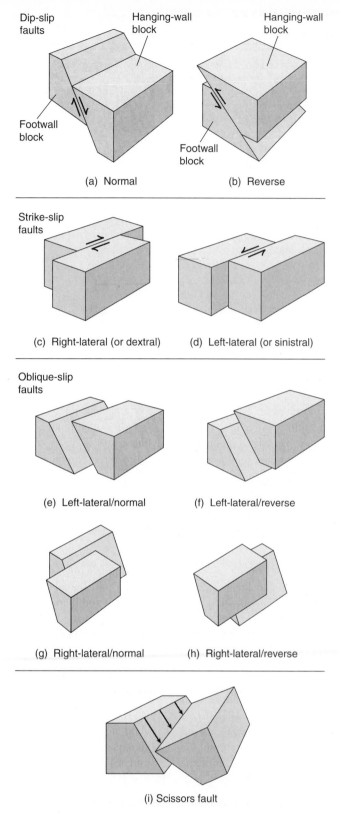

(a) Normal   (b) Reverse

(c) Right-lateral (or dextral)   (d) Left-lateral (or sinistral)

(e) Left-lateral/normal   (f) Left-lateral/reverse

(g) Right-lateral/normal   (h) Right-lateral/reverse

(i) Scissors fault

Figure 8.4   Block diagram sketches showing the different types of faults.

Table 8.2   Types of Faults

Allochthon	The thrust sheet above a detachment is the allochthon (meaning that it is composed of *allochthonous* rock; i.e., rock that has moved substantially from its place of origin).
Autochthon	The footwall below a detachment is the autochthon; it is composed of *autochthonous* rock, or rock that is still in its place of origin.
Contractional fault	A contractional fault is one whose displacement results in shortening of the layers that the fault cuts, regardless of the orientation of the fault with respect to horizontal.
Décollement	The French word for detachment.
Detachment (fault)	This term is used for faults that initiate as a horizontal or subhorizontal surface along which the hanging-wall sheet of rock moved relative to the footwall. An older term, "overthrust," is a regional detachment fault on which there has been a thrust sense of movement. Some detachments are listric, and on some detachments regional normal-sense displacement occurs.
Dip-slip fault	The slip direction on a dip-slip fault is approximately parallel to the dip of the fault (i.e., has a rake between ~80° and 90°).
Extensional fault	An extensional fault is one whose displacement results in extension of the layers that the fault cuts, regardless of the orientation of the fault with respect to horizontal.
Normal fault	A normal fault is a dip-slip fault on which the hanging wall has slipped down relative to the footwall.
Oblique-slip fault	The slip direction on an oblique-slip fault has a rake that is not parallel to the strike or dip of the fault. In the field, faults with a slip direction between ~10° and ~80° are generally called oblique-slip.
Overthrust fault	You may find this term in older papers on faults, but it is no longer used much today. The term is used for thrust faults of regional extent. In this context, "regional extent" means that the thrust sheet has an area measured in tens to hundreds of square km, and the amount of slip on the fault is measured in km or tens of km. Today, such faults are generally called regional detachments.
Par-autochthonous	If a fault block has moved only a small distance from its original position, the sheet is *par-autochthonous* (literally, "relatively in place").
Reverse fault	A reverse fault is a dip-slip fault on which the hanging wall has slipped up relative to the footwall.
Scissors fault	On a scissors fault, the amount of slip changes along strike so that the hanging-wall block rotates around an axis that is perpendicular to the fault surface (Figure 8.4h).
Strike-slip fault	The slip direction on a strike-slip fault is approximately parallel to the fault strike (i.e., the line representing slip direction has a rake (pitch) in the fault plane of less than ~10°). Strike-slip faults are generally steeply dipping to vertical.
Transfer fault	A transfer fault accommodates the relative motion between blocks of rock that move because of the displacement on other faults.
Transform fault	In the context of plate tectonics, transform faults are plate boundaries at which lithosphere is neither created nor destroyed. In a general sense, a transform fault links two other faults and accommodates the relative motion between the blocks of rock that move because of the displacement on the other two faults. The term *transfer fault* is also used for this general type of displacement, independent of scale.

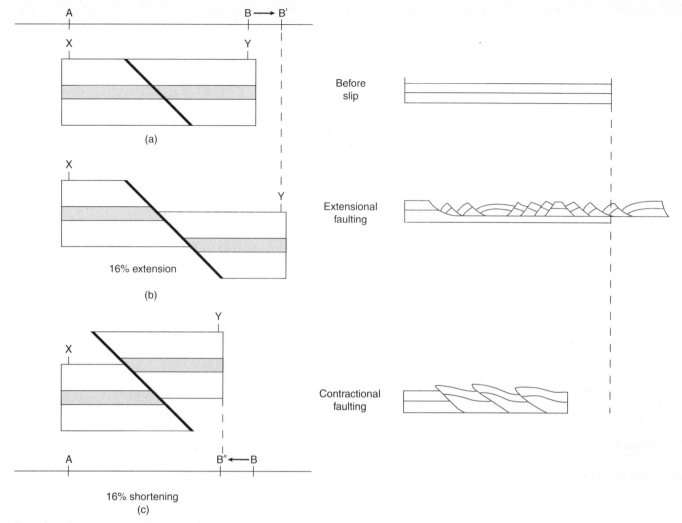

Figure 8.5 Concepts of extensional and contractional faulting. (a) Starting condition, (b) extension, and (c) contraction.

movement was considered to be the norm. It is a safe guess that geologists came up with the name "reverse fault" to describe faults that are the opposite of normal.

You can also distinguish among faults based on whether faulting causes shortening or lengthening of the layers that it cuts. Imagine that a fault cuts and displaces a horizontal bed marked with points X and Y (Figure 8.5a). Before movement, X and Y project to points A and B on an imaginary plane above the bedding plane. If the hanging wall moves down, then points X and Y project to A and B'. The length AB is greater than the length AB' (Figure 8.5b). In other words, movement on this fault effectively lengthens the layer. We call a fault that results in lengthening of a layer an *extensional fault*. By contrast, the faulting shown in Figure 8.5c resulted in a decrease in the distance between points X and Y (AB > AB''). We call a fault that results in shortening of a layer a *contractional fault*. Contractional faults result in duplication of section, as measured along a line that crosses the fault and is perpendicular to stratigraphic boundaries, whereas extensional faults result in loss of section. Generally, you can use the term "normal fault" as a synonym for an extensional fault, and the term "reverse fault" as a synonym for contractional faults. But

such usage is not always correct. Consider a normal fault that rotates during later deformation. In outcrop, this fault may have the orientation and sense of slip you would expect on a reverse fault, but in fact, its displacement represents extensional strain parallel to layering.

8.2.2 Representation of Faults on Maps and Cross Sections

Remember that a fault is a type of geologic contact (i.e., a fault forms the boundary between two bodies of rock), so like other contacts, faults are portrayed as a (thick) line on geologic maps. You can distinguish among different types of faults on maps through the use of symbols (Figure 8.6). For example, thrust faults are decorated with triangular teeth placed on the hanging-wall side of the trace (but note that the teeth do not necessarily indicate the direction of movement!). Where erosion cuts a hole through a thrust sheet, exposing rocks of the footwall, the hole is a *window* and the teeth point outward from the hole (Figure 8.7). An isolated remnant of a thrust sheet surrounded by exposures of the footwall is a *klippe*. The thrust-fault symbol with the teeth pointing inward encircles a klippe. Normal faults, regardless of dip, are commonly portrayed by placing barbs

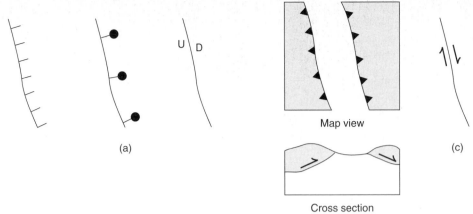

(a)

Map view

Cross section

(c)

(b)

Figure 8.6 Basic map symbols for (a) normal fault, (b) thrust fault, and (c) strike-slip fault.

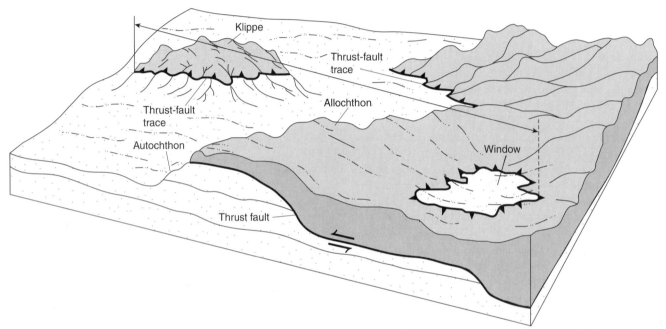

Figure 8.7 Block diagram illustrating klippe, window, allochthon (gray), and autochthon (stippled) in thrust-faulted region. The minimum displacement is defined by the farthest distance between thrust outcrops in klippe and window.

on the hanging-wall block. You can represent strike-slip faults on a map by placing half arrows on either side of the fault that indicate the sense of slip (Figure 8.6).

In cross sections, you also represent faults by a thick line (Figure 8.8). If the slip direction on the fault roughly lies in the plane of the cross section, then you indicate the sense of slip on the fault by oppositely facing half-arrows drawn on either side of the fault. If the movement on the fault is into or out of the plane of the section for a strike-slip fault, you indicate the sense of slip by drawing the head of an arrow (a circle with a dot in it) on the block moving toward you, and tail of an arrow (a circle with an X in it) on the block moving into the plane. If the movement is into or out of the page for a dip-slip fault, you place the map symbol (teeth for thrust faults and barbs for normal faults) for the fault on the hanging-wall block.

If a fault cuts across the contact between two geologic units, it must displace this contact unless the net-slip vector happens to be exactly parallel to the intersection line be-

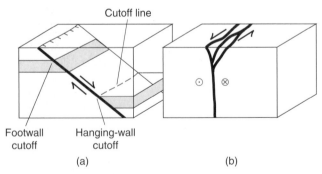

Figure 8.8 Block diagrams with cross-sectional fronts, showing the different symbols for representing (a) dip-slip faults and (b) strike-slip faults; in (a) we also show cutoffs.

tween the fault and the contact. The point on a map or cross section where a fault intersects a preexisting contact is called a *cutoff;* and in three dimensions (Figure 8.8), the intersection between a fault and a preexisting contact is a *cutoff line*. If the truncated contact lies in the hanging-wall

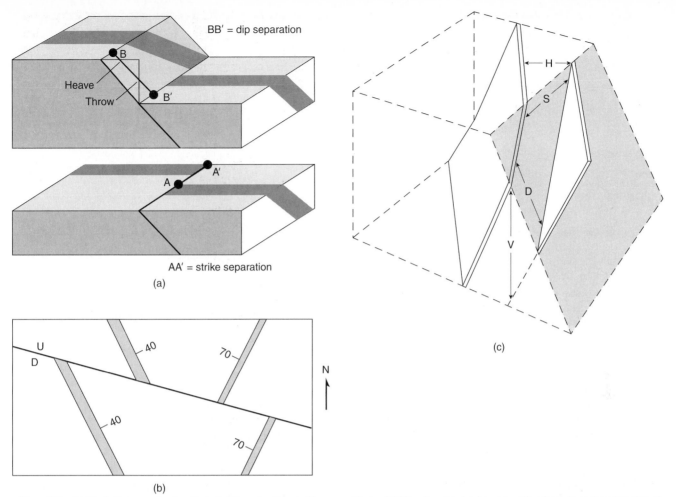

Figure 8.9 (a) Block diagrams showing dip and strike separation, and heave and throw. (b) Map view showing how separation depends on the orientation of the offset layer. The two dikes shown in this figure dip in different directions and have, therefore, different strike separations. (c) Block diagram illustrating horizontal (H) and vertical (V) separation, as well as the dip (D) and strike (S) separation.

block, the truncation is a *hanging-wall cutoff,* and if the truncated contact lies in the footwall, it is a *footwall cutoff.*

8.2.3 Fault Separation and Determination of Net Slip

Imagine a *marker horizon* (a distinctive surface or layer in a body of rock; e.g., a bed) that has been cut and offset by slip on a fault (Figure 8.9a). We define *separation* as the distance between the displaced parts of the marker horizon as measured along a specified line. "Separation" and "net slip" are not synonymous, unless the line along which we measure the separation parallels the net-slip vector. The separation for a given fault along a specified line depends on the attitude of the offset marker horizon. Therefore, separation along a specified line is not the same for two nonparallel marker horizons (Figure 8.9b). Fault separation is a little difficult to visualize, so we will describe different types of fault separation with reference to Figure 8.9c, which shows an oblique-slip fault that cuts a steeply dipping bed. We define the types of separation illustrated in this figure in Table 8.3.

With the terms of Table 8.3 in mind, note that horizontal beds cut by a strike-slip fault show no strike separation and vertical beds cut by a perpendicular dip-slip fault show

no dip separation. If the fault cuts ground surface, this surface itself is a marker horizon for defining vertical separation, and linear features on the ground (e.g., fences, rows of trees, roads, railroads, or river beds) serve as markers for defining horizontal separation. Note that Table 8.3 also defines *heave* and *throw,* which are old terms describing components of dip separation (Figure 8.9a).

In order to completely define the net-slip vector, you must specify its magnitude (e.g., in meters), the direction of displacement (as a plunge and bearing), and the sense of slip. If you are lucky enough to recognize two points now on opposite walls of the fault that were adjacent prior to displacement,[2] then you can measure net slip directly in the field. For example, if you observe a fence on the ground surface that has been offset by a fault, then you can define net slip, because the intersection of the fence with the ground defines a line, and the intersection of this line with the walls of the fault defines two previously adjacent points.

More commonly, however, you won't be lucky enough to observe an offset linear feature, and you must calculate the net-slip vector from other information. This can be done

---

[2]Sometimes referred to as *piercing points.*

Table 8.3 Fault Separation and Fault-Separation Components (Figure 8.9)

Dip separation	The distance between the two bed/fault intersection points as measured along a line parallel to the dip direction.
Heave	The horizontal component of dip separation.
Horizontal separation (H)	The offset measured in the horizontal direction along a line perpendicular to the offset surface.
Stratigraphic separation	The offset measured in a line perpendicular to bedding.
Strike separation (S)	The distance between the two bed/fault intersection points as measured along the strike of the fault.
Throw	The vertical component of dip separation.
Vertical separation (V)	The distance between two points on the offset bed as measured in the vertical direction. Vertical separation is the separation measured in vertical boreholes that penetrate through a fault.

by measurement of: (a) separation, along a specified line, of the intersection between a single marker horizon and the fault, plus information on the direction of slip; (b) separation, along two nonparallel lines, of the intersection between a single plane and the fault; or (c) separation, along a specified line, of two nonparallel marker horizons. Look at any standard structural geology methods book for an explanation of how to carry out such calculations.

If you do not have sufficient information to determine the net slip, you can at least provide valuable information about displacement on the fault by searching for slip lineations and shear-sense indicators. *Slip lineations* are structures on the fault that form parallel to the net-slip vector for at least the last increment of movement on the fault, and possibly for accumulated movement during a progressive deformation event. *Shear-sense indicators* are structures on the fault surface or adjacent to the fault surface that define the direction in which one block of the fault moved with respect to the other. Slip lineations alone define the plunge and bearing of the net-slip vector, and with kinematic indicators they define the direction that the vector points. Such information can help you interpret the tectonic significance of a fault, even if you don't know the magnitude of displacement across it. We'll discuss types of slip lineations and kinematic indicators, and how to interpret them, later in this chapter, after we have reviewed the process that leads to their formation.

The magnitude of the net-slip vector (i.e., fault displacement) on real faults ranges from millimeters to thousands of kilometers. For example, about 600 km of net slip occurred on the oldest part of the San Andreas Fault in California. In discussion, geologists refer to faults with large net slip as *major faults* and faults with small net slip as *minor faults* or *mesoscopic faults*. Keep in mind that such adjectives are relative, and depend on context; a major fault on the scale of an outcrop may be a minor fault on the scale of a continent. In the relatively rare cases where an earthquake-generating fault cuts ground surface, you can directly measure the increment of displacement accompanying a single earthquake. Generally, however, the displacement that you measure when studying ancient faults in outcrop is a *cumulative displacement* representing the sum

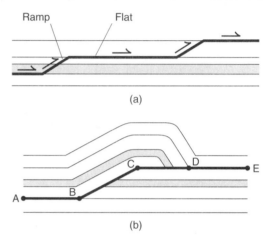

Figure 8.10 (a) Cross section showing the geometry of ramps and flats along a thrust fault. The fault geometry is shown prior to displacement on the fault. (b) Cross section illustrating how to describe hanging-wall and footwall flats and ramps. Segment AB is a hanging-wall flat on a footwall flat. Segment BC is a hanging-wall flat on a footwall ramp. Segment CD is a hanging-wall ramp on a footwall flat. Segment DE is a hanging-wall flat on a footwall flat.

of many incremental offsets that occurred over a long period of time.

8.2.4 Changes in Fault Attitude (Fault Bends)

As we mentioned earlier, fault surfaces are not necessarily planar. It is quite common, in fact, for the attitude of a fault to change down-dip or along-strike. In some cases, the change is gradual. For example, the dip of a 'down-to-the-Gulf' fault (see Chapter 2) along the coastal plain of Texas typically decreases with depth, so that the fault overall has a concave-up shape, making it a *listric fault*. Other faults have wavy traces because their attitude changes back and forth.

If the dip and/or strike of a fault abruptly changes, the location of the change is called a *fault bend*. Dip-slip faults that cut across a stratigraphic sequence in which layers have different mechanical properties typically contain numerous stratigraphically controlled bends that make the trace of the fault in cross section resemble a staircase. Some fault segments run parallel to bedding, and are called *flats*, and some cut across bedding, and are called *ramps* (Figure 8.10a). If the fault has not been folded subsequent to its formation,

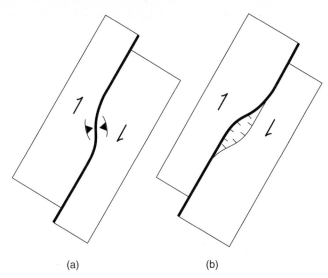

Figure 8.11 Map-view illustrations of (a) a restraining bend, and (b) a releasing bend along a right-lateral strike-slip fault.

(a)   (b)

flats are (sub)horizontal, whereas ramps have dips of about 30–45°. Note that, as shown in Figure 8.10b, a segment of a fault may parallel bedding in the footwall, but cut across bedding in the hanging wall. Thus, when describing stairstep faults in a stratified sequence, you need to specify whether a fault segment is a ramp or flat with respect to the strata of the hanging wall, footwall, or both.

Fault bends (or *steps*) along strike-slip faults cause abrupt changes in the strike of the fault. To describe the orientation of such fault bends, imagine that you are straddling the fault and are looking along its strike; if the bend moves the fault plane to the left, you say the fault steps to the left, and if the bend moves the fault plane to the right, you say that the fault steps to the right. Note that the presence of bends along a strike-slip fault results in either contraction or extension across the step, depending on its geometry. Locations where the bend is oriented such that blocks on opposite sides of the fault are squeezed together are *restraining bends,* whereas locations where the bend is oriented such that blocks on opposite sides of the fault pull away from each other are *releasing bends* (Figure 8.11).[3] Where movement across a segment of a strike-slip fault results in some compression, we say that *transpression* is occurring across the fault; where movement results in some extension, we say that *transtension* is occurring across the fault. Note that a step to the left on a right-lateral fault yields a restraining bend, whereas a step to the right on a

right-lateral fault yields a releasing bend. You can think of the rules for a left-lateral fault yourself! Natural examples of these structures, such as the San Andreas Fault of California, are discussed in Chapter 18.

8.2.5 Fault Terminations and Fault Length

Faults develop at all scales (Figure 8.2): microscopic faults offset the boundaries of a single grain, whereas, megascopic faults cut across thousands of kilometers of crust. But even the longest faults do not extend infinitely in all directions. Faults terminate in several ways.

Some faults terminate where cut by younger structures, such as another fault, an unconformity, or an intrusion (Figure 8.12a). Application of the principle of cross-cutting relations allows you to determine the relative age of faults with respect to the structures that cut them. Faults that you can map today in the field terminate at the ground surface either because the fault intersected the ground when it moved or has been subsequently exposed by erosion. If the fault intersected the ground surface while it was still active, it is an *emergent fault* (e.g., the San Andreas Fault); but if it intersects the ground surface only because the present surface of erosion has exposed an ancient, inactive fault, it is an *exhumed fault.*

Some faults link to other faults while both are active (Figure 8.12b). For example, fault splays diverge from a larger fault, and faults in an anastamosing array merge and diverge along their length. Where a fault does not terminate against another structure, it must die out, meaning that the magnitude of displacement decreases along the trace of the fault, becoming zero at its tips. The boundary between the slipped and unslipped region at the end of a fault is the *tip line* of the fault (Figure 8.13a). Exhumed and emergent faults must die out along their strike, unless they terminate at another structure; their tip line intersects the ground surface as a point. A fault that dies out in the subsurface before intersecting the ground surface is a *blind fault* (Figure 8.13b). In some cases, a fault splits into numerous splays near its tip line, thereby creating a fan of small fractures called a *horsetail* (at B in Figure 8.12b), or it may die in an array of pinnate fractures. Alternatively, the deformation associated with the fault dies out in a zone of ductile deformation (e.g., folding or penetrative strain) that surrounds the tip line (at C in Figure 8.12b).

As we just pointed out, the length of some faults is limited by their intersection with, or truncation by, other structures. For many faults the trace length changes with time as the fault evolves. To visualize this process, imagine a fault that initiates at one point and grows outward. At a given instant of time, slip has occurred where the fault surface already exists, but there is no slip beyond the tip of

---

[3]The terms "restraining bends" and "releasing bends" can also be applied to steps along dip-slip faults that connect two parallel fault segments.

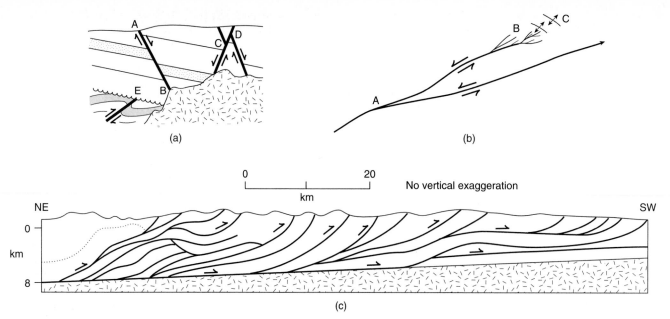

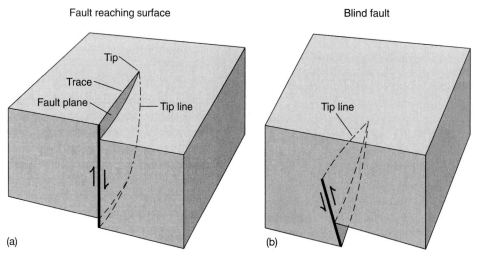

Figure 8.12 (a) Cross-sectional sketch showing various types of fault terminations. At point A, the fault terminates at the ground surface; at B the fault has been cut by a pluton; at C and D, the one fault cuts another; at E, the fault was eroded at an unconformity. (b) Termination of a fault by merging with another fault (at point A), or by horsetailing (at point B) and dying out into a zone of ductile deformation (at point C). (c) A series of ramps merging at depth with a basal detachment.

(c) Modified from R. A. Price, 1981, "The Cordilleran foreland thrust and fold belt in the Southern Canadian Rocky Mountains" in M. P. Coward and K. R. McClay, (eds.) *Thrust and Nappe Tectonics,* Geological Society of London Special Publication 9, pp. 427–428. Used by permission.

Figure 8.13 Tip lines for (a) an emergent fault and (b) blind fault.

the fault (Figure 8.14a). A little later, after more fault-tip propagation, there is increased slip in the center of the fault (Figure 8.14b). As a consequence, the magnitude of displacement changes along the length of the fault, and the magnitude of displacement on the fault must be less than the length of the fault. Considering this relation, we might expect a general relationship between fault length and displacement: the longer the fault trace, the greater the displacement. Recent work supports this idea, though the details remain controversial. Faults that are meters long display offsets typically on the order of centimeters or less, whereas faults that are tens of kilometers long

show typical offsets on the order of several hundreds of meters. Figure 8.14c shows a log-log plot of length versus offset, based on examination of thousands of faults. The faults shown on this diagram occur in a variety of lithologies and range in length from centimeters to hundreds of kilometers. Nonetheless, you will notice that the points plot between two 45° lines, suggesting that, given knowledge of fault length, we can predict its displacement (or vice versa), independent of the properties of the material and scale. However, debate continues whether a single relationship describes the length versus offset relationship for both regional and mesoscopic faults, for, as evident by

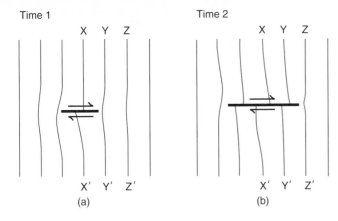

(a) (b)

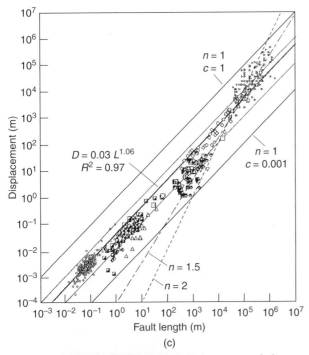

(c)

Figure 8.14 (a, b) Map-view showing that displacement on a fault grows as the fault length increases. At time 1, the short fault offsets marker line XX' by only a small amount. At time 2, the fault has grown in length, and marker line XX' has been offset by a greater amount. Note that the displacement decreases toward the ends of the fault. (c) Log-log plot showing the apparent relationship between fault length (L) and displacement (D): $D = C \cdot L^n$; R^2 is the correlation coefficient.

the figure, the best-fitting straight line slightly overestimates displacement for shorter faults and gives underestimates for longer faults.[4]

---

[4]The single, best-fitting power-law relationship is: $D = 0.03 \cdot L^{1.06}$; the exponent (here $\simeq 1$) is called the *fractal dimension.*

8.3 CHARACTERISTICS OF FAULTS AND FAULT ZONES

8.3.1 Brittle Fault Rocks

Faulting involves either shear fracturing of a previously intact rock, or slip on a previously formed fracture. Shear fracturing occurs when a multitude of cracks coalesce (Chapter 6), and shear displacement on a preexisting fracture may lead to formation of new off-plane fractures and fault splays. Thus, the process of brittle faulting tends to break up rock into fragments, which is called *brittle fault rock.* We classify brittle fault rock based on whether it is *cohesive* or not (i.e., whether the fragments composing the fault rock remain stuck together to form a coherent mass without subsequent cementation or alteration) and on the *size* of the fragments making up the fault rock. Table 8.4 summarizes the principal terms that we use to describe brittle fault rocks.

A random array of nonsystematic mesoscopic fractures that surround angular blocks of rock creates *fault breccia* (Figure 8.15a). Continued displacement across the fault zone may crush and further fragment breccia, and/or may break off microscopic asperities protruding from slip surfaces in the fault zone, thereby creating a fine-grained rock flour that we call *fault gouge* (Figure 8.15b). Gouge and (micro)breccia are noncohesive fault rocks, meaning that they easily fall apart when collected at a fault zone or hit with a hammer. In general, noncohesive fault rocks have *random fabrics,* meaning they do not contain a distinctive foliation.

The network of fractures between fragments in breccia and gouge allows groundwater to pass through the fault zone. Minerals like quartz or calcite may precipitate out of the groundwater, thereby cementing together rock fragments in the fault zone. As a result, breccia and gouge become *indurated,* meaning that the fragments are cemented together. In coarse breccia, the cement typically also fills veins of euhedral or blocky crystals in the open spaces between rock fragments, resulting in formation of a *vein-filled breccia* (Figure 6.1b). Circulating groundwater may also have the effect of causing intense alteration of minerals in the gouge or breccia zone. In fact, alteration rates in fault zones tend to be greater than in intact rock, because fragmentation of rock increases the net area of reactive surfaces. As a consequence of alteration, some minerals (e.g., feldspar) transform into clay. The layer of clay that develops in some fault zones can act as an impermeable barrier to further fluid movement. In olivine-rich rocks, such as basalt and peridotite, reaction of fault rock with water yields serpentine. As we mentioned earlier, clay and serpentine are groups of weak minerals, so their presence along a fault allows it to slip at lower differential stresses than it would if the original minerals were present.

Table 8.4   Classification of Fault Rock

Noncohesive Fault Rocks

Fault gouge	Composed of material whose grain size has been mechanically reduced by pulverization. Grains in fault gouge are less than about 1 mm in diameter. Like breccia, gouge is noncohesive. Shearing of gouge along a fault surface during progressive movement may create foliation within the gouge. Clay formed by alteration of silicate minerals in fault zones may be difficult to distinguish from true gouge.
Indurated gouge	Fault gouge that has been cemented together by minerals precipitated from circulating groundwater.
Fault breccia	Composed of angular fragments of rock greater than about 1 mm and as much as several meters across; fault breccia is noncohesive.
Vein-filled breccia	Fault-breccia blocks that are cemented together by vein material. Another term, *indurated breccia,* is synonymous.

Cohesive Fault rocks

Pseudotachylyte	A glass or microcrystalline material that forms when frictional heating melts rock during slip on a fault. Pseudotachylyte commonly flows into cracks between breccia fragments or into cracks penetrating the walls of the fault. In special cases, pseudotachylyte may be several m thick (e.g., impact sites), but generally it occurs in lenses or sheets that are mm to cm in thickness.
Argille scagliose	A fault rock that forms in very fine grained clay- or mica-rich rock (e.g., shale or slate) and is characterized by the presence of a very strong wavy anastamosing foliation. As a consequence, the rock breaks into little scales or platy flakes.
Cataclasite	A cohesive fault rock composed of broken, crushed, or rolled grains. Unlike breccia, it is a solid rock that does not disintegrate when struck with a hammer.

(a)

(b)

Figure 8.15   (a) Fault breccia along the Buckskin detachment. (Battleship Peak, Arizona). (b) Fault gorge from the Lewis thrust (Southern Alberta).

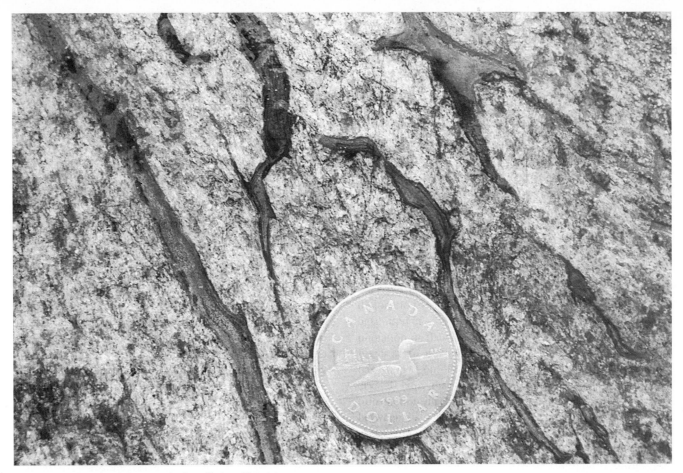

Figure 8.16 Pseudotachylyte near the Grenville front (Ontario).

Cohesive brittle fault rock is known as *cataclasite* Cataclasite differs from gouge or breccia in that the fragments interlock and allow the fragmental rock to remain coherent even without later cementation. Cataclasites generally have random fabrics (i.e., no strong foliation or lineation).

Table 8.4 also lists two less common types of fault rock, pseudotachylyte and argille scagliose. *Pseudotachylyte* (from the prefix *pseudo-*, which means "like," and the noun *tachylyte,* which is a type of volcanic glass) is glass or very fine crystalline material that forms when frictional sliding generates enough heat to melt the rock adjacent to the fault (Figure 8.16). Such conditions can occur during earthquakes. Because rock is not a good conductor, the heat generated by frictional sliding cannot flow away from the fault, and temperatures in the fault zone become very high (>1000°C). Melt formed in such a setting squirts into cracks and pores in the fault walls where it cools so quickly that it solidifies into a glass. *Argille scagliose* refers to a strongly foliated fault rock formed by pervasive shearing of a clay-rich or very fine-grained mica-rich lithology such as shale or slate. In argille scagliose (which means "scaly clay" in Italian), foliation planes, which are microscopic shear surfaces, are anastomosing and very shiny, yielding a rock that has the overall appearance of a pile of oyster shells. Argille scagliose can develop under nonmetamorphic conditions (e.g., at the base of a sedimentary mélange) or in low-grade metamorphic conditions. Scaly fabrics also form in other fine-grained lithologies, such as coal and serpentinite.

8.3.2 Slickensides and Slip Lineations

Displacement on a fault in the brittle field involves frictional sliding and/or pressure-solution slip. Each process yields distinctive structures (slickensides and slip lineations) on the fault surface, which may provide information about the direction of net slip and even the shear sense of slip during faulting.

If slip on a fault takes place by frictional sliding, asperities on the walls of the fault break off and/or plow into the opposing surface and wear down. As a result, the two walls of the fault may become smoother and, in some cases, attain a high polish. Fault surfaces that have been polished by the process of frictional sliding are called *slickensides* (Figure 8.17). Slickensides form either on the original wall rock of the fault, or on the surface of a thin layer of gouge. During frictional sliding, asperities on one

Figure 8.17 Shiny slickensided surface in Paleozoic strata (Maryland Appalachians).

wall of the fault plow into the surface of the other wall, thereby creating *groove lineations* (striations or scratches) on the slickenside, resulting in formation of a *lineated slickenside* (Figure 8.17).[5] Groove lineations resemble the glacial striations created when rocks entrained in the base of a moving glacier scratch across bedrock, though they are much smaller.

Groove lineations are not the only type of lineation that forms on faults in the brittle regime. On fault surfaces coated with gouge, fault slip may mold gouge into microscopic linear ridges that, along with grooves, create a lineation visible on the fault surface. Also, some fault surfaces initiate with small lateral steps whose presence gives the fault a corrugated appearance. These *corrugations* resemble grooves, but can be longer than the total displacement on the fault. The origin of corrugations is not well understood.

On a fault developed by pressure-solution slip, the fault surface becomes coated by elongate fibers of vein minerals (typically quartz, calcite, or chlorite), whose long axes lie subparallel to the fault surface (Figure 8.18). These *fibers* grow incrementally by the "crack-seal" deformation mechanism, or by solution mass transfer through a fluid film along the fault surface. Thus, fiber formation does not necessarily involve brittle rupture along the fault surface. We discuss fibers here because they form under shallow crustal conditions, and thus occur in association with structures formed by frictional sliding. Whether frictional sliding or pressure-

solution slip takes place on a given fault reflects the strain rate and fluid conditions during faulting. Fibers form at smaller strain rates, require the presence of water films, and typically form in imbricate sheets (Figure 8.18). As movement continues, multiple sheets of fibers may develop on top of one another, so that a vein up to several centimeters thick eventually develops along the fault plane. In relatively thick veins (>2 or 3 cm), the internal portion of the vein may consist of blocky spar, which forms either by recrystallization of earlier formed fibers, or by precipitation of euhedral crystals in gaps along the fault surface.

Slip lineations trend parallel to an increment of displacement on a fault, and thus allow you to determine whether the increment resulted in strike-slip, dip-slip, and oblique-slip offset. If all movement on the fault has been in the same direction, then the slip lineation defines the orientation of the net-slip vector. You must use caution when interpreting slip lineations, however, because they may define only one increment of displacement on a fault surface on which slip in a range of directions occurred. Moreover, the last increment of frictional sliding sometimes erases grooves formed during earlier increments, and formation of one sheet of fiber lineations may cover and obscure preexisting sheets of fiber lineations. If more than one set of lineations is preserved on a fault surface, you can study them to distinguish among multiple nonparallel slip increments.

[5]Also called *slickenlines*.

(a)

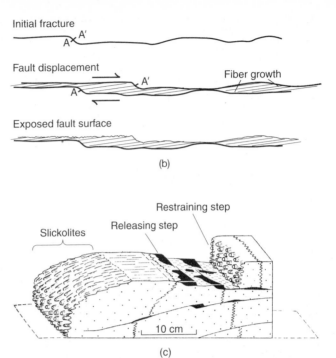

Initial fracture

Fault displacement

Fiber growth

Exposed fault surface

(b)

Restraining step

Releasing step

Slickolites

10 cm

(c)

Figure 8.18 (a) Slip fibers on a fault showing steps that indicate sense of shear; compass; notebook for scale. (b) Illustration of the growth of slip fibers along a fault, and (c) block diagram illustrating steps along a fiber-coated fault surface. Restraining steps become pitted by pressure solution, releasing steps become the locus of vein growth, and oblique restraining steps become slickolites.

(b) *Basic Methods of Structural Geology* by Marshak/Mitra, © 1988. Reprinted by permission of Prentice-Hall, Inc. Upper Saddle River, NJ.

Structures on a fault surface may provide constraints on the shear sense. For example, as frictional sliding occurs, an original irregularity on a slickenside becomes polished more smoothly on the upslip side, and remains rougher on the downslip side, so that subtle steps may develop on the fault surface. These features create an anisotropy on a slickenside; the surface feels smoother as you slide your hand in the shear direction. However, anisotropy on a slickenside is subtle and may be ambiguous, because the intersection of small fractures with the fault surface may create steps that face in the opposite direction to the steps formed by differential polishing.

Sheets of mineral fibers formed by pressure-solution slip on a fault provide a more reliable indication of shear sense. Because the fibers composing the sheets form at a low angle to the fault surface, fiber sheets tend to overlap one another like shingles, and tilt away from the direction of shear. Features developed on mesoscopic steps along a fault

that moved by pressure-solution slip may also define shear sense. Restraining steps oppose movement on the fault, and therefore become pitted by pressure solution. Pit axes on the steps roughly parallel the net-slip vector, and thus are subparallel to the fault surface. If the restraining-step face is not perpendicular to the fault surface, its pit axes are oblique to the step face. Surfaces containing such oblique pits look like a cross between a stylolite and a slip lineation, and thus some authors call them *slickolites* (Figure 8.18c). Releasing steps, at which the opposing walls of the fault pull away from each other, typically become coated with fibrous veins. The long axes of the fibers of these veins parallel the pit axes on releasing steps along the same fault.

8.3.3 Subsidiary Fault and Fracture Geometries

As mentioned previously, some fault zones consist of a major fault along with an array of subsidiary faults, including both discrete smaller faults that occur within a larger fault zone (and that may anastomose with one another) and fault splays that branch off. Such subsidiary faults may initiate when the primary rupture splits into more than one surface during its formation in intact rock; when numerous

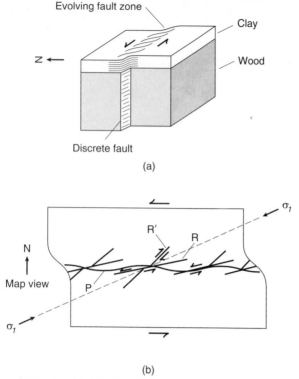

(a)

(b)

Figure 8.19 Growth of Riedel and P shears. (a) Block diagram illustrating a layer of clay that deforms when underlying blocks of wood slide past each other. (b) Map view of the top surface of the clay cake, illustrating the orientation of Riedel (R), conjugate Riedel (R′), and P shears. The acute bisectrix of the R and R′ shears is parallel to the remote σ_1 direction.

subparallel faults initiate simultaneously in a fault zone; when conjugate shear fractures develop at an angle to the principal fault; or when numerous preexisting surfaces in the fault zone reactivate during a deformation event.

In the case of emergent strike-slip faults, like the San Andreas Fault, a particularly interesting array of subsidiary faults, known as Riedel and P shears, develops. To picture how these develop, imagine an experiment in which you place a layer of clay over two wooden blocks, and then shear one block horizontally past the other (Figure 8.19a). In such an experiment, the clay accommodates strain before any fractures appear, but when fracturing in the clay begins to develop, the first fractures to appear are short shear fractures inclined at an angle to the trace of the throughgoing fault that eventually forms (Figure 8.19b). These short fractures are called *Riedel shears*.[6] Generally, you will find two distinct sets of Riedel shears (R and R′) that together define a conjugate pair. The bisectrix of the acute angle between conjugate R and R′ shears reflects the local orientation of σ_1 adjacent to the future fault. As shear continues, a third set of fractures, P shears, develops. P shears link the previously formed Riedel shears, and, eventually, a throughgoing fault zone consisting of linked R, R′, and P shears develops. Whether such subsidiary shears also form at

---

[6]After W. Reidel who first discussed this relationship in the igzo's.

depth in the Earth remains debatable. In fact, some geologists suggest that they form only when a weak layer is sheared by relative displacement of two stronger blocks on either side or below, much as in our experiment.

Tensile cracks can also develop in association with faulting. These cracks may occur in the wall rock, or in the fault zone itself. A series of parallel tensile fractures that form within a fault zone defines an *en echelon* array that tends to dilate and become veins. Typically, *en echelon* veins in a fault zone initiate at an angle of about 45° to the zone boundary and tilt toward the direction of shear; thus their orientation provides a shear-sense indicator. With progressive displacement of the fault-zone walls, the earlier formed parts of the veins rotate, but each new increment of vein initiates at an angle of 45° to the walls. Eventually, *en echelon* veins evolve into a sigmoidal shape whose sense of rotation defines the sense of shear on the fault (see Figure 7.22).

8.3.4 Fault-Related Folding

Faults and folds commonly occur in the same outcrop. The spatial juxtaposition of these structures, one brittle and one ductile, may seem paradoxical at first. How can rock break brittlely at the same time that it distorts ductilely? One explanation for the juxtaposition of faults and folds in an outcrop is that the structures formed at different times under different pressure and temperature conditions. For example, imagine that the folds you see in an outcrop formed 500 million years ago at a depth of 15 km in the crust, but, once formed, the rock body containing the folds was unroofed and moved to shallower depths, where a later and separate deformation event created faults in the body. Typically, in such examples, the faults cut across the preexisting folds and are not geometrically related to the folds. But in many examples, it is clear from the spatial and geometric relation between folds and faults that the two structures formed in association with each other. Folds that form in association with faults are called *fault-related folds*.

From earlier experiments we learned that the transition between brittle and ductile deformation depends on strain rate (Chapter 5). If ductile strain (e.g., folding) in a local region in a rock body develops fast enough to accommodate regional deformation, then differential stress does not get very high in the rock and faulting need not occur. But if regional deformation cannot be accommodated sufficiently fast by ductile strain, then differential stress builds up in the rock until it exceeds the failure strength of the rock or the frictional sliding strength of a preexisting fracture in the rock, and faulting occurs. In fact, strain rate may vary with position in a rock body, so that while one region in a rock body faults, another region in the rock body folds. Also, strain rate and differential stress magnitude can change with time at a given location in a rock body during the same overall deformation event, so that episodes of faulting and folding may alternate at the location.

In this section, we introduce several types of folds that develop in association with faults. We also return to the

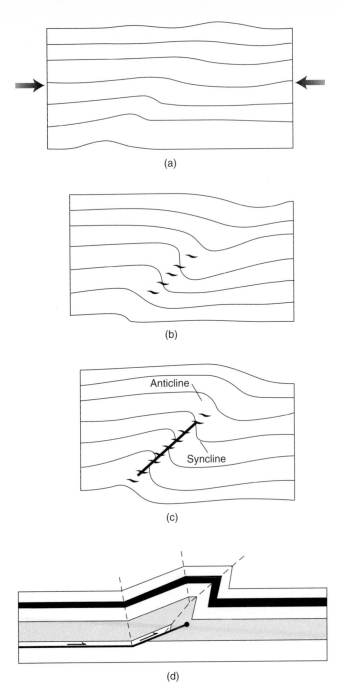

Figure 8.20 Progressive development of a fault-inception fold in a stratified sequence. (a) A small flexure develops during shortening of the layers, and a pronounced anticline-syncline pair develops. (b) *En echelon* gashes form in the fold. (c) A fault breaks through the fold. (d) Geometry of a fault-propagation fold.

(d) *Principles of Structural Geology* by Suppe, John, © 1985. Adapted by permission of Prentice-Hall, Inc., Upper Saddle River, NJ.

layers fold. Such folding may yield merely a gentle flexure of adjacent beds, or a pronounced asymmetric anticline-syncline pair (Figure 8.20).[7] If, at a later time, folding can no longer accommodate the internal displacement of the block, the differential stress magnitude in the block increases and a fault initiates. Stress buildup leading to faulting may reflect a change in the regional strain rate or may reflect locking up of the fold. By "locking up," we mean that the change in the geometry of the beds resulting from fold formation makes continued folding more difficult (i.e., like a stick, the layers can be bent only so far before they break). We call folds that develop during ductile deformation preceding faulting *fault-inception folds*. Note that the sense of asymmetry of a fault-inception fold defines the sense of shear on the fault.

The spatial relation between fault-inception folds and associated faults is variable. In some cases, a fault cuts through the fold along the hinge of the fold; in other cases, the fault breaks through the limbs of the folds between the adjacent anticline and syncline (i.e., at the inflection surface of the fold). In Chapter 17, we will see that fault-inception folds breaking through at the syncline hinge are called *fault-propagation folds* (Figure 8.20d). After an increment of faulting occurs, fault-inception folds adjacent to the fault may continue to develop again until stress on the fault exceeds the failure strength for frictional sliding and a new increment of displacement occurs.

The presence of bends along a fault passively causes folding of strata that move past the bend during displacement on the fault, because the moving layers must accommodate the bends in the fault without gaps or overlaps developing along the fault. Folds that form in this manner are called *fault-bend folds* (Figure 8.21a). Fault-bend folds form in association with all types of faults, but most of the literature concerning them pertains to examples developed along dip-slip faults (see Chapter 17 for more detail).

A third situation in which folding accompanies faulting occurs where a sequence of interlayered weak and strong rock layers (e.g., interbedded shale and sandstone) is caught between the opposing walls of a fault zone (Figure 8.21b). In such a circumstance, displacement of the rigid fault-zone walls with respect to each other causes the intervening rock to fold, much like a carpet that is caught between the floor and a sliding piece of furniture. *Folds in fault zones* formed by such a process tend to be very asymmetric. Typically, the hinges of folds within the fault zone are initially perpendicular to the overall shear direction, but with continued displacement, the hinges themselves may be bent, eventually curving into

subject of fault-related folding when we discuss aspects of tectonics (Part D).

Imagine a body of stratified rock with horizontal bedding that is undergoing shortening due to regional compression. Initially, the ends of the body move toward each other very slowly. If, during this stage, differential stress in the body is lower than the shear failure strength of the rock, the

---

[7]An anticline is an archlike fold, and a syncline is a troughlike fold. The hinge of a fold is the line on the folded surface where the curvature is a maximum. The axial surface of a fold is the surface containing the hinges of successive layers. Inclined folds are folds whose axial surface is dipping (i.e., it is not vertical or horizontal). Asymmetric folds are folds whose limbs are not equal in length, so that the portion of the fold on one side of the axial surface is not a mirror image of the portion of the fold on the other side. We'll discuss fold terminology in Chapter 10.

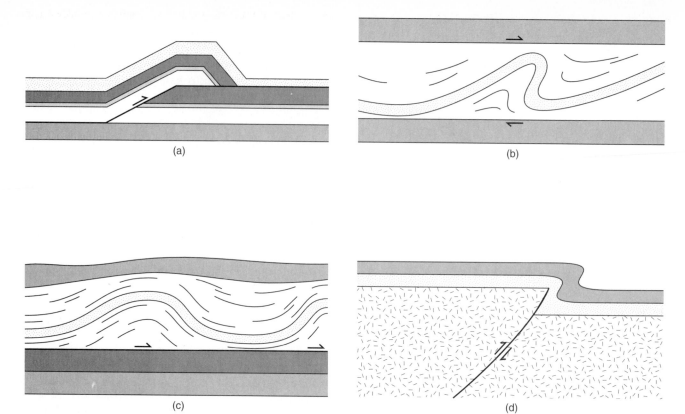

Figure 8.21 Fault-related folds. (a) Fault-bend fold. (b) Fold in fault zone. (c) Detachment fold. (d) Drape fold.

parallelism with the shear direction. Note that fault zones in which folding occurs are effectively ductile shear zones. We mention them here because such ductile deformation can occur in association with brittle sliding surfaces.

As is the case with fault-inception folds, the asymmetry of fault-zone folds provides a clue to the sense of shear on the associated fault. However, because the hinges of fault-zone folds change during progressive shear, the geometry of a single fold may not necessarily provide the trend of the net-slip vector. You can define the trend if you measure a slip lineation on the fault surface; but if such lineations are not present, you may be able to obtain this information by measuring numerous folds with a range of hinge trends using a technique called the *Hanson slip-line method.*[8] We leave a description of this method for laboratory texts on structural geology.

The sheet of rock above a detachment fault may deform independently of the rock below. The resulting folds are called *detachment folds* (Figure 8.21c). Shear on the detachment fault accommodates the contrast in strain between the folded hanging wall and the unfolded footwall.

In continental-interior platform regions (e.g., the Midcontinent of the United States), a thin and relatively weak veneer of Phanerozoic sedimentary rock was deposited over relatively rigid Precambrian crystalline basement. At various times during the Phanerozoic, steeply dipping faults in the basement reactivated, causing differential movement of basement blocks. This movement forces the overlying layer of sedimentary rocks to passively bend into a fold that drapes over the edge of the basement block. Fault-related folds that form in this way are called *drape folds* or *forced folds* (Figure 8.21d). Kinematically, they are fault-propagation folds, but they have been given a separate name to emphasize their unique tectonic setting.

In closing, a final comment on the term *drag fold,* which is used to refer to many fault-related folds in older literature. This general application of the term is misleading, because "drag" implies that the fold formed because shear resistance on the fault retarded movement of the hanging wall with respect to the footwall. Rather, fault-related folds form in a number of different ways, most of which do not involve such "drag." Thus we avoid using this term.

8.3.5 Shear-Sense Indicators of Brittle Faults—A Summary

So far, we have mentioned how various features of fault zones can be used to determine the shear sense. In Table 8.5, we provide a concise summary of the various shear-sense indicators.

8.4 RECOGNIZING AND INTERPRETING FAULTS

Displacement on a fault disrupts rock bodies, but it can also alter the landscape. The nature of the disruption depends on the type of fault, on how much displacement oc-

---

[8]Named after its originator.

Table 8.5   Shear-Sense Indicators for Brittle Faults and Fault Zones

Carrot-shaped grooves	Grooves on slickensides tend to be deeper and wider at one end and taper to a point at the other, thus resembling half a carrot. The direction that the 'carrot' points defines the direction of shear.
Chatter marks	As one fault block moves past another, small wedge-shaped blocks may be plucked out of the opposing surface. The resulting indentations on the fault surface are known as chatter marks.
***En echelon* veins**	*En echelon* veins tilt toward the direction of shear. If the veins are sigmoidal, the sense of rotation defines the shear sense.
Fault-related folds	The sense of asymmetry of fault-related folds defines the shear sense. Typically, fault-inception folds verge in the direction of shear (see Chapter 10 for a definition of fold vergence). If the hinges of folds in a fault zone occur in a range of orientations, you may need to use the Hanson slip-line method to determine shear sense. Note that the asymmetry of rollover folds relative to shear sense is opposite to that of other fault-related folds.
Fiber-sheet imbrication	The imbrication of slip fiber sheets on a fault provides a clear indication of shear sense. Fiber sheets tilt away from the direction of shear.
Offset markers	You can define shear sense if you are able to define the relative displacement of two piercing points on opposite walls of the fault, or if you can calculate the net-slip vector based on field study of the separation of marker horizons.
Pinnate fractures	The inclination of pinnate fractures with respect to the fault surface defines the shear sense.
Steps on slickensides	Microscopic steps develop along slickensided surfaces. Typically, the face of the step is rougher than the flat surface. However, slickenside steps may be confused with the intersection between pinnate fractures and the fault, giving an opposite shear sense.

curred, on the rate of displacement, on how recently the faulting occurred, on the climate in the zone of faulting, and on whether the fault is emergent or blind. We include climate and age of the faulting in our list of factors because they determine the extent to which erosion can erase the effects of faulting. For example, in a humid climate with abundant rainfall, a fault's geomorphologic manifestation may disappear within months or years, whereas in a desert climate, fault's geomorphologic manifestation may last for millennia.

Seismic faulting along an emergent strike-slip fault typically creates a *surface rupture,* manifested by broken ground and fissures. Displacement on an emergent dip-slip fault creates a step in the ground surface, called a *fault scarp* (Figure 8.22). Both dip-slip and strike-slip faulting offset topographic features such as ridges or river beds, and if the fault is very recent, it may even offset roads and fences. You can identify the trace of the San Andreas Fault in the wine district of California, for example, by where it offset rows of casks in a wine cellar. Erosion typically destroys surface ruptures and fault scarps within tens to thousands of years. However, the abrupt change in elevation of the ground surface caused by the faulting can be maintained even after the original scarp has eroded away. Recent movement on a blind fault, by definition, does not result in a surface rupture, but the trace of the fault may be evident by differential movement of the ground surface. Such movement can be detected by using ground-based

surveying equipment and, more recently, by using GPS (global positioning system) equipment.

Displacement across regional strike-slip faults is not necessarily exactly parallel to the fault surface, in part due to the existence of restraining and releasing bends along the fault, and in part due to changes in the relative motion of plates after the fault formed. Transpression across a large strike-slip fault results in development of *fault-parallel ridges* tens to hundreds of meters high (Figure 8.23), whereas transtension along a large strike-slip fault results in the development of *sag ponds.* Because of such features, the traces of major strike-slip faults, like the San Andreas Fault, are so obvious, that they can be seen from space. If transpression occurs along a fault for millions of years, a mountain range develops adjacent to the fault; if transtension occurs for millions of years, a sizable sedimentary basin develops adjacent to the fault. As we will see in Chapter 18, displacement in a regional-scale strike-slip fault zone may result in the formation of *en echelon* folds and faults near the ground surface. These structures may have surface manifestations in the form of local ridges or troughs oriented at an angle to the main fault trace.

Inactive emergent or exhumed faults also have a topographic manifestation, for two reasons. First, the fault zone itself may have a different resistance to erosion than the surrounding intact rock. If the fault zone consists of noncohesive breccia or gouge, it tends to be weaker than the surrounding intact rock and thus preferentially erodes. As a

Figure 8.22 Fault scarp in the Basin and Range Province of Nevada. Note the normal-sense of offset.

Figure 8.23 Ridges formed along a transpressional zone of the San Andreas Fault. The rocks forming the ridges have been thrust up in response to the component of shortening along the fault.

consequence, the fault trace on the ground surface evolves into a linear trough that may control surface drainage. Alternatively, if the fault becomes indurated, it may become more resistant to erosion than the surrounding region and will, therefore, stand out in relief. Second, if the fault juxtaposes two rock units with different resistances to erosion, then a topographic scarp develops along the trace of the fault because the weaker unit erodes more rapidly and the land surface underlain by the weaker unit becomes topographically lower (Figure 8.24). Such a *fault-line scarp* differs from a fault scarp, in that it is not the plane of the fault itself. In fact, the slope direction of a fault-line scarp may be opposite to the dip direction of the underlying fault.

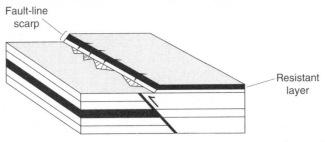

Figure 8.24 Block diagram illustrating a fault-line scarp caused by the occurrence of a resistant stratigraphic layer (in black) that has been uplifted on one side.

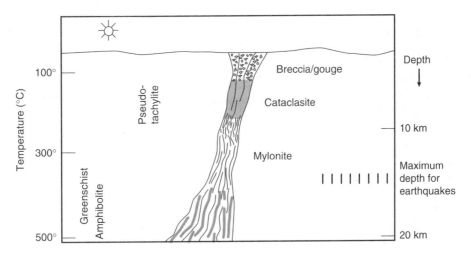

Figure 8.25 Change in fault character with depth for a steeply dipping fault.

Fault zones are sometimes manifested by subtle features of the landscape. For example, fault zones containing abundant fractures may control local groundwater movement and may preferentially drain the surface. This effect can cause changes in vegetation that appear as lineaments in remote-sensing images. Similarly, faults may offset structurally controlled topographic ridges, and thus may cause an alignment of ridge terminations.

If you are lucky enough to find an exposure of a fault surface or zone in the field, it is an exciting experience if the discovery helps resolve problems on your map. Ancient and inactive faults can be identified in the field, even if they do not disrupt the landscape and/or fault-related structures have been covered or removed. Features indicative of faulting include: juxtaposition of rock units that were not in contact when first formed, offset of marker horizons, loss or duplication of section, and the presence of slickensided surfaces, slip lineations, fault rocks, and fault-related folds. Doing some of your own fieldwork in faulted regions is simply the best way to obtain an appreciation of the many manifestations of faulting.

8.4.1 Recognition of Faults from Subsurface Data

To many geologists, particularly those working in economic geology, petroleum geology, or hydrogeology, the "field" consists of the subsurface region of the crust. The

"field data" include *drill-hole logs* (i.e., downhole records of lithology based on analysis of cores or cuttings, as well as on electrical-conductivity measurements or gamma-ray measurements), geophysical measurements (e.g., maps showing regional variation in the Earth's gravity and magnetic fields), and seismic-reflection data. Analyses of drill-hole logs and geophysical data provide a basis for identification of faults.

Subsurface evidence for faulting includes: (1) abrupt steps on structure-contour maps; (2) excess section (i.e., repetition of stratigraphy) or loss of section in a drill core; (3) zones of brecciated rock in drill core, although weak fault rocks typically do not survive drilling intact (and thus fault zones may appear as gaps in a core); (4) seismic-reflection profiles, on which faults appear either as reflectors themselves, or as zones that offset known reflectors; and (5) linear anomalies or an abrupt change in the wavelength of gravity and/or magnetic anomalies, suggesting the occurrence of an abrupt change in depth to a particular horizon.

8.4.2 Changes in Fault Character with Depth

The physical appearance of a fault depends on the magnitude of displacement on the fault, on whether or not faulting ruptures a previously intact rock or follows a preexisting surface, and on the pressure and temperature conditions (i.e., burial depth) at which faulting occurs (Figure 8.25).

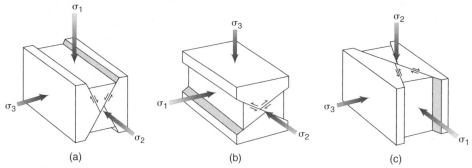

Figure 8.26 Anderson's theory of faulting. (a) Normal faults, (b) thrust faults, and (c) strike-slip faults.

At very shallow depths in the earth (< ~5 km), mesoscopic faults that form by reactivation of a preexisting joint or bedding surface typically result in discrete slickensided or fiber-coated surfaces. Mesoscopic shallow-level faults that break through previously intact rock tend to be bordered by thin *breccia* or *gouge zones*. Macroscopic faults, which inevitably break through a variety of rock units and across contacts, tend to be bordered by wider breccia and gouge zones and subsidiary fault splays.

As we discussed in Chapter 6, rocks tend to become progressively more ductile with depth in the crust because of the increase in temperature and pressure that occurs with depth. Consequently, at depths of between about 3–5 and 10–15 km, faulting tends to yield a fault zone composed of *cataclasite*. Recall that cataclasite forms by brittle deformation processes on a grain scale, but that mesoscopically, movement in the zone resembles viscous flow and that strain is distributed across the zone (hence, ductile). The *brittle-plastic transition* for typical crustal rocks lies at a depth of 10–15 km in the crust. Note that we specify the transition as a range, because rocks consist of many different minerals each of which begins to behave plastically under different conditions, and because the depth of transition strongly depends on the geothermal gradient that varies with location. At temperature conditions achieved in the 10–15 km depth interval (~250° to 350°C; i.e., lowermost greenschist facies of metamorphism), plastic deformation mechanisms become the dominant contributor to strain in typical (quartz-rich) crustal rocks. The activity of plastic deformation mechanisms below this brittle-plastic transition yields a fine-grained and strongly foliated shear-zone rock called *mylonite,* which we discuss in more detail later.

The degree of ductile deformation that accompanies faulting also depends on the strain rate and on the fluid pressure in the rock. At slower strain rates, a given rock type is weaker and tends to behave more ductilely. Thus, rocks can deform ductilely even at shallow crustal levels if strain rates are slow, whereas rocks below the brittle-plastic transition can deform brittlely if strain rates are high. A given shear-zone interval may contain both mylonite and cataclasite, either as a consequence of variations in strain rate at same depth or because progressive displacement eventually transported the interval across the brittle-plastic boundary.

The width of a given fault zone typically varies as a function of original rock strength; fault zones tend to be narrower in stronger rock. Thus, the width of a transcrustal fault zone may vary with depth. Very near the surface (within a few km), the fault diverges into numerous spays, because the near-surface rock is weakened by jointing and by formation of alteration minerals (e.g., clay). At somewhat greater depths, rock is stronger and the fault zone may be narrower. At still greater depths, where cataclastic flow dominates, the fault zone widens. Similarly, we might expect to find mylonite widening with greater depth when the rocks become weaker as a whole.

8.5 RELATION OF FAULTING TO STRESS

Faulting represents a response of rock to shear stress, and therefore occurs only when the differential stress ($\sigma_d = \sigma_1 - \sigma_3$) does not equal zero. Given that shear stress (σ_s) causes faulting, you might expect a relationship between the orientation of faults formed during a tectonic event and the trajectories of principal stresses during that event, because the shear stress magnitude on a plane changes as a function of the orientation of the plane with respect to the principal stresses. Indeed, if faults initiate as Coulomb shear fractures, they will form at about 30° to the σ_1 direction and contain the σ_2 direction. This relationship is generally called *Anderson's theory of faulting*.[9] Recall that the ratio of shear stress to normal stress on planes orientated at about 30° to σ_1 is at a maximum (see Chapter 6).

The Earth's surface is a "free surface" (the contact between ground and air) that cannot transmit a shear stress. Therefore, principal stresses are parallel or perpendicular to the surface of the Earth near the surface. Considering that gravitational body force is a major contributor to the stress state, and this force acts vertically, stress trajectories in homogeneous, isotropic crust can maintain this geometry at depth. Anderson's theory of faulting states that in the Earth-surface reference frame, normal faulting occurs where σ_2 and σ_3 are horizontal and σ_1 is vertical; thrust faulting occurs where σ_1 and σ_2 are horizontal and σ_3 is vertical, and strike-slip faulting occurs where σ_1 and σ_3 are horizontal and σ_2 is vertical (Figure 8.26). Moreover, the

[9]After the British geologist E. M. Anderson.

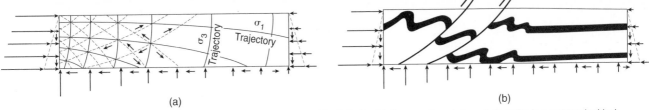

Figure 8.27 Curving principal stress trajectories with depth and the origin of listric faults. (a) Cross section of stress trajectories in a rectangular block bounded on the top by a free surface and on the bottom by a frictional sliding surface. (b) Predicted pattern of reverse faulting, assuming that faults form at ~30° to the σ_1 trajectory (only one set of reverse faults is illustrated).

dip of thrust faults should be ~30°, the dip of normal faults should be ~60°, and the dip of strike-slip faults should be about vertical. For example, the σ_1 orientation is horizontal beneath collisional orogens. Anderson's theory predicts that thrust faults should form in this environment, and indeed belts of thrust faults do form in such a setting.

But we cannot use Anderson's theory to predict all fault geometries in the Earth's crust, for several reasons. Faults do not necessarily initiate in intact rock. Frictional sliding strength of a preexisting surface is less than the shear failure strength of intact rock; thus, preexisting joint surfaces or faults may be reactivated before new faults initiate, even if the preexisting surfaces are not inclined at 30° to σ_1 and do not contain the σ_2 trajectory. Preexisting fractures that are not ideally oriented with respect to the principal stresses become oblique-slip faults. Second, a fault surface is a material feature in a rock body whose orientation may change as the rock body containing the fault undergoes progressive deformation. Thus, the fault may rotate into an orientation not predicted by Anderson's theory. Finally, local stress trajectories may be different from regional stress trajectories because of local inhomogeneities and weaknesses (e.g., contacts between contrasting lithologies, and preexisting faults) in the Earth's crust. As a consequence, local fault geometry might not be geometrically related to regional stresses. Systematic changes in stress trajectories probably occur with depth in mountain belts (e.g., along a regional detachment), but Anderson's theory assumes that the stress field is homogeneous and that the principal stresses are either horizontal or vertical, regardless of depth.

8.5.1 Formation of Listric Faults

Let us consider changes in stress field with depth in more detail, because it explains the geometry of *listric faults*. Deformation intensity, as manifested by strain magnitude, decreases from the interior to the margin of an orogenic belt. We can infer from this observation that the magnitude of horizontal tectonic stress decreases from the interior to the margin of the belt. Now, envision a sheet of rock above a detachment that is subjected to a greater horizontal σ_1 at the hinterland end than at the foreland end (Figure 8.27). You might expect that such an apparent violation of equilibrium would cause the block to translate toward the foreland, but in our model friction inhibits such movement.

You can picture the frictional resistance as a shear stress acting on the base of the sheet. The top of the sheet, meanwhile, is a free surface (the contact between rock and atmosphere), so at the top of the sheet, principal stresses must be parallel or perpendicular to the surface. However, σ_1 near the bottom of the sheet cannot parallel σ_1 at the top surface, because there is shear stress at the bottom of the sheet but not at the top. Calculating stress trajectories given these conditions, we find that they curve into the bottom of the sheet. If we take Anderson's assumption that faults in the sheet initiate as Coulomb shears at about 30° to the σ_1 trajectory, then the faults must also curve; that is, their dips decrease with depth. Certainly this is a reasonable explanation for the occurrence of listric faulting, but note that this theory does not apply where the detachment is ductile.

8.5.2 Fluids and Faulting

There is no doubt that fluids play a role in fault zones, for commonly you will find that fault rocks are altered by reaction with a fluid phase (e.g., clay in fault zones forms by reaction of feldspar with water), and that fault zones contain abundant veins composed of minerals that precipitated from a fluid (e.g., quartz, calcite, chlorite, numerous economic minerals). Apparently, the fracturing that accompanies displacement creates open space within the fault zone for fluid to enter. In fact, because of the increase in open space, fluid pressure in the fault zone temporarily drops relative to the surrounding rock. The resulting fluid-pressure gradient actually drives groundwater into the fault zone until a new equilibrium is established. Such faulting-triggered fluid movement is known as *fault valving.*[10]

The presence of water in fault zones affects the shear stress at which faulting occurs in three ways. First, alteration minerals formed by reaction with water in the fault zone tend to have lower shear strength than minerals in the unaltered rock, and thus their presence may permit the fault to slip at a lower differential stress than it would otherwise. Second, the presence of water in a rock may cause hydrolytic weakening of silicate minerals, and therefore allow deformation to occur at lower stresses. Third, the pore pressure (P_{fluid}) of water in the fault zone decreases

---

[10]Or *seismic pumping.*

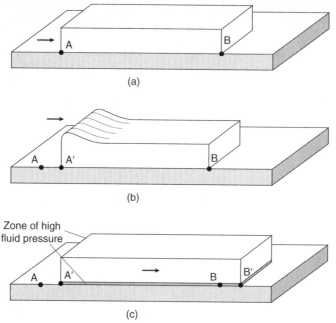

Figure 8.28 The thrust sheet paradox. (a) Block and sectional view of a thrust sheet on a frictional surface. (b) Because of frictional resistance, the shear stress necessary to initiate sliding exceeds the yield strength at one end of the thrust sheet. (c) Thrust sheet moves easily if it rides on a 'cushion' of fluid; A′ and B′ are displaced positions of points A and B.

the effective normal stress in a rock body (Chapter 6), and thus decreases the magnitude of the shear stress necessary to initiate a shear rupture in intact rock or to initiate frictional sliding on a preexisting surface.

The concept that increasing fluid pressure leads to faulting at a lower regional differential stress is illustrated by the history of earthquake activity in the Rocky Mountain Arsenal near Denver, Colorado. In the early 1960s, the U.S. military chose to dispose of large quantities (sometimes as much as 30 million liters per month) of liquid toxic waste by pumping it into the groundwater reservoir via a 4 km deep well at the arsenal. Geophysicists noticed that when the waste was injected, dozens of small earthquakes occurred near the bottom of the well. Evidently, injection of the waste increased P_{fluid} at the bottom of the well. This decreased the effective normal stress on faults such that the local σ_d was sufficient to cause preexisting faults near the well to slip until the elevated fluid pressure dissipated by fluid flow out of the well.

The concept that an increase in fluid pressure decreases effective stress across a fault helps to resolve one of the great paradoxes of structural geology, namely the movement of thrust sheets on regional-scale detachments. Look at the cross-sectional image of a thrust sheet depicted in Figure 8.28a. The sheet looks like a large rectangle slipping on a detachment at its base. If you assume that the shear resistance at the base of the sheet is comparable to the frictional sliding strength of rock observed in the laboratory, then in order to move the sheet, the magnitude of the horizontal stress applied at the end of the sheet must be very large. In fact, and herein lies the paradox,

the horizontal stress must be so large that it would exceed the strength of the thrust sheet itself. Thus, the thrust sheet would deform internally (by faulting and folding) close to where the stress was applied before the whole sheet would move (Figure 8.28b). As an analogy of this paradox, picture a large Persian rug lying on a floor. If you push at one end of the rug, it simply wrinkles at that end, but it does not slip across the floor, because the shear resistance to sliding is too great. So, how do large thrust sheets move great distances on detachment faults?

Fluid pressure offered the first reasonable solution to this paradox. If fluid pressure (P_{fluid}) in the detachment zone approaches lithostatic pressure (i.e., the magnitude of fluid pressure approaches the weight of the overlying rock), then the effective normal stress across the fault plane approaches zero (see Chapter 6) and the shear stress necessary to induce sliding on the detachment would become very small so that thrust sheets could move before deforming internally (Figure 8.28c). This clever idea, known as the *Hubbert-and-Rubey hypothesis*,[11] emphasizes the importance of elevated fluid pressure during movement on detachments, and has wide applicability. Indeed, modern measurements confirm that in regions where detachment faults move, P_{fluid} near the detachment interval approaches lithostatic values.

Initially, geologists thought that all regional thrust sheets slide down gentle foreland-dipping slopes in response

---

[11]Named after the American geologists M. King Hubbert and William Rubey who presented this idea in the 1950s.

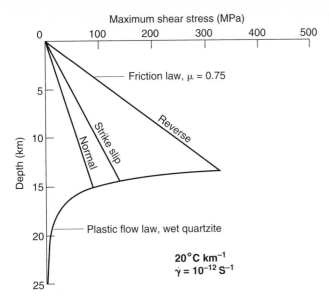

Figure 8.29 Graph showing variation in maximum shear stress necessary to initiate sliding for reverse, strike-slip, and normal faults as a function of depth.

to gravity, effectively gliding on a cushion of water. But gravity sliding cannot explain most examples of thrusting in orogenic belts, because thrust sheets typically dip away from the direction in which they move. We will explore this intriguing problem in Chapter 17, where we discuss newer concepts of thrust-sheet movement, and the role of the Hubbert-and-Rubey hypothesis in these models.

8.5.3 Magnitude of Stress Required for Faulting— A Continuing Debate

The issue of how large the shear stress (σ_s) must be in order both to initiate faults and to reactivate preexisting faults remains very controversial. Experimental rock mechanics provides one approach to this problem. From laboratory triaxial loading experiments, geologists determined that, all other factors being equal, the differential stress necessary to cause failure increases as confining pressure increases, and further that σ_d is greatest for contractional faulting and least for extensional faulting (Figure 8.29). The actual magnitude of σ_s necessary to trigger faulting depends on fluid pressure, lithology, strain rate, and temperature. If faulting occurs by frictional sliding on a preexisting fault, then the σ_s necessary to initiate sliding also depends on the orientation of the fault with respect to the principal stresses. Recall that for a given range of orientations, the σ_d necessary to initiate frictional sliding on a preexisting fault is less than that necessary to initiate a new fault.

Many geologists have questioned the validity of stress estimates based on laboratory studies because of uncertainty about how such estimates scale up to crustal dimensions. Thus, geologists have used alternative approaches to determine stress magnitudes needed to cause crustal-scale faulting, including analysis of heat generation during faulting and direct measurement of stresses near faults.

If movement on a fault surface in the brittle regime of the crust involves frictional sliding, some of the work done during fault movement is transformed into heat. This process, called *shear heating*, obeys the equation:

$$\sigma_s \cdot u = E_e + E_s + Q \qquad \text{Eq. 8.1}$$

where σ_s is the shear stress across the fault, u is the amount of slip on the fault, E_e is the energy radiated by earthquakes, E_s is the energy used to create new surfaces (i.e., break chemical bonds), and Q is the heat generated. This equation suggests that $\sigma_s \cdot u > Q$. Thus if u is known, the value of Q should provide a minimum estimate of σ_s. With this concept in mind, geologists have studied metamorphism near faults in order to calculate the amount of heat (Q) needed to cause the metamorphism and thereby provide an estimate of σ_s during faulting. If the metamorphism is assumed to be entirely a consequence of shear heating, these studies conclude that shear stresses across faults are quite large. These studies, however, have been criticized, because not all geologists accept the assumption that the observed metamorphism is due to shear heating, and because observed heat flow adjacent to active faults is not greater than heat flow at a distance from the fault. For example, even though the San Andreas Fault is a huge active fault, it is not bordered by the zone of high heat flow predicted by the shear-heating model.

In recent years, geologists have attempted to define the stress state during crustal faulting by *in-situ* measurements of stress fields in the vicinity of faults. Recall from Chapter 3 that hydrofracture measurements in drill holes, strain-release measurements, and borehole breakouts all provide an estimate of stress orientation, and in some cases magnitude, in the crust. In general, these measurements suggest that stresses during faulting are relatively low. For example, direct measurement of stress in the crust around the San Andreas Fault suggests that the remote σ_1 trajectory bends so that it is nearly perpendicular in the immediate vicinity of the fault. Such a change in stress trajectory orientation suggests that the fault is behaving like a surface of very low friction, and therefore could not support a large σ_d. Based on these observations, some geologists have concluded that the σ_s needed to cause movement on the San Andreas Fault, and by inference other faults, is relatively low.

Another way in which stresses necessary to initiate faulting have been estimated comes from studies of seismicity. When an earthquake occurs, energy is released and the value of σ_s across the fault decreases. This decrease is called the *stress drop*, and its value provides a minimum estimate of the value of σ_s that triggered the earthquake. Stress drops estimated for earthquakes range between about 0.1 and 150 MPa, with an average of about 3 MPa. Finally, estimates of differential stress magnitudes based on the tensile strength of rocks and grain-scale deformation mechanisms indicate values on the order of 20–100 MPa.

In summary, geologists do not yet agree about the stress state necessary to cause crustal faulting, for estimates

Table 8.6 Geometric Classification of Fault Arrays (Figure 8.30)

Anastamosing array	A group of wavy faults that merge and diverge along strike, thereby creating a braided pattern in map view or cross section.
Conjugate fault array	Composed of two sets of faults that are inclined to one another at an angle of about 60°. Conjugate fault arrays can consist of dip-slip faults or strike-slip faults. If the faults in the array are strike-slip, then one set must be dextral and other sinistral.
En echelon array	A group of parallel fault segments that lie between two enveloping surfaces and are inclined at an angle to the enveloping surfaces.
Parallel fault array	As the name suggests, a parallel fault array includes a number of fault surfaces that roughly parallel one another.
Random fault array	In some locations, faulting occurs on preexisting fractures. If the fracture array initially had a wide range of orientations, then slip on the fractures will yield faults in a wide range of orientations. Such an array is called a random or nonsystematic array.
Relay array	In map view, a relay array is a group of parallel or subparallel noncoplanar faults that are spaced at a distance from one another across strike, but whose traces overlap with one another along strike. As displacement dies out along the strike of one fault in the array, displacement increases along an adjacent fault. Thus, displacement is effectively "relayed" (transferred) from fault to fault. In a thrust belt containing a relay array of faults, regional shortening can be constant along the strike of the belt, even though the magnitude of displacement along individual faults dies out along strike.

derived from different approaches do not agree with one another. Most likely, there is truly a range of stress conditions that can cause faulting, which reflects parameters such as the geometry of the fault, the nature of the faulted material, and the fluid pressure in the fault zone.

8.6 FAULT SYSTEMS

Faults are not just geologic curiosities that complicate mapping projects. Rather, they are, first and foremost, the manifestation of a fundamental way in which stress causes deformation in the Earth's upper crust. A fault typically does not occur in isolation, but rather occurs as part of a group of associated faults that develop during the same interval of deformation and in response to the same regional stress field. We classify groups of related faults either by their geometric arrangement or their tectonic significance (i.e., the type of regional deformation resulting from their movement). A group of related faults is a *fault system* or *fault array;* the terms "system" and "array" can be used interchangeably, but commonly geologists use the word "array" when talking about geometric classifications and "system" in the context of tectonic classifications. Below, we'll first describe a geometric classification for fault arrays, and then we'll introduce the three tectonically defined types of fault systems: normal, thrust, and strike-slip systems. In many areas, faults in normal and thrust systems merge with a detachment in sedimentary rocks at

shallow depth (~10–15 km), in which case the system may be referred to as a *thin-skinned system.* If faulting involves deeper crustal rocks (i.e., basement), we call the system *thick skinned.* Generally, however, this distinction is artificial as these systems are laterally related.

Our description of fault systems in this chapter is meant to be only a brief introduction. We provide detailed descriptions of fault systems later in this book (Chapters 15–18), where you will also find ample illustration of the associated structures.

8.6.1 Geometric Classification of Fault Arrays

Groups of faults may be classified as parallel, anastamosing, *en echelon,* relay, conjugate, or random arrays, depending on the orientation of the faults in the array with respect to one another, and on how the faults link with one another along strike. We define these terms in Table 8.6 and illustrate them in Figure 8.30.

Typically, faults in parallel arrays of normal or thrust faults dip in the same direction over a broad region. Subsidiary faults in the array that parallel the major faults are called *synthetic faults,* whereas subsidiary faults whose dip is opposite to that of the major faults are called *antithetic faults.*

8.6.2 Normal Fault Systems

Regional normal fault systems form in *rifts,* which are belts in which the lithosphere is undergoing extension,

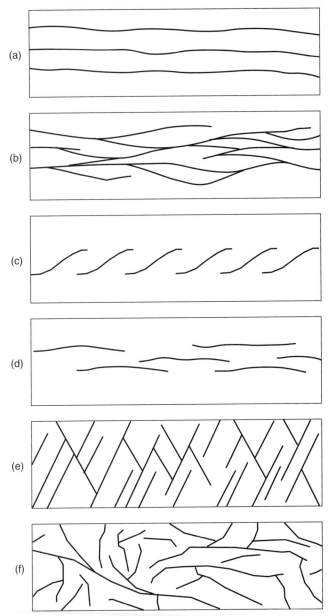

Figure 8.30 Map-view sketches of various types of fault arrays. (a) Parallel array, (b) anastamosing array, (c) *en echelon* array, (d) relay array, (e) conjugate array, and (f) random array. See Table 8.6.

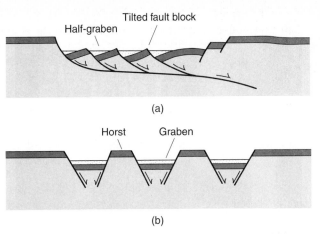

Figure 8.31 Normal fault systems. (a) Half-graben system. (b) Horst-and-graben system.

along *passive margins,* which are continental margins that are not currently plate margins, and along *mid-ocean ridges.* Typically, faults in a normal fault system compose relay or parallel arrays, and they can be listric, planar, or contain distinct fault bends. Movement on both planar and listric normal fault systems generally results in rotation of hanging-wall blocks around a horizontal axis, and therefore causes tilting of overlying fault blocks and/or formation of rollover anticlines and synclines. The geometry of tilted blocks and rollover folds developed over normal faults depends on the shape of the fault and on whether or not synthetic or antithetic faults cut the hanging-wall block.

As a consequence of the rotation accompanying displacement on a normal fault, the original top surface of the hanging-wall block tilts toward the fault, creating a depression called a *half graben* (Figure 8.31a). Note that a half graben is bounded by a fault on only one side. Most of the basins in the Basin-and-Range Province of the Western USA are half grabens, and the ranges consist of the exposed tips of tilted fault blocks. In places where two adjacent normal faults dip towards each other, the fault-bounded block between them drops down, creating a *graben* (from the German word for "trough"). In places where two adjacent normal faults dip away from one another, the relatively high footwall block between the faults is a *horst* (Figure 8.31b). Horsts and grabens commonly form because of the interaction between synthetic and antithetic faults in rift systems. Chapter 15 discusses the tectonic setting of normal faults in detail.

8.6.3 Thrust-Fault Systems

Thrust-fault systems are arrays of thrust faults that form to accommodate regional shortening of the lithosphere. Not surprisingly, thrust systems are common along the margins of convergent plate boundaries and along the margins of collisional orogens. In such tectonic settings, thrusting occurs in conjunction with formation of folds, resulting in tectonic provinces called *fold-thrust belts* (Chapter 17).

To a first approximation, fold-thrust belts resemble the wedge of snow that is scraped off a road by a plow. Typically, the numerous thrusts in the belt merge at depth with a shallowly dipping detachment. At a crustal scale, major thrust faults are listric; but in detail, where thrusts cut upwards through sequences of contrasting strata, they have a stair-step profile, with *flats* following weak horizons and *ramps* cutting across rigid beds. Ramp-flat geometry generally develops best in sequences of well-stratified sedimentary rock, such as occurs at passive-margins caught in the vice between colliding continents, or in the

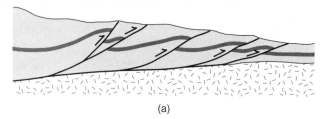

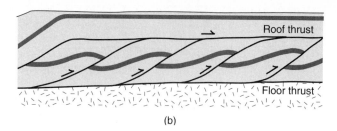

Figure 8.32 Thrust-fault systems. (a) Imbricate fan. (b) Duplex system.

sequence of sediment, called a foreland basin on the craton side of the orogen, derived by erosion of the recently uplifted orogen.

As is the case with normal fault systems, faults in a thrust-fault system tend to compose relay or parallel arrays. An *imbricate fan* of thrust faults consists of thrusts that either intersect the syntectonic ground surface or die out up dip (Figure 8.32a); whereas a *duplex* consists of thrusts that span the interval of rock between a higher-level detachment called a *roof thrust* and a lower level detachment called a *floor thrust* (Figure 8.32). The fault-bounded bodies of rock in a duplex are *horses*. We can't say much more about thrust systems at this point without introducing a lot of new terminology, so we'll delay further discussion of the subject until Chapter 17.

8.6.4   Strike-Slip Fault Systems

Strike-slip fault systems occur both at transform plate boundaries (boundaries where two plates slide past each other without the creation or subduction of lithosphere), within plates, and as components of collisional or convergent orogens. Major continental strike-slip fault systems are complicated structures. Typically, in the near surface they splay into many separate faults, which, in cross section, resembles the head of a flower. Because of this geometry, such arrays are commonly called *flower structures* (Figure 8.33). We present further description of these complexities and of the tectonic settings in which strike-slip fault occurs in Chapter 18.

8.6.5   Inversion of Fault Systems

Once formed, a fault is a material discontinuity in the lithosphere that may remain weaker than surrounding regions for long periods of geologic time. Thus, faults can be reactivated during successive pulses of tectonism at different times during Earth's history. If the stress field during successive pulses is different, the kinematics of movement on the fault during the different episodes will not likely be the same, and the displacement resulting from one event may be opposite to displacement resulting from another event. For example, a normal fault formed

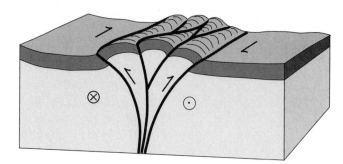

Figure 8.33 Strike-slip related flower structure.

during rifting of a continental margin may be reactivated as a thrust fault if that margin is later caught in the vice of continental collision. Likewise, the border faults of a half graben or unsuccessful rift may be reactivated later as thrust or oblique-slip faults if the region is later subject to compression. The reversal of displacement on a fault is called *fault inversion*. When inversion results in contraction of a previously formed basin, the process is called *basin inversion*.

8.6.6   Fault Systems and Paleostress

With the Andersonian concept of faulting in mind, the geometry of a fault system that is active at a given time is a clue to the regional stress field that caused the faulting. A thrust system reflects conditions where regional σ_1 is horizontal and trends roughly perpendicular to the trace of the system. A normal fault system reflects conditions where regional σ_3 is horizontal and trends roughly perpendicular to the strike of the system. There is no simple rule defining the relationship between strike-slip fault systems and stress trajectories. Oceanic transforms, for example, typically parallel the σ_3 direction, whereas continental strike-slip faults commonly trend oblique to the σ_1 direction.

In recent years, geologists have studied the possibility that the stress field can be derived by studying slip on a nonsystematic array of faults in a region, assuming that all the faults moved in response to the same stress field. Most places in the Earth's crust are fractured, and though there may be dominant systematic arrays of fractures in the area, there are likely to be many nonsystematic fractures in all orientations. If a body of rock containing abundant preexisting fractures in a range of orientations is subjected to a regional homogeneous stress field, a shear stress will exist on all fractures that are not principal planes (i.e., not perpendicular to one of the principal stresses). The orientation of the resolved shear stress on each fracture is determined by the orientation and relative magnitudes of the principal stresses defining the regional stress state. Fractures whose resistance to sliding is exceeded will move, with the direction of movement being approximately parallel to the maximum resolved shear stress on the fracture surface. Fractures whose dip direction happens to be parallel to the shear stress become dip-slip faults, and fractures whose strike direction happens to be parallel to the shear stress become strike-slip faults. All other fractures that slip become oblique-slip faults.

During the past few decades, methods have been developed that permit the principal stress directions to be derived from measurements of slip trajectories on a nonsystematic array of faults, which were assumed to have moved in response to a regionally homogeneous stress state. In principle, you need measurements of the shear sense and the trend of the net-slip vector on only four nonparallel faults to complete such *paleostress analysis*. But in practice, geologists use measurements from numerous faults and employ statistics to obtain the best-fit solution. It is beyond the scope of this book to provide the details of paleostress analysis from slip data on fault arrays, but we provide some references on the subject at the end of Chapter 4.

8.7 FAULTING AND SOCIETY

After all this terminology and theory, it is easy to forget that the study of faulting is not just an academic avocation. Faults must be studied carefully by oil and mineral exploration geologists, because faulting controls the distribution of these resources. Similarly, faults and fractures play an important role in groundwater mobility and the general availability of water. Regional fault analysis is required for the localization and building of large human-made structures, such as dams and nuclear power plants. Such structures, which may have devastating impact when they fail, should clearly not be built near potentially active faults. But perhaps the most dramatic effect of fault activity on society is in the form of earthquakes, which can be responsible for great loss of life (in some cases measured into the 100,000s) and can destroy the economic stability of industrialized countries (damages may exceed 100 billion dollars when urban areas are involved). In this final section we

briefly look at these perhaps more immediate consequences of faulting.

8.7.1 Faulting and Resources

Faults contribute to the development of oil traps by juxtaposing an impermeable seal composed of packed gouge or fault-parallel veining against a permeable reservoir rock, or by juxtaposing an impermeable unit, like shale, against a reservoir bed. Alternatively, faulting may affect oil migration, by providing a highly fractured zone that can serve as a fluid conduit through which oil migrates. Finally, syn-depositional faulting (growth faults) may affect the distribution of oil reservoir units.

Valuable ore minerals (e.g., gold) commonly occur in veins or are precipitated from hydrothermal fluids that passed preferentially through fault zones because fracturing provides enhanced permeability. Thus, fault breccias are commonly targets for mineral exploration. As we pointed out earlier, displacement on faults may control the distribution of ore-bearing horizons, and thus mining geologists must map faults in a mine area in great detail.

Hydrogeologists also are cognizant of faults, because of their effect on the migration of groundwater. A fault zone may act as a permeable zone through which fluids migrate if it contains unfilled fractures, or the fault zone may act as a seal if it has been filled with vein material or includes impermeable gouge. In addition, faults may truncate aquifers, and/or juxtapose an aquitard against an aquifer, thereby blocking fluid migration paths.

8.7.2 Faulting and Earthquakes

Nongeologists may panic if they hear that a fault has been discovered near their home, because of the common misperception that all faults eventually slip and cause earthquakes. Faulting is very widespread in the Earth's crust, but fortunately, most faults are *inactive*, meaning that they haven't slipped in a long time, and are probably permanently stuck. Relatively few faults are *active*, meaning that they have slipped recently or have the potential to slip in the near future. Even when slip occurs, not all movement on active faults results in seismicity. If an increment of faulting causes an earthquake, we say that the fault is *seismic*, but if the offset occurs without generating an earthquake, we say that it is *aseismic*. Aseismic faulting is commonly called *fault creep* in the geophysical literature.

Why do earthquakes occur during movement on faults? Earthquakes represent the sudden release of elastic strain energy that is stored in a rock, and can be generated when an intact rock ruptures, or when asperities on a preexisting fault snap off or suddenly plow. Rubbing two bricks while applying some pressure is a good analogy of this process.

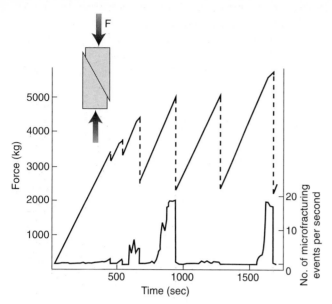

Figure 8.34 Laboratory frictional sliding experiment on granite, showing stick-slip behavior. The stress drops (dashed lines) correspond to slip events. The microfracturing activity is also indicated.

The fault (bricks) moves until the indentation of asperities once again anchors it. This start-stop behavior of faults is called *stick-slip behavior* (Figure 8.34). During the stick phase, stress builds up (as illustrated by the solid line); during the slip phase, the fault moves and the stress at the site of faulting drops (dashed line). Typically, the stress drop is not complete, meaning that the differential stress does not decrease to zero. *Fault creep* occurs where a fault zone is very weak due to the presence of weak material (e.g., clay or serpentine), to hydrolytic weakening in the fault zone, or to high fluid pressures in the fault zone.

Geoscientists have struggled for decades to delineate regions that have the potential to be seismic. This work involves study of faults to determine if they are active or inactive, and whether active faults are seismic or aseismic. But fault studies alone do not provide a complete image of seismicity, for not all earthquakes occur on recognized faults. Some represent development of new faults, some represent slip on blind faults, and some represent nonfault-related seismicity (e.g., volcanic explosions). Delineation of seismically active regions plays a major role in land-use planning. Obviously, the potential for seismicity must be taken into account when designing building codes or when siting special facilities like nuclear power plants, schools, hospitals, dams and pipelines, and your own home.

The primary criterion for delineating a seismically active region comes from direct measurements of seismicity.

The underlying idea is that places with a potential for earthquakes in the near future probably have suffered earthquakes in the recent past. Networks of seismographs record earthquakes and provide the data needed to pinpoint the *focus* of each earthquake, that is, the region in the Earth where the seismic energy was released. As you know from earlier geology courses, maps of earthquake *epicenters* (the point on the Earth's surface that lies directly above the focus) emphasize that most earthquakes lie along plate boundaries, but that dangerous earthquakes also occur within plate interiors. Thus, we distinguish between *plate-boundary seismicity* and *intraplate seismicity*. Cross sections showing the distribution of earthquake foci emphasize that, with the exception of convergent-margin seismicity, most earthquakes occur at depths shallower than ~15 km, which defines the lower boundary of the brittle upper crust.[12] Convergent-margin earthquakes occur along subducted slabs down to depths of about 650 km, defining the *Wadati-Benioff zone* (see Chapter 14). The deep earthquakes in a Wadati-Benioff zone occur well below the expected depth for brittle faulting. So why do these deep-focus earthquakes occur? One suggestion is that they represent movement on brittle faults in the cool interior of the downgoing slab, but modeling suggests that at such great depth the interior of the downgoing slab should be quite warm also. An alternative is that deep-focus earthquakes represent the stress release associated with sudden mineral phase changes in the downgoing slab (e.g., olivine to spinel). Since different mineral phases occupy different volumes, the sudden phase change causes a movement in the rock body that could result in the generation of an earthquake. The question is not resolved.

A reliable and detailed record of seismicity is only a few decades old, because a worldwide network of seismograph stations was not installed until after World War II. The information from these stations proved critical for the formulation of plate tectonic theory in the 1960s (Chapter 14), but governments actually funded this network to monitor underground nuclear testing in the Cold War era. However, seismic studies cannot delineate the potential for seismicity in areas that have only infrequent earthquake activity. To identify such cryptic seismic zones, geologists have to rely on field data. If the fault cuts a very young sequence of sediment or a very young volcanic flow or ash, then the fault must itself be very young. Similarly, if the fault cuts a young landform, such as an alluvial fan or a

---

[12]This seismically active region of the crust is also called the *schizosphere*, whose lower boundary is called *brittle-ductile transition* by some seismologists; we prefer to use *brittle-plastic transition* for this boundary.

glacial moraine, then the fault must be young. Fault scarps and triangular facets suggest that the faulting occurred so recently that erosion has not had time to erase its surface manifestation. The presence of uncemented gouge suggests that a fault was active while the rock it cuts was fairly close to the Earth's surface, a situation that implies movement on the fault occurred subsequent to uplift and exhumation of rock. Moreover, the presence of pseudotachylyte may indicate that the fault was seismic. Finally, accurate surveying of the landscape may indicate otherwise undetectable ground movements that could be a precursor to seismicity.

In special cases, it may be possible to determine the *recurrence interval* on a fault (meaning the average time between successive faulting events). This is done by studying the detailed stratigraphy of sediments deposited in marshes or ponds along the fault trace. Seismic events are recorded by layers in which sediment has been liquefied by shaking, which causes disruption of bedding and the formation of sand volcanoes. By collecting organic material (e.g., wood) from the liquefied interval, it may be possible to date the timing of liquefaction. Once we know the recurrence interval and the size of the earthquakes, we can estimate the seismic risk for an area.

The past few decades have seen intense study of earthquakes and related processes. As a result we have become reasonably successful in predicting *where* earthquakes will occur (if not at least the general area). But *when* they occur remains an imprecise science at best. Error margins of 50–100 years are simply inadequate for modern society, but improving significantly on this with our current understanding of faulting seems unlikely. Perhaps preparation remains our best bet when it comes to earthquake hazards.

8.8 CLOSING REMARKS

Chapters 6, 7, and 8 have provided an overview of brittle deformation processes and structures. At the end of Chapter 8, we also discussed briefly society's need for better understanding of fault processes. But brittle deformation is only part of what contributes to the development of Earth structures and deformation. We have alluded, for example, to the existence of shear zones in which displacement is not accommodated by brittle failure, and we have mentioned folding, which also deforms rocks without loss of cohesion.

In the next section of this book, we turn our attention to ductile processes and resulting structures. After we have introduced these structures, we will be able to explore the relationship between faults and other geologic structures in greater detail, and describe the tectonic settings in which they form.

ADDITIONAL READING

Anderson, E. M., 1951, *Dynamics of faulting and dyke formation,* Edinburgh, Oliver and Boyd, 191 p.

Boyer, S. E., and Elliot D., 1982, Thrust systems, *American Association of Petroleum Geologists Bulletin,* v. 66, p. 1196–1230.

Chester, F. M., and Logan, J. M., 1987, Composite planar fabric of gouge from the Punchbowl Fault, California, *Journal of Structural Geology,* v. 9, p. 621–634.

Hubbert, M. K, and Rubey, W. W., 1954, Role of fluid pressure in mechanics of overthrust faulting, *Bulletin of the Geological Society of America,* v. 70, p. 115–205.

Keller, E., and N. Pinter, 1996, *Active tectonics: Earthquakes, uplift, and landscape,* Englewood Cliffs, NJ, Prentice-Hall.

Mandl, G., 1988, *Mechanics of tectonic faulting, models and basic concepts,* Amsterdam, Elsevier, 407 p.

Petit, J. P., 1987, Criteria for the sense of movement on fault surfaces in brittle rocks, *Journal of Structural Geology,* v. 9, p. 597–608.

Scholz, C. H., 1990, *The mechanics of earthquakes and faulting,* Cambridge University Press, Cambridge, 439 p.

Sibson, R. H., 1977, Fault rocks and fault mechanisms, *Journal of the Geological Society of London,* v. 133, p. 190–213.

Sylvester, A. G., 1988, Strike-slip faults, *Geological Society of America Bulletin,* v. 100, p. 1666–1703.

Wise, D. U., Dunn, D. E., Engelder, J. T., Geiser, P. A., Hatcher, R. D., Kish, S. A., Odom, A. L., and Schamel., S., 1984, Fault-related rocks: Suggestions for terminology, *Geology,* v. 12, p. 391–394.

C
Ductile Structures

Chapter 9
Ductile Deformation Processes and Microstructures

9.1 INTRODUCTION

Why is it that a layer of rock can be permanently bent into a fold, or that a material such as ice can be permanently contorted (or flow) over a period of time while it remains a solid? Ice is a particularly instructive example of flow in crystalline solids because it moves on human time scales (Figure 9.1). At first you might think that this is accomplished by bending and stretching of atomic bonds in the crystal lattice. But bending and stretching is *elastic deformation,* and, as described in Chapter 5, elastic deformation is recoverable (i.e., nonpermanent). The movement of glaciers, however, is a permanent feature and they retreat only by melting. Such permanent bends and contortions represent *ductile deformation,* which is not recoverable. If we were able to remove a folded layer from an outcrop or a deformed mineral from a hand specimen, they would not jump back to their original forms. So these distortions that must be a result of permanent changes in the material. Understanding some of the principles that underlie the ability of rocks to accumulate permanent strain is contained in a vast and ever growing body of materials science literature. Structural geologists have applied many of the concepts from materials science to geologic conditions; use of the associated terminology, however, is not always consistent between these fields. In trying to keep the terms and concepts to a minimal, we invariably have to make some choices about what to include in this chapter; otherwise we will lose sight of our ultimate goal: understanding the way rocks deform in the ductile regime.

Figure 9.1 Malaspina Glacier in Alaska showing moraines (dark bands) that became folded during differential flow of the ice.

In Chapter 5 we first introduced the concept of *flow.* At that point we described this topic merely in terms of stress and strain rate. We contrasted linear viscous (Newtonian) and nonlinear viscous (non-Newtonian) behavior using analogs and simple mechanical models. In this chapter we will examine the physical processes that underlie this ability of materials to undergo appreciable, permanent distortions that are distributed through a volume of rock. Just to refresh your memory, strain that is distributed over the body is what distinguishes *ductile behavior* from *brittle behavior.* But strain that appears homogeneous on one scale may represent heterogeneity on another, so we again have to note the scale of our observation. The scales of observation in structural geology range from nanometers (10^{-9} m) to kilometers, and single grains to mountain ranges. For practical reasons, therefore, we define ductile behavior as uniform flow down to the scale of the hand specimen, that is, down to a mesoscopic-scale phenomenon.

We distinguish three fundamental mechanisms that produce ductile behavior in rocks and minerals: (1) *cata-* *clastic flow,* (2) *diffusional mass transfer,* and (3) *crystal plasticity.* Which of these processes actually occurs at a given time in a rock is primarily a function of temperature, stress, strain rate grain size, rock composition, and fluid content. At the onset we mention that temperature is a particularly important parameter, but since different minerals behave ductilely at different temperatures, what is considered high-temperature behavior for one mineral may be low-temperature behavior for another mineral. Thus, when talking about the relationship between temperature and deformation, we need to use a normalized parameter that is called the *homologous temperature,* T_m. The homologous temperature is a dimensionless parameter that is defined as the absolute temperature divided by the absolute melting temperature of the material:

$$T_h = T/T_m \qquad\qquad \text{Eq. 9.1}$$

with T for temperature and T_m for the melting temperature of the material, both in K. We loosely define 'low-temperature conditions' as $0 > T_h > 0.3$, 'medium-

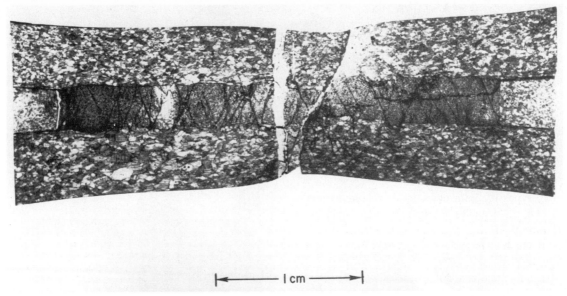

|←——— I cm ———→|

Figure 9.2 Extension experiment showing cataclastic flow in Luning dolomite (Italy), which is surrounded by marble that deformed by crystal plastic processes.

temperature conditions' as $0.3 > T_m > 0.7$, and 'high-temperature conditions' as $0.7 > T_h > 1$.

After discussing the fundamental mechanisms and the *microstructures* (i.e., mineral geometries on the microscopic scale) that result from the activity of these mechanisms, we will close this chapter by examining the interrelationship between the various rheologic parameters (such as stress and strain rate; see Chapter 5), and by introducing the powerful concept of *deformation-mechanism maps*. Now let us first turn to the three mechanisms of ductile behavior—cataclastic flow, dislocation movement, and diffusion.

9.2 CATACLASTIC FLOW

We start our discussion of cataclastic flow with a simple experiment. Consider a bean bag that is shaped like a ball. Now squash the bean bag so it fits into a small box (i.e., distort the bag into a brick shape). In order for the bag to change shape, the beans have to slide past one another. Now imagine that the surface of the bag is very strong and you attach the bag to a winch that pulls it through an opening that is smaller than a single bean. For the bag to pass through the small opening, all the individual beans must fracture into smaller pieces (brittle deformation), but the bag as a whole remains coherent. Such a process, where a mesoscopic body (the bean bag) changes shape without breaking into separate pieces, but microscopically the grains (the beans) fracture into smaller pieces and/or slide past one another, is called *cataclastic flow*. In rocks the tiny fractures are called *microcracks,* and the pieces move past one another by the process of *frictional sliding* (see Chapter 6).

During cataclastic flow, a rock deforms without obvious strain localization on the scale of the hand specimen, yet the *mechanism* of deformation is (micro)fracturing and frictional sliding (Figure 9.2). You may now better understand the source of confusion surrounding the terms brittle and ductile (Section 5.5). Cataclasis is mesoscopic ductile behavior, yet the process by which it occurs is microscopic brittle fracturing and sliding!

In rocks, microfractures may occur at grain boundaries (intergranular) or within individual grains (intragranular). In both cases the process occurs by breaking many atomic bonds at the same time. The crystal structure away from the fracture, however, remains unaffected. Frictional sliding is strongly dependent on pressure: with increasing pressure the ability of fracturing to occur is reduced (see Chapter 6). Therefore, we expect to find cataclastic flow in rocks at relatively low lithostatic pressures. This condition is met in the upper several km of the crust, and indeed we typically find cataclastic flow in shallow-crustal rocks. This strong stress-dependence of cataclasis is one practical characteristic that distinguishes it from ductile mechanisms involving crystal defects, which are discussed in the next section.

9.3 CRYSTAL DEFECTS

Ductile behavior of materials at elevated temperatures is most commonly achieved by the motion of crystal defects. In simple terms, a *crystal defect* is an error in the crystal lattice. There are three basic types: (1) point defects, (2) line defects, or dislocations, and (3) planar defects, or stacking faults. The motion of defects that occurs under specific conditions of differential stress gives rise to permanent strain without the material losing cohesion (i.e., without fracturing). Point and line defects are most important for the deformation of rocks; planar defects, which arise from errors in the stacking of atomic layers, play little role in deformation. In order to understand diffusional mass transfer and crystal plasticity, we first need to take a more detailed look at point and line defects.

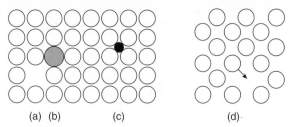

Figure 9.3 Point defects: (a) vacancy, (b) substitutional impurity, (c) interstitial impurity, (d) vacancy migration.

9.3.1 Point Defects

There are two types of point defects: (1) *vacancies* and (2) *impurity atoms.* Vacancies are empty sites in the crystal lattice (Figure 9.3a). Impurity atoms either are (a) *substitutionals,* in which an atom in a lattice site of the crystal is replaced by a different atom (Figure 9.3b), or (b) *interstitials,* in which an atom is at a nonlattice site of the crystal (Figure 9.3c). Vacancies can migrate by exchange with atoms in neighboring sites (Figure 9.3d). At first glance the concept of migrating vacancies sounds a bit odd. But consider an atom that moves to a vacant site; you can equally say that the vacancy moves to the atom. The general term for this process of atom or vacancy migration is *diffusion.* It is discussed later in this chapter (Section 9.4). Application of stress to a crystal causes a gradient in the vacancy concentration. Vacancies migrate down these concentration gradients, which causes material to flow.

9.3.2 Line Defects, or Dislocations

A line defect, usually called a *dislocation,* is a linear array of lattice imperfections. More formally stated: a dislocation is the linear array of atoms that bounds an area in the crystal that has slipped relative to the rest of the crystal (Figure 9.4). We realize that this definition is hardly informative at stage, until we look at the geometry of two endmember configurations, the edge dislocation and the screw dislocation, and before turning to the concept of slip in crystals.

An *edge dislocation* occurs where there is an extra half-plane of atoms in the crystal lattice (Figure 9.5). As illustrated in Figure 9.5a, there are seven vertical planes of atoms at the top half of the crystal and only six vertical planes of atoms at the bottom half. The termination of the *extra half-plane* (the plane that ends halfway in the crystal) is the dislocation. It extends into the crystal as the dislocation line, **l** (line CD in Figure 9.5a). The symbol for an edge dislocation is ⊥ or ⊤, depending on the location of the extra half-plane in the upper or lower half of the crystal, respectively. The presence of a dislocation causes a distortion of the crystal structure, just as a wedge distorts the log that is being split. In *screw dislocations,* the atoms are arranged in a corkscrew-like fashion (Figure 9.5b); the axis of the screw marks the dislocation line.

In a deformed crystal, an atom-by-atom circuit around the dislocation fails to close by one or more atomic distances, whereas a similar circuit around atoms in a perfect crystal would be complete. The arrow connecting the two ends of the incomplete circuit is called the *Burgers vector,*

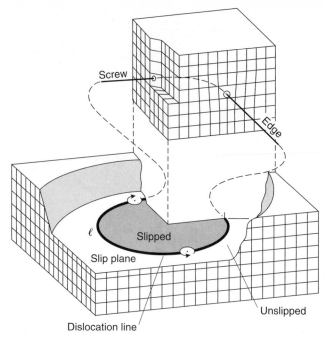

Figure 9.4 Geometry of a dislocation showing the edge- and screw-types and their relationship. The boundary between the unslipped and slipped portion of the crystal is the dislocation line **l**.

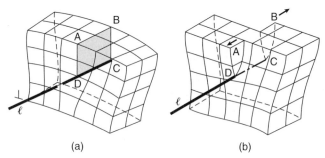

Figure 9.5 Types of dislocations. (a) The extra half-plane of atoms in an edge dislocation. (b) The corkscrew-like displacement of a screw dislocation.

b. The length of the Burgers vector in most minerals is on the order of nanometers (1 nm = 1×10^{-9} m). For an edge dislocation, the Burgers circuit remains in the same plane (Figure 9.6a), while for a screw dislocation the circuit steps up or down to another plane (Figure 9.6b). Edge and screw dislocations are therefore distinguished on the basis of the relationship between Burgers vector and dislocation line. For edge dislocations the Burgers vector is *perpendicular* to the dislocation line, while for screw dislocations the Burgers vector is *parallel* to the dislocation line. In fact these properties are used to characterize the nature of dislocations using the electron microscope (see Section 9.8). Beside very large magnifications, crystal-defect features may be seen by using a decoration technique that is described in the appendix at the end of this chapter. Figure 9.7 shows dislocations in the mineral olivine using an etched sample under the optical microscope. Edge and screw dislocations are only endmember geometries, and dislocations that consist of part edge and part screw components are called *mixed dislocations.*

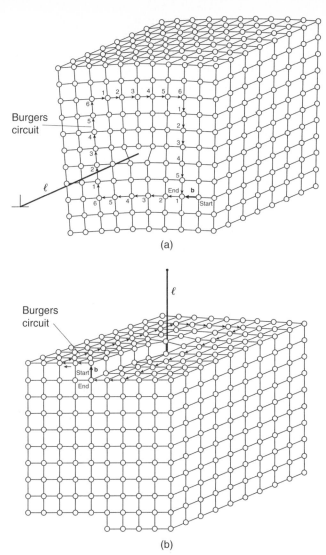

(a)

(b)

Figure 9.6 The determination of the Burgers vector, **b**, of a dislocation in a cubic lattice using a Burgers circuit. (a) The Burgers circuit around an edge dislocation marked by **l**. (b) The Burgers circuit in a screw dislocation. The closure mismatch for both edge and screw dislocations is the Burgers vector. In the edge dislocation **b**⊥**l**, and in the screw dislocation **b**//**l**.
Principles of Structural Biology by Suppe, John, © 1985. Adapted by permission of Prentice-Hall, Upper Saddle River, NJ.

The presence of a dislocation distorts the crystal lattice (Figure 9.6), which gives rise to a local stress field around the dislocation. In an edge dislocation, there is compressive stress on the side of the extra half-plane of atoms and tension on the opposite side (Figure 9.8a). The ax analog serves to illustrate this point rather well. The ax forces the wood away, giving rise to compression, which may result in the ax becoming stuck. Just beyond the tip of the ax, however, there is tension, which is why you are able to split wood without the ax going all the way through. Similarly, in a screw dislocation we find shear stresses (Figure 9.8b). What is the effect of these local stresses? The role of compressive and tensile stresses associated with dislocations can be understood intuitively by considering the behavior of magnets and charged particles. The compressive stress fields of two edge dislocations repel, and the compressive and ten-

sile fields of two edge dislocations attract (Figure 9.9), just as the poles of two magnets attract and repel if their polarities are opposite and reverse, respectively. Analogously, two screw dislocations with the same sense of shear repel each other, and those with opposite sense of shear attract. In a crude way you can say that dislocations are able to 'see' each other by the stress fields that arise from the distortion of the crystal lattice. Remember that this internal stress is not the same as the remote stress arising from, for example, squeezing the crystal.

Edge and screw dislocations are endmember configurations, called *perfect dislocations,* because the Burgers vector has a length of one unit lattice distance (i.e., the length of one atomic bond or multiples thereof). However, studies of some minerals (e.g., calcite) have shown Burgers vectors that differ from one unit lattice distance; these are called *partial dislocations.* Partial dislocations may be formed by splitting a long Burgers vector into two or more components by the process of *dissociation.* Dissociation is energetically more favorable in these cases, and arrays of partial dislocations may, for example, produce twinning in crystals (see Section 9.5.3).

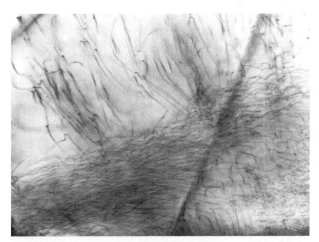

Figure 9.7 Dislocations in the mineral olivine from a Hawaiian mantle nodule. The dislocation appears by a decoration technique that is described in the appendix, which allows for optical inspection. Width of view is 200 μm.

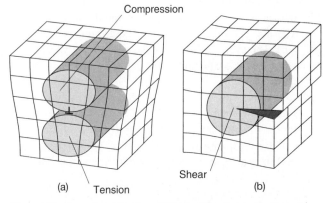

Figure 9.8 Geometry of the stress field (shaded regions) around an edge dislocation (a) and around a screw dislocation (b).

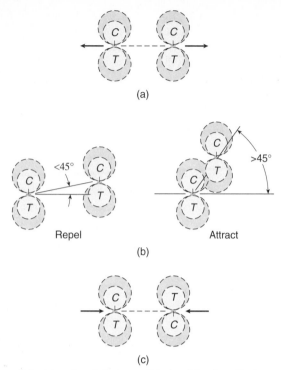

Figure 9.9 Interactions between parallel edge dislocations. Regions labeled C and T are areas of compression and tension, respectively, associated with the dislocations. (a) Like dislocations on the same or nearby planes repel. (b) Like dislocations on widely separated planes may attract or repel depending on the angle between the lines joining the dislocations. (c) Unlike dislocations on the same or nearby planes attract.

9.4 DIFFUSIONAL MASS TRANSFER

Flow of rocks can occur by the transfer of material through the process of diffusion. Three diffusion-related deformation mechanisms are important for natural rocks: (1) *pressure solution,* (2) *grain-boundary diffusion,* and (3) *volume diffusion.* Diffusion occurs when an atom or a point defect migrates through the crystal. Diffusion is temperature dependent, because thermal energy causes atoms to vibrate more, so that bonds can break and reattach more easily. When we increase the temperature of a material, the ability of individual atoms to jump to neighboring vacant sites increases. For example, at the melting temperature of Fe (i.e., $T_h = 1$) the *jump frequency,* Γ, of vacancies is on the order of 10^{10} per second. The jump distance, r (the distance between atoms in the crystal structure), for each jump is 10^{-10} m (0.1 nm). The average distance traveled, R, for a vacancy is given by Einstein's equation:[1]

$$(\Gamma t r)^{\frac{1}{2}} \qquad \text{Eq. 9.2}$$

If we use t = 1 second, then R = 0.001 mm at $T_h = 1$ for Fe-metal. This distance is very small; you might even think that it is insignificant. However, considering that geologic time is measured in millions of years, the value of R becomes quite large. For example, at t = 1 m.y. (3.1×10^{13} sec), the value of R is 56 m. However, such values are representative for minerals in rocks only near their melting temperature, which is not the most typical condition during deformation. At lower T_h, diffusion distances are on the order of cm to m. One aspect that may not be immediately appreciated is that R does not represent the linear distance between the original position of the vacancy or an atom and its position after time t. Diffusion is nondirectional in an isotropic field, and it is what we call a *random-walk process;* the linear distance after time t may lie anywhere between 0 and R.

Theoretical arguments, which we will not discuss here, allow us to define a diffusion coefficient, D, for a given mineral, which describes movement of a species down a concentration gradient:[2]

$$D = \Gamma/6\ r^2 \qquad \text{Eq. 9.3}$$

Without giving the derivation here, we can rewrite Equation 9.3 in a form that shows the temperature dependence for diffusion and a minimum energy for migration to occur:

$$D = D_o \exp(-E^*/RT) \qquad \text{Eq. 9.4}$$

where D_o is a material constant that is empirically determined, E^* is the activation energy for migration (kJmol^{-1}), R is the gas constant (8.31 Jmol^{-1}K^{-1}), and T is temperature (in °K).[3] We present the equation in this particular form because it compares with the constitutive equations for flow that were given in Section 5.3.6 and those later in this chapter (Section 9.6).

Solid diffusion in crystals may take place in two manners: (1) grain-boundary diffusion, or Coble creep and (2) volume diffusion, or Nabarro-Herring creep. Diffusional mass transfer that involves a fluid phase is called pressure solution. Each of these three mechanisms is discussed below.

9.4.1 Volume Diffusion and Grain-Boundary Diffusion
Given sufficient time, diffusing vacancies reach the surface of the crystal where they disappear. To see how this causes deformation, consider a crystal that is being subjected to a differential stress (Figure 9.10). The vacancies migrate toward the site where stress is greatest, and the atoms move to the sides where the stress is least. This results in an overall change in the distribution of mass, which in turn changes the shape of the crystal, but remember that this occurs without significant distortion of the crystal lattice. Thus, in the presence of a nonisotropic stress field, diffusion is directional. The diffusion of vacancies can occur through the entire crystal or can be concentrated along a

---

[1]Another Albert Einstein (1879–1955) equation.

[2]Strictly, these equations are for vacancy movement, and define D_{vac}.
[3]We may also write this equation involving Boltzmann's constant (k), in which case E^* is given in a different form; k and R are related as: $R = kN_A$, with N_A being Avogadro's number (6.02×10^{23}mol^{-1}), which gives k = 1.38×10^{-23}. Note that "exp(a)" means e^a in this and subsequent equations.

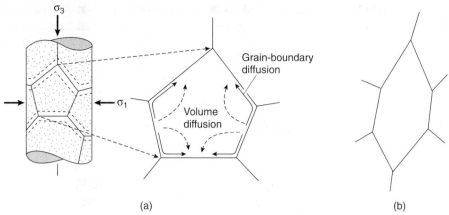

Figure 9.10 Diffusional flow by material transport through (volume diffusion, or Nabarro-Herring creep) and around grains (grain-boundary diffusion, or Coble creep) in the presence of a differential stress (a) produces a shape change (b).

narrow region at the grain boundary (Figure 9.10a), resulting in a permanent shape change as shown in Figure 9.10b. These deformation mechanisms are called *volume diffusion,* or *Nabarro-Herring creep,* and *grain-boundary diffusion,* or *Coble creep,* respectively.

Both Nabarro-Herring creep and Coble creep achieve strain by the diffusion of vacancies. The strain rate for each mechanism is, therefore, a function of the diffusion coefficients (volume diffusion [D_v] and grain-boundary diffusion coefficients [D_b], respectively), but also of the grain size (d):

$$\dot{e}_{Coble} \cong D_b/d^3 \qquad \text{Eq. 9.5}$$

$$\dot{e}_{Nabarro-Herring} \cong D_v/d^2 \qquad \text{Eq. 9.6}$$

These simplified relationships emphasize the critical importance of grain size in diffusional creep: the larger the grain size, the smaller the strain rate.

The activation energy for grain-boundary diffusion (included in D_b) is less than that for volume diffusion (included in D_v), and grain-size dependence of volume diffusion is larger. Thus, Coble creep is a more significant process in crustal rocks than Nabarro-Herring creep; the latter is restricted to very high temperatures (e.g., temperatures in the mantle) and/or very small grain sizes.

9.4.2 Pressure Solution

Pressure solution is the third diffusional mass transfer process in natural rocks. Its occurrence requires the presence of a fluid film on grain boundaries. The process is important in crustal rocks because material transfer can occur at temperatures well below those required for vacancy diffusion, because a chemically active fluid film dissolves the crystal. The dissolved ions then move along a *chemical gradient* to regions of new growth. In rocks this chemical gradient arises from differential solubility in the presence of a differential (nonisotropic) stress. Recall that fluids do not support shear stresses (Section 3.9), so pressure solution works only if the fluid film is 'attached' to the grain boundary by chemical bonds; thus, when the stress is applied, the fluid does not move. Areas of high stress, say, surfaces perpendicular to the maximum principal stress, exhibit enhanced solubility, and the dissolved material is transported to regions under lower stress (surfaces perpendicular to the minimum principal stress). The geometric consequences sound very similar to our previous description of grain-boundary diffusion. Indeed, pressure solution produces shape changes like that in Figure 9.10, except that it can occur at such low temperatures that it can be active even at the Earth's surface. We distinguish between these diffusional deformation mechanisms by referring to pressure solution as *fluid-assisted diffusion* and Nabarro-Herring creep and Coble creep as *solid-state diffusion;* in colloquial terms we sometimes call them 'wet' and 'dry' diffusion, respectively. Fluids are mostly present in shallow crustal rocks, and this, combined with low ambient temperatures, makes pressure solution a characteristic deformation mechanism in low-grade, upper-crustal rocks.

We can observe the past activity of pressure-solution diffusion in rocks by the presence of, for example, stylolites in limestones, grain overgrowths in sandstones, and cleavage and pressure shadows in some slates (Figure 9.11; see also Chapter 10). In contrast to 'dry' diffusion, the distance over which material may be transported by fluid-assisted diffusion is not limited to individual grains, but can be substantial, particularly if the dissolved ions migrate into the pore fluid of the rock. Movement of pore fluid (i.e., groundwater flow) can flush the dissolved ions completely out of the rock, resulting in substantial volume loss. On the mesoscale, pressure solution may result in the formation of alternating layers of different composition, such as quartz- versus mica-rich layers, in a process called *differentiation* (Section 11.3.6). Alternatively, the dissolved ions may precipitate as vein fillings in cracks. The widespread occurrence of these rock

Figure 9.11 Bedding-perpendicular pressure solution seams (stylolites) in argillaceous limestone. The middle bed is pure carbonate and does not contain as many seams (Pennsylvania, Appalachians); The white stripes are veins. Note that the stylolites cut across bedding.

structures formed by pressure solution shows the great geological importance of this deformation mechanism.

The strain rate associated with pressure solution is a function of the rate of atoms that go into dissolution (i.e., the solubility of a material) in the fluid:

$$\dot{e}_{\text{pressure solution}} \cong D_f/d^2 \qquad \text{Eq. 9.7}$$

where D_f is the diffusion coefficient of a phase in a fluid and d the grain size.

9.5 CRYSTAL PLASTICITY

The conditions for vacancies and atoms to move are not always sufficient to cause significant distortion, especially at low temperatures or in the absence of a reactive fluid film. Yet from direct observations, we know that crustal rocks deform quite readily under these conditions. They are able to do so by the motion of dislocations. We last left our description of dislocations by distinguishing between edge and screw dislocations, and between perfect and partial dislocations (Section 9.3.2). Under certain stress conditions, dislocations are able to migrate through the crystal lattice if the activation energy for movement is exceeded. Strain produced by this movement represents our third ductile deformation mechanism: crystal plasticity. Dislocation movement may occur by *glide* and by a combination of glide and climb (*creep*), depending mainly on temperature. A special case of crystal-plastic behavior, *twinning,* occurs at low temperature in some minerals.

9.5.1 Dislocation Glide

Deformation and temperature introduce energy into the crystal, which allows dislocations to move. However, dislocations are not free to move in any direction through the crystal. At low temperatures they are restricted to *glide planes* (or *slip planes*). The glide plane of a dislocation is the plane that contains the Burgers vector, **b,** and the dislocations line, **l.** Because a plane is defined by two nonparallel lines, each edge dislocation has one slip plane because **b** and **l** are perpendicular. A screw dislocation, on the other hand, has many potential slip planes, because **b** and **l** are parallel. Physically, a glide plane is a crystallographic plane across which bonds are relatively weak. Some crystals have only one crystallographic plane that is an easy glide plane; others may have many. Table 9.1 lists the dominant slip systems for some of the more common rock-forming minerals. Note that in many crystals more than one slip systems may be active at the same time.

What is the physical process that allows the movement of dislocations? Nature has devised an energetically clever way for dislocations to move: rather than breaking all atomic bonds across a plane simultaneously, such as occurs during fracturing and which requires considerable energy, only bonds along the dislocation line are broken during an increment of movement. Let us again turn to an analogy to illustrate the process. The movement of dislocations is comparable to moving a large carpet across a room that contains heavy pieces of furniture. The easiest way to move the rug is to ruck up one end and propagate the ruck across the room. Energy is only needed to lift up selected furniture legs to propagate the ruck past these obstacles rather than lift all the furniture simultaneously. In a natural example, caterpillars or snakes move by displacing one segment of their body at a time, instead of moving their entire body simultaneously. Edge dislocations move by successive breaking of bonds under the influence of a minimum stress acting on the

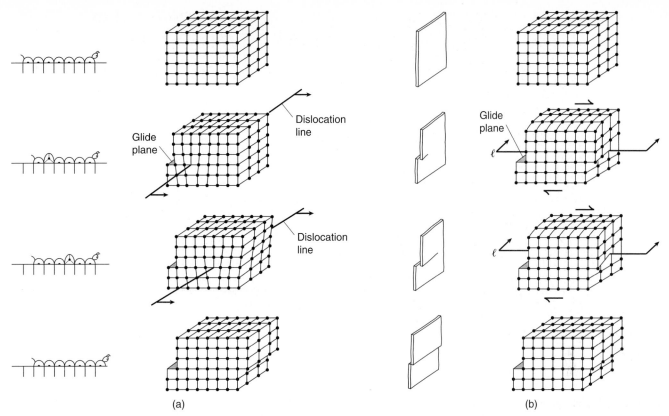

(a) (b)

Figure 9.12 Dislocation glide. (a) Movement of an edge dislocation in a cubic lattice; this process may be thought of as the movement of a caterpillar. (b) Movement of a screw dislocation; this is analogous to tearing a sheet of paper, with the screw dislocation at the tip of the tear. After the dislocation passes through the lattice, it leaves behind a strained crystal with a perfect lattice structure. The dislocation lines and the glide planes are shown.

Table 9.1 Dominant Slip Systems in Common Rock-Forming Minerals

Mineral	Glide Plane and Slip Direction[a]	Comments
Calcite	$\{\bar{1}018\}<40\bar{4}1>$	e-twinning
	$\{10\bar{1}4\}<\bar{2}021>$	r-twinning
	$\{10\bar{1}4\}<\bar{2}021>$	r-glide
	$\{01\bar{1}2\}<2\bar{2}01>$ or $<\bar{2}021>$	f-glide
Dolomite	$\{\bar{1}012\}<10\bar{1}1>$	f-twinning
	$(0001)<2\bar{1}\bar{1}0>$	c-glide
	$\{01\bar{1}2\}<2\bar{2}01>$ or $<\bar{2}021>$	f-glide
Mica	$(001)<110>$	basal (c) slip
Olivine	$(001)[100]$	
	$\{110\}[001]$	
Quartz	$(0001)<11\bar{2}0>$	basal (c) slip
	$\{10\bar{1}0\}[0001]$	prism (m) slip, along c
	$\{10\bar{1}0\}<11\bar{2}0>$	prism (m) slip, along a
	$\{10\bar{1}1\}<11\bar{2}0>$	rhomb (z) slip

[a]Miller indices for equivalent glide planes from crystal symmetry are indicated by { }; specific glide planes are indicated by ( ); equivalent slip directions from crystal symmetry are indicated by < >; individual slip directions are indicated by [ ].

Source: data from Wenk, (1985).

glide plane. This stress is called the *critical resolved shear stress* (*CRSS*). Recognizing this, it is easy to see why some minerals may deform more easily than others at a given temperature. If a crystal has many nonparallel glide planes, it is more likely that, for a given applied stress, the CRSS is exceeded on at least one and sometimes more than one glide plane, than when the crystal has only one glide plane. An edge dislocation moves when the unattached atoms at the bottom of the extra half-plane bond to the next lower and forward atoms that are located directly below the glide plane. Thus the position of the extra half-plane changes as the dislocation moves (Figure 9.12a). A screw dislocation moves by shearing one atomic distance forward (Figure 9.12b), similar to tearing a piece of paper. In spite of the fact that atomic bonds are broken and reattached when both types of dislocations move toward the edge of a crystal, they leave a perfect crystal lattice behind. When a dislocation reaches the edge of the crystal, there are no more atoms below to attach to and the crystal edge becomes offset. This offset of the crystal edge produces on the surface of the crystal a stair-step structure known as a *slip band*. Importantly, the process of dislocation movement produces permanent strain in rocks and minerals without causing them to lose coherency.

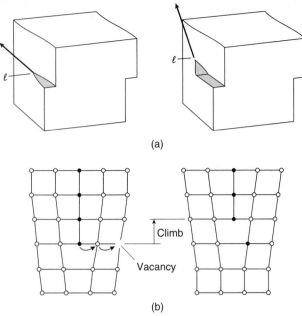

Figure 9.13 (a) Cross-slip of a screw dislocation, and (b) climb of an edge dislocation.

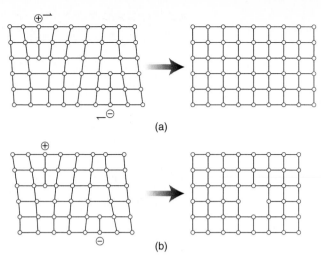

Figure 9.14 (a) Two edge dislocations with opposing extra half-planes that share a glide plane move in opposite directions to meet and form a perfect crystal. (b) When they move in different glide planes, a vacancy may be formed where they meet.

9.5.2 Cross-Slip and Climb

It is not always possible for dislocations to propagate to the edge of the crystal. Point defects, such as impurity atoms bonded tightly to their neighbors, can resist the breaking of bonds that is required for dislocation glide. Unfavorable stress fields of the dislocations themselves can also resist their motion (see Figure 9.8), especially when many dislocations are present. Just consider trying to work your way past a one-lane car accident. Obstacles that result from the presence of a great many immobile dislocations are therefore (morbidly) called *pile-ups*. In order to overcome these obstacles, edge and screw dislocations must move out of their current glide plane. They do so by the processes of climb and cross-slip. The activation of these mechanisms requires additional energy beyond that for dislocation glide. Screw dislocations, unlike edge dislocations, are not confined to a single glide plane, because the dislocation line and Burgers vector are parallel. They can leave one glide plane and simply move to another glide plane, a process called *cross-slip* (Figure 9.13a). If it is so easy, why does cross-slip not occur all the time? Typically, cross-slip requires that the dislocation leaves a favored glide plane (one with short Burgers vectors) for a less-favored one. Thus cross-slip takes place only if the CRSS on the less-favored plane is increased. Alternatively, raising the temperature (which means increasing the energy state of the system) lowers the CRSS that is needed for cross-slip because chemical bonds are weakened, and cross-slip will occur more easily.

Edge dislocations cannot cross-slip because they have only one glide plane. However, they can *climb* to a different, but parallel glide plane if there is a vacancy to accept the lowest atom of the extra half-plane (this is shown two-dimensionally in Figure 9.13b). Climb, therefore, involves

diffusion. Because the rate of vacancy production increases with rising temperature, the efficiency of dislocation climb is temperature dependent. Both cross-slip and particularly climb become activated at temperature conditions that exceed that for dislocation glide in a single mineral given the same stress conditions. Therefore, they typically occur at deeper (= hotter!) levels in the Earth. It is not possible to put an absolute number on the necessary depth in the Earth, because this is a function of the mineral in question as well as the Earth's thermal structure. Nevertheless, as a general guide, we can say that glide and climb occur at temperatures greater than 300°C for quartzitic rocks and carbonates, and at higher temperatures (>500°C) for such common minerals as dolomite, feldspar, and olivine. In the literature you will find that the term *dislocation creep* is used for the combined activity of glide and climb, which should not be confused with the term *dislocation glide*.

Although interacting dislocations often repel each other, unlike dislocations may attract and annihilate. *Dislocation annihilation* is one way of reducing the internal strain energy that arises from lattice distortion of a crystal containing dislocations. For example, two edge dislocations are lying in the same glide plane, with the extra half-plane of one dislocation inserted upwards (positive edge dislocation) and the other downward (negative edge dislocation). When these meet, they annihilate each other (Figure 9.14a). Similarly, convergence of screw dislocations with Burgers vectors in opposite directions will also result in annihilation. Two dislocations of opposite sign but on different glide planes may still attract, but they cannot fully annihilate each other. In such cases a defect remains (e.g., a vacancy; Figure 9.14b). However, allowing climb and cross-slip increases the probability of dislocation annihilation, so the rate of dislocation annihilation is also temperature dependent.

Figure 9.15 Calcite e-twins in a marble from southern Ontario (Canada). Width of view is ~4 mm.

Table 9.2 Crystal Systems

System	Symmetry	Crystal Axes
Triclinic	1 one-fold axis or center of symmetry	$a \neq b \neq c, \alpha \neq \beta \neq \gamma \neq 90°$
Monoclinic	1 two-fold axis or 1 symmetry plane	$a \neq b \neq c, \alpha = \gamma = 90°, \beta \neq 90°$
Orthorhombic	3 two-fold axes or 3 symmetry planes	$a \neq b \neq c, \alpha = \beta = \gamma = 90°$
Trigonal	1 three-fold axis	$a_1 = a_2 = a_3 \neq c, \beta = 90°$
Hexagonal	1 six-fold axis	$a_1 = a_2 = a_3 \neq c, \beta = 90°$
Tetragonal	1 four-fold axis	$a = b \neq c, \alpha = \beta = \gamma = 90°$
Cubic	4 three-fold axes	$a = b = c, \alpha = \beta = \gamma = 90°$

[1]a, b, c describes length of crystal axes; α is angle between b and c; β is angle between a and c; γ is angle between a and b.

9.5.3 Mechanical Twinning

Twins are common in many minerals. You have probably seen them with the handlens in the minerals plagioclase and calcite. In thin section, under crossed polarizers, they are easily recognized by their extinction behavior as you rotate the stage. Twins that develop during the growth of a crystal are called *growth twins,* but these twins say little or nothing about the conditions of deformation (stress, strain). In contrast, minerals, such as calcite, may form twins in response to an applied stress, which are called *mechanical twins* (Figure 9.15). Let's first have a look at twinning in general and then see what information mechanical twins can provide for deformation studies.

Mechanical twinning is a type of crystal plastic process that involves glide of *partial dislocations.* A surface imperfection, the *twin boundary,* separates two regions of a twinned crystal. The lattices in these two portions are mirror images of each other; in other words, a twin boundary is

a mirror plane with a specific crystallographic orientation. As a rule, twinning planes cannot already be mirror planes in the untwinned crystal. Therefore, mechanical twinning is most common in low-symmetry minerals such as trigonal calcite and dolomite, and triclinic feldspar. Recall that crystal symmetry is a geometric operation that repeats a crystal plane in another position. For reference, Table 9.2 lists the seven crystal systems and their symmetry.

Mechanical twins are produced when the resolved shear stress acting on the future twin boundary exceeds some critical value (the CRSS for twinning). During twinning, the crystal lattice rotates in the direction that produces the shortest movement of atoms, with a unique rotation angle. As such, mechanical twinning has similarities with dislocation glide, but differs in two respects. First, atoms are not moved an integral atomic distance as in glide, but rather some fraction; consequently, twinning involves partial dislocations. Secondly, the twinned portion of a grain is

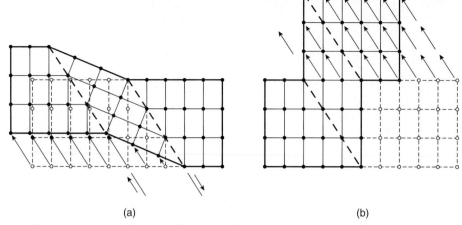

Figure 9.16 (a) Schematic illustration of mechanical twinning. The heavy outline marks a twinned grain, in which the twin boundaries (heavy dashes) are mirror planes. The atomic displacements are of unequal length and generally do not coincide with one atomic distance. Closed circles are atoms in an orthogonal crystal structure, and open circles give the original positions of displaced atoms. Twinning contrasts with dislocation glide (b), in which atoms move one or more atomic distances in the glide plane (heavy dashed line).

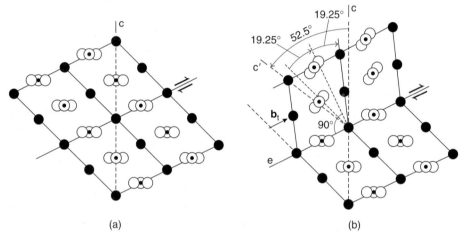

Figure 9.17 Calcite crystal lattice showing layers of Ca (large black dots) and CO_3 groups (C is small dots, O is large open circles); the crystallographic c-axis is shown (a). The twinned calcite lattice in (b) shows the partial dislocation (b_t) and angular rotations of the c-axis and the crystal face.

a mirror image of the original lattice (Figure 9.16a), whereas the slipped portion of a grain has the same crystallographic orientation as the unslipped portion of the grain (Figure 9.16b). For deformation studies, we are mainly interested in mechanical twins, that is, those that are produced by stress. We digress briefly to illustrate these in calcite.

The fact that twinning takes place along specific crystallographic planes in a calcite crystal,[4] and that rotation occurs over a specific angle and in a specific sense, allows us to use twinning as a measure of finite strain. The atomic structure of calcite twins is illustrated in Figure 9.17, showing the specified rotation angle of the crystallographic c-axis, which is perpendicular to the planes containing the CO_3 groups, and that of the crystal face. In Figure 9.18a, a deformed grain A′B′CD with one twin is shown; the original grain outline is ABCD with its sides parallel to calcite crystal planes. From this figure, you can see that the shear strain for the twinned grain is:

$$\gamma = \tan\psi = q/T \qquad \text{Eq. 9.8}$$

For one twin, q = p, so

$$\gamma = (2t \tan(\phi/2))/T \qquad \text{Eq. 9.9}$$

where T is the grain thickness and t is the twin thickness. For a grain containing several twins (Figure 9.18b), the shear strain is obtained by adding the strain due to each twin, or:

$$\gamma = \frac{2}{T} \sum_{i=1}^{n} t_i \tan (\phi/2) \qquad \text{Eq. 9.10}$$

where n is the number of twins in the grain. Given that the angle ϕ is constant in the case of calcite (~38°; Figure 9.17b), Equation 9.10 simplifies to:

$$\gamma = \frac{0.7}{T} \pi \sum_{i=1}^{n} t_i \qquad \text{Eq. 9.11}$$

---

[4]We will consider only e-twins ($\{1018\}\langle4041\rangle$) with a rotation angle for the c-axis of 52.5°, and a CRSS of 10MPa.

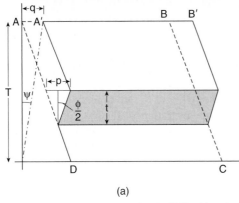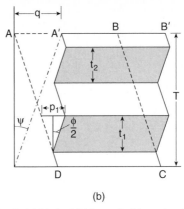

Figure 9.18 Calcite strain-gauge technique. An original grain ABCD with a single (shaded region) (a) twin of thickness. In (b) a grain with multiple twins is shown.

So, if we measure the total width of twins and the grain size perpendicular to the twin plane, we can obtain the total shear strain for a single twinned grain. In an aggregate of grains, the shear strains will vary as a function of the crystallographic orientation of individual grains relative to the bulk strain ellipsoid. We use this variation to determine the orientation of the principal strain axes by determining the orientations for which the shear strains are zero[5] and maximum. This strain analysis technique is called the *calcite strain-gauge method*. Looking again at Figure 9.18 and Equation 9.11, we can now determine the maximum amount of shear strain that can be accumulated using twinning: γ_{max} occurs when the entire grain is twinned, so t = T; thus, $\gamma_{max} = 0.7$ or X/Z $\approx$ 2. This maximum contrasts with the amount of strain that can accumulate during dislocation glide, which is unrestricted.

The calcite strain-gauge technique has proven to be very useful in studying strain fields in limestones subjected to small strains. Such work helps elucidate the kinematics of folding, the formation of veins, the early deformation history of fold-and-thrust belts, and the strain patterns in continental interiors. The great advantage of this method lies in the fact that twinning occurs at low T_h and low σ_d, and that the orientation and magnitude of very small finite strains may be recorded. An example of the application of this method showing a regionally consistent strain field in the cratonic interior of eastern North America was shown in Figure 4.29.

9.5.4 Strain-Producing and Rate-Controlling Mechanisms

Dislocations are not stationary in a crystal. They are able to move (slip or glide) while leaving behind what is called the *slipped portion* of the crystal. This slipped portion shows no crystallographic distortion even though the dislocation once passed through this part of the crystal. The ability of dislocation to move finally brings us back to the earlier definition of a dislocation that was given without much explanation at the time (Section 9.3.2): a dislocation is the linear array of atoms that bounds an area in the crystal that has slipped one Burgers vector more than the rest of the crystal. Now that we have discussed the various dislocation motion mechanisms, there is an important distinction between dislocation glide, on the one hand, and dislocation cross-slip and climb, on the other hand, that needs to be recognized. Dislocation glide is the process that produces a change in the shape of grains; it is therefore the main *strain-producing mechanism* of crystal plasticity. Cross-slip and climb facilitate glide, but by themselves they produce little strain; they mostly allow a dislocation to leave its original glide plane, for example, to bypass an impurity. Cross-slip and climb are therefore the *rate-controlling mechanisms* of crystal plasticity. Climb occurs at temperatures that are higher than those required for glide in the same mineral. Sometimes you will therefore see the terms *low-temperature creep* and *high-temperature creep* for glide (and twinning), and glide plus climb, respectively.

9.5.5 Where Do Dislocations Come From?

Nothing in life is perfect! You will have undoubtedly heard and probably experienced this yourself. The same goes for minerals. Defects, such as dislocations, are a part of all minerals, but not without reason. For example, the small offsets that occur at the edges of crystals containing dislocations (on the order of nanometers) are used as nucleation sites during mineral growth. For deformation, dislocations are necessary to enable the shape change during crystal plasticity. So far we have talked about only a couple of dislocations at the same time. Actually, the number of dislocations in a mineral, the *dislocation density, N,* is quite large. For example, 'perfect' grains that have grown from a melt have a dislocation density of 10^6 cm^{-2}; for deforming grains, N is even several orders of magnitude larger. What is this strange unit "cm^{-2}" for dislocation density? Dislocation density is the total length of dislocations

---

[5]A computer routine is normally used for the analysis. For details of the method see Additional Reading. A method has also been developed that allows the determination of the differential stress from twinned aggregates.

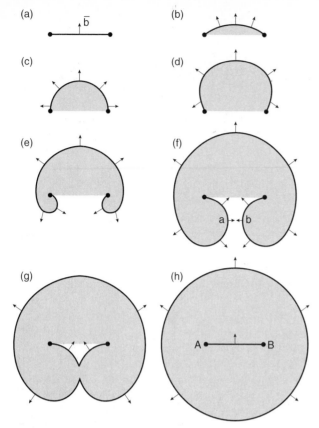

(a) $\bar{b}$

(b)

(c)

(d)

(e)

(f)

a → ← b

(g)

(h)

A ————— B

Figure 9.19 Dislocation multiplication by a Frank-Read source. A pinned dislocation with Burgers vector **b** bows out during glide to form a new dislocation. The slipped portion of the grain is shaded.

per volume of crystal; thus N = length/volume, so the unit of N is $[l]/[l^3] = [l^{-2}]$. Measuring dislocation length per volume is not a very convenient way to determine N. Practically, we measure the number of dislocations (scalar) that intersect an area (l^2), which also gives the unit $[l^{-2}]$. Later, in Section 9.9.1, we will give an example of a dislocation density calculation.

If we wish to obtain appreciable strains using dislocation movement, we will need a great many dislocations. At the same time we already learned that strain is produced by dislocations moving to the edge of the crystal (Figure 9.12), after which they leave a perfect lattice behind. So in order for crystal plastic processes to occur, we actually need to generate dislocations. In fact we already mentioned that dislocation density is greater in strained grains than in unstrained, 'perfect' grains. This suggests that dislocations can be generated during deformation. One important mechanism for dislocation generation (or multiplication) is by *Frank-Read sources* (Figure 9.19). Consider a dislocation that is anchored at two points, A and B; this *pinning* may arise from impurities, climb, or the interaction with other dislocations (not shown in the figure). During glide, the A-B dislocation will bow out because it is pinned at its edges (Figure 9.19b–d). Eventually this produces the kidney-shaped loop in Figures 9.19e and f. Note that the dislocation segments at a and b in

Figure 9.19f are opposite in sign because their Burgers vectors are opposite. So as a and b come together, they annihilate each other (Figure 9.19g), forming a new A-B dislocation line, while leaving the old loop present (Figure 9.19h). The process starts again for the new A-B dislocation line, while the old loop continues to glide. There is no restriction to number of cycles and therefore the number of dislocation loops that are generated in this manner, and the mechanism occurs for both edge and screw dislocations. This and other dislocation multiplication mechanisms collectively produce the high dislocation densities that are required in grains to deform by crystal plastic processes.

9.6 FLOW LAWS

Previously we defined the rate at which shape change occurs as the strain rate, $\dot{e}$ (Section 5.1.1). Since dislocation movement is a function of the differential stress (either arising from internal distortion or imposed on the system from outside), temperature, and the activation energy for breaking bonds, the rate at which strain occurs by dislocation movement is of course also a function of these parameters. This relationship is described by a *constitutive equation* or *flow law,* with the general form:

$$\dot{e} \equiv A\, f(\sigma_d)\, \exp(-E^*/RT) \qquad \text{Eq. 9.12}$$

where A is a material constant, E^* is the activation energy, R is the gas constant, T is the absolute temperature (in K), and $f(\sigma_d)$ simply means "function of differential stress"; characteristic values for these parameters were given in Table 5.6. In this chapter we will focus only on the stress function, $f(\sigma_d)$, which is determined from experiments on natural rocks and minerals. For dislocation glide (low-temperature creep), the function of stress is *exponential,* so the flow law is of the form:

$$\dot{e} = A\, \exp(\sigma_d)\, \exp(-E^*/RT) \qquad \text{Eq. 9.13}$$

Because of the form of this equation, dislocation glide is also called *exponential creep.*

For dislocation glide and climb (high-temperature creep), which is typical for many crustal and mantle rocks, the stress is raised to the *power* n. This flow law takes the general form:

$$\dot{e} = A\, \sigma_d^n\, \exp(-E^*/RT) \qquad \text{Eq. 9.14}$$

Analogously, climb-assisted glide is therefore also called *power-law creep,* with the power n called the *stress exponent.*

In Section 9.4 we presented the diffusion coefficient for point defects (Equations 9.4–9.6). The motion of vacancies is also driven by differential stress, and requires an activation energy. The flow law for diffusion has the form:

$$\dot{e} = D_o\, \sigma_d\, \exp(-E^*/RT)\, d^r \qquad \text{Eq. 9.15}$$

You will notice that the stress function of Equation 9.15 has exactly the same form as that in Equation 9.14, except

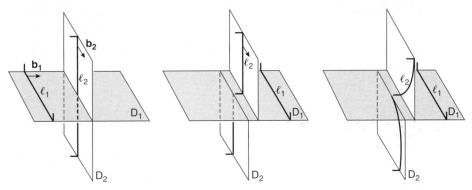

Figure 9.20 The formation of a jog from the interaction of two edge dislocations. For simplicity, dislocation D_2 is initially kept stationary while dislocation D_1 moves; the glide planes, Burgers vectors **(b)**, and dislocation lines **(l)** for each edge dislocation are shown (a). As D_1 passes through dislocation line l_2, a small step of one Burgers vector (b_1) length is created; this small step is a jog, with a differently oriented dislocation line segment but with the same b_2 (b). As a consequence, the glide plane, containing l_2 and b_2, is different along l_2. In fact, for the jog, the glide plane is the same as that for D_1, but with a different Burgers vector. Now assuming that the CRSS for glide differs in different directions, the ability of D_2 to move is no longer the same along l_2, and the jog pins the dislocation anchoring a segment of l_2 (c).

that the stress exponent, n, equals 1. This means that diffusion is *linearly* related to the strain rate and therefore that diffusional creep is a linear viscous process (or Newtonian viscous; Section 5.3.2). Note, however, that the strain rate for diffusional creep is nonlinearly related to the grain size, with r = 2 or 3 (Equations 9.5, 9.6, and 9.15).

We will see later that these various creep regimes produce characteristic microstructures. But let us first return to the deformation experiments of Chapter 5 and interpret the behavior in light of what we now have learned about crystal defects, and dislocations in particular.

9.7 A MICROSTRUCTURAL VIEW OF LABORATORY BEHAVIOR

So far our discussion of ductile behavior has been pretty theoretical. The reason we first discussed all this material is that defect microstructures allow us to understand how materials respond to stress, that is, their rheology. Remember that the deformation experiments discussed in Chapter 5 showed much the same general behavior: after an initial elastic stage, permanent (ductile) strain occurs. The elastic component is recoverable and does not involve crystal plastic processes, but the ductile component of the curve is generally achieved by the motion of dislocations. Strain accumulates at constant stress or will require increasingly higher stress at constant ė, which we previously called steady-state flow and work (or strain) hardening, respectively. From the microstructural perspective, *steady-state flow* simply implies that the generation, motion, and removal of dislocations is sufficiently fast to achieve strain at a constant rate for a certain level of stress. But what is the microstructural explanation for *work hardening* (Section 5.4.6)? The lack of climb and cross-slip at relatively low temperatures prevents dislocations from slipping past inclusions and other dislocations. This, combined with a decreased frequency of dislocation annihilation,

causes the dislocation density in a crystal to increase, which affects the ability of dislocations to glide because they tangle with one another. Remember that it is primarily the ability of dislocations to glide that produces the strain, so when dislocations tangle, this restricts their motion and the rate of strain accumulation decreases (unless, of course, the differential stress is increased).

Let's look at dislocation interaction. Figure 9.20 shows a situation where one edge dislocation (D_1) moves relative to another edge dislocation (D_2) with a different slip plane (initially we keep dislocation D_2 stationary merely for the convenience of illustrating our point). As D_1 passes through D_2, the dislocation line l_2 is offset (Figure 9.20b). This offset is called a *jog* and has an important implication. The Burgers vector b_2 for dislocation D_2 remains the same along the dislocation line, but the slip plane has changed at the jog. For dislocation D_2 to move, it not only needs a critical resolved shear stress (CRSS) that allows movement on the initial slip plane, but also movement on a second slip plane for that same dislocation. Typically, the values of the CRSS differ for crystal planes in different orientation. Thus, the ability for glide along the dislocation line varies when a jog is present, and the result is that the dislocation is held back by the jog (Figure 9.20c). This reduced ability of dislocation to move is what causes the material to strengthen and results in work hardening. Diffusion of vacancies to the dislocation can overcome the restriction, so work hardening is much less important in the high-temperature creep regime. The presence of impurities that pin dislocations or high dislocation densities that restrict dislocation motion (tangles) result in work hardening of the material, which can be overcome by the activity of dislocation climb. We also noticed *work softening* in some of our previous experiments (Section 5.4.6). But we wait until Section 9.9.2, in which we discuss effective grain-size reduction of material during deformation, to give an explanation of this effect.

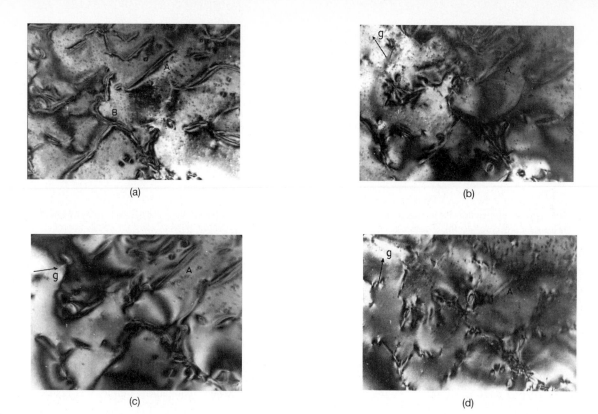

Figure 9.21 Dislocations in calcite (a) and determination of the Burgers vector using TEM. View of the same area for different diffracting lattice planes: (0006) (b), (10$\bar{1}$2) (c), (10$\bar{1}$4) (d); the orientation of the lattice planes in each image is indicated by its pole (g). The presence of dislocation A in (b) and (c) rules out all possible Burgers vectors in calcite with the exception of ($\bar{2}$021); this is confirmed by the absence of contrast from A in (d). This (time-consuming) procedure is called the *invisibility criterion*. The Burgers vector is not perpendicular nor parallel to the dislocation line, so dislocation A is a mixed dislocation. Width of view in each TEM image is ~ 1.7 μm.

9.8 IMAGING DISLOCATIONS

In Section 9.5.6 we learned that the dislocation density of an unstrained crystal is on the order of 10^6 cm^{-2}, and that this number is orders of magnitude higher in strained crystals. So, a 1 cm^3 volume of a strained quartz crystal with a dislocation density of 10^{10} cm^{-2} will have a total dislocation line length of 10^{10} cm, or 100,000 km (or 2.5 × Earth's circumference!). Obviously, dislocations must be quite small to fit so many in this small a volume. Thus we need very large magnifications to image them. Transmission electron microscopy (TEM) is generally used for imaging and analysis of crystal defects. This technique permits imaging of microstructures at magnifications of up to 500,000×, with a resolution better than 1 nm (1 nanometer = 10^{-9} m[6]). However, such very high resolution is not usually necessary for the examination of defect microstructures, and more conventional TEM examination is done at magnifications of 10,000–100,000× (Figure 9.21). TEM samples are prepared by processes that provide sufficient thinning of the material so that it is transparent to the electron beam; generally, the thickness of the thin foil is a few hundred nm. Crystal defects in thin foils are revealed by diffraction contrasts that result from lattice distortions surrounding the defect. Analysis of such images allows us to determine both the Burgers vector of a dislocation and the crystallographic orientation of the dislocation line (Figure 9.21). Once these are established, it is possible to determine the nature of a dislocation, as being an edge, screw, or mixed, by the angular relationship between **b** and **l**, recalling that for edge dislocations **b** and **l** are perpendicular, and for screw dislocations **b** and **l** are parallel.

In Figure 9.7 we were able to see dislocations in olivine using the standard petrographic microscope and a decoration technique. For comparison, a variety of dislocation geometries in olivine in the transmission electron micrograph are shown in Figure 9.22. We can distinguish arrays of parallel dislocations (lower right), straight dislocations (upper half), and dislocation loops (lower left). The terminations of the dislocations that you see in this photomicrograph arise from the intersection of the dislocation line with the lower and upper boundaries of the thin foil; the thicker the foil, the longer the dislocation will appear until it intersects the crystal edge. Note the geometric similarity between the optical and transmission electron micrographs of Figures 9.7 and 9.22, which both contain straight dislocation lines with sharp angular bends, as predicted by slip systems in olivine (Table 9.1).

---

[6]In TEM study, the non-SI unit Ångstrom continues to be used: 1Å = 10^{-10}m = 10 nm.

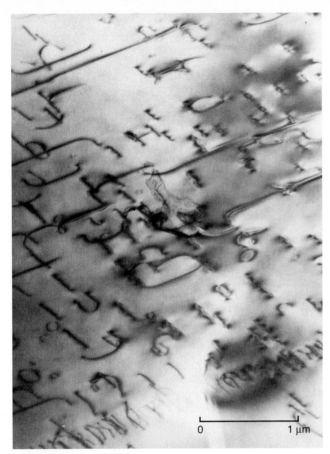

Figure 9.22 Transmission electron micrograph showing dislocation lines, loops, and arrays in experimentally deformed olivine.

9.9 DEFORMATION MICROSTRUCTURES

Now that we have established the various deformation mechanisms by which rocks deform, the critical question arises: can we recognize the (past) activity of a particular deformation mechanism in a rock and, by inference, determine the rheologic conditions during deformation? The answer to this question is often yes, because most deformation mechanisms produce a relatively characteristic microstructure that can be observed in thin section. However, this 'memory' may be incomplete; only the latest deformation mechanism is preserved. Once we have established the operative deformation mechanism from the characteristic microstructure, we can then proceed to make predictions about the past conditions of temperature, stress, and strain rate during rock deformation, which is the ultimate reason to examine these structures. Throughout the book we use the term *microstructure* to describe geometric characteristics of rocks on the scale of the thin section. For example, twins are a microstructural element. To avoid confusion, we will use the term *(micro)fabric,* which means different things to different people, only with an appropriate modifier (e.g., as dimensional-preferred fabric for geometric alignments). In Chapter 12 we will discuss another type of fabric, *crystallographic-preferred fabric,* that describes the degree of crystal lattice orientation of a mineral aggregate.

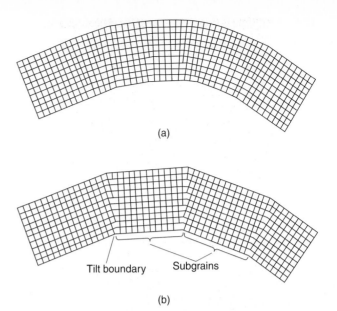

Figure 9.23 Irregularly distributed dislocations (a) are rearranged by glide and climb to form a dislocation wall or tilt boundary that separates subgrains (b).

For the next several pages we will look at the most characteristic microstructures in deformed rocks that arise from three mechanisms: recovery, recrystallization, and superplastic creep. Mechanical twinning, a fourth mechanism, was already discussed (Section 9.5.3). To guide you through the many new concepts that will be introduced here, brief descriptions of the processes, characteristic microstructures, and some related terms of crystal plastic and diffusional creep are given in Table 9.3.

9.9.1 Recovery

The presence of crystal defects such as dislocations and twins increases the *internal strain energy* of a grain, because the crystal lattice surrounding a defect is distorted. This means that the atomic bonds are bent and stretched (explaining the "strain" in strain energy), and thus that the crystal lattice is not in its lowest energy state. This internal strain energy is also called *stored strain energy.* The process of dislocation creep lowers the internal strain energy by annihilation and/or moving the dislocations to the edge of crystals, so the internal strain can be minimized. However, this does not mean that internal strain is recoverable (as in elastic), because permanent distortions are produced when dislocations move through the crystal (recall Figure 9.12). Another way to reduce the overall internal strain energy of a grain is by the localization of crystal defects. As a result of climb, cross-slip, and glide, dislocations can be arranged into a zone of dislocation called a *dislocation wall* or *tilt boundary.* Such tilt boundaries represent a lower strain energy state than if the dislocations were more evenly distributed across the grain (Figure 9.23). An individual dislocation produces only a very small crystallographic distortion that is not visible optically, but an array of dislocations in a tilt boundary makes this mismatch

Table 9.3 Some Terms and Concepts Related to Crystal Plasticity and Diffusional Creep

Annealing	Loosely used term for high temperature grain adjustments, including static recrystallization and grain growth.
Bulge nucleation	A type of migration recrystallization in which a grain boundary bulges into a grain with higher internal strain energy, forming a recrystallized grain.
Dislocation wall	Concentration of dislocations in a planar array.
Dynamic recrystallization	Formation of relatively low-strain grains under an applied differential stress.
Foam structure	Recrystallized grain structure characterized by the presence of energetically favorable grain-boundary triple junction (at $\approx 120°$ angles).
High-angle boundary	Boundary across which the crystallographic mismatch exceeds 10°; characteristic of recrystallization.
Low-angle boundary	Tilt boundary across which the crystallographic mismatch is less than 10°; characteristic of recovery.
Migration recrystallization	Recrystallization mechanism by which grain boundaries move driven by a contrast in strain energy between neighboring grains.
Polygonized microstructure	Elongate to blocky subgrains within grains (mostly used for phyllosilicates).
Recovery	Process that forms low-angle grain boundaries by the temperature-activated rearrangement of dislocations.
Recrystallization	Mechanism that removes internal strain energy of grains remaining after recovery, producing high-angle grain boundaries that separate relatively strain-free (recrystallized) grains.
Recrystallized grains	Relatively low-strain grains that are formed by recrystallization.
Rotation recrystallization	Recrystallization mechanism by which dislocations pile up in a tilt boundary, thereby 'rotating' the crystal lattice of the area that is enclosed by the tilt boundary.
Static recrystallization	Formation of strain-free grains after deformation has stopped (i.e., differential stress is removed).
Subgrain	Area of crystallographic mismatch with host grain that is less than 10°.
Subgrain rotation	Rotation recrystallization mechanism by which dislocations continue to move into a low-angle tilt boundary surrounding a subgrain, thereby increasing the crystallographic mismatch and forming a high-angle grain boundary.
Superplastic creep	Grain-size sensitive deformation mechanism by which grains are able to slide past one another without friction because of the activity of diffusion (as opposed to frictional sliding or cataclasis).
Tilt boundary	Concentration of dislocations in a planar array.
Twinning	Deformation mechanism that rotates the crystal lattice over a discreet angle such that the twin boundary becomes a crystallographic mirror plane. Such a planar defect is produced by the motion of partial dislocations.
Undulose extinction	Irregular distribution of dislocations in a grain, producing small crystallographic mismatches or lattice bending that is visible under crossed polarizers.

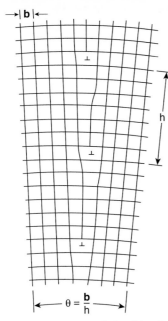

Figure 9.24 A tilt boundary composed of edge dislocations at a distance h apart in a simple cubic lattice. The crystal lattice across the boundary does not have the same orientation, but is rotated over an angle θ (in radians) = **b**/h, with **b** the Burgers vector and h the spacing of dislocations in the tilt wall.

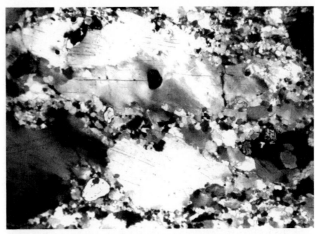

Figure 9.25 Subgrain microstructure and undulose extinction in a marble mylonite from southern Ontario. Width of view is 4 mm.

Figure 9.26 Recrystallization microstructure, showing relatively strain-free grains with straight grain boundaries. This image represents the most deformed stage in a mylonite from southern Ontario. Width of view is 2 mm.

optically visible (Figure 9.24). The greater the number of dislocations in the wall, that is, the closer their spacing, the greater the mismatch. The angular mismatch, θ, arising from a tilt boundary is a function of the length of the Burgers vector (**b**) of a dislocation and the spacing (h) of individual dislocations in the wall:

$$2\sin(\theta/2) = \mathbf{b}/h \qquad \text{Eq. 9.16}$$

or for small angles of θ (in radians):

$$\theta = \mathbf{b}/h \qquad \text{Eq. 9.17}$$

The area of a crystal that is enclosed by a low-angle tilt boundary is called a *subgrain*. The convention to distinguish between low-angle (subgrains) and high-angle boundaries (recrystallized grains; discussed below) is that the angular difference across the tilt boundary is less than 10°. We can estimate the number of dislocations in a tilt wall 500 μm long by 2 nm wide by considering Equations 9.12 and 9.13 and assuming a Burgers vector of 0.5 nm length and an angular mismatch θ of 10°. This says that the dislocation spacing is approximately 2.9 nm and thus that there are more than 170,000 (!) dislocations in the 500-μm-long low-angle tilt boundary. This represents a dislocation density in the 1×10^{-8} cm^2 area of the low-angle tilt wall of 1.7×10^{13} cm^{-2}.

In thin section, under cross-polarized light, undulatory extinction is a manifestation of the crystallographic mismatch that results from subgrain formation. It is particularly common in the minerals calcite, quartz, olivine, and pyroxene (Figure 9.25). *Recovery* is the name of the process that forms low-angle grain boundaries by the temperature-activated rearrangement of dislocations, producing the characteristic deformation microstructure of subgrains. In the case of phyllosilicates such as micas, the presence of subgrains is sometimes called a *polygonized microstructure,* which describes the characteristic archlike geometry where each segment is oriented slightly different from the next. Experiments in which recovery dominates have shown that the stress function of the associated flow law is exponential (Equation 9.13). Materials scientists, therefore, also use the term *exponential-law creep* for this regime of deformation.

9.9.2 Recrystallization

The process that removes the internal strain energy of grains that remains after recovery is called *recrystallization.* During recrystallization, *high-angle grain boundaries* are formed that separate relatively strain-free grains from each other. In rocks, a recrystallized microstructure is characterized by grains without undulous extinction and with relatively straight grain boundaries that meet at angles of approximately 120° (Figure 9.26). An everyday example of

Figure 9.27 Microstructure of a mylonite. Note the fine-grained, quartz-rich matrix that surrounds relatively rigid feldspar clasts. Width of view is ~1 cm.

this structure is found in the foam of soap. Looking closely at foam when you are doing the dishes or washing your hair, you will see all the characteristics of a recrystallized microstructure. In fact, similar energy considerations are involved with the formation of the microstructure of foam;[7] perhaps appropriately, we call the geometrically similar structure in rock a *foam microstructure.*

Recrystallization that occurs in the presence of a differential stress (i.e., during deformation) is called *dynamic recrystallization.* Dynamic recrystallization generally produces grain-size reduction, which is well known in sheared rocks (e.g., mylonites; Figure 9.27). We will return in much more detail to mylonites in Chapter 11, but at this point we note that they are characterized by a grain size that is smaller than that of the host rock from which they were formed. In fact, the term *mylonite* is unfortunate for these microstructures, as it derives from the Greek word "*mylos,*" meaning milling. At the time of their discovery in northern Scotland by Sir Charles Lapworth (late nineteenth century; Chapter 12), it was thought that they were formed by a grinding process (which we now call *cataclasis;* Section 6.5.3). Since their original discovery we have learned that this is incorrect, and that dynamic recrystallization is responsible for the grain-size reduction; nonetheless the name "mylonite" has persisted. Dynamic recrystallization can be used as a qualitative indicator of the temperature conditions during deformation, for example, >300°C for calcite, >350°C for quartz, and >450°C for feldspar. These estimates are based on experimental work, but they agree well with field observations.

Recrystallization that occurs after the differential stress is removed (i.e., after deformation has stopped) is called *static recrystallization,* or more loosely, *annealing.*

From a microstructural perspective, its only distinction from dynamic recrystallization is a relatively larger recrystallized grain size. Static recrystallization reduces the internal strain energy by the formation of relatively large, strain-free grains that grow to decrease the total free energy of the rock.[8]

A final comment about the use of the term *recrystallization* before we turn to the operative mechanisms. Recrystallization as we have discussed so far involves changes in the strain energy of a *single phase,* whereas recrystallization in petrology typically involves a *phase change.* Thus, petrologists consider the process to be governed by chemical potentials rather than by the strain potentials that are implied in our usage. Be sure not to confuse these different meanings of the word recrystallization in geology.

9.9.3 Mechanisms of Recrystallization

There are two main mechanisms by which recrystallization occurs: (1) rotation recrystallization and (2) migration recrystallization. *Rotation recrystallization* results from the progressive misorientation of a subgrain as more dislocations move into the tilt boundary, thereby increasing the crystallographic mismatch across the boundary. This produces a high-angle grain boundary without appreciable migration of the original subgrain boundary (Figure 9.28a). Eventually, the subgrains become so misoriented that individual grains can be recognized. Remember that the progressive rotation of the subgrain occurs only by adding more dislocations in the boundary and that there is never loss of cohesion with the crystal lattice of the host grain. The convention that we previously used to distinguish subgrains (low-angle grain boundaries) from recrystallized grains (high-angle grain boundaries) is an angle of 10°. This is a somewhat arbitrary convention, and indeed we often see a progression from low-angle grain boundaries to high-angle grain boundaries in rocks. Recrystallized grains are best developed where large strain gradients occur, such as at grain boundaries. The common microstructure of a change from relatively deformation-free grain interiors to subgrains to recrystallized grains at grain boundaries (Figure 9.29) is called a *core-mantle structure,* or *mortar structure.* Rotation recrystallization has been observed in such common rock-forming minerals as calcite, quartz, halite, and olivine.

Migration recrystallization is a process by which a grain grows, at the expense of its neighbor, when the grain's boundary effectively moves through its neighbor. The grain that grows has a lower dislocation density and the grain that is consumed has a higher dislocation density. Let's consider an example in which the boundary of grain A migrates into grain B. Keep in mind that a grain boundary separates two crystals whose lattices are not parallel. Migration happens when the atoms in grain B near the

---

[7]In foam, however, surface energy is most important, whereas internal strain energy is more important in rocks.

[8]This is sometimes called *secondary* or *exaggerated grain growth.*

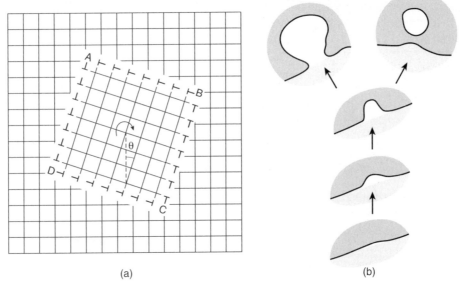

(a)   (b)

Figure 9.28 (a) Recrystallization by subgrain rotation and (b) bulge nucleation. In (a), a portion of a crystal is bounded by four tilt boundaries (ABCD) and rotation by adding more dislocations of the same sign leads to a progressively greater misorientation, that is, a recrystallized grain. In (b), growth of dynamically recrystallized grains occurs by bulge nucleation of the grain boundary into a grain with higher internal strain energy, leaving behind a relatively strain-free region that eventually develops into a recrystallized grain.

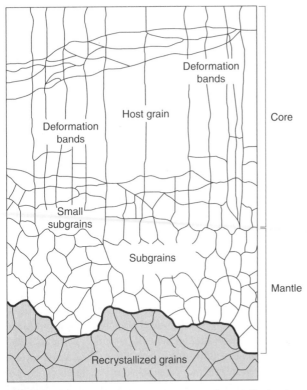

Figure 9.29 Core-and-mantle or mortar structure. Recrystallized grains occur at the edge of the mantle by the progressive misorientation of subgrains (mantle). The internal portion of the host grain (core) shows weak deformation features such as undulatory extinction or deformation bands, or may even be strain free. The microstructure is based on observations in quartz aggregated, with the recrystallized grain size on the order of tens μm.

grain boundary rearrange so they fit into the lattice of the crystal with low dislocation density. As soon as this happens, the atoms are considered to be part of grain A. It's easier to rearrange atoms in grain B; because of its higher dislocation density, bonds in B are already stretched and misoriented. As the grain with lower dislocation density, grows at the expense of the grain with higher dislocation density, the overall strain energy of the system decreases. Typically, where migration recrystallization occurs, the boundary of the grain with lower dislocation density bulges into the grain with higher dislocation density (Figure 9.28b). Thus, this process has been called *bulge nucleation*. If the new grains become deformed as they grow, their growth may cease. In nature, quartz, halite, and feldspar commonly recrystallize by bulge nucleation.

Materials may recrystallize by both rotation recrystallization (subgrain rotation) and migration recrystallization (bulge nucleation). The dominance of a particular mechanism is largely a function of strain rate. Consider this: if you are in a hurry to get somewhere, you will try to take the fastest means of transportation. Similarly, nature will use the mechanism that produces the fastest strain rate to reduce the internal strain energy of the system. For recrystallization, bulge nucleation is favored at higher strain rates; moreover, bulge nucleation is favored at high temperatures. For both these recrystallization mechanisms, the recrystallized grain size is inversely proportional to the strain rate. For example, the smaller recrystallized grain size in rocks in mylonite zones is an expression of strain rate increase. Previously we called strain-rate increases at constant stress *work softening* (Chapter 5), a process which may be caused by grain-size reduction during deformation.

Table 9.4 Empirically-Derived Parameters for Recrystallized Grain Size–Differential Stress Relationships

Mineral	A (in MPa)	i (with d in mm)
Calcite	467	1.01
Quartz	381	0.71
Quartz ('wet')	4090	1.11
Olivine	4808	0.79

Sources: Mercier et al., (1977); Ross et al., (1980); Schmid et al., (1980); Ord and Christie, (1984).

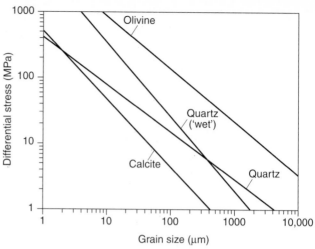

Figure 9.30 Empirically derived recrystallized grain size versus differential stress relationships for calcite, quartz, 'wet' quartz, and olivine using the parameters listed in Table 9.4. Note that we plot log σ_b versus log d, so that small shifts in the position of each curve reflect large changes in ambient conditions.

The formation of recrystallized grains is driven by the generation and motion of dislocations, which in turn is driven by differential stress. So you may expect that a relationship exists between recrystallized grain size and differential stress magnitude. Indeed, experiments have shown that a characteristic grain size occurs for a specific condition of stress and mechanism of recrystallization. This means that we can estimate paleostress during deformation if we know the operative recrystallization mechanisms; that is, recrystallized grain size can be a *paleopiezometer* (derived from the Greek "piezo," meaning "to press").[9] This is potentially a very powerful tool for understanding deformation, because paleostress is a difficult parameter to extract from rocks. The debate about this relationship is not settled, but it is clear that recrystallized grain size is inversely proportional to differential stress magnitude:

$$\sigma_d = Ad^{-i} \qquad \text{Eq. 9.18}$$

where A and i are empirically derived parameters for each mineral, and d is grain size in μm. To give you a rough idea of these relationships and the differential magnitudes stress, we give some representative parameters for three common minerals; calcite, quartz, and olivine, in Table 9.4. These data are plotted in Figure 9.30, but remember that considerable uncertainty surrounds these paleopiezometers.

Overall, a small recrystallized grain size in a deformed rock, say a quartzite, reflects a high strain rate, a high differential stress magnitude, or a combination of both. So returning to the flow laws, rock experiments have shown that the corresponding stress function during recrystallization has the form: $f(\sigma) = \sigma^n$ (Equation 9.14). The value of n, the *stress component,* varies, but typically lies in the range 2–5 for common minerals in monomineralic rocks (see Table 5.6). In materials science, this deformation regime dominated by recrystallization is therefore also called the *power-law creep* regime.

9.9.4 Superplastic Creep

Superplastic creep, which is more completely described by the somewhat cumbersome name *grain-boundary sliding superplasticity* (GBSS), may at first seem a bit out of place after the description of dislocation creep mechanisms, because it returns us to the topic of diffusion. This mechanism is kept to the last because it is the least understood deformation mechanism and it occurs at relatively high temperatures. We look at its characteristics first. Most importantly, superplastic creep is a grain-size sensitive deformation mechanism in which grains literally slide past one another during deformation. Aside from the part on grain-size dependence, this description sounds a lot like that for cataclastic flow, but with the important difference that volume and grain boundary diffusion are efficient enough to keep gaps or voids from forming between moving grains, and therefore grains are able to slide *without friction*. Deformation is accommodated by neighbor switching, as illustrated in Figure 9.31, and superplastic creep can produce very large strains (>1000%) without showing appreciable deformation of the grains. The original definition of superplasticity is, in fact, this ability of rocks to accumulate very large strains without mesoscopic breaking, but we emphasize the mechanism by which it occurs rather than the phenomenon only. Even after large finite strains, grains are equiaxial and 'fresh-looking' and show no preferred elongation. This mechanism is only active in materials with very small grain sizes (5 to 15 μm) to facilitate diffusion. In this context, recall Equations 9.5 and 9.6, which showed the inverse exponential proportionality of diffusion to grain size.

---

[9]Piezometers using free dislocation density and subgrain size also exist, but these data are more difficult to obtain and the methods appear less reliable.

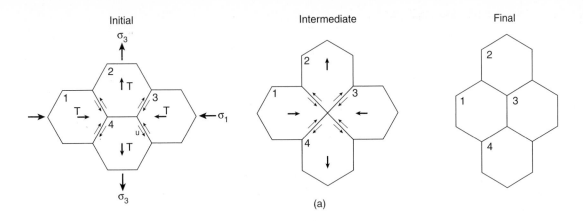

Initial · Intermediate · Final

(a)

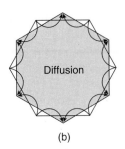

Diffusion

(b)

Figure 9.31 Grain boundary sliding superplasticity or superplastic creep. In (a), neighbor switching in the superplastic regime is shown; a group of four grains suffers approximately 55% strain without appreciable deformation of the grains except locally to accommodate grain boundary sliding (small arrows). The required accommodation of local strain by diffusion is shown in (b).

So far, superplastic creep has been proposed as a natural deformation mechanism for fine-grained calcite- and quartz-rich rocks. The very high temperatures that occur in the upper mantle may allow this mechanism to be active in coarse-grained olivine. Superplastic creep takes place at lower differential stresses than dislocation creep, but it occurs mainly in rocks with a small grain size. Thus, a rock may initially deform by dynamic recrystallization until the grain size is sufficiently reduced for superplastic creep to occur. When this happens, the rock becomes much weaker; i.e., the stress necessary to produce strain decreases. This weakening, known as strain softening, is common in ductile shear zones.

The stress function of the corresponding flow law for superplasticity approaches linearity between strain rate and stress; that is, the stress component n of Equation 9.15 approaches 1. Consequently, the strain rate is inversely proportional to the grain size:

$$\dot{e} \propto d^{-r}$$

where r is 2 or 3 based on experimental work. Now, recalling that a linear relationship between strain rate and stress is linear viscous or Newtonian rheology (Chapter 5), superplastic creep with n = 1 is viscous behavior. This contrasts with dislocation creep, which typically has nonlinear rheology (n ≠ 1). To emphasize the grain-size dependence of superplastic creep, we sometimes call it *grain-size sensitive creep*.

9.10 DEFORMATION-MECHANISM MAPS

Quite an array of concepts and terms have passed before your eyes in this chapter. We therefore close by attempting to create some order in all this information, which may help you to remember the most important elements. The activity of the various ductile deformation mechanisms can be summarized in a single diagram that shows over what ranges of stress, strain rate, temperature, and grain size each mechanism dominates for a given material. Such diagrams are called *deformation mechanism maps* or *Ashby diagrams*.[10] The plotted parameters may be stress (e.g., differential stress), temperature, and grain size, but for comparison between different materials we generally use normalized parameters. A normalized physical quantity is the ratio between a variable and a material constant measured in the same units. In this case, stress is normalized to an elastic

---

[10]After the British material scientist Michael F. Ashby, who proposed these constructions in the early 1970s.

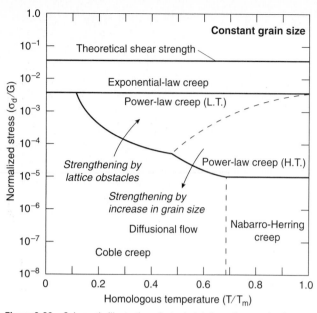

Figure 9.32 Schematic illustration of a typical deformation mechanism map showing normalized stress versus homologous temperature at constant grain size.

modulus of the material (typically the shear modulus, G), and temperature T (absolute temperature, in °K) is normalized to the absolute melting temperature T_m of the material (Figure 9.32), which we previously called the homologous temperature, T_h. On the deformation mechanism map, we plot lines of constant strain rate, as shown in Figures 9.33 and 9.34. Only a part of the diagram space is constructed from experimental results, and because most natural conditions are outside the range of laboratory experiments, we must extrapolate. This is comparatively easy where an essentially linear (Newtonian) relationship exists between ė and σ, such as for diffusional flow. For other regimes, such as those dominated by dislocation glide (exponential-law creep) and dislocation glide and climb (power-law creep), the extrapolation of nonlinear relationships derived from experimental conditions is more tenuous. Figures 9.33a and 9.34a show examples of deformation mechanism maps for two common crustal minerals, calcite and quartz; in Figures 9.33b and 9.34b, the pressure solution field ('wet' diffusion) has been added.

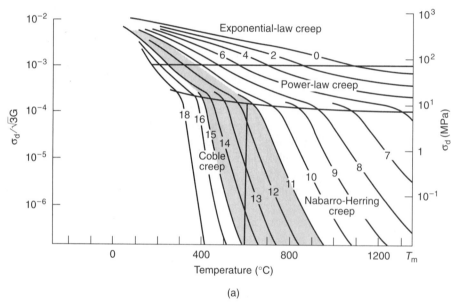

(a)

Figure 9.33 Deformation mechanism maps for calcite (a) without and (b) with a pressure solution field for a grain size of 100 μm. Contours of −log strain rate are shown; σ_d is differential stress; G is shear modulus; σ_d-scale at right is for a shear modulus G at 500°C. The undulation of strain rate contours in the pressure solution field arises from the competition between change of solubility of calcite and fluid concentration with pressure as temperature increases. The range of reasonable geologic strain rates (10^{-11}–10^{-15} sec^{-1} is shaded).

(b)

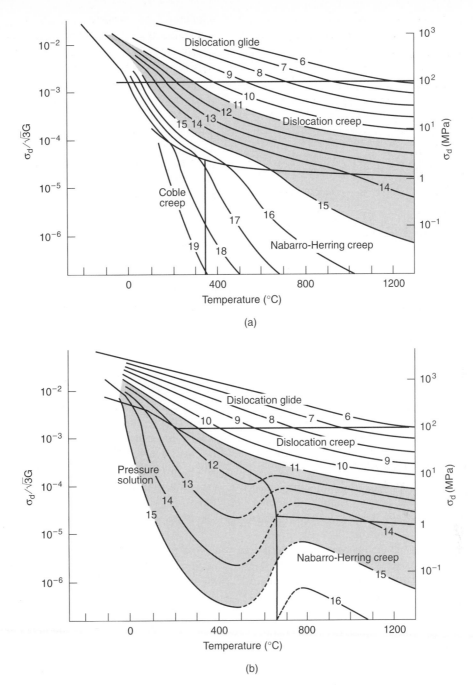

Figure 9.34 Deformation mechanism maps for (a) quartz without and (b) with a pressure solution field, for a grain size of 100 μm. Contours of −log strain rate are shown; σ_d is differential stress; G is shear modulus; σ_d-scale at right is for a shear modulus G at 900°C. The stress exponent, n, in the power law field is 4. The region of dashed strain-rate contours represents the inhibition of pressure solution through decrease in pore water concentration.

The meaning of the boundaries between the different fields on the deformation mechanism map are not all that easy to interpret, because deformation mechanisms do not change abruptly at this boundary. Several mechanisms will operate simultaneously, but the one that generates the highest strain rate will be the dominant deformation mechanism. Fields in deformation mechanism maps are defined by first calculating the strain rate for each flow law. Then the mechanism giving the fastest strain rate is considered representative for the field (i.e., the mechanism that domi-

nates the flow). For example, the field for dislocation creep represents the range of conditions for which dislocation motion creates a strain rate faster than any other mechanism, even though these other mechanisms may be operating. This means that at a boundary the two adjacent deformation mechanisms are equally important.[11] Let's

---

[11]To emphasize this aspect, deformation mechanism maps are also called *deformation-regime maps*.

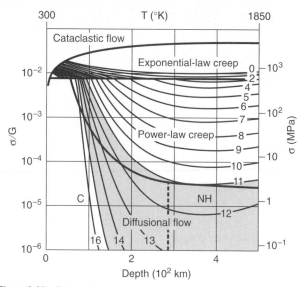

Figure 9.35 Deformation mechanism map for olivine with a grain size of 100 μm. Variables are the same as in Figure 9.32, except that depth is substituted for temperature given an exponentially decreasing geothermal gradient with 300°C at the surface and 1850°C at 500 km depth.

give a practical example. During mylonitization of a quartzite, dynamic recrystallization may dominate; yet some diffusional flow may occur simultaneously if, for example, the grain size and the strain rate are sufficiently small. Consequently, the map may indicate that we are in the power-law field, but we may also see some microstructural evidence for diffusional flow.

A general pattern is common to all deformation mechanism maps. We will take the classic example of the mineral olivine to illustrate this (Figure 9.35). Instead of the homologous temperature, we plot depth in the Earth in Figure 9.35, based on a thermal gradient that exponentially decreases from 300°K at the surface to 1850°K at 500-km depth. This enables us also to take into account the effects of pressure, which plays a role in the mantle by increasing the flow strength and decreasing the strain rate. From the map you see that cataclastic flow and even exponential-law creep are restricted to relatively large differential stresses (here ≈8 × 10² MPa), which means that these mechanisms are restricted to shallow crustal levels. As depth increases, we pass from exponential-law creep to power-law creep to diffusional creep given a constant geologic strain rate (say, 10⁻¹⁴/sec). In the latter regime, we may pass from grain-boundary diffusion (Coble creep) to volume diffusion (Nabarro-Herring creep), given further temperature or strain rate increase. The value of deformation mechanism maps lies in their ability to predict the mechanism that dominates under various natural conditions. In the absence of direct observations, we may predict the mechanism dominating the flow of rocks if the conditions of stress, T, and grain size are known. For example, if we assume that the Earth's upper mantle consists mainly of olivine, we can predict that at strain rates greater than 10⁻¹¹/sec, dislocation glide and climb dominate the

flow in the upper 100 km, given a grain size of 100 μm. If the strain rate is lower, diffusional creep will become increasingly important, especially if the grain size is small. The latter point, the effect of grain-size variation, is not clear from any of the deformation mechanism maps so far, because grain size was taken as a constant value in each (d = 100 μm). So how do we know what the role of grain-size variation is? Again consider the flow laws for diffusional creep (Equation 9.15), which states that strain rate is inversely proportional to the square or cube of the grain size. Thus, reducing grain size by, say, one order of magnitude (d = 10 μm) increases the strain rate by two to three orders of magnitude, which will move the field of geologic strain rates further into the regime of diffusional flow. Similarly, if we construct a map for a grain size of, say, 1 mm (1000 μm) or larger, the field of geologic strain rates moves into the regime of power-law creep. An estimate for the grain size of olivine in the upper mantle lies between 100 and 1000 μm. Observed microstructures of mantle rocks generally support the predictions based on olivine deformation mechanism maps.

9.10.1 How to Construct a Deformation Mechanism Map

The concept of deformation mechanism maps tends to be overwhelming, until you construct your own. So in Table 9.5 we list constitutive equations for various deformation mechanisms, based on experiments on natural limestones and marbles, to make our own map.

First, choose the axes for the plot. Let's plot differential stress versus temperature. Now calculate the corresponding strain rate from each of the four constitutive equations for a specific stress and temperature condition (i.e., a point in the diagram). You recall that the mechanism producing the fastest strain rate is dominant; so from the four solutions, the one giving the highest strain rate is dominant for that particular point. Using individual points is an unnecessarily slow and cumbersome approach. Instead, we calculate the stress-temperature curves at a given strain rate for each equation and plot these four curves in the diagram. However, some of these curves will intersect and the final strain rate curve is composed of the curves for which the differential stress is the smallest. When you use different strain rates, you will see that the position of intersection points changes. Also, the dominant deformation mechanism may change. So connecting these intersection points defines the boundary between fields. A worked-out example is given in Figure 9.36, in which differential stress is plotted as a function of grain size for T = 475°C. Obviously, you can vary any condition and calculate the corresponding map. Fairly simple spreadsheet calculations allow you to do this on a personal computer. This returns us to the main reason for construction deformation mechanism maps: they give a first-order understanding of the most likely deformation mechanism under given conditions of stress, temperature, grain size, and strain rate.

Table 9.5 Experimentally Derived Constitutive Equations Used for the Construction of Figure 9.36

Exponential-law creep:	$\dot{e} = 10^{5.8}\, e^{(-62,000/RT\, +\, \sigma/114)}$
Power-law creep regime a:	$\dot{e} = 10^{-5.5}\, e^{(-75,000/RT)}\, \sigma^{6.0}$
Power-law creep regime b:	$\dot{e} = 10^{3.8}\, e^{(-86,000/RT)}\, \sigma^{2.9}$
Superplastic creep:	$\dot{e} = 10^{5.0}\, e^{(-51,000/RT)}\, \sigma^{1.7}\, d^{-3}$

$\dot{e}$ = strain rate (sec^{-1})

σ = differential stress (bar)

T = absolute temperature (°K)

R = gas constant

d = grain size (μm)

Source: Rutter, (1974); Schmid et al., (1977); Schmid, (1982).

Figure 9.37 Marble bench that sagged with time under the influence of gravity.

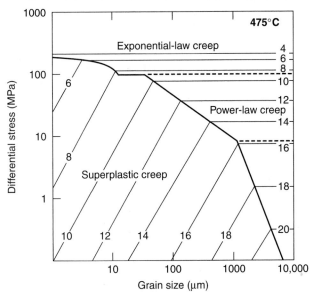

Figure 9.36 Deformation mechanism map for calcite at T = 475°C, constructed from the constitutive equations listed in Table 9.5. The thin lines represent strain rates (given as −log) whereas the thick lines separate deformation mechanism fields.

9.10.2 A Note of Caution

Deformation mechanism maps allow us to evaluate material behavior that is not just restricted to rocks. They also enable predictions about the creep of metal in a nuclear reactor and thereby aid their safe design, they determine the life span of lightbulb filaments, they explain the behavior of ice sheets, and provide such pivotal information as the sagging rate of marble benches (Figure 9.37). Clearly, deformation mechanism maps are a powerful tool to understand the rheology of materials and to apply this information to significant problems. However, they are not without their limitations. First of all, the extrapolation over several orders of magnitude that is needed to go from experimentally derived flow laws ($\dot{e} > 10^{-8}$/sec) to geologic conditions ($\dot{e} < 10^{-11}$/sec) is a major source of uncertainty. Secondly, the maps assume steady-state flow (i.e., stress is strain independent), which may or may not be relevant to geologic conditions. Moreover, as a microstructure evolves, it may affect the operative deformation mechanism. For example, dynamic recrystallization tends to reduce the grain size, which in turn may enhance the importance of diffusional creep and weaken the material. In spite of these considerations, deformation mechanism maps are a handy and powerful way to evaluate and predict deformation mechanism or ambient conditions for materials. The best test is a comparison of natural deformation structures with predictions based on deformation mechanism maps. Such tests indicate that these maps indeed provide reasonable estimates about the conditions of deformation. Now return to Figures 9.33–9.35, showing deformation mechanism maps for several common minerals, and use these maps to think about the interplay between deformation mechanism, strain rate, temperature, and stress and their corresponding microstructures.

9.11 CLOSING REMARKS

Modern structural geology interpretations are relying increasingly on a synthesis of microscopic and mesoscopic observations. Microstructures especially play a growing role in unraveling the deformation histories of rocks and regions. For example, mechanical twinning is used to unravel the early stress and strain history of fold-and-thrust belts, and other microstructures allow us to estimate the conditions of T, σ_d, $\dot{e}$ during deformation. One important product of crystal plastic deformation mechanisms, crystallographic-preferred fabrics, has not been addressed here. This will be introduced in Chapter 12, after we explore three common ductile structures in outcrop, namely, folds, foliations, and lineations.

ADDITIONAL READING

Ashby, M. F., and Verrall, R. A., 1978, *Micromechanisms of flow and fracture, and their relevance to the rheology of the upper mantle:* Philosophical Transactions of the Royal Society of London, series A, v. 288, p. 59–95.

Frost, H. J., and Ashby, M. F., 1982, *Deformation-mechanism maps. The plasticity and creep of metals and ceramics:* Oxford, Pergamon Press, 166 p.

Groshong, R. J., Jr., 1972, Strain calculated from twinning in calcite: *Geological Society of America Bulletin,* v. 83, p. 2025–2038.

Hayden, H. W., Moffatt, W. G., and Wulff, J., 1965, *Mechanical behavior:* The structure and properties of materials, v. III, New York, John Wiley and Sons, 247 p.

Hull, D., and Bacon, D. J., 1984, *Introduction to dislocations:* Oxford, Pergamon Press, 257 p.

Loretto, M. H., and Smallman, R. E., 1975, *Defect analysis in electron microscopy:* London, Chapman and Hall, 134 p.

Nicolas, A., and Poirier, J.-P., 1976, *Crystalline plasticity and solid state flow in metamorphic rocks:* London, John Wiley and Sons, 444 p.

Passchier, C. W., and Trouw, R. A. J., 1996, *Microtectonics:* Berlin, Springer, 289 p.

Poirier, J.-P., 1985, Creep in crystals: *High-temperature deformation processes in metals, ceramics and minerals:* Cambridge, Cambridge University Press, 260 p.

Rutter, E. H., 1974, The influence of temperature, strain rate and interstitial water in the experimental deformation of calcite rocks: *Tectonophysics,* v. 22, p. 311–334.

Schmid, S. M., Boland, J. N., and Paterson, M. S., 1977, Superplastic flow in finegrained limestone: *Tectonophysics,* v. 43, p. 257–291.

Schmid, S. M., Paterson, M. S., and Boland, J. N., 1980, High temperature flow and dynamic recrystallization in Carrara marble: *Tectonophysics* v. 65, p. 245–280.

Twiss, R. J., 1986, Variable sensitivity piezometric equations for dislocation density and subgrain diameter and their relevance to quartz and olivine, in Hobbs, B. E., and Heard, H. C., eds., *Mineral and rock deformation: laboratory studies* (The Paterson Volume), American Geophysical Union, Geophysical Monograph 36, p. 247–261 .

Urai, J. L., Means, W. D., and Lister, G. S., 1986, Dynamic recrystallization of minerals, in Hobbs, B. E., and Heard, H. C., eds., *Mineral and rock deformation: laboratory studies* (The Paterson Volume), American Geophysical Union, Geophysical Monograph 36, p. 161–200.

Wenk, H.-R., (ed.), 1985, *Preferred orientation in deformed metals and rocks: An introduction to modern texture analysis:* Orlando, Academic Press, 610 p.

White, S., 1976, *The effects of strain on the microstructures, fabrics, and deformation mechanisms in quartzites:* Philosophical Transactions of the Royal Society of London, series A, v. 283, p. 69–86.

APPENDIX: DISLOCATION DECORATION

The principle behind the dislocation decoration technique of olivine is that iron oxides preferentially precipitate along lattice dislocations in Fe-rich olivines. In order to decorate dislocations in olivine, a sample with one polished surface is heated in air for approximately one hour at 900°C. A standard petrographic thin section is prepared, with the previously polished surface in contact with the glass slide. For most crystallographic directions, dislocation lines as far as 50 μm from the polished surface are decorated. Under the optical microscope, screw dislocations generally appear as long and straight lines. Using a Universal Stage the crystallographic relationship can be determined. Decoration is most effective in samples with relatively low dislocation densities ($<10^8$/cm^2), such as occurs in mantle xenoliths in volcanic flows (Figure 9.7).

Chapter 10
Folds and Folding

10.1 INTRODUCTION

Ask structural geologists, or other geologists for that matter, about their favorite structure and chances are that they will choose folds. If you have seen a fold in the field, you will have marveled at its appearance. Let's face it, it is pretty unbelievable that 'hard rocks' are able to change shape in such a dramatic way (Figure 10.1). In simple terms, a fold is a structural feature that is formed when planar surfaces are bent or curved. In fact, if such surfaces are

not available (e.g., bedding, cleavage, inclusions), you will not see folds even though the rock is deformed. Folding is a manifestation of ductile deformation because it can develop without fracturing, and the deformation is distributed over the entire structure. Some fracturing may occur in places, but processes such as grain-boundary sliding, kinking, dissolution, and crystal plasticity are the dominant deformation mechanisms yielding folding. Looking at a fold from a kinematic perspective, you may realize that the strain in this structure cannot be the same everywhere. We recognize distinct sections in a fold, such as the hinge area and the limbs, the inner arc, and the outer arc, each of which reflect different strain histories. Although much of our discussion in this chapter focuses on mesoscale folds, you will find that folds occur on all scales, from mm-size or smaller microfolds in thin sections to single large folds that occupy entire mountains (Figure 10.1).

Why do folds exist, how do folds develop, and what does folding mean for regional analysis? Well, these and many other questions were first asked quite some time ago, and much of what we know today about folds and folding was established well before the 1980s. The geometry of folds tells us something, for example, about the conditions of stress and strain, which in turn provides critical information about the deformation history of a deformed area. Much of the work on folds in recent years presents refinements of some of this earlier work. In particular, we use more sophisticated numerical and experimental approaches. Yet the fundamental observations remain essentially intact. Therefore we will mainly look at some of these first principles of folding in this chapter, and their application in

Figure 10.1 Large-scale recumbent fold in the Caledonides of northeast Greenland. The height of the cliff is ca. 800 m and the view is to the northwest.

structural analysis. First, however, we discuss the basic vocabulary needed to communicate information about folds and fold systems.

10.2 ANATOMY OF A FOLDED SURFACE

The schematic illustration in Figure 10.2 shows the basic geometric elements of a fold. First, we recognize the hinge area and the limbs of a fold. The *hinge area* is the region of greatest curvature and separates the two limbs. The line on the folded surface when curvature is greatest is called the *hinge line*. You may think of a *limb* as the less curved portion of a fold. In a limb the point where the sense of curvature changes is called the *inflection point.* In Figure 10.3a the hinge line of the fold is straight. Folds with a straight hinge line are called *cylindrical folds* when the folded surface can wrap partway around a cylinder. If this is not the case and the hinge line curves, the folds are called *noncylindrical* (Figure 10.3b). In reality, the extent of

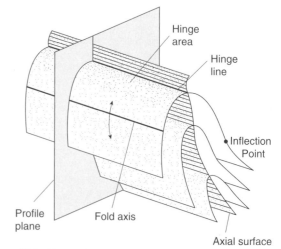

Figure 10.2 The terminology of a fold.

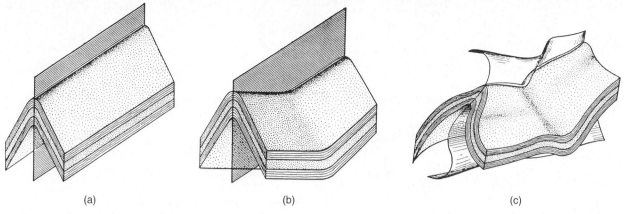

Figure 10.3 A cylindrical fold (a) is characterized by a straight hinge line and a noncylindrical fold, and (b) characterized by a curved hinge line. The axial surface is planar in (a) and (b), but it may also be curved (c).

cylindrical folds is restricted to the outcrop scale or even less. Over any distance, the hinge line of folds typically curves. Nevertheless, you will find that we may often treat natural folds as cylindrical by dividing them into segments with straight hinge lines.

A cylindrical surface consists of an infinite number of lines that are parallel to an imaginary generator line. This generator line is called the *fold axis*, which, when moved parallel to itself through space, outlines the folded surface. In the case of cylindrical folds, the fold axis is of course parallel to the hinge line.[1] The topographically highest and lowest points of a fold are called the *crest* and *trough*, respectively, and do not always coincide with hinge lines. The surface containing the hinge lines from consecutive folded surfaces in a fold is the *axial surface* (Figures 10.2 and 10.3). The term *axial plane* is loosely used by some, but the surface is not necessarily planar, as seen in Figure 10.3c. Moreover, the axial surface does not always divide the fold into equal halves that are mirror images of each other (see Section 10.4.2). The reference plane used to describe fold shape is called the *fold profile plane*, which is perpendicular to the hinge line (Figure 10.2). Note that the profile plane is not necessarily the same as a cross section through the fold; a cross section is any vertical plane through a body, much like the sides of a slice of layered cake. Therefore, if the hinge line is not horizontal, the profile plane is not a cross-section plane.

The angle between fold limbs in the profile plane is called the *interlimb angle* (Figure 10.4). Intuitively you may realize that the interlimb angle gives a qualitative estimate of the degree or intensity of folding; the smaller the

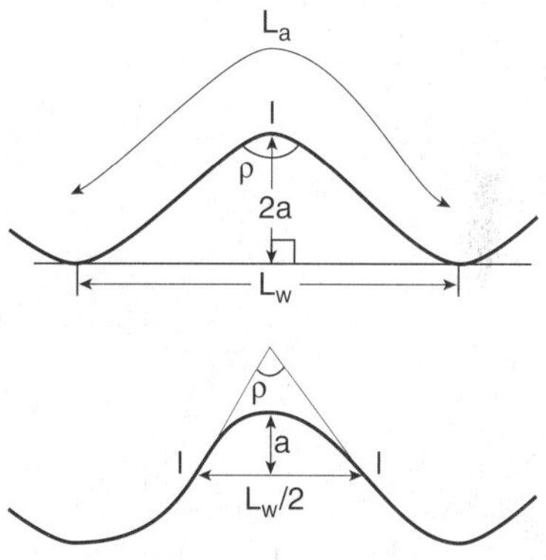

Figure 10.4 The interlimb angle (ρ), the wavelength (L_w), the amplitude (a), and the arc length (L_a) of a fold system in profile.

interlimb angle, the greater the intensity of folding. Finally, we recognize the *amplitude, wavelength,* and *arc length* of a fold system in profile. These terms are used in the same manner as they are in wave physics, because to some extent, fold systems look a bit like harmonic functions such as a sine curve. The wavelength is defined as the distance between the hinges of successive folds whereas the arc length is this distance measured over the folded surface. The amplitude is half the height of the structure measured from crest to trough (Figure 10.4). These and some other terms associated with folds are summarized in Table 10.1.

When successive layers in a folded sequence have approximately the same wavelength and amplitude, the folds are called *harmonic*. If some layers have distinctly different wavelengths and/or amplitudes, the folds are *disharmonic* (Figure 10.5). In extreme circumstances, a series of folded layers may be totally decoupled from unfolded layers above or below. When this happens, a *detachment* horizon exists between folded and unfolded layers.

---

[1]Sometimes "fold axis" is used as a synonym for hinge line, but this is not strictly correct.

Figure 10.5 Small-scale disharmonic folds in anhydrite of the Permian Castile Formation in the Delaware Basin of Texas. White layers are anhydrite; dark layers consist of calcite that is rich in organic material (hence the dark color). Note that detachments occur in the organic-rich calcite layers and that the different fold shapes in the anhydrite vary as a function of layer thickness.

Table 10.1   Vocabulary of a Fold

Amplitude	The height of the structure measured from crest to trough.
Arc length	The distance between two adjacent fold hinges of the same orientation as measured along the folded surface.
Axial surface	The surface containing the hinge lines from consecutive folded surfaces.
Crest	The topographically highest line on a folded surface, which need not coincide with the fold hinge.
Culmination	High point of the hinge line in a noncylindrical fold.
Cylindrical fold	Fold in which a straight hinge line parallels the fold axis; in other words, the folded surface wraps partway around a cylinder. Cylindrical folds can be generated by moving the fold axis parallel to itself.
Depression	Low point of the hinge line in a noncylindrical fold.
Fold axis	Fold generator in cylindrical folds.
Hinge	The region of greatest curvature in a fold.
Hinge line	The line of greatest curvature.
Inflection line	The line on a limb where the sense of curvature changes.
Limb	Lesser curved portion of a fold.
Median surface	The surface that passes through the inflection lines on opposing limbs.
Noncylindrical fold	Fold with a curved hinge line.
Profile plane	The surface perpendicular to the hinge line.
Trough	The topographically lowest point of a fold, which need not coincide with the fold hinge.
Wavelength	The distance between two hinges of the same orientation.

10.2.1   Fold Facing

Take a deep breath. We have already sprung a sizable array of terms on you, and before you can understand the significance of folding, we have to add a few more. Maybe you will find comfort in the knowledge that generations of students before you have plowed their way through this terminology, happily discovering that in the end it really is useful for the description and interpretation of regional deformation. Having said this, now draw a fold profile on a piece of paper. Chances are that you place the hinge area at the top of the structure, outlining something of an arch. In fact, a psychological study among geologists found this invariably to be true (just look at your neighbor's sketch).[2] This particular fold geometry is called an *antiform*. The opposite geometry, when the hinge zone is at the bottom

(outlining a valley), is called a *synform*. The explanation for the modifiers *anti* and *syn* is that the limbs dip away or toward the center of the fold, respectively. You will find that some geologists use anticline and syncline as synonyms for antiform and synform, respectively but this is incorrect. "Anticline" and "syncline" infer that the stratigraphic top or younging direction in the folded beds is known. This is an important distinction for regional analysis, so let's look at this in some detail.

Imagine a sequence of beds that is laid down in a basin over a period of 50 m.y. Obviously, the youngest bed lies at the top, while the oldest bed is at the bottom of the pile (this is Steno's Law of Superposition). When we fold this sequence into a series of antiforms and synforms, we see that the oldest bed lies in the core of the antiform and the

---

[2]This is not (yet) linked to any criminal behavior.

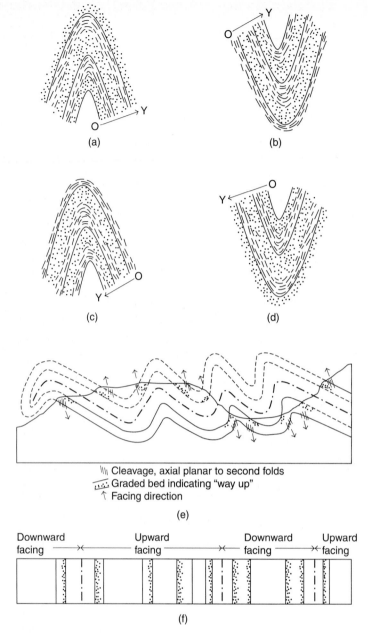

Cleavage, axial planar to second folds
Graded bed indicating "way up"
Facing direction

(e)

| Downward facing | Upward facing | Downward facing | Upward facing |

(f)

Figure 10.6 Antiforms, synforms, and fold facing. Upward-facing antiform (a) is also called an anticline, and an upward-facing synform (b) is called a syncline. Downward-facing antiforms (c) and downward-facing synforms reflect an early history that placed the beds upside down prior to folding, which may occur in a region containing two generations of folding (e). The corresponding facing in map view across this area is shown in (f).

youngest bed in the core of the synform (Figure 10.6a and b). Under these circumstances we call them anticlines and synclines, respectively. In an *anticline* the beds young away from the core (Figure 10.7); vice versa, in a *syncline* they young toward the core. In both cases the younging direction points (or faces) upward, so collectively we call them *upward-facing folds*. Now turn the original sequence upside down: the oldest bed is now at the top and the youngest bed at the bottom (Figure 10.6c and d). We generate the same geometry of antiforms and synforms, but in this case the younging direction is opposite to what we had before. In this antiform, the beds young toward the core, while in the synform the beds

young away from the core. In both cases they are *downward-facing folds*. Therefore, an antiform with this younging characteristic is called *downward-facing antiform;* analogously, we recognize a *downward-facing synform.*[3] Remember that when you find downward-facing folds in the field, you immediately know that some secondary process has inverted the normal stratigraphic sequence; that is, we cannot violate

---

[3]Note that this antiform and synform have the younging characteristics of a syncline and anticline, respectively, and are therefore also called *antiformal syncline* and *synformal anticline,* respectively.

Figure 10.7 An asymmetric, plunging fold (The Sheep Mountain Anticline in Wyoming).

Steno's Law of Superposition. Such downward-facing folds are not as uncommon as you might think at first. They are often associated with areas containing an early 'generation' of regional folds with horizontal axial surfaces, which are quite common in mountain belts. Subsequent folding of such early structures will generate a series of upward- and downward-facing folds, as shown in Figure 10.6e. But we must not get ahead of ourselves; the principle of superposed folding is discussed in Section 10.6.

10.3 FOLD CLASSIFICATION

Now that we have established a basic vocabulary, we can classify folds, which is based on four components:

1. Fold shape in three dimensions, primarily distinguishing between cylindrical folds and noncylindrical folds (Figure 10.3).

2. Fold facing, separating upward-facing folds and downward-facing folds (Figure 10.6).
3. Fold orientation.
4. Fold shape in the profile plane.

The first two components, three-dimensional fold shape and fold facing, were discussed above (Figures 10.3 and 10.6). In this section we concentrate on the other two components that are used in fold classification: fold orientation and fold shape in profile.

10.3.1 Fold Orientation
Looking at the surface of a natural fold makes one wonder if there is any representative measurement for this structure. Taking your compass to various locations on the folded surface will give you a large number of different readings for dip and strike (or dip direction), especially if the fold is noncylindrical. In some folds the limbs are relatively planar, and you will find that all the measurements in

Figure 10.8 Chevron folds in Franciscan chert of California (Marin County).

a single limb are pretty much alike (Figure 10.8). However, in folds with curving limbs, this will not be the
case. So what do we measure if we want to describe the
orientation of a fold to another geologist? The first measurement we take is the orientation of the hinge line (Figure 10.9). On the scale of an outcrop, the hinge line typically is fairly straight, and we determine its plunge (20°)
and direction of plunge (190°). We can now say, for example, that the fold is shallowly plunging to the south.
Secondly, we measure the orientation of the axial surface. We measure a dip direction/dip of 270°/70° for the
axial surface, which then completes our description of the
fold: *a shallowly south-plunging, upright fold.* Remember
that the hinge line always lies in the axial surface. Test
your measurements in a spherical projection to see
whether this relationship holds. What constrains these
terms *shallow* and *upright*? As a practical convention, we
use the angular ranges given in Table 10.2.

In Figure 10.10 some representative combinations of
hinge line and axial surface orientations are shown with
their terminology. A fold with a horizontal axial surface (as

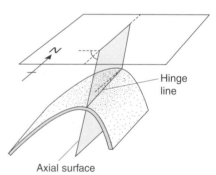

Figure 10.9 Fold orientation. Note that the axial surface is a plane, whose
orientation is given by a strike and dip, whereas the hinge is a line whose
orientation is given by a plunge and bearing.

in Section 10.2.3) by definition will have a horizontal hinge
line, and is called a *recumbent fold* (Figure 10.1). In the
Alps of Europe are large-scale recumbent folds often associated with thrust faulting; they are called *nappes* (see
Chapter 17). Another term that is used for a steeply plunging, inclined fold is *reclined fold*. In all cases, remember

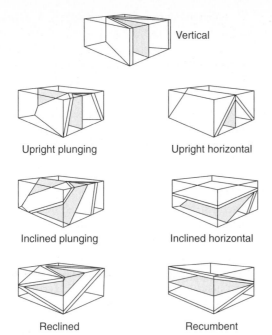

Vertical

Upright plunging   Upright horizontal

Inclined plunging   Inclined horizontal

Reclined   Recumbent

Figure 10.10 Fold classification based on the orientation of the hinge line and the axial surface (shaded).

Table 10.2 Fold Classification by Orientation

Plunge of Hinge Line	Dip of Axial Surface
Horizontal: 0–10°	Recumbent: 0–10°
Shallow: 10–30°	Inclined: 10–70°
Intermediate: 30–60°	Upright: 70–90°
Steep: 60–80°	
Vertical: 80–90°	

that your field measurements may be accurate within 2° (compass accuracy), but that the feature you measure will probably vary over an angle of 5–10°. Thus, the values in Table 10.2 can serve as only a rough guide and should not be applied too literally.

10.3.2 Fold Shape in Profile

The profile plane of a fold is defined as the plane perpendicular to the hinge line (Figure 10.2). The fold shape in profile (as viewed, by convention, down the plunge) allows further classification of folds. Fold classification by its shape in profile involves specification of the interlimb angle and of the change in bed thickness. The *interlimb*

angle of a fold is the angle between the limbs to define this angle, we assume that the limbs are relatively planar. If they are not, we measure the angle between the tangents at the inflection points (Figure 10.4). The values corresponding to the various terms are listed in Table 10.3. As with those in Table 10.2, the values serve only as a rough guide.

The second characteristic of fold shape in profile is the change in *bed thickness* across the structure. If you look at Figure 10.11a, you will notice that the bed thickness does not change appreciably as we go from one limb of the fold to the other. In contrast, the fold in Figure 10.11b has thin limbs and a relatively thick hinge area. We can quantify these observations by using a method called *dip-isogon analysis*. Dip isogons connect points on the upper and lower boundary of a folded layer where the layers have the same dip relative to a reference frame (Figure 10.12). The construction method is explained step by step in all structural geology laboratory manuals. Three classes are recognized, based on isogon patterns: *convergent dip isogons* (Class 1), *parallel dip isogons* (Class 2), and *divergent dip isogons* (Class 3). The terms "convergence" and "divergence" are used with respect to the core of the fold[4]; when

---
[4]The same terminology returns with cleavage (Chapter 11).

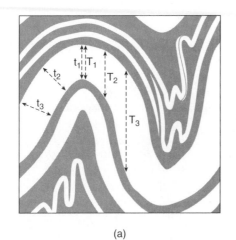

(a)

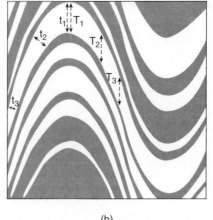

(b)

Figure 10.11 Parallel folds (a) maintain constant layer thickness across the folded surface, that is, $t_1 = t_2 = t_3$. Note that parallelism must eventually break down in the cores of folds because of space limitation, which is illustrated by the small disharmonic folds. In similar folds (b), the layer thickness parallel to the axial surface remains constant, that is, $T_1 = T_2 = T_3$. Similar folds do not produce the space problem inherent in parallel folds.

Table 10.3   Fold Classification by Interlimb Angle

Isoclinal	0–10°
Tight	10–60°
Open	60–120°
Gentle	120–180°

the dip isogons intersect in a point at the core of the fold, it is called convergent, and vice versa. The two geometries of Figure 10.11 return as special cases. Dip isogons that are perpendicular to bedding throughout the fold define a *parallel fold,* whereas dip isogons that are parallel to each other characterize a *similar fold.* This terminology (especially the use of the word "parallel") may be confusing unless you remember that parallel and similar describe the geometric relationship between the top and bottom surfaces of a folded layer; they do *not* describe the relationship between individual dip isogons in a fold. The positions of parallel and similar folds anchor the finer five-fold subdivision that is used mainly for detailed analysis. In the field, using the terms similar (representing Classes 2 and 3) and parallel (representing Classes 1A, 1B, and 1C) is generally sufficient to describe the shape of folds in profile.[5]

So we have added two more components to our description of a fold. Now as test, sketch a shallowly plunging, upright, tight, similar, downward-facing synform in the margin of the text. Hopefully these terms have become sufficiently clear that this is a relatively simple task. The only parameter

we have excluded in our description is *fold size.* We may use microfold (microscopic size; i.e., <1 mm), mesofold (hand specimen to small outcrop size; i.e., cm–m), and macrofold (mountain size and larger; i.e., hundreds to thousands of m) for this. The above lengthy description is certainly not pretty, but it is informative. Remember that the goal of any good description is first to remember the characteristics of a feature for yourself, and secondly to relay this information in an understandable and unequivocal fashion to someone else.

10.4 FOLD SYSTEMS

Our treatment of folds so far has concentrated mostly on single antiforms and synforms. If we have a series of antiforms and synforms, we call this a *fold system.* The information we can obtain from fold systems provide some of the most powerful information for the analysis of deformed regions, and involves such elements as fold symmetry, fold vergence, and the enveloping surface. We will start with the last.

10.4.1 The Enveloping Surface

Let's draw an imaginary plane that is tangential to the hinge zones of a series of small folds (surface A, Figure 10.13). We call this curved surface the *enveloping surface.* It contains all the antiformal hinges or synformal hinges.[6] Figure 10.13 also shows that we can draw an additional enveloping surface (surface B) when we connect the hinges of the regularly folded enveloping surface A. We call the enveloping surface (ES) for the largest folds the first-order ES (here

---

[5]Graphically, the dip isogon classification plots angle α versus normalized distance between the two tangents defining a dip isogon.

[6]In the case of folds with horizontal or shallowly dipping axial surfaces, we use crests and troughs, respectively.

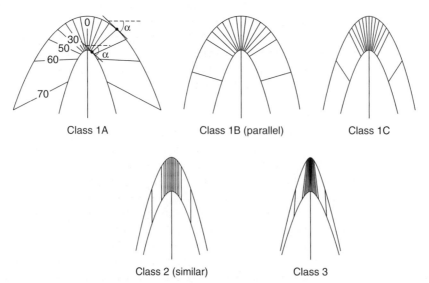

Class 1A   **Class 1B (parallel)**   **Class 1C**

Class 2 (similar)   **Class 3**

Figure 10.12   Fold classes based on dip-isogon analysis. In Class 1A the construction of a single dip isogon is shown, which connects the tangents to the upper and lower boundary of the folded layer with equal angle (α) relative to a reference frame; dip isogons at 10° intervals are shown for each class. Class 1 folds have convergent dip isogon patterns; dip isogons in Class 2 folds are parallel; Class 3 folds have divergent dip isogon patterns. In this classification, parallel and similar folds are labeled as Class 1B and Class 2, respectively.

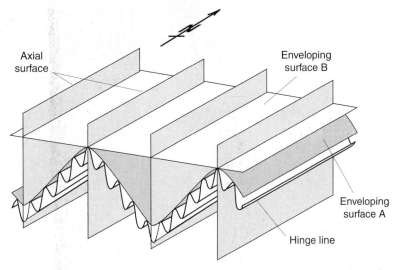

Figure 10.13 The enveloping surface connects the antiform (or synform) hinges of consecutive folds (surface A). If this imaginary surface appears to be folded itself, we may construct yet a higher-order enveloping surface (surface B). Note that the orientations (hinge line and axial surface) of the small folds and the large-scale folds are very similar.

From *An Outline of Structural Geology* by Hobbs, Means, and Williams. Copyright © 1976, John Wiley & Sons, New York. Reprinted by permission of John Wiley & Sons, Inc.

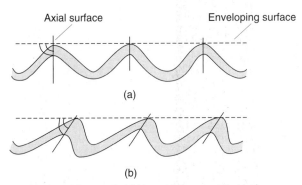

Figure 10.14 (a) Symmetrical folds and (b) asymmetrical folds are defined by the angular relationship between the axial surface and the enveloping surface.

surface B). The enveloping surfaces of successively smaller structures have higher order (second-order ES, third-order ES, etc.; here surface A). The first-order enveloping surface is typically of regional scale, whereas higher-order enveloping surfaces may go down as low as the thin-section scale. But what is the point of determining the enveloping surface? With decreasing order, the enveloping surfaces reduce the structural information of a folded area into increasingly simpler patterns. For example, the second-order ES in Figure 10.13 shows that the small-scale folds define a larger-scale fold pattern consisting of antiforms and synforms. Such large-scale structures have traditionally been called an *anticlinorium* and a *synclinorium,* respectively. Unfortunately these terms suggest upward-facing structures, which may not always be the case. The presence of anticlinoria and synclinoria implies that somehow the many small folds that we see are related, even though they vary in shape and position in the larger structure. It is important to note, however, that the *orientations* of these small folds (both hinge line and axial surface) are all approximately the same, and also that they are approximately equal to that of the anticlinorium and

synclinorium. For that reason, the small folds are sometimes called *parasitic folds,* because they are closely related to a larger structure.

The geometric relationship between parasitic folds and regional structures gives us a powerful concept in structural analysis, which states that the orientation of small (high-order) structures is representative of the orientation of regional (low-order) structures. This concept is known as *Pumpelly's rule,* after the nineteenth-century American geologist Raphael Pumpelly. Thus, the orientations of the hinge line and the axial surface of a small fold predict these elements for a regional fold; in Figure 10.13 you will see that this is indeed the case. Obviously, this 'rule' requires testing in the field and serves only as a convenient working hypothesis, but it has proven to be remarkably robust in regional analyses. Use Pumpelly's rule on a field trip and surprise friends and foes with your great insight into regional structures.

10.4.2   Fold Symmetry and Fold Vergence

The relationship between the enveloping surface and the axial surface of folds also enables us to describe the symmetry of folds. If the enveloping surface and the axial surface are approximately perpendicular (+/–10°), we have *symmetrical* folds (Figure 10.14a); otherwise, the folds are *asymmetrical* (Figure 10.14b). In the case of an isolated fold, an enveloping surface cannot be defined. Thus, to determine if the fold is symmetric or asymmetric, we use the median surface (i.e., the surface that passes through the inflection lines on opposing limbs). If the axial surface is perpendicular to the median surface, then the fold is symmetric, otherwise the fold is asymmetric. There are other definitions of fold symmetry that involve, for example, the relative steepness of limbs, but the above definition of fold symmetry are unambiguous because they are independent of fold orientation.

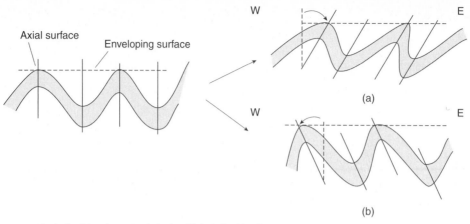

Axial surface   Enveloping surface

(a)

(b)

Figure 10.15   Fold vergence, clockwise (a) and counterclockwise (b), is defined by the apparent rotation of the axial surface from a hypothetical symmetrical fold into the observed asymmetrical fold, without changing the orientation of the enveloping surface. In a given geographic coordinate system, we may also say east-verging (a) and west-verging (b) folds for clockwise and counterclockwise folds, respectively (for this example). In all descriptions we are looking down the plunge of the fold.

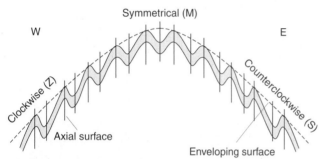

Figure 10.16   Characteristic fold vergence of parasitic folds across a large-scale antiform. Looking down the plunge, the parasitic fold change from clockwise asymmetry (east-verging in the geographic coordinate system) to symmetrical to counterclockwise asymmetry (west-verging) when going from west to east. You may also find that some geologists use the terms Z-, M-, and S-folds for this progression (for obvious reasons).

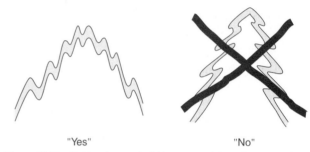

"Yes"   "No"

Figure 10.17   Right and wrong in fold vergence. It is useful to copy the images in Figures 10.16 and 10.17 in your notebook for reference.

Now let's look in some detail at the practical significance of fold symmetry. The second-order enveloping surface defined by the small folds of Figure 10.13 outlines a large antiform-synform pair. The small folds (i.e., the parasitic folds) show distinct shapes and asymmetries as we move along the second-order enveloping surface. On the west limb of the large antiform, the minor folds are asymmetric and have what we call a clockwise asymmetry when looking down the plunge of the fold. In the hinge area, the folds are symmetrical, because the axial surface is perpendicular to the enveloping surface. In fact, as we move from the limb toward the hinge area, the clockwise asymmetry becomes progressively less. As we move into the east limb of the antiform, the fold asymmetry returns but now with opposite sense to that in the west limb; in the east limb the asymmetry is counterclockwise. Clockwise and counterclockwise are defined by the rotation of the axial surface relative to a hypothetical symmetrical fold (Figure 10.15). In the past, parasitic folds were erroneously given the genetic name 'drag folds', because it was assumed that the apparent rotation of the axial surfaces reflects drag between the layers during folding. In fact, the small folds are probably more symmetrical in the incipient stages of regional

folding and become more asymmetrical when the large folds tighten. So, across a fold the *vergence*[7] of parasitic folds changes in a characteristic manner that allows us to predict the location of the hinge area of large antiforms and synforms (Figure 10.16). Using Pumpelly's rule, we may even predict the orientation of these large regional folds.

There is need for some caution when the folds are not plunging, that is, in horizontal upright folds. If we view the structure in Figure 10.16 from the opposite side (just place the face of the page holding Figure 10.16 on a window, viewing it from the back), the clockwise folds become counterclockwise! This situation is not as paradoxical as it may sound at first. Imagine an antiform that is cut by a road perpendicular to the axial surface. Depending on which side of the road you look, the asymmetry of parasitic folds in this large structure is clockwise or counterclockwise *in the same limb*. However, in either case they make the same prediction for the location of the hinge area of the antiform. So long as you define the direction in which you view the structure, there is no problem using

---

[7]Do not confuse fold vergence with fold facing.

Figure 10.18 Monocline in the Big Horn Mountains (Wyoming).

fold vergence as a mapping tool in an area. A useful tip is to copy the geometry of Figure 10.16 in the back of your field notebook. Matching field observation with asymmetries in your sketch, which may require some rotation of your notebook, will ensure a reliable application. In any case, a pattern of fold vergence opposite to that in Figure 10.16 ('Christmas tree' geometry) cannot be produced in one single fold generation (Figure 10.17).[8]

10.5 SOME SPECIAL FOLD GEOMETRIES

We end the descriptive part of this chapter with a few special fold geometries that you may encounter in your field career. *Monoclines* are fold structures with only one tilted limb; the beds on either side of the tilted limb are horizontal. Monoclines typically result from a vertical offset on a steeply dipping fault in the subsurface near the tilted portion of the structure. The fault uplifts a block of relatively rigid igneous or metamorphic rock, and the overlying sedimentary layers drape over the edge of the uplifted block to form the monocline. Spectacular monoclines are found, for example, in the western United States (Figure 10.18). *Kink folds* are small folds (less than a meter down to microscale) that are characterized by straight limbs and sharp hinges. Typically they occur in finely laminated (i.e., strongly anisotropic) rocks such as shales and slates (Figure 10.19). Sharply bending a deck of cards is a good illustration of the kinking process, because kink folds are formed by

Figure 10.19 Kink folds in the mica-rich portion of the Yuso Formation of the Cantabrian Mountains (northern Spain).

---

[8]This geometry is, in fact, diagnostic of the presence of at least two fold generations (see Section 10.6).

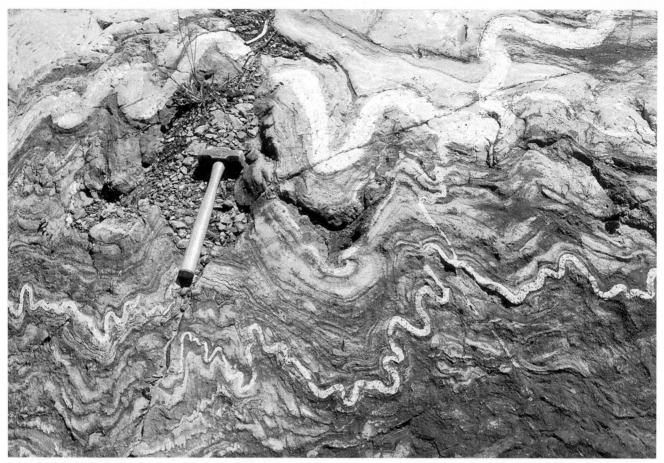

Figure 10.20 Ptygmatic folds in the Grenville Supergroup (Ontario, Canada). Note the wavelength variation as a function of layer thickness. Hammer for scale (Author provided).

displacements between individual laminae (the individual cards in the analogy; see Section 10.8.2). *Chevron folds* (Figure 10.8) are the larger scale equivalent of kink folds.

A *box fold* describes a geometry that is pretty self-explanatory (Figure 10.5)—in order to form, the layer must be detached from the underlying and overlying layers. *Ptygmatic folds* are irregular and isolated fold structures that typically occur as tightly folded veins or thin layers of strongly contrasting lithology (and, thus, contrasting competency; Figure 10.20). They look like intestines.

Doubly-plunging folds are structures with hinge lines that change curvature along their length. Along the trend of plunge the folds may die out or even change from antiforms to synforms. The high point of the hinge line in a doubly-plunging fold is called the *culmination,* and the low point along the same hinge line is called a *depression.* The typical change in plunge angle is less than 50°. If additional folds are present, this may result in *en echelon folds,* in which a gradually opening fold is replaced by a neighboring, gradually tightening fold of opposite form. Such a geometry may occur on all scales, ranging from the hand specimen (Figure 10.21) to the size of whole mountain belts. Note that doubly-plunging folds are by definition

Figure 10.21 *En echelon* folds on the scale of cm.

noncylindrical. *Sheath folds*[9] show extreme hinge line curvature, to the extent that it approaches parallelism (change in plunge up to 180°!). What is typically visible in outcrop is the elliptical cross section of the nose of the fold (see Figure 12.28), but this pattern itself does not necessarily imply the presence of a sheath fold. Any doubly-plunging fold may give the same outcrop pattern, but only if a highly

---

[9]So named because they resemble the sheath of a sword, in shape.

220 Ductile Structures

curved hinge line is visible can sheath folds be recognized as such. You can produce sheath folds by taking a mildly doubly-plunging fold and 'pulling' at its crest, which occurs in zones of high shear strain. We return to the formation and significance of these structures in Section 12.6.1.

Finally, two fold types that have been studied extensively in recent years because of their association with hydrocarbon potential in fold-thrust belts are fault-bend folds and fault-propagation folds. Because of their intimate association with thrusting, we will examine their formation and significance in Chapter 17 (see also Section 8.3.4). At this point we merely include them for completeness: *fault-bend folds* are formed as thrust sheets move over irregularities in the thrust plane, whereas *fault-propagation folds* are accommodation structures above the frontal tip of a thrust.

10.6 SUPERPOSED FOLDING

Structural geologists use the term *fold generation* to refer to groups of folds that formed at approximately the same time interval and under similar kinematic conditions. However, commonly we find several fold generations in an area, which are labeled by the letter F (for *F*old) and a number reflecting the relative order of their formation: F_1 folds form first, followed by F_2 folds, F_3 folds, and so on, at any given outcrop. In turn, several fold generations may form during an *orogenic phase* (such as the Siluro-Devonian 'Acadian' phase in the Appalachians or the Cretaceous-Tertiary 'Laramide' phase in the North American Cordillera), which is given the letter D (for *D*eformation). In any mountain belt, several phases may be present, which are labeled D_1, D_2, and so on, each containing one or more generations of folds. For example, the Appalachian Orogeny of the eastern United States contains three main deformation phases (Taconic, Acadian, and Alleghenian; Chapter 19). From the onset it is important to realize that neither a deformation phase nor each individual fold generation has to be present everywhere along the orogen, nor do they occur everywhere at the same time. On a regional scale, deformation is irregularly distributed and commonly diachronous. You can imagine that fold generations and deformation can rapidly become pretty complex and confusing. So we'll stick to fold generations to examine the principles of superposed folding, which enable us to unravel the sequence (i.e., relative timing) of folding.

Generation is a relative time concept and implies only "older than" or "younger than." There are methods to determine the absolute ages of folds, such as dating of minerals that formed during folding (e.g., micas in cleavage), but we will not go into them here. The principle of *superposed folding*[10] is simple: folds of a later generation are superimposed on folds of an earlier generation. The determination of this temporal sequence, however, is not straightforward and requires careful geometric analysis. In the field we find

that superposed folding is a widespread phenomenon, which is not restricted to high-grade metamorphic areas. Even in regions down to subgreenschist facies, superposed folding is common, and one can determine the temporal relationships by studying the patterns that arise from the superimposition of one fold generation on another. After discussing fold superposition and criteria for its recognition, we close with the concept of fold style, which is used to place our findings on fold generations in a regional context.

10.6.1 The Principle of Fold Superposition

Figure 10.22 is a field photograph of a complex fold geometry that contains two fold generations. How do we know this from looking at the picture, and how do we separate F_1 and F_2 folds in this pattern? Fold superposition has a simple rule at its root, yet the concept requires the ability to visualize and analyze sometimes very complex three-dimensional geometries. Let us first start with the rule: a superposed fold must be younger than the structure it folds. This is simply restating the Law of Superposition so that it applies to folding. Unless a fold was present previously, it cannot be modified by a younger fold. Now what is left is to determine the criteria by which we obtain this temporal relationship. We will begin with an example.

Figure 10.23a shows a sequence of recumbent folds that we will call F_A; the associated axial surface is called S_A. We now superimpose a series of upright folds of approximately the same scale (F_B with S_B axial surface; Figure 10.23c). The superimposition of F_B on F_A produces the interference pattern shown in Figure 10.23b. Elements of both fold generations are preserved; for example, the recumbent nature of F_A is still there, but its limbs are now folded. Similarly, the upright F_B folds remain clearly visible, but they are superposed on a pattern that repeats and inverts bedding (from the recumbent F_A folds). The way to determine the temporal relationship from our interference pattern is to invoke the rule of superposition. Both the bedding and S_A are folded, but S_B is essentially planar. So, SS and S_A were already present before S_B; consequently, the upright folds must be F_2, which are younger than the recumbent F_1 folds. The axial surfaces may not be always visible in the field (although axial plane foliations are common; Chapter 10), but you can always draw an imaginary axial surface to evaluate these complex folding patterns. Now let's examine this pattern with an analog. Take a piece of paper and fold it in two (our F_1 fold) and orient it into a recumbent orientation. Then fold the paper again to create an upright fold whose hinge parallels the hinge of the recumbent fold (our F_2 fold). Voila, you get the pattern of Figure 10.23b. If you are comfortable with this explanation, we will proceed with the four basic fold interference patterns.

10.6.2 Fold Interference Patterns

Four basic patterns can be recognized arising from the superimposition of upright F_2 folds on F_1 folds of variable orientation (Figure 10.24). Looking at these *fold interference types,* you will notice that we earlier examined Type 3

---

[10]Also called "fold superimposition" or "fold overprinting."

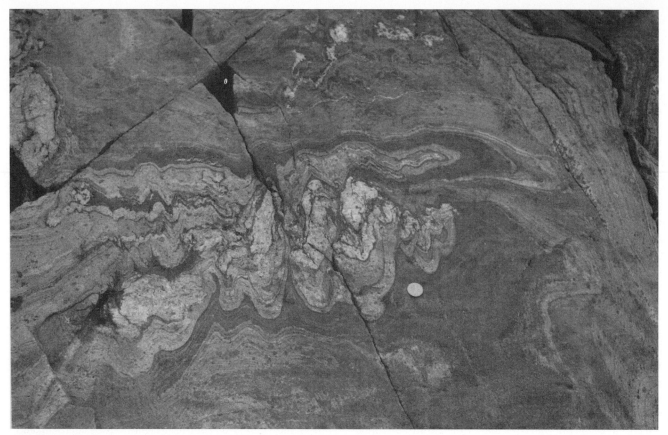

Figure 10.22 Fold interference pattern of Type 3 ('refolded fold') geometry.

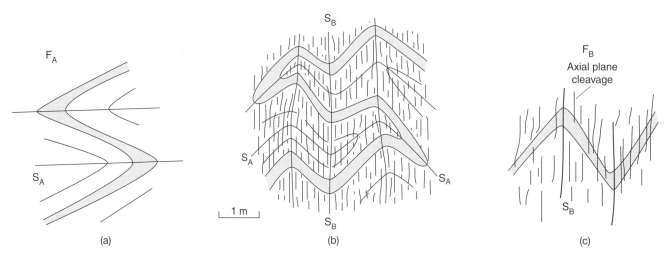

Figure 10.23 Dissecting a fold interference pattern. The cross-sections show F_A recumbent folds (a) overprinted by F_B upright folds (c), producing the fold interference pattern in (b).

using a piece of folded paper. Types 0 and 2 can equally well be examined by folding a piece of paper, but Type 1 requires some additional crumpling. Instead of describing these patterns in confusing detail, look at Figure 10.24, and experiment with a piece of paper and your two hands. One approach that may offer further insight is to make fold interference patterns using a few thin layers of colored modeling clay, which can then be cut with a knife to see the geometry of intersecting surfaces.

Welcome back from spending time with Figure 10.24. We will now look at some important properties of the various fold interference types. Type 0 is a very special condition, because the hinge lines and the axial surfaces of both fold generations are parallel. As a consequence, F_1 is merely tightened by the superimposition of F_2. You realize that, practically, Type 0 can therefore not be recognized in the field as an interference type by geometry alone (that's why we use the number 0). Type 1 is also called 'dome-

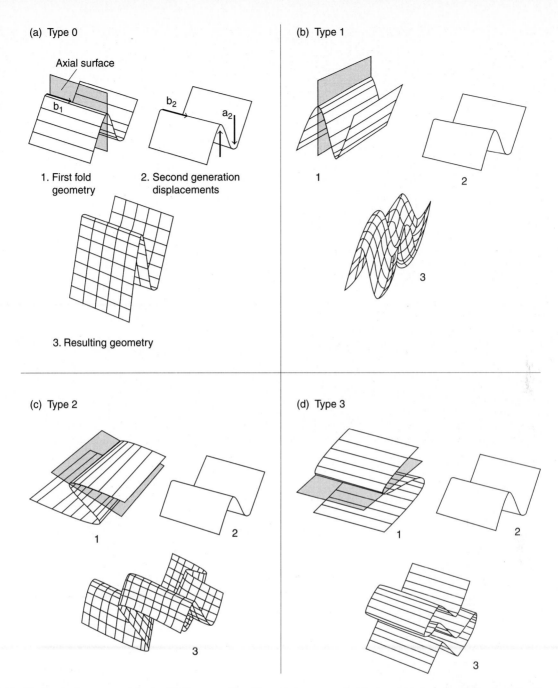

(a) Type 0

Axial surface

b_1

1. First fold geometry

b_2

a_2

2. Second generation displacements

3. Resulting geometry

(b) Type 1

1

2

3

(c) Type 2

1

2

3

(d) Type 3

1

2

3

Figure 10.24 The four basic patterns arising from fold superposition. The analysis assumes that F_2 shear folds (a_2 is the relative shear direction and b_2 is the hinge line) are superimposed on a preexisting F_1 fold of variable orientation. Shear folds are modeled by moving a deck of cards. The shaded surface is the S_1 axial surface. It's as easy as 1, 2, 3.

and-basin' structure and resembles an egg carton. Both the axial surfaces and the hinge lines of the two generations are perpendicular, producing this characteristic geometry (Figure 10.24a). Type 2 is perhaps the most difficult geometry to visualize, but folding a piece of paper helps enormously. In outcrop, we typically see a section through this geometry, which commonly gives what we call the 'mushroom' pattern (Figure 10.24b). Note that this outcrop pattern is generated only in the horizontal surface that intersects Type 2; if we take another cut, say vertical, the outcrop pattern is quite different. Finally, Type 3 (Figure 10.24c) is

known sometimes as the 'refolded fold' pattern, which is really a misnomer because all four types are refolded folds. We use the name just so that you have heard it, and because very few people are able to remember the corresponding numbers of the types. In fact, we recommend that you use the descriptive terms 'dome-and-basin,' 'mushroom,' and 'refolded-fold,' however flawed, instead of the uninformative Type 1, Type 2, and Type 3, respectively.

Interference patterns are a function of the spatial relationship between hinge lines and axial surfaces of the fold generations, as well as the surface in which we view the

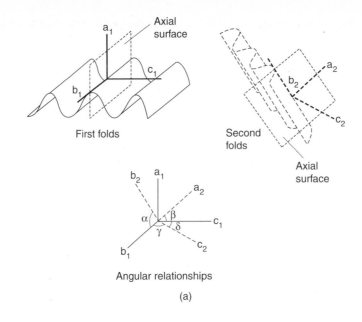

First folds

Second folds

Axial surface

Angular relationships

(a)

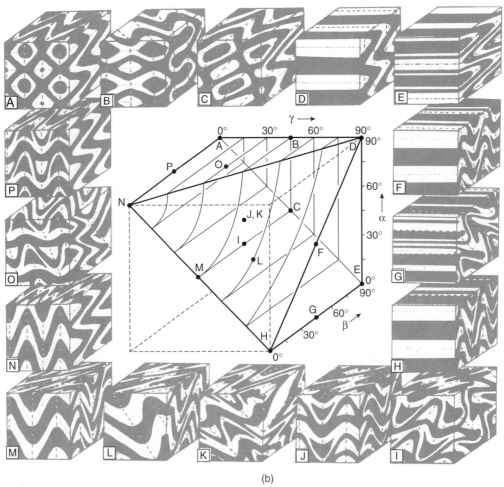

(b)

Figure 10.25 Geometric Axes describing the orientation of F_1 and F_2 (a), and representative interference patterns (b). In all patterns, the layering was initially parallel to the front face of the cube. S_1 is shown with dotted lines and S_2 with dashed lines.

pattern. Thus, the analysis of fold superposition is a distinctly three-dimensional problem. The four types that are shown in Figure 10.24 are only endmember configurations in an infinite array of possibilities. Figure 10.25 is a summary diagram that shows the patterns we observe as we vary the spatial relationships as well as the observation surface (i.e., an outcrop surface). Even more so than before, understanding these sectional patterns requires self-study. Ultimately, such multisurface patterns reward you with complete information on the sequence and orientation of fold generations. So, take your time.

Table 10.4   The Characteristic Elements of Fold Style

- In profile plane, is the fold classified as parallel or similar (or a further refinement)?
- What is the interlimb angle in profile?
- In three dimensions, is the fold cylindrical or noncylindrical?
- Is there an associated axial plane foliation and/or lineation present, and of what type are they (these will be discussed in Chapter 11)?

Note that orientation and symmetry are not style criteria.

The fold interference patterns we have seen are produced when fold generations of similar scale are superimposed. If the scales are very different, there may be no interference pattern visible on the outcrop scale, and only through regional structural analysis does the large-scale structure appear. After some field work, it is therefore not uncommon to find one or more additional fold generations that show only on the map scale. A reexamination of some puzzling notes and outcrop sketches may all of a sudden be easily explained by recognizing this extra fold generation.

The occurrence of several fold generations has major implications for the interpretation of the deformation history of your area. First, it implies that the kinematic conditions have changed to produce a fold generation in a different orientation than before (except Type 0); so the deformation regime must have changed somehow. Secondly, folds of the first generation will have variable orientations depending on where they are measured in the fold superposition pattern. Orientation, therefore, is *not* a characteristic of fold generations in multiply-deformed areas and should be used carefully as a mapping tool (see below). That leaves the final question: how may we recognize folds of a certain generation in the absence of interference patterns at each and every locality in our area? For this we turn to the powerful concept of fold style.

10.6.3   Fold Style

In our field area we encounter a number of folds. Of course the question of their significance arises. Are they part of the same generation, or do they represent one or more generations? At one locality in our area we are actually able to determine a sequence of F_1 and F_2 folds, so we know that there are at least two generations. From our experience with superposed folding, we are also aware that only F_2 folds have an orientation that may persist over any distance, and that the orientation of F_1 folds depends entirely on their position in the fold interference pattern (we measure their orientation nonetheless because the distribution should 'fit' this pattern). We now are at an outcrop where we find only one fold, which is not in the exact same orientation as either F_1 or F_2 in the previous outcrop. Nonetheless, we wish to predict to which generation it belongs. For that, we use several characteristics of each fold generation, which are

grouped under the term *fold style*. What are these fold style characteristics (Table 10.4)?

The four elements of Table 10.4, parallel/similar, interlimb angle, cylindrical/noncylindrical, and foliations/lineations, are used to describe the style of a fold. The first three have been discussed in detail and need no further clarification. Foliations and lineations will have more meaning after you read Chapter 11, but this fourth characteristic is included nonetheless because of its discriminatory ability. For example, an axial plane crenulation cleavage may be a characteristic of F_2 folds, and the presence of a mineral lineation may reflect special metamorphic conditions that occurred only during the first fold generation. Notably absent in our list are fold orientation and fold symmetry, which are *not* style criteria. Discriminating a fold generation on its orientation may work only for the youngest fold generation; the older ones most likely have become variably oriented. Secondly, we already learned that fold symmetry may change within a single-generation, large fold (Figure 10.16). So, just like orientation, symmetry is not a style criterion.

10.6.4   A Few Philosophical Points

We close the section on superposed folding with a few considerations. You will find that often it is not possible in any single outcrop to determine the complete sequence of fold generations because discriminatory interference patterns may not be exposed, or one or more generations may not be present at all. However, by combining information from several outcrops, as well as using fold style, you should eventually be able to obtain a reasonable folding sequence. As you map, you should continue to test this hypothesis, after which you can place your findings in a regional kinematic picture. For example, you may find that the first generation of folds is recumbent, a geometry that we quite commonly associate with thrusting. The second fold generation may be upright, reflecting folding of the thrust sequence. Maybe a third fold generation represents a very different shortening direction, possibly related to a different orogenic phase. Very commonly we find small kink folds in well-foliated rocks that complete the folding sequence. Whereas the possibilities may seem limitless, reasonable interpretations are not. There is sometimes a tendency to

Figure 10.26 Fold interference pattern of Type 1 (dome-and-basin) geometry.

recognize too many fold generations. In the end, the number of fold generations should be based *only* on interference patterns, either on local or regional scale. Practically, in any one outcrop you may be able to see three or maybe four fold generations, and regionally perhaps a couple more. Always remember that structural analysis is not helped by proposing an unnecessarily long and complex sequence of fold generations, because each generation reflects a corresponding deformation regime, and one can reasonably expect only so many different patterns. With these musings and a field example of a Type 1 fold interference pattern (Figure 10.26), we leave the descriptive part of folds and their field analysis to turn our attention to the mechanics of folding.

10.7 THE MECHANICS OF FOLDING

Why do folds occur? As we have spent so much time on their description, this is a question that has probably surfaced several times. Well, if you ever were involved in a minor car collision, you do not need to be reminded that forces cause folding.[11] Similarly, natural forces applied to rocks cause folding. But we will see that forces alone are not sufficient to form folds. Consider a block of plasticine that is reshaped by external forces (from your hands). The block will change form, but in doing so the internal structure does not show any folds (Figure 10.27a). However, when we add slightly wavy layers of different color, but otherwise the same material properties to the block, we see folds appear (Figure 10.27b). In a third experiment we add

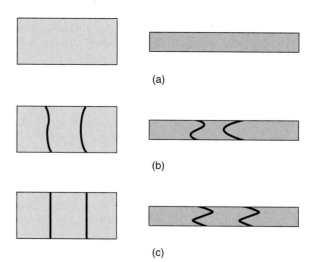

Figure 10.27 Plasticine experiments without (a) and with colored bands (b), and containing rubber bands (c).

thin sheets of rubber to our plasticine block, and when force is applied, folds appear (Figure 10.27c). Thus, to see folds, we must have some visible layering to define the fold shape. These simple experiments provide us with a fundamental subdivision of folding based on the mechanical role of layers: passive folding and active folding.

10.7.1 Passive versus Active Folding

During *passive folding,* the layering has no mechanical significance. In the color-banded plasticine block, folds are formed by the amplification of small perturbations in the color bands, but the strain pattern in the block is unaffected by these layers. You can see the same result by squeezing multicolor toothpaste on a counter top. The complex folding patterns that you see in the blob are visible only be-

---

[11]The obsession with this analogy does not reflect the authors' personal experiences.

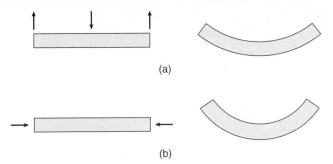

Figure 10.28 Bending (a) and buckling (b) of a layer.

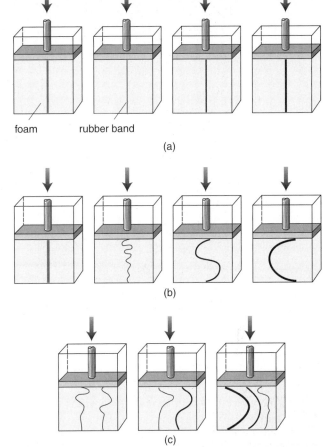

foam   rubber band

(a)

(b)

(c)

Figure 10.29 Line drawings of deformation experiments with transparent boxes containing foam and rubber bands. In (a), four setups are shown that contain foam only (with a passive marker line added), thin, medium, and thick rubber bands, from left to right. When we apply the same displacement (b), the setups respond differently. The foam only box shows only thickening of the marker line, whereas the boxes with rubber bands shows folds with different arc lengths depending on the thickness of the bands. When we use more than one rubber band (c), the behavior depends on the combination of bands and their thicknesses, with the thicker band dominating the behavior.

cause of the color contrast; if you do the same experiment with single-color toothpaste, you will not see folds even though the internal structure of the two blobs is similar. To reproduce such toothpaste-like behavior in nature, we need conditions in rocks that have little or no competency contrast between the layers. Elevated temperatures can produce the right conditions for passive folding, and it is not uncommon to find toothpaste-like structures in deformed metamorphic rocks. Rocks that were deformed at or near their melting temperature (i.e., high homologous temperature, $T_h = T/T_m$), called *migmatites,* can give wonderfully complex structures (Figure 2.20). Similarly, passive folding occurs in glaciers that deform close to their melting temperature (Figure 9.1). Passive folds are merely the amplification of natural imperfections in the layers or are a consequence of differential flow in a volume of rock. To picture such differential flow, think of the fold-like pattern that develops in the scum on the surface of a slowly flowing river. But don't think that passive folds have to be chaotic in appearance because of this. Sheath folds in a shear zone are a natural example of passive folding, and they commonly show very consistent orientation and style.

During *active folding,* also called *flexural folding,* the layering has mechanical significance. This means that the presence of layers with different competency directly affects the strain pattern in the deforming body and that there is contrasting behavior between the layers. There are two dynamic conditions that we distinguish for active folding: bending and buckling. In *bending,* the applied force is oriented at an oblique angle to the layering (Figure 10.28a). In nature this may occur during basin formation or loading of a lithospheric plate (also called flexural loading), or during the development of monoclines over fault blocks. In buckling, the force is oriented parallel to the mechanical anisotropy (Figure 10.28b).

Let's see what happens in a series of analog experiments of active folding, in this case buckling (Figure 10.29). We surround a thin band of rubber with foam in a plastic box that is open at one end, and place a plunger at the opening. As we apply force to the plunger, the band of rubber forms a series of folds. The surrounding foam (the 'matrix') accommodates the shape of the rubber band by filling the gaps that would otherwise be present. We repeat the experiment with a thinner band of rubber and the same foam. Now

we produce several more folds and the weaker foam again accommodates this new pattern. So, in spite of applying the same force and producing the same bulk shortening strain, the folding patterns are different as a function of the thickness of the rubber band. In other words, the rubber band introduces *mechanical anisotropy* during folding.

10.7.2 Buckle Folds

We will look at some physical properties of buckle folds in this section. First we return to our rubber band-foam experiments, where we saw that the thickness of the layer somehow affects the fold shape. Now we will carefully monitor the conditions that result in these differences. The applied force (and thus stress), the bulk strain (the distance the plunger moves into our box), the strain rate (the speed at which we plunge), as well as the metamorphic conditions (room temperature and pressure), are assumed to be the same in all experiments; only the thickness, t, of the layer

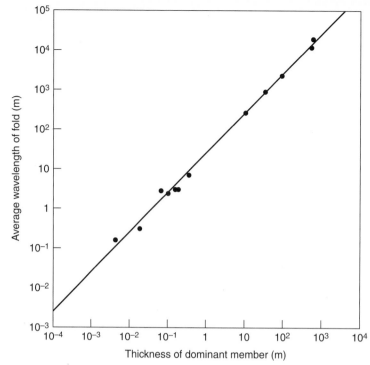

Figure 10.30 Log-log plot of wavelength versus layer thickness in competent sandstone layers.

varies. We notice that with increasing thickness the wavelength and arc length become larger. Because arc length can directly be measured in the field, we generally use this parameter in fold analysis. Already in the early part of this century this proportionality was known (Figure 10.30).[12] Secondly, if we compare bands of equal thickness but with different stiffness, we find that the arc length of the stiffer layer is larger than that of the weaker layer. So, we observe that both thickness and the parameter 'stiffness' increase the wavelength. In nature, we are not really interested in the stiffness of elastic rubber bands. Folds are permanent strain features, so it is more useful to consider this problem in terms of viscous behavior or more complex rheologic models (e.g., elastico-viscous, nonlinear viscosity; Chapter 5). For simplicity we assume Newtonian (i.e., linear) viscosity, but eventually we need to examine this problem using nonlinear rheologies; the latter will be mentioned only in passing and will not be explored here.

Using Newtonian viscosity, the theoretical arc length-thickness relationship for a layer with viscosity μ_L surrounded by a matrix with viscosity μ_M was determined in the early 1960s:

$$L = 2\pi t \, (\mu_L/6\mu_M)^{\frac{1}{3}} \qquad \text{Eq. 10.1}$$

This equation, which is known as the *Biot-Ramberg equation,*[13] tells us that arc length is directly proportional

to thickness and to the cube root of the viscosity ratio. Therefore, if we know the arc length/thickness ratio of a layer, we can obtain the viscosity ratio. Reorganizing Equation 10.1:

$$\mu_L/\mu_M = 0.024 \, (L/t)^3 \qquad \text{Eq. 10.2}$$

This states that the viscosity ratio is proportional to the cube of the L/t ratio. The measurements shown in Figure 10.30 give a viscosity ratio on the order of 475. We intentionally say "on the order of," because Figure 10.30 is a log-log plot, meaning that a small change in L/t ratio will result in a large change in viscosity ratio; therefore this μ_L/μ_M ratio is only a first-order estimate.

However, there is an additional element that we need to consider when the viscosity ratio of layer and matrix is small (say, <100). Again we return to our box experiment. This time we place only foam in our plastic box, and draw a layer with a marker pen (Figure 10.29), which represents a very small viscosity ratio (in this case, of course, $\mu_L/\mu_M = 1$). As we shorten the foam we do not get any folds, but we see that the layer thickens. So, low viscosity contrasts result in a component of strain-induced *layer thickening* during folding. We simplify this effect in our analysis by inferring that a component of layer thickening occurs before folding instabilities arise. Recalling the effect of layer thickness on L (Equation 10.1), we therefore include a strain component in Equation 10.1:

$$L = 2\pi t \, [(\mu_L(R-1))/(6\mu_M \cdot 2R^2)]^{\frac{1}{3}} \qquad \text{Eq. 10.3}$$

[12]The arc length is confusingly referred to as the wavelength in some older studies.

[13]After the two persons who simultaneously, but independently, carried out this analysis.

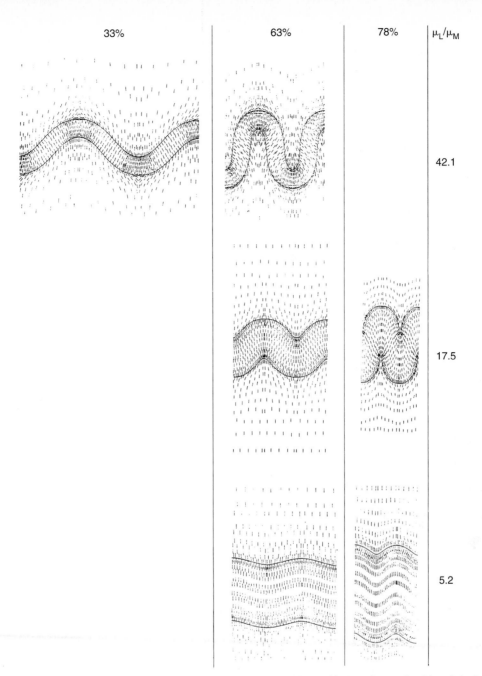

Figure 10.31 Finite-element modeling of single-layer buckling for various viscosity contrasts (μ_L/μ_M) between layer and matrix and shortening strains (%). The tick marks represent the orientation of the long axis of the strain ellipse in the fold profile plane.

where R is the strain ratio X/Z. This *modified Biot-Ramberg equation*,[14] gives quite a reasonable prediction for the shape of natural folds in rocks with low viscosity contrast, such as occur in many metamorphic rocks.

These roles of viscosity contrast and layer thickening during shortening are nicely illustrated in numerical models of folding. Figure 10.31 shows the results of a series of computer simulations of single-layer folding for three viscosity ratios using a finite-element method. As we saw earlier, with decreasing viscosity ratio the arc length becomes less and the layer increasingly thickens. One advantage of computer modeling is that we can track the strain

field in our system. Note, for example, that the strain pattern in and immediately surrounding the layer, which is indicated by the ticks in Figure 10.31, becomes increasingly homogeneous with decreasing viscosity contrast. This is obvious when you realize that no viscosity contrast (i.e., $\mu_L/\mu_M = 1$) reflects the situation in which there is no mechanical significance to the layer (see the foam-only box experiment in Figure 10.29).

These analyses of folding are entirely based on linear-viscous behavior of both the layer and the matrix. However, it is likely under elevated conditions of P and T that folding involves nonlinear rheologies, in which the viscosity is stress dependent (Section 5.3.6). Without going into the details, we mention that nonlinear rheologies result in lower

---

[14]Also called the *Sherwin-Chapple equation*.

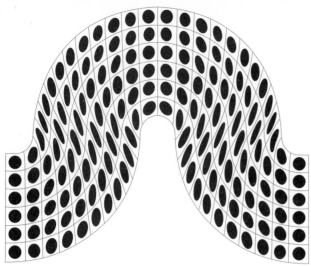

Figure 10.32 The characteristic strain pattern of flexural folding in the fold profile plane.

values of L/t as well as a small to negligible component of layer-parallel shortening. In fact, in addition to a low viscosity contrast, natural folds in metamorphic terranes typically have low L/t ratios (L/t < 10), indicating that power-law rheology and therefore crystal-plastic processes are indeed important during folding under these conditions.

10.7.3 Multilayers

Our discussion so far has focused on folding of a single layer in a weaker matrix. Although this situation occurs in nature, it is hardly representative. So, what happens if several layers are present? Again we turn to our simple experimental setup. We now take two rubber bands of equal thickness, and we see that folds develop as we strain the system. Can you explain the difference from our previous single-layer experiments (Figure 10.29)? The wavelength in the multilayer experiment is greater and seems to act as a thicker single layer. In our next experiment we combine a thick and a thin layer (Figure 10.32). In this case the thin layer does not at all give the fold shape predicted from single-layer theory, but the thick layer behaves pretty much the same as that predicted from a single-layer experiment (compare with Figure 10.29). In fact, the thin layer mimics the shape of the thick layer, indicating that the behavior of the thin layer is dominated by that of the thick layer. Obviously, we can try many other combinations, which in the end show that the behavior of a multilayer system reflects the interaction between layers. In general, the thicker and more competent ('stiffer') layers will dominate in a multilayer system. Therefore, the previous Biot-Ramberg buckling theory applies only when layers in natural rocks are

sufficiently removed from one another. We can theoretically determine the region over which the effect of a buckled layer dies off to negligible values in a weaker matrix, which is called the *contact strain zone* (CSZ). This width of the contact strain zone $(2 \cdot d_{CSZ})$, with d_{CSZ} measured from the midpoint of the buckled layer, is a function of the arc length:

$$d_{CSZ} = 2/\pi \, L \qquad \text{Eq. 10.4}$$

Practically, this means that the width of the CSZ is slightly greater than the arc length of a fold. Again, the predictions from this relationship are supported by field observations on folds.

When layers interact, there is a relatively simple extension of the theory for multilayer folding, if we assume a stack consisting of several superposed thin layers that are free to move relative to one another (i.e., no coupling). The corresponding equation for this multilayer case is:

$$L_m = 2\pi t \, (N\mu_L/6\mu_M)^{\frac{1}{3}} \qquad \text{Eq. 10.5}$$

where N is the number of superposed layers and the space between the layers is infinitely small. Note that this is not the same as the equation for a single layer with thickness N·t! In fact, if we calculate the arc-length ratio of a system with N superposed multilayers (L_m) and a system with a single layer of thickness Nt (L_s) by dividing Equation 10.1 (with t = Nt) and Equation 10.5, we find:

$$L_s/L_m = N^{\frac{2}{3}} \qquad \text{Eq. 10.6}$$

This shows that the arc length of N multilayers is less than that of a single layer with thickness N·t. This case is certainly not applicable to all geologic conditions. Commonly we find that layers of one viscosity alternate with layers of another viscosity. For example, a turbidite sequence contains alternating layers of sandstone and shale. In this case the analysis is considerably more complex and, because increasingly restrictive assumptions have to be made (which also includes the spacing between the layers), we stop here. The general point is that multilayers behave like thick single layers, but that the resulting arc length in a multilayer system will be less than that for a single layer.

10.8 KINEMATIC MODELS OF FOLDING

The distinction made earlier between active and passive folds describes the mechanical role of layers under an imposed stress. But this says little about the inner workings of the folded layer and, in turn, the associated strain pattern. To this end, we contrast four fundamental models, flexural folding, neutral-surface folding, shear folding, and modification of folds by superimposed strain. Folds resulting from each model have distinct properties and characteristics. Included in these descriptions are the associated strain patterns. The models are compared with natural examples in the final section.

10.8.1 Flexural Slip/Flow Folding

Take a phone book, or better yet a deck of cards, and bend it into a fold. The ability to produce a fold is achieved by slip on the surfaces of the cards relative to one another, without appreciable distortion within the surface of any individual card. If we use small circles on the top surface as well as on the two sides of the card stack as a strain gauge and we fold the stack, we indeed notice that strain accumulates only in the surfaces that are at an angle to the individual cards (i.e., the sides of the deck). The circles become ellipses in the profile plane (Figure 10.32) as well as on the other side of our folded stack. Within the plane of the cards, however, there is no strain. Folds that form in association with slip between layers (i.e., beds) are called *flexural slip folds*. The amount of slip between the layers increases away from the hinge zone and has a maximum at the inflection point. Moreover, the amount of slip is proportional to the limb dip; slip increases with increasing limb. The card deck analogy highlights three important characteristics of flexural slip folding. First, at each point in the profile plane the strain ratio and orientation differ (Figure 10.32). Secondly, in three dimensions the strain state of the fold is plane strain ($X > Y = 1 > Z$), with the orientation of the intermediate strain axis (Y) parallel to the hinge line. Thirdly, a geometrical consequence of the flexural slip model is that the fold is cylindrical and parallel (Class 1B); the bed thickness in flexural slip folds does not change. The geometrical consequences of flexural slip folding, however, are not diagnostic of the model, because they also occur in other models (see Section 10.8.2 on neutral-surface folding). Chevron folds (Figure 10.8) and kink folds (Figure 10.19) are good examples of flexural slip folding in natural rocks because of their well-developed layer anisotropy.

Slip may also occur on individual grains within a layer, without the presence of visible slip surfaces, in which case we refer to the folding as *flexural flow folding*.[15] Whereas a few details differ, the geometrical and kinematic consequences of both flexural slip and flexural flow folding are alike.

Another diagnostic feature for flexural slip folding that can be used in field analysis is that any original angular relationship in the slip surface (say, flute casts in the bedding surfaces of turbidites) before folding will maintain this an-

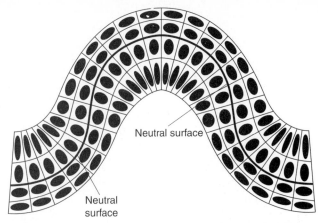

Figure 10.33 The strain pattern of neutral-surface folding in the fold profile plane.

gular relationship across the fold, because there is no strain on the folded surface. Consequently, a lineation at an angle α to the future hinge line will distribute in a cone with angle α (or small-circle pattern in spherical projection) around that line.

10.8.2 Neutral-Surface Folding

If we bend a layer of plasticine or a metal bar, we obtain a fold geometry that in many aspects is identical to that produced by flexural folding, but with a distinctly different strain pattern. This is illustrated by the distortion of circles drawn on the sides of the undeformed layer. In all three surfaces we find that circles have become ellipses. In the top of the folded layer, the long axis of each ellipse is perpendicular to the hinge line, but on the bottom of the layer the long axis is parallel to the hinge line. In the profile plane, the long axis is either parallel or perpendicular to the top and bottom portions of the folded layer, depending on where we are in that plane (Figure 10.33). This implies, of course, that there must be a surface in the fold where there is no strain. This zero-strain surface gives the model its name, *neutral-surface folding*. If you try to mimic this behavior with the deck of cards used in the previous model, you will need quite some muscle power, because the cards in the outer arc need to stretch while those in the inner arc are compressed.

The fold shape from neutral-surface folding is parallel and cylindrical, and the intermediate strain axis is parallel to the fold axis. These characteristics and plane strain conditions also hold for flexural slip folding, so they are not diagnostic for either model; the strain patterns, however, are. Because strain also accumulates in the folded surfaces during neutral-surface folding, the orientation of any feature in these surfaces changes with its position in the fold. In the outer arc an initial angle α with the fold axis

---

[15]"Flexural flow folding" has also been used to describe the migration of material from the limbs to the hinge area of folds, but we do not adopt this usage here.

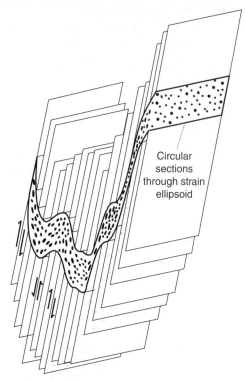

Circular
sections
through strain
ellipsoid

Figure 10.34   The strain pattern of shear folding.
From *An Outline of Structural Geology* by Hobbs, Means, & Williams. Copyright ©
1976, John Wiley & Sons, New York. Reprinted by permission of John Wiley
& Sons, Inc.

increases, while in the inner arc it decreases. So, the angle
between flute marks and the hinge line on the top surface of
the folded layer will have increased, while in the bottom
surface this angle has decreased. In both cases the orientations in any individual surface describe neither a cone nor a
plane (or small circle and great circle in spherical projection, respectively). Only in the neutral surface does the
angle remain undisturbed. In this plane the linear feature
describes a cone around the hinge line. It is not easy to use
this criterion as a field tool unless great care is taken to
group lineations from individual layers across a fold. The
strain pattern in the profile plane, however, is characteristic, and is sometimes (cumbersomely) called *tangential longitudinal strain*. Note that the position of the neutral surface is not restricted to the middle of the fold, nor does it
necessarily occur at the same relative position across fold.
In extreme cases the neutral surface may coincide with the
outer arc, in which case the long axis of each strain ellipse
in the profile plane is perpendicular to the folded surface. Is
it possible to have the neutral surface at the inner arc? This
is a good question to test your understanding of neutral-
surface folding before moving on to the third model.

10.8.3   Shear Folding

To represent sheer folding, we again turn to a deck of cards,
but now we draw a layer oblique to the sides of the deck.
When we move the cards relative to one another, we produce

a fold by a mechanism called *shear folding* (Figure 10.34).
The fold shape varies with the amount and sense of displacement between individual cards, and the layer has no
mechanical significance; that is, shear folds are passive
folds. While slip occurs on individual cards, the slip surface
and the slip vector are not parallel to the folded surface, as
they are in flexural folding. Circles drawn on the deck before we move the cards show that there is no strain in the
surface of individual cards, but there is strain in the other
surfaces. The overall strain state is plane strain, but the
hinge line is not by definition perpendicular to the displacement vector, nor parallel to the intermediate strain axis.
Most notably, the folds produced have a distinct shape because the trace of the layer on the card remains equal in
length after shearing. As a consequence, we produce similar folds (Class 2). It is about time that we produced similar
folds, because so far we have generated only parallel folds
with our models, and any field geologist will testify to the
common occurrence of similar folds. In fact, much of the
charm of shear folding is its ability to form similar folds,
which cannot be explained by the other two mechanisms
without additional modification (see Section 10.8.4).

In nature, there are no playing cards slipping past one
another. Rather, the rock flows as a continuum.[16] Shear
folds may be formed in regions where the flow field is heterogeneous, such as in glaciers (Figure 9.1). In shear zones,
relatively narrow regions that experienced high shear
strains (Chapter 12), a mylonitic foliation is commonly
present, which may act as the shear plane for folding.
Sheath folds are a spectacular example of the development
to such passive folds.

10.8.4   Fold Shape Modification

The appeal of shear folding is that it allows the formation
of similar (Class 2) folds. Recall that both flexural folding
and neutral-surface folding produce only parallel (Class
1B) folds. Now let's look at the latter two mechanisms
again, but this time we allow some additional modification
to the fold shape.

Experiments and geometrical arguments place a limit
on the degree of strain that can be accumulated in parallel
folds due to fold tightening. You may have noticed this
when folding the card deck, during which the inner-arc region increasingly experiences space problems as the fold
tightens (Figure 10.11a). Layers that occupy the inner-arc
region may be able to accommodate this space crunch,
much as the foam does in our earlier box experiments.
However, if there is no competency contrast during shortening, the later strain increments will increasingly affect
the entire fold structure (i.e., limbs and hinge), resulting in
superimposed homogeneous strain. This strain component

---

[16]Axial plane cleavage is close to the orientation of the finite XY principal
plane of strain; however, when strictly parallel, cleavage cannot be a shear
plane.

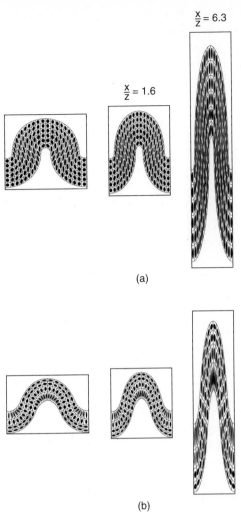

$$\frac{X}{Z} = 6.3$$

$$\frac{X}{Z} = 1.6$$

(a)

(b)

Figure 10.35 The effect of superimposed homogeneous strain on (a) a flexural and (b) a neutral-surface fold. Constant volume, plane strain with X/Z = 1.6 (20% shortening), and X/Z = 6.3 (60% shortening) are shown.

has a marked effect on the fold shape. Figure 10.35 shows the shape change and corresponding strain distributions in a flexural fold and a neutral-surface fold after superimposed homogeneous strains (constant volume, plane strain) of 20% (X/Z = 1.6) and 60% (X/Z = 6.3) shortening. You see that the initially parallel fold changes shape by thinning the limbs relative to the hinge area, resulting in a geometry approaching that of similar folds (Class 1C). However, perfectly similar (Class 2) folds are achieved only at an infinite X/Z ratio, which is obviously unrealistic. A note of caution regarding strain superimposition. In Chapter 4 you learned that strain is a second-rank tensor, and one property of tensors is that they are not commutative; that is, $a_{ij} \cdot b_{ij} \neq b_{ij} \cdot a_{ij}$. This implies that we obtain different finite strains if we reverse the sequence of parallel folding and homogeneous strain (i.e., begin with a component of layer-parallel shortening), or simultaneously add homogeneous strain during folding.

Regardless of the details of superimposed homogeneous strain, we are not able to produce Class 3 folds from initially parallel or similar folds, which therefore poses the final challenge. One likely explanation for Class 3 folds lies in the interaction between layers of different competency in a multilayered system. Consider a sequence of shale and sandstone layers (Figure 10.36). As we shorten the sequence, the strong (competent) sandstone layers form Class 1B to 1C folds (as outlined above). The weaker (less competent) intervening shale layers accommodate the shape of the folded sandstone layers when they lie within the contact strain zone, much like the thin rubber bands in our box experiments. This means that folds of Class 3 are formed when the sandstone layers are closely spaced, and folds of Class 2 when the sandstone layers are farther

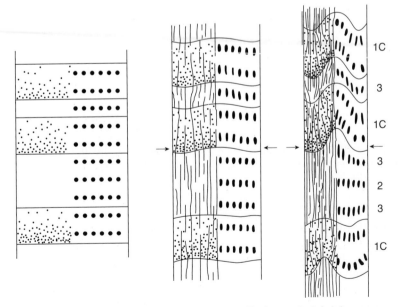

Figure 10.36 Folding of a multilayer consisting of sandstone (stippled) and shale layers. The incompetent shale layers accommodate the strong sandstone layers. This results in Class 3 and Class 2 folds when the sandstone layers are closely and more widely spaced, respectively.

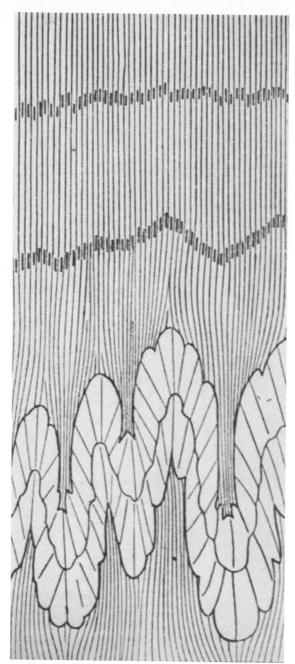

Figure 10.37 Sketch of a folded sandstone layer surrounded by slate by H. C. Sorby in the 19th century. Note that the sandstone folds are Class 1C and that the strain in the slate appears to be much more homogeneous except in the vicinity of the sandy layer.

removed. Such widely different fold geometries and thus strain states were already recognized nearly 150 years ago in the classic work of H. C. Sorby (Figure 10.37).

10.8.5   A Natural Example

All this theory begs the question of its application to nature. Do these folding models apply to real rocks? To answer this, we look at an example of an essentially parallel fold in a limestone-pebble conglomerate. Strain measured from pebbles across the folded layer gives an estimate of the overall strain pattern (Figure 10.38a). When we compare this pattern with those predicted for flexural folding (Figure 10.38c) and neutral-surface folding (Figure 10.38d), we see that the measured pattern most resembles that predicted for neutral-surface folding. The X-axis is parallel to the folded layer in the outer arc and perpendicular to the layer in the inner arc. Yet the magnitude of the strain ratios predicted by neutral surface folding is too low in the natural pattern. So, while

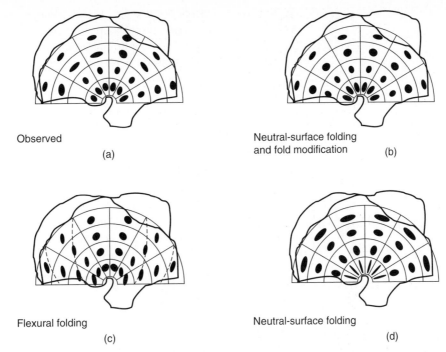

Observed
(a)

Neutral-surface folding
and fold modification
(b)

Flexural folding
(c)

Neutral-surface folding
(d)

Figure 10.38 Strain pattern in a natural fold specimen of limestone-pebble conglomerate (a). This pattern better resembles the strain predicted in neutral-surface folding (c) than in flexural folding (d). With further modification, which consists of initial compaction and material transport away from the inner-arc region, a strain pattern much like that observed in the natural sample can be produced (b).

neutral-surface folding appears to be a reasonable model, the pattern requires additional modification to match the natural fold. One solution is shown in Figure 10.38b, where prefolding compaction (nonconstant volume, layer-perpendicular shortening) is followed by neutral-surface folding, during which material is preferentially removed from the inner arc. Dissolution and material transport during folding are quite common (see Chapter 11). Other studies of natural folds show the importance of flexural folding and shear folding, so all of our folding models have proven to be reasonable first-order approximations of the inner workings of a folded layer.

Whereas the example in Figure 10.38 again emphasizes that strain in folds is highly heterogeneous, we can nonetheless estimate the bulk shortening strain from the fold shape by adapting Equation 4.1. The bulk strain is given by:

$$\mathbf{e} = (W - L)/L \qquad \text{Eq. 10.7}$$

where L is the arc length and W is the wavelength. Applying this to our example, we obtain approximately 35% bulk shortening strain. Finally, remember that the strain pattern is representative of the clasts in the folded layer and not necessarily of the matrix or the rock as a whole (Section 4.11.1). In the likely circumstance that the clasts

are relatively competent compared to the rest of the rock, the calculated strain will be an underestimate of the actual strain value.[16]

10.9 A POSSIBLE SEQUENCE OF EVENTS

We close the chapter on folding by an interpretive sequence of stages that may be involved in the formation of a single-layer fold (or a multilayer system with sufficient spacing between the layers). As before, we concentrate on buckle folds (Figure 10.39).

Immediately following deposition, compaction reduces the bed thickness. We use an intermediate value of 20% layer-perpendicular shortening strain for this first component in our example (i.e., $X_c/Z_c = 1.25$), which represents 20% area loss (Figure 10.39b); in nature, compactional strains range from 0 to 50%. During the first stage of buckling, the competent layer changes dimensions by layer-parallel shortening (lps). You recall that layer thickening is increasingly important at low viscosity ratios, and in our example we assume 20% layer-parallel shorten-

---

[16]Compaction will not lengthen the layer, so the longitudinal strain estimate from Equation 10.7 is a 'tectonic' strain.

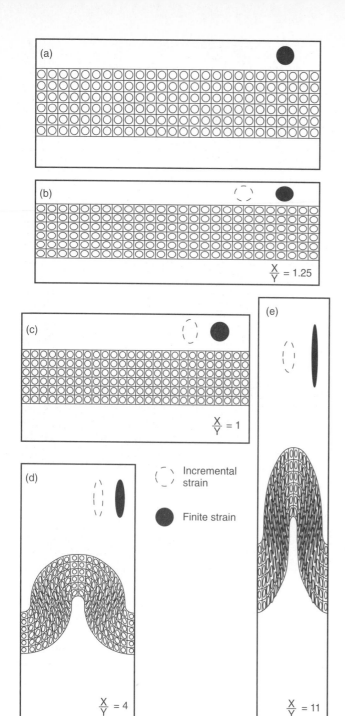

Figure 10.39 Folding scenario with the corresponding strain states at each step. An initial layer (a) undergoes 20% compaction (b). This is followed by layer-parallel shortening (c) and buckling (d). The final stage is a homogeneous shortening strain that transforms the parallel fold into a similar fold. The finite strain in the layer is indicated at each step; the incremental strain (i) and finite strain (f) ellipses of the system are shown in the upper right. Strain ratios are calculated for each step. It is assumed that volume loss (represented by Δ) occurs only at the compaction stage.

In the figure: (a), (b) $\frac{X}{Y} = 1.25$, (c) $\frac{X}{Y} = 1$, (d) $\frac{X}{Y} = 4$, (e) $\frac{X}{Y} = 11$.

Legend: Incremental strain; Finite strain.

ing. This constant-volume, homogeneous strain component (conveniently) restores the finite strain ellipse to a circle ($X_f/Z_f = 1$), but the corresponding lps strain ratio X_{lps}/Z_{lps} is 1.25 (Figure 10.39c). Continued shortening results in the initiation and growth of a parallel fold by flexural folding, with a characteristic arc length (L) as a function of the thickness (t) of the layer and the linear viscosity ratio (μ_L/μ_M). We can estimate the viscosity ratio of the system by measuring the ratio L/t and the strain ratio X_{lps}/Z_{lps} of the layer. Until this buckling stage, the strain has been homogeneous; but after fold initiation the strain becomes heterogeneous, with coaxial strain dominating the hinge area and noncoaxial strains in the limbs of the fold (Figure 10.39d). Consequently, we can calculate only the bulk finite strain of the system, represented by the ratio $X_b/Z_b(= 4)$. At this stage the resistance to folding has reached a critical value, and further shortening is achieved by superimposed homogeneous strain (Figure 10.39e). The end result is a similar fold with a strain pattern that will vary as function of the degree of compactional strain, the operative fold mechanism, the viscosity ratio, and the degree of superimposed homogeneous strain. The finite strain ratio X_f/Z_f for our complete history is 11, which corresponds to a total layer-parallel shortening strain of 70%.

Obviously you can vary this scenario in many ways by simply changing the value of the strain ratio at each step, but also by introducing volume loss during the stages of buckling and superimposed homogeneous strain. Examining these alternative scenarios will give you a good idea of folding and strain distributions under different conditions. For example, in metamorphic rocks there is little competency contrast between the layers, and layer-parallel thickening may be much more important than in low-grade sedimentary beds. Moreover, elevated temperature conditions promote grain dissolution and transport of material. It is therefore an instructive exercise to determine the strain history in a system experiencing 50% volume loss (Δ = –0.5) during the buckling and superimposed homogeneous strain stages, which represents a geologically reasonable condition that will return in the chapter on rock fabrics (Chapter 11).

10.10 CLOSING REMARKS

Hopefully you did not lose sight of the natural beauty of folds after learning so much about description, terminology, and mechanics. Folds are simply fascinating to look at (Figure 10.40). Ultimately, we study them to understand the conditions and significance of deformation of rocks and regions. Folding patterns give us a good estimate of the orientation of regional strain, as well as its change with time through the analysis of fold superposition. Superposed

Figure 10.40 Large recumbent fold in Nagelhorn (Switzerland), showing the characteristic nappe style of deformation in this part of the European Alps (see Section 19.2).

folding presents a particular challenge that awakens the puzzler in us. Recognition of fold generations should be based only on fold superimposition patterns, aided by style for correlation between outcrops, if necessary. As a 'tongue-in-cheek' rule, the number of fold generations should always be less than the number of folds you have encountered in the field.

Folds are good strain markers, but the mechanical contrast between neighboring layers or with the matrix implies that strain data is representative only of the folded layer, and not necessarily that of the bulk rock. This should not be a crippling limitation for regional analysis, nor is the problem unique to folds; it holds essentially for all strain markers. Strain within a fold is markedly heterogeneous and the pattern may be completely different from the regional conditions. The relationship of folds to regional stresses is even less straightforward because of the mechanical interaction between layers with contrasting competency. Nonetheless, in many circumstances, folds will tell us a lot about these and other properties of rocks (e.g., temperature).

Much of the material in this chapter focuses on single-layer folding rather than on more complex multilayer fold-

ing, but the material should give you sufficient insight to tackle the literature on this and other advanced topics (such as nonlinear material rheologies). You must be itching to go out into the field and test some of these concepts and ideas; otherwise, a few laboratory experiments may satisfy your appetite until the field summer comes around. When you are in the field, you will find that folds are commonly associated with other structural features, such as foliations and lineations; these features form the topic of the next chapter.

ADDITIONAL READING

Biot, M. A., 1961, Theory of folding of stratified viscoelastic media and its implication in tectonics and orogenesis: Geological Society of America Bulletin, v. 72, p. 1595–1620.

Currie, J. B., Patnode, H. W., and Trump, R. P., 1962, Development of folds in sedimentary strata: Geological Society of America Bulletin, v. 73, p. 655–674.

Dietrich, J. H., 1970, Computer experiments on mechanics of finite amplitude folds: *Canadian Journal of Earth Sciences,* v. 7, p. 467–476.

Hudleston, P. J., and Lan, L., 1993, Information from fold shapes: *Journal of Structural Geology,* v. 15, p. 253–264.

Latham, J., 1985, A numerical investigation and geological discussion of the relationship between folding, kinking and faulting: *Journal of Structural Geology,* v. 7, p. 237–249.

Price, N. J., and Cosgrove, J. W., 1990, *Analysis of geological structures:* Cambridge, Cambridge University Press, 502 p.

Ramberg, H., 1963, Strain distribution and geometry of folds: Bulletin of the Geological Institute, University of Uppsala, v. 42, p. 1–20.

Ramsay, J. G., and Huber, M. I., 1987, *Folds and Fractures:* The techniques of modern structural geology, v. 2, New York, Academic Press, 700 p.

Sherwin, J., and Chapple, W. M., 1968, Wavelengths of single layer folds: A comparison between theory and observation: *American Journal of Science,* v. 266, p. 167–179.

Thiessen, R. L., and Means, W. D., 1980, Classification of fold interference patterns: A reexamination: *Journal of Structural Geology,* v. 2, p. 311–316.

Chapter 11
Fabrics: Foliations and Lineations

11.1 INTRODUCTION

In everyday language, we use the word "fabric" frequently; we even wear it. When talking about fabrics that are used to make curtains, we mean a cloth made by weaving fibers in some geometric arrangement. In a philosophical moment we might wonder about the "fabric of life," by which we mean the underlying organization of life. But, as we have found with many terms, the word "fabric" has a related yet somewhat different meaning in geology. To a structural geologist, the *fabric* of a rock is the geometrical arrangement of component features in the rock, seen on a scale large enough to include many samples of each feature. The features themselves are called *fabric elements*.

Examples of fabric elements include mineral grains, clasts, compositional layers, fold hinges, and planes of parting. Fabrics that form as a consequence of tectonic deformation of rock are called *tectonic fabrics,* and fabrics that form during the formation of the rock are called *primary fabrics*. It may sound picky, but some structural geologists also make a distinction between 'fabric' and 'texture.' Whereas *texture* is sometimes used as a synonym for *microstructure,* for example, in igneous texture, we restrict the term *texture* to crystallographic orientation patterns in an aggregate of grains (see Chapter 12).

Tectonic fabrics provide clues to the strain state of the rock, the geometry of associated folding, the processes involved in deformation, the kinematics of deformation, the timing of deformation (if the fabric is defined by an arrangement of datable minerals), and ultimately about the tectonic evolution of a region. The purpose of this chapter is to describe two common fabric elements in rocks, *foliations* and *lineations,* and to introduce you to the outcrop characteristics and interpretation of these elements.

11.2 FABRIC TERMINOLOGY

Let's start by developing the inevitable vocabulary to discuss tectonic fabrics (see Table 11.1). If there is no *preferred orientation* (i.e., alignment) of the fabric elements, then we say that the rock has a *random fabric* (Figure 11.1a). Undeformed sandstone, granite, or basalt are rocks with random fabrics. If the fabric is the same throughout the rock, we say that the rock has a *homogeneous fabric*. Deformed rocks typically have a *nonrandom fabric* or a *preferred fabric,* in

Table 11.1 Tectonic Fabric Terminology

Axial plane cleavage Cleavage that is parallel or subparallel to the axial plane of a fold; it is generally assumed that the cleavage formed roughly synchronous with folding and is subparallel to the XY-plane of the bulk finite strain ellipsoid.

Cleavage A secondary fabric element, formed under low temperature conditions, that imparts on the rock a tendency to split along planes.

Crosscutting cleavage Cleavage that is not parallel to the axial plane of a fold (also *nonaxial plane cleavage*); *transecting cleavage* is used when cleavage and folding are considered roughly synchronous in a transpressional regime.

Fabric The geometrical arrangement of component features in a rock, seen on a scale large enough to include many samples of each feature.

Foliation The general term for any type of 'planar' fabric in a rock (e.g., bedding, cleavage, schistosity).

Gneissosity Foliation in feldspar-rich metamorphic rock, formed at intermediate to high temperatures, that is defined by compositional banding; the adjective *ortho* and *para* are used for igneous and sedimentary protoliths, respectively.

Lineation A fabric element that is best represented by a line, meaning that one of its dimensions is much longer than the other two.

Migmatite Semi-chaotic mixture of layers formed by partial melting and deformation.

Mylonitic foliation A foliation in ductile shear zones that is defined by the dimensional preferred orientation of flattened grains; the foliation tracks the XY-flattening plane of the finite strain ellipsoid and is therefore at a low angle to the shear zone boundary.

Phyllitic cleavage Foliation that is composed of strongly aligned fine-grained white mica and/or chlorite; the mineralogy and fabric of phyllites give the rock a distinctive silky appearance, called *phyllitic luster*.

Schistosity Foliation in metamorphic rock, formed at intermediate temperatures, that is defined by mica (primarily muscovite and biotite) giving the rock a shiny appearance.

Texture The pattern of crystallographic axes in an aggregate of grains; also *crystallographic fabric*.

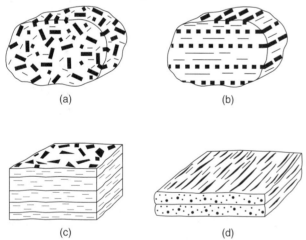

Figure 11.1 Basic categories of fabrics. (a) A random fabric. The fabric elements are dark, elongate crystals. The long axes of these crystals are not parallel to one another. (b) A preferred fabric, in which the long axes of the elongate crystals are aligned with one another. (c) A foliation. The fabric elements are planar, and the elements are essentially parallel to one another. (d) A lineation. The fabric elements are linear; in this example, we show the alignment of the fabric elements in a single plane.

which the fabric elements are aligned in some manner and/or are repeated at an approximately regular spacing (Figure 11.1b). There are two main classes of preferred fabrics in rock. A planar fabric, or *foliation* (Figure 11.1c), is one in which the fabric element is a planar or tabular feature (i.e., it is shorter in one dimension than in the other two), and a linear fabric, or *lineation* (Figure 11.1d), is one in which the fabric element is effectively a linear feature (i.e., it is long in one dimension relative to the other dimensions). In conversation, structural geologists may use the word "fabric" alone to imply the existence of a preferred fabric (as in, "that rock has a strong fabric"), but you should use appropriate modifiers if your meaning is not clear from context alone.

We're not quite done with terminology yet! Fabrics are complicated features, and there are lots of different adjectives used by structural geologists to describe them. For example, if you can keep splitting the rock into smaller and smaller pieces, right down to the size of the component grains, and can still identify a preferred fabric, then we say that the fabric is *continuous* (Figure 11.2a). In practice, if the fabric elements are closer than 1 mm (i.e., below the resolution of the eye), the fabric is continuous. When there is an obvious spacing between fabric elements, we say that the fabric is *spaced* (Figure 11.2b).

Rocks with a *penetrative* tectonic fabric are called *tectonites*. In this context, the term penetrative refers to the repetition of a fabric elements throughout the rock at the scale of your observation. Rocks with dominantly linear fabric elements are called *L-tectonites;* rocks with dominantly planar fabrics are called *S-tectonites;* and, not surprisingly, rocks with both types of fabric elements are called *LS-tectonites* (Figure 11.3).[1] Why create such jargon? Simply to highlight rocks whose internal structure has been substantially

---

[1]"L" stands for "lineation"; "S" stands for "surface" in English, "schistosité" in French, or "Schieferung" in German.

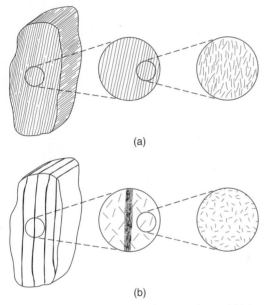

(a)

(b)

Figure 11.2 The distinction between continuous and spaced fabrics. (a) A continuous fabric. The lines represent a planar fabric element that is still visible, no matter how small your field of view (at least down to the scale of individual grains). (b) A spaced fabric. The rock between the fabric elements does not contain the fabric. The circled areas represent enlarged views.

changed during deformation. Typically, the deformation that leads to the formation of a tectonite is accompanied by metamorphism, so the fabric is defined by grains that have been partially or totally recrystallized, and/or by new minerals that have grown during deformation. Examples of tectonites include slates, schists, and mylonites.

11.3 FOLIATIONS

A *foliation* is any type of planar fabric in a rock. We are admittedly a bit loose in our use of 'planar.' Strictly speaking, a plane does not contain any curves or changes in orientation; "curviplanar" or "surface" would be more appropriate terms. Although foliations are generally not perfectly planar, structural geologists nonetheless are in the habit of talking about 'planar' fabrics. Thus, bedding, cleavage, schistosity, and gneissosity (topics of this chapter) all qualify as foliations (Figure 11.4). Fractures, however, are not considered to be foliations, because fractures are breaks through a rock and are not a part of the rock itself. A rock may contain several foliations, especially if it has been deformed more than once. To keep track of different foliations, geologists add subscripts to the foliations: S_0, S_1, S_2, and so on. S_0 is used to refer to bedding, S_1 is the first foliation formed after bedding, S_2 forms after S_1, and so on. The temporal sequence of foliation development is defined by crosscutting relationships, but in complexly deformed areas it may be quite a challenge to determine which foliation is which unless independent constraints on time are available.

There are many types of tectonic foliations that are distinguished from one another simply by their appearance. The physical appearance of a foliation reflects the process(es) by which it formed. The processes, in turn, are controlled partly by the composition and grain size of the original lithology, and partly by the metamorphic conditions at which the foliation formed. Different names are used for the different types of foliations. In Table 11.2 and the following discussion, we will introduce different types of foliation roughly in sequence of increasing metamorphic conditions—cleavage first, then schistosity, then gneissic layering.

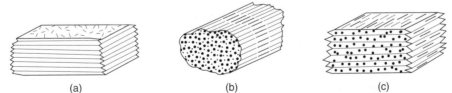

(a)   (b)   (c)

Figure 11.3 The nature of tectonites. (a) An S-tectonite. This fabric is dominantly a foliation, and the rock may tend to split into sheets parallel to the foliation. Within the planes of foliation, linear fabrics are not aligned, or are not present at all. (b) An L-tectonite. The alignment of linear fabric elements creates the dominant fabric, so the rock may split into rodlike shapes. In L-tectonites, there is not a strong foliation. (c) An L/S-tectonite. The rock possesses a strong foliation and a strong lineation.

Figure 11.4 Disjunctive cleavage in mica-rich rocks (Rhode Island). Note the variation in cleavage spacing between steeply clipping bedding layers. Width of view is ~4 m.

Table 11.2   Foliation Classification Scheme

Continuous cleavage	(a) coarse cleavage (e.g., pencil cleavage)
	(b) fine cleavage (e.g., slaty cleavage, phyllitic cleavage)
Spaced cleavage	(a) disjunctive cleavage (e.g., stylolitic cleavage)
	(b) crenulation cleavage
Schistosity	
Gneissic layering	

11.3.1   What Is Cleavage?

Cleavage in rocks has been defined in different ways by different people, and thus use of this term in the literature is confusing. We favor using a nongenetic definition in which cleavage is defined as a secondary fabric element, formed under relatively low temperature metamorphic to non-metamorphic conditions, that imparts on the rock a tendency to split along planes. The point of this definition is to emphasize: (1) that cleavage is a component of the rock that forms subsequent to the origin of the rock; (2) that the term 'cleavage' is used, in practice, for tectonic planar fabrics formed at or below lower greenschist facies conditions (i.e., <350°C). 'Cleavage' is not used when referring to the fabric in coarse-grained schists or in gneiss; and (3) that in a rock with cleavage, there are planes of weakness across which the rock may later break when uplifted and exposed at the surface of the earth), even though cleavage itself forms without loss of cohesion. By this definition, an array of closely spaced fractures is not a cleavage.

We recognize four main categories of cleavage, which are differentiated from one another by their morphological characteristics (i.e., by how they look in outcrop). These are disjunctive cleavage, pencil cleavage, slaty cleavage, and crenulation cleavage.

Figure 11.5 Spaced disjunctive cleavage (or solution cleavage) in steeply-dipping limestone from the Harz Mountains of Germany. The cleavage is marked by the narrow dark bands that cut across the original lighter-colored argillaceous limestone. (a) Outcrop view; (b) close-up view of central portion. In (c) the cleavage domain and microlithon of spaced cleavage are illustrated. Note that in (c), bedding dips gently to the left.

11.3.2   Disjunctive Cleavage

Disjunctive cleavage is a foliation that forms primarily in sedimentary rocks[2] that have been subjected to a tectonic differential stress under subgreenschist facies metamorphic conditions. It is defined by an array of subparallel fabric elements, called *cleavage domains,* in which the original rock fabric and composition have been markedly changed by the process of *pressure solution.* Domains are separated from one another by intervals, called *microlithons,* in which the original rock fabric and composition are more or less preserved (Figure 11.5). The adjective "disjunctive" implies that the cleavage domains cut across a preexisting foliation (usually bedding) in the rock without affecting the orientation of the preexisting fabric in the microlithons.

---

[2]There are descriptions of disjunctive cleavage in igneous rocks, but it is most typical of sedimentary rocks.

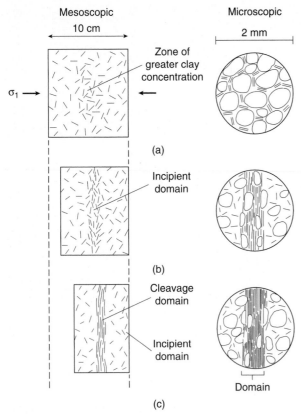

Zone of
greater clay
concentration

$\sigma_1 \rightarrow$

(a)

Incipient
domain

(b)

Cleavage
domain

Incipient
domain

Domain

(c)

Figure 11.6 Evolution of spaced disjunctive cleavage. (a) Pre-cleavage fabric of the rock. In the area indicated by the arrow in the mesoscopic image, there happens to be a greater initial concentration of clay. The microscopic image indicates that the clay flakes are randomly oriented. (b) As shortening and pressure solution occur, the zone in which there had initially been a greater clay concentration evolves into an incipient cleavage domain. At this stage, grains are being preferentially dissolved on the faces perpendicular to S_1, and the clay flakes are collapsing together. (c) Ultimately, a clearly defined cleavage domain is visible. In the domain, the clay flakes are packed tightly together and only small relicts of the soluble mineral grains are visible.

Because pressure solution is always involved in the formation of a disjunctive cleavage, other terms such as (*pressure-solution cleavage*) and *stylolitic cleavage* have been used for this structure. If the context is clear, some authors may refer to the structure simply as *spaced cleavage*, though as we see later, there are other types of spaced cleavages (e.g., crenulation cleavage). Earlier in this century, many geologists incorrectly considered cleavage domains to be brittle fractures formed by loss of cohesion. The term fracture cleavage should be avoided when referring to solution cleavage, because cleavage cannot be composed of fractures. Such arrays of closely spaced fractures should be called fracture or joint arrays.

Now that we've gotten through the jargon, let's discuss how disjunctive cleavage forms. Consider a horizontal bed of argillaceous (clay-rich) limestone or sandstone that is subjected to a compressive stress (σ_1 in Figure 11.6). Dissolved ions produced by pressure solution diffuse away from the site of dissolution through a fluid film that adheres to the grain surfaces. The ions then either precipitate at crystal faces where compressive stress is less, precipitate nearby in the *pressure shadows* adjacent to rigid grains, or enter the pore fluid system and are carried out of the local rock environment entirely. Remember that in order for pressure solution to occur, a thin layer of water molecules must be chemically bonded to grain surfaces. If the water is not bonded, a grain in the water can sustain only isotropic stress and pressure solution won't occur. Thus, pressure solution is not just the dissolution of minerals in the free fluid of pore spaces.

The distribution of clay in rock is not uniform. Pressure solution occurs more rapidly where the initial concentration of clay is higher. The exact role of clay on pressure solution rates remains enigmatic. Perhaps swelling clay, which contains interlayers of bonded water, multiplies the number of diffusion pathways available for ions and thus multiplies the diffusion rate. Alternatively, the highly charged surfaces of clay grains may act as a chemical catalyst for the dissolution reaction. Field observations suggest that a rock must contain >10% clay in order for solution cleavage to develop. Where it occurs, the process preferentially removes more soluble grains. Thus, in an argillaceous limestone, calcite is removed and clay and quartz get progressively concentrated. In an argillaceous sandstone, the process is effectively the same as that in argillaceous limestone, except that quartz is the mineral that preferentially dissolves, and clay alone is concentrated. As the framework grains of carbonate or quartz are removed, the platy clay grains collapse together like a house of cards. Concentration of clay in the domain further enhances the solubility of the rock in the incipient domain; that is, there is positive feedback. Eventually, a discrete domain develops in which there is a *selvage,* the material filling the domain, composed of mostly clay (and quartz) with some relict corroded framework grains. In the selvage, the clay flakes are packed together with a dimensional preferred orientation. If deformation continues, the domain continues to thicken as pressure solution continues along its edges. As a result, compositional contrast between cleavage domains and microlithons becomes so pronounced that it defines a new stratification in the rock that is mechanically more significant than original bedding. The process of spaced solution cleavage formation is identical to the processes by which bedding-parallel stylolites form as a consequence of compaction loading.

The *spacing of cleavage domains* in a rock depends on the initial clay content and on strain. If the clay content is higher, the domains are more closely spaced (Figure 11.4). Spacing also changes with progressive deformation. As strain increases, more cleavage domains nucleate and thus the domain spacing decreases. Domain spacing may also be related to the magnitude of differential stress during deformation, for experiments suggest that the rate of pressure solution is proportional to the magnitude of differential stress. During the process of domain formation, the fabric

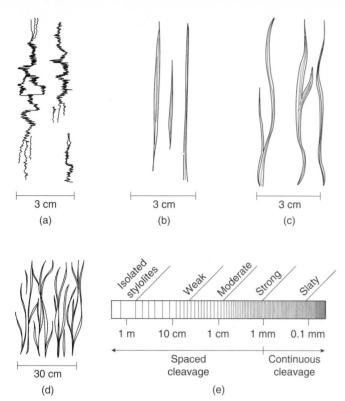

3 cm   3 cm   3 cm
(a)   (b)   (c)

30 cm
(d)   (e)

Isolated stylolites   Weak   Moderate   Strong   Slaty

1 m   10 cm   1 cm   1 mm   0.1 mm

Spaced cleavage   Continuous cleavage

Figure 11.7 Cross-sectional sketches of morphological characteristics used for cleavage description and classification. (a) Sutured domains; (b) planar domains; (c) wavy domains; (d) an anastamosing array of wavy domains. In (e) the description of spaced cleavage based on domain spacing is shown.

Figure 11.8 Pencil cleavage in shale.

of the microlithons is relatively unaffected, though microscopic analysis of microlithons indicates the presence of incipient pressure solution features or newly crystallized grains.

Before we leave the topic of disjunctive cleavage, we need to introduce you to the vocabulary that geologists use to describe cleavage in the field. Classification and description of disjunctive cleavage is based on characteristics of surface morphology of domains, and on domain spacing (Figure 11.7). If the initial clay content in the host rock is low, domain surfaces are severely pitted. In cross section, such domains resemble toothlike or jagged sutures on a skull; this type of cleavage domain is called a *sutured domain.* If the clay content is high, cleavage domains tend to have thick selvages that have smooth borders; these domains are called *nonsutured.* When viewed under high magnification with a microscope, you will see that thick domains, in some cases, are composed of a dense braid of threadlike sutured domains. Cleavage domains can be called either *wavy,* if the domain undulates, or *planar,* if the domain does not. If wavy domains are closely spaced, such that the domains merge and bifurcate and give the fabric the appearance of braided hair, the cleavage is called an *anastamosing cleavage.*

The average spacing between domains is also a useful criterion for classification of cleavage. Figure 11.7e shows a simple field classification based on spacing. If the spacing between domains is greater than about 1 m, then the rock really doesn't have an obvious fabric. Such isolated pressure-solution domains, which formed in response to tectonic stress, are called *tectonic stylolites.* The prefix "tectonic" distinguishes these structures from bedding-parallel stylolites formed by compaction. In general practice, if domains are spaced between about 10 cm and 1 m, it is *weak cleavage;* 1 cm to 10 cm, it is *moderate cleavage;* and less than 1 cm, it is *strong cleavage.* When cleavage domains are less than 1 mm apart, the cleavage is *continuous* (e.g., slaty cleavage). These distinctions are only approximate, and we caution you again that different authors may use the adjectives differently.

11.3.3 Pencil Cleavage

If a fine-grained sedimentary rock (shale or mudstone) breaks into elongate pencil-like shards because of its internal fabric, we say that it has a *pencil cleavage* (Figure 11.8). Typically, pencils are 5–10 cm long and half a centimeter to a centimeter in diameter. In outcrop, pencil cleavage looks like it results from the interaction of two fracture sets (and in some locations, it is indeed merely a consequence of the intersection between a fracture set and bedding), but in a rock with a true pencil cleavage, the parting reflects the internal alignment of clay grains in the rock.

Pencil cleavage forms because of the special characteristics of clay. Because of their strong shape anisotropy, clay flakes attain a preferred orientation parallel to bedding when they settle out of water and are compacted together. This preferred orientation imparts the tendency for clay to break on bedding planes that is displayed by *shale*. After compaction, a strain ellipsoid representing the state of strain in the shale relative to a mass of uncompacted clay would look like a pancake parallel to bedding (Figure 11.9). Now, imagine that the shale is subjected to layer-parallel shortening. Cleavage formation processes begin to take place: large detrital phyllosilicates fold and rotate, whereas clay flakes undergo pressure solution along domains perpendicular to the shortening direction, and new clay crystallizes out of solution. The plane of new flakes or parts of rotated grains is roughly perpendicular to the shortening direction (i.e., at a high angle to bedding). In addition, quartz may begin to dissolve, and as these framework grains are removed, clay flakes progressively rotate so that their planes are at a high angle to the shortening direction. Microfolding may occur during this stage of the process, but because of the fine grain size, the effects of microfolding are hard to see, except under the microscope. In any case, a tectonic foliation forms at high angles to original bedding. At an early stage during this process, the new tectonic fabric is comparable in degree of development to the initial bedding-parallel fabric. At this stage, the strain ellipsoid representing this state would look something like a big cigar or a French fry, and the rock displays pencil cleavage. Thus, pencil cleavage is a fabric found in weakly deformed shale in which the tendency of the shale to part on bedding planes is about the same as the tendency for it to part on an incipient tectonic cleavage that is at a high angle to bedding.

11.3.4 Slaty Cleavage

Pencil cleavage may be considered a snapshot of an early stage in the process by which slaty cleavage develops. If shortening perpendicular to cleavage planes continues, clay throughout the rock displays a preferred orientation at an angle to the primary sedimentary fabric. Eventually, the tectonic fabric dominates over the primary fabric (Figure 11.9d). The finite strain ellipsoid representing cleavage development at this stage is a pancake parallel to the tectonic fabric. Formation of this fabric occurs by much the same process as the formation of disjunctive cleavage in argillaceous sandstone or limestone, but the resulting domains are so closely spaced that there are effectively no uncleaved microlithons, and the clay in the entire rock mass displays the tectonically induced preferred orientation. When a rock has this type of fabric, we say that it has *slaty cleavage* (Figure 11.10). In other words, slaty cleavage is defined by a strong dimensional-preferred orientation of phyllosilicates in a very clay-rich sedimentary rock. The resulting rock, which is a low-grade metamorphic rock, is called a *slate*. Slaty cleavage tends to be smooth and planar. This characteristic, coupled with the penetrative nature of slaty cleavage, means that slates can be split into thin

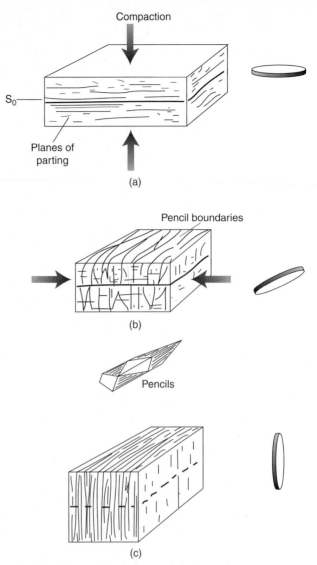

Figure 11.9 Sketches illustrating the progressive development of slaty cleavage via formation of pencil structure. (a) Compaction during burial of a sedimentary rock produces a weak preferred orientation of clay parallel to bedding. A representative strain ellipsoid is like a pancake in the plane of bedding. (b) Shortening parallel to layering creates an incipient tectonic fabric. Superposition of this fabric on the primary compaction fabric leads to the formation of pencils. The representative strain ellipsoid is elongate, with the long axis parallel to the pencils. (c) Continued tectonic shortening leads to formation of slaty cleavage at a high angle to bedding. The phyllosilicates are now dominantly aligned with the direction of cleavage, and a representative strain ellipsoid is oblate and parallel to cleavage.

sheets, which made them popular roofing material in the nineteenth and early part of the twentieth century.

Not all slaty cleavage looks the same. If we say that a rock has *weak slaty cleavage,* then the degree of preferred orientation is not perfect and the rock may still contain thin microlithons in which cleavage fabric is not completely formed. If we say that a rock has *strong slaty cleavage,* then the cleavage is completely penetrative or continuous. To emphasize this contrast with spaced cleavage, strong

Figure 11.10 An overturned syncline with well-developed axial plane slaty cleavage (Southern Appalachians).

slaty cleavage is sometimes called *continuous cleavage*. However, at high magnifications the domainal character of slaty cleavage becomes apparent (Figure 11.11). Slaty cleavage forms under temperature conditions that mark the onset of metamorphism and new growth of phyllosilicates. At lower-grade conditions, the clay mineralogy of slate tends to resemble that of shale, though there is a notable decrease in the amount of interlayered water (i.e., smectite, the water-bearing clay, transforms to illite). But as metamorphic temperature increases, the illite becomes more micalike in structure, though it is still very fine grained. Thus, rocks with higher-grade slaty cleavage display a distinct

sheen on cleavage planes, the slate becomes significantly harder than shale, and may ring when you hit it. Excellent material for the construction of pool tables.

11.3.5 *Phyllitic Cleavage and Schistosity*

If metamorphic conditions reach the greenschist facies, that is, higher than the metamorphic conditions that lead to formation of slate, the clay and illite in a pelitic rock react to form white mica[3] and chlorite. If reaction occurs in an

---

[3]White mica is essentially very fine-grained muscovite. It contains more potassium and aluminum than illite, and is better ordered than illite.

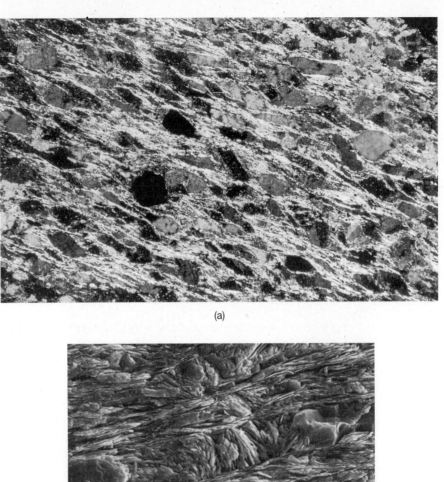

(a)

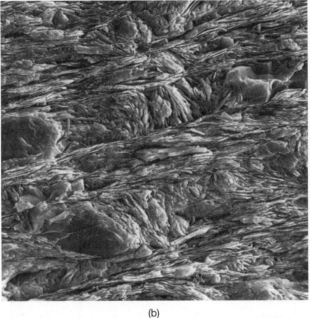

(b)

Figure 11.11 (a) Photomicrograph of a continuous cleavage (Newfoundland); width of view is 2 mm. (b) Scanning electron micrograph from a slate in the Rheinische Schiefergebirge (Germany). Note the microcrenulation of bedding-parallel micas, creating distinct microlithons with mica roughly parallel to bedding and cleavage domains with mica parallel to cleavage. Width of view is 100 μm.

anisotropic stress field, these phyllosilicates grow with a strong preferred orientation. Rock that is composed of strongly aligned fine-grained white mica and/or chlorite is called *phyllite,* and the foliation that it contains is called *phyllitic cleavage.* The mineralogy and fabric of phyllites give the rock a distinctive silky luster.

When metamorphic conditions get into the middle greenschist facies or higher, the minerals in a pelitic lithology react to form coarser-grained mica and other minerals. Again, if the reactions take place in an anisotropic stress field, the mica has a strong preferred

orientation. The resulting rock is a *schist,* and the foliation it displays is called *schistosity.* The specific assemblage of minerals that forms depends not only on the specific pressure and temperature conditions of the rock, but also on the chemical composition of the protolith and on the degree to which chemicals are added or removed from the rock by migrating fluids. Field geologists name a schist by the assemblage of metamorphic minerals that it contains (e.g., a garnet-biotite schist). If the schist contains *porphyroclasts* (relict large non-platy crystals) or *porphyroblasts* (newly grown large non-platy crystals),

Figure 11.12 Photomicrograph of crenulation cleavage in a phyllite (Vermont). The spaced crenulation cleavage deforms an earlier continuous cleavage. Note that cleavage spacing and intensity are less in the sandy layers.

3 mm

the schistosity tends to be wavy, for the micas curve around the large, nonplaty crystals. In some literature, you may find that schists containing large relict grains of feldspar are called *augen schists,* because the feldspar grains look like eyes ("Augen" in German).

11.3.6   *Crenulation Cleavage*

If a lithology containing a closely and evenly spaced foliation is shortened in a direction at a low angle to this foliation, it will crinkle like the baffles in an accordion (Figure 11.12). In fine-grained lithologies like slate or phyllite, the microfolds are closely spaced and the spacing tends to be very uniform. In such cases, the axial planes of the crenulations defines a new foliation called *crenulation cleavage* (Figure 11.13). Crenulations that look like the baffles of an accordion are called *symmetric crenulations.* If the shortening direction during formation of a crenulation cleavage is not parallel to the preexisting foliation in

the rock, there will be a component of shear along the crenulation cleavage planes. As a result, the microfolds that define the cleavage are shaped like little sigmoids (i.e., they are S- or Z-shaped). The resulting fabric is called *asymmetric crenulation cleavage.* In a given area, both symmetric and asymmetric crenulation cleavages may occur. For example, on the limbs of a fold, where there has been shear during folding, asymmetric crenulation occurs, whereas in the hinge zone of the fold, the crenulation is symmetric (Figure 11.13).

We remind you again that a requirement for the formation of crenulation cleavage is the existence of a preexisting strong lamination or foliation. Crenulation cleavage won't form in a sandstone, but may form in a micaceous shale with a strong bedding-plane foliation or in a rock that already contains a slaty or phyllitic cleavage. Crenulation cleavage that is visible in the field generally forms from rocks with a preexisting tectonic cleavage. Thus, a

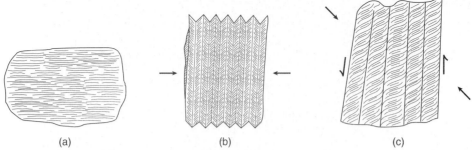

(a)   (b)   (c)

Figure 11.13 The two basic categories of crenulation cleavage. (a) Symmetric crenulation cleavage; (b) asymmetric (sigmoidal) crenulation cleavage. The arrows indicate a possible component of shear associated with this crenulation geometry.

crenulation is typically an S_2 foliation, which has been superimposed on an earlier S_1 foliation.

Crenulation cleavage forms under conditions that are also amenable for the occurrence of pressure solution. If the starting rock contains a mixture of quartz and clay or fine-grained mica, then as the crenulations form, the quartz is preferentially removed from the limbs of the microfolds and is reprecipitated at the hinges, where the differential stress is less (Figure 11.14). Gradually, phyllosilicate minerals concentrate on the limbs and quartz is concentrated at the hinges. As the process continues, the phyllosilicate minerals recrystallize and progressively rotate into parallelism with the new foliation. The mineralogical *differentiation* can be so complete, that the old foliation disappears entirely and is replaced by a new foliation, which is defined not only by preferred orientation of the phyllosilicates, but by microcompositional layering (Figure 11.15). This process is a type of *transposition,* by which a preexisting foliation is transposed into a new orientation. If quartz is largely removed by progressive pressure solution, a new slaty or phyllitic cleavage eventually develops and the crenulated appearance of the rock fades (Figure 11.15c). When this happens, all traces of the original slaty or phyllitic cleavage, the one predating the crenulation, might be destroyed. Thus a crenulation cleavage can evolve from a rock with a preexisting slaty or phyllitic cleavage, but, paradoxically, it can also evolve into a rock with a slaty or phyllitic cleavage.[4]

---

[4]To some extent, crenulation cleavage is a matter of observation scale; many 'good' slates show evidence of microfolds (Figure 11.11b).

Figure 11.14 Incipient differentiation in a spaced crenulation cleavage (Vermont). Width of view is 2 cm.

11.3.7 Gneissic Layering and Migmatization

Gneiss is a metamorphic rock in which the foliation is defined by compositional banding (Figure 11.16). Gneiss can be derived from a sedimentary protolith, in which case it is called a *paragneiss,* or from an igneous protolith, in which case it is called an *orthogneiss.* It is often difficult to decide whether a particular rock is an orthogneiss or a paragneiss; this requires careful field and petrologic analysis. Low-grade (greenschist facies) rocks may contain mica and have a schistosity. However, in high-grade gneiss (amphibolite to granulite facies), muscovite reacts to form feldspar, so

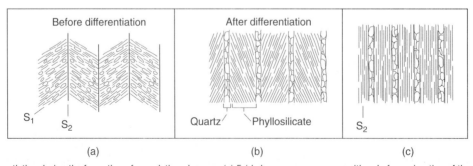

(a)   (b)   (c)

Figure 11.15 Differentiation during the formation of crenulation cleavage. (a) Fairly homogeneous composition, before migration of the quartz. (b) Quartz accumulates in the hinges of the crenulations, and the phyllosilicates are concentrated in the limbs; the result is the formation of compositionally distinct bands in the rock. (c) Complete transposition of the S_1 foliation into a new S_2 foliation.

Figure 11.16 Note the transposed gneiss near Parry Sound, Grenville Orogen (Canada).

the rock contains no good schistosity. Typically, the layers in a gneiss have different colors. For example, a gneiss formed from a rock of granitic composition consists of alternating felsic layers and mafic layers; the light-colored layers are rich in feldspar and quartz and/or muscovite, whereas darker layers contain more biotite and amphibole. This color banding in gneiss is called compositional banding or gneissis. A special kind of gneiss, called *augen gneiss,* contains relatively large feldspar clasts surrounded by a finer-grained matrix.

How does the compositional banding in gneiss form? There are actually several processes involved in the formation of gneissic foliation (Figure 11.17). First, it may occur by *inheritance* from original compositional contrasts. If the protolith (the rock from which the metamorphic rock formed) was a stratified sequence in which layers had different compositions (e.g., alternating sandstone and shale), then metamorphism will transform this sequence into a compositionally banded metamorphic rock. Secondly, it may occur by *transposition* via folding of an earlier layering. *Transposition* is an important process during deformation under metamorphic conditions, and appears at several places in this book (see also Chapter 12). When we discussed crenulation, we already introduced one type of transposition. Another type of transposition involves intense folding. If a package of rock containing compositional layering is subjected to intense deformation, it may develop isoclinal folds. If the hinges of the fold are detached and a new sequence of compositional layers has formed, then we say that the new layering is a type of *transposed foliation.* The apparent stratigraphy in a rock with a transposed foliation does not represent the original stratigraphy of the rock, though it may have been derived from the original stratigraphy. In other words, the sequence of compositional layers in a transposed rock is not the original stratigraphic succession. In practice, the resulting rock is considered to be a gneiss if it is a high-grade metamorphic rock, and the transposition involved passive folding. Thirdly, gneissic layering may be formed by *metamorphic differentiation.* The thermodynamics governing diffusion

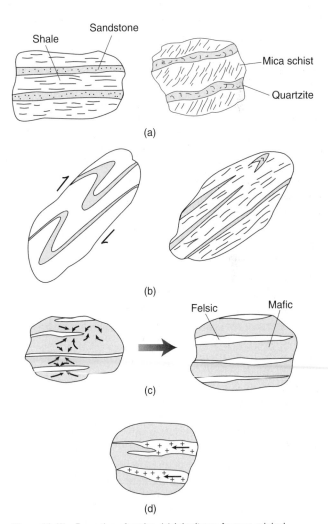

Figure 11.17 Formation of gneiss. (a) Inheritance from an original lithology; (b) creation of new compositional banding via transposition; (c) metamorphic differentiation; (d) *lit-par-lit* intrusion.

during metamorphism causes certain ions to be excluded from the formation of new metamorphic minerals in a specific layer. The excluded ions accumulate to form different minerals in an adjacent layer. Thus, minor differences in original composition between layers may be amplified into

Figure 11.18 Photograph of reduction spots in slate (Vermont).

major compositional changes during metamorphism. The resulting rock with alternating layers of different composition is a gneiss. Finally, gneissic banding may originate from an igneous process called *lit-par-lit intrusion* (French for "layer-by-layer"). Melts inject as thin sills along many weak planes in the protolith, and this interlayering of sills and host rock defines the gneiss. Usually the process of injection is accompanied by passive folding, so the igneous nature of the contacts is often not evident.

If metamorphic temperatures get high enough, a rock begins to melt, but not all minerals melt at the same temperature. Quartz, some feldspar, and muscovite melt at lower temperatures than mafic minerals like amphibole, pyroxene, and olivine. Therefore, when an intermediate composition rock begins to melt, certain minerals become liquid, while others remain solid. The minerals that stay solid until extremely high temperatures are achieved are called *refractory minerals*. When only part of a rock melts, we say that it has undergone *partial melting,* or *anatexis*. A rock that is undergoing partial melting is a mixture of pockets of melt and lenses of solid, both of which are quite soft. Thus, shortening will cause the mass to flow, and the contrasting zones of melt and solid fold and refold much like chocolate and vanilla batter that you mix together to make marble cake. When this happens in rock, the resulting semi-chaotic mixture of light and dark layers is called a *migmatite*. Because of the chaotic nature of structures in migmatite their analysis may provide little or no information about the kinematics of regional deformation.

11.3.8 Mylonitic Foliation

Mylonitic foliation will be discussed in detail in the chapter on ductile shear zones (Chapter 12), but we mention it here so that you will also think of this fabric element in the context of foliation. Mylonites are very fine grained, strongly foliated, and lineated rocks that form by crystal plastic deformation mechanisms. The foliation in mylonites is defined

by dimensional-preferred orientation of flattened grains (usually quartz, but also feldspar and olivine) and mica as a consequence of the deformation process. Typically, transposition via folding occurs during the progressive deformation that characterizes mylonites, so the lithologic banding in a mylonite is not the original layering of the rock. Mylonitic foliation is axial planar to these folds, but the fold hinges may vary widely in orientation or be strongly curved in the plane of foliation, as a consequence of differential flow in the mylonite. Similar to other fine laminations, a mylonitic foliation is susceptible to the formation of crenulations at a later stage in the deformation.

11.4   CLEAVAGE AND STRAIN

Does the study of a foliation in the field provide any constraints on the nature of strain in the region? Unfortunately, we can offer only a wishy-washy answer: sometimes yes, sometimes no. In order to determine the relationship of cleavage to strain, we need to look at strain markers in a cleaved rock. Red slate is a good lithology for such strain studies, because it may contain reduction spots. *Reduction spots* (Figure 11.18) are small zones where the iron in the rock is reduced and therefore is greenish in color. If these spots start out as spheres around an inclusion, they make ideal strain markers because they behave totally passive, meaning that they have exactly the same competency as the bulk rock. In studies of reduction spots in slate, geologists find that the deformed spots are flattened ellipsoids, and that the plane of flattening (i.e., the XY principal plane of the finite strain ellipsoid) is essentially parallel to the slaty cleavage domains. In these cases, therefore, cleavage appears to define a principal plane of strain, with the shortening direction being perpendicular to the plane. But cleavage is probably not strictly parallel to the XY principal plane of strain in all situations. For example, when cleavage occurs in flexural slip folds, or adjacent to fault zones, there may be a component of shear on the cleavage domains themselves, and when this happens the cleavage, by definition, is not a principal plane. Moreover, a spaced cleavage may initiate as a principal plane, but subsequent folding of the bed in which the cleavage occurs may rotate it away from the bulk principal strain directions. Reduction spot studies give estimates of a total shortening strain (e_1) associated with slaty cleavage formation in excess of 60%.

Perhaps the most controversial issue about low-grade cleavage formation is the question of whether strain resulting from cleavage formation is a volume-constant strain or a volume-loss strain. Imagine a cube of rock that is 10 cm long on each edge. If it is shortened in one direction, but does not stretch in the other directions (i.e., X = Y = 1 > Z), then it must lose volume during deformation (Figure 11.19). Alternatively, if shortening in one direction causes it to expand by an equal proportion in one other direction (i.e., X > Y = 1 > Z or plane strain) or in two other directions (X > Y > Z), then the strain may be volume constant.

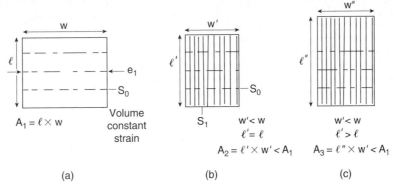

(a) (b) (c)

Figure 11.19 To lose or not to lose volume; that's the question. (a) A block of height I and width w is shortened and forms a cleavage. If there is volume loss strain, then w′ < w, but l′ = l as in (b), (we are assuming no change in the third dimension). However, if there is volume constant strain, then l′ > l, as in (c).

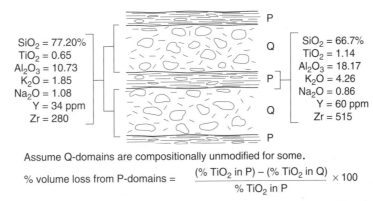

Assume Q-domains are compositionally unmodified for some.

$$\% \text{ volume loss from P-domains} = \frac{(\% \ TiO_2 \text{ in P}) - (\% \ TiO_2 \text{ in Q})}{\% \ TiO_2 \text{ in P}} \times 100$$

Figure 11.20 Volume loss may occur by preferential removal of certain elements. If the composition of the microlithons is assumed to be constant, then the amount of volume loss can be calculated (see equation); using the relatively immobile elements TiO_2, Y, and Zr, the volume loss is ~45%.

Remember that formation of low-grade cleavage involves significant activity of dissolution and new growth of grains. If the ions of soluble minerals enter the pore water system, they can be carried out of the local rock system by movement of groundwater. Thus, volume loss during cleavage formation is a distinct possibility. In fact, studies of deformed markers (such as the fossil graptolite whose constant protrusion spacing can be used to measure finite strain) and geochemical studies demonstrate that rock volume may have decreased by up to 50% during cleavage formation (Figure 11.20). But not all strain analyses of cleaved rocks yield clear indications of volume loss. Probably the degree of volume loss is affected by the degree to which the rock is an open or closed geochemical system during deformation, and whether the fluids passing through the rock during deformation are saturated or undersaturated with respect to the soluble mineral phase.

Volume-loss studies have an important implication concerning the nature of *rock-water interactions* during deformation. If we assume that most of the volume loss reflects dissolution and removal of certain minerals, then we can calculate the volume of fluid required for a given amount of volume loss if we know the solubility of the mineral and the amount that can be held in solution (saturation). Such calculations suggest that the amount of fluid needed for observed amounts of volume loss to occur are very large. The results of these calculations are commonly expressed in terms of the ratio of fluid volume to rock volume (called the *fluid/rock ratio*). In the example of Figure 11.21,[5] this ratio is much greater than 100, but whether such large ratios are reasonable is a matter of debate.

11.5 FOLIATIONS IN FOLDS AND FAULT ZONES

So far we have discussed tectonic foliations mainly from a descriptive point of view, without mentioning how they are related to other structures. Tectonic foliations do not occur

---

[5] Given 0.019 kg/l solubility of quartz and a 20% undersaturated fluid.

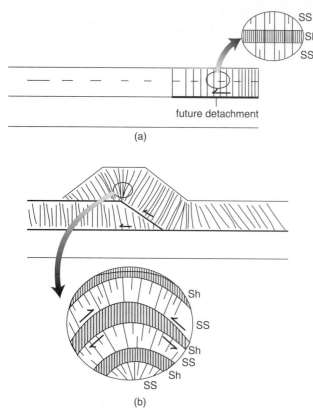

Figure 11.21 Evolution of cleavage in a foreland fold-thrust belt. (a) Initiation of cleavage during layer-parallel shortening. Cleavage starts by being nearly perpendicular to bedding. Inset shows that cleavage is parallel in sandstone (SS) and shale (Sh) beds, but that the initial spacing is different. After formation of a ramp anticline (b), regional shear has caused cleavage to be inclined with respect to bedding. In the fold itself, cleavage refracts as it passes from lithology to lithology. Cleavage in the shale beds is axial planar with respect to the folds, but fans around the folds in sandstone layers.

in isolation; rather, they are integral components of the suite of structures that represent the manifestation of deformation in a region. In this section we will look at how cleavage develops during *progressive deformation*, and use this description to introduce a number of *field characteristics* of cleavage.

First, let us consider the geometric relationship between low-grade cleavage (slaty, spaced disjunctive, and crenulation cleavage) and folds. Low-grade cleavage forms in regions where metamorphic conditions are low (i.e., subgreenschist facies), a condition that is prevalent in the foreland region of a collisional orogen or a convergent plate margin. Such regions display fold-thrust belt style deformation in which the uppermost part of the crust is shortened above a basal detachment (see Chapter 17). This shortening is partitioned among several strain mechanisms: thrust

faulting, folding (which is generally upright to inclined), and formation of cleavage and intragranular strains (calcite twining, quartz deformation bands, etc.). All of these different strain mechanisms are geometrically related because they all accommodate the same regional shortening event.

Imagine that a region of flat-lying strata is subjected to layer-parallel shortening (Figure 11.21a). At stresses that are lower than necessary to initiate faults, pressure solution is activated. As a consequence of pressure solution, spaced disjunctive cleavage domains initiate in argillaceous sandstone and limestone, and pencil cleavage develops in shale. Because of the orientation of the shortening direction, the cleavage domains are oriented approximately perpendicular to bedding. The initial spacing of the domains, you will recall, depends on the original clay content, so in more clay-rich units, the cleavage domains tend to be more closely spaced (Figure 11.21a inset). With continued deformation, individual domains get thicker, and new domains initiate. As a consequence, cleavage gets stronger (i.e., domains are more closely spaced). In shales, a weak slaty cleavage develops. If a detachment fault exists at depth, deformation of the strata by cleavage formation may be decoupled from the deformation of rock below the detachment. In fact, there may be movement on a detachment simply to accommodate the required shortening of strata that occurs during cleavage formation.

Eventually, if stresses get high enough for faulting to initiate, the package of strata containing the cleavage is carried in the hanging wall of the thrust and is folded. During folding, flexural slip occurs between layers. Weaker shale layers are caught in between the more competent sandstone or limestone layers. As a consequence, the cleavage within the shale layers rotates and becomes inclined to bedding (Figure 11.21b). In the more rigid layers, the cleavage maintains its original orientation at a high angle to bedding. Thus, cleavage in the more rigid layers *fans* around the folds, whereas cleavage in the weaker (shaly) horizons is roughly parallel to the axial plane of the fold (Figure 11.21b inset). In a graded bed, the change in cleavage orientation occurs gradually across the bed. Cleavage that is parallel or subparallel to the axial plane of the fold is called an *axial plane cleavage*. During the final phases of deformation, the fold as a whole is flattened and the limbs are squeezed together. In the vertical to overturned limb of the fold, late-stage tectonic cleavage forms, and this cleavage is just about parallel to the steeply dipping beds.

Because cleavage fans in some beds and is axial planar in others, cleavage changes orientation from bed to bed, which is called *cleavage refraction* (Figures 11.21b inset and Figure 11.22). In other words, cleavage refraction is the change in cleavage attitude that occurs where cleavage do-

Figure 11.22 Cleavage refraction in a phyllite-quartzite sequence (Scotland).

mains cross from one lithology into another of different *competency,* and reflects variation in the local strain field between beds. Usually, changes in domain spacing accompany cleavage refraction, as domain spacing is also controlled by lithology.

If the entire unit being deformed is dominantly shale, and conditions are appropriate for slaty cleavage to form, then the regional slaty cleavage forms approximately parallel to the axial plane of the regional folds. Usually, in such tectonic settings, folds are overturned in the direction of regional tectonic transport. Therefore, the regional cleavage dips toward the hinterland. If the cleavage is not exactly parallel to the axial plane, but cuts obliquely across the axial planes of the folds, then we say that the cleavage is *crosscutting.* The occurrence of crosscutting cleavage may indicate that the cleavage was superimposed on preexisting folds, or that there were local complexities in the strain field. For example, rotation of a thrust sheet during folding

may cause fold hinges to be oblique to the regional shortening direction. *Transecting cleavage* (Figure 11.23) is crosscutting cleavage that forms in transpressional environments (i.e., containing components of both pure and simple shear). Counterclockwise transecting cleavage, in which the cleavage cuts counterclockwise relative to roughly synchronous fold hinges is used to indicate a component of dextral shear (dextral transpression). Similarly, a component of sinistral shear may give a clockwise transecting cleavage, but the use of transecting cleavages as a shear-sense indicator remains controversial.

There is no controversy among structural geologists on cleavage-bedding geometry as a powerful clue to the position of rocks with respect to the axial surface of a fold, and that cleavage refraction can help determining the facing direction of folds. Let's take the example of an initially recumbent fold (F_1), in which the S_1 axial surface is folded by a second, upright folding generation (F_2). As a

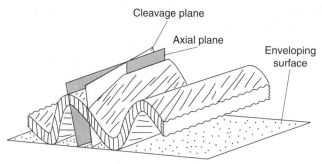

Figure 11.23 The relationship between cleavage, axial plane, and enveloping surface in folds with transecting cleavage. The counterclockwise cleavage transection that is illustrated may be indicative of dextral transpression. Note the obliquity of the bedding-cleavage intersection lineation to the hinge line.

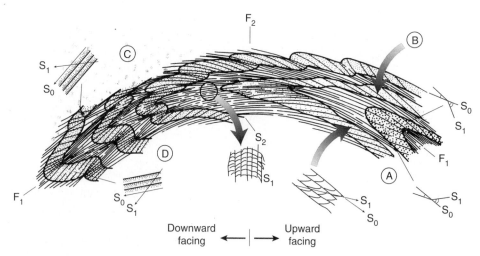

Figure 11.24 Cleavage-bedding relationships and cleavage refraction on upright and overturned limbs of both upward-facing and downward-facing folds.

consequence of this history, a portion of the large fold faces up and a portion of the fold faces down (Figure 11.24). If the strata in this fold still contain the S_1 cleavage, and if this cleavage is axial planar to the F_1 fold in mica-rich horizons, then on the overturned limb of the upward-facing part of the F_2 fold (at A), bedding dips more steeply than cleavage, and on the upright limb (at B) cleavage dips more steeply than bedding. On the upright limb of the downward-facing part of the large fold (at C), bedding dips more steeply than cleavage, whereas on the overturned limb cleavage is steeper (at D). You may also find a crenulation cleavage (S_2) that is axial-planar to the F_2 fold, especially in the hinge region. Notice that cleavage refraction indicates the younging direction (i.e., upright and overturned) in each region of the fold. Figure 11.24 is a diagram that lets you explore quite a variety of geometries and relationships that will prove to be quite helpful in the field; that is, once you have figured it out.

11.6 LINEATIONS

A *lineation* is a fabric element that can be represented by a line, meaning that one of its dimensions is much longer than the other two. There are many types of lineations. Some are manifestations of the occurrence of other structures (e.g., folds or boudins), some are visible only on specific surfaces in a rock body, whereas others reflect the arrangement and shape of mineral grains or clasts within the rock. Interpretation of lineations is not straightforward, because some lineations reflect strain axes or kinematic trajectories, while others appear to have no kinematic significance. We broadly group lineations into three categories: form lineations, surface lineations, and mineral lineations.

11.6.1 Form Lineations

The hinge of any cylindrical fold is a linear feature. If the folds are closely spaced, the fold hinges effectively define a rock fabric that we can measure as a *fold-hinge lineation*

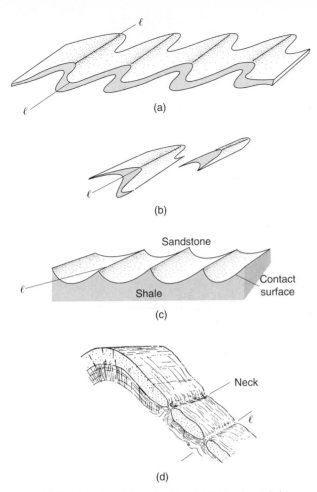

Figure 11.25 Examples of form lineations. (a) Fold and crenulation hinges, (b) rods, (c) mullions, and (d) boudins.

(Figure 11.25a). Imagine a multilayered sequence in which alternating layers are composed of different lithologies. Because of differences in compositions, the different layers have different mechanical behavior. If the multilayered sequence is deformed, it may develop folds. Similarly, a *crenulation lineation* is defined by the hinge lines of the microfolds in a crenulated rock. Why bother measuring crenulation lineations if a crenulation cleavage is present in the rock? The reason is that sometimes the interval in which the crenulation occurs is not thick enough for the cleavage domains to be measurable. It is common that phyllitic beds in which a crenulation has formed occur in between layers of uncrenulated rock. For example, deformation and metamorphism of a protolith composed of thick sandstone beds separated by thin interbeds of shale will create quartzite beds separated by thin layers of phyllite. Even if the phyllite layer is so thin that a crenulation cleavage plane cannot be measured, it may still be possible for you to see and measure the crenulation lineation.

When intense deformation detaches the limbs of folds, such as occurs during fold transposition (Chapter 12), only fold hinges may be left in the rock. Such hinges are called *rods,* and are a measurable lineation (Figure 11.25b). Rodding typically occurs in a multilayer composed of phyllite (or schist) and quartzite; the quartz layers are relatively rigid and define visible folds, the limbs of which may be thinned so severely that they pinch out, and the quartz flows into the hinge zone, where it is preserved as a rod. Rodding may also occur in mylonites, because the progressive folding in mylonites may generate rootless isoclinal folds whose limbs are detached and whose hinges (the rods) have rotated into parallelism with the shear direction (see Chapter 12). *Mullions* are cusplike corrugations that form at the contact between units of different competencies in a deformed multilayered sequence (Figure 11.25c); the axes of the mullions are a lineation. Typically, the more rigid lithology occurs in convex bulges that protrude into the ductile lithology, and the bulges connect in pointed troughs. Because of their mechanical origin, mullions cannot be used as a facing indicator.

Boudins are sausage-shaped lenses of a relatively rigid layer, embedded in a more ductile matrix, in a rock that has undergone layer-parallel stretching (Figure 11.25d). In the third dimension, they are long tabular bodies separated by *boudin necks* that can be measured as linear objects. Other elongate objects in rock that are useful lineations include *elongate pebbles* and elongate pumice fragments. If the long axes of elongate objects in a rock are aligned, then they define a lineation (Figure 11.26). The elongation of the object is generally a manifestation of deformation in some form. But when using such objects for structural analysis, it is important to make sure that their alignment is not a primary feature (e.g., alignment of pebbles during deposition in flowing water).

11.6.2 Surface Lineations

An *intersection lineation* is a linear fabric element defined by, as the name suggests, the intersection of two planar fabric elements (Figure 11.27a). An intersection lineation that structural geologists commonly use in the field is the *bedding-cleavage intersection,* which is manifested by the traces of cleavage domains on a bedding plane or vice versa. If the cleavage is parallel to the axial plane of a fold, the bedding-cleavage intersection parallels the hinge line in cylindrical folds.

Slip lineations form on surfaces on which there has been sliding and trend parallel to the sliding directions (Figure 11.27b). This may occur, for example, on fault surfaces, but also at the interface between beds in flexural slip folds. There are two basic types of slip lineations: *groove lineations,* formed by plowing of surface irregularities, and *fiber lineations,* which are formed when vein mineral fibers precipitate along a sliding surface via the crack-seal deformation process (see Chapter 7). Slip lineations, as the name suggests, are parallel to the slip direction.

Figure 11.26 Elongate pebbles in a stretched conglomerate (Rhode Island).

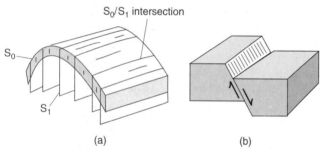

Figure 11.27 Examples of surface lineations. (a) Intersection lineation of bedding, S_0, and (axial plane) cleavage, S_1, in a fold, and (b) slip lineation.

11.6.3 Mineral Lineations

When we say that a rock contains a *mineral lineation,* we mean that the fabric element defining the lineation is the size of a mineral grain or a cluster of mineral grains (Figure 11.28). Mineral lineations commonly occur in the plane of metamorphic foliation, on shear surfaces, or on the plane of mylonitic foliation. There are many types of mineral lineations, which do not all have the same tectonic significance. So when using lineations in a structural analysis, it is important to determine what type of mineral lineation is present in the rock.

Some minerals, such as kyanite and amphibole, grow such that they are very long in one direction relative to the other two directions. If the long axes of the crystals are aligned in a rock, they create a mineral lineation. The alignment may be due either to growth of the crystal in a preferred direction (controlled by differential stress or by flow-controlled diffusion) or because elongate grains are

rotated toward a principal strain direction during progressive deformation. In addition to aligned, elongate single crystals, rocks can contain several other types of mineral lineations. For example, during development of slate, pyrite grains behave as rigid inclusions, so that when the rock matrix stretches, a space develops on either side of the grain. These spaces, called *pressure shadows,* fill with elongate fibers of quartz or calcite, defining a lineation parallel to the direction of stretching. In higher-grade rocks, porphyroclasts may undergo metamorphic reactions that create a rim of alteration minerals (e.g., fine-grained mica and quartz around a feldspar porphyroclast). During shear, these mantles smear out in the direction of shear, creating long streaks of small grains on opposite ends of the porphyroclast. Such streaks define a lineation. Similar mica streaks may form in tectonites even in the absence of porphyroclasts. Finally, during plastic deformation, quartz gains can become elongate lenses, contributing to defining a mesoscopic lineation (comparable in origin to stretched pebbles).

11.6.4 Tectonic Interpretation of Lineations

In structural analysis, the *bedding-cleavage intersection lineation* provides a clue to the orientation of fold hinges in a region where the hinges themselves may not be exposed. Other types of lineations, however, can be quite difficult to interpret, because there are at least two alternative interpretations for their origin. First, a lineation can parallel a principal strain; specifically, the direction of stretching or elongation. When we talk about *stretching lineations,* as defined by elongate mineral grains or pebbles, we are implying that the lineation is roughly parallel to the direction of maximum elongation (the X-axis of the finite strain ellipsoid). Other lineations, like boudins, are roughly

Figure 11.28 A mineral lineation (hornblende) in the Gotthard Massif (Switzerland).

perpendicular to the stretching direction. Second, a lineation can be parallel to a shear direction, meaning that it is parallel to a vector defining the motion of one part of a rock with respect to another. Slip lineations (fibers or grooves) and mica streaks are good examples of shear-direction lineations. Whereas, the shear direction is not parallel to the stretching direction, in zones of high shear strain it is nearly parallel to the finite elongation direction; grains are stretched and mineral clusters are smeared out in the direction of shear.

11.7 OTHER PHYSICAL PROPERTIES OF FABRICS

We focused in this chapter primarily on dimensional and crystallographic aspects of rock fabrics, but their presence also imparts an anisotropy that may be manifested by other physical properties. It is beyond the scope of this text to explain the underlying principles of these other properties and discuss the details of the methods, so we merely mention two: seismic and magnetic anisotropy. *Seismic velocity* varies as a function of material characteristics; a property that we use to understand the deep structure of the Earth (see Chapter 14). It also varies with direction in a single crystal or aggregate of crystals, because the spacing of atoms in the crystal lattice is different in different directions. Thus, when a fabric is defined by the preferred orientation of crystals, the seismic velocity of the sample will vary (by a few percent) in different directions. In recent years, seismic-wave splitting is being used increasingly to study the structure of the mantle (e.g., shear-wave splitting).

The *magnetic properties* of rocks also vary as a function of direction, because the ability of a rock to be magnetized similarly depends on the geometric arrangement of atoms in individual crystals and/or the orientation of its mineral constituents. Because of the physical anisotropy of a foliated rock, its magnetic fabric is also anisotropic. *Anisotropy of magnetic susceptibility* (AMS) is one measure of the magnetic fabric in rocks that is easy to obtain in the laboratory, and is a recorder of even the weakest fabric element. AMS is often touted as a finite strain gauge, but whereas the orientations of the strain ellipsoid and the ellipsoid representing directional variability of the magnetic fabric are often parallel, the relative magnitudes of the ellipsoids depend strongly on the magnetic phases involved. Finite strain-magnetic fabric relationships are therefore not straightforward. Nevertheless, magnetic fabrics offer a sensitive measure of the orientation and intensity of fabric elements in natural rocks.

11.8 CLOSING REMARKS

You may find it surprising that after nearly 150 years of research on foliations and lineations (dating from the early studies of Sorby, Darwin, and their contemporaries in the nineteenth century), many questions about these fabric elements remain unresolved. Perhaps one reason for this uncertainty is our inability to create fabrics in the laboratory under conditions and with processes similar to those in nature; especially foliation formation involving fluid migration and material transfer poses experimental limitations. So, many of our ideas about tectonic fabrics are based on field evidence, which by its nature is often circumstantial. However, in the past few decades, great

Figure 11.29 Transmission electron micrograph of microfolds in mica from a slate in southern Wales. Width of view is 0.4 μm.

progress has been made in our understanding of fabric formation by using modern approaches such as electron microscopy. These studies reveal the detailed behavior of detrital and newly grown grains (e.g., Figure 11.29), suggesting their respective role during deformation. Complementary data obtained from X-ray techniques (texture goniometry) further support a general view of fabric formation. And all structural geologists agree? Well, it is still quite common for a group of professionals to stand on an outcrop and argue about the origin and the meaning, and in some cases the very existence, of a fabric. In this chapter, we have attempted to de-emphasize our own biases with regard to fabrics, but they have crept into the text in various places and may be criticized by your instructor. No matter, debate is what keeps structure and tectonics a vibrant and exciting field!

ADDITIONAL READING

Borradaile, G. J., 1988, Magnetic susceptibility, petrofabrics and strain: *Tectonophysics,* v. 156, p. 1–20.

Engelder, T. E., and Marshak, S., 1985, Disjunctive cleavage formation at shallow depths in sedimentary rocks: *Journal of Structural Geology,* v. 7, p. 327–343.

Etheridge, M. A., Wall, V. J., and Vernon, R. H., 1983, The role of the fluid phase during regional metamorphism and deformation: *Journal of Metamorphic Geology,* v. 1, p. 205–226.

Gray, D. R., 1979, Microstructure of crenulation cleavages: An indicator of cleavage origin: *American Journal of Science,* v. 279, p. 97–128.

Hobbs, B. E., Means, W. D., and Williams, P. F., 1982, The relationship between foliation and strain: an experimental investigation, *Journal of Structural Geology,* v. 4, p. 411–428.

Knipe, R. J., 1981, The interaction of deformation and metamorphism in slates, *Tectonophysics,* v. 78, p. 249–272.

Powell, C. M. A., 1979, A morphological classification of rock cleavage, *Tectonophysics,* v. 58, p. 21–34.

Williams, P. F., 1972, Development of metamorphic layering and cleavage in low grade metamorphic rocks at Bermagui, Australia, *American Journal of Science,* v. 272, p. 1–47.

Wood, D. S., Oertel, G., Singh, J., and Bennett, H. F., 1976, Strain and anisotropy in rocks: *Philosophical Transactions of the Royal Society of London,* v. 283, p. 27–42.

Wright, T. O., and Platt, L. B., 1982, Pressure dissolution and cleavage in the Martinsburg shale, *American Journal of Science,* v. 282, p. 122–135.

Chapter 12

Ductile Shear Zones, Textures, and Transposition

12.1   INTRODUCTION

Imagine a cold and wet day in northern Scotland, which may not be a far stretch of the imagination if you've ever visited the area. As you are mapping a part of the Scottish Highlands, you are struck by the presence of a layer of highly deformed rocks that overlie relatively undeformed, flat-lying, fossiliferous sediments. This relationship is even more odd because the overlying unit is unfossiliferous and has experienced much higher grade metamorphism than the underlying sediments. Where you find the contact between these two characteristic rock suites, you notice that they are separated by a distinctive layer of par-

ticularly fine-grained rock. The regional relationships of these two suites and their superposition already suggest that the contact is a low-angle reverse (i.e., thrust) fault. So, what is the distinctive fine-grained rock at the contact? In your mind you envision the incredible forces associated with the emplacement of one unit over the other, and you surmise that the rock at the contact was crushed and milled, like what happens when you rub two bricks against each other. Using your class in ancient Greek, you decide to coin the name *mylonite* for this fine-grained rock unit, because mylos is Greek for 'milling.'

Something like this happened over a hundred years ago in Scotland where the late Precambrian Moine Series ('crystalline basement') overlie a Cambro-Ordovician quartzite and limestone ('platform') sequence along a Middle Paleozoic low-angle reverse fault zone, called the Moine Thrust. This area was mapped by Sir Charles Lapworth of the British Geological Survey in the late nineteenth century. Anecdote has it that Lapworth became convinced that the Moine Thrust was an active fault and that it would ultimately destroy his nearby cottage and maybe take his life; Lapworth's later years were spent in great emotional distress.

In many areas you will find similar zones in which the deformation is markedly concentrated (Figure 12.1). Deformation in these zones is manifested by a variety of structures, which may include isoclinal folds, disrupted layering, and well-developed foliations and lineations. These zones, called ductile shear zones, may contain some of the most

Figure 12.1 The Parry Sound shear zone in the Grenville Orogen displays many of the characteristics that are common in ductile shear zones, including mylonitic foliation, mineral lineation ('stretching lineation'), and rotated clasts, (Southern Ontario, Canada).

important information about the deformation history of an area, so they deserve special treatment. A *ductile shear zone* is a tabular band of definable width in which there is considerably higher strain than in the surrounding rock. The total strain within a shear zone typically has a large component of simple shear, and as a consequence, rocks on one side of the zone are displaced relative to those on the other side. In its ideal form, a shear zone is bounded by two parallel boundaries outside of which there is no strain. In reality, however, shear zone boundaries are typically gradational. The adjective "ductile" is used because the strain accumulated by ductile processes which may range from cataclastic flow to crystal-plastic processes and diffusional flow (Chapter 9). So, a shear zone is like a fault in the sense that it accumulates relative displacement of rock bodies. But unlike a fault, displacement in a ductile shear zone occurs by ductile deformation mechanisms and no throughgoing fracture is formed. This absence of a single fracture is largely a consequence of movement under relatively high temperature and pressure conditions.

Consider a major discontinuity that cuts through the crust and breaches the surface (Figure 12.2). In the first few km below Earth's surface, *brittle processes* will occur along the discontinuity, which result in earthquakes if the frictional resistance on discreet fracture planes is overcome abruptly. Displacement may also occur by the movement on many microscopic fractures, a ductile process called *cataclastic flow* (Chapter 9). In either case, frictional processes dominate the deformation at the higher crustal levels along the discontinuity, and this crustal segment is therefore called the *frictional regime*.

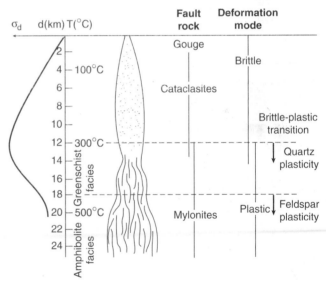

Figure 12.2 Model for a discontinuity that cuts deep into the crust, showing the frictional (FR) and plastic regimes (PR), the frictional-plastic (or brittle-plastic) transition, and a relative crustal strength curve. Fault rocks typically associated with each level are indicated.

With depth, crystal plastic and diffusional processes, such as recrystallization and superplastic creep, become increasingly important mainly because temperature increases. Where these mechanisms are dominant, typically below 10–15 km depth for normal geothermal gradients (20–30°C/km) in quartz-dominated rocks, we say that displacement on the discontinuity occurs in the *plastic regime*. The transitional zone between a dominantly frictional and

Figure 12.3 Clastomylonite containing relatively rigid clasts of varied lithologies in a very fine-grained, crystal-plastically deformed marble matrix (Grenville Orogen, Ontario).

dominantly plastic regime is called the *frictional-plastic transition.*[1] Technically it is not correct to call this the brittle-ductile transition, because ductile processes (such as cataclasis) can occur in the frictional regime as well. A crustal discontinuity that is a brittle fault at the surface is a ductile shear zone with depth. Associated with this contrast in deformation processes the zone of shear changes from a relatively narrow fault zone to a broader ductile shear zone because the host rock weakens (Figure 12.2).

Mylonites are dominated by the activity of crystal-plastic processes, which produces yet another characteristic feature of deformed rocks: *crystallographic-preferred fabrics,* or *textures.* The topic of textures is introduced in this chapter, although some of the theory logically follows the material presented in Chapter 9. Secondly, rocks within ductile shear zones typically are intensely folded, and the original layering is transposed into a tectonic foliation. Therefore, a discussion of transposition will close the chapter, but we emphasize that transposition is not unique to ductile shear zones. In this chapter you will see that shear zones may have more than one foliation, a strong lineation, and that shear zone rocks commonly contain rotated fabric elements, grain-shape fabrics, and in particular a smaller grain size than that of the host rock. Arguably, ductile shear zones are the most varied structural feature, and perhaps the most informative. Enough hype; let's look at these things.

---

[1]We favor "frictional-plastic transition" to avoid confusion surrounding the terms brittle, ductile, and plastic (see Chapter 5), but "brittle-plastic transition" is more widely used.

12.2 MYLONITES

The inference made by Lapworth for the formation of mylonites at the Moine Thrust is incomplete and, as a general rule, incorrect. As mentioned earlier, the derivation of the name "mylonite" suggests that the process by which grain-size reduction occurs in mylonites is cataclastic flow. In fact, study of microstructures in samples that represent the change from relatively undeformed to mylonitized rock shows that *crystal plastic processes,* and *dynamic recrystallization* in particular, are mainly responsible for the observed grain-size reduction (Figure 12.3). Therefore, we restrict the use of the term "mylonite" to a fault rock type with a relatively fine grain size as compared to the host rock, resulting from crystal-plastic processes. Brittle fracturing may also produce a reduced grain size. Fault rock types formed by brittle processes are "fault rocks." *Cataclasite* is a fault rock in which grain size has been mechanically reduced, but the rock has not lost cohesion (the grains interlock). The field terminology of fault rocks and the processes by which they primarily form are summarized in Table 12.1.

Mylonites are associated with all kinds of ductile shear zones, whether they result in reverse displacement (such as the Moine Thrust in Scotland), normal displacement (such as those in core complexes and the Basin-and-Range Province of the western USA), or strike-slip displacement (such as the Alpine fault in New Zealand). In each of these structural settings, mylonites have one element in common: conditions that promote crystal-plastic deformation mechanisms. Such conditions are reached at various conditions of temperature, strain rate, and stress depending on the dominant mineral in the rock (Chapter 9). For example, marble

Table 12.1   Fault Rocks and Processes

Breccia	Incohesive fault rock with randomly oriented fragments that make up >30% of the rock mass, formed by brittle processes.
Cataclasite	Cohesive fault rock generally with randomly oriented fabric, formed by brittle processes.
Gouge	Incohesive fault rock with randomly oriented fragments that make up <30% of the rock mass, formed by brittle processes.
Mylonite	Cohesive, foliated fault rock, formed dominantly by crystal-plastic processes.
Pseudotachylite	Glassy fault rock along fractures, formed by melting of the host rock from heat generated during frictional sliding possibly associated with paleo-earthquake activity.

mylonites and quartzite mylonites form at temperature conditions that are lower than those under which feldspar mylonites form, based on the temperature for the onset of dynamic recrystallization in calcite (>250°C), quartz (>300°C), and feldspar (>450°C). Rocks that contain a variety of minerals show mixed behavior. For example, quartz grains in a sheared granite may dynamically recrystallize, while feldspar grains deform predominantly by fracturing. If the rock displays this contrast, we surmise that it was sheared at temperatures greater than 300°C, but less than 500°C (i.e., greenschist facies of metamorphism). However, mylonitic microstructures should not be used as a quantitative indicator of the conditions of temperature, stress, or strain rate, because of the uncertainties surrounding these parameters and their mutual dependence; at best we can make semiquantitative estimates such as those above. Quantitative metamorphic petrology and isotope geology in many instances provide reliable methods to determine the past conditions of temperature and pressure of deformed rocks (see Chapter 13), so we do not have to rely on microstructures only.

12.2.1   Types of Mylonites

The widespread occurrence of mylonites in crustal rocks and their study has led to use of several prefixes to distinguish among different types (Table 12.2). *Protomylonite* and *ultramylonite,* which are used to describe mylonites in which the proportion of matrix is <50% and 90–100%, respectively. In a protomylonite, only part of the rock was mylonitized, whereas pervasive mylonitization occurred in an ultramylonite. Mylonites containing 50–90% matrix are known simply as *mylonite. Blastomylonite* and *clastomylonite* are used to describe mylonites containing large grains surrounded by a fine-grained matrix that grew during mylonitization or that remained from the original rock, respectively. The terms derive from the Greek words "blastos," meaning growth, and "klastos," meaning broken. The prefixes are often used for microstructural descriptions of mylonites; for example, Figure 9.27 is a photomicrograph of a clastomylonite. However, they are also used as field terms; for example, clastomylonite is used to describe mylonites that contain coarse fragments of less deformed

Table 12.2   Types of Mylonites

Blastomylonite	Mylonite that contains relatively large grains that grew during mylonitization (e.g., from metamorphic reactions or secondary grain growth).
Clastomylonite	Mylonite that contains relatively large grains or aggregates that remain after mylonitization reduced the grain size of most of the host rock (e.g., relatively undeformed feldspar grains or clumps of mafic minerals).
Mylonite	In a strict sense, mylonite contains 50–90% matrix.
Phyllonite	Mica-rich mylonite (derived from phyllite).
Protomylonite	Mylonite in which the proportion of matrix is <50% (i.e., rocks in which only a minor portion of the minerals underwent grain-size reduction).
Ultramylonite	Mylonite in which the proportion of matrix is 90–100% (i.e., rocks in which mylonitization was nearly complete).

host rock or other rock types, such as igneous clasts in the marble mylonite shown in Figure 12.3.

12.3 SHEAR-SENSE INDICATORS

Ductile shear zones record displacements between blocks at deeper levels in the crust, where recognizable markers such as bedding, allowing us to determine offset, are generally absent. Consider a greenschist facies shear zone in a large granitic body. The granitic rocks on either side of the mylonite are indistinguishable, so there is nothing at first glance to predict the sense of displacement, let alone the magnitude of displacement. *Sense of displacement* describes the relative motion of opposite sides of the zone (left-lateral or right-lateral, up or down), whereas *magnitude of displacement* is the distance over which one side moves relative to the other. This poses an interesting challenge to the structural geologist. The solution is to look for *shear-sense indicators*[2] in ductile shear zones. Before we start looking, however, it is critical to identify the optimal frame of reference for our analysis.

12.3.1 The Plane of Observation

In order to use shear-sense indicators, we need to examine our shear zone in the most informative orientation. Many mylonites contain one or more foliations and a lineation, which we will use as an *internal reference frame* for our analysis of ductile shear zones. In the field we look for outcrop surfaces (or cut our sample in the lab) that are perpendicular to the mylonitic foliation *and* parallel to the lineation (Figure 12.4a). We make the (reasonable) assumption that the lineation coincides with the movement direction of the shear zone. When two foliations are present, this surface is also generally perpendicular to their intersection. This plane maximizes the expression of the rotational component of the deformation (Chapter 4); in any other surface this component is less. Secondly, we must place the orientation of our surface in the context of the region. Say we find that the displacement is right-lateral in our cut surface. If we look at this same surface from the opposite side, the displacement would appear to be left-lateral (Figure 12.4b). Of course this is not a paradox, but simply a matter of reference frame; we encounter the same situation with fold vergence (Chapter 9). The displacement sense is the same in geographic coordinates, and it is therefore a good habit to analyze surfaces in the same orientation across the field area to avoid confusion. If this is not possible, make careful field notes of the orientation of the surface in which you determine the shear sense. Back in the laboratory, the rock saw offers complete control as long as you orient the sample prior to removing it from the field.

[2]The term *kinematic indicators* is also used.

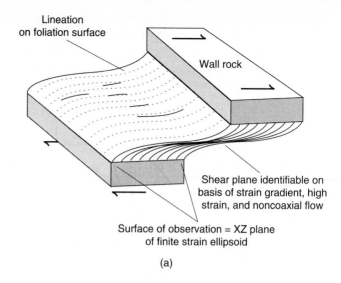

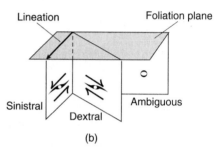

Figure 12.4 (a) Schematic ductile shear zone and optimal surface for analysis. (b) Apparent difference in shear sense, which results from observing the same indicator in different surfaces. Note that the surface perpendicular to the lineation and the foliation gives no shear-sense information.

Having cautioned you sufficiently about orientation, let us look at the most informative shear-sense indicators. They fall into five main groups: (1) grain-tail complexes, (2) disrupted grains, (3) foliations, (4) textures (or crystallographic fabrics), and (5) folds.

12.3.2 Grain-Tail Complexes

Mylonites commonly contain large grains or aggregates of grains that are surrounded by a fine matrix; for convenience, we use the term *grain* in a general sense to describe both large single grains and grain aggregates. These grains in turn may have *tails* of material with a composition and/or grain shape and size that differ from the matrix, such that they are distinguishable. For example, large feldspar grains connected by thin layers of finer-grained feldspathic material are common in gneisses (Figure 12.5). The tail may represent highly attenuated, preexisting mineral grains, dynamic recrystallization of material at the rim of the grain, or material formed by synkinematic metamorphic reactions (neocrystallization). During deformation, the grains act as

(a)

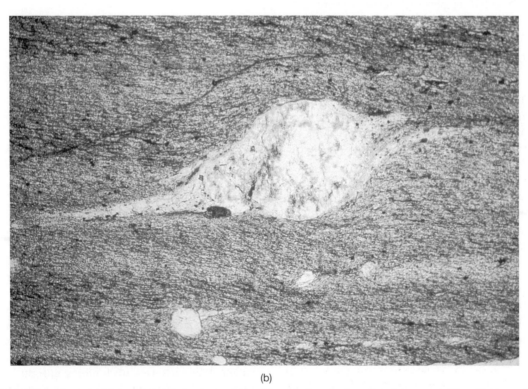

(b)

Figure 12.5 Grain-tail complexes. (a) A δ-type grain-tail complex in a feldspathic gneiss from the Parry Sound shear zone of Grenville Orogen (Canada). (b) A K-feldspar clast with a tail of fine-grained plagioclase forming a σ-type complex (California). Width of view is ~10 cm.

rigid bodies, and from the tails we may be able to determine the sense of displacement. In order to use grain-tail complexes as shear-sense indicators we use an internal reference frame (Figure 12.4). Based on their relationship with the shear zone foliation, we recognize two end-member types of grain-tail complexes: σ-type and δ-type.[3]

Grain-tail complexes of the σ-type are characterized by wedge-shaped tails that do not cross the reference plane when tracing the tail away from the grain (Figure 12.6a). Sometimes the tail is flat at the top and the other side is curved toward the reference plane. This grain-tail geometry looks like the Greek letter σ, at least in the case of right-lateral displacement, hence the name "σ-type complex." Figure 12.5a is an example of a σ-type complex in a quartzo-feldspathic shear zone. In the case of left-lateral displacement, the geometry is the mirror image. Overall it has a stair-stepping geometry. In δ-type grain-tail complexes, tails wrap around the grain such that they crosscut the reference plane when tracing the tail away from the grain (Figure 12.6b). If you rotate the Greek letter δ over 90°, you will see why we use this symbol. Figure 12.5b is an example of a δ-type complex in a feldspathic shear zone. Note that rotation in the δ-type complex is counterclockwise for left-handed and clockwise for right-handed displacement.

The two grain-tail complexes are related (Figure 12.6c). It is quite common, in fact, to find both σ- and δ-types in one surface and even a combination of the two may occur. A possible reason for this mixed occurrence is a varying relationship between the rate of recrystallization or neocrystallization, and rotation of the grain. If tail formation is fast relative to rotation, the tails are of the σ-type. If, on the other hand, the rotation of the grain is fast, the tail will simply be dragged along and wrap around the grain (δ-type). The case of preexisting tails, which quite often occurs with stretched pegmatites that are incorporated in shear zones, falls in the latter category.

The presence of both σ- and δ-type grain-tail complexes may indicate different rates of tail growth, different initial grain shape, different times of tail formation, or different degrees of coupling (see Section 12.4.1). From these many uncertainties, it is clear that we should use grain-tail complexes with some caution for strain quantification purposes, but there is no doubt about their importance as shear-sense indicators. In the field, geometries of the σ-type may sometimes be difficult to interpret, but δ-type complexes offer intuitively obvious and unequivocal information on the shear sense.

12.3.3 Fractured Grains and Mica Fish

Some minerals, such as feldspar, may deform by fracturing. Fracturing can occur along crystallographic directions or parallel to the shortening direction (extension cracks). As

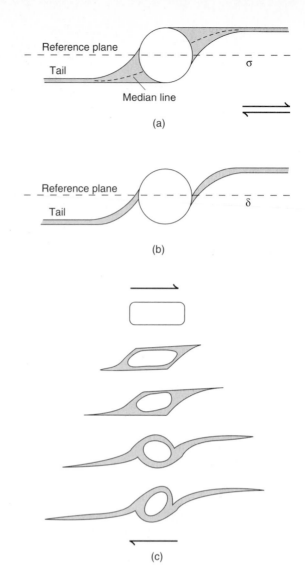

Figure 12.6 Grain-tail complexes as shear-sense indicators. (a) σ-type complex, (b) δ-type complex, and (c) the development of a δ-type grain-tail complex from σ-type complex.

long as the approximate orientation of these fractures before shear is known, we can determine the shear sense from their displacement (Figure 12.7). Fractures oriented at low angles to the mylonitic foliation show a displacement sense that is consistent with the overall shear sense of the zone; these fractures are called *synthetic* fractures (Figure 12.7a). Fractures at angles greater than 45° to the foliation show an opposite sense of movement; these are called *antithetic* fractures (Figure 12.7b). These opposite motions are not contradictory, which we can examine with a simple experiment. Place a series of dominos upright between your hands, and move your hands in opposite directions. You will notice that as long as the angle of the dominos with your hands remains greater than approximately 45°, the displacement between individual dominos is opposite (antithetic) to the relative movement direction of your hands. At lower angles the experiment is a bit harder; but with some care you will find that displacement of the dominos at very low angles reverses to a motion that is the same (synthetic)

---

[3]A third type, θ-complexes, occurs around large grains with alteration rims, but these are equivocal shear-sense indicators.

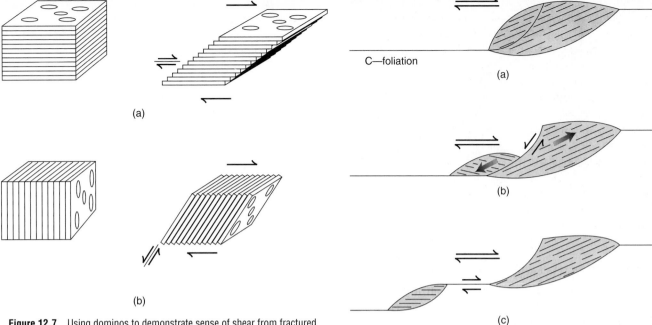

Figure 12.7 Using dominos to demonstrate sense of shear from fractured grains if the fractures are at a low angle (a) or high angle (b) to the shear plane. Note that the rotation according to the 'domino model' of individual segments of fractured grains in (a) is the same as that for rotated grains.

Figure 12.8 The formation of mica fish. C-foliation marks the shear plane. (a), (b), and (c) show successive stages in fish development.

as the motion of your hands. In fact, you know that it is working when the dominos slide away from your hands. Fractured minerals and clasts behave similarly in shear zones, and we therefore call this the *domino model*[4] for shear sense.

Feldspar and quartz are not the only minerals useful for determining shear sense in mylonites. It is quite common to find large phyllosilicate grains, such as mica and biotite in quartzo-feldspathic rocks and phlogopites in marbles, that display a useful geometry. The micas are connected by a mylonitic foliation, and their basal (0001) planes are typically oriented at an oblique angle to the mylonitic foliation, such that they point in the direction of the instantaneous elongation axis. In this orientation they show a stair-stepping geometry in the direction of shear (Figure 12.8), which is similar to σ-type grain-tail complexes that also step up in the shear direction. When phyllosilicates are large enough to be seen in hand specimen they look like scales on a fish (hence they are called 'mica fish'). You can use a simple field test to determine their approximate orientation. The basal planes of phyllosilicates are excellent reflectors of light, so when you turn the foliated shear zone sample in the sun you encounter an orientation that is particularly reflective. When this happens and the sun is behind you, you are looking in the direction of shear. The method is affectionately known as the 'fish flash,' and was obviously developed by those fortunate geologists who work in sunnier parts of the world.

---

[4]More literate, less playful geologists prefer the term *bookshelf model.*

12.3.4 Foliations: C-S and C-C' Structures

Most mylonites show at least one well-developed foliation, which is generally at an angle less than 45° to the boundaries of the shear zone. This foliation is what we previously called the *mylonitic foliation,* otherwise known as the *S-foliation;* S is derived from the French word for foliation, "schistosité." Its angle with the shear-zone boundary may be as little as a few degrees, at which point it is hard to distinguish it from another foliation that parallels the shear-zone boundary, called the *C-foliation;* C stems from "cisaillement," which is French for shear. A third foliation showing discrete shear displacements, but which is oriented oblique to the shear-zone boundary, is called the *C'-foliation.* These are the three most common foliations, and each one has distinct kinematic significance; they may reflect grain-shape fabrics or discrete shear surfaces. At first, these many foliations will seem confusing, but, when correctly identified, they become very powerful shear-sense indicators.

Thin-section study of well-developed mylonites in quartzites, granites, and marbles often shows the presence of a foliation that is defined by elongate grains (Figure 12.9). This foliation, the S-foliation, reflects the activity of crystal-plastic processes that tend to elongate grains toward the extension axis of the finite strain ellipsoid. It is uncertain whether the S-foliation exactly tracks the XY-plane of the finite strain ellipsoid; nor is it certain whether S and C form simultaneously or sequentially. These distinctions are important when C-S structures are used to determine the degree of noncoaxiality (i.e., the kinematic vorticity number;

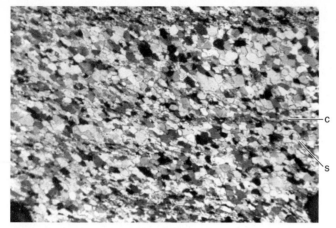

Figure 12.9 Photomicrograph of a C-S structure in a marble mylonite (Grenville Orogen, Canada). Note that the S-orientation is defined by elongate calcite grains, and that the C-foliation is marked by smaller grains and a concentration of dark material (predominantly graphite). Width of view is 1.5 mm.

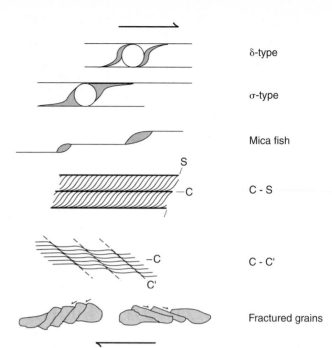

Figure 12.11 Summary diagram of shear-sense indicators in dextral shear zone. A copy of this figure on a transparency (for left- and right-lateral shear zones) is a handy inclusion for your field notebook.

see Chapter 4). For their purpose as a shear-sense indicator, however, these questions are academic, because in all cases the geometry of a *C-S structure* gives the same shear sense. The long axis of elongate grains in the S-surface points up in the direction of shear, and the shear direction lies in the C-plane, perpendicular to the intersection line of the S- and C-foliations.

Another common shear-sense indicator is a series of oblique, discrete shears that occur in strongly foliated mylonites. These small shears, called C′-surfaces because they also accumulate shear strain, are particularly common in phyllosilicate-rich mylonites (Figure 12.10), and crenulate or offset the mylonitic foliation. C′-foliations are therefore also called *shear bands,* or *extensional crenulations.* The offset on C′-surfaces is in the same direction as the overall displacement of the shear zone (i.e., displacement on C). The C′-surfaces contrast with S-surfaces that do not appear to displace the C-surface, suggesting that C′-surfaces form late in the mylonitic evolution. As with C-S structures, the kinematic significance of *C-C′ structures* is incompletely understood, but it appears that their

formation reflects a component of continued extension along the main anisotropy (the C-surface) of the mylonite. Thus, the shear sense on C-C′ structures is *synthetic* to the sense of shear in the zone as a whole. Figure 12.10 summarizes the characteristic displacement sense associated with both C-S and C-C′ structures.

12.3.5   A Summary of Shear-Sense Indicators

Shear sense in ductile shear zones is reliably determined only when a variety of indicators give a consistent sense. To give you an overview we close this section with a summary diagram (Figure 12.11) showing the various shear-sense indicators that may be encountered in a ductile shear zone. The presence of a particular indicator may vary within the same zone and even within the same outcrop as a function of the dominance of grains or foliations. Note that, except for fractured grains, the shear sense can be determined from any of these indicators without knowing their (original) relationship to the shear-zone boundary.

12.4   STRAIN IN SHEAR ZONES

In Chapter 4 we introduced several methods to determine strain in deformed rocks; in particular we made use of grain shapes. But do any of these methods work in shear zones? Well, even if we accept that grain shape or degree of grain alignment reflects strain, the activity of dynamic recrystallization in mylonites tells us that, at best, we may be able to measure an incremental part of the strain history. However, we get little or no idea how much of the finite strain magnitude this represents, or even to what

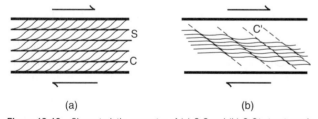

Figure 12.10 Characteristic geometry of (a) C-S and (b) C-C′ structures in a dextral shear zone. The C-surface is parallel to the shear-zone boundary and is a surface shear accumulation (i.e., nonparallel to a plane of principal finite strain). The S-foliation is oblique to the shear-zone boundary and may be close to the XY-plane of the finite strain ellipsoid. The C′-foliation displaces an earlier foliation.

Figure 12.12 Snowball garnet (Sweden); width of view is ~7 mm.

extent this increment coincides with the orientation of the finite strain ellipsoid. Two approaches give us at least a reasonable estimate of the finite strain amount in shear zones: rotation of grains, and deflection of foliations. But as you will see, they are not without their limitations either. It is because of these uncertainties surrounding strain analysis in shear zones that we have left a description of these methods until now.

12.4.1 Rotated Grains

The formation of ductile shear zones involves an internal rotation component of strain (the internal vorticity; Section 4.5) that may be recorded in rotated grains. One example we saw earlier are δ-type grain-tail complexes; others include minerals that grow during rotation, resulting in the progressive incorporation of matrix grains. In particular the mineral garnet shows this behavior, in which 'trapped' matrix grains produce a spiraling trail. Such garnets are called *snowball garnets,* for obvious reasons (Figure 12.12). We return to the process of mineral growth and deformation in Chapter 13, but for now we use a simple analog experiment. Place a ball bearing or a marble between oppositely sliding hands and you will see that its rotation is directly related to the motion of your hands (Figure 12.13). In fact, the amount of rotation of the marble is proportional to the relative displacement of your hands

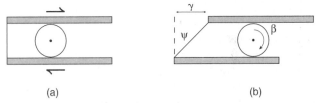

Figure 12.13 A simple ball-bearing experiment.

(i.e., the amount of simple shear); mathematically, this relationship is quite simple:

$$\beta = \tan\psi = \gamma \qquad \text{Eq. 12.1}$$

with β the rotation angle in radians (1 radian is 180°), ψ the angular shear, and γ the shear strain (see Section 5.7.3). However, if the ball bearing is greasy, the rotation angle may be less, because there is some slip between your hands and the ball bearing. This requires that we add a factor that describes the *coupling* between matrix and grain to Equation 12.1:

$$\beta = \Omega \tan\psi = A\,\gamma \qquad \text{Eq. 12.2}$$

where the parameter Ω describes the mechanical coupling between the ball bearing and your hands. The value of Ω is equal to 1 for full coupling (clean ball bearing), less than 1 for partial coupling (greasy ball bearing), and 0 for

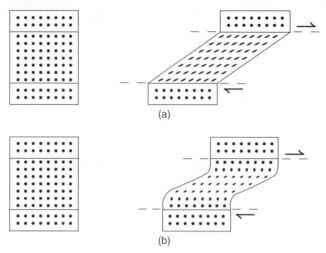

(a)

(b)

Figure 12.14   Strain in shear zones. The equal ellipses in (a) show that strain in the shear zone is homogeneous, whereas the strain in (b) is heterogeneous as shown by ellipse shapes that vary across the zone.

no coupling.[5] There is no single value for Ω that is unique for natural rocks. But if we assume that grains rotate in a linear viscous (Newtonian) fluid, where considerable slippage will occur at the contact between matrix and grain, we obtain a theoretical value for Ω of 0.5. Thus, by measuring the rotation angle of the spiraling snowball garnet, we can determine the shear strain given an assumption about coupling. Regardless of this assumption, $\Omega = 1$ gives us an estimate of the *minimum* shear strain.

---

[5]Kinematically, coupling describes the degree by which internal vorticity is converted to spin.

12.4.2   Deflected Foliations

Now let's take a closer look at the strain distribution in a shear zone by deforming the central portion of a block (Figure 12.14) under conditions of noncoaxial strain (simple shear). In Figure 12.14a the strain in the sheared portion of the block is homogeneous because the initial circles become ellipses with the same axial ratio and orientation. The strain pattern in Figure 12.14b, on the other hand, is clearly heterogeneous as shown by the ellipses of different

Figure 12.15   Small-scale, left-lateral shear zone, showing deflection of mylonitic foliation of an anorthosite (Grenville Orogen, Canada). Width of view is ∼20 cm.

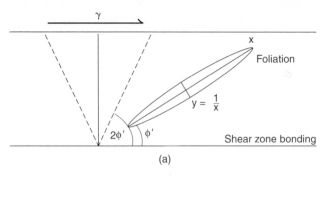

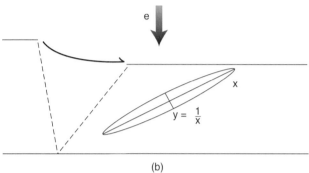

Figure 12.16 Angular relationship ϕ′ between foliation and shear-zone boundary and shear strain γ in (a) a perfect shear zone ($W_k = 1$), and (b) a shear zone with a component of shortening perpendicular to the shear-zone boundary ($0 < W_k < 1$). Dashed lines are the traces of circular sections of the strain ellipse.

axial ratio and orientation. To compare these two theoretical patterns with realistic examples, we look at a discrete shear zone in natural rock (Figure 12.15). The shear zone is characterized by a mylonitic foliation (S-foliation) that is at approximately 45° to the shear-zone boundary at the edges of the zone. This foliation becomes increasingly parallel to the shear-zone boundary as we approach the center of the shear zone. The progressive deflection of the foliation is similar to the pattern in Figure 12.14b, which indicates that the strain in natural shear zones is heterogeneous. Now, if we assume that the trace of the foliation tracks the X-axis of the finite strain ellipsoid (Figure 12.16a), the shear strain is readily determined by the equation:

$$\gamma = 2/\tan 2\phi' \qquad \text{Eq. 12.3}$$

Unfortunately, small differences in the value of ϕ′, such as occur at the center of the zone, significantly affect the calculated magnitude of the shear strain. Just compare the results for values for ϕ′ of, say 10° and 5°. Practically, therefore, we obtain a minimum shear strain with this method.

But there is yet a bigger problem. Strain determination using Equation 12.3 is complicated when a component of shortening or extension perpendicular to the shear-zone boundary is present (Figure 12.16b); that is, the strain is nonperfect simple shear (or general shear; see Section 4.5). General shear with a shortening component is called *transpression,* and with an extensional component it is called *transtension.* This pure shear component will change the

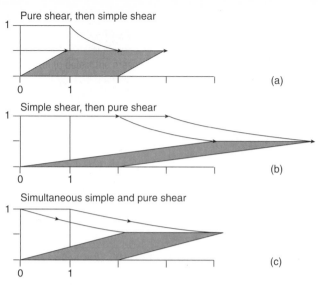

Figure 12.17 Difference between (a) superimposing simple shear on pure shear, (b) pure shear on simple shear, and (c) simultaneously adding simple and pure shear. The magnitudes of the simple shear and pure shear components are equal in all three examples, but they produce distinctly different finite strains because of the noncommutative nature of tensors.

angle of ϕ′, and thus invalidate Equation 12.3. The combination of coaxial and noncoaxial strain cannot be modeled by simply superimposing the two components, because the order of addition matters (recall that tensors are noncommutative; Figure 12.17a and 12.17b). Moreover, simultaneously adding these components will produce yet another finite strain tensor (Figure 12.17c). Deriving the associated equations requires an understanding of strain tensors and matrix operations that is beyond the scope of this book. (We refer you to the Additional Reading section at the end of the chapter.) We leave you with the observation that the angle ϕ′ may be small even for low shear strains in the presence of a shear zone-perpendicular shortening component and, analogously, that the angle may be large with a shear zone-perpendicular extensional component. In these cases shear strains obtained from Equation 12.3 will be overestimates and underestimates, respectively. Clearly, strain analysis in shear zones is a complex problem because of the many assumptions it requires, and the results are therefore rough estimates at best.

12.5 TEXTURES

Microstructures describe the geometric relationship between the various constituents of a rock and can give a good indication of the operative deformation mechanism, and even an estimate of the ambient conditions (Chapter 9). Crystal-plastic deformation mechanisms provide us with an additional tool for the analysis of rock deformation, which will prove to be particularly useful in the analysis of shear zones. Hence we waited till this point to bring it up. Recall that dislocation creep is the strain-producing crystal-plastic mechanism and that glide occurs on specific crystallographic planes in a crystal (Section 9.5; Table 9.1 lists

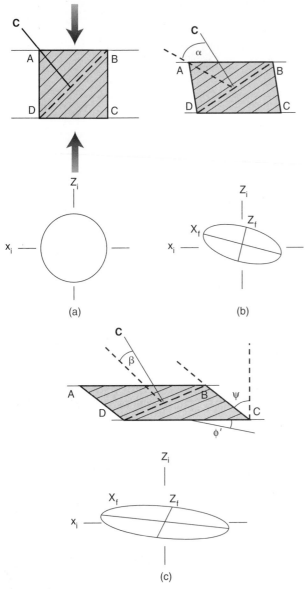

Figure 12.18 The development of crystallographic-preferred fabric by dislocation glide. The thin lines in grain ABCD are crystallographic glide planes, which in our example coincide with the basal plane (0001). The c-axis is indicated with the large C and the shortening direction by the heavy arrows. Strain arises from glide on the crystal planes; note that the length of individual planes remains the same, which requires that the glide planes rotate to accommodate the deformation. A number of different angular relationships are indicated: (α) angle of shear along glide plane; (β) rotation angle of the c-axis with respect to an external reference system (e.g., shear-zone boundary); (ψ) rotation angle of material line BC with respect to an external reference system (angular shear); (ϕ') angle of finite extension axis with respect to external reference system. Note that the instantaneous strain axes define the external reference system, and that they do not change in orientation during progressive deformation (i.e., no spin).

these glide planes in common rock-forming minerals). The process of dislocation movement holds the key to understanding *textures* or crystallographic-preferred fabrics in rocks. In fact, the particular type of texture formed may be indicative of the dominant deformation mechanisms, which in turn provide valuable information on the rheologic conditions of rocks.

What, yet another type of fabric in deformed rocks? You have already heard about dimensional-preferred fabric; and now we add crystallographic-preferred fabric to our structural vocabulary. Let's first see what their difference is, because dimensional- and crystallographic-preferred fabrics describe different properties of a rock. A *dimensional-preferred fabric* is the quantification of grain shapes in a rock; it is in essence a geometric parameter. Elongate quartz crystals that are all aligned in the same direction is an example of a dimensional-preferred fabric. A *crystallographic-preferred fabric,* on the other hand, describes the collective crystallographic orientation of all the individual grains that make up the rock. In other words, crystallographic-preferred fabrics represent the degree of alignment of crystallographic axes. We start by looking at the principles that govern the development of crystallographic-preferred orientation by intracrystalline slip.

The square ABCD in Figure 12.18a marks the cross section of a single grain that is deformed by homogeneous shortening perpendicular to the top and bottom sides of the square. Stated more formally, the infinitesimal shortening strain (Z_i) is parallel to AD (heavy arrows). The grain can deform only by dislocation glide along specific crystallographic planes, which are shown by the thin lines (parallel to the diagonal BD). Let us assume that these planes coincide with the basal plane of the crystal with the indices (0001) for hexagonal minerals, which means that the crystallographic c-axis is oriented perpendicular to these glide planes. We add one other restriction to the deformation: the faces AB and CD and the infinitesimal strain axes are held in a constant orientation relative to the external reference frame, which makes this a nonspinning deformation. Because dislocation glide is a volume constant mechanism, the dimension of the grain measured parallel to the glide plane cannot change during progressive deformation. At first glance this may not appear so, but measure the length of the same glide plane in each step to prove this point to yourself.

During deformation, shortening is accomplished by slip on the glide planes; obviously, this is accompanied by simultaneous extension of the grain, as shown by the finite strain axis (X_f) in Figure 12.18b. In order not to change their length or spacing, the glide planes have to rotate to accommodate the strain. As a consequence, the c-axis (perpendicular to the glide plane) rotates toward the infinitesimal shortening direction. Continued shortening produces further rotation of the c-axis toward the infinitesimal shortening axis. Meanwhile, the grain continues to elongate, with the long axis of the finite strain ellipsoid approaching AB. Note that the finite strain axes and infinitesimal strain axes rotate relative to each other, so that we have a nonspinning, noncoaxial mode of deformation. In this simplified scenario, we therefore observe two important characteristics: (1) the c-axis rotates toward the instantaneous shortening axis (Z_i), and (2) the strain ellipsoid elongates progressively toward the instantaneous shortening direction (X_i). In nature matters are more complex, because we are dealing with three-dimensional space in which several glide

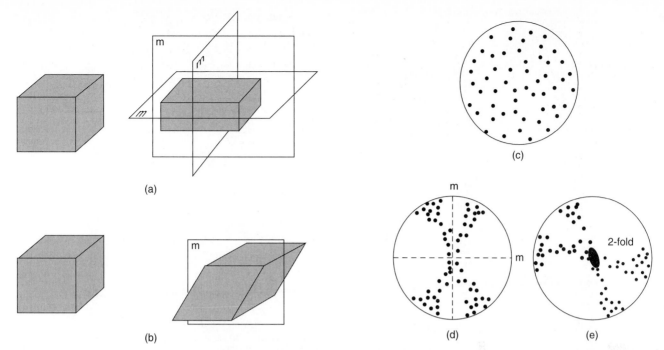

Figure 12.19 The Symmetry Principle. A cube is distorted into a body with (a) orthorhombic and (b) monoclinic symmetry. (c) A random distribution of c-axes reorganizes in response to deformation into (d) a high-symmetry and (e) a low-symmetry pattern. The symmetry of the c-axes patterns enables us to predict the strain path.

Table 12.3 Crystal Systems in Order of Increasing Symmetry

System	Symmetry	Crystal Axes[a]
Triclinic	1 onefold axis or center of symmetry	$a \neq b \neq c, \alpha \neq \beta \neq \gamma \neq 90°$
Monoclinic	1 twofold axis or 1 symmetry plane	$a \neq b \neq c, \alpha = \gamma = 90°, \beta \neq 90°$
Orthorhombic	3 twofold axes or 3 symmetry planes	$a \neq b \neq c, \alpha = \beta = \gamma = 90°$

[a]a, b, and c describe the lengths of the crystal axes; α is the angle between b and c; β is the angle between a and c; γ is the angle between a and b.

planes may be active simultaneously. Computer modeling of the development of crystallographic fabrics in these more complex situations, however, shows overall behavior similar to that in our simple model. Perhaps the most practical application of crystallographic-preferred fabrics lies in the determination of the *shear sense* in mylonites, but for this we must become familiar with the Symmetry Principle.

12.5.1 The Symmetry Principle

Symmetry is all around us. If you look at your neighbor, you will see that a person's face is symmetrical (at least in general) across a single plane; this *symmetry plane* cuts between the eyes and divides the nose in halves. A cube also has symmetry, but unlike the human face it contains three mutually perpendicular symmetry planes, or *mirror planes.* We say that a cube has a higher symmetry than a human face because it contains more mirror planes. In mineralogy you will have learned, for example, about triclinic (only a

center of symmetry or a twofold axis), monoclinic (one mirror plane), and orthorhombic (three mutually perpendicular planes with intersections of unequal length) symmetry. Any introductory mineralogy textbook will help you to refresh your memory on crystal symmetry. In Table 12.3 we list the systems that are most pertinent to our discussion.

Before formally stating the *Symmetry Principle* (or Curie Principle),[6] we illustrate it with a simple example (Figure 12.19). You recall that we can apply strain to a cube such that the resulting shape is a rectangular block. The symmetry of the rectangular block is orthorhombic (three perpendicular symmetry planes or three perpendicular twofold axes; Figure 12.19a). The symmetry of the simplest strain path causing this shape change is also orthorhombic: the incremental and finite strain ellipsoids differ only in shape, not in orientation (i.e., coaxial strain). Now we restate this relationship by distinguishing the cause (the strain path) and the effect (the rectangular block): the

---

[6]After the French scientist Pierre Curie (1859–1906).

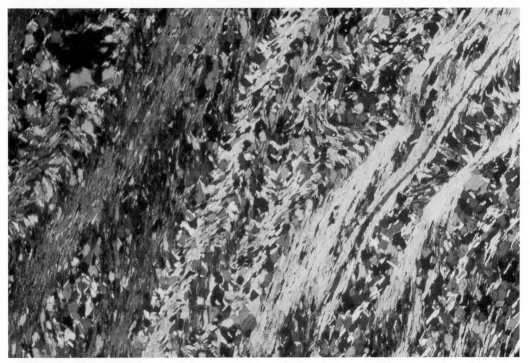

Figure 12.20 Photomicrograph showing mica fabric in differentiated crenulation cleavage (Pyreneus). Width of view is 1.5 mm.

effect has a symmetry that is equal to or greater than the cause. This is the Symmetry Principle. Now we distort our cube such that it becomes a block with a lower symmetry, say by shear (Figure 12.19b). In this case the resulting symmetry is monoclinic, because we can recognize only one mirror plane or one twofold axis. Using the Symmetry Principle, we therefore predict that the strain path (i.e., the cause) must have monoclinic or lower symmetry in the second experiment; thus, the strain must have been noncoaxial.

So, if the Symmetry Principle says that the effect is of equal or greater symmetry than the cause, is the reverse also true? No, the cause cannot have a symmetry that is higher than the effect. This is all fine and well, but what does this have to do with crystallographic-preferred fabrics in rocks? Figure 12.19c shows a random pattern of crystallographic c-axes in lower hemisphere projection, say, from a statically recrystallized quartzite. In two separate deformation experiments, we form two c-axis patterns (Figure 12.19d and e). The symmetry of the two patterns is distinctly different: the pattern in Figure 12.19d has orthorhombic symmetry, whereas the pattern in Figure 12.19e is monoclinic. What does this say about the strain that produced these patterns? Using the Symmetry Principle, the pattern in Figure 12.19d must have been caused by a strain path with a symmetry equal to or less than orthorhombic; analogously, the symmetry of the strain path that produced the pattern in Figure 12.19e is monoclinic or lower. Thus, the pattern in Figure 12.19e ('the effect') can have been formed only by non-coaxial strain ('the cause'). However, the pattern in Figure 12.19d cannot be uniquely interpreted; it can have formed either by coaxial or noncoaxial strain, according to the Symmetry Principle. So, determining the c-axis fabric pro-

vides information on the strain history of the rock. In particular we use low-symmetry patterns to indicate that the rocks were deformed in a noncoaxial strain ('general shear') regime.

12.5.2 Textures as Shear-Sense Indicator

The asymmetrical shape of crystallographic patterns can offer an additional piece of information about ductile shear zone (Figure 12.20). Let us again use a simple geometric experiment. Imagine an aggregate of grains that allow slip on the same single glide plane, which is initially randomly oriented (Figure 12.21). This situation is more complex than our previous single-grain condition (Figure 12.18), because neighboring grains will affect the ability of a grain to slip.[7] Nevertheless, as we shear the aggregate a pattern

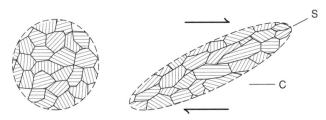

Figure 12.21 Relationship between shape, crystallographic fabric, mylonitic foliation (S), and shear plane (C) in an aggregate with a single glide plane.

---

[7]This is called *strain compatibility*.

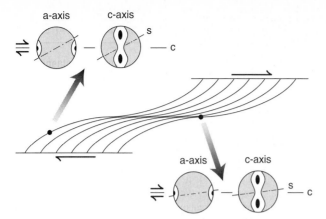

Figure 12.22 Schematic illustration of foliation pattern in shear zone showing angular relationships with increasing shear strain between finite strain (as recorded by foliation deflection) and quartz c-axes.

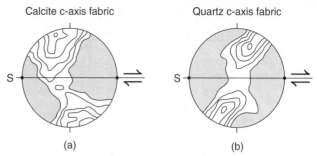

Figure 12.23 Asymmetrical c-axis fabrics for (a) calcite and (b) quartz from shear zones, showing contrasting c-axis patterns resulting from the operation of different glide planes. In quartz, basal slip occurred, whereas e-twinning dominated calcite deformation. Note that the mylonitic foliation plane (S) is used for reference.

emerges in which the majority of c-axes rotate toward an orientation perpendicular to the shear plane (C). At the same time a dimensional-preferred fabric is formed, defining a mylonitic foliation (S-foliation) that is oblique to the shear-zone boundary (C-surface). Thus, the c-axis fabric is oblique to the S-foliation in a way similar to that of the shear sense in the zone (Figure 12.22); in other words, the c-axis girdle tilts in the direction of shear. With increasing shear, this obliquity becomes less because the S- and C-surfaces approach parallelism, which means that in the center of shear zones this use of textures may be difficult (Figure 12.22).

The natural c-axis fabric for quartz shown in Figure 12.23 is consistent with our model, but is this pattern a rule that can be applied to all minerals? Unfortunately, the answer is no. Look at the natural c-axis fabric for calcite (Figure 12.23), in which slip occurred on the e-plane of grains (calcite twinning; Table 9.1). The displacement is again right-lateral, but in this case the c-axis pattern lies in the opposite quadrant. This is not a paradox; rather, the explanation lies in the operative slip system. Quartz in the first example deformed predominantly by crystal slip on the basal plane ((0001)-plane), whereas calcite slip occurred by e-twinning. However, calcite crystallographic fabrics that are formed by slip on other glide planes can produce c-axis patterns that are similar to those for quartz. This highlights the importance of knowing the operative slip system before crystallographic patterns are used as a shear-sense indicator. Such information may be obtained by measuring the orientation of other crystallographic axes in the sample (such as the a-axis; Figure 12.22).

The analysis of crystallographic fabrics requires careful preparation of oriented thin sections and tedious measurement of crystallographic orientations with an optical microscope equipped with a universal stage or other techniques (e.g., X-ray methods). Crystallographic fabric analysis is most commonly applied in monomineralic rocks consisting of relatively simple minerals such as quartz, calcite, and olivine, but other minerals, such as

feldspar, can also be used. For each grain the orientation of a particular crystallographic axis is measured and plotted in spherical projection. Typically, the orientation of the mylonitic foliation (S) or, if known, the shear-zone boundary (C) is taken as E-W and vertical; it is critical for the interpretations to label these foliations carefully (recall their relationship as a function of shear sense; see Figure 12.10). A reliable crystallographic-preferred fabric analysis involves at least 100–150 grains unless a very simple and well-defined pattern is present. Before starting your measurements, you can get a crude estimate of the degree of preferred orientation by inserting the gypsum plate (1/4-wavelength) in the optical microscope under crossed polarizers and by rotating the sample. If there is an obvious crystallographic fabric, the color of most grains will change at the same time (e.g., from blue to yellow in quartz); otherwise, you will continue to see an irregular pattern of colors (a mix of red [= extinction color], yellow, and blue).

As a final note, crystallographic fabrics mainly reflect the *latest stages* of the deformation history, because recrystallization processes tend to erase the early history. Nonetheless, crystallographic fabrics can offer a valuable tool to unravel the deformation history of rocks and regions, as you will see by looking at the modern structural geology literature.

12.6 FOLD TRANSPOSITION

At first glance, folds seem not so common in ductile shear zones, but upon closer inspection they have a peculiar appearance. In Figure 12.24 an asymmetrical fold is formed from a small instability by shear parallel to a well-foliated rock. With increasing shear, the oblique (short) limb of the asymmetrical fold rotates back into a layer-parallel orientation (Figure 12.24a–d). The resulting perturbation from this fold gives rise to a new fold that is superimposed on the original structure. Continued shear reorients the fold pattern back into a layer-parallel orientation (Figure 12.24e–f), leaving behind a complex geometry that can be interpreted only by carefully tracing the layering (Figure 12.25). This scenario highlights two aspects of folding in shear zones:

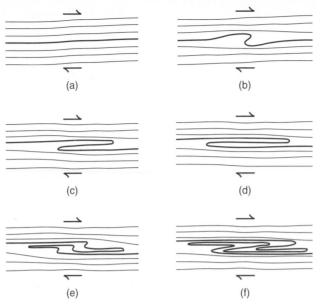

Figure 12.24 Fold transposition in a layered rock that undergoes noncoaxial, layer-parallel strain. An asymmetrical fold develops at a perturbation (a–d), which in turn becomes refolded (e–f). The end result is a superimposed fold structure that is essentially parallel to the dominant layering. Note that if the second fold formed at a place other than the location of the first fold, the two would have been indistinguishable, yet formed at different times.

Figure 12.25 Transposed mafic layer in granitic gneiss (Grenville Orogen, Canada). The dark layer can be traced as a single bed that is folded numerous times in the outcrop. This structure is affectionately known as the "snake outcrop."

(1) Fold asymmetry may be representative for the sense of shear; that is, Z-vergence occurs in right-lateral shear zones, while S-vergence occurs in left-lateral shear zones. We purposely state this as a possibility and not as a rule, because at very high shear strains the vergence of small folds may reverse (Figure 12.26). (2) Folding is a progressive process, resulting in complex patterns of folding and refolding. Such complex patterns in which limbs of isoclinal folds are oriented parallel to the dominant layering are called *fold transposition*. Folds in areas of high shear are sometimes disrupted, preserving only a few isolated fold hinges or 'fold hooks' (because of their resemblance to fishing hooks) and otherwise consisting mostly of layers representing the parallel limbs. Fold transposition occurs at all scales, from microfolds to km-scale folds, and is not restricted to zones of noncoaxial strain, but may occur in any type of high strain.

East-west shortening parallel to competent layers in a less competent matrix, say, sandstone layers in a shale matrix, results in folds with an upright axial surface (Figure 12.27a); the *fold enveloping surface* is drawn for reference. Continued shortening produces some thinning of the layers, and locally the limbs and hinges become detached (Figure 12.27b; a natural example of this stage is shown in Figure 12.27d). Eventually, the folds become isoclinal and we are left with a rock texture that is dominated by a vertical layering (Figure 12.27c). Without knowledge of the history, the fabric in Figure 12.27c shows little or no indication of the original geometry, especially if the structures are very large (say kilometers) and you have only a limited view of the area (imagine mapping only the square in Figure 12.27c). A cursory examination would suggest that overall bedding is vertical, until you notice the presence of the small fold hooks and maybe some minor

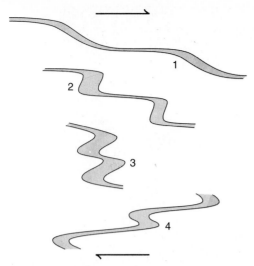

Figure 12.26 Reversal in fold vergence (from 'S-shape' to 'Z-shape'), with increasing shear strain in a right-lateral shear zone.

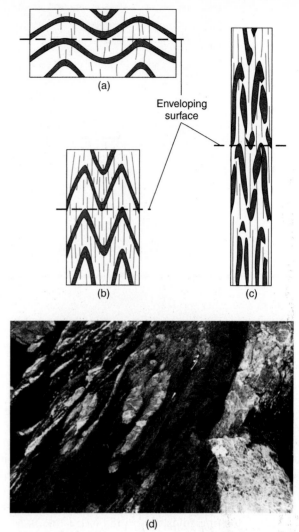

Figure 12.27 Fold transposition. Progressive coaxial strain (a–c) results in isoclinal folds that become disrupted by limb attenuation and boudinage. (d) A natural example of an early stage of transposition (New World Island, Newfoundland).

isoclinal folds. Then you realize that fold transposition may have occurred. Transposition fabrics are even more pronounced when an axial-plane foliation is formed early on, which will be nearly parallel to transposed layering (Figure 12.27c). As a consequence, pretty much all layering in the rock seems parallel, with the exception of a few, partly preserved fold hinges. Similarly, transposition occurs in ductile shear zones, resulting in a geometry with all layering oriented roughly parallel to the shear-zone boundary. Try to sketch the evolution of folds in a noncoaxial environment and track the orientation of the enveloping surface for comparison with the coaxial strain scenario in Figure 12.27.

If transposition sounds a bit contrived, it helps to realize that it is more likely the rule than the exception in deformed metamorphic areas, where it is facilitated by the overall weakness of layers at elevated pressures and temperatures (e.g., Figure 12.25). Unfortunately, in many instances we are not alerted to its presence until a lot of detailed study is carried out. Yet, identifying transposition is important to erect a regional stratigraphy for an area, because it poses the question of whether repetition of a particular lithology is primary or structural in origin, and to unravel the deformation history of an area. So, are there *criteria* to recognize transposition? One indicator is a regular repetition of lithologies in an area; for example, repeated marble layers in an area of mafic gneisses may, on a regional scale, outline large isoclinal folds. When a sequence of lithologies exists, say gabbro-felsic gneiss-marble, its repetition in reversing order is suggestive of transposition. Parallelism between foliation and bedding is not diagnostic, but it certainly should make you suspicious. The occurrence of minor isoclinal folds and fold hooks (especially if the foliation is axial-planar to these structures) adds considerable weight to the case. You will wrap it up if you can show that the direction of younging in layers across the area changes back and forth, but younging

evidence may be hard to find in medium- and high-grade metamorphic areas. However, in low-grade metamorphic areas (greenschist facies and down) the observant geologist often finds sufficient evidence for younging. In the end, your efforts proving a case of transposition may be large, but the fruits from its recognition are even larger. Transposition usually results in (1) a radical reinterpretation of stratigraphic relationships by showing that the 'stratigraphy' seen in the area is in fact of structural origin and that the real stratigraphy is much simpler, and (2) a more complete structural history by recognizing a previously unknown isoclinal stage of folding that has major implications for the geologic evolution of the area. In Chapter 19 you will see that many mountain belts have experienced early stages of large-scale isoclinal folding, typically associated with thrusting, which is often manifested in outcrop by fold transposition.

Figure 12.28 Sheath fold in amphibolite gneiss (Grenville Front, Ontario, Canada).

12.6.1 Sheath Folds

In closing, we turn to one particular type of fold that is restricted to regions of high shear, *sheath folds* (Figure 12.28; see also Section 10.5). In contrast to the ambiguity from fold vergence (Figure 12.26), sheath folds may define shear sense in ductile shear zones. They are typically recognized by an eye-shaped outcrop pattern that represents a section through the nose of the fold. Sheath folds are a special case of doubly-plunging folds, where the hinge line is bent around by as much as 180° (Figure 12.29). Thus, layering in a sheath fold is everywhere at a high angle to the profile plane, giving the characteristic eye-shaped pattern in section A. Sheath fold is formed when the hinge line of a fold rotates passively into the direction of shear and the axial surface rotates toward the shear plane (Figure 12.29a). Since such rotations require high amounts of shear strain, the occurrence of sheath folds is limited mainly to shear zones. The degree of hinge rotation is a direct function of the shear strain and the angular relationship of the hinge line with the shear plane. However, sheath folds cannot be used as a gauge of strain beyond the general comment of "large amount of shear," unless the initial angular relationships are known.

The location of the nose of sheath folds points in the sense of movement, but this can be determined only when the fold is fully exposed (i.e., including the nose!). In the less-rotated portions of a sheath fold, the shear sense may also be derived from fold asymmetry (Figure 12.29a). More typically, sheath folds define the direction of shear rather than direction and sense, with the hinge line approximately parallel to the shear direction. Note that sheath folds differ from 'dome-and-basin'-type fold interference patterns by

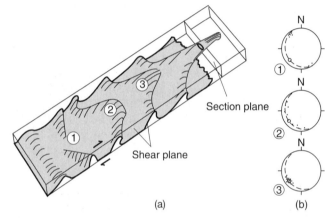

Figure 12.29 Sheath folds form when initially weakly doubly-plunging folds are modified in a zone of high shear strain. Several stages of fold modification occur, with the most evolved stage producing the characteristic conical geometry of a sheath fold. In (a), the lowest amount of shear occurs at left and the highest shear strain at right. Note the curvature of the hinge line in fold 3, which typically is associated with a strong (stretching) lineation in outcrop. In (b), the corresponding lower hemisphere projections of folds 1, 2, and 3 are shown. The great circle represents the shear plane, whereas the open circle represents the shear direction and the small dots are hinge line measurements. When fully exposed, sheath folds may be used as a shear-sense indicator, with the nose pointing in the direction of transport; otherwise, sheath folds indicate the direction of shear only (but not the sense!).

the fact that they form by progressive strain in the same overall deformation regime, rather than reflecting two discrete deformation regimes (i.e., superposed fold generations; Section 10.6).

12.7 CLOSING REMARKS

It is fairly certain that you will encounter shear zones and transposition on a field trip or perhaps in your study area (and possibly you already have). There is no better way to learn about the variety of deformation features in shear zones than by such direct observation. In many instances you will be unable to convince yourself of aspects such as sense of shear or 'stratigraphy' unless you allow yourself ample time for outcrop examination. Remember that it is more useful to understand one outcrop well than many outcrops in a cursory fashion; one single outcrop may, in fact, hold the key to the interpretation of an entire area. Unfortunately, it is an unwritten rule that this special outcrop occurs in the most remote part of the area and on the highest peak. However, any well-studied outcrop will allow you to erect a working hypothesis that can be tested in other places. Only later, after many outcrops, will you learn which of them is your area's 'Rosetta stone.'

Once mylonitization begins, the associated microstructures weaken the rock relative to the host rock. As a consequence, further strain will continue to be localized and the mylonite will evolve further (perhaps into an ultramylonite). In other words, shear zones represent zones of strain softening. If they didn't, strain would be distributed throughout the rock body, and would not be concentrated in the shear zone. Within ductile shear zones quite a large variety of shear-sense indicators may develop, but you will have noticed that their kinematic significance is often not fully understood. Many of these indicators are based only on empirical observations. Therefore it is recommended not to interpret the shear sense based only on any one single indicator, no matter how often it occurs or how "classic" an example it is. Reliable predictions are based on several different indicators that each show the same shear sense. Of course caution should be exercised with any interpretation, but because shear zones often are central to unraveling the deformation history of an area, we wish to emphasize this point here. Finally, when you are adequately convinced of the significance of a shear zone in your area, you should keep in mind that the zone may be part of a larger system, and that the shear sense is representative only for the zone in which you carried out the analysis (remember the analogous problem with changing vergence of minor folds in a single large fold; Figure 10.16). To end this sequence of chapters on ductile deformation, we discuss some tools from related Earth science disciplines that aid the modern structural geologist in unraveling the history of deformed rocks and regions.

ADDITIONAL READING

Fossen, H., and Tikoff, B., 1993, The deformation matrix for simultaneous simple shearing, pure shearing and volume change, and its application to transpression and transtension tectonics: *Journal of Structural Geology*, v. 15, p. 413–422.

Hanmer, S., and Passchier, C., 1991, Shear-sense indicators: a review, *Geological Survey of Canada*, Paper 90-17, 72 p.

Law, R. D., 1990, Crystallographic fabrics: a selective review of their applications to research in structural geology, in Knipe, R. J., and Rutter. E. H., eds., Deformation mechanisms, rheology and tectonics: Geological Society Special Publication, v. 54, p. 335–352.

Lister, G. S., and Snoke, A. W., 1984, S-C mylonites: *Journal of Structural Geology*, v. 6, p. 617–638.

Passchier, C. W., and Simpson, C., 1986, Porphyroclast systems as kinematic indicators, *Journal of Structural Geology*, v. 8, p. 831–843.

Ramsay, J. G., and Graham, R. H., 1970, Strain variation in shear belts, *Canadian Journal of Earth Sciences*, v. 7, p. 786–813.

Ramsay, J. G., and Huber, M. I., 1983/1987, The techniques of modern structural geology, Volumes 1 and 2: New York: Academic Press, 700 p.

Sibson, R. H., 1977, Fault rocks and fault mechanisms: *Journal of the Geological Society of London*, v. 133, p. 191–213.

Simpson, C., 1986, Determination of movement sense in mylonites: *Journal of Geological Education*, v. 34, p. 246–261.

Simpson, C., and De Paor, D. G., 1993, Strain and kinematic analysis in general shear zones: *Journal of Structural Geology*, v. 15, p. 1–20.

Chapter 13

Deformation, Metamorphism, and Time: D-P-T-t Paths—An Essay

13.1 INTRODUCTION

Unraveling the deformation history of a region can be like a detective story. Well, maybe solving how an area became deformed isn't quite as exciting as some of the stuff in the movies, but it's still a mystery. Just like any modern sleuth, you have many fancy tools available in the forensic lab to solve the geologic 'crime.' It is impossible to be expert in all the methods (which range from physics to chemistry), but a basic knowledge is needed to help you decide what may be useful for your particular case. In this chapter we look at some approaches, mainly from the fields of metamorphic petrology and isotope geochemistry, that are in-

creasingly used in the study of deformed regions. We concentrate on medium- to high-grade metamorphic areas, which are representative of processes active in the deeper levels of the crust, to complement the emphasis on relatively shallow-level deformation in much of the book. We will use a hypothetical area with hypothetical (= perfect?) rocks,[1] with the purpose of showing the complementary nature of these methods to 'traditional' structural analysis. As we go along, you will see that this chapter is far from comprehensive; so we call it an essay.

First we need to collect, organize, and examine the evidence, and then chart our course of action. Some of the associated terminology and concepts that follow are listed in Table 13.1.

13.2 FIELD OBSERVATIONS AND OUR GOAL

Let's assume that we are studying the tectonic evolution of a metamorphic terrane containing rocks in greenschist to granulite facies. Our first field summer was used to map the lithologies and structure of the area. Now that we have a good understanding of the structural geometry (e.g., locations of shear zones, nature of fabrics, etc.), we select

---

[1]Although real names and natural examples are used in this dramatization, any resemblance to reality is purely coincidental.

Table 13.1   Some Terms and Concepts Related to Deformation, Metamorphism, and Time

(Geo)barometry	Determination of pressure during metamorphism.
(Geo)chronology	Isotopic age determination of a mineral or rock; these ages are given in years, sometimes called "absolute" age.
Clockwise P-T-t path	A schematic plot in P-T space showing changing conditions in a rock subjected to burial followed by uplift and exhumation, such that during burial, temperature increases relatively rapidly while pressure increases relatively slowly, and during uplift temperature decreases relatively slowly while pressure increases relatively rapidly. (*Note:* clockwise is defined in a P-T plot, with P increasing downward; petrologists use the opposite terminology because they plot P increasing upward.)
Closure temperature	Temperature at which a system becomes closed to loss of the radiogenic daughter isotope.
Cooling rate	The rate at which temperature decreases (with time).
Counterclockwise P-T-t path	A schematic plot in P-T space showing changing conditions in a rock subjected to burial followed by uplift and exhumation, such that during burial pressure increases relatively rapidly while temperature increases relatively slowly, and during uplift, temperature decreases relatively rapidly while pressure decreases relatively slowly. (*Note:* Counterclockwise is defined with P increasing downward.)
Exhumation (rate)	Displacement relative to the Earth's surface (with time); also *denudation, unroofing.*
Geothermal gradient	Temperature change per depth unit (typically °C/km or K/km).
Paragneiss	Mineral assemblage in a metamorphic rock.
Peak metamorphism	Condition of peak temperature and corresponding pressure. Note that peak temperature generally does *not* coincide with peak pressure (and vice versa).
Porphyroblastesis	Timing of mineral growth relative to deformation (strictly, mineral growth).
Postkinematic growth	Metamorphic mineral growth after deformation.
Prekinematic growth	Metamorphic mineral growth before deformation.
Prograde metamorphism	Metamorphic history before peak temperature, characterized by dehydration reactions and density increase.
P-T-t path	History of rock or region in pressure(P)-temperature(T)-time(t) space.
Retrograde metamorphism	Metamorphic history after peak temperature condition, which may involve hydration reactions in the presence of an H_2O-rich fluid.
Synkinematic growth	Metamorphic mineral growth during deformation.
Tectonite	General term for a deformed rock; the adjectives L, S, and L-S indicate dominance of a lineation, a foliation, or a combination, respectively (see Chapter 11).
(Geo)thermobarometry	Temperature-pressure determination using equilibrium reactions.
(Geo)thermochronology	Temperature-time history of rocks.
(Geo)thermometry	Determination of metamorphic temperature.
Uplift (rate)	Displacement relative to a fixed reference frame, such as the geoid (with time).
(Mineral) zonation	Presence of compositionally distinct regions in a mineral grain; this records changing conditions of chemical equilibrium during growth or subsequent modification at times of chemical disequilibrium. Typically zonation is found from core to rim in such minerals as garnet and feldspar.

(a)

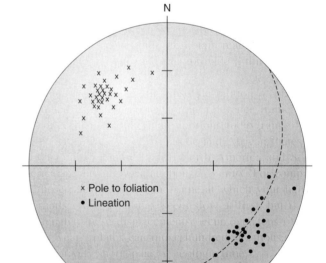

Figure 13.1 Field photograph of mylonitic gneiss (a) and lower-hemisphere, equal-area projection of spatial data (b).

(b)

samples from outcrops for further analysis (Figure 13.1a). The goal of this aspect of our study is to determine the spatial and temporal conditions at which these rocks were deformed. Specifically, we try to determine the crustal depth at which these rocks were deformed, the timing of metamorphic and deformational events, and the overall burial and exhumation history.

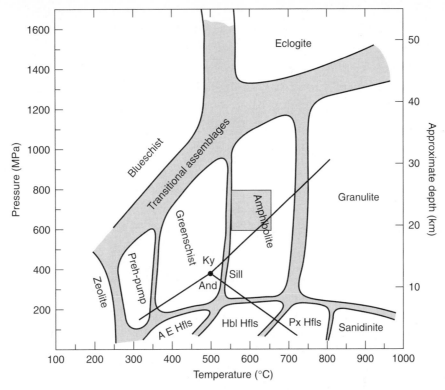

Figure 13.2 Pressure-temperature diagram showing the fields of the various metamorphic facies and the Kyanite (Ky)-Andalusite (And)-Sillimanite (Sill) triple point. The approximate crustal depth is also shown. Abbreviations used are: A E = albite-epidote, Hbl = hornblende, Hfls = hornfels, Preh-pump = prehnite-pumpellyite, Px = pyroxene.

Lithologically, our samples are *paragneiss* and *marble,* which are metamorphic rocks of sedimentary origin (as opposed to *orthogneiss,* which is a rock of igneous origin). Prior to removing samples, we characterize the outcrop as a whole and carefully orient our specimens by marking dip and strike or dip direction on the samples. These *tectonites* have several distinct field characteristics. Most obvious is a well-developed foliation that is defined by compositional variation or color banding, and a lineation that is defined by oriented large hornblendes in the paragneiss and black graphitic stripes in the marble. The orientation of these fabric elements in outcrop is shallowly SE-dipping and SE-plunging, respectively. Upon closer inspection of the gneiss, we see a fine matrix that surrounds mm-size grains with distinct shapes and textures. In particular the asymmetry of tails on feldspar grains in the gneiss draws our attention, because it indicates a left-lateral sense of displacement along the foliation when looking to the NE (δ-type porphyroclasts; Section 12.3). Based on the presence of a strongly foliated and lineated fabric, relatively fine grain size of the sample relative to neighboring outcrops, and the occurrence of rotated porphyroclasts, we conclude that the rock is a mylonite; that is, the rock developed significant shear strain by plastic deformation mechanisms. We plot the spatial information in spherical projection (Figure 13.1b), and based on this we conclude that the outcrop is part of a low-angle, SE-dipping shear zone with a reverse sense of displacement.

The paragneiss contains the minerals hornblende, garnet, quartz, feldspar, biotite, muscovite, and kyanite, as well as accessory phases that cannot be readily identified with the handlens. This mineral assemblage indicates that the rock was metamorphosed in the amphibolite facies, suggesting a temperature range of 500–700°C and a pressure range of 400–1200 MPa (Figure 13.2). Staining of the marble with a solution of HCl and red dye indicates the presence of dolomite in addition to calcite; the main accessory phase is graphite. The assemblage of minerals in the marble does not allow us to determine the metamorphic grade in the field, but the coexistence of calcite and dolomite will be useful in the laboratory as a geothermometer. After taking some photographs of the scene and some sketches of geometrical relationships, the samples are removed, labeled, and taken to the laboratory for further analysis.

Upon returning to the bug-free environment of the laboratory, we cut oriented samples and prepare two mutually perpendicular thin sections. The sections are oriented such that one is perpendicular to the foliation and parallel to the lineation, and the other is perpendicular to both the foliation and the lineation. These two surfaces will provide a three-dimensional description of the microscopic characteristics of our samples. Thin sections of the gneiss reveal that our field conclusions about metamorphic grade were generally correct. The main minerals in the rock are plagioclase and alkali feldspar, quartz, hornblende, biotite, muscovite, garnet, and kyanite, and the accessory phases

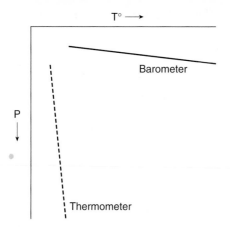

Figure 13.3 Reliable geothermometers (dashed line) are largely pressure insensitive, whereas good geobarometers (solid line) are largely temperature independent; they are schematically plotted in P-T space.

are rutile, ilmenite, and titanite. The foliation is defined by alternating bands of light and dark color, which is a consequence of the relative proportion of light-colored quartz and feldspar, and dark-colored hornblende and biotite. The marble consists of calcite, dolomite, and minor amounts of graphite and titanite. The foliation in the marble is defined by alternating layers of different grain size and by concentrations of opaques (mainly graphite) and other accessory phases (primarily, monazite and sphene).

13.3 PRESSURE AND TEMPERATURE: GEOTHERMOBAROMETRY

Geothermobarometry is the quantification of temperature and pressure conditions to which a rock was subjected during its metamorphic history. It contributes three important pieces to our geologic puzzle: (1) the peak temperature condition, (2) the approximate depth at which the rocks were buried, and (3) (part of) the retrograde history of the rock. We will look at each piece below.

Experimental and thermodynamic work in metamorphic petrology, supported by empirical observations, define a large number of mineral reactions that can be used to estimate the P and T conditions of metamorphism. In making such estimates, it is assumed that the mineral assemblage or parts of individual minerals represent *equilibrium conditions,* which means that at a given condition of pressure and temperature, certain minerals have specific compositions. Mineral assemblages make good geobarometers if the equilibrium reaction is relatively insensitive to temperature (Figure 13.3). Similarly, a good geothermometer is largely pressure independent (Figure 13.3); thermodynamically, there is a large entropy[2] change, but little volume change in

---

[2]Entropy, S, is the thermodynamic parameter describing the degree of randomness or disorder in a system.

geothermometers. For example, consider the following reaction involving the exchange of ions:

$$A(i) + B(j) = A(j) + B(i) \qquad \text{Eq. 13.1}$$

where A and B are two minerals in equilibrium (say, garnet and biotite) and i and j are two cations (say, Fe^{2+} and Mg). Because only cation exchange occurs in this reaction, there is little or no change in volume involved, and consequently many such reactions are pressure insensitive.

Now let's look at a geobarometer. The mineral garnet is common in amphibolite facies rocks, and is formed when the rocks contain sufficient amounts of the respective elemental constituents of that mineral. An example of a common garnet-forming reaction in metasediments is:

$$3CaAl_2Si_2O_8 = Ca_3Al_2Si_3O_{12} + 2Al_2SiO_5 + SiO_2$$
plagioclase (anorthite) = garnet (grossular) +
aluminosilicate (e.g., kyanite) + quartz   Eq. 13.2

Garnet is formed by the breakdown of plagioclase with increasing pressure. This particular reaction (known by the acronym GASP) is a geobarometer because it involves significant volume change and is relatively insensitive to temperature. Thus, in a plot of pressure versus temperature (P-T diagram; Figure 13.3), we find that lines with a shallow slope are good barometers and those with steep slopes are good geothermometers. An intermediate slope indicates a sensitivity to both pressure and temperature (one reaction, two unknowns), and the reaction can be used only if one parameter is determined independently by another geothermobarometer.

A single mineral reaction describes a line in the P-T diagram; for example the kyanite-silliminate boundary in Figure 13.2. Two mineral reactions define a single P-T point provided that the lines intersect, and the greater the angle between the two intersecting lines the more reliable the P-T estimate. Examples of some widely used geothermometers and geobarometers are listed in Table 13.2; for details of these systems and their applications you are referred to the petrologic literature (see Additional Readings at the end of the chapter).

In theory, the application of thermobarometric systems should give single, internally consistent values for pressure and temperature, but in practice this is rarely the case. Nevertheless, using a few different systems, one is generally able to constrain the pressure conditions within ±100 MPa (= ±1 kbar) and temperature conditions within ±50°C. In all our future results we simply assume these error estimates.

After considering the mineral assemblages in our samples, we decide to apply the geobarometers GASP and GRIPS, and two-feldspar and garnet-biotite geothermometry to the paragneiss (Table 13.2). Remember that the most reliable results are obtained when more than one

Table 13.2 Selected Geobarometers and Geothermometers

Geobarometers:

Garnet-plagioclase (Ca exchange between garnet and plagioclase, for example: garnet-plagioclase-quartz-Al_2SiO_5 (GASP); garnet-plagioclase-rutile-ilmenite-quartz (GRIPS); garnet-rutile-ilmenite-Al_2SiO_5-quartz (GRAIL))

Pyroxene-plagioclase-quartz (Na or Ca exchange)

Phengite (phengite content in muscovite)

Hornblende (Al content)

Geothermometers:

Garnet-biotite (Fe-Mg exchange)

Garnet-pyroxene (Fe-Mg exchange)

Two-feldspar (Na exchange between alkali feldspar and plagioclase)

Two-pyroxene (Ca-Mg exchange)

Calcite-dolomite (Mg exchange)

Calcite-graphite (^{13}C isotopic fractionation)

In part after Essene, 1989.

---

system is used to determine P or T. The marble from the same outcrop does not contain an assemblage that allows a pressure determination, but the calcite-dolomite and calcite-graphite systems are available as geothermometers. After many hours in a darkened microprobe room and some stable isotopic work using a mass spectrometer (which we will not go into here), we obtain an internally consistent pressure of 700 MPa (7 kbar) and a temperature of 600°C for our samples. These results place the rocks firmly in the amphibolite facies, just above the kyanite-sillimanite boundary (Figure 13.2). This refines our rough field estimate.

What do these values of temperature (T) and pressure (P) really mean? In general, thermobarometry measures the conditions at the time of mineral formation or when chemical exchange between phases ceases to occur. If the mineral formed during *prograde metamorphism* (i.e., a history of increasing metamorphic temperature), we assume that temperature represents the peak temperature achieved during metamorphism, because at those conditions the rate of mineral reactions is highest. However, the chemical composition of minerals formed at peak temperature conditions may be modified during cooling of the rock, which is called *retrograde metamorphism*. Retrograde metamorphism may lead to zoned minerals when only part of the mineral reequilibrates, or to complete compositional resetting. This is like erasing with one hand part or all that you wrote with

the other. The rock achieves equilibrium for certain T conditions until this thermally activated process is too slow to allow complete exchange.

The pressures that we derive from geobarometry do not describe the stress state that caused deformation of the rock. Rather, metamorphic pressures are *lithostatic pressures* and are a measure of the weight of the overlying rock column (recall Chapter 3). So, if we know the density of the rock column, we can directly translate a metamorphic pressure into a value for depth. For every kilometer in the crust, the pressure increases by about 27 MPa (assuming an average rock density of 2700 kgm^{-3}; Section 3.9). Thus, the 700 MPa pressure estimate of our sample implies that these rocks were buried to a depth of approximately 26 km.

Zoned minerals are particularly useful for deciphering P-T history, because zonation records changes in conditions during prograde and/or retrograde metamorphism. Typically, the rims of minerals reset to retrograde P-T conditions. Zonation is not always visible in the optical microscope; rather, zonation patterns are determined using an electron microprobe, which allows us to recognize small compositional differences. Mineral zonation stems from the *incomplete* equilibration of previously formed minerals. Garnets in particular are prone to zonation. They are large and relatively fast growing minerals that may act as the memory of the metamorphic history. Figure 13.4 is an example of a large garnet that preserves a complex history of growth based on the irregular Ca pattern, as well as a retrograde exchange based on decreasing Mg/Fe ratio (i.e., the Mg decreases and Fe increases) and increasing Mn content toward the rim. In our garnet sample, we are able to determine P-T for both core and rim of the mineral. The results from the analysis of garnet cores and inclusions were established earlier (700 MPa, 600°C), and for the rims we obtain values of 500 MPa and 550°C.

13.3.1 Status Report I

What have we learned so far? After the gneiss and marble were deposited at the surface (they are metasediments!), they were buried by approximately 26 km where they became exposed to a peak temperature of 600°C ('core' P-T). Today they are again at the surface, because we are able to sample them. They recorded lower temperature and pressure during their ascent as the 'rim' P-T data (550°C at ca. 19 km depth). Shallower conditions are not recorded because the kinetics of the geothermobarometric reactions became too slow. Thus, we are beginning to unravel the burial and exhumation history or the *P-T-t path* of our rocks, which eventually helps us to provide information on the structural evolution of our area. But before we can address this, we need to establish the temporal relationship (t) between deformation and metamorphism, as well as obtain absolute ages.

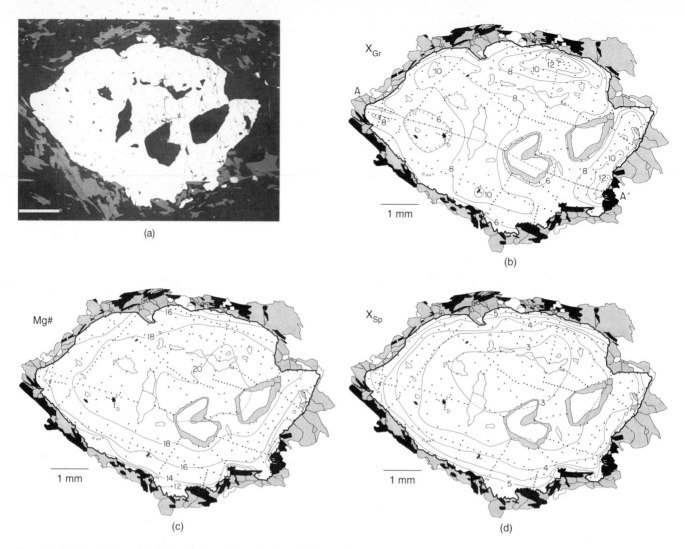

Figure 13.4 Complexly zoned, inclusion-rich garnet from the Grenville Orogen, showing (a) back-scattered electron image, (b) Ca ($X_{Gr} = 100 \times$ Ca/(Ca + Mn + Mg + Fe)), (c) Mg# (= $100 \times$ Mg/(Mg + Fe)), and (d) Mn ($X_{Sp} = 100 \times$ Mn/(Ca + Mn + Fe)) composition based on detailed electron microprobe analysis. Each dot in garnet represents an analytical point; black is biotite, gray is plagioclase, and white is quartz. Scale bar is 1 mm; zonation is given in mol%.

13.4 DEFORMATION AND METAMORPHISM: PORPHYROBLASTESIS

Because of our interest in the conditions during deformation, we must first determine the relative timing of metamorphism with respect to deformation. *Prekinematic growth* means that mineral growth occurs before deformation, *synkinematic growth* means that growth occurs during deformation, and *postkinematic growth,* you guessed it, implies that minerals grow after deformation. The shape and internal geometry of minerals that grow during metamorphic reactions and their relationship to external fabric elements, in particular foliations, will help us to determine this temporal relationship. As newly grown minerals are called *porphyroblasts,* derived from the Greek word "blastos" for growth, the relationship between growth and deformation is called *porphyroblastesis.*

When thinking of mineral growth, what perhaps comes to mind is those nice crystals that decorate geodes. But rocks that are buried a few km in the earth rarely have any

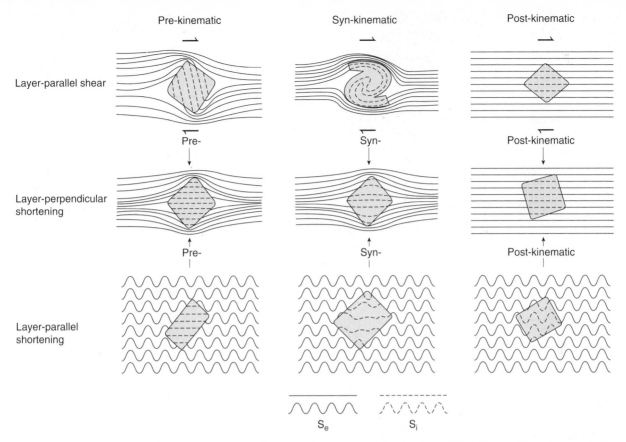

Figure 13.5 Schematic diagram showing the diagnostic forms of porphyroblasts that grow before (pre-kinematic), during (syn-kinematic), and after (post-kinematic) layer-parallel shear, layer-perpendicular shortening, and layer-parallel shortening. The temporal relationship is based largely on the relationship between the internal foliation (S_i; dashed lines) and the external foliation (S_e; solid lines).

open spaces, and mineral growth under these conditions generally occurs by the replacement of preexisting phases. For example, garnet grows by the consumption of another mineral (e.g., the GASP reaction, Equation 13.2). As garnet grows, phases that are not involved in the garnet reaction (e.g., accessory phases such as zircon, monazite), or phases that are left over because the rock does not contain the right mix of ingredients, will become *inclusions*. These inclusions may form ordered trails that define an *internal foliation* (S_i) in the porphyroblast; the foliation outside the blast is called the *external foliation* (S_e). The relationship between S_i and S_e allows us to determine the relative timing of mineral growth and deformation (Figure 13.5). Pre-kinematic growth is characterized by an internal foliation (S_i) whose orientation is unrelated to S_e, and typically the

internal and external foliations do not connect. At the other end of the spectrum lies postkinematic growth, which shows an external foliation that continues into the grain without any disruption. Figure 13.6 shows some examples. In the intermediate case, synkinematic growth, growth and deformation coincide. The evidence for synkinematic growth is an external foliation that can be traced into a blast containing an internal foliation with a pattern different from S_e. A classic example is that of *snowball garnets* (Figure 12.12), in which S_i spirals around the core of the garnet until it connects with S_e, which generally has a simpler geometry.

How do we get such spiral patterns in garnet? Imagine a rock that becomes metamorphosed in an active shear zone. After a small porphyroblast nucleates, it rapidly

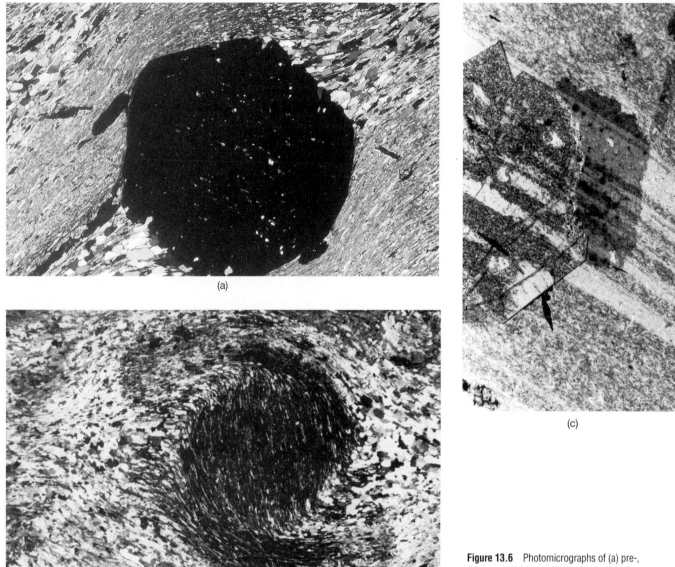

(a)

(b)

(c)

Figure 13.6 Photomicrographs of (a) pre-, (b) syn-, and (c) post-kinematic growth of garnet (a and b) and staurolite (c). Long dimension of view is ca. 1.5 mm.

grows by consumption and inclusion of other minerals (Figure 13.7). At the same time, a shear zone foliation is formed that becomes incorporated into the blast as an inclusion trail. However, because the blast itself is affected by the shearing motion, like ball bearings rotating between your hands (Section 12.4.1), continued mineral growth produces an internal foliation that appears to be rolled up. This process forms the characteristic snowball geometry. But consider the alternative: the same pattern can be generated if the matrix rotates around a stationary, growing clast (Figure 13.8). The relative rotation of blast versus matrix is a matter of some debate, and it has been suggested that blasts are located in nondeforming pods surrounded by bands that concentrate the deformation.[3] Most geologists, however, believe that the blasts rotate, not the matrix. In our sample, we find that garnets display snowball inclusion trails, so we conclude that the metamorphism is synkinematic. Furthermore, the counterclockwise rotation sense supports a reverse motion in our mylonite zone that was also determined from shear-sense indicators in the field (see Section 13.2).

---

[3]Matrix rotation is fiercely advocated by a few structural geologists.

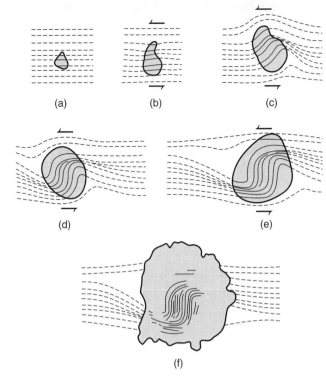

Figure 13.7 The progressive development of snowball textures in a counterclockwise rotating, growing porphyroblast (a–f).

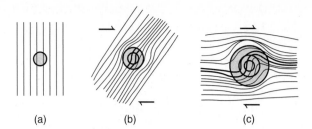

Figure 13.8 The development of snowball textures in a stationary, growing porphyroblast with a progressively rotating matrix (a–c). Note that the relative rotation sense of the matrix is opposite that of the blast in Figure 13.7.

In many instances we do not find such compelling evidence for synkinematic growth as snowball garnets. However, it seems generally true that mineral reactions are triggered by deformation, because the stress gradients that exist during deformation provide avenues for material transport and grain growth (for example, by fluid-assisted or 'wet' diffusion, Chapter 9). Stated otherwise, mineral assemblages and mineral compositions may be metastable until deformation triggers the reactions that produce equilibrium assemblages for the ambient P-T conditions. The suggestion that most mineral growth occurs during deformation is a rule that should be applied with caution. Environments of contact metamorphism immediately come to mind as an exception (post-kinematic growth).

13.4.1 *Status Report II*

From thin-section study we learned that previously determined peak P and T values represent the metamorphic conditions during deformation. In other words, deformation in the shear zone occurred under amphibolite facies conditions. Much of the necessary information to solve our geologic mystery has been obtained; what remains is: *when?* This has long been a difficult question to answer, but progress in the field of radiogenic isotope geochemistry comes to the rescue.

13.5   TIME: GEOCHRONOLOGY

You will remember that a mineral consists of a specific arrangement of atoms. Atoms, in turn, consist of protons and neutrons that make up the nucleus, which is surrounded by a cloud of electrons. The number of protons, the *atomic number* (symbol Z), defines an element; for example, Si contains 14 protons and its atomic number is therefore 14: $_{14}$Si. The atomic number specifies the order in the Periodic Table of the Elements. The protons and neutrons combined define the mass of an atom, which is called the *mass number* or *isotopic number* (symbol A). The isotopic number is the second value that you find in the Periodic Table; for example, Si has a mass of 28, which we write as: $_{14}^{28}$Si. In contrast to the atomic number, a single element can have several isotopic numbers, that is, different numbers of neutrons. For example, the element rubidium has the isotopes $_{37}^{85}$Rb and $_{37}^{87}$Rb. Some isotopes are unstable and fall apart spontaneously. Of the two Rb isotopes, $_{37}^{87}$Rb is unstable and decays to $_{38}^{87}$Sr; rubidium is called the *parent isotope* and strontium the *daughter isotope*.[4]

The number of isotopes that decay per unit time is proportional to the total number of parent isotopes present; this proportionality factor is a constant for each unstable isotope and is called the *decay constant,* λ. Isotope geologists use the same symbol that we previously defined as the quadratic elongation (Chapter 4); they are unrelated. Alternatively we can define the *half-life* ($t_{1/2}$) of an isotope, which is the time required for half of a given number of parent isotopes to decay. In the case of Rb to Sr, the decay constant is $1.4 \times 10^{-11}\text{y}^{-1}$, which implies a half-life of 48.8×10^{9}y based on the relationship:

$$t_{1/2} = 0.693/\lambda \qquad \text{Eq. 13.3}$$

---

[4]There are no sons in this dating game.

Table 13.3 Commonly Used Long-Lived Isotopes in Geochronology

Radioactive Parent (P)	Radiogenic Daughter (D)	Stable Reference (S)	Half-Life, $t_{1/2}$ (10^9y)	Decay Constant, λ (y^{-1})
^{40}K	^{40}Ar	^{36}Ar	1.25	0.58×10^{-10}
^{87}Rb	^{87}Sr	^{86}Sr	48.8	1.42×10^{-11}
^{147}Sm	^{143}Nd	^{144}Nd	106	6.54×10^{-12}
^{232}Th	^{208}Pb	^{204}Pb	14.01	4.95×10^{-11}
^{235}U	^{207}Pb	^{204}Pb*	0.704	9.85×10^{-10}
^{238}U	^{206}Pb	^{204}Pb*	4.468	1.55×10^{-10}

*^{204}Pb is not stable, but has an extremely long half-life of ca. 10^{17} years.

Source: Faure (1986).

Table 13.3 lists common radiogenic systems and their corresponding half-life and decay constant.

The principle behind isotopic dating is quite simple. If we measure the amount of parent (P) and daughter (D) present in a mineral now, their sum gives the original amount of parent (P_o). The ratio D/P_o is proportional to the age of the mineral. A useful object to illustrate the fundamentals of geochronology is an hourglass. If we start with one side of the hourglass full (containing the 'parent') and the other side empty (containing the 'daughter'), we need to know only the rate at which the sand moves from one chamber to the other (the 'decay constant') and the amount of sand in the daughter chamber, or the amount of parent remaining, to determine how much time has passed. In mathematical terms:

$$-dP/dt = \lambda P \qquad \text{Eq. 13.4}$$

where dP/dt is the rate of change of the parent, with a negative sign because the parent decreases with time. However, in reality the process is more complex.

13.5.1 The Isochron Equation

The first complication occurs because at the time the radiogenic clock starts, the rock already contains some daughter; in other words, some sand already exists in the daughter chamber before we start to measure time. This amount of daughter is referred to as the *initial daughter*. When we measure the total amount of daughter product in our specimen, we are combining the amounts of daughter from decay of the parent and initial daughter. The latter amount, however, needs to be subtracted for age determination. The solution to this problem involves using rocks and/or minerals of the same suite containing different isotopic ratios, which allows us to define an *isochron*. The details of this approach lie outside the scope of this chapter, but you will find them explained in every textbook on

geochronology. We merely give you the *isochron equation* without the theory:

$$D/S = D_o/S + P/S \, (e^{\lambda t} - 1) \qquad \text{Eq. 13.5}$$

where D is total daughter present at time t (radiogenic and common daughter), D_o is common daughter, P is parent present at time t, S is reference isotope (stable, nondaughter isotope), λ is the decay constant, and t is time. You may notice that this equation has the general form $y = a + bx$, which is the equation of a line. So, when we determine the values of D/S and P/S for several minerals in a single rock or for a suite of rocks, and plot these in a diagram of D/S versus P/S, the slope of this line (the isochron) represents the age of the sample; the slope is defined by $(e^{\lambda t} - 1)$ (Equation 13.5). An example of this procedure is given in Figure 13.9 for the Rb-Sr system.

The situation is more complex for the U/Pb system, because here we are dealing with two parents and two daughters (Table 13.3); however, this added complexity also gives great reliability to U/Pb ages. We simultaneously solve two isochron equations that give equal or different ages. If the ages are equal within errors (typically within 2–3 m.y.), the age is called *concordant;* otherwise we have a *discordant* age (Figure 13.10). Discordant mineral ages are quite common and reflect loss of Pb after the mineral was formed, but they are not useless. Techniques to extract the original age of the mineral as well as the time of Pb loss have been developed. These properties and the high closure temperature for some minerals (see Section 13.5.2) make U/Pb dating a very sensitive (standard error around 2 m.y.) and informative method, which has become widely used in regional geology.

The ^{40}Ar/^{39}Ar method of dating is the second geochronologic approach that we use in our study. The advantage of this method over traditional K-Ar determinations is that all measurements are carried out on the same

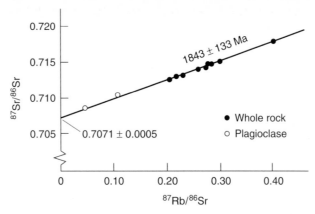

Figure 13.9 Rb-Sr isochron based on a combination of whole-rock analyses and plagioclase concentrates; each point represents an analysis. The different ratios lie on a straight line (the isochron) whose slope gives the age of the rocks.

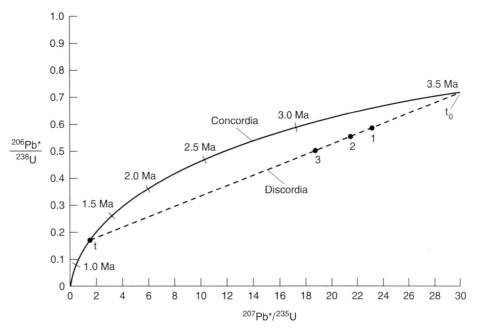

Figure 13.10 U-Pb concordia diagram showing three discordant zircon ages that reflect Pb loss at time t = 1250 Ma; the original zircon growth age is derived from the intersection between the line passing through the discordant zircon ages and concordia (t_0 = 3500 Ma). Modern U/Pb analysis of zircon uses portions of individual grains to determine discordia. The grain is progressively abraded to remove reset rims.

sample (after neutron irradiation), allowing us to use the *incremental heating technique* during which Ar is progressively released with increasing temperature steps. In theory, the age at each step (derived from the $^{40}Ar/^{39}Ar$ ratio) should be the same, but in practice this is generally not the case. Some Ar may have left a mineral (called Ar-loss) or have been gained (called excess-Ar), which is examined by stepwise release patterns (Figure 13.11). A reliable age for the mineral is based on the recognition of a 'plateau' comprising several steps of equal age within some error limits. Finally we need to know the closure temperature for all geochronometers, which is critical for the interpretation of age data.

13.5.2 The Closure Temperature of Minerals

What does an isotopic age mean? We need to know when the radiogenic clock starts to tick before we use isotopic ages in regional analysis. Previously you saw that the age of a mineral is a function of the ratio D/P, but remember that we measure only the radiogenic daughter that remains in the mineral. Now, consider that the daughter chamber in our hourglass has no bottom; so, the sand will not stay until we are able to close this chamber. Similarly in nature, closure of the system for daughter product does not necessarily coincide with the time of mineral growth, nor does it occur at the same temperature for each isotopic system. This important characteristic is addressed by the concept of

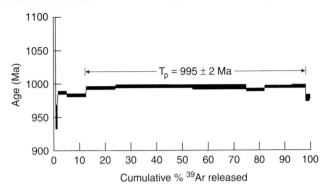

Figure 13.11 ^{40}Ar/^{39}Ar stepwise release spectrum of hornblende, with a plateau age, T_p, of 995±2 Ma that is indistinguishable from the total gas age (or fusion age).

Table 13.4 Selected Chronologic Systems and Approximate Closure Temperatures[a]

System-Mineral	T_c
U-Pb-zircon	>900°C
U/Pb-garnet	>800°C
U/Pb-monazite	725°C
U-Pb-titanite	600°C
U-Pb-rutile	400°C
^{40}Ar/^{39}Ar-hornblende	480°C
^{40}Ar/^{39}Ar-muscovite	350°C
^{40}Ar/^{39}Ar-biotite	300°C
^{40}Ar/^{39}Ar-alkali feldspar	200°C
Fission track-apatite	100°C

[a]For a cooling rate of 1–10°C/m.y. and a common grain size.

Source: Mezger (1990).

the *closure temperature*, T_c. Below its closure temperature, a mineral retains the amount of daughter produced; in other words, the radiogenic clock starts to tick at T_c. Above the closure temperature the daughter can escape the system. Table 13.4 lists closure temperatures for some common minerals and their isotopic systems, which shows that T_c differs greatly for different minerals and isotopic systems. Note that the closure temperature of a mineral is not a set value, but changes as a function of grain size (larger grains have higher T_c), cooling rate, and other parameters. The values listed in Table 13.4 are based on common grain sizes as well as on the range of cooling rates that are typical for deformed metamorphic rocks.

A high T_c value implies that the radiogenic clock in a mineral is not easily reset by regional metamorphism, which explains, for example, the popularity of the mineral zircon for U/Pb dating of the time of formation of igneous rocks. Thus, dating porphyroblasts with a closure temperature that exceeds the metamorphic temperature (e.g., monazite and titanite in greenschist facies rocks) gives a mineral *growth age*. Note that minerals can grow at temperatures higher than, equal to, or lower than T_c! In general, minerals with low T_c give growth ages only if the metamorphic temperature is less than T_c; otherwise they give *cooling ages*, the time at which the mineral cooled below the closure temperature after having achieved peak metamorphic temperature. If the closure temperature is above peak metamorphic temperature, then the date gives the time at which the mineral reached the peak metamorphic temperature. The large difference in T_c for various minerals has advantages that are used in various ways to understand the history of a rock or region, as we'll see later.

13.5.3 Dating Deformation

Now we can turn to the question of when our rocks were deformed. It is more difficult to obtain absolute timing constraints on deformation than relative timing; the latter may

be determined from crosscutting relationships and stratigraphic constraints. Most 'dateable' minerals are not involved in microstructures that give the time of their growth relative to deformation. One way to constrain the absolute age of deformation is to bracket the deformation by dating rocks that intrude the area. U/Pb ages on zircons from granites and pegmatites are typically used to this end. Deformed intrusive rocks give ages that predate the deformation, whereas undeformed intrusives are postkinematic. With some luck you can tightly constrain the deformation within, say, 10 m.y. Unfortunately, in older rocks these constraints are often very loose, resulting in a broad age bracket. Careful field study may offer a second approach, permitting identification of intrusives that are restricted to deformation zones. This suggests that intrusion occurred due to deformation, as such, these intrusives are synkinematic and they date the deformation. Quite commonly, U/Pb zircon ages from pegmatites in Precambrian rocks are used for this purpose.

In our area we seem to have no dateable intrusions to bracket deformation, but we have a range of minerals present that can be dated with great precision. Monazite and titanite in our samples show no evidence that they were derived elsewhere, that is, that they were eroded and transported. The crystals are euhedral and look 'fresh', and have distinctly different morphologies (color, size) from those in less-sheared samples elsewhere in the area. This suggests that they grew synkinematically in our shear zone. Dating monazite and titanite using the U/Pb method gives us concordant ages of 1010±2 Ma and 1008±2 Ma, respectively. The closure temperature for monazite (725°C; Table 13.4) is well above the peak metamorphic temperature determined from thermometry (600°C), so this mineral grew below its closure temperature and thus dates the deformation. The titanite age is slightly less, but the difference is not statistically significant (although it might reflect some resetting). As the monazite represents a growth age, we conclude that the age of deformation is 1010 Ma.

Both approaches (dating of intrusives and dating of accessory phases) have some drawbacks: (1) they do not date the mineral(s) involved in the mineral reactions on which we base the P and T conditions, and (2) they do not date minerals that are syn-kinematic based directly on microstructural relationships (as in Figure 13.5). In greenschist facies rocks, this can be overcome by looking at mineral reactions involving hornblende, muscovite, and other phases that can be dated with the $^{40}Ar/^{39}Ar$ technique. In amphibolite facies rocks, however, this method yields only cooling ages because T_c for these minerals (Table 13.4) is less than T_{peak}. As an option for high-grade rocks, garnet may be used for dating. This mineral is involved in many geothermobarometric reactions (Table 13.1) and commonly allows determination of the relative timing of deformation and metamorphism (Figure 13.6). After considerable effort, U/Pb dating of synkinematic garnets gives an age of 1011 ± 2 Ma. This age is within error equal to the monazite age and supports our previous determination of the age of deformation.

With our geochronologic work we have answered the question of when deformation occurred, but we are also able to learn something about the subsequent history by dating other minerals. The range in T_c values for various systems enables us to determine the cooling history of the area, or its *thermochronology*. If we date several minerals within a single rock, we obtain the times at which the rock passed through the respective closure temperatures, provided that the minerals grew above these closure temperatures. In other words, we are able to calculate the temperature change per time unit for our rocks (Figure 13.12). We apply the $^{40}Ar/^{39}Ar$ method to our gneiss sample, which yields ages for hornblende, muscovite, and biotite of 1000 Ma, 984 Ma, and 976 Ma, respectively. The *cooling rate* is the temperature difference divided by the age difference:

$$\text{cooling rate} = \delta T / \delta t \qquad \text{Eq. 13.6}$$

In our sample, the cooling rate from the time of peak metamorphic temperature to the closure temperature for hornblende is:

$$(T_{peak} - T_{c,hornblende})/(age_{monazite} - age_{hornblende}) = 120°C/10 \text{ m.y.} = 12°C/\text{m.y.}$$

Furthermore, the cooling rate seems to change with time based on the ages from other minerals:

$$(T_{c,hornblende} - T_{c,muscovite})/(age_{hornblende} - age_{muscovite}) = 130°C/16 \text{ m.y.} = 8°C/\text{m.y.}$$

$$(T_{c,muscovite} - T_{c,biotite})/(age_{muscovite} - age_{biotite}) = 50°C/8 \text{ m.y.} = 6°C/\text{m.y.}$$

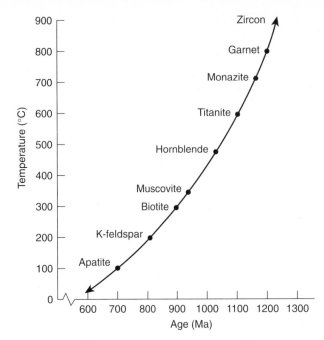

Figure 13.12 Hypothetical cooling history of a rock based on determining the ages of various minerals with different closure temperatures (see Table 13.4).

The decreasing cooling rate of our rocks places constraints on the uplift history of our area, as discussed after the third status report.

13.5.4 Status Report III

Our data base has expanded considerably in size and scope (as has our bill for analytical work) since we hammered our samples in the field. Let's summarize the key information:

- low-angle mylonite zone with reverse sense of displacement; based on field data and microfabrics
- metamorphosed sediments; thus, originally deposited at the surface
- peak temperature of 600°C (middle amphibolite facies)
- peak pressure of 700 MPa and burial depth of 26 km; thus, metamorphosed in the middle crust
- 1010 Ma age of peak temperature and deformation (i.e., mid-Proterozoic orogenic activity), based on U/Pb ages of monazite and garnet
- retrograde P-T conditions of 500 MPa and 550°C, based on garnet rim analyses
- cooling ages of 1000 Ma at 480°C, 984 Ma at 350°C, and 976 Ma at 300°C, based on hornblende, muscovite, and biotite $^{40}Ar/^{39}Ar$ dating, respectively
- cooling rate of 12°C/m.y. during early retrogression that slows down to 6°C/m.y.

13.6 DEFORMATION, PRESSURE, TEMPERATURE, AND TIME: D-P-T-t PATHS

Status report III admittedly provides quite a lot of information to digest, but we are now able to construct a remarkably detailed history of the rocks from deposition to burial and back to surface exposure. This history is described by the pressure-temperature-time (*P-T-t*) path, which in turn makes predictions about the deformational setting of the area. In Figure 13.13, the P, T, and t data that we have collected (solid dots) and inferences based upon them (open dots) are summarized in three related diagrams: a temperature-time diagram, a pressure (or depth)-temperature diagram, and a pressure (or depth)-time diagram. Each is discussed separately below.

13.6.1 Temperature-Time History

Our geochronologic and thermochronologic work give quite a few data points from which we may construct a reliable T-t diagram. Each mineral age is plotted against the corresponding peak T or T_c, depending on whether the age is a growth or cooling age, respectively. From this plot we see that the curve is not linear, but that it decreases in slope toward the surface temperature (for convenience, we place the mean surface temperature at 0°C). Each segment of the curve represents a different cooling rate that changes from 12°C/m.y. to 8°C/m.y. to 6°C/m.y.; the youngest segment is unconstrained other than that it eventually intersects the origin of the T-axis. We choose a gradient of 5°C/m.y., so that the rocks are at the surface around 920 Ma (dashed line in Figure 13.13a).

13.6.2 Pressure-Temperature History

In P-T space we have determined two independent points from thermobarometry: 700 MPa and 600°C, and 500 MPa and 550°C. We also know from the occurrence of retrograde rims on garnet that the change in P and T trends toward the lower values. Thus, we have determined the sense of part of the P-T path. In addition, we know that the path should intersect the origin (heavy dashed line in Figure 13.13b) after 976 Ma because the rocks are now at the surface. With the temporal information, we can place ages at points where the T-t curve meets the P-T curve. This highlights one important characteristic of P-T paths: there is no simple relationship between length of a P-T path segment and length of time. P-T segments of equal length early in the retrograde history represent only a few m.y., whereas in the late history they require tens of m.y. The time at which we reach the P-T point at 500 MPa and 550°C is predicted from the T-t curve at 1006 Ma. The path from peak temperature to surface exposure, the *retrograde path*, is more likely smooth rather than kinked, as in Figure 13.13b, which we mark by the dashed line.

We have no direct information on the *prograde path* of our rocks before they reached peak metamorphic conditions, other than that the path starts at the origin (recall that the rocks were deposited as sediments). Thus, we connect the high-temperature end of our retrograde path with the origin (Figure 13.13b).

13.6.3 Pressure-Time History

Only one point is constrained by barometric and geochronologic work in P-t space: 700 MPa at 1010 Ma. Nevertheless, we are able to construct a P-t path by combining the T-t and P-T curves (Figure 13.13c). The resulting path is a pattern of decreasing pressure change with time, which means decreasing rate of removal of the overlying rock column with time, called *exhumation rate*.[5] For our purposes, it is most informative to convert pressure to depth (1 km = 27 MPa), which shows that the exhumation rate changes by at least one order of magnitude, from approximately 2 km/m.y. to <0.2 km/m.y.

13.6.4 The Geothermal Gradient

The diagrams in Figure 13.13 highlight a second important property: the geothermal gradient is not a constant. The geothermal gradient, temperature change with depth, equals the ratio of cooling rate (Equation 13.6; Figure 13.13a) over exhumation rate (Figure 13.13b):

$$\text{geothermal gradient} = \delta T / \delta d \qquad \text{Eq. 13.7}$$

Calculating Equation 13.7 for various time intervals, we get:

1010–1000 Ma	12°C/km
1000–984 Ma	26°C/km
984–976 Ma	30°C/km
976–(920) Ma	30°C/km

In other words, the gradient changes with time, especially during the early retrograde history. This shows that the geothermal gradient changes during the process of sediment burial to a depth of 26 km and subsequent return to the surface. Thus, no single geothermal gradient is representative for the entire rock's history. Using the metamorphic P and T peak estimates to determine the geothermal gradient in deforming areas is therefore incomplete. In our example, this would give a gradient of 600°C/26 km, or 23°C/km, which lies somewhere in between the range of values determined above.

Why does the geothermal gradient change with time? Consider a slice of continental crust with an undisturbed geothermal gradient of 25°C/km (Figure 13.14a). In reality,

---

[5]*Uplift* is sometimes used, but this describes displacement relative to a fixed coordinate system. Here we uplift the rock and remove part of the overlying column, hence *exhumation*.

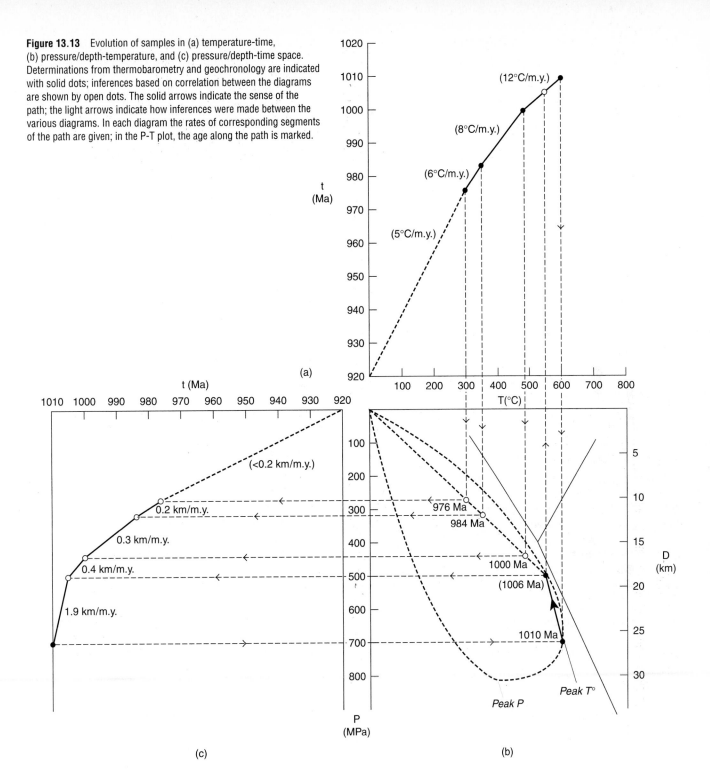

Figure 13.13 Evolution of samples in (a) temperature-time, (b) pressure/depth-temperature, and (c) pressure/depth-time space. Determinations from thermobarometry and geochronology are indicated with solid dots; inferences based on correlation between the diagrams are shown by open dots. The solid arrows indicate the sense of the path; the light arrows indicate how inferences were made between the various diagrams. In each diagram the rates of corresponding segments of the path are given; in the P-T plot, the age along the path is marked.

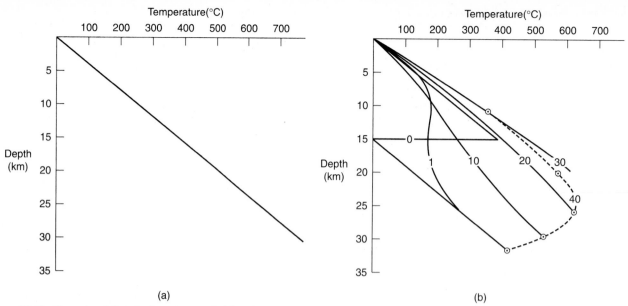

Figure 13.14 Thermal evolution after instantaneous doubling of a 15 km thick crustal section. A simple geothermal gradient of 25°C/km is assumed before thickening of the crust (a). In (b) we show the geotherms at the time of instantaneous thrusting (t_o), and after 1 (t_1), 10 (t_{10}), 20 (t_{20}), and 30 (t_{30}) m.y. After a few m.y. the irregularity in the thermal perturbation is largely removed, but the geothermal gradient remains depressed. At t_{10} the gradient is <20°C/km; at t_{20} it is approximately back to the original value (ca. 25°C/km); and after 30 m.y. the gradient is 30°C/km. These diagrams are schematic, but the shapes of the curves are based on thermal modeling.

this gradient is not linear because heat-producing elements are preferentially concentrated in the upper crust, but for our purposes this simplification will do (in Chapter 14 we return to the thermal structure of the earth). At time t_o, the top 15 km of the crust becomes doubled by thrusting. This disturbs the geothermal gradient, making it increase toward the thrust boundary. At the thrust boundary it sharply jumps back to the surface temperature, after which it again increases; this pattern is called the *sawtooth model*. Note that the development of the sawtooth shape implies that deformation (here thrusting) is instantaneous relative to thermal equilibration. How can we evaluate this?

The time over which a thermal perturbation decays by conduction is given by heat flow's "rule of thumb" equation:

$$t = h^2/K \qquad \text{Eq. 13.8}$$

where h is the distance over which thermal conduction is occurring and K is the thermal diffusivity (5×10^{-7} m^2 sec^{-1} for average crustal rock). Substituting h = 15 km gives an equilibration time on the order of 14 m.y. Thrusting, on the other hand, proceeds by many tens of km per m.y. Thus, whereas deformation may not be really instantaneous, it does occur at much faster rates than changes in thermal structure, permitting the simplification above. Equation 13.8 quantifies the change of a sawtooth thermal gradient with time, but let's use a comforting analog. When placing a cold blanket over your warm body, there is a strong temperature gradient at the contact between blanket

and body. Pretty soon the heat of your body affects the blanket and the temperature difference becomes less. After a while there will be no temperature difference between blanket and body and you are comfortably warm. Before dozing off, we return to the cold reality of Figure 13.14b, where you see that after approximately 20 m.y. the original gradient is restored, and that after 30 m.y. it even exceeds the original value (ca. 30°C/km). Thus, we conclude that deformation can significantly affect the thermal structure of the crust for time periods of up to tens of m.y.

13.6.5 The Deformational Setting

Finally, at the end of a chapter that seems dominated by metamorphic petrology and geochronology, we return to the D in D-P-T-t: the deformation history. The patterns in Figures 13.13 and 13.14 can be analyzed using numerical models involving input parameters describing heat production and heat transfer, and familiar parameters such as density, time, and depth. This theoretical approach, the details of which remain outside the scope of our discussion, can give some critical insights into orogenic evolution. A combination of observations on natural rocks, such as our samples, and modeling has produced characteristic relationships between P-T-t paths and the deformation history. For example, crustal thickening by thrusting results in rapid burial of rocks (deformation is relatively instantaneous), a thermal structure that evolves from having a sawtooth shape to elevated temperatures, and peak pressures that precede peak temperatures (Figure 13.15a). As a consequence of the

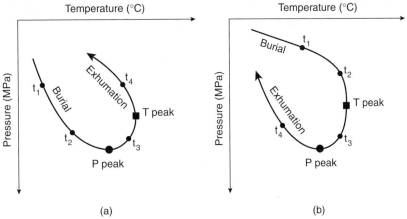

Figure 13.15 (a) Counterclockwise and (b) clockwise P-T-t paths may reflect characteristic deformation histories. Note that peak temperature (square) and peak pressure (circle) do not occur at the same time. Because metamorphic reactions proceed faster at peak temperatures, metamorphic pressures obtained from geothermobarometry are generally less than peak pressures. The time steps (t_i) do not represent equal time increments.

latter, rocks record pressure at the time of peak temperature, which is equal or less than the time of peak pressure. This combined burial (prograde) and exhumation (retrograde) history describes a *counterclockwise path*[6] (Figure 13.15a), which fits well with the analysis of our own field observations and P-T-t data.

Paths from other tectonic environments also have characteristic geometries. For example, a clockwise path (Figure 13.15b) may reflect increasing heat from below due to extension and crustal thinning, or adding a heat source at the base of the crust (a process called magmatic underplating). Clearly, numerical models are a powerful way to analyze orogenic evolution, but they do not substitute for observations. The main advantage of numerical modeling is that it can constrain part of the history that is not preserved in natural rocks; but it requires a critical approach because the input parameters vary considerably and the equations are quite sensitive to uncertainties.

13.7 CLOSING REMARKS

Modern structural analysis requires the integration of a variety of approaches. This essay shows only a glimpse of the possibilities that move the fields of structure, petrology, and geochronology into the full realm of orogenic evolution and tectonic processes of the deeper crust. Each individual approach has its strengths and weaknesses, but combined they provide enough information to determine a remarkably complete scenario. Central to all this, however, remain good field observations. There is no sense in analyzing a sample with the most sophisticated techniques and

numerical models if the results cannot be related to the area or if the sample is not representative. If we have met these requirements, an integrated approach such as used in this chapter will greatly expand our understanding of the deformation history of a region and provide fundamental information on lithospheric processes. With these remarks we close the part on ductile structures and start dealing with the fundamentals of plate tectonics and lithospheric structure.

ADDITIONAL READING

Barker, A. J., 1990, *Introduction to metamorphic textures and microstructures,* Glasgow, Blackie, 162 p.

Bell, T. H., Johnson, S. E., Davis, B., Forde, A., Hayward, N., and Wilkins, C., 1992, Porphyroblast inclusion-tail orientation data, eppure non son girate! *Journal of Metamorphic Geology,* v. 10.

England, P. C., and Thompson, A. B., 1984, Pressure-temperature-time paths of regional metamorphism I. Heat transfer during the evolution of regions of thickened continental crust, *Journal of Petrology,* v. 25, p. 894–928.

Essene, E. J., 1989, The current status of thermobarometry in metamorphic rocks, in Daly, J. S., Cliff, R. A., and Yardley, B. W. D., eds., Geological Society Special Publication, v. 43, p. 1–44.

Faure, G., 1986, *Principles of Isotope Geology* (second edition): New York, John Wiley and Sons, 589 p.

Hodges, K. V., 1991, Pressure-temperature-time paths: *Annual Reviews of Earth and Planetary Sciences,* v. 19, p. 207–236.

Mezger, K., 1990, Geochronology in granulites, in Vielzeuf, D., and Vidal, P., eds. Granulites and crustal evolution: Kluwer, p. 451–470.

---

[6]Petrologists often plot pressure up, thereby creating a clockwise path and some confusion.

Passchier, C. W., Meyers, J. S., and Kroner, A., 1990, *Field geology of high-grade gneiss terranes,* Springer Verlag, 150 p.

Passchier, C. W., Trouw, R. A. J., Zwart, H. J., and Vissers, R. L. M., 1992, Porphyroblast rotation: eppur si muove? *Journal of Metamorphic Geology,* v. 10, p. 283–294.

Spear, F. S., and Peacock, S. M., 1989, Metamorphic pressure-temperature-time paths, American Geophysical Union, Short Course in Geology, v. 7, 102 p.

Spry, A., 1969, *Metamorphic textures,* Oxford, Pergamon, 350 p.

Vernon, R. H., 1976, Metamorphic processes: London, Murby, 247 p.

Wood, B. J., and Fraser, D. G., 1977, *Elementary thermodynamics for geologists*, Oxford, Oxford University Press, 303 p.

Zwart, H. J., 1962, On the determination of polymetamorphic mineral associations, and its application to the Bosost area (central Pyrenees): *Geologische Rundschau,* v. 52, p. 38–65.

D

Tectonics and Regional Deformation

Chapter 14
Whole Earth Structure and Plate Tectonics

14.1 INTRODUCTION

Tectonics is the study of the origin and geologic evolution of large regions of the Earth's lithosphere. In this part of the book we will focus on regional geology and tectonics. Structural studies are used in tectonic analysis to provide information about the conditions during regional deformation (Figure 14.1). Likewise, structural geologists explore the tectonic framework in which deformation occurred because they want to know why the structures they've been studying formed in the first place. The speculations that characterize tectonic debates can be good fun, but be forewarned that the conclusions of such debates are sometimes not well constrained! Tectonics concerns most of the fundamental

Figure 14.1 Satellite image of the Appalachian Valley and Ridge Province in Pennsylvania. In this region, Paleozoic strata were folded into large amplitude folds in the fold-thrust belt of the Alleghanian Orogeny (see Chapters 17 and 19). Ridge-forming sandstone units outline the folds.

features of our evolving Earth. The origin of the continents, the building of mountain belts, the formation of ocean floor, and plate movement are only a few of the many topics that will be addressed under this heading.

Before we can discuss *how* and *why* the upper layers of the Earth were, and continue to be mobile, we must

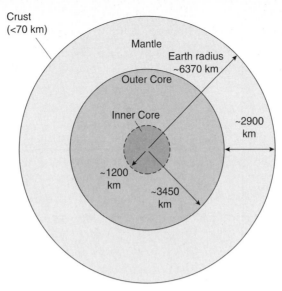

Crust
(<70 km)

Mantle

Earth radius
~6370 km

Outer Core

Inner Core

~2900
km

~1200
km

~3450
km

Figure 14.2 Traditional image of the layered Earth and the approximate thickness of each layer. The boundaries between these layers are defined by seismic discontinuities.

first have a sense of what we are working with. So, we need to answer two fundamental questions. First, what are the entities that move and deform? To answer this question, we briefly look at information about the internal structure of the Earth. Second, how do the outer layers of the Earth move? To answer this question, we examine the basic tenets of plate tectonics theory.

14.2 EARTH LAYERING

Remember the image of the Earth's interior that your introductory geology book presented? Probably you were left with the impression that the Earth resembles an onion, composed of concentric shells of rock or liquid with contrasting physical properties (Figure 14.2). This image of the Earth, on the scale of human history, is actually quite young. Before the nineteenth century, little was known of the Earth's interior except that it must be hot enough locally to generate volcanic lavas and gases. But whether the interior was solid or liquid, homogeneous or inhomogeneous, was quite a mystery. The first insights into the nature of the interior came from astronomers. With knowledge of the gravitational constant and the radius of the Earth (which has been known for more than 2000 years!), they were able to calculate the mass, and therefore the mean density of the Earth, and came up with a value of about 5500 kg/m³. Earth scientists realized that this value is more than twice the density of granite or sedimentary rocks, which are the principal rock types exposed on continents. Thus, a *density gradient* (i.e., a change in density as a function of depth) must exist in the Earth. As a starting point, it was assumed that this gradient was smooth, and was entirely a consequence of compression caused by the increase in lithostatic pressure with depth. Based on this assumption, it was estimated that the density of whatever composed the center of the Earth

was 10,000–12,000 kg/m³, a result that proved to be pretty close to the presently accepted value.

It was not until the end of the nineteenth century that another breakthrough occurred, which ultimately led to our present layered image of the Earth's interior. That breakthrough was the proposal that seismic waves generated during a large earthquake could travel through the whole Earth and be measured elsewhere on the globe. When this proposal was confirmed, the science of seismology was born. Using seismic data, we gradually developed an understanding of the basic distinctions between crust, mantle, and core. By the time of World War II, we had the image of the Earth's interior as divided into discrete shells separated from each other by seismic discontinuities (i.e., places where there is an abrupt change in velocity of the seismic waves), which correspond to relatively abrupt changes in density.

Analysis of thousands of earthquake records has yielded a fairly refined model of the Earth's interior, illustrated in Figure 14.3, which is a representation of an average profile of seismic velocity versus depth. Note the seismic discontinuities that define the crust, the upper mantle, the transition zone, the lower mantle, the outer core, and the inner core. Keep in mind, though, that this image is only a model; it is a best-fit approximation to explain existing seismic measurements. There is still disagreement about the exact position of many discontinuities, and velocity versus depth curves do vary with location. For example, the layering of the Earth beneath oceans differs from that beneath continental shields, which differs in turn from that beneath an orogenic belt.

What causes this pattern of density variation in the Earth? Changes in density with depth can reflect changes in chemical composition (*compositional changes*) or changes in mineral structure not requiring chemical change (*phase*

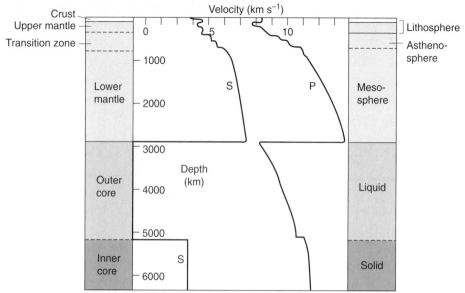

Figure 14.3 Variation of P- and S-wave velocities with depth in the Earth and their correlation with compositional and rheological layering. P-waves are "primary" waves, the fastest waves to travel through the Earth. S-waves are "secondary" waves, the next fastest. Also P-waves are compressional waves, whereas S-waves are shear waves.

changes). Exactly what changes occur at what depths is a question that continues to challenge researchers, because we are not able to make direct observations. Our understanding developed out of consideration of data from the study of geochemistry, mineral physics, meteoritics, and igneous petrology, among other fields.

Geologists are now confident that the average crust is composed of rocks ranging from an average *felsic* (granitic) composition, at shallower levels, to an average *mafic* (basaltic) composition, that the mantle is composed of *ultramafic* rock (peridotite), that the outer core is composed of *liquid iron alloy* (because it is unable to transmit S-waves;[1] Figure 14.3), and that the inner core is composed of *solid iron alloy*. Thus, the boundary between crust and mantle, known as the Mohorovičić discontinuity (or Moho), after its discoverer, and the boundary between mantle and core, called the Gutenberg discontinuity after its discoverer or CMB (core-mantle boundary), are *compositional boundaries*. Other major boundaries between layers of the mantle, however, probably reflect *phase changes*. For example, at the 400-km discontinuity, *olivine* (the principal Mg-silicate in peridotite) may transform into the more densely packed *spinel* structure; and at the 670-km discontinuity, the spinel structure may transform into the even denser *perovskite* structure. The boundary between outer and inner core is also probably a phase boundary, in this case between liquid and solid. Perhaps the main significance of the core for tectonics is that the Earth's magnetic field originates from inner core-outer core interactions.

In summary, we see that the familiar image of an Earth with a distinct crust, mantle, and core is an image based on analysis of variation in seismic velocity with depth. This

variation is caused by changes in physical properties (primarily density and elastic moduli), which in turn are caused by compositional and/or phase changes. The distinction between *lithosphere* and *asthenosphere*, terms that you probably remember from introductory geology and that we will clarify later in this chapter, is something quite different. They are based on the way in which these layers respond to stress; in other words, they describe rheologic properties of the Earth. Below, we first provide some details of the structure of the crust and the mantle, and then discuss the distinction between lithosphere and asthenosphere.

14.3 THE CRUST AND THE MANTLE

Much of this book so far has been devoted to teaching you how to examine and interpret geologic structures that you can see, if not on a map or in outcrop, at least with a microscope. But how do we know what is in the subsurface, say, 20, 200, or even 2000 km down? We can't generally walk around on the Moho or tunnel through the mantle, but we can infer what these places are like by interpreting a variety of data. What we know about the outer layers of the Earth comes from many sources, briefly described in Table 14.1.

14.3.1 Oceanic Crust

It is obvious, especially from space, that there are two fundamentally different physiographic features on planet Earth: oceans and continents. Approximately 29% of the Earth's surface is continents,[2] and the rest is oceans. Yet, even without the oceans, the Earth has a very characteristic physiography. This is graphically illustrated by a *cumula-*

---

[1]S-waves are shear waves, which cannot be transmitted in fluids.

[2]If we add the continental shelves, this value becomes ca. 40%.

Table 14.1   Studying the Outer Layers of the Earth

Electrical conductivity	These studies are particularly useful to detect the presence of partial melts.
Exposed deep crust	Particularly in mountain belts there are localities where faulting, folding, and exhumation have exposed very deep crustal levels, so that the rocks now at the surface were once at depths as much as 50 km in the crust (commonly granulites); a few localities even expose rocks from approximately 100 km depth (e.g., coesite-bearing rocks). Study of these localities gives a direct image of the geology at depth.
Geochemistry and elemental abundances	These help define the range of possible compositions of rocks in the Earth.
Gravity anomalies	These investigations use density differences between rocks and can show the shape of basins, the distribution of subsurface rock units, and the existence of isostatically uncompensated loads.
Lithospheric flexure	The lithosphere bends (flexes) in response to surface loads or load removal, such as the rebound of crust after glaciation and basins formed by loading the crust with thrust sheets or sediment. These studies gain insight into crust and mantle response to stress (e.g., is it elastic, viscous, or viscoelastic).
Magnetic anomalies	These can show the distribution of subsurface rock units, from variation in magnetic properties.
Ophiolites	An ophiolite is a rock suite representing oceanic lithosphere. These rocks may have been thrust over continental crust during collisional orogeny, and are now exposed subareally so that they can be examined in detail. Ophiolites as old as 2 Ga are known.
Petrology of xenoliths in volcanic rocks	Xenoliths provide direct observations of rocks from the deep crust and/or upper mantle (*xeno* = "strange" or "foreign"), though they are altered during their journey to the surface.
Seismic-reflection	These define locations in the crust where there are abrupt changes in lithology, creating a horizon where seismic energy is reflected. Such zones may be interpreted in terms of compositional or textural layering, geologic contacts, or structures.
Seismic-refraction	These define the velocity of seismic waves at depth in the Earth.
Seismic tomography	New techniques allow seismologists to create a 3-D picture of seismic velocity as a function of location in the crust and mantle. These images can be interpreted in terms of variation in density (primarily controlled by temperature) and chemistry.
Seismic velocity as a function of lithology	Laboratory measurements of velocities in rock samples allow interpretation of seismic-refraction studies.

tive frequency curve or *hypsometric curve,* which displays the percentage of surface area as a function of elevation (Figure 14.4). The high mountains of the Earth, peaked by Mt. Everest in the Himalayas, and the deep trenches, such as the Mariana Trench near the Philippines, comprise only a very small fraction of the Earth's surface. In contrast, we find two distinct plateaus on the curve, with little transition between them. Most of the continental surface lies within 1000 m of sea level, and most of the ocean floor lies at depths around 4 km below sea level. Note that sea level has varied in geologic time, but only within a few hundred m of today's value. This characteristic elevation distribution reflects fundamental differences in the composition and thickness of oceanic and continental crust,

parameters that reflect contrast in the process of formation of these two types of crust (Figure 14.5).

Oceanic crust forms at mid-ocean ridges by the process of seafloor spreading. Partial melting of underlying mantle peridotite, an ultramafic rock that is very high in magnesium and iron-bearing minerals, occurs because of depressurization as it rises beneath the ocean ridges. Silica is preferentially concentrated in the melt, so the resulting magma is mafic (basaltic) in composition. At fast-spreading ridges, the magma enters a large magma chamber in the crust below the ridge axis, whereas at slow ridges, the crustal magma chamber is temporary, and it freezes solid between pulses of spreading. Oceanic crust is created both by igneous extrusion and intrusion—part of

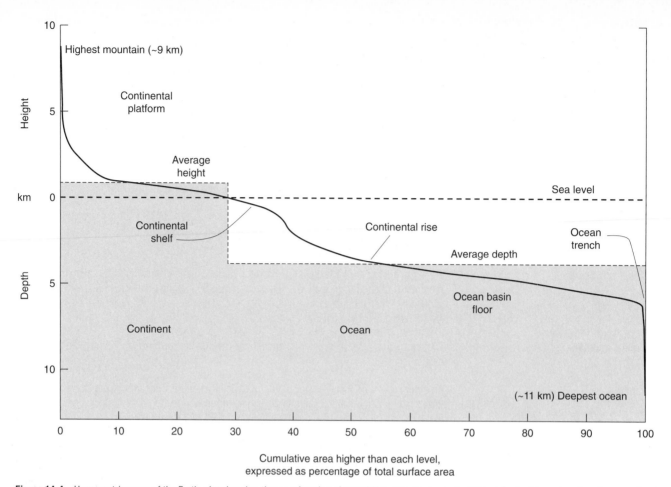

Figure 14.4 Hypsometric curve of the Earth, showing elevation as a function of cumulative area. For example, 29% of the Earth's surface lies above sea level; the deepest oceans and highest mountains comprise only a small fraction of the total area. The total surface area of the Earth is 510 × 10⁶ km².
From *The Way the Earth Works* by Peter J. Wyllie. Copyright 1976 © John Wiley & Sons, Inc. Reprinted by permission of John Wiley & Sons, Inc.

the magma is extruded as *pillow basalt* on the seafloor, part is intruded as vertical dikes (called *sheeted dikes*) at shallow depths below the ridge axis, and part cools more slowly to form *gabbro* at greater depth. Cumulates of olivine and other dense minerals settle to the base of the magma chamber. As seafloor spreading progresses, newly formed crust moves away from the ridge axis and is eventually buried by deep-marine sediment (siliceous ooze that eventually turns into chert, clay that eventually turns into shale, and carbonate ooze that eventually turns into limestone).

Because the process of seafloor spreading is similar everywhere, oceanic crust has distinctive layers that show up clearly in seismic-refraction studies (Figure 14.5). These layers have been named, simply, Layer 1, Layer 2a, Layer 2b, and Layer 3. Geologists studying *ophiolites,* which are slices of oceanic lithosphere, and marine geologists who have drilled holes into the oceanic crust, have confirmed that Layer 1 is the sediment layer, Layer 2a is pillow basalt, Layer 2b is sheeted-dike complex, and Layer 3 is massive gabbro. The lowest layer of an ophiolite is ultramafic rock, representing cumulates at the base of the crust or upper mantle (Figure 14.5). The total thickness of oceanic crust is about 6–10 km.

Recall from introductory earth science that oceanic lithosphere is subducted at trenches and descends into the mantle. As a consequence, all of today's oceanic crust is less than 200 Ma, except in a few places on the continents where ophiolites as old as 2 Ga have been preserved in ancient orogens. This young age of today's ocean basins is only one of many remarkable consequences of plate tectonics.

14.3.2 Continental Crust

Continental crust is five to ten times thicker than oceanic crust (Figure 14.5). Most continental crust is about 40 km thick, but crust beneath orogenic belts can attain thickness of up to 70 km, and crust beneath active rifts is thinned to less than 25 km. The average chemical composition of the upper portion of the continental crust (upper crust) is similar to that of *granodiorite* (rock of silicic to intermediate composition), whereas the average chemical composition of the lower portion of the crust (lower crust) is chemically approaching *basalt* (rock of mafic composition). The contrast in composition between upper and lower crust probably reflects the early differentiation of the continental crust during the Precambrian. Partial melting of the lower crust created intermediate to silicic magmas that rose to shallow

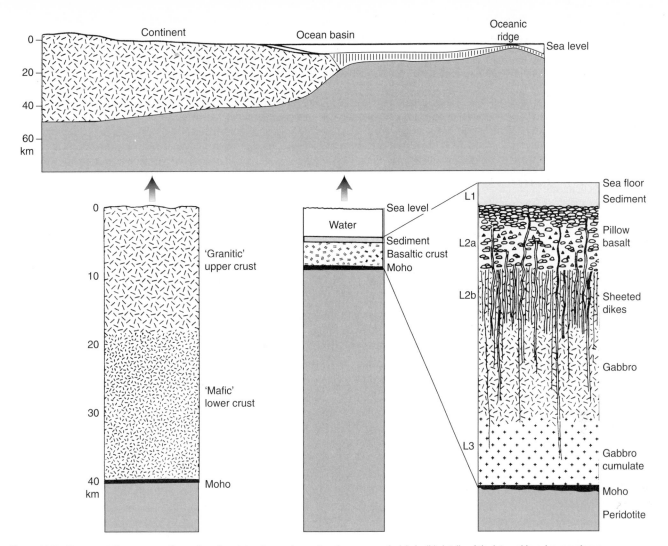

Figure 14.5 Representative cross sections of continental and oceanic crust and upper mantle (a). In (b) details of the internal layering are shown.

levels to make the upper crust, leaving a more mafic rock forming the lower crust. Keep in mind that even though a gross layering exists in the continental crust, represented by the contrast between upper and lower crust, the continental crust is very heterogeneous and varies laterally in both lithology and structure.

This heterogeneity of the continental crust reflects its long and complicated history. Most of the elements forming the present continental crust migrated out of the Earth's interior in Archean time, between about 3.8 and 2.5 Ga, when Earth was much hotter than today. This migration represents *differentiation* of the Earth. The present crust did not exist when Earth first formed. In fact, in one popular model of crustal evolution, the Earth was so hot during its very early history (4.5–3.8 Ga), that its outer layer was simply a convecting ultramafic mantle with a scumlike ultramafic crust at the surface. This crust, which formed because of cooling at the interface between the mantle and cold space, resembles the 'scum crust' that forms on the surface of a lava lake in a volcanic crater. Convective overturn caused the scum crust to be 'subducted,' such that it eventually sank back into the underlying mantle; but in contrast to the present process of subduction, the scum totally remelted because the mantle was so hot. Therefore, no permanent differentiation occurred, and no permanent crust formed during the early history of the Earth. Superimposed on this process was the relatively frequent disruption of the early Earth's surface by major meteor or planetesimal impacts, which pulverized the protocrust and tapped magma sources at depth.

After about 3.8 Ga, the interior of the Earth became sufficiently cool that when the scum crust was subducted, it underwent only *partial melting*, meaning that only minerals that melted at lower temperatures went into a liquid state. The resulting magmas, therefore, did not have the same composition as the subducting crust and differentiation occurred. Magmas generated by this partial melting rose, fractionated, and eventually crystallized at the surface to form a range of igneous rocks of mafic to silicic composition, which constituted the first continental crust. Meanwhile, at zones of upwelling (proto-oceanic ridges), decreasing temperatures also meant that only partial melting could occur at the ridges, and the new oceanic crust that formed was therefore mafic (basalt) rather than ultramafic in composition.

Table 14.2 Contrasts between Oceanic and Continental Crust

- Continental crust has a mean composition equivalent to granodiorite, whereas oceanic crust has a more mafic composition (approaching basalt).
- Continental crust is an amalgamation of rock that has not been subducted, and most of which was formed in Precambrian times. Oceanic crust, in contrast, continues to form today at mid-ocean ridges by the process of seafloor spreading.
- Continental crust is 5 to 10 times thicker than oceanic crust. Most oceanic crust is between 6 and 10 km thick, and can be subdivided into distinct layers, which are recognized from seismic data, from drilling data, and from field studies of ophiolites (fragments of oceanic lithosphere that have become exposed above sea level). From top to bottom, oceanic crust consists of a sediment layer on top, a layer of pillow basalt, a layer of diabase dikes ("sheeted dikes"), a layer of massive gabbro; the bottom layer of peridotite in an ophiolite represents cumulate or ultramafic rock of the upper mantle. Continental crust ranges in thickness from 25 to 70 km. It has a gross layering of a felsic to intermediate upper crust and an intermediate to mafic lower crust; it is much more heterogeneous and complicated than oceanic crust.
- Continental crust, which is buoyant relative to the upper mantle, is normally not subducted, and thus portions of the continental crust are very old. Oceanic crust is mostly subducted, so there is no oceanic crust on Earth older than about 200 Ma, with the exception of small fragments (ophiolites) that have been emplaced on continents.
- The Moho at the base of the oceanic crust is very sharp, suggesting that the boundary between crust and mantle is sharp. The continental Moho is less distinct.

When this crust subducted, it generated progressively less mafic magmas, which in turn created progressively more silicic magmas that rose to the surface.

The above model suggests that a primitive type of plate tectonics was taking place very early on in Earth's history. As a consequence, crustal rocks gradually separated from the mantle and accumulated in little volcanic chains at the surface. The resulting continental crust could not be permanently returned to the mantle because it was too buoyant. Consumption of intervening oceanic crust caused these blocks to collide and amalgamate, eventually forming large fragments, that is, the *protocontinents.* Most of this process appears to have been completed by about 2.5 Ga, so that the rate of *continental growth* since then has greatly diminished. Since Proterozoic and Phanerozoic time, the crust has grown primarily by the creation of new crust at volcanic chains above subduction zones, but the rate of growth appears to be far less than during the Archean. Continental crust, being buoyant, cannot be easily subducted, so, with the exception of sediments deposited in trenches and carried into the mantle with descending oceanic lithosphere, continental crust is relatively permanent. In other words, the differentiation of the Earth that separated the continental crust from the mantle cannot be reversed by plate tectonics. Evidence for this accumulation is that Paleozoic to Archean rocks are widely present on almost all continents (the oldest known rock is ca. 4 Ga), whereas the oceanic crust is no older than Early Mesozoic (ca. 200 Ma).

Since its initial origin, continental crust has been further differentiated, reheated, sometimes remelted, deformed, and deformed yet again. Ultimately, this process yielded the *cratons,* which, because of their mafic lower layer, have acted as relatively strong blocks during subse-

quent Earth history. The cratons collided and were sutured together along orogenic belts into *supercontinents* (such as Rodinia in the Late Proterozoic and Pangaea in the Late Paleozoic), which later broke apart. This break up, or rifting, typically occurred in the relatively warm and weak orogens bounding the cratons, and thus Phanerozoic orogens have not (yet) had the opportunity to transform into craton. Effectively, since orogenic belts are the weaker portions of the crust, they tend to be the sites of rifting and collision again and again, leaving cratonic regions in the interior relatively unaffected. This opening and closing of oceans at roughly similar positions is now called the *Wilson cycle.*[3]

In summary, the continental crust is an amalgamation of fragments that formed at different times, and joined at different times. Continental crust is generally weaker than oceanic crust, because it contains so much quartz, which is able to undergo plastic deformation at relatively low temperatures (see Chapter 5). Because of these contrasts, continental crust and oceanic crust behave quite differently when involved in deformation. The contrasts between these two types of crust are summarized in Table 14.2.

14.3.3   Types of Continental Crust

As we mentioned, continental crust is quite variable, so it is not realistic to draw a type cross section of the continental crust in the way that we do with oceanic crust. Nevertheless, we can generalize by dividing the continental crust into several fundamental types (Table 14.3; Figure 14.6).

---

[3]Named after the Canadian geophysicist J. Tuzo Wilson, who envisioned many fundamental tectonic concepts.

Table 14.3 Principal Types of Continental Crust

Continental platforms	These are broad regions in which predominantly crystalline Precambrian rocks are covered by a relatively thin veneer of Phanerozoic sedimentary strata. Strata thickness is variable and may range to depths of >5 km in intracratonic basins. The central United States and eastern Russia are situated on platforms.
Cratons	A general term for stable continental regions that have not undergone orogenic activity since latest Precambrian (~600 Ma) and in many cases well before that. Cratons include shields, regions without a cover sequence of sedimentary strata, as well as platforms.
Phanerozoic orogenic belts	These are mountain belts (a belt is a region that is significantly longer than it is wide) that have been affected by convergent orogeny since the end of the Precambrian. Examples are the Alps, Himalayas, North American Cordillera, and the Appalachians-Caledonides.
Precambrian shields	These are broad regions in which Precambrian crystalline rocks and sedimentary strata are exposed. At some point in time these rocks may have formed parts of ancient, now deeply eroded mountain belts. Some shields (e.g., Canadian, Baltic, Siberian) have relatively low elevations and subdued topography, whereas others (Brazilian, southern African) attain elevations of up to two kilometers above sea level. Small outcrops of Precambrian rocks that occur within orogenic belts are not considered to be shields.
Rifts and passive margins	These are regions where the crust has been tectonically thinned and then buried by sediment. The largest rift in the United States is the Basin-and-Range Province. The eastern seaboard and the Gulf of Mexico coast are passive margins.

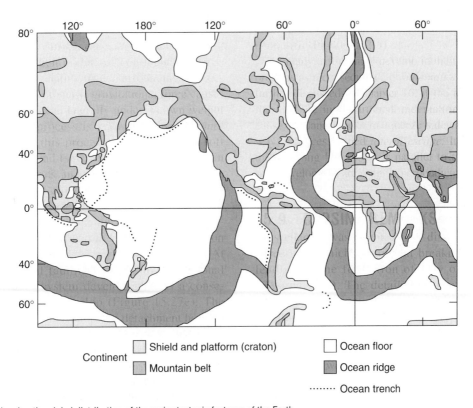

Figure 14.6 Map showing the global distribution of the major tectonic features of the Earth.
From *The Way the Earth Works* by Peter J. Wyllie. Copyright 1976 © John Wiley & Sons, Inc. Reprinted by permission of John Wiley & Sons, Inc.

In Table 14.3 we include *craton*, which is used as a general term for a relatively stable portion of continental crust, meaning a region that has not undergone orogenic activity for several hundred million years. Cratons are differentiated such that at depth they have a distinct lower crust of mafic rock, and they have relatively low heat flow. Under the thermal conditions found at depth in cratons, the mafic lower crust is very strong, so for this reason cratons tend to behave as rigid blocks, in contrast with orogenic belts, which are hotter and significantly weaker. An example of

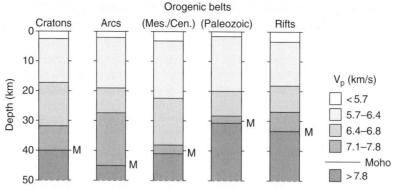

Figure 14.7 Vertical sections through continental crust of different tectonic settings, based primarily on seismic data. The P-wave velocities are shown; M is Moho. Note especially the varying thickness of the highest velocity layer in the lower crust, which ranges from 20 km in arc regions to only a few km in young orogenic belts.

the contrast between craton and orogen, in terms of mechanical behavior, is illustrated by the Cenozoic collision of India with Asia. The interior of India, an old cold craton, is largely unaffected by the collision, but southern Asia has evolved into a very broad mobile belt in which the crust is greatly thickened (Figure 14.6). The interaction somewhat resembles a collision between an armored bank truck and a mountain of jelly.

The word *craton* is used differently in different regions. In North American usage, the craton includes the platform and shield together, a region that is also called the *continental interior,* and has not been subjected to Phanerozoic orogeny. In South American usage, the term craton is used to refer to residual fragments of Archean continental crust that have not been substantially affected by Late Proterozoic orogeny. Does this sound confusing? Unfortunately, the terminology is not used consistently in the geologic literature; we will follow the North American usage.

In continental crust, composition varies laterally and vertically, for it consists of crustal blocks of different origin and age that have been sutured together. Thus intensity and orientation of fabrics and folds, the distribution and composition of intrusions, and thickness vary with location. A series of vertical sections through various tectonic settings (Figure 14.7) emphasizes this heterogeneity. Inhomogeneities may cause stress concentrations and differential movements when the stress state in the crust changes.

14.3.4 The Moho

The Moho, short for *Mohorovičić discontinuity,* is a seismic boundary, above which seismic P-wave velocities are about 5–8 km/s and below which seismic velocities are >8

km/s (Figure 14.7; see also Figure 14.3). Beneath the ocean floor, this discontinuity is very distinct and probably represents a very abrupt contact between two lithologies (gabbro above peridotite), and therefore cleanly defines the boundary between crust and mantle. Beneath continents, the Moho is less distinct, meaning the zone over which there is a change in P-wave velocity is broader. In some locations, it appears to coincide with a distinct reflector in seismic-reflection profiles, but this is not always the case. Possibly the Moho beneath continents is a contact between mantle and crustal lithologies in some localities, but it may locally be a zone of sill-like mafic intrusions (emplaced by a process called *magmatic underplating*) or a shear zone. In contrast to the impression that you may have developed from your first course in geology, the crust-mantle boundary is a variable and sometimes enigmatic entity. Because of this uncertainty, you should indicate whether you are referring to a seismic discontinuity, a reflector, or a petrologic boundary when discussing continental Moho.

14.3.5 The Mantle

In general, seismic velocities of the mantle, observations of mantle-derived xenoliths, and other data are compatible with the mantle having the general composition of peridotite. *Seismic tomography* (a sort of CAT scan of the Earth's interior; Figure 14.8) suggests major inhomogeneities in three dimensions, which can represent variations in temperature, composition, or both. Compositional variation can be caused by depletion of basaltic components during generation of seafloor or volcanic arc rocks, or by incorporating subducted slabs. Mantle from which a basaltic magma has been extracted is called *depleted mantle* (composing much of the

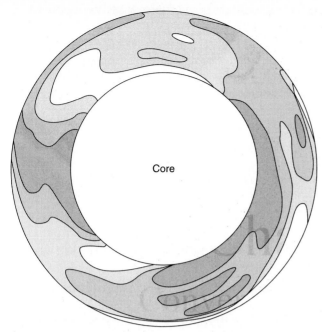

Figure 14.8 Seismic tomographic image of the mantle. The pattern reflects differences in velocity of seismic waves, with the highest velocity (colder material) indicated by the darkest color and the lowest velocity (hotter material) by white. Note that in places these patterns suggest mantle convection where relatively hot material rises from the core-mantle boundary region to the upper mantle.

lithospheric part of the upper mantle), and portions that still contain the basaltic fraction are called *undepleted mantle* or "fertile mantle" (composing much of the astheno-sphere). Temperature variations in the mantle can reflect patterns of convective flow, the distribution of materials with contrasting thermal conductivities, and the distribution of mantle plumes that originate near the core-mantle boundary (Figure 14.8).

14.4 RHEOLOGY OF THE EARTH

We have tacitly assumed that you already have a sense of the meaning of the words *lithosphere* and *asthenosphere* from earlier courses in geology, so we have used the terms already. But to progress with our discussion of plate dynamics, we now need to look at these layers in a bit more detail. Lithosphere and asthenosphere are distinguished from one another primarily by their response to stress; that is, their *strength*, not by seismic discontinuities. In other words, the terms lithosphere and asthenosphere describe *rheological layering* in the Earth.

14.4.1 The Lithosphere

The existence of a lithosphere was first proposed during research on the phenomenon of *isostasy*[4] and on the response of the Earth's surface to imposed vertical loads (e.g., growth of a glacier and formation of ocean islands). This work was done long before the proposal of modern plate tectonics. However, the word "lithosphere" came into much wider usage after the plate tectonics revolution, because geologists began to think of the lithosphere as the layer that was divided into plates, and everyone began talking about lithospheric plates moving around.

But what is lithosphere? Derived from the Greek word "lithos" for rock (implying that it has strength; i.e., resistance to stress), the *lithosphere* is the uppermost rheological layer of the Earth. It is distinguished from the underlying layer because, overall, it exhibits flexural rigidity on geological time scales. *Flexural rigidity* is the resistance to bending (flexure) of a material, and is a manifestation of the elastic component of the material's behavior. For example, a steel beam has a relatively high flexural rigidity, whereas a thick band of rubber has a relatively low flexural rigidity. Totally nonelastic materials, like honey or lava, have no flexural rigidity at all.

Lithosphere is not totally elastic, meaning that over very long periods of time it does flow, but the rate of viscous flow tends to be less than typical geologic strain rates (10^{-11} to 10^{-15} sec^{-1}). The response of the lithosphere to a load is time dependent. A load will be supported for some time, but then it gradually sinks by flow in the lower portion of the plate. Thus, to be accurate, the rheology of lithosphere is best described as *visco-elastic* (see Chapter 5). You can picture the division between the elastic and viscous components by representing lithosphere as having two imaginary layers, an upper elastic layer and a lower viscous layer. The thickness of this imaginary elastic layer is called the *effective elastic thickness* of a plate. Lithosphere with a greater elastic thickness has greater flexural rigidity than lithosphere with less elastic thickness. Remember that this distinction is imaginary, because no physical boundary exists between such layers in the subsurface. As an analogy, imagine green paint to be composed of a blue component and a yellow component. The shade of green depends on the proportion of these two components. Not all lithosphere on the Earth has the same effective elastic thickness (just as not all paint is the same shade of green). It is greater in old, cold cratons than in younger and warmer regions (young mountain belts), and it can be affected by stress.

---

[4]Derived from the Greek words "isos" and "stasis," meaning equal and standing.

Table 14.4 Types of Heat Transport

Conduction	The phenomenon that occurs when you place one end of an iron bar in a fire and the other end eventually gets hot too. The iron atoms do not physically move from the hot end of the bar to the cool end. Rather, atoms nearest the fire vibrate faster, which in turn cause adjacent atoms to vibrate, and so on. This vibration increase transfers along the bar, so eventually the whole bar gets hot.
Convection	The phenomenon that occurs when you place a pot of soup on a hot stove. Conduction through the base of the pot causes the soup at the bottom to heat up. The heated soup becomes less dense than the overlying (cold) soup and this density inversion is unstable in a gravity field. Consequently, the hot soup starts to rise and replaces the cold soup that sinks. Thus, convection is driven by *density gradients* that generate *buoyancy forces.* Convection occurs when two conditions are met: (1) the rate at which heat is added at the bottom must exceed the rate at which heat can be conducted upward through the layer, and (2) the material that gets heated must be able to flow on some time scale.
Advection	Heat may also be transported in a material by a secondary phase, such as a moving fluid through pore spaces or fractures in rock. The heat conducts from the fluid into the host rock. Consider our example of heating a metal bar: running cold water over the bar will prevent heating the bar's other end, because heat is transported away by advection.

Some geologists suggest that another characteristic of lithosphere is that it transmits heat primarily by conduction; Table 14.4 provides descriptions of the modes of heat transport in the Earth. Buoyancy contrasts between adjacent parts of the lithosphere, either due to compositional or thermal contrasts, cannot cause the lithosphere to convect. Therefore, the lithosphere can support loads that are not fully isostatically compensated. In other words, the lithosphere moves as a coherent entity, which is called the *plate.*

The lithosphere includes all of the crust and the uppermost part of the mantle. For practical purposes, the base of the lithosphere is defined as the 1280°C isotherm in the mantle (there is continuing debate about this temperature). At approximately this temperature, olivine, the dominant silicate in the peridotite that composes the upper mantle, becomes very weak because dislocation glide and climb become efficient deformation mechanisms (see the deformation mechanism map for olivine, Figure 9.35). Because the depth of this isotherm is not fixed in the Earth, but varies with the thermal structure of the mantle, the base of the lithosphere is not at a fixed depth everywhere around the Earth. Moreover, at a given location, the depth of the base of the lithosphere can vary with time if the region is heated or cooled. For example, directly beneath the axis of mid-ocean ridges, the lithosphere is probably only a few kilometers thick, because the 1280°C isotherm comes right to the base of the newly formed crust. Yet beneath oceanic abyssal plains, the lithosphere is probably around 100 km thick, and beneath cratons the lithosphere may be more than 150 km thick. In summary, the boundary between the lithosphere and underlying asthenosphere may be viewed as a *thermal boundary,* whose position may vary in space as well as in time. In contrast, the boundary between the crust and the upper mantle (which collectively make up the lithosphere) is a compositional boundary (Figure 14.9).

14.4.2 The Asthenosphere

The *asthenosphere,*[5] which is the layer of the mantle that underlies the lithosphere, behaves much like a viscous fluid without flexural rigidity on geologic time scales. Recall that over very short periods of time, however, it behaves elastically, because it is able to transmit seismic waves. If we were to represent the rheology of the asthenosphere by a two-layer model similar to that used for the lithosphere (an elastic layer and a viscous layer), the elastic layer would be so thin that it would not be able to hold up a load, and any load placed on it would sink (e.g., a subducted lithospheric slab sinks through the asthenosphere). The lack of flexural rigidity in the asthenosphere and lateral density differences from variations in temperature and/or composition cause material flow of the asthenosphere. In other words, heat transfer in the asthenosphere can occur by *convection.* Keep in mind, of course, that the asthenosphere is composed predominantly of solid, though weak, rock (except where partial melting has occurred), so convective flow in the asthenosphere refers to crystal plastic and diffusional processes and not the motion of fluids.

The definition of the *base of the asthenosphere* is problematic. Some geologists think that convection in the upper mantle is largely independent of convection in the lower mantle, or transition zone, and thus the base of the

---

[5]Derived from "stenos," which is Greek for strength, the prefix "*a*" changes the meaning to no strength.

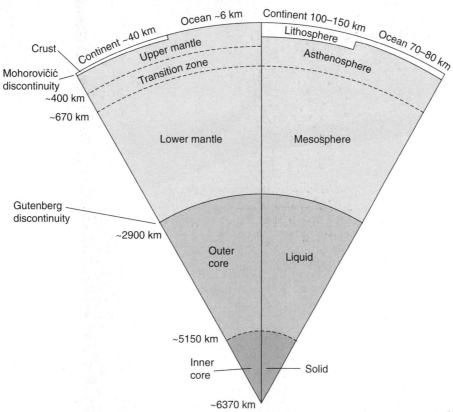

Figure 14.9 Comparison of compositional and rheological layering of the Earth. The base of the asthenosphere in this diagram is placed at the base of the transition zone. This definition is controversial. Some geologists use the term "mesosphere" for all mantle below the asthenosphere, but this term is not universally accepted.

asthenosphere is represented by the 400 km discontinuity. Others believe that there is a significant rheological contrast between the transition zone and the lower mantle, and place the base of the asthenosphere at the top of the lower mantle (ca. 670 km). Still others believe that whole mantle convection occurs and that there is no reason to distinguish a base to the asthenosphere. Evidence for this view comes from *mantle plumes* that originate at the core-mantle boundary and are observed at the surface as *hot spots,* and from tomographic studies that track subducted slabs into the lower mantle. This has led to the concept of *plume tectonics* as a complementary model to plate tectonics. Consensus is lacking on this issue, but for purposes of understanding isostasy, it doesn't really matter where you put the base of the asthenosphere, as long as it is well below the base of the lithosphere.

Beneath oceanic regions, seismic waves travel anomalously slow in the upper part of the asthenosphere. This portion of the asthenosphere is called the *low-velocity zone* (LVZ), and it extends to a depth of approximately 150 km (i.e., it does not correspond to the entire asthenosphere). The existence of the LVZ may be a consequence of the occur-

rence of 1–6% partial melt, or may reflect the rheology of olivine under the P-T conditions found in this interval. The LVZ is not readily apparent beneath continental lithosphere.

14.5 THE TENETS OF PLATE TECTONICS

Plate tectonics is a *geotectonic theory.* It is a comprehensive set of ideas that explain the development of regional geologic features, such as the distinction between oceans and continents, the origin of mountain belts, and the nature and distribution of earthquakes, volcanoes, and rock types. It is not the first geotectonic theory to be proposed. Ancient cultures already had concepts of tectonics embedded in their creation legends. The history of geotectonic theories is an interesting one, but its details are beyond the scope of this book. Suffice to say, that when plate tectonics was proposed in the 1960s, after several decades of critical discoveries in paleomagnetism, marine geology, and seismology, it replaced its immediate predecessor (geosynclinal theory). It is supported so well by observations that it is likely to remain the dominant geotectonic theory.

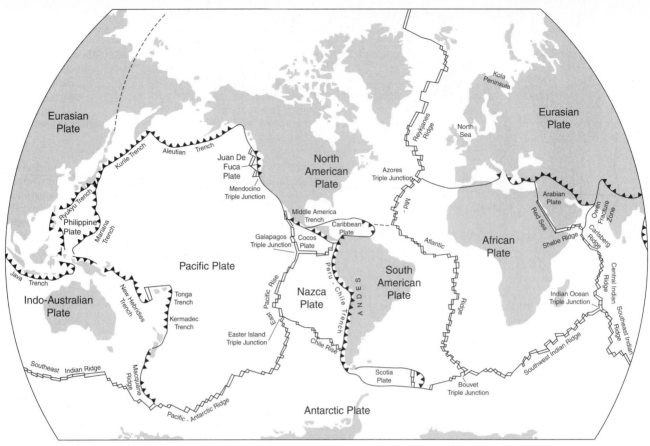

Figure 14.10 The seven major and several minor plates of the Earth. Ocean ridges (double lines), trenches (barbed lines), and transform faults (continuous lines) mark the plate boundaries, but in a few places these plate boundaries are ill defined (such diffuse boundaries are marked by dashed lines).

When someone talks about "plate tectonics," they are generally referring to a series of related ideas, which we briefly review in this section. We don't have the space to provide a detailed consideration of all the ins and outs of the theory, but we want to cover key terms and concepts that will return in subsequent chapters.

14.5.1 Mobility of the Outer Layer of the Earth

According to the theory of plate tectonics, the outer layer of the Earth is divided into discrete entities called *plates*. Geometrically, a plate can be viewed as a cap on the surface of a sphere. Plates move relative to one another as a consequence of dynamic processes in the Earth. The seven major plates are the Pacific, North American, South American, Eurasian, African, Indo-Australian, and Antarctic plates. In addition there are a dozen or so smaller plates (e.g., Cocos and Nazca Plates; Figure 14.10). When plate boundaries are not completely obvious; they are called *diffuse plate boundaries*. For example, the boundaries between the African and Eurasian plates are diffuse.

Traditionally, the lithosphere is considered to be the entity that moved during plate tectonics. This definition works well in oceanic regions, where the lithosphere is decoupled from the asthenosphere along the very weak LVZ.

But beneath continents the LVZ is not so obvious, and in the last decade it has been suggested that part of the asthenosphere moves with the lithosphere. In this latest concept, continents have very deep roots that include part of the asthenosphere. If so, then the entity that moves during plate tectonics is not simply the lithosphere, but is a thicker layer that has been referred to as the *tectosphere*. The base of the continental tectosphere may be as deep as 250 km, whereas beneath oceans the tectosphere equals the lithosphere. Whether continental plates really have asthenospheric roots, and hence whether the term "tectosphere" is needed, remains a matter of debate. For the sake of simplicity, therefore, we will continue the tradition of equating plate with lithosphere.

A plate can be entirely oceanic lithosphere (such as the Pacific Plate), but more generally is a combination of oceanic and continental lithosphere (Figure 14.10). Thus, not all of today's continental margins are plate boundaries. We make the distinction between *active continental margins,* which are plate boundaries, and *passive continental margins,* which are not plate boundaries. The present eastern margin of the North American continent is a passive margin, whereas the western margins of the South and North American continents are examples of active margins.

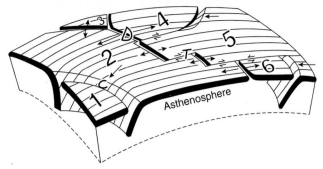

Figure 14.11 Types of plate boundaries and various triple junctions on the surface of the Earth. Relative displacement rates (linear velocities) are proportional to arrow length. D is divergent boundary (characterized by ridges), C is convergent boundary (characterized by trenches), and T is transform boundary. RRR triple junction where plates 2-3-4 meet; RRT junction where 2-4-5 meet; RTT junction where 2-5-6 meet.
From *An Outline of Structural Geology* by Hobbs, Means & Williams. Copyright © 1976 John Wiley & Sons, Inc., NY. Reprinted by permission of John Wiley & Sons, Inc.

Table 14.5 Types of Plate Boundaries

Divergent plate boundaries	Boundaries where two plates are moving apart. Such movement occurs at ocean ridges, where new oceanic lithosphere is created. This process is called seafloor spreading, and thus, divergent boundaries are also called *spreading centers* or *constructive boundaries.*
Convergent plate boundaries	Boundaries where two plates are moving together. Such movement occurs where, for example, one oceanic plate bends and descends beneath another plate (either continental or oceanic). At a convergent plate boundary, oceanic lithosphere is consumed. This process is called *subduction,* so such boundaries are also called *subduction zones* or *destructive boundaries.* Typically, the edge of a subduction zone is topographically defined by the presence of a trench. Only oceanic lithosphere can be subducted, because oceanic lithosphere is denser than the continental lithosphere and can sink into the asthenosphere. Continental lithosphere cannot subduct, so when two pieces of continental lithosphere converge, they collide and a *mountain belt* develops. Therefore, the oldest ocean lithosphere on Earth is less than 200 Ma, whereas there are many regions of continental lithosphere that are over 3 Ga.
Transform plate boundaries	Boundaries where two plates slide past each other, and lithosphere is neither created nor destroyed; therefore, they are also called *conservative boundaries.* By definition, a perfect transform plate boundary is a pure strike-slip fault, but in reality the boundary may have a component of compression or extension across it. In the former case, we say that the boundary is *transpressive;* in the latter case, we say that the boundary is *transtensile.*

14.5.2 *Plate Boundaries versus Plate Interiors*

The location of plate boundaries is based on the distribution of earthquakes, volcanoes, and topographic features (primarily ocean ridges and deep-sea trenches). The three basic types of plate boundaries, as defined by the relative motion of the two plates in contact at the boundaries are *divergent, convergent,* and *transform boundaries* (Figure 14.11, Table 14.5). These boundaries will be discussed in detail in the next chapters. Plate tectonic theory concludes that tectonic activity is largely confined to plate boundaries and that the interiors of plates are relatively stable. This assumption works quite well for oceanic plates, which are composed of strong mafic lithosphere, but does not work perfectly in the case of continental lithosphere, which contains weaker quartz-rich crust that deforms internally. The internal deformation of continents is manifested at zones of continental rifting (where continents are stretching, such as the African rift zone), by regional deformation

in plate interiors from plate margin collision (such as east-central Asia), and by the presence of large basins in the middle of continental interiors (i.e., intracratonic basins such as the Michigan Basin in North America).

It is the processes of *seafloor spreading* and *subduction* that cause continents to move around as envisioned in Wegener's[6] original concept of *continental drift.* In the modern concept, continents move relative to one another as seafloor between them is either created or destroyed. At various times in the past, continental movements and collisions have produced large continental masses called *supercontinents,* which lasted for tens of millions of years. Examples appear in the essays of Chapter 19, and include the Late Precambrian supercontinent Rodinia and

---

[6]German explorer Alfred Wegener (1880–1930) is considered the 'father' of continental drift, but suggestions of continental movement go back as far as the seventeenth century.

the Permo-Triassic supercontinent Pangaea (meaning "all land"). These supercontinents are not stable and break apart again into pieces that move relative to one another.

Plate tectonics in the form that we now observe has probably been active at least since the Early Proterozoic (2.5 Ga). In some form, however, dynamic interactions among continental masses have occurred since the first continents were created in the Archean.

14.6 BASIC PLATE KINEMATICS

Plate kinematics refers to the description of plate motion on the surface of the Earth. With the movement and interaction of plates responsible for so much of the structure that we have described, it is important to have an image of rates and directions of plate motion. Description of plate motion is essentially a geometric exercise, with, however, the extra complexity of motion on a sphere. Using *spherical geometry* to understand plate kinematics requires us to make three assumptions. The first assumption, obviously, is that the Earth is a sphere. In reality, we know that the Earth is not perfectly spherical, but the slight flattening at the poles is not sufficiently large for us to worry about here.[7] The second assumption is that the circumference of the (spherical) Earth is constant; in other words, the Earth is neither expanding nor contracting. This assumption implies that the amount of seafloor spreading at the surface of the whole Earth is equivalent to the amount of subduction. An expanding Earth was once proposed as a mechanism, but is not supported by physical evidence (e.g., from paleomagnetism). Note that this assumption does not mean that the rate of subduction on one side of a specific plate is the same as the amount of spreading on the other side of that plate, but only that growth and consumption of plates are balanced for all plates combined. The third assumption is that the plates are rigid. As we said earlier, this assumption is reasonable for oceanic plates but less for continental plates. This produces a slight error in plate motion calculations, but this error is only a few percent.

To describe any motion, including plate motion, we must have a reference frame. The reference frames that we use to describe plate motions are the absolute and the relative reference frame. In the *absolute reference frame,* we describe plate motions with respect to a fixed point in the Earth's interior. In the *relative reference frame,* we describe the motion of one plate with respect to another, so no point is fixed.

14.6.1   Absolute Plate Motion

What can we use as a fixed point to provide the reference frame in which we describe absolute plate motion? Perhaps the 'fixed points' that are easiest to visualize when describing absolute plate motion are hot spots (Figure 14.12). A

hot spot is a narrow plume of upwelling mantle originating at great depth in the Earth, probably at or near the core-mantle boundary, that impinges on the bottom of the lithosphere. In essence, plumes rise much like salt diapirs in the crust due to density contrasts, except that the density inversion in plumes originates from temperature contrasts. The relatively high temperature of plumes triggers partial melting of mantle peridotite at shallow levels in the Earth where pressure is less. The resulting basaltic magmas rise to the surface and create volcanic chains on overlying plate as it moves over to the plume. The age of the volcanoes in the chain increases progressively along the chain, and only one end of the chain is active at a given time. These chains are called *hot-spot tracks;* the Hawaiian-Emperor chain in the Pacific is the classic example of this process. For a simple experiment, place a stationary flame under a moving sheet of plastic. The heat of the flame melts the plastic, leaving a track on the sheet as it moves relative to the flame. Some plumes are long lived (i.e., 100 m.y. or more), whereas others are short lived (i.e., less than 10 m.y.), probably as a function of how much material flowed into the plume at depth. Note that hot spots are, in fact, not strictly stationary. They slowly move relative to one another, but at a rate one or more orders of magnitude less than the rate of plate motion.

Another absolute frame of reference is provided by a *no net torque calculation.* This calculation simply assumes that the sum of all the plate movements shearing against the underlying asthenosphere creates zero torque on the mantle; otherwise, the rotation of the Earth would change. Finally, modern global positioning systems (GPS) are rapidly improving our ability to obtain real-time measurements of plate motions and thus provide a third frame of reference.

Figure 14.12 shows the absolute motion of plates as determined using the no net torque reference frame. You will notice that absolute plate motions today range from less than 1 cm/year (e.g., Antarctic and African Plates) to about 10 cm/y (Pacific Plate). North America and South America are moving west at about 2–3 cm/y. For comparison, these rates are about the same at which your hair or fingernails grow. This may seem slow at first, but if you consider that these displacements occur over tens of millions of years, plates can move hundreds to thousands of km in relatively short spans of geologic time.[8]

14.6.2   Relative Plate Motion

The motion of any plate with respect to another can be defined by fixing one plate and moving the other. For example, if we want to describe the motion of plate A with respect to plate B, we fix plate B on the surface of the Earth and see how plate A moves (Figure 14.13). In plate kinematic calculations, the Earth is considered to be a

---

[7]The Earth's radius to the poles is approximately 20 km less; this flattening at the poles may give rise to *membrane stresses* in plates that move north-south, as their radius of curvature changes.

[8]. . . and your nails would grow quite long.

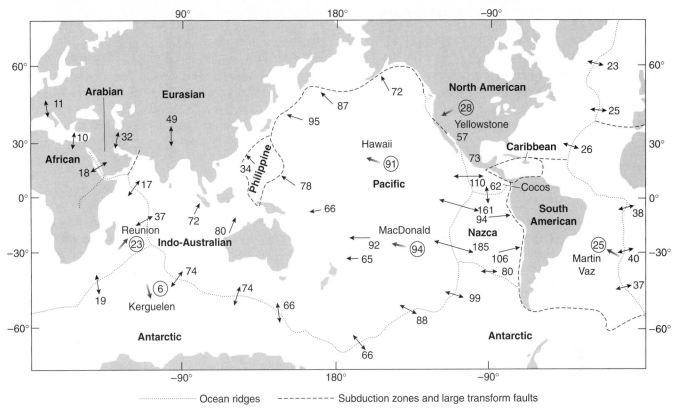

Figure 14.12 Directions and rates of plate motion in mm/y. Absolute motions of plates relative to a hotspot reference frame are shown by circled numbers and heavy arrows. Relative motions of the major plates are given for selected points along their boundaries.

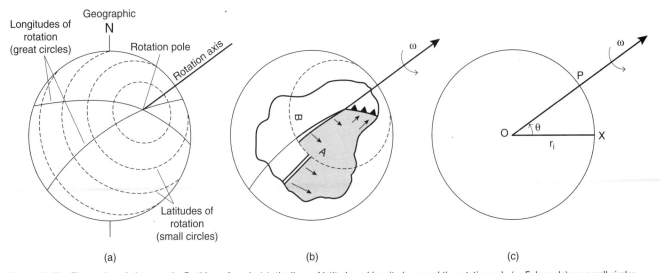

Figure 14.13 The motion of plates on the Earth's surface. In (a), the lines of latitude and longitude around the rotation pole (or Euler pole) are small circles and great circles, respectively. In (b), three types of plate boundaries are present between plates A and B. We assume that plate B is fixed, so that only plate A moves relative to plate B. The spreading and subduction velocities increase away from the rotation pole, and the transform fault is an arc of a small circle around the rotation pole. (c) A cross section through the center of the earth (O) with radius r_i. P is the rotation pole and X is a point on the plate boundary.

sphere, so such a movement is described by a rotation at a specified angular velocity around an imaginary axis that passes through the center of the Earth. The intersection between this imaginary rotation axis and the surface of the Earth is called the *rotation pole,* or *Euler pole,*[9] between

two plates. There are two types of Euler poles. An *instantaneous Euler pole* describes relative motion at an instant in geologic time, whereas a *finite Euler pole* describes net relative motion over a long period of geologic time. It is important to realize that Euler poles are merely geometric elements, and that they are not related to the Earth's

---

[9]Named after the eighteenth-century Swiss mathematician Leonhard Euler.

geographic poles (i.e., points where the Earth's spin axis intersects the surface); nor are they related to the Earth's magnetic poles (i.e., points where the magnetic dipole vector is perpendicular to the Earth's surface).

To visualize the movement of a plate on a sphere, let's consider an example (Figure 14.13). Using spherical geometry, the relative motion of plate A with respect to plate B is defined by a vector $_A\Omega_B$:

$$_A\Omega_B = \omega\mathbf{k} \qquad \text{Eq. 14.1}$$

where ω is the angular velocity and $\mathbf{k}$ is a unit vector parallel to the rotation axis. Usually, it is easier for us to think of plate motion in terms of the linear velocity v as measured in cm/y at a point on a plate. However, we can describe v only if we specify the point at which v is to be measured, because we are dealing with spherical geometry (see below). The value of v is the vector cross product of $_A\Omega_B$ and the radius vector (r_i) drawn from the center of the Earth to the point in question. This relation can be represented by the equation:

$$v = {_A\Omega_B} \times r_i \qquad \text{Eq. 14.2}$$

As in any vector cross product (check a math book to remind yourself, if necessary), this equation can be rewritten as:

$$v = \omega r_i \sin\theta \qquad \text{Eq. 14.3}$$

where θ is the angle between r_i and $\mathbf{k}$ (Figure 14.13c). We see from these relations that v is a function of the distance between the point at which v is determined and the Euler pole. As you get closer to the Euler pole, the value of θ becomes progressively smaller, and at the pole itself, $\theta = 0°$. Since $\sin 0° = 0$, the relative linear velocity (v) at the Euler pole is zero. This relation is perhaps easy to picture if you think of an old Beatles record playing on a stereo. The linear velocity at the center of the turntable is zero and increases toward the edge of the record, whereas angular velocity is the same at all positions. Similarly, the maximum relative linear velocity on a sphere occurs where $\sin\theta = 1$; that is, at 90° from the Euler pole.

In our simple model, we drew the Euler pole to be on the plate boundary, but this is not necessary. In fact, for many plates the Euler pole is not on the plate boundary.

Because the relative velocity between two plates can be described by a vector, plate velocity calculations obey the *closure relationship,* namely:

$$_A\Omega_C = {_A\Omega_B} + {_B\Omega_C} \qquad \text{Eq. 14.4}$$

Using the closure relationship, we can calculate the relative velocity of two plates that do not share a common boundary. For example, to calculate the relative motion of the African Plate with respect to the Pacific Plate, we use the equation:

$$_{\text{Africa}}\Omega_{\text{Pacific}} = {_{\text{Africa}}\Omega_{\text{S. America}}} + {_{\text{S. America}}\Omega_{\text{Nasca}}}$$
$$+ {_{\text{Nasca}}\Omega_{\text{Pacific}}} \qquad \text{Eq. 14.5}$$

An equation like this is called a *vector circuit.*

At this point, you may be asking yourself the question, how do we determine a value for $_A\Omega_B$ in the first place? Actually, we can't measure $_A\Omega_B$ directly. We must calculate it by knowing ω, which we determine, in turn, from a knowledge of v at various locations in the two plates or along their plate boundary. Values for v are directly measurable along divergent and transform boundaries. Let's take the example of a point P on the Mid-Atlantic Ridge, the boundary between the African and the North American Plates. We want to determine the instantaneous Euler pole describing the motion between these two plates, and the value for v at point P.

First, as v is a vector, we first need to specify the *orientation of v.* To a first approximation, the spreading direction is given by the orientation of the transform faults that connect segments of the ridge. On a transform fault, plates slip past one another with no divergence or convergence, so, geometrically, a transform fault must be a small circle around the Euler pole (Figure 14.13), just as a line of latitude is a small circle around the Earth's geographic pole. Therefore, the direction of v at point P is parallel to the nearest transform fault. Secondly, we determine the *location of the Euler pole,* describing the motion of Africa with respect to North America. Considering that transform faults are small circles, a great circle drawn perpendicular to a transform fault must pass through the Euler pole (Figure 14.13b), just as lines of longitude must pass through the geographic pole. So, to find the position of the Euler pole, we draw great circles perpendicular to a series of transforms along the ridge; where these great circles intersect is the Euler pole.

Finally, we need to determine the *magnitude of v.* To do this, we look at the magnetic anomalies of the oceanic crust on either side of point P. A *magnetic anomaly* is a place where the magnetic field strength is higher or lower than the Earth's average magnetic field predicted for that location. The oceanic crust is composed of basalt, which contains small amounts of the magnetic mineral magnetite (Fe_3O_4). When new basaltic magma is extruded or intruded during the process of oceanic crust formation at the ridge, the magnetite grains lock in the Earth's magnetic field when the basalt cools down below a certain temperature.[10] Thus, the seafloor records information about the Earth's past magnetic field just like a stereo tape recorder. You will recall that the polarity of the Earth's magnetic field reverses every now and then. Thus, the magnetization of basalt formed when the polarity is normal (i.e., the same as today) will add to today's Earth's magnetic field and create a *positive anomaly.* However, the magnetization of basalt

---

[10]This is called the Curie temperature (ca. 580°C for magnetite):
Conceptually, the Curie temperature (named after French Scientist Pierre Curie) is much like the *closure temperature* in geochronology.

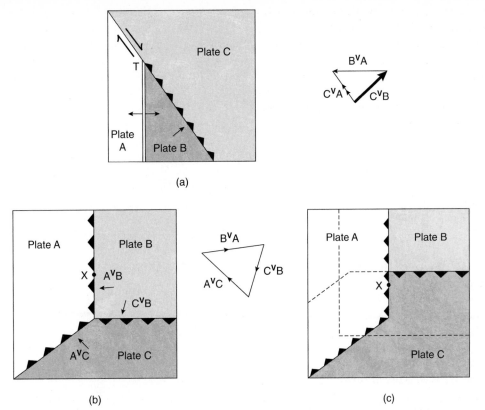

Figure 14.14 Triple junctions. The ridge-trench-transform triple junction in (a) is stable. Although the absolute position of the triple junction moves with time, the basic geometry remains the same. The rate of convergence between plate B and C can be determined from the vector circuit. Given the velocities between plates A and B (divergent boundary), and plates A and C (transform boundary), the velocity between plates B and C is obtained by completing the vector circuit (heavy line). In contrast, the trench-trench-trench triple junction in (b) is unstable. In (c), the positions of plates B and C after some time is shown if no subduction had occurred and plate A remained fixed; the associated vector circuit is also shown. Point X originally on the AB plate boundary (in b) is now located on the boundary between plates A and C (in c). Because the geometry changes, the trench-trench-trench junction in (b) is an unstable triple junction. Note that the new geometry in (c) created a stable triple junction, whose geometry will not change with time.

formed when the polarity was reversed subtracts from today's Earth's magnetic field and creates a *negative anomaly*. Thus, we see alternating stripes of positive and negative anomalies that roughly parallel the ocean ridges, and that are symmetrical with respect to the ridge. The age of reversals can be dated radiometrically and/or by fossils, which determines the age of the stripes on either side of the ridge. Velocity is simply distance divided by time, so we measure the distance between two magnetic anomalies of equal age on either side of the ridge, and we calculate the spreading velocity. This gives us the magnitude of v at point P.

The magnitude of v for oceanic transform faults is determined using magnetic anomalies if the direction of motion is parallel to the transform. On continental transform boundaries, like the San Andreas Fault in California, the magnitude of v is determined by the age of features that crosscut the fault after it formed, and were then offset by the fault. It is more difficult to determine relative motion across a convergent boundary, in part because the convergence direction is often oblique (in this case we talk about *oblique convergence*). Note that the orientation of trans-

forms or ridges that are being subducted cannot tell us about the direction of v, and anomalies tell us nothing about the magnitude of v. Therefore, the value of v across convergent boundaries must be determined using vector circuits or using modern GPS.

14.6.3   Triple Junctions

The point where three plates meet is called a *triple junction*. We define specific types of triple junction by the nature of plate boundaries that intersect. For example, a ridge-ridge-ridge triple junction is one where three ridges join (e.g., in the Indian Ocean; Figure 14.9), and a ridge-trench-transform triple junction is a place where a ridge, trench, and transform intersect. A *stable triple junction* is one whose basic geometry can exist for a long time, even though the absolute position of the triple junction changes with time. For example, the ridge-trench-transform junction in Figure 14.14a is a stable configuration. An *unstable triple junction* is one whose geometry must change to create a new arrangement of plate boundaries (Figure 14.14b and c).

The migration of a triple junction along a plate boundary can lead to the transformation of the plate boundary from one type to another. Perhaps the best known example is the San Andreas System in California (Figure 14.15). The western North American coast was a convergent plate boundary until the Farallon-Pacific Ridge was first subducted at around 30 Ma. When this occurred, the Pacific Plate came into contact with the North American Plate, and two triple junctions formed. One junction is moving north and the other is moving south. As the triple junctions move, the plate boundary changes from a convergent boundary into a transform boundary. This transform segment of the western North American Plate boundary is the origin of the San Andreas Fault. Thus, migration of triple junctions is one reason that tectonic styles along a given plate boundary may change with time.

14.7 THE PLATE TECTONIC ENGINE

We cannot close a discussion of the fundamentals of plate tectonics without some mention of its driving forces. You may recall from introductory geology that the energy driving the plate tectonic engine is heat transport from the core and mantle to the Earth's surface. This is nicely illustrated by the global heat flow map (Figure 14.16), which shows high heat flow associated with ocean ridges and relatively low heat flow in cratons. At first, heat transfer by *convection* was considered the main plate driving mechanism (called *mantle drag*). But calculations of the relative magnitudes of plate forces have emphasized the importance of plate edge forces: *ridge push* and *slab pull* (Figure 14.17). Slab pull arises from the negative buoyancy of relatively cold oceanic lithosphere. Sinking of the cold subducting slab into the mantle pulls oceanic plate along with it. This sinking is further facilitated by density-increasing phase transformations in the plate at depth (basalt to eclogite). Ridge push is caused by the gravitational potential energy of the elevated lithosphere at ridges. This potential energy causes plates to spread laterally away from the ridge axis, much like a block of cheese spreads laterally when heated by the sun. So, whereas a cooling core and mantle fuel the engine, it appears that the plates themselves are at least equally important in the driving process. Today's question is no longer what drives the process, but rather what controls the throttle and steers!

14.8 THE SUPERCONTINENT CYCLE

And for dessert, we serve some intriguing speculation. The geologic record shows that a supercontinent (Pangaea) formed by the end of the Paleozoic (~250 Ma), only to disperse in the Mesozoic. Similarly, a supercontinent formed at the end of the Precambrian (1.1 Ga, called Rodinia), only to disperse at the beginning of the Paleozoic. This long-term cycle of supercontinent assembly and breakup is known as the *supercontinent cycle*. Whether supercontinents occurred earlier in the Precambrian is less certain (proposed candidates are at ~1.7, 2.1, and 2.6 Ga), but this possible cyclicity warrants further examination.

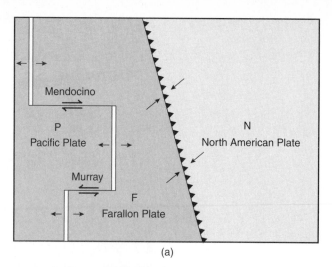

(a)

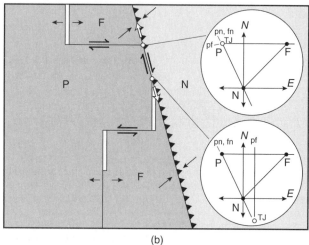

(b)

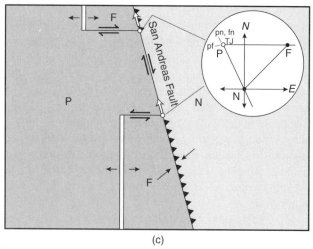

(c)

Figure 14.15 Schematic evolution of the San Andreas Fault System and triple junction motion. As the ridge and Mendocino transform between the Farallon (F) and Pacific (P) Plates get subducted under the North American (N) Plate (b), the San Andreas transform fault grows by the opposite motion of two triple junctions (white arrows). The geometry of the southern triple junction changes from ridge-trench-transform to transform-ridge-transform when the Murray transform arrives at the trench (c). Then, it begins to move in a northerly direction until it again changes into ridge-trench-transform (not shown). The corresponding vector circuits for each triple junction are given.

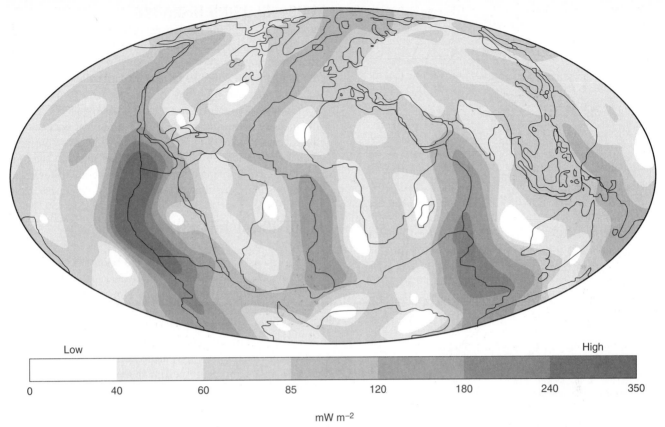

Figure 14.16 Today's pattern of global heat flow in mW m^{-2}, showing that ridges coincide with areas of high heat flow, whereas cratons are regions of relatively low heat flow.

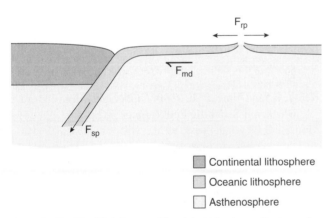

Figure 14.17 Simplified diagram of the plate driving forces. F_{md} is mantle drag, F_{rp} is ridge push, and F_{sp} is slab pull.

Recent geodynamic models suggest that the supercontinent cycle may be related to long-term convection patterns in the mantle. In these models, relative motion between plates at any given time is determined by ridge-push and slab-pull forces (see above). But over long periods of time (about 400 million years), continents tend to accumulate over a zone of major mantle downwelling to form a super-continent (Figure 14.18). When this happens, the thermal structure of the mantle changes dramatically, because the supercontinent acts as a giant insulator that does not allow heat to escape from the mantle. Over 80% of the Earth's internal heat escapes at ocean spreading centers, where seawater circulating through the hot crust and upper mantle cools the lithosphere much like a coolant cools an automobile engine. However, in a supercontinent configuration there are no ridges, so heat is not easily lost. As a consequence, the mantle beneath the supercontinent heats up, and can no longer be a region of downwelling. Thus, a new downwelling zone develops elsewhere, and hot mantle begins to upwell beneath the supercontinent. This causes the continental lithosphere to heat up and weaken, so that ultimately it rifts apart into smaller continents separated by new oceans. These ideas predict that the current cycle will end in about 250 million years when the Pacific and Atlantic Oceans close to form a new supercontinent.[11]

---

[11]Based on a collision between the Americas and Asia, this future supercontinent has been called Amasia.

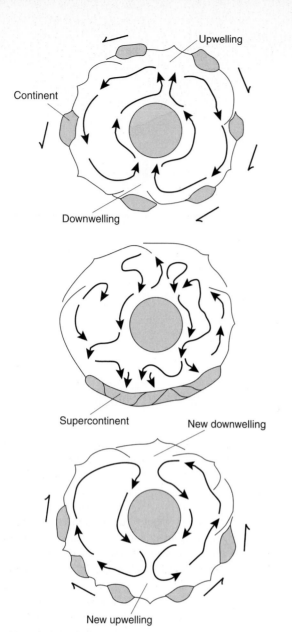

Figure 14.18 Stages in the supercontinent cycle. (a) Continents move relative to one another, but gradually accumulate over a mantle downwelling zone. (b) While the supercontinent exists, large-scale convection in the mantle reorganizes (b), and the resulting upwelling beneath the supercontinent weakens it and results in breakup (c).

14.9 CLOSING REMARKS

A complete understanding of tectonics fills an entire book, and, in fact, many have been written (see, and Additional Reading). For now, you have become familiar with some of the fundamental concepts of tectonics that allow you to understand why and what type of structures are formed in response to plate motions. In the next chapters we will examine in some detail the nature of structures that occur at the three types of plate boundaries (convergent, divergent, and transform boundaries). You will find that many of the topics discussed in this and subsequent chapters will return in the essays that close this part of the book (Chapter 19).

ADDITIONAL READING

An enormous number of articles and books have appeared on plate tectonics over the past three decades, and this body of literature keeps growing at a dazzling rate. It is difficult to make a selection of even the most important articles without offering a long list. Therefore, we include only a short list of material that is primarily instructive as well as accessible for the neophyte tectonicist. In the books you will be able to find the pertinent references, which will allow you to dive into any topic that sparked your interest.

Condie, K. C., 1989, *Plate tectonics and crustal evolution* (3rd edition), Pergamon Press, Oxford.

Cox, A., and Hart, R. B., 1986, *Plate tectonics. How it works:* Oxford, Blackwell Scientific Publications, 392 p.

Davies, G. F., 1992, Plates and plumes: dynamos of the Earth's mantle: *Science,* v. 257, p. 493–494.

Fowler, C. M. R., 1990. *The solid earth. An introduction to global geophysics:* Cambridge, Cambridge University Press, 472 p.

Keary, P., and Vine, F. J., 1990, *Global tectonics:* Oxford, Blackwell Scientific Publications, 302 p.

Moores, E. M., and Twiss, R. Y., 1995, *Tectonics:* W. H. Freeman, New York.

Nance, R. D., Worsley, T. R., and Moody, J. B., 1986, Post-Archean biogeochemical cycles and long-term episodicity in tectonic processes: *Geology,* v. 14, p. 514–518.

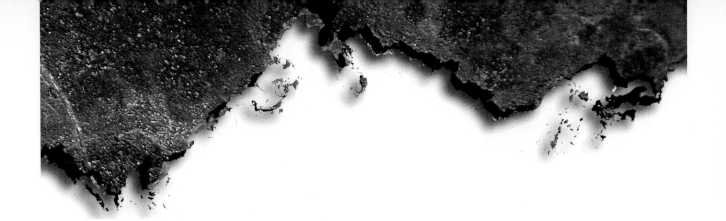

Chapter 15
Extensional Tectonics

15.1 INTRODUCTION

Two hundred and fifty million years ago, a dinosaur could have walked from North America to Africa. At that time, most of the continental crust on the planet was sutured together to form the supercontinent Pangaea. Beginning in Late Triassic, however, Pangaea began to stretch and thin along discrete belts. One of these belts roughly followed the trace of the Appalachian Orogen, a collisional mountain range in eastern North America, that formed during the Paleozoic assembly of Pangaea. By Cretaceous time, North America had separated from Africa and Europe along this belt, and seafloor spreading at the new Mid-Atlantic Ridge began to form the Atlantic Ocean (Figure 15.1). African dinosaurs bade farewell to their American cousins.

The drama that we just described is an example of rifting. Simply put, *rifting* is the process by which continental lithosphere stretches. A *rift* or *rift system* is the belt of continental lithosphere that is stretching. In *active rifts,* which are places where extensional deformation is currently taking place (Figure 15.2), the ground surface is cut by faults, earthquakes rumble with unnerving frequency, and volcanic eruptions occasionally bury the countryside. Typically, active rifts display a topography that is characterized by the occurrence of linear ridges or mountain ranges that separate nonmarine or shallow-marine sedimentary basins (Figure 15.3). In *inactive rifts,* meaning rifts at which extensional deformation has ceased, the record of rifting is manifested by the occurrence of normal faults in association with thick sequences of redbeds, conglomerates, evaporites, and volcanics. When extension continues until a continent separates into two pieces and a new oceanic

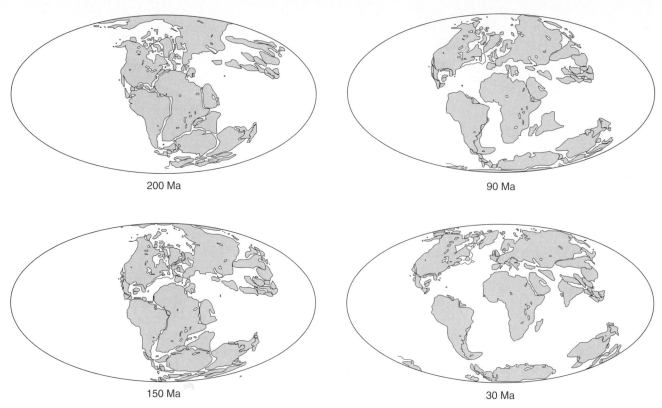

Figure 15.1 Maps showing opening of the Atlantic Ocean, between Lower Jurassic and Oligocene time.

200 Ma

90 Ma

150 Ma

30 Ma

Figure 15.2 Edge of the Gregory Rift in southwest Kenya.

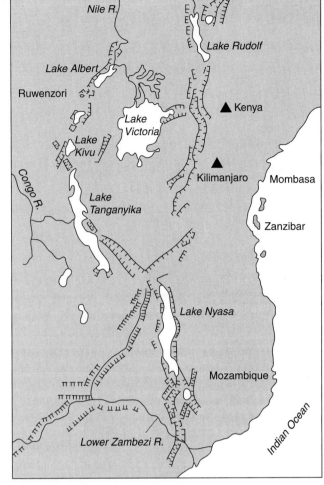

Figure 15.3 Map of the East African rift system. The filled triangles are prominent volcanoes.

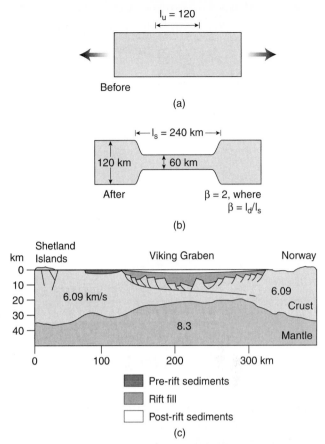

Figure 15.4 Concept of stretching factor. (a) Unstretched lithosphere of rift is 120 km wide; (b) stretched lithosphere is 240 km wide. The stretching factor (β) for this example is 2; the lithosphere thins by 50% as a consequence. (c) Simplified sketch through the Viking Graben (North Sea) based on seismic refraction, illustrating stretched crust. Note that the depression has filled with sediment.

spreading ridge forms, we say that the rifting was *successful*. But not all rifts are successful. If extension ceases before the continent separates into two pieces, we say that the rifting was unsuccessful. *Unsuccessful rifts* that cut into cratonic areas of continents at a high angle to the continental margin are also known as *aulacogens* (from the Greek for furrow). These and other terms related to rifting and extension are summarized in Table 15.1.

Typically, 20 to 60 million years pass between inception of a rift and the time (called the *rift-drift transition*) at which successful rift faulting ceases and seafloor spreading begins. The amount of lithospheric stretching that takes place prior to the rift-drift transition is variable. Typically, continental lithosphere stretches by a factor of two to four times before separation, meaning that the lithosphere of the rift region eventually becomes two to four times its original width and therefore about one-half to one-quarter of its original thickness (Figure 15.4). The amount of stretching prior to rifting at a given locality, as well as the overall width of the rift, depends on such variables as pre-rift strength of the lithosphere and syn-rift heat flow at the base of the lithosphere.

During the rift-drift transition, movement on rift-related faults ceases, and the thinned continental lithosphere underlying the newly formed continental margins begins to *subside*. As subsidence occurs, relicts of the now inactive rift sink relative to the Earth's surface and are progressively buried beneath a thick accumulation of marine sedimentary rock. Continental margins at which this process is occurring are known as *passive margins,* to distinguish them from *active continental margins* at which you find either strike-slip motion or convergence. The trough containing the generally thick sedimentary accumulation is a *passive-margin basin.* Passive-margin basins, which underlie the continental shelves of many continents, are not really static, but rather are the locus of gravity-driven deformation causing the sedimentary wedge to gradually slide seaward on an array of normal faults (see Section 2.3).

Rifts and passive margins are fascinating regions, not only because of the complex structural and stratigraphic assemblages that occur within them, but also because of the petroleum resources that they contain. In this chapter, we survey the principal structural, petrologic, and topographic features of rifts and passive margins, and introduce some of the current speculations about how and why rifting occurs.

Table 15.1 Terminology of Rifting and Extensional Tectonics

Accommodation zones	Regions where faults interact and connect along strike.
Active margins	Continental margins that coincide with either strike-slip or convergent plate boundaries, and thus are seismically active.
Aulacogens	Unsuccessful rifts that cut into continental interiors, but did not evolve into an ocean.
Drifting	The separation of continental plates by seafloor spreading.
Listric normal faults	Normal faults whose dip decreases with depth, thereby making the fault surface concave upward.
Passive margins	Continental margins characterized by thick accumulations of marine sediment deposited in response to subsidence. Passive margins are not plate boundaries and thus are not seismically active.
Planar normal faults	Normal faults whose dip is constant with depth.
Rift (system)	Generally linear belt in which continental lithosphere is stretching.
Rift-drift transition	Time at which active rift faulting ceases and seafloor spreading begins.
Rifting	The process by which continental lithosphere stretches.
Subsidence	The process of sinking (generally to maintain isostatic equilibrium).

15.2 CROSS-SECTIONAL STRUCTURE OF RIFTS

Let's start our discussion of rifts by developing a cross-sectional image of an active rift. Rifts evolve with time, so the geometry of a rift at a very early stage in development may differ from that of a rift just prior to the rift-drift transition. We'll examine such changes later. For now, we'll focus on the geometry of a rift that is at an intermediate stage in its development; examples of rifts at this stage include the active Basin-and-Range Province of the southwestern United States, the inactive North Sea Rift of northwestern Europe, and the active East African Rift.

On pre-1970s cross sections, geologists suggested that rifts were symmetric arrays of horsts and grabens (see Chapter 8), with the horsts being the ranges and the grabens being the basins in the *basin-and-range topography* (Figure 15.5a). According to these cross sections, rift faults are planes that simply die out at depth, either in a zone of distributed strain, or in a cluster of question marks. Modern seismic-reflection data, however, are not compatible with this image. In contemporary models of rifts, upper crustal extension is accommodated by displacement on arrays of subparallel faults, which merge with a regionally extensive subhorizontal *detachment fault* at depth (Figure 15.5b). Note that in the context of these models, the ranges in rifts are the unburied tips of tilted hanging-wall fault blocks, and the basins are *half grabens*. Moreover, because faults across the extended region dominantly dip in the same direction, the structure as a whole is markedly asymmetric.

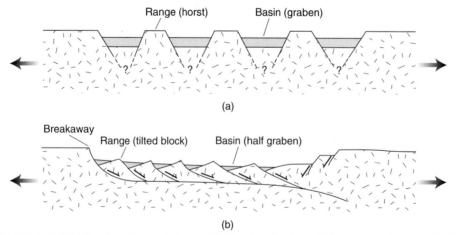

Figure 15.5 Contrasting models of rift faulting. (a) Old concept of symmetrical horsts and grabens. (b) Contemporary image of tilted fault blocks and half grabens, all above a detachment.

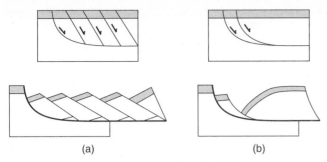

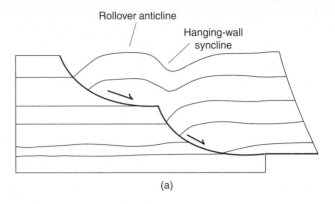

(a)

Figure 15.6 Normal fault array geometry. (a) Parallel rotational faults (note that the tilt of the fault blocks increases as the slip on the fault blocks increases). (b) Listric faults and associated unfaulted rollover.

Detailed studies of rift structure show that rifts contain both *planar normal faults,* meaning faults whose dip is constant with depth (Figure 15.6a), and *listric normal faults,* meaning faults whose dip decreases with depth, thereby making the fault surface concave upward (Figure 15.6b). Movement on these faults typically causes the fault blocks above them to tilt progressively during regional extension in proportion to the amount of displacement on the fault. When this happens, the faults are *rotational normal faults.* In such geometries, a system of planar rotational normal faults forms a set of tilted fault blocks that resembles a suite of tilted books. In fact, the book analogy offers a simple classroom experiment to examine the relationship between amount of fault displacement and tilt.

Normal faults may consist of subhorizontal segments (*flats*) linked by steeply dipping segments (*ramps*), producing a staircase geometry in profile (compare this geometry with that of thrust faults in Chapter 17; Figure 15.7a). During displacement on a listric or stairstep normal fault, strata of the hanging wall either bend to form a rollover fold, and/or are cut by *synthetic* or *antithetic* faults (Figure 15.7b). In places, many adjacent ramps merge with the same upper and lower flats, creating an *extensional duplex* (Figure 15.7c). The net result can be a very complex geometry.

We've concentrated, so far, on the structural geometry of rifts at a very shallow level in the continental crust. What do rifts look like at greater depths in the lithosphere? Currently, two endmember models provide insight into deeper-level rift geometry. In the *pure-shear model* (Figure 15.8a), the detachment defining the base of upper-crustal normal faulting lies at or near the brittle-plastic transition in the crust. Beneath this detachment, the crust accommodates stretching across a broad zone, either by development of penetrative ductile strain, or by movement on an array of anastamosing shear zones. The name "pure-shear model" comes from the recognition that a square superimposed on a cross section of the crust prior to extension becomes a rectangle after extension. By contrast, in the *simple-shear model* (Figure 15.8b), the basal detachment cuts down all the way through the deeper

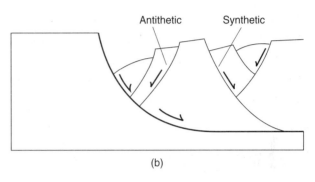

(a)

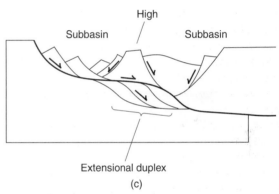

(b)

(c)

Figure 15.7 (a) Rollover anticline and hanging-wall syncline. (b) Rollover cut by antithetic (dipping toward the main fault) and synthetic (dipping in the same direction as the main fault) faults. (c) Complex fault array underlain by an extensional duplex.

lithosphere as a discrete shear zone. In some versions of this model, the detachment may be subhorizontal for a substantial distance before ramping down through the lower crust, so that the region of upper crustal extension does not lie directly over the region of deeper extension (Figure 15.8c). The detachment in a simple-shear model is the fundamental boundary between the two pieces of lithosphere (upper plate and lower plate) that will separate from each other if the rift is successful. Simple-shear models have the advantage of readily explaining the asymmetry of many rifts. Quite likely, some combination of these models represents the true geometry of rifts. For example, the crustal-scale shear zone of the simple shear model may merge at depth with a zone of distributed shear (Figure 15.8d).

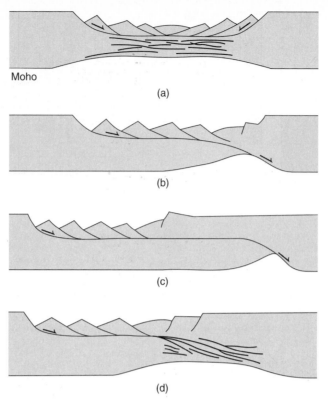

Figure 15.8 Models of rifting at the crustal scale. (a) Pure shear model; (b) simple-shear model; (c) delamination model; (d) hybrid model (simple shear plus broad zone of distributed strain.

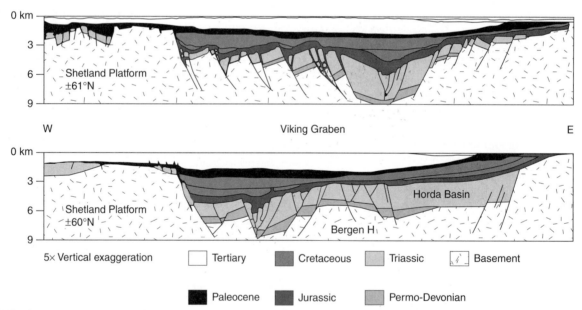

Figure 15.9 Cross sections of the Viking Graben (North Sea Rift), showing the relation of the basin to the position of faults. See Figure 15.4c for a model showing the linkage of the faults to a deep detachment.

To conclude our discussion of rift structure in cross section, let's look at the North Sea Rift system in northwestern Europe, which formed during the Mesozoic and Early Cenozoic, and has been studied intensively because of its petroleum reserves. According to a cross-sectional interpretation of the northern portion of the basin (Viking Graben; Figure 15.9), the rift had a long history of activity, and it changes geometry along the strike. Our second example is the Gulf of Suez, which formed in the Cenozoic during a now terminated northward propagation of the Red Sea Rift. According

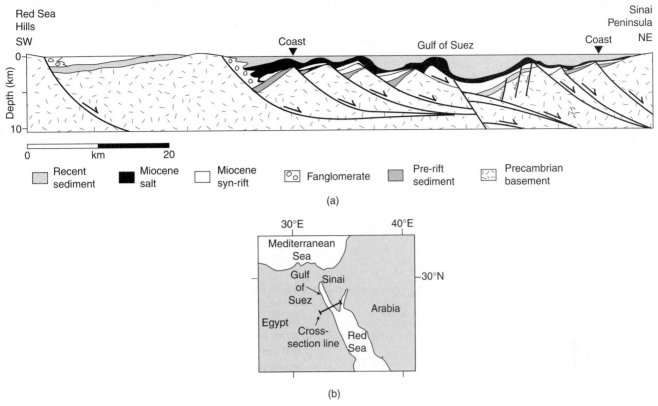

Red Sea Hills SW

Coast

Gulf of Suez

Sinai Peninsula

Coast NE

Depth (km)

| Recent sediment | Miocene salt | Miocene syn-rift | Fanglomerate | Pre-rift sediment | Precambrian basement |

(a)

30°E   40°E

Mediterranean Sea

Gulf of Suez   Sinai

30°N

Egypt   Arabia

Cross-section line   Red Sea

(b)

Figure 15.10 Cross section of the Gulf of Suez (a) and location map (b).

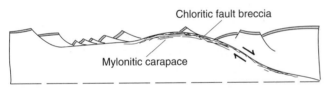

Chloritic fault breccia

Mylonitic carapace

Figure 15.11 Generalized geometry of a metamorphic core complex.

to the cross-sectional interpretation of Figure 15.10, the rift is very asymmetric, for most of the faults dip eastward at the latitude of the cross section, and extension lies to the east of a distinct breakaway fault, which is the boundary between the stretched crust of the rift and the unstretched crust outside the rift. In the Gulf of Suez, it appears that after the original breakaway had been active for some time, a new breakaway developed to the east, and this younger breakaway cuts down to a deeper crustal level.

15.2.1   Metamorphic Core Complexes

The simple-shear model of rifts provides a possible explanation for the origin of enigmatic terranes now known as metamorphic core complexes. *Metamorphic core complexes* were first recognized in a belt that rims the eastern edge of the Cenozoic Basin-and-Range Province of the North American Cordillera (Section 19.4). They were so named because they are distinct bodies of metamorphic rock that occur in the core (or hinterland) of the orogen (Figure 15.11). But what distinguishes core complexes from other regions of metamorphic rock is a characteristic zonal arrangement of features. Specifically, the interior of

an idealized cordilleran core complex is composed of Cenozoic or Mesozoic granite which cuts nonmylonitic pre-Cenozoic country rock (gneiss, sedimentary strata, volcanics). This interior zone is surrounded by a carapace of mylonite (Figure 15.12). The basal contact of the mylonite is gradational, such that the fabric intensity diminishes progressively downward into the nonmylonitic rocks of the complex's interior. Regionally, the foliation of the mylonite is arched into a gentle dome that is shaped somewhat like a turtle shell. But regardless of dip direction of the mylonitic foliation, it has essentially the same lineation direction, and shear-sense indicators in the mylonite provide the same regional displacement sense. The top of the mylonite carapace is an abrupt contact typically delineated by a chloritic fault breccia; rocks above this breccia zone are nonmylonitic. Movement on rotational normal faults that cut the hanging wall above the chlorite breccia zone, but do not cut the breccia zone itself, resulted in large extensional strains in the hanging wall. In fact, locally, faulting has isolated individual blocks of the hanging wall, so that they appear as islands of unmetamorphosed strata floating in a 'sea' of mylonite. Because of the rotation of the fault blocks, strata in these blocks locally intersects the mylonite/breccia zone at almost a right angle.

An explanation of how core complexes formed remained elusive for many years, until the process of mylonitization, the tools for interpreting shear-sense indicators, the implications of asymmetric rifting, and the possibility that the basal detachment of rifts could penetrate below the brittle/plastic transition became clear. In the

Figure 15.12 Exposure of the detachment fault in the Whipple Mountains. The dark rocks are tilted fault blocks composed of Cenozoic sedimentary and volcanic rocks. The light colored rocks are footwall mylonites.

context of these new concepts, metamorphic core complexes are considered exposures of the regional detachment at the base of the normal fault system in a rift (Figure 15.13). Core complexes occur where the extension of the hanging wall above this shear zone has been so extreme that the detachment and the footwall beneath it become exposed. In this interpretation, the nonmylonitic core is the rock of the footwall below the detachment, and the mylonite zone is the footwall portion of the crustal-scale shear zone. The footwall rock is mylonitized at depth in the crust and then moved up to shallower crustal levels by normal motion. The chlorite breccia zone represents the boundary between fault rocks deformed in the brittle field and the mylonite. Chlorite in this breccia is due to retrograde metamorphism accompanying fluid circulation.

In summary, core complexes represent regions where crustal-scale normal-sense shear on a detachment brings mylonitized footwall rocks up from depth and juxtaposes them beneath brittlely deformed rocks of the hanging wall. The arching of the mylonitic shear zone is a consequence of isostatic uplift in response to unloading when the hanging wall is tectonically thinned and removed.

15.3 FORMATION OF A RIFT SYSTEM

Extension along the entire length of a continent-scale rift system does not all begin at the same time. In a given region, the rift system begins as a series of unconnected normal faults each of which grows in length until many faults link to form a continuous zone of extension, and the

zone of extension itself then propagates along strike. The initial faults that ultimately link to form the rift are not necessarily aligned end to end, but may be offset from one another. Some faults are planar in three dimensions, and thus have a straight-line trace. Others, however, are spoon shaped in 3-D, and thus their map trace resembles a 'C' (on the order of 10 km from end to end). At the center of the C-shaped faults, slip is dominantly down-dip, whereas at the ends of the C, displacement is strike-slip or oblique-slip (Figure 15.14). Correspondingly, the thickness of sediment in the half graben developed above the down-dropped, hanging-wall block diminishes toward the ends of the C. With continued growth, the individual faults begin to overlap and interact.

Regions where rift segments interact and connect are *accommodation zones*. In these zones, you'll find complex deformation involving strike-slip, dip-slip, and oblique-slip faulting (Figure 15.15). The geometry of an accommodation-zone depends on whether the faults that it links dip in the same or different directions, on whether the faults were initially aligned end to end, and on how much the fault traces overlap along strike. As a consequence, accommodation-zone geometry is quite variable. Accommodation zones may evolve into strike-slip faults (called *transfer faults*) oriented at a high angle to the regional trend of the rift, which, in turn, evolve into *transform faults* that connect segments of ocean ridges if the rift is successful (see Chapter 18 for the specific terminology of strike-slip faults).

A large rift system typically consists of several smaller *rift segments,* 200–700 km long, that are linked by

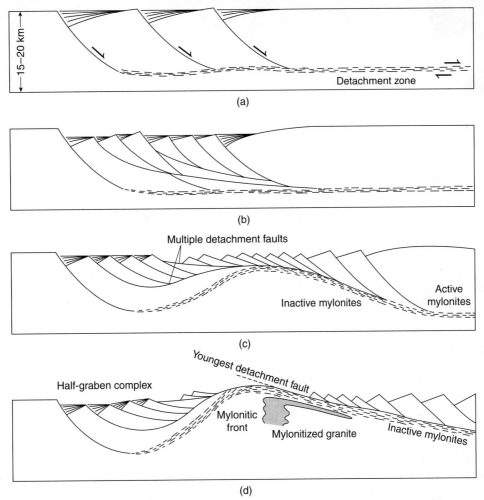

(a)

15–20 km

Detachment zone

(b)

(c)

Multiple detachment faults

Inactive mylonites

Active mylonites

(d)

Youngest detachment fault

Half-graben complex

Mylonitic front

Mylonitized granite

Inactive mylonites

Figure 15.13 Model of the development of a metamorphic core complex. (a) An initially subhorizontal, midcrustal ductile detachment zone beneath an array of steeply dipping normal faults in the upper plate are formed; in (b) additional normal faults have formed, increasing the geometric complexity. (c) As a result of unloading and isostatic compensation, the lower plate bows upward, until the metamorphic core is exposed underneath the detachment fault (d). The steeply dipping mylonitic front is a tilted section of the original detachment zone.

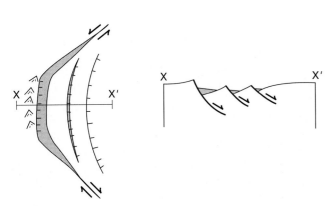

X   X'

X   X'

Figure 15.14 Map view and cross section of a C-shaped half graben. Note that the basin tapers towards its ends as the displacement on the fault changes from dip-slip to strike or oblique-slip. In this example, the hanging-wall block is cut by synthetic faults.

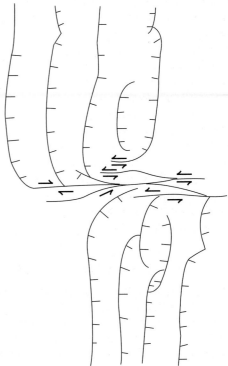

Figure 15.15 Map-view geometry of an idealized accommodation zone.

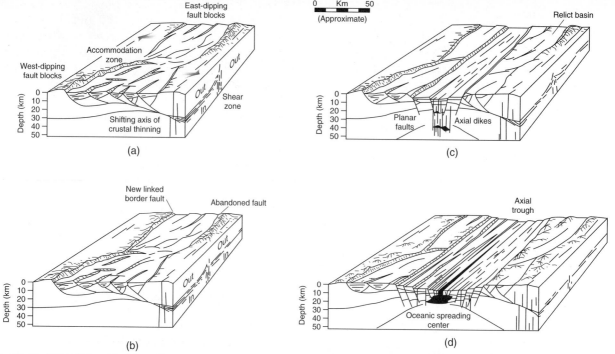

Figure 15.16 Rift evolution. After initial rifting, fault propagation results in linked subbasins (a). After some fault reorganization (b), the rift reaches the rift-drift transition stage (c), which eventually leads to a new ocean basin (d).

accommodation zones. The numerous normal faults that compose an individual segment dip dominantly in the same direction, but different segments may have different dominant dip directions. Figure 15.16 illustrates the process by which uniform dip direction may develop a segment. At an early stage in rifting, normal faults are isolated and are separated from one another by unfaulted crust. As displacement on the faults increases, the length of the faults also increases. Eventually, the individual blocks interact along strike, with one dip direction dominating, at which time the opposite-dipping faults die out or link with antithetic upper-plate faults (Figure 15.16b). Presumably, the underlying detachment also grows during this time, and propagates to greater depths in the crust. Perhaps at a later stage, the original breakaway becomes inactive and a new breakaway is established.

With continued extension, a rift passes through the rift-drift transition stage (Figure 15.16c). The region is underlain predominantly by stretched continental crust, but a discrete elongate 'deep' is present in which heat flow is particularly high. These deeps are sites of nascent seafloor spreading. As stretching continues, the deep widens until the continental lithosphere breaks apart and an oceanic spreading center initiates (Figure 15.16d), and a new ocean is born.

15.4 CONTROLS ON RIFT ORIENTATION

Two fundamental factors, *preexisting structure* and *syn-rift stress field,* appear to control the regional orientation of a rift. Examination of rift geometry worldwide suggests a correlation between the rift-axis strike and/or transfer-zone strike and the orientation of basement foliations, faults, and shear zones. In the Appalachian Mountains, for example, the trend of Mesozoic rift basins (Figure 15.17) closely follows the trend of Paleozoic fabrics and faults, and in the South Atlantic Ocean, the Romanche fracture zone aligns with a major Precambrian shear zone in the Brazilian Shield. Your intuition probably anticipates such relationships, because planes of foliation are weak relative to other directions in a rock. Preexisting orogens are particularly favored sites for later rifting, for not only do orogens contain preexisting faults that can be reactivated, but orogens are warmer than cratonic areas of continents, and thus are weaker. Orogens are, effectively, the weak link between continental blocks that had been connected to form a larger continent.

The geometric relation between the orientation of preexisting structure and the orientation of rifts or transfer faults is not universal. For example, in the interior of Africa, contemporary structures of the East African Rift system obliquely cut across a preexisting rift system. Thus, at least locally, another factor must control rift geometry, and this factor is the syn-rifting stress field. In the case of East Africa, stress measurements suggest that faults in the rift are perpendicular to contemporary σ_3, which is predicted from basic theory (Chapter 8). Evidently, at some localities where the faults crosscut older structures, stress control is more influential in affecting rift geometry than the older structures themselves.

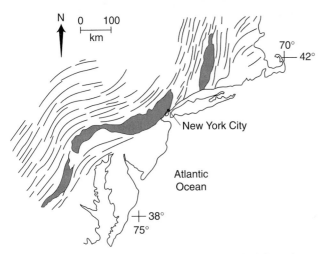

Figure 15.17 Location of Triassic/Jurassic rift basins (shaded areas) along the east coast of the USA, showing how these relatively young basins parallel the Paleozoic structural grain of the Appalachians. Lines represent trends of major faults and folds.

15.5 LITHOLOGIC AND TOPOGRAPHIC FEATURES OF RIFTS

Perhaps the first geologic feature that comes to mind when you think about a rift is a system of normal faults and associated folding. But extension-related structures are not the only features of a rift. Rifts are also characterized by development of distinct sedimentary- and igneous-rock assemblages, and a distinct topography.

15.5.1 Sedimentary Rock Assemblage of Rifts

Formation of a rift results in the development of a depression surrounded by highlands. This depression is a *rift basin,* which traps the sediments derived by erosion of the adjacent highs. Subsidence of the rift floor can be so rapid that several kilometers of sediment accumulate in the basin in just a few million years. The depositional sequence in a rift basin reflects the evolution of the basin.

During initial stages of rift formation (Figure 15.18a), the floor of the rift basin is above sea level, so the sediment carried into the basin is composed of nonmarine clastic debris that is deposited in alluvial fans. Intermittent lakes locally form along the axis of the basin, and fine-grained mud settles out in the quiet water of these lakes. Some of these lakes may be saline. Thus, the sedimentary sequence deposited during the early stages of rifting contains coarse gravels interstratified with red sandstone and siltstone, lacustrine shale and thin evaporite.

As a rift basin continues to widen, its floor further subsides and eventually sinks below sea level (Figure 15.18b). At this stage, the sea that covers the basin floor is very shallow, so evaporation rates are high; in fact, at times, the sea may dry up entirely. In such an environment, seawater becomes supersaturated and salt (dominantly halite ± gypsum ± anhydrite) precipitates. The resulting evaporite sequence, which may be quite thick, buries the continental clastic deposits that had formed earlier. With continued extension, the rift basin broadens and deepens, thereby becoming a narrow ocean (Figure 15.18c). At this stage, marine strata (e.g., carbonates, sandstones, and shales) bury the evaporites.

As we will see in our later discussion of passive margins, if the rift is successful, the syn-rifting sequence forms the base of a passive-margin wedge that may become many km thick. If the rift is unsuccessful, then as the lithosphere beneath the central part of the rift cools, thickens, and subsides, the rift acts as a load that bends down the rift margins. As a result, a thin tapering wedge of sediment covers the basin, and a vertically exaggerated profile of the rift basin and its margins resembles the head of a longhorn bull (Figure 15.19).

15.5.2 Igneous Rock Assemblage of Rifts

Stretching of the lithosphere during rifting decreases its thickness, and thus decreases the lithostatic pressure in the underlying asthenosphere. The asthenosphere is so hot (>1280°C) that, when pressure decreases, it undergoes *partial melting.* Partial melting of the asthenosphere, which is composed of ultramafic rock (peridotite), yields a mafic magma. By partial melting, we mean that only the minerals with lower melting temperatures in the rock melt, leaving behind a residue of unmelted minerals. Partial melting always results in a magma that is less mafic than its source, because mafic minerals (e.g., olivine and pyroxene) have higher melting temperatures than felsic minerals (e.g., quartz and K-feldspar). The amount of partial melting at a given pressure depends on temperature, so if there is a mantle plume beneath the rift making the asthenosphere hotter than normal, large amounts of magma may form.

Magma is less dense than the overlying lithosphere, and thus magma generated by partial melting of the asthenosphere rises into the lithosphere. Locally, migrating magma follows cracks and faults formed during stretching of the lithosphere. Some of the magma is trapped at the base of the crust, a process called *magmatic underplating;* elsewhere magma intrudes the crust or even erupts in volcanoes at the Earth's surface. Magmatic underplating is an important process that effectively thickens the crust, but it is difficult to document because the product is not exposed. In contrast, igneous rocks that intrude the crust are readily observed. Because rifts are zones of horizontal extension, σ_3 is horizontal in the crust; so basalt intruded at these levels creates subvertical *dike swarms* (i.e., arrays of many parallel dikes) that strike parallel to the rift axis. At very shallow crustal levels, however, σ_3 may be vertical, and basalt intruded at these levels forms *sills,* such as the Palisades sill that forms the cliffs along the Hudson River opposite New York City. Basalt erupting at the ground surface has low viscosity and thus spreads laterally to create aerially extensive flows. In a few locations, such as the

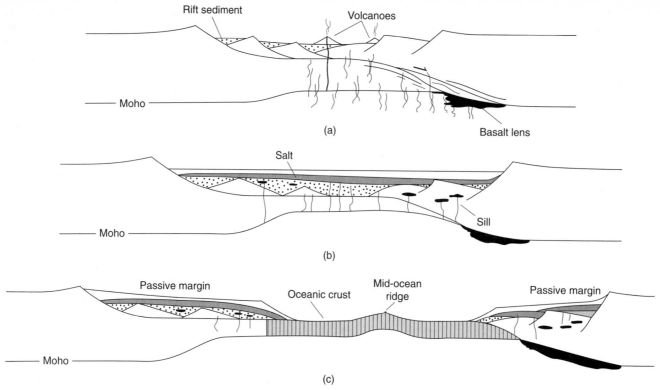

Figure 15.18 Evolution of a rift system into paired passive margins and sedimentary environment. (a) Rift stage with nonmarine basins; (b) rift-drift transition with evaporite deposition; (c) drift stage (seafloor spreading is occurring and passive-margin basins are evolving) with marine deposition.

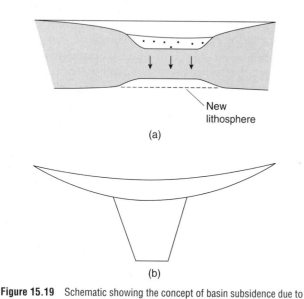

Figure 15.19 Schematic showing the concept of basin subsidence due to thermal cooling and consequent thickening of the lithosphere.
(a) Development of a 'steerhead' basin after instantaneous stretching.
(b) Schematic drawing of the basin fill, showing the resemblance to a Texas longhorn.

Columbia Plateau in Washington and the Parana Basin in Brazil, so much basaltic lava erupted that vast floods of lava buried the landscape in immense sheets; the resulting rock is aptly named *flood basalt.* Recently, it has been suggested that formation of flood basalt occurs specifically where there are mantle plumes, leading to generation of particularly large volumes of magma.

Mafic magma is so hot (1100°–1200°C) that when it is trapped in the continental crust, it triggers partial melting of the adjacent continental crust that consists of minerals with melting temperatures <1000°C. Partial melting of continental crust yields felsic magma that then rises to form granite plutons and rhyolite dikes at depth, or rhyolite domes and ignimbrites (sheets of welded tuff formed when hot ash flows are erupted from volcanoes) at the surface. This association of felsic and mafic volcanism that is characteristic of rifts is called a *bimodal volcanic suite.*

15.5.3 Active Rift Topography and Rift-Margin Uplifts

Figure 15.20 shows profiles across two active rifts, the Basin-and-Range Province of Utah and Nevada, and the Red Sea Rift east of Egypt. Note that in both of these examples, the axis of the rift is a low area relative to the rift margin, and that within the rift, there are numerous elongate asymmetric ridges separated from each other by narrow plains. The highlands bordering the rift are known as rift-margin uplifts, and they can reach an elevation of 2 km above the axis of the rift. The asymmetric ridges within the rift are the tips of tilted fault blocks, and the plains are the surfaces of sediment-filled basins formed above the down-dropped fault blocks. In the first example, which is an early to intermediate stage rift, the axis of the basin is itself substantially above sea level, whereas in the Red Sea, which is a late-stage rift, the axis of the basin is below sea level.

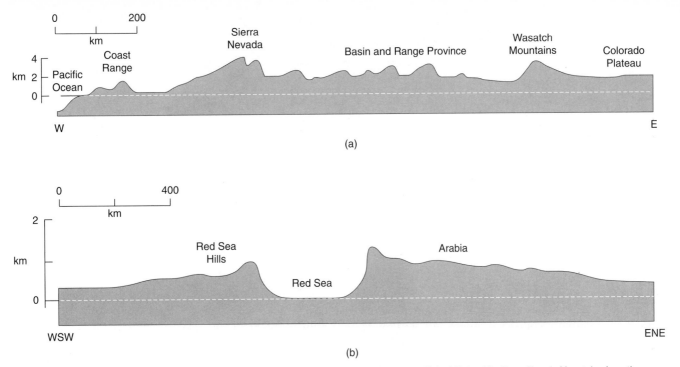

Figure 15.20 Topographic profiles across two active rifts. (a) The Basin and Range of the western United States. The Sierra Nevada Mountains form the western edge of the rift, and the Wasatch Mountains form the eastern edge. (b) The Red Sea rift along the eastern edge of Egypt.

Clearly, the topography of rifts evolves through time in concert with the development of strain in the rift. During most of their evolution, rifts are actually relatively high. It is only during their old age, after they have ceased being active or when they are entering the rift-drift transition, that rift floors subside to sea level or deeper.

What causes the overall topographic elevation of many rifts? One cause may be preexisting elevation of the region undergoing extension. Rifts forming in regions that were orogenic belts previously are high to start with. For example, rifting is currently occurring in Tibet, a region that is several kilometers above sea level (Chapter 19). But even if rifting begins in a region at low elevation, the process of rifting commonly causes an increase in ground elevation. In other words, rifting itself causes uplift. Let's see how this works.

Heating in rifts occurs both because the magma that rises through the lithosphere during rifting carries heat that is transferred to the surrounding rock, and because thinning of the lithosphere that accompanies rifting causes the 1280°C isotherm defining the lithosphere-asthenosphere boundary to move closer to the Earth's surface. Bringing the isotherm closer to the surface is much like placing a hot plate beneath the lithosphere: at shallower depths the once cool lithosphere gets warmer. Heat-flow studies demonstrate that heating of lithosphere does occur in rifts. For example, the average heat flow for the Canadian Shield is ~1 HFU (Heat Flow Unit is a measure of the heat passing through an area in a given time; 1 HFU = 40mW m^{-2}), whereas the average heat flow in the East African Rift is about 2.5 HFU (see Figure 14.16). Heating the lithosphere

causes it to become less dense; so to maintain regional isostatic compensation, the surface topography rises. Put in simpler terms, as the lithosphere heats and becomes less dense, it floats higher, just like a block of balsa wood floats higher than a block of oak when placed in a bathtub.

At the same time that uplift is occurring because of high temperatures, extensional faulting and graben or half-graben formation is occurring due to lithosphere stretching. Thus, the central axis of the rift is a down-dropped trough, bordered by a series of fault-bounded blocks that step upward to the highest elevations on either side of the trough. Unloading of the lithosphere forming the rift margin causes the margin to rebound and move up, much like a trampoline surface moves up when a gymnast steps off. Why then do rifts ultimately subside? When rifting ceases, meaning that lithospheric stretching ceases, the rift zone begins to cool. As a consequence, the 1280°C isotherm migrates to lower depth (i.e., the lithosphere thickens), so there is a greater thickness of cooler, denser lithosphere. To maintain isostatic balance, the lithosphere sinks.

Rift margins may remain high long after rifting has ceased. Examples of contemporary passive-margin uplifts that reach elevations of several km include the Serra do Mar along the southeast coast of Brazil, the Blue Mountains of Australia, and the Transantarctic Mountains of Antarctica. It is important to realize that the topography of these uplifts is not a relict of collisional mountain belts that deformed and metamorphosed the crust of these regions, but that it is a much younger feature related to rifting of adjacent regions. For example, the rocks that compose the famous high peaks in and around Rio de Janeiro are

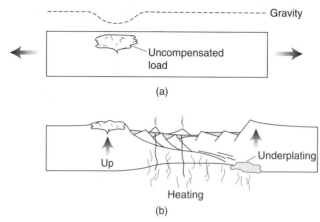

Figure 15.21 Other causes for rift-margin uplifts. (a) Before rifting, an uncompensated load is held in place by the strength of the lithosphere. (b) After rifting, the uncompensated load rises. Underplating and heating may also cause rise.

Proterozoic in age, but the present uplift is Late Mesozoic to Cenozoic. The basement of the Transantarctic Mountains formed during a collisional orogen in Late Proterozoic to Early Paleozoic times; this orogen was beveled flat and was covered by Jurassic strata and volcanics before uplift in Late Mesozoic and Cenozoic time, in association with development of the Ross Sea Rift. The Jurassic rocks are still nearly flat lying, even though they are now exposed at elevations of over 4 km.

The cause of long-lived rift-margin uplift remains a matter of debate. Certainly, the initial uplift is, in part, related to heating of the lithosphere by the emplacement of hot asthenosphere beneath the axis of the adjacent rift. However, calculations suggest that this effect cannot explain all the uplift, and it also cannot explain why the margins of some rifts remain high tens of millions of years after rifting has ceased. Other factors may contribute to rift-margin uplift. Flexural rebound accompanying unloading of the lithosphere when the hanging wall above a normal fault moves off the footwall causes uplift (analogous to the proposed scenario for metamorphic core complexes; Figure 15.13). Also, the process of underplating causes thickening of less-dense crust relative to denser mantle lithosphere, which results in isostatic uplift (Figure 15.21). Finally, the process of rifting decreases the strength of the lithosphere, and may allow isostatically uncompensated loads (e.g., large granitic plutons from an earlier episode of intrusion that are no longer held in place by the strength of the lithosphere) to rise to higher elevations.

15.6 OCEAN RIDGES

You will recall that *ocean ridges* are plate boundaries at which new oceanic lithosphere is created. Once formed, the new lithosphere moves away from the ridge axis, from which came the term *seafloor spreading*. At the ridge axis,

this process involves extensional tectonics. Considering that the net length of the current ocean ridges on this planet is over 40,000 km, the ridge system is an important region of extension. To understand how seafloor spreading occurs, we return to the structure of oceanic lithosphere, and then the bathymetry of the seafloor.

Oceanic lithosphere consists of two components (Chapter 14). The upper part is the *oceanic crust,* which is about 6–10 km thick, and is almost uniform in thickness across the width of the ocean. Oceanic crust is subdivided into four layers, which are from top to bottom: pelagic sediment, pillow basalt, sheeted dikes, and massive gabbro. The lower part of oceanic lithosphere is upper mantle, which varies in thickness from essentially 0 km beneath the ridge to over 70 km at the margins of a large ocean.

The *pelagic sediment* layer of oceanic crust forms from snowlike settling of dust and plankton shells through seawater. Brown clay deposits form where the major component of the sediment is dust; siliceous ooze forms where the major component is diatom tests (skeletons); and carbonate ooze forms where the dominant sediment is foram tests. The composition of the sediment at a specific location depends on factors like water temperature, sunlight, and nutrient content, which determine the volume and identity of plankton. Water depth is also an important factor, because below a certain depth most calcite and silica dissolves in seawater before reaching the seafloor. Thus the composition of sediment being deposited over a particular spot on the seafloor changes as the position of the spot changes due to subsidence of the plate as it gets older, and due to the position of the site relative to the equator.

The *pillow basalt* of oceanic crust forms by submarine extrusion at the axis of spreading ridges. Observations suggest that extrusion occurs in a zone that is only a few kilometers wide. Large volumes of hot water are also expelled at the ridge axis, at vents called black smokers, because of

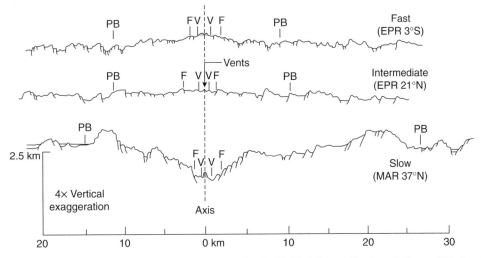

Figure 15.22 Bathymetric profiles of ocean ridges with fast (>9 cm/year; e.g., East Pacific Rise), intermediate (e.g., Galápagos Ridge), and slow (<5 cm/year; e.g., Mid-Atlantic Ridge) spreading rates. V marks areas of volcanic activity flanked by zones of fissuring (F), and PB (plate boundary) marks the edge of active faulting.

the cloud of fine sulfide minerals that precipitates upon cooling of the water when it mixes with seawater above the vent. The *sheeted dike complex* consists of dikes that intrude dikes that intrude dikes, all of basaltic composition. Dike intrusion occurs in a hypabyssal setting below the ridge axis; the dikes strike parallel to the ridge axis. *Gabbro* formation represents slow cooling of a crystal mush ± magma at depth. Settling of crystals in the magma may locally give rise to layering in the gabbro and an ultramafic cumulate at the base of the crust.

The bathymetry of oceanic spreading ridges depends on the rate of spreading at the ridge. Fast ridges (>9cm/year), such as the East Pacific Rise, are broad and fairly smooth swells with no major axial trough. In contrast, slow-spreading ridges (<5 cm/year), like the Mid-Atlantic Ridge, are narrower, with axial regions that are defined by deep troughs bordered by steplike escarpments (Figure 15.22). Why are fast ridges wider than slow ridges? The answer comes from a consideration of the relationship between the depth of the seafloor and its age. Ocean above young seafloor is shallower than ocean above older seafloor, due to isostasy. Beneath young seafloor, the crust is warm and hot asthenosphere lies just below the crust; whereas beneath older seafloor, the lithosphere is thicker, cooler,[1] and therefore denser, so it sinks to a greater depth to maintain isostatic compensation. The contrast in morphology of the ridge axis area of fast ocean ridges versus slow ocean ridges probably reflects a balance between the amount of magma rising at the axis and the rate at which

stretching of the crust occurs. At slow ridges, seafloor spreading stretches the crust so that a graben develops over the axis; whereas at fast ridges, the crust at the axis is held high by magma beneath. As the new oceanic crust moves away from the ridge axis, slip on faults may reverse, so the crustal blocks rise back up to higher elevation along the margins of the central trough. However, the nature of this movement is not well understood.

Keeping these concepts in mind, we can create a composite image of the process of seafloor spreading (Figure 15.23). Let's imagine that an increment of spreading occurs. This spreading stretches the crust, resulting in the formation of normal faults parallel to the ridge axis. As spreading occurs, hot asthenosphere rises from below and is depressurized, which produces partial melting. The resulting mafic magma rises into the crust. At fast-spreading ridges, it creates a zone of crystal mush, at the top of which is a small melt lens. Denser minerals accumulate at the floor of this mush and magma zone, creating a cumulate

<hr>

[1]Remember that lithosphere is defined merely as the rheologic layer above the 1280°C isotherm; so, as it moves away from the axis and cools, the 1280°C isotherm sinks (i.e., rock that was part of the asthenosphere becomes part of the lithosphere).

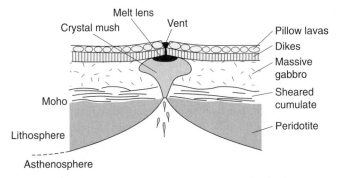

Figure 15.23 Schematic section through the oceanic crust and upper mantle at a spreading ridge.

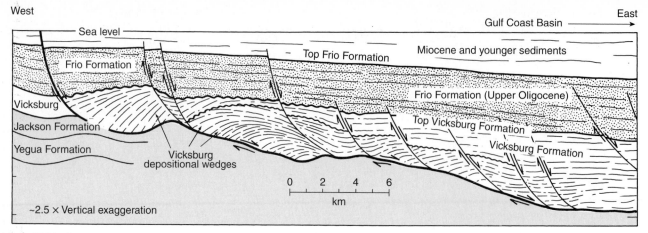

Figure 15.24 Extensional geometry of the southern U.S. passive margin. Note syndepositional listric faulting, and the formation of rollover anticlines. Modified from Erxleben and Carnahan (1983).

that is sheared by plate movement. Cooling at the margins of the magma body creates massive gabbro. Magma from the melt lens injects up as the crust rifts. Some of this melt cools within the crust, forming new dikes; some extrudes at the seafloor, creating pillow basalt. At slow ridges, the melt lens does not appear to exist; magma may inject directly from the mantle below, and have time to cool completely before the next increment of spreading occurs.

15.7 PASSIVE MARGINS

There are two basic types of continental margins. *Active margins* are continental margins that coincide with either transform or convergent plate boundaries, and thus are seismically active. In contrast, *passive margins* are not plate boundaries and thus are not seismically active. As we already pointed out, passive margins develop over the edge of a rift after the rift-drift transition. Examples of present-day passive margins include the eastern and Gulf coast margins of North America, the eastern margin of South America, both the eastern and western margins of Africa, the western margin of Europe, the western, southern, and eastern margins of Australia, and all margins of Antarctica.

The subsidence of passive margins is a direct consequence of the rifting process from which they formed. During rifting, the lithosphere is thinned. Once rifting ceases, the thinned, hot lithosphere begins to cool and thicken. To maintain isostatic equilibrium, the surface of the stretched lithosphere sinks. This process of sinking to maintain isostatic equilibrium during cooling is called *thermal subsidence*. Effectively, thermal subsidence creates a 'hole' that fills with sediment. This hole is the *passive-margin basin,* and the sediment pile filling the basin is the *passive-margin sedimentary wedge.* Typically, the portion of the passive-margin wedge that underlies the continental shelf contains shallow-water marine strata, whereas the part that underlies the continental slope or rise contains deep-

water strata.[2] Passive-margin basins are typically segmented along strike into discrete basins separated by basement highs where sediment is thinner. Individual basins probably correspond to discrete rift basins, and the highs to accommodation zones. Loading by sediment causes the floor of the basin to sink even more (the weight of 1 km of sediment causes the floor of the basin to sink by about 1/3 km) than it would if no sediments were present!

We noted earlier that active tectonic extension ceases after the rift-drift transition. By *tectonic extension,* we mean the faulting that leads to stretching of pre-rift continental crust. Extensional faulting, however, does occur along passive margins long after the drift stage has begun. This postdrift faulting of passive margins involves the strata deposited above the rifted crust. The basal detachment during this stage of faulting is typically the evaporite horizon at the base of the passive-margin basin deposits (Figure 15.24). Gravity is the force that drives postdrift extensional faulting of strata in a passive-margin basin. The seaward face of the passive margin is a sloping surface separating denser sediment below from water above. Much like a slump in the sediment of a hill slope, the strata of the passive margin slip seaward along the weak underlying salt horizon. This movement leads to development of a system of half grabens and rollover folds in the passive-margin wedge, and a system of submarine thrusts at the toe of the passive-margin wedge. These structures grow as deposition continues, and may be modified by salt movement.

Considering the evolutionary history of margins from rifts, and considering that rifts tend to be asymmetric, we

---

[2]In pre-plate tectonic literature, this package of strata was referred to as a *geosyncline,* which was subdivided into a miogeosyncline, which is the part with shallow-water facies, and a eugeosyncline, which is the part with deep-water facies.

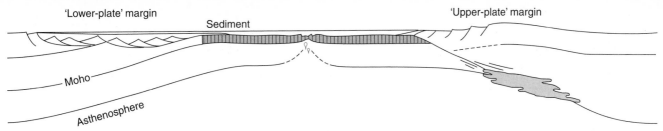

'Lower-plate' margin

Sediment

'Upper-plate' margin

Moho

Asthenosphere

Figure 15.25 Illustration of the concept of upper-plate and lower-plate margins.

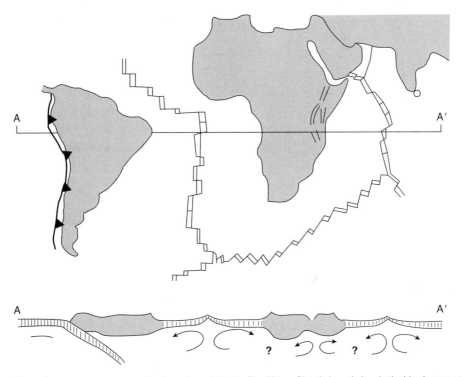

A   A'

A   A'

? ?

Figure 15.26 The convection cell concept cannot explain the formation of all rifts. The African Plate is bounded on both sides by spreading ridges, so the site of the East African Rift should be a downwelling zone, assuming a simple convective cell geometry.

might expect two different endmember classes of passive margins. These classes depend on whether the margin evolved from the upper-plate side of the rift or the lower-plate side of the rift, respectively. In margins originating as upper plates (*upper-plate margins*), the lithosphere has not been stretched substantially, and there is relatively little subsidence. In margins that started as the lower plates (*lower-plate margins*), however, the lithosphere has been stretched substantially and there is more subsidence. Figure 15.25 schematically illustrates the basic characteristics of these two classes, but their presence remains speculative at this point.

15.8 CAUSES OF RIFTING

Finally, we address the fundamental question of why rifting occurs in the first place. Ultimately, rifting, and all plate motion for that matter, takes place because of flow in the mantle. If the mantle did not flow, then Earth's lithosphere

would not move. The asthenosphere flows because buoyancy forces resulting from lateral density gradients exceed the mantle's strength and cause it to convect. Lateral density variations in the mantle are caused either by temperature gradients (in a rock of uniform composition density decreases as temperature increases), or by compositional differences. Given that convection occurs in the asthenosphere, is it correct to picture all rifts as places where the continent lies above the upwelling part of a convective cell? Actually, no! First, it is impossible to devise a geometry of simple convective cells that is compatible with the present-day geometry of rifts. For example, if upwelling is occurring at the South Atlantic Ridge and at the Indian Ocean Ridge, how can there also be upwelling along the East African Rift (Figure 15.26)? Second, calculations suggest that tractions applied to the base of plates by moving asthenosphere are too small to drive plates apart (see Chapter 14). Several more likely reasons for rift formation are explored below.

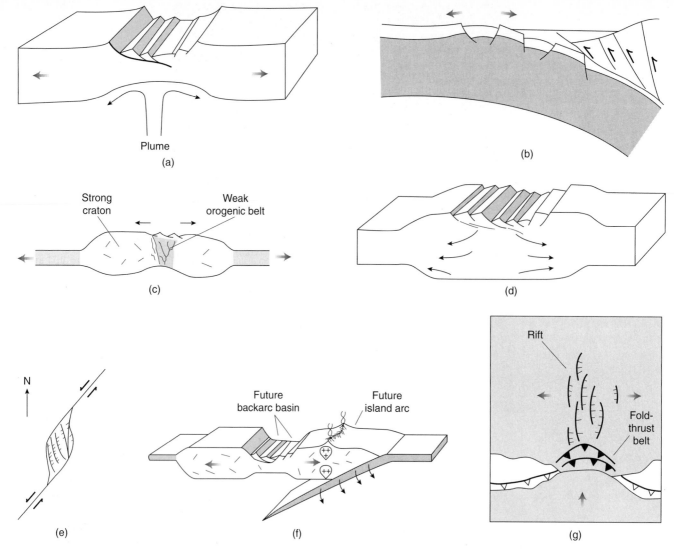

Figure 15.27 Causes of rifting. (a) Rifting above a thermal plume; (b) outer-arc extension of a bending slab at a subduction zone; (c) rifting due to stretching apart of continental lithosphere containing a weak zone (old orogen); (d) gravitational collapse of thickened crust at an orogen; (e) pull-apart basin at a releasing bend along a strike-slip fault; (f) backarc extension associated with convergence; (g) rifting in the foreland of an orogen due to the collision of irregular continental margins.

In places where *plumes* of hot mantle rise (i.e., under *hot spots*), the overlying lithosphere heats up and stretches (Figure 15.27a). As a consequence of this stretching, the upper part of the lithosphere is broken by normal faulting. Rifts that form in response to the rise of hot mantle are *thermally activated rifts*. Whether thermally activated rifts can be successful is not clear.

Rifting may also be caused by changing the radius of curvature of a plate. Such a change occurs where lithosphere bends just prior to descending beneath a collision zone (Figure 15.27b). As a consequence of stretching around a smaller radius of curvature, a series of normal faults that parallel the trench develop seaward of the trench. Similarly, because the Earth is not a perfect sphere, the radius of curvature of an elastic plate must change as the

plate moves from one latitude to another. Whether such *membrane stresses* are sufficiently large to break plates is controversial, because overall plate rheology is still poorly understood.

As first discussed in Chapter 14, plates are subject to ridge-push and slab-pull forces. Thus, it is likely that some rifts form when the two ends of a continent are pulled in opposite directions by plate-driving forces (Figure 15.27c). If the continent contains a weak zone (e.g., a recent orogen), these forces may be sufficient to pull the continent apart. Such rifting may be called *drift activated*.

Some rifts appear to develop in regions of thickened and elevated crust, even while contractional deformation continues, suggesting a relationship between zones of crustal thickening and zones of extension. This relationship

Figure 15.28 Space Shuttle image looking south along the Red Sea. The Gulf of Suez (at the bottom of the image) and the northern portion of the Red Sea are underlain by extended continental crust, whereas oceanic lithosphere has already begun to form in the southern half of the Red Sea.

occurs because the quartz-rich rocks composing the continental crust are not very strong, especially where heated in an orogenic belt. So when a zone of continental crust is thickened and elevated relative to its surroundings, as occurs during a collisional orogeny, gravitational force causes the thickened zone to spread laterally under its own weight (Figure 15.27d). This process is known as *gravitational collapse.* To visualize this process, take a block of soft cheese that has a hard rind (say, Brie). When you put it in the sun, the cheese warms and attempts to flow laterally. Eventually the rind splits along discrete 'faults' to accommodate the overall displacement.[3]

Rifting also occurs in association with large continental strike-slip faults. As described in Chapter 18, extension must occur at releasing bends along strike-slip faults. At these locations, normal faulting oblique to the regional trend of the strike-slip system develops, and as a consequence *pull-apart basins* develop (Figure 15.27e). The faults in pull-apart basins may stop at a detachment horizon at depth in the crust.

As we will see in the next chapter, a zone of extension may develop in *backarc regions* at convergent margins. If the volcanic arc was originally developed on continental crust, this backarc extensional zone is, effectively, a continental rift (Figure 15.27f).

Finally, rifting may develop in the foreland of collisional orogens as a consequence of *indentation* (Figure 15.27g). As further discussed in Chapter 16, if the lateral margins of the foreland are unconstrained, indentation by the orogen may cause foreland blocks to squeeze sideways, to get out of the way, a process called *lateral escape.* In the region between the escaping blocks, σ_3 is parallel to the orogen front, and rifts develop that are roughly perpendicular to the orogen.

15.9 CLOSING REMARKS

This chapter was devoted to a discussion of rifting, the process by which continental breakup occurs, ultimately leading to the formation of new oceanic lithosphere (Figure 15.28). The details of rifting remain relatively uncertain, and many of the concepts presented here were developed only recently. Especially the nature of continental extension has given rise to a lively debate in the structural geology community since the early 1980s. In this chapter we did not discuss the process of seafloor spreading in detail, because this topic is more appropriate for a marine geology course. In the next chapter, we move to convergent tectonics, which is the process by which oceanic lithosphere is consumed, and collisional tectonics, which is the process by which nonsubductable blocks of lithosphere merge.

---

[3]According to some connoisseurs, this is the exact time to consume the experimental material.

ADDITIONAL READING

Dewey, J. F., 1988, Extensional collapse of orogens, *Tectonics*, v. 7, p. 1123–1139.

Gibbs, A. D., 1984, Structural evolution of extensional basin margins, *Journal of the Geological Society of London,* v. 141, p. 609–620.

Lister, G. S., and Davis, G. A., 1989, The origin of metamorphic core complexes and detachment faults during Tertiary continental extension in the northern Colorado River region, *Journal of Structural Geology,* v. 11, p. 65–94.

Macdonald, K. C., 1982, Mid-ocean ridges: Fine-scale tectonic, volcanic and hydrothermal processes within the plate boundary zone, *Annual Review of Earth and Planetary Sciences,* v. 10, p. 155–190.

McKenzie, D. P., 1978, Some remarks on the development of sedimentary basins, *Earth and Planetary Science Letters,* v. 40, p. 15–32.

Rosendahl, B., 1987, Architecture of continental rifts with special reference to East Africa, *Annual Review of Earth and Planetary Sciences,* v. 15, p. 445–503.

Wernicke, B., 1985, Uniform-sense normal simple shear of the continental lithosphere, *Canadian Journal of Earth Sciences,* v. 22, p. 108–125.

Wernicke, B., and Burchfiel, B. C., 1982, Modes of extensional tectonic, *Journal of Structural Geology,* v. 4, p. 105–115.

Chapter 16
Convergence and Collision

16.1   INTRODUCTION

Plate convergence and collision cause a variety of highly visible tectonic phenomena, including mountain belts (Figure 16.1), volcanism and (major) earthquakes. Besides their mere beauty, the vertical relief of mountain ranges provides a detailed view of the structural consequences of plate interactions and lithospheric evolution.

In the previous chapter, we examined the process by which continental lithosphere rifts apart, and we noted that an oceanic spreading ridge forms between the once adjacent blocks if rifting is successful. At first, the oceanic lithosphere that forms at the ridge is warm, and relatively buoyant. This is the reason that oceanic ridges are, in fact, ridges, and rise to shallower depths than the surrounding abyssal plains. But as the new lithosphere moves away from the ridge axis, it cools, the mantle portion of the lithosphere thickens, and lithospheric density increases. Eventually, oceanic lithosphere actually becomes negatively buoyant; in other words, it becomes denser than the underlying hot asthenosphere. Once oceanic lithosphere has become negatively buoyant, it can *subduct,* meaning it can sink into the asthenosphere. Exactly how the subduction process initiates remains somewhat of a puzzle, but possibly it begins in response to compression across an existing weakness in the oceanic lithosphere (e.g., a transform fault or the contact between oceanic and continental lithosphere). Compression causes thrusting of more buoyant lithosphere over less buoyant lithosphere, which in turn results in insertion of the negatively buoyant oceanic lithosphere into the asthenosphere. Once there, this lithosphere begins to sink, pulling the rest of the oceanic plate with it (Figure 16.2).

Figure 16.1 The majestic Matterhorn of the Swiss Alps, a product of glacial erosion of continental crust uplifted in a collisional orogen.

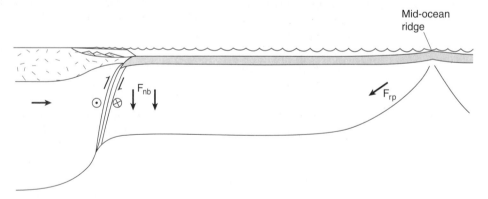

Figure 16.2 Cross section that illustrates how a subduction zone may initiate from a strike-slip continental margin. As the stress field changes across this plate boundary, the negatively buoyant oceanic lithosphere (F_{nb}) underthrusts the buoyant continent.

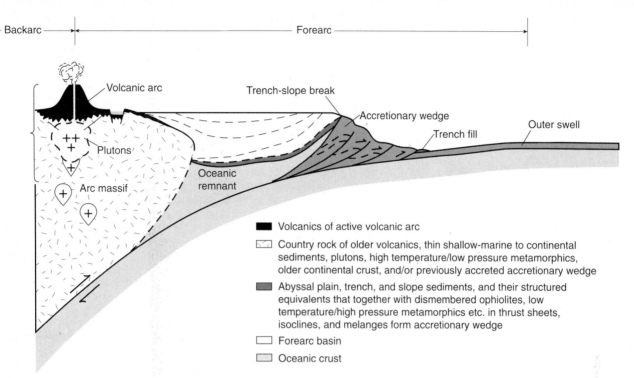

Figure 16.3 Idealized cross section of a convergent plate margin. Modified from Dickinson and Seely (1979).

With the onset of subduction, a new convergent plate margin develops (Figure 16.3). At a convergent plate margin, oceanic lithosphere of the *downgoing slab* (or downgoing plate) bends and descends into the mantle beneath the *overriding slab* (or overriding plate). A deep, locally sediment-filled trough, the *trench,* forms at the boundary between the downgoing and overriding slabs, and an *accretionary wedge,* which is a package of intensely deformed sediment and oceanic basalt scraped off the downgoing slab, may form at the edge of the overriding slab. The top of the accretionary wedge is buried by undeformed strata of the *forearc basin.* This basin, which may also bury trapped oceanic lithosphere, lies between the accretionary wedge and a chain of volcanoes known as the *volcanic arc.* For purposes of directional reference, the region on the trench side of a volcanic arc is the *forearc region;* the accretionary wedge, forearc basin, and trench are all in the forearc region. The other side of the overriding plate with respect to the arc is the *backarc region,* where contraction, extension, and/or strike-slip deformation may occur. These and related terms are summarized in Table 16.1.

As subduction consumes an oceanic plate, pieces of buoyant lithosphere attached to the downgoing slab may eventually be brought to the convergent boundary. Examples of buoyant lithosphere (Figure 16.4) include large continents, small continental fragments, *island arcs* (volcanic arcs built on seafloor), *oceanic plateaus* (broad regions of anomalously thick crust), and some spreading ridges. Re-

gardless of type, buoyant lithosphere cannot be subducted, and when it merges with the overriding slab, the boundary becomes a *collision zone.*

Both convergent plate boundaries and collision zones are sites of contractional tectonics, meaning that at these locations lithosphere shortens in the horizontal direction, and consequently thickens in the vertical direction. Contractional tectonics causes thrust faults, folds, tectonic foliations, metamorphism, and igneous activity. It produces great mountain ranges, like the present-day Himalayas and Andes, and the Paleozoic Appalachians-Caledonides. You will learn, however, that extension and strike-slip may occur coevally with this contraction. In this chapter, we discuss the principal structural and lithologic features formed at convergent boundaries and in collision zones.

16.2 CONVERGENT PLATE MARGINS

Convergent plate boundaries presently rim much of the Pacific Ocean, occur at the east edge of the Caribbean and Scotia Seas, along the western and southern margin of southeast Asia, and locally in the Mediterranean Sea. During most of the Mesozoic, the west coast of North America and southern margins of Europe and Asia were convergent plate margins (see Chapter 19). Some convergent plate margins mark localities where oceanic lithosphere is being subducted under oceanic lithosphere (e.g., Marianas, Aleutians), and others mark localities where oceanic lithosphere

Table 16.1 Elements of Convergent Plate Boundaries

Accretionary wedge/prism	A wedge of deformed sediment and sometimes oceanic basalt that forms at the edge of the overriding slab.
Arc-trench gap	Horizontal distance between arc and trench.
Backarc basin	Basin that is located behind the volcanic arc.
Backarc region	Region that is on the opposite side of the volcanic arc from the forearc basin and trench.
Continental arc	A chain of subduction-related volcanoes along the margin of a continent.
Coupled subduction	A subduction system in which the balance between the downgoing plate velocity, the rollback velocity, and the overriding plate velocity yields a condition that the entire subduction system is in compression and the overriding plate is pushing against the downgoing plate.
Décollement	French term used in some literature for a detachment fault.
Detachment	Basal fault zone; in accretionary wedges it marks the top of the descending slab.
Downgoing slab/plate	Oceanic lithosphere that bends and descends into the mantle beneath the *overriding slab.*
Forearc basin	A depression containing undeformed strata that bury the top of the accretionary wedge, adjacent to the volcanic arc.
Forearc region	Region on the trench side of a volcanic arc.
Island arcs	Volcanic arcs built on oceanic lithosphere. They consist of a chain of volcanic islands.
Marginal sea	Term used for a backarc basin that is underlain by oceanic lithosphere.
Mélange	Chaotic mixture of lithogies of variable origin and metamorphic grade, typically in a fine-grained matrix.
Oceanic plateau	Broad region of anomalously thick crust.
Offscraping	Decoupling mass from the downgoing slab that is accreted at the toe of the wedge.
Outer swell	Broad topographic arch formed by flexure of lithosphere outboard of a subduction system.
Peripheral bulge	*See* Outer swell.
Tectonic erosion	Removal of rock from the base of the overriding plate as a consequence of subduction.
Tectonic underplating	Mass from the downgoing slab that is accreted to the base of the accretionary prism.
Trench	Trough that forms at the boundary between the downgoing and overriding slabs.
Trench-slope break	Topographic ridge marking a sudden change in slope at the top of the accretionary wedge.
Uncoupled subduction	In this subduction system, the balance of velocities yields a condition in which the subduction system is effectively under tension.
Volcanic arc	Chain of subduction-related volcanoes.

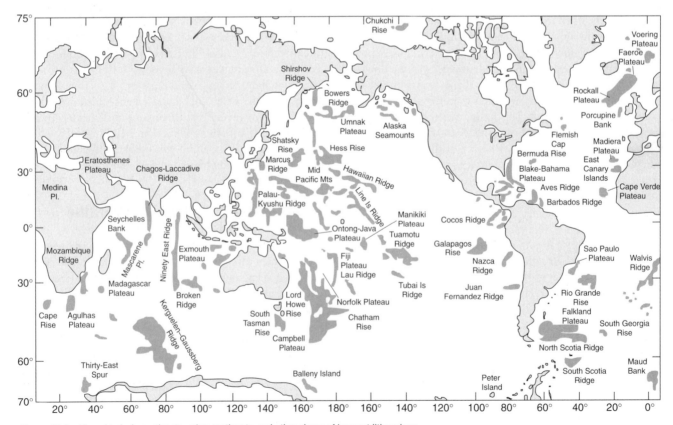

Figure 16.4 Map of today's continents, microcontinents, and other pieces of buoyant lithosphere.

Reprinted with permission from Z. Ben-Avraham, et al., "Continental accretion and orogeny," *Science* 213:47–54. Copyright 1981 American Association for the Advancement of Science.

is being subducted under continental lithosphere (e.g., Andes). To get an overview of what a convergent plate margin looks like, we take you on a tour from the ocean basin across a convergent plate margin. We start on the downgoing slab, cross the trench, climb the accretionary wedge, and trundle across the forearc basin and frontal arc into the volcanic arc itself. We conclude our journey by visiting a variety of backarc regions.

16.2.1 The Downgoing Slab

The first hint that oceanic lithosphere is approaching a subduction zone occurs about 250 km outboard of the trench. Here, the surface of the lithosphere begins to rise and bend around a broad arch called the *outer swell,* or *peripheral bulge* (Figure 16.5a). The elevation difference between the abyssal plain surface seaward of the outer swell, and the crest of the swell itself, is less than 1 km. Outer swells exist because of the flexural rigidity of the lithosphere. Downward bending of the lithosphere below the trench levers up the lithosphere outboard of the trench, a phenomenon that you can duplicate by bending a sheet of stiff rubber or plastic over the edge of a table (Figure 16.5b). The upper layer of the lithosphere stretches to accommodate the bending and is typically cut by an array of trench-parallel normal faults in the vicinity of the outer swell (see Figure 16.5c). This gives rise to a sawblade geometry of the descending slab near the trench.

The location of the downgoing slab as it descends into the mantle is defined by seismicity. Earthquakes occur along an inclined belt (the *Wadati-Benioff zone*)[1] that reaches a maximum depth of around 670 km (Figure 16.6). Recall that a depth of ~670 km is the boundary between the transition zone and lower mantle (see Figure 14.9). Focal mechanisms of earthquakes change in character along the Wadati-Benioff zone. Beneath the outer swell, earthquakes are associated with stretching and normal faulting, whereas at the base of the accretionary wedge, earthquakes are associated with thrust movements due to shear interaction between the overriding and downgoing slab. It is in this zone that the destructive earthquakes of the Pacific Rim region occur. At depths of about 150–300 km, earthquakes in the downgoing slab indicate extensional motions, suggesting that this portion of the plate is being stretched by the pull of the deepest part of the slab. At very deep levels, earthquakes in the slab show contractional motions. There is uncertainty about the significance of such deep-focus earthquakes, because the downgoing slab should not be brittle at such depths. Geophysicists now believe that these deep earthquakes may represent stresses accompanying phase changes and/or dehydration reactions in the rock composing the slab.

The depth to which the downgoing slab descends is also controversial. Until recently, it was thought that the

---

[1]Named after its discoverers, who first recognized the feature in the 1940s and 50s.

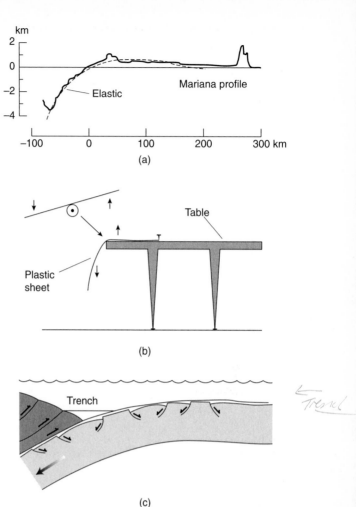

Figure 16.5 ─Trench

Figure 16.5 The peripheral bulge. (a) Example of the peripheral bulge in the Mariana region. (b) Table-top model of a peripheral bulge, with the inset showing lever concept. (c) Stretching along the outer swell produces normal faulting (a sawblade geometry) in the descending slab.

slab mixes with the upper mantle and is no longer a distinct mechanical entity at depths greater than 670 km. But increasingly, geophysicists have come to believe that slabs descend deeper, but are no longer seismic. Recent interpretations of seismic tomographic data show slab graveyards at great depth in the mantle. Perhaps it is in this region that the *refractory* (high-melting temperature) parts of slabs accumulate. If this image of the deep Earth is correct, it suggests that the mantle is not stratified into layers that do not mix, and that upper mantle is carried by subduction into the lower mantle. The presence of whole-mantle versus layered convection is a fiercely debated topic among geophysicists and geochemists (see Section 14.4).

16.2.2 The Trench

Trenches are linear or curvilinear troughs that mark the boundary, at the Earth's surface, between the downgoing slab and the accretionary wedge (Figure 16.3). Trenches exist because the subducted portion of the downgoing slab pulls the slab downward. The top of the slab is depressed

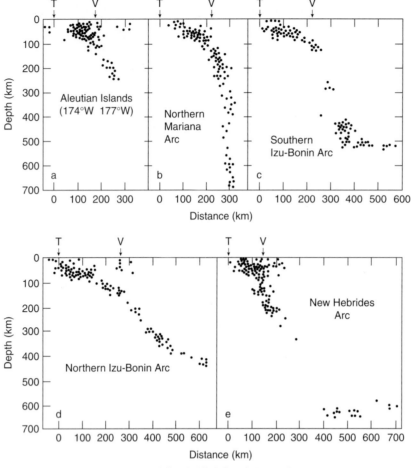

Figure 16.6 Cross sections showing earthquake hypocenters that define the Wadati-Benioff zone. T is the location of the trench and V is the volcanic arc; the distance between T and V is the arc-trench gap. Note that the patterns are quite different for different subduction systems.

by this load, forming a surface (the trench) at a depth greater than it would be if the lithospheric plate were isostatically compensated. The resulting mass deficit from this depression at trenches produces a large *negative gravity anomaly,* which is a signature of subduction zones.

The deepest locations in the oceans occur at trenches. For example, the floor of the Mariana Trench in the western Pacific (Figure 16.7) is nearly 12 km deep, deep enough to swallow Mt. Everest (nearly 9 km high) without a trace. But not all trenches are so deep. In some subduction systems, such as the one off the coast of Oregon and Washington, the trench as a distinct troughlike topographic feature is almost nonexistent. The depth of trenches reflects: (1) the age of the downgoing slab (old oceanic lithosphere is colder and denser than young oceanic lithosphere, and thus sinks to greater depth), and (2) the sediment supply into the trench (if a major river system from a continent spills into a trench, the trench fills with sediment). To see the effect of these parameters, we can compare the very deep trench bordering the Mariana Island arc to the very shallow trench

along the northwest coast of the USA adjacent to the Cascade volcanic arc. The great depth of the Mariana Trench occurs because the trench is not near a continental sediment supply, and because the plate that is being subducted at the Mariana Trench is relatively old (Middle Mesozoic). In contrast, the shallow trench along the Pacific northwest margin of the United States is being filled by sediments carried by the Columbia River, and the downgoing slab beneath the trench is quite young (Late Tertiary).

Even though the thickness of sediments in trenches is variable, all trenches contain some sediment, called *trench fill* (Figure 16.3). Typically, trench fill consists of turbidites that are derived from the volcanic arc and its basement, from the forearc basin, and from older parts of the accretionary wedge. Turbidites spill into the trench mostly down submarine canyons. Once in the trench, the sediment spreads out laterally. Eventually, the flat-lying strata composing the trench fill will be incorporated into the accretionary wedge, where they become deformed (Figure 16.8).

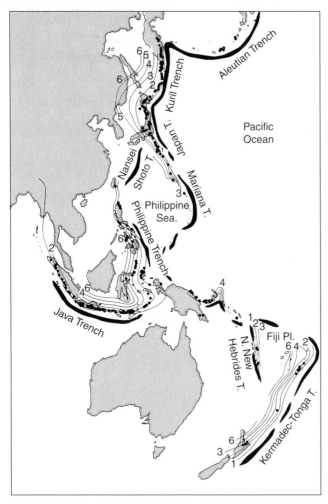

Figure 16.7 Trenches (heavy lines), Wadati-Benioff zones, and volcanicity (black dots) for the western Pacific region. Depth to Wadati-Benioff zones is shown by multiples of 50 km contours.

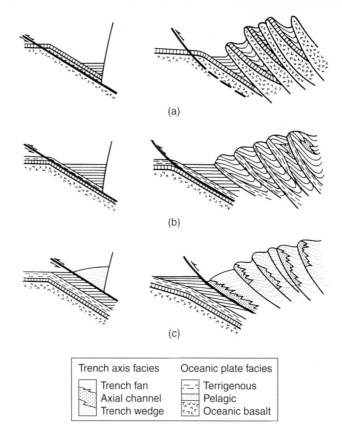

Figure 16.8 Accretion of trench deposits and location of basal detachment. Sediment supply varies from meager (a) to abundant (c) along the margin, resulting in subduction complexes (right) that include (a) slivers of oceanic lithosphere with thin stratal packages, (b) mixtures of trench and oceanic facies, or (c) predominantly trench facies.

16.2.3 The Accretionary Wedge

The accretionary wedge is the most structurally complex part of a subduction system. It is composed both of sediment and rock scraped off of the downgoing slab during subduction, and of trench-fill sediments (Figure 16.3). Because the material composing wedges is so heterogeneous, the internal structure of wedges is highly variable. In places, accretionary wedges contain coherent sequences of strata that have been folded and faulted (Figure 16.9); in some cases a tectonic foliation is present. Elsewhere, wedges contain a chaotic mixture of different rock types in a fine-grained matrix (Figure 16.10), which is commonly referred to as *mélange* (the French word for mixture). Rocks in accretionary wedges include: graywacke and slate (trench-fill turbidites), ribbon chert (pelagic-silicic sediments), micrite (pelagic carbonates), greenstone (altered seafloor basalt), and blueschist (high pressure-low temperature metamorphic rocks). In mélange, bedding cannot be traced very far, and rocks of radically different lithology and grade may be spatially juxtaposed.

How do accretionary wedges develop? An accretionary wedge is a type of fold-thrust belt that forms because oceanic lithosphere underthrusts the overriding plate. The wedge is underlain by a *detachment fault*[2] on the top of the descending slab (Figure 16.9). This fault ramps to progressively shallower levels toward the trench. Internally, the wedge is penetratively deformed. The detachment separates the downgoing slab from the overriding slab and is the actual plate boundary. Note that the seaward propagation of the detachment decouples mass from the downgoing slab. This mass accretes to the overriding slab. Beneath the base of the wedge this process is called *tectonic underplating*[3] (Figure 16.11). While this happens, the trench fill continues to prograde over the subducting ocean floor. At depth, a blind detachment propagates seaward in the pelagic strata beneath the trench fill. As a consequence, some sediment is incorporated in the accretionary prism at the toe of the wedge. This process is called *offscraping*. With continued deformation, faults in the internal part of the wedge progressively steepen and

---

[2]The synonym *décollement* continues the French lesson of this section.
[3]This is different from magmatic underplating (Section 18.5.2).

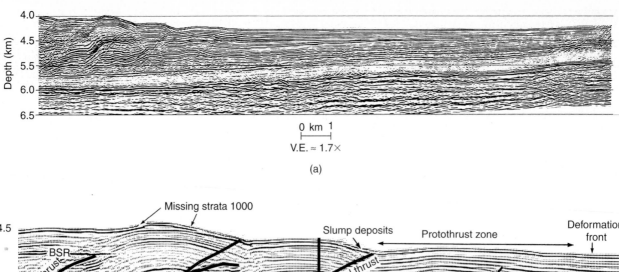

(a)

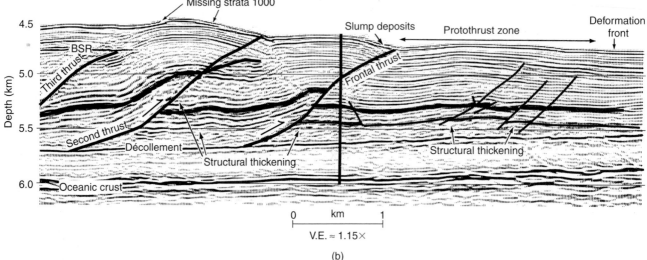

(b)

Figure 16.9 Multichannel seismic-reflection profile of seaward part of the accretionary wedge in the Nankai trough region, Japan (a), and structural interpretation of the toe of the wedge (b).

rocks become more highly strained. Not all thrusting in an accretionary wedge is necessarily toward the trench; arc-verging thrusts occur close to the arc in some wedges (Figure 16.12).

Movement on thrust faults is not all that characterizes an accretionary wedge. Part of the mixing process that leads to the formation of mélange is a consequence of slumping down the slope of the wedge (Figure 16.13). Huge slump blocks, locally tens of kilometers across, carry already deformed rocks down the slope of the wedge. If they reach the toe of the wedge, these blocks may recycle back into the wedge again. Moreover, when wedges reach a substantial thickness, the upper part may collapse by the development of *normal faults* (Figure 16.13).

We'll see in Chapter 17 that the accretion of material in an accretionary wedge resembles the process by which snow piles up in front of a moving snow plow. Briefly summarizing the essentials, a plow blade acts as a rigid backstop that transmits stress into the snow. In a subduction system, the already accreted portions of the wedge act as a backstop with respect to new material being added at the toe of the wedge. As the wedge thickens, gravitational

potential energy also causes stress. In actively developing wedges, a balance is maintained between new material added by offscraping and underplating, internal shortening within the wedge by thrusting, foliation, and fold development, slumping down the slope of the wedge, and extension by normal faulting. The manifestation of this balance is the maintenance of a *critical-taper angle,* which is the angle between the surface of the wedge and the surface of the downgoing slab (Figure 16.14). If internal shortening and underplating cause the internal part of the wedge to thicken, so that the surface slope becomes steeper, then the taper angle of the wedge becomes too large. At this time, the wedge slides seaward and new material is added to the toe of the wedge by offscraping, the net result being that the taper angle decreases below the critical value again. The taper angle of the wedge can also be decreased as a consequence of slumping and extensional faulting in higher parts of the wedge. Alternatively, if the surface slope of the wedge becomes so small, so that the taper angle becomes less than the critical value, then internal strain of the wedge and underplating occur, resulting in an increase of the taper angle (Figure 16.14). The dynamic

Figure 16.10 Mélange from northcentral Newfoundland representing deformed wedge rocks associated with the Lower Paleozoic Iapetus Ocean.

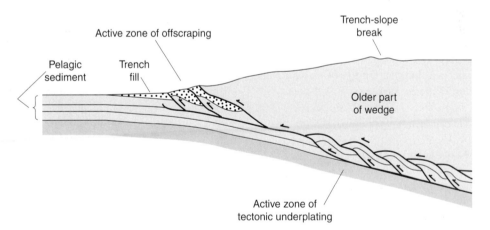

Figure 16.11 Geometry of offscraping and underplating in an accretionary wedge.

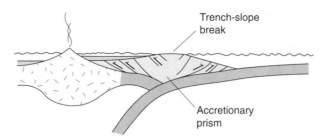

Figure 16.12 Geometry of doubly-vergent thrusting in forearc regions.

interplay between gravitational forces, tectonic compression, and the strength of wedge material is examined further in Chapter 17.

The overall consequence of deformation and mass-wasting processes in accretionary wedges is a long-term internal circulation of rock within the wedge. For example, sediment that accreted to the base of the wedge by subduction may later be brought back to the surface because of internal thrusting in the wedge, and because of removal of the upper levels of the wedge by normal faulting and slumping (Figure 16.15). *Blueschist,* a glaucophane-bearing metamorphic rock, forms at the base of large accretionary wedges (>20 km), where pressures are very high because of the substantial lithostatic overburden, but temperatures are relatively low because the wedge is underlain by 'cold' downgoing slab. The net material flow within an accretionary wedge sometimes brings blueschist up to the surface of the wedge. If, however, the subducting slab is not cool, as is the case where ridges are subducted, blueschist does not form, and the wet sediment in the wedge may actually melt, creating small granitic plutons.

16.2.4   The Forearc Basin and the Volcanic Arc

As we continue our tour up the slope of the wedge and toward the volcanic arc, we find that the top of the wedge is defined by an abrupt decrease in slope. This topographic ridge is the *trench-slope break* (Figures 16.3 and 16.11). At many convergent margins, a broad shallow basin on the arc

side of the wedge fills with nearly flat-lying strata, derived principally by erosion of the arc or its substrate. Such *fore-arc basins* are underlain by older portions of the wedge, which subsided in response to the progressively increasing depth of the underlying downgoing slab, and/or by a slice of trapped ocean lithosphere that was left between the arc axis and the trench when subduction initiated, and/or by older parts of the volcanic arc and its basement that have eroded and subsided.

A *volcanic arc* is the chain of volcanoes that forms at the Earth's surface, about 100–150 km above the surface of the subducted oceanic lithosphere (Figure 16.3). The magmas that rise in an arc form at or above the surface of the downgoing slab as it enters the asthenosphere. At this depth, the slab and the mantle above the slab begins to partially melt, probably in response either to the release of volatiles like H_2O or CO_2 from the downgoing slab as it is heated, or to the flow of hot asthenosphere against the surface of the slab. Once melting has occurred, the resulting magma coalesces into diapirs that rise and fuel the volcanoes.

Island arcs form where one oceanic plate is subducted beneath another, and *continental arcs* grow where an oceanic plate is subducted beneath continental lithosphere. Volcanism at island arcs tends to produce mostly mafic and intermediate igneous rocks, whereas volcanism at continental arcs produces intermediate and felsic igneous rocks, including massive 'granitic' batholiths. The large volumes of felsic magmas in continental arcs form because

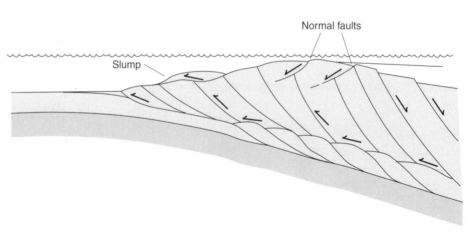

Figure 16.13 Formation of slumps and extensional structures in an accretionary wedge.

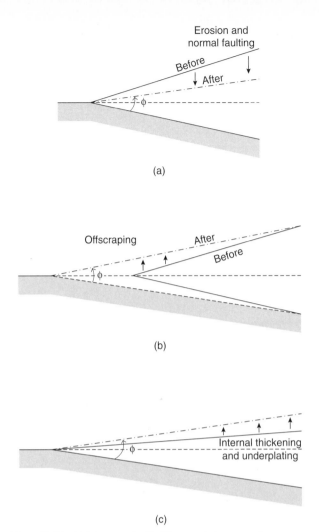

(a)

(b)

(c)

Figure 16.14 Maintenance of critical-taper angle in an accretionary wedge. (a) Wedge slope is too steep and steepness is decreased by erosion and normal faulting. (b) Slope is too steep and steepness is decreased by offscraping. (c) Slope is not steep enough and steepness is increased by underplating and internal thickening. See Chapter 17 for a quantitative approach to the force balance in wedges.

mafic magmas rising from the mantle are so hot that they cause partial melting of the preexisting continental crust, which yields felsic magma.

Whereas volcanic arcs are always about 100–150 km above the top of the downgoing slab, the distance between the arc axis and the trench axis, called the *arc-trench gap,* varies (Figure 16.6). The arc-trench gap is affected by the dip of the downgoing slab. If the downgoing slab is dipping very steeply, then the arc-trench gap is smaller than if the downgoing slab is dipping gently. Slab dips vary from nearly 0°, meaning that the slab is shearing along the base of the overriding slab, to 90°. Why we see such variation remains uncertain. The dip is controlled, in part, by the age of the subducting lithosphere, for older oceanic plate is denser and may sink more rapidly. It may also be controlled by *convergence rate,* which is the horizontal rate at which plates are converging across the trench. A faster convergence would presumably decrease the angle of subduc-

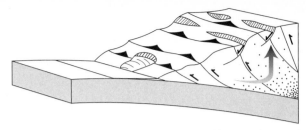

Figure 16.15 Bulk flow path of sediment in an evolving accretionary wedge.

tion for a constant sinking velocity. Perhaps the angle is also affected by the flow direction and velocity of the asthenosphere into which the lithosphere is sinking. Secondly, the width of the accretionary wedge may play a role in the dimension of the arc-trench gap. If subduction has continued for a long time or is occurring where there is a large sediment supply into the trench, the accretionary wedge becomes very large. In effect, the wedge acts as a weight that flexurally depresses the downgoing slab. As a result, the wedge builds seaward, causing a broad, shallowly dipping segment in the downgoing slab, and thus an increase in the width of the arc-trench gap.

16.2.5 The Backarc Region

The backarc region refers to the area on the opposite side of the arc from the forearc basin. The structural character of backarc regions is highly variable. For the purpose of discussion, we'll define five types of backarc regions: (1) extensional, (2) contractional, (3) stable, (4) strike-slip, and (5) trapped ocean lithosphere.

In *extensional backarcs,* we find active rifting and/or seafloor spreading (Figure 16.16a). The result of this movement is to create an active *backarc basin* that, as the name suggests, is a small ocean basin underlain by continental/oceanic lithosphere. The Mariana Arc in the western Pacific (Figure 16.7) is an example of an island arc with an extensional backarc region (Philippine Sea); thus, extensional backarcs are sometimes called 'Mariana-type' backarc basins. Seafloor spreading in backarc basins is not as well organized as that along major mid-ocean ridges. Frequent shifts in the locus of spreading create a muddled pattern of magnetic anomalies. In some active backarcs, distinct and separate episodes of rifting break off pieces of the volcanic arc and separate it from the active arc by a new segment of seafloor. These extinct arc slices are called *remnant arcs.*

In *contractional backarcs* (Figure 16.16b), a backarc basin does not form, but rather contractional deformation generates a fold-thrust belt and/or a belt of basement-cored uplifts (see also Chapter 17). Which of these two styles of contractional deformation occurs may depend on the angle of subduction and on the nature of the crustal section in the overriding plate that is being deformed. If subduction angles are moderate to steep, and the backarc region contains thick strata of a former passive margin, a fold-thrust belt forms with a subhorizontal basal detachment; examples are

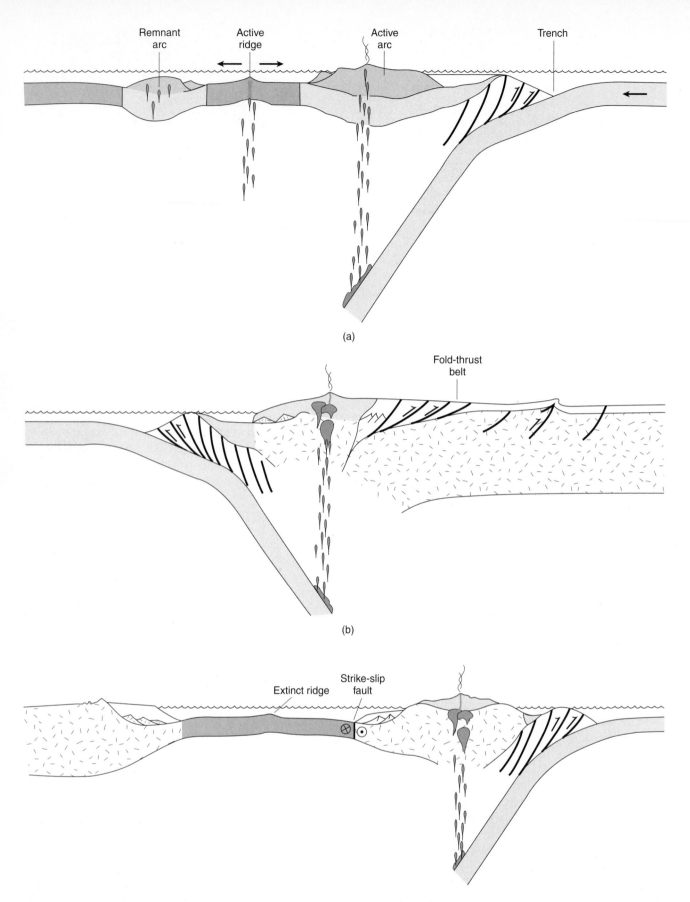

Figure 16.16 Different kinds of backarc regions: (a) extensional (Mariana-type) backarc; (b) contractional (Andean-type) backarc; (c) stable (± strike-slip; Japan-type) backarc. Backarcs underlain by oceanic lithosphere are also called marginal seas.

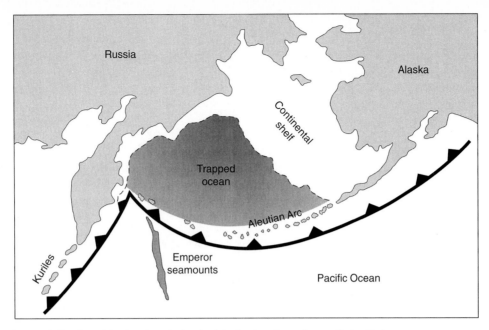

Figure 16.17 Oceanic backarc (marginal sea) originating from trapped ocean (Bering Sea).

the continental side of the Andes Mountains in South America (the 'Andean-type' backarc basin) and by the Mesozoic Sevier Orogen of the western United States. If, alternatively, subduction angles are shallow and stress is transferred into the interior of the plate, then basement-penetrating thrusts develop in the continental interior. These faults uplift basement blocks that develop monoclinal folds, as exemplified by the Rocky Mountain region of the North American Cordillera, where numerous basement-cored uplifts formed during the Laramide Orogeny (Late Cretaceous to Early Tertiary), and by an actively deforming region on the continental margin of the Andes today.

In *stable backarcs,* no motion is currently occurring (Figure 16.16c), whereas in *strike-slip backarcs,* significant strike-slip faulting is occurring. Of course, strike-slip deformation may also be concurrent with overall contraction (transpression) or extension (transtension) in backarcs. Finally, not all backarc basins are sites of deformation and seafloor spreading. Some are merely composed of oceanic lithosphere that was trapped when a convergent margin developed within an oceanic plate. For example, the Bering Sea is largely underlain by Mesozoic-age trapped ocean floor that was isolated from the rest of the Pacific Plate when the Aleutian volcanic arc formed (Figure 16.17).

Why is there such a range of behavior for backarc regions? The answer comes from examining the relative motions of the backarc region and the volcanic arc. As subduction progresses, the location of the bend in the downgoing slab *rolls back,* in the direction seaward of the arc (Figure 16.18). It is impossible for an open space to form, so the arc moves with the rollback. If the overriding plate is stationary or is moving away from the arc, then rifting and

a backarc basin develop. If the overriding plate is moving in the direction of rollback at the same rate as rollback, then the backarc region is stable. Analogously, the other types of backarc basins can be defined in terms of relative motion with respect to the arc. If the overriding plate is moving in the same direction but at a rate faster than rollback, the result is a contractional backarc. If the overriding plate is moving obliquely to the plate boundary, then there will be a component of strike-slip displacement.

It is important to realize that plate boundaries evolve with time. In some backarc regions, therefore, the style of deformation may change. For example, the Japan Sea initiated as an extensional backarc. Japan is underlain by basement that was originally part of eastern Asia, but was separated when the Japan Sea backarc basin developed. Currently, however, the Japan Sea is the locus of shortening and strike-slip deformation, which may ultimately result in closure of the basin.

16.2.6 Curvature of Island Arcs

If you look in map view at many island arcs (e.g., Figure 16.7), you will notice that they are curved, which is why they are called "arcs"! No one is totally sure why the curvature exists, but there are a few interesting ideas. One is that the curves reflect the natural shape of an intersection line where the surface of a sphere is indented and pushed inward (Figure 16.19a). To visualize this geometry, take a ping-pong ball and push one side in with your thumb. Another idea is that the curves reflect indentation of an originally straight arc by subduction of chains of seamounts. This concept is supported by the observation that in the western Pacific, some of the major cusps in subduction zones coincide with sites of seamount subduction (Figure 16.19b). For example, the Emperor Seamount Chain is subducted at the

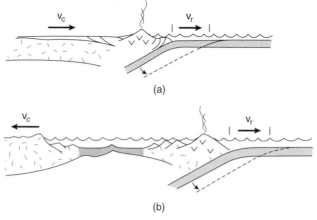

Figure 16.18 Coupled (a) versus uncoupled (b) subduction offers an explanation for the difference between Andean-type and Mariana-type convergent margins, respectively; v_r is rollback velocity and v_c is convergence velocity.

cusp between the Aleutian Trench and the Kurile Trench (Figure 16.17). Perhaps subduction of seamounts inhibits propagation of the accretionary wedge, and/or the buoyancy of seamounts decreases the rate of rollback. A third idea is that strike-slip faulting and drag at either end of an arc causes the curvature (Figure 16.19c).

16.2.7 Coupled versus Uncoupled Subduction Systems

Putting the above descriptions together, convergent boundaries consist of basically two endmember types of subduction systems, which yield contrasting suites of structures. The first type is a *coupled subduction system.* In such a system, the balance between the downgoing plate velocity, the rollback velocity, and the overriding plate velocity yields a condition that the entire subduction system is in compression and the overriding plate is pushing against the downgoing plate (Figure 16.18a). As a consequence, there are large shear stresses across the contact, causing efficient offscraping and tectonic underplating, and therefore buildup of an accretionary wedge, as well as generating large earthquakes. The compression also causes development of a contractional backarc region, and slows the rise of mantle-derived magmas. Thus, the rising magmas have more time

to fractionate, and heat brought into the lithosphere by these magmas is able to cause partial melting of the lithosphere. As a result, the ultimate intrusive rocks that are produced tend to be intermediate to felsic.

The second endmember type is an *uncoupled subduction system,* where the balance of velocities yields a condition in which the subduction system is effectively under tension (Figure 16.18b). As a consequence, shear stresses across the plate boundary are relatively low, thrust earthquakes tend to be smaller, and relatively little offscraping and underplating occurs. In uncoupled systems we find extensional backarcs, and mantle-derived magmas are able to rise directly to the surface before significant fractionation or crustal contamination occurs.

16.3 THE BASIC STAGES OF COLLISIONAL TECTONICS

When two buoyant pieces of lithosphere converge, a *collision* ensues, because neither piece can subduct beneath the other. As we pointed out earlier in this chapter, examples of buoyant lithosphere include continents, island arcs, oceanic plateaus, microcontinents, and some ridges. Collision is always preceded by convergent tectonics, because it is

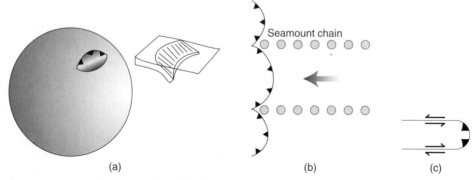

Figure 16.19 Reasons for arc curvature: (a) Ping-Pong ball model for island-arc curvature; (b) subducting seamount origin for curvature; (c) strike-slip and drag origin for island-arc curvature.

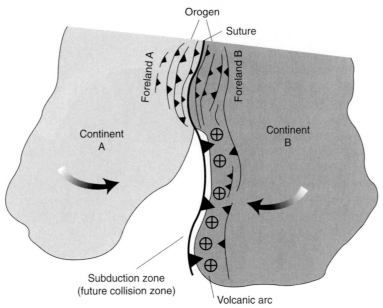

Figure 16.20 Time-transgressive collision between two major continents as a result of zipperlike closure of the intervening ocean basin.

Table 16.2   Terminology of Collision

Accreted terrane	Piece of exotic crust that has been attached to the margin of a larger continent as a consequence of collision (note spelling of "terrane"; as opposed to the geographic term for a tract of land, "terrain"). See Exotic terrane.
Basin inversion	A region that has undergone extension during basin formation subsequently telescopes back together by reverse slip reactivation of faults.
Delamination	The removal of layers; in the context of collisional tectonics, the removal of part of the lower lithosphere.
Exotic terrane	Independent pieces of buoyant crust that are swept into a convergent margin during subduction and dock against the continent. Synonym for accreted terrane.
Lateral escape	Tectonic style in which blocks of the overriding plate are squeezed out sideways (at high angles to convergence vector) in response to plate collision.
Orogenic collapse	Extensional processes in orogens; in some cases extension is coeval with thrusting (*synorogenic collapse*); in other cases it occurs after thrusting ceased (*postorogenic collapse*).
Suspect terrane	A particular block whose origin is unclear (i.e., exotic or not?).
Suture	The irregularly shaped surface that marks the boundary between once separate tectonic plates; commonly this boundary is marked by the occurrence of ophiolite slivers.
Tectonic collage	A term to describe the geology of regions that are characterized by the presence of several exotic terranes.

subduction of intervening ocean floor that brings the two pieces together. As long as the buoyant block remains connected to the downgoing slab along strike, collision continues (e.g., India is still pushing into Asia). But when the forces driving collision cease, the relative motion ceases, and when this happens, the once separate plates have become one. The surface that marks the boundary between these once separate plates is a *suture,* which is often lined with remnants of ancient oceanic lithosphere (*ophiolites*). These and other terms related to collisional tectonics are summarized in Table 16.2.

The nature of a particular collisional orogeny depends on numerous variables, such as the relative motion between the colliding pieces, the rate of collision, whether the colliding pieces are bordered to the sides by continental lithosphere or by oceanic lithosphere, and on the physical characteristics (e.g., age, temperature, size, shape, and composition) of the colliding pieces. For example, if an old ('cold') craton collides with a young ('warm') orogenic belt, the craton will act as a rigid indenter that pushes into the relatively soft, preexisting orogenic belt, so that a broad band of the orogenic belt is deformed while the craton remains largely undeformed. If pieces collide obliquely, the timing of collision is diachronous along strike (Figure 16.20), and the pieces merge together like the two sides of a zipper. Because of all the variables, no two collisional

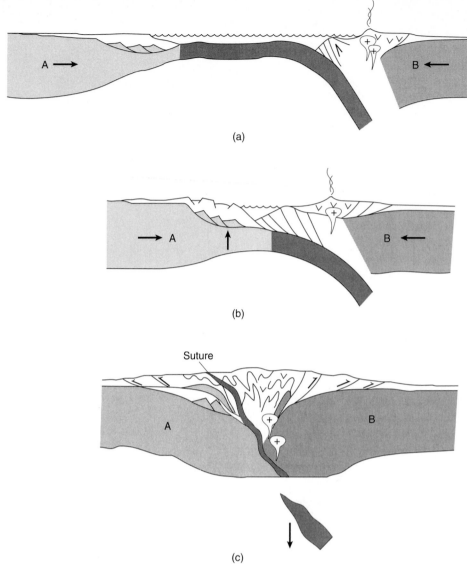

Figure 16.21 Stages in continent-continent collision: (a) pre-collision; (b) abortive subduction; (c) formation of a suture, and breakaway of the downgoing slab.

orogenies are exactly the same. However, we can introduce the basic concepts of what is involved by outlining the various stages in an idealized collision between two continents (A and B; Figure 16.20). We refer to the portion of the orogen that is on the craton side of the collision as the *foreland,* and the internal part of the orogen as the *hinterland.*

Stage 1, Pre-Collision

Before collision, Continent A is connected to oceanic lithosphere that is being subducted beneath Continent B (Figure 16.21a). The margin of Continent A is a passive margin, along which a sedimentary basin has developed. The margin of Continent B is an Andean-type convergent margin, where the velocity of the overriding plate exceeds the rollback velocity.

Stage 2, Initial Interaction

The first indication that a collision is imminent occurs when the edge of Continent A begins to rise and stretch as it bends

around the outer swell just prior to being pulled into the subduction system by the downgoing slab (Figure 16.21b). Uplift may be sufficient to raise the shelf of the passive-margin basin above sea level. Stretching of the continental edge is accommodated by the formation of normal faults that strike parallel to the margin.

Stage 3, Abortive Subduction Stage/Collision

With continued convergence, the passive margin is carried into the trench (Figure 16.21b). Turbidites derived from the arc bury the margin.[4] The contact between the top surface of the passive-margin sequence and the turbidites is, therefore, a major *unconformity.* Collision causes a fold-

---

[4]In older literature, this sequence of turbidites is called *flysch,* which was defined as 'synorogenic' strata. Deposition was thought to signal the onset of orogenic activity (see also Chapter 19.2). However, turbidites form in other settings as well, so *flysch* as a tectonic term is confusing and we discourage its use.

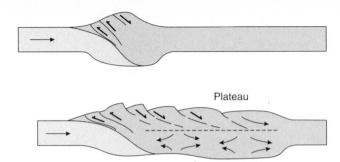

Figure 16.22 Sketches showing orogenic collapse and progressive growth of a Tibet-like plateau.

thrust belt to initiate in the well-stratified sequence of the passive margin, which, with time, grows in the direction of the continental interior (i.e., toward the foreland) of Continent A. The stack of thrust slices acts as a load that depresses the surface of the craton and results in the formation of a foreland sedimentary basin on the craton of Continent A (Figure 16.21c).

Eventually, shortening during the collision causes inversion of the normal faults that bound basement slices at the base of the passive margin, and slices closest to the hinterland may thrust over strata of the former passive margin (Figure 16.21c). This process is called *basin inversion,* because a region that has undergone extension during basin formation now telescopes back together by reverse slip reactivation of these preexisting faults. In the hinterland, a slice of oceanic lithosphere (ophiolite) derived from the continent/ocean boundary area thrusts over rock of Continent A. This slice, which is preserved in the orogen as a band of highly sheared mafic and ultramafic rock, defines the *suture.* In the internal part of the orogen, the crust thickens considerably, with ductile folding and shearing at depth. This deformation involves the roots of the now extinct volcanic arc. Large isoclinal folds may develop in this region in association with crustal shear zones. Regional metamorphism affects the rocks that were once near the surface of the Earth but are now at depth, where they are heated and sheared. As a result, broad regions of foliated metamorphic rock (gneiss and schist) form in the internal part of the orogen. On Continent B, deformation styles are the same, but the vergence of structures is opposite to those that form on the edge of Continent A. Also, the foreland fold-thrust belt of Continent B may be much older than that of Continent A, because it initiated during the pre-collision stage.

The depth in the crust to which faults penetrate during collision is not clear. It is possible that much of the deformation in a collisional orogen is confined to crust above a *detachment,* which lies at midcrustal levels. Pressure, temperature, and differential stress conditions below this detachment may allow the crust to behave like a viscous fluid. However, a translithospheric boundary between the two plates remains at least as long as subduction continues.

16.4 OTHER CONSEQUENCES OF COLLISIONAL TECTONICS

The stages of collision that we just described are only a simple representation of the collision process. We kept all motion within the two-dimensional plane of the cross section, and deformation primarily involved shortening of the lithosphere. In this section, we relax these conditions and describe additional tectonic phenomena that may occur during collision in specific situations.

16.4.1 Synorogenic Collapse

If collision produces a large degree of crustal shortening, then substantial crustal thickening occurs. Igneous activity and deep burial due to thickening heat up the crust, so locally it becomes relatively weak. As a consequence, the differential stress developed in the orogen due to lithostatic overburden exceeds the yield strength of the rock at depth, and it begins to flow and extend laterally under its own weight. As we pointed out in Chapter 15, you can picture this process by imagining a block of cheese that is heated in the sun. Eventually, the cheese gets so soft that it spreads out, and the thickness of the block diminishes. Lateral spreading at depth in an orogen causes stretching of the upper crust, and in this region, rocks respond by developing normal faults. Therefore, even while shortening and thrusting continue in a collisional orogen, extensional structures may also develop (i.e., normal-sense faults and shear zones). This process of coeval thrusting and extension is called *synorogenic collapse.* It keeps mountain ranges from exceeding elevations of about 10 km, and contributes to the development of broad, high plateaus like the Tibetan Plateau of Asia (Figure 16.22; see also Chapter 19.3).

16.4.2 Synorogenic Deformation of Continental Interiors

Collisional orogeny along a continental margin may lead to the development of major strike-slip faults in the orogenic foreland that propagate into the interior of the overriding plate. A series of experiments involving a rigid indenter and plasticine nicely illustrate the resulting fault pattern and regional kinematics of indentation (Figure 16.23a). In the case of one unconstrained margin in the experimental setup (as opposed to two constrained margins), a sequence of sideways moving blocks form. The geometric similarity between experimentally produced faults and real faults in southern Asia is quite striking (Figure 16.23b). This pattern of continental strike-slip faults and regional tectonics is called *lateral escape.* Effectively, during lateral escape, blocks of the overriding plate are squeezed out of the orogen, much like a watermelon seed that you squeeze between your fingers. Another example of lateral escape is currently occurring in the eastern Mediterranean region, where Turkey is being squeezed westward along the Northern and Southern Anatolian faults as the Saudi Arabian Plate moves north. Note that where lateral escape occurs, the strain resulting from collision cannot be depicted in a cross section, because of movement into or out of the plane of the section.

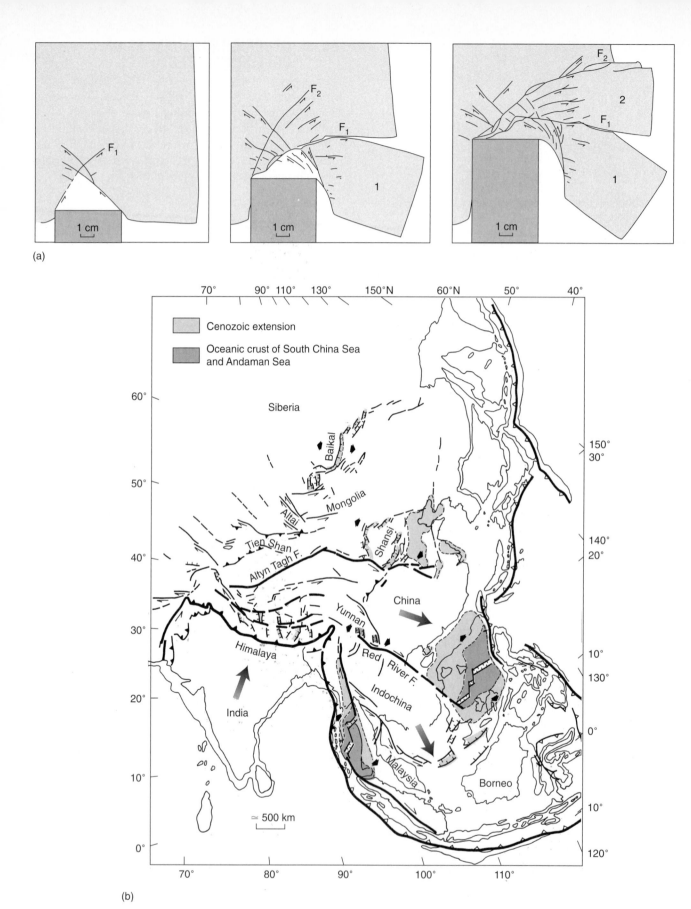

Figure 16.23 Lateral escape tectonics as illustrated by progressive indentation experiments using a rigid indenter and plasticine (a), and its application to the southeastern Asia-India collision (b).

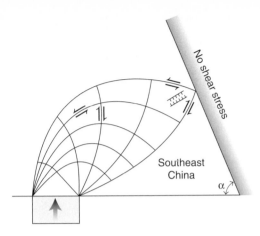

Figure 16.24 Theoretical model of a slip-line field caused by indentation of a rigid block into an elastic half-space.

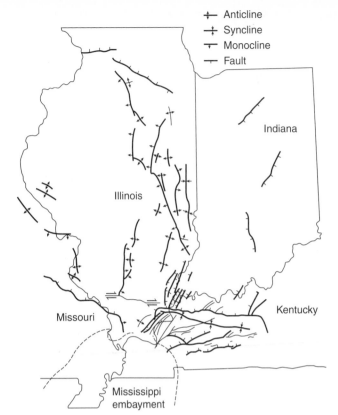

Figure 16.25 Structures in the continental interior part of the eastern USA that were reactivated by Paleozoic Orogeny.

The pattern of continental-scale strike-slip faults illustrated in Figure 16.23 can also be examined from the theory of elasticity. If a rigid indenter pushes into an elastic plate, trajectories of maximum shear stress define an array of curved lines that is called a *slip-line field* (Figure 16.24). Substantial debate continues about the angle and amount of slip on faults in southern Asia, and thus the applicability of slip-line field theory to these faults remains uncertain. It is possible that the strike-slip faults are merely reactivated preexisting basement faults.

In the previous paragraphs we alluded to the possibility that collision causes differential stresses in the continental interior that are sufficient to reactivate preexisting faults. For example, in the Midcontinent region of the United States, numerous basement-penetrating faults reactivated during Late Paleozoic orogenic activity in the Appalachians (Figure 16.25). Movement on these faults displaced stratigraphic markers at depth and led to the development of folds at shallower crustal levels. In addition, intragranular strain on the order of a few percent developed at this time in Midcontinent limestones (see Figure 4.29). The change in differential stress in continental interiors may also trigger *epeirogenic movements,* meaning the gentle vertical displacement of broad regions in continental interiors. These movements include the uplift of regional domes and arches as recorded by development of unconformities, and subsidence of intracratonic basins as recorded by the sudden increases in the rate of basin subsidence. We return to these types of continental interior movements in Section 19.10.

16.4.3 Accretion Tectonics

If you look at the present-day Pacific Basin on a topographic/bathymetric map of the Earth, you will notice many bits and pieces of crust that are neither pieces of large continents nor simply oceanic crust (Figure 16.4). Some of these fragments (e.g., Japan) were once a part of a larger continent, but were rifted off; some (e.g., Borneo in the Indonesian archipelago) are microcontinents that have been independent for a long time; some (e.g., the Mariana Islands) are volcanic island arcs formed on oceanic litho-

sphere; some (e.g., the Hawaiian Islands) are the tracks of hot spots; and some are oceanic plateaus (enigmatic features that may be underlain by a stack of flood basalts extruded at a particularly productive hot spot). In any case, all of these pieces are relatively buoyant and cannot be subducted. If subduction along the eastern margin of Asia continues, these individual pieces eventually will collide with Asia, creating collisional orogenies that will suture them to Asia. A new convergent margin will form on the outboard (oceanic) side of each sutured piece, so that it becomes a fault-bounded block within a larger accretionary complex. Each block has a geologic record that is independent of neighboring pieces. Furthermore, paleomagnetic and paleontologic data may indicate that neighboring blocks originated at paleolatitudes that were quite different from each other, and from their current location along the continent to which they are now attached. Looking again at Figure 16.4, but now imagining that the Pacific Ocean has closed, an interpretation like that above is quite reasonable. Indeed, the geologic record shows such *tectonic collages* for almost all of the world's collisional mountain ranges.

The North American Cordillera is a classic example of an orogen with a history of *terrane accretion.* Small independent pieces of buoyant crust that are swept into a convergent margin during subduction and dock against the continent are called *exotic terranes*[5] (emphasizing that

---

[5]Note the spelling of *terrane* (a 3-D geologic block) as opposed to *terrain* (geographic term for an area of land).

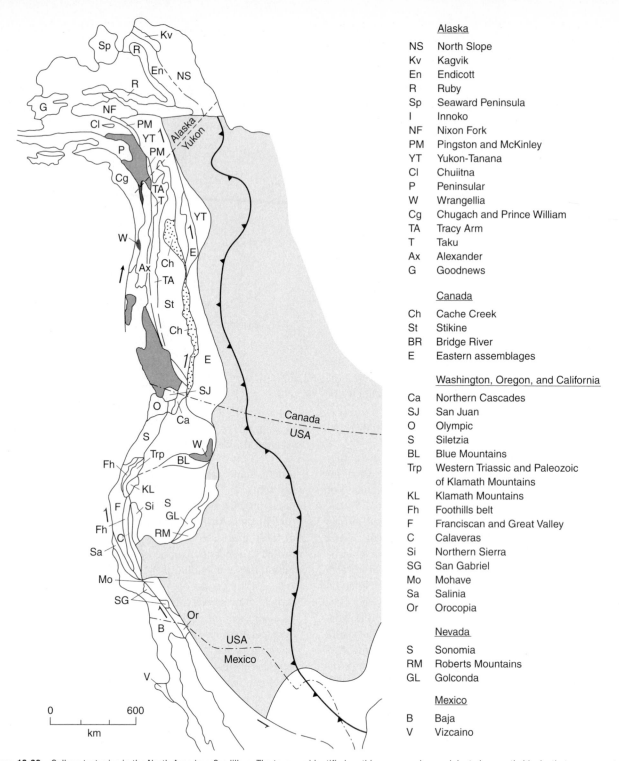

Figure 16.26 Collage tectonics in the North American Cordillera. The terranes identified on this map may have originated as exotic blocks that were accreted and dissected along the margin of North America or as Phanerozoic igneous provinces. Lines are faults that outline distinct litho-tectonic blocks (terranes); barbed line is the eastern extent of Mesozoic-Cenozoic deformation in western North America. Light shading is the Precambrian core of North America; darkly shaded blocks are part of the accreted Wrangellia terrane that originated at an equatorial position.

they came from somewhere else) or *accreted terranes* (emphasizing that they have been attached to the margin of a larger continent). If it's not clear whether a particular block of crust is exotic or not, it is called a *suspect terrane* (meaning that its origin is unclear). In the North American Cordillera, a vast tract including most of

California, Oregon, Washington, and Alaska in the USA, and Canada's British Columbia and Northwest Territories, either originated outboard of North America and was accreted to the continental margin or formed by igneous intrusion (Figure 16.26; see also Section 19.4).

During oblique convergence, the strike-slip component will move blocks along the strike of the orogen. Strike-slip faulting may also dice initially coherent terranes into several pieces that are subsequently dispersed along the length of the orogen. In western North America, for example, an exotic terrane known as Wrangellia has been found in bits and pieces from Washington (maybe even Idaho) to Alaska (Figure 16.26).

When two major continents collide along a margin that previously was the locus of terrane accretion, you can imagine that the resulting collisional orogen will be very complex. It will contain several sutures separating different blocks, and each block will have its own unique geologic history. To our knowledge, all major collisional orogens involve accretion of exotic terranes prior to collision of the larger continents and final closure of the intervening ocean, so such complexity is the norm rather than the exception. Therefore, you should not assume that the history in one particular region is representative for the orogen as a whole, nor should you be surprised to find radically different geologic histories when examining a large orogen along its trend.

16.4.4 Lithosphere Delamination

So far, we have mainly focused our attention on the crust, but it is important to keep in mind that the lithospheric mantle also thickens during collisional orogeny. What happens to this part of the lithosphere is yet another topic of ongoing debate. Some geologists suggest that the lithospheric mantle separates from the crust during collision and sinks into the deeper mantle. Others suggest that the lithospheric mantle is itself shortened and thickened during collision, which leads to interesting consequences. If lithospheric mantle is thickened by the same proportion as the crust, the lower part of the lithosphere will reach a depth of 200 or more km. Calculations show that this relatively cool mantle portion of the lithosphere is denser than the surrounding asthenosphere, and therefore is negatively buoyant. Under such conditions, the lower part of the lithosphere may peel off and sink into the mantle after the collision (Figure 16.27). As hot asthenosphere flows in to replace delaminated lithosphere, heating may cause partial melting of the lithosphere and magma generation. This magma rises, is contaminated, and intrudes upper crustal layers as postorogenic plutons. Delamination also leads to postorogenic isostatic uplift of the orogen, because the heavy lower lithosphere is being replaced with hotter, less dense asthenosphere.

16.5 CLOSING REMARKS

In the descriptions provided in Chapters 15 and 16, we've illustrated the start and finish of one loop in the *Wilson cycle*, the history of opening and closing of ocean basins. Geologic mapping around the world emphasizes that pat-

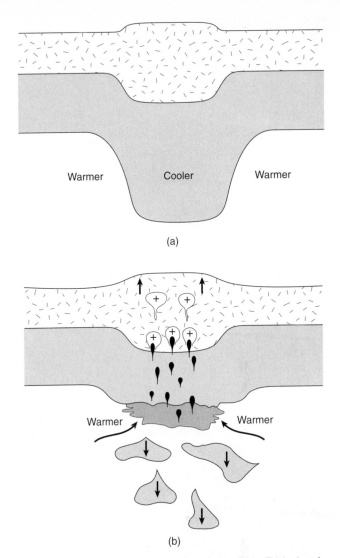

Figure 16.27 The concept of lithosphere delamination. (a) Thickening of lithosphere causes deep keel of cool lithosphere; (b) the keel drops off and is replaced by warm asthenosphere, causing partial melting and formation of anorogenic (postcollisional) plutons.

terns of structures created during continental breakup influence patterns of structures developed during collision, and vice versa. Rifts are superimposed on ancient collisional orogens, and in turn, rifts are the likely sites of later collision (so rift basins ultimately get inverted). Because rifts generally control the location of collisional orogens, and vice versa, rocks in a single orogen tend to record the effects of multiple phases of contraction and extension. For example, the eastern United States records a history that involves two principal phases of rifting (Late Precambrian and Middle Mesozoic) and two principal phases of continental collision (Late Precambrian and Paleozoic, each involving several subphases). Multiple reactivations have kept Phanerozoic collisional orogens weak relative to continental interiors (cratons). In a way, they are like weak scars that never heal, and they protect cratons from major deformation.

Figure 16.28 Glarner Thrust of the Swiss Alps, emplacing Permian Verrucano (coarse clastics) over Tertiary Flysch (turbidites). Highly deformed Jurassic limestone marks the contact.

Our discussion of contractional tectonics is not yet complete. In the next chapter (Chapter 17), we narrow our focus and look into the world of fold-thrust belts (Figure 16.28). In Chapter 19 we will look at the geology of several collisional orogens in detail, where many of the concepts introduced in Chapters 15 through 18 will be placed in a regional context.

ADDITIONAL READING

Ben-Avraham, Z., Nur, A., Jones, D., and Cox, A., 1981, Continental accretion: from oceanic plateaus to allochthonous terranes, *Science,* v. 213, p. 47–54.

Coney, P. J., 1989, Structural aspects of suspect terranes and accretionary tectonics in western North America: *Journal of Structural Geology,* v. 11, p. 107–125.

Dewey, J. F., 1980, Episodicity, sequence, and style at convergent plate boundaries, *in* Strangway, D. W., ed., The continental crust and its mineral deposits, *Geological Association of Canada Special Paper,* v. 20, p. 553–574.

Dickinson, W. R., and Seely, D. R., 1979, Structure and stratigraphy of forearc regions: *American Association of Petroleum Geologists Bulletin,* v. 63, p. 2–31.

Malavieille, J., 1993, Late orogenic extension in mountain belts: insights from the Basin and Range and the Late Palezoic Variscan belt, *Tectonics,* v. 12, p. 1115–1130.

Moore, J. C., 1989, Tectonics and hydrogeology of accretionary prisms: Role of the decollement zone, *Journal of Structural Geology,* v. 11, p. 95–106.

Oldow, J. S., Bally, A. W., Avé Lallemant, H. G., 1990, Transpression, orogenic float, and lithospheric balance, *Geology,* v. 18, p. 991–994.

Tapponnier, P., Peltzer, G., and Armijo, R., 1986, On the mechanics of the collision between India and Asia, *in* Coward, M. P., and Ries, A. C., eds., Collision tectonics, *Geological Society Special Publication,* v. 19, p. 115–157.

Chapter 17
Fold-Thrust Belts—An Essay[1]

17.1 INTRODUCTION

Picture yourself hiking among glaciated spires in the Canadian Rockies or on an Alpine meadow in Switzerland (Figure 17.1). The strata forming the massive mountains around you were not deposited at their present location; in fact, most were deposited in a marine environment and then they were buried deeply. In the Cenozoic, these rocks moved tens of kilometers horizontally and several kilometers vertically to arrive at their present position. This movement involved slip on a series of *low-angle reverse* or *thrust faults,* which in turn led to the development of spectacular panels of tilted rocks and complex folds. Geologic areas like the Canadian Rockies, in which regional upper-crustal shortening resulted in formation of a distinctive suite of thrust faults and folds, are called *fold-thrust belts.* Fold-thrust belts occur worldwide. Principal locations include the foreland side of collisional orogenic belts, restraining bends of large strike-slip faults, the seaward toes of passive-margin sedimentary basins, and the backarc region of an Andean-type convergent margin.

Generations of geologists have struggled to decipher the character and origin of fold-thrust belts. Thrust faults were first recognized in 1826 near Dresden, Germany, where Precambrian granites overlie Cretaceous strata across a fault contact. Geologists in the 1840s mapped and described allochthonous strata in the Swiss Alps (Figure 17.2) and proposed the then radical idea that sheets of rocks could travel long distances along subhorizontal faults that we now call *detachments.* This proposal was so radical, in fact, that it was not until careful mapping and documentation of the Moine Thrust in northwestern Scotland forty years later that geologists came to accept the concept of detachments and regional-scale thrust faults. Documentation of thrust faults solved a number of previously intractable stratigraphic problems. Imagine how confused early geologists must have been when they climbed uphill, crossed a covered thrust, and abruptly entered a sequence of strata that was older than the sequence at the base of the mountain!

In the nineteenth century, geologists also began to investigate the relationship of faults to nearby folds. The Rogers brothers first proposed such a relationship in 1843 based on field work in the Appalachians. Subsequent authors followed up on these pioneering observations and recognized that fold formation may be an inherent consequence of faulting. It was proposed that in areas where regional-scale folds crop out on the surface, genetically related thrusts lie beneath them in the subsurface. Later, in the twentieth century, geologists refined these fundamental observations and developed geometric and kinematic rules to help constrain the fold geometries portrayed in cross sections of fold-thrust belts, and grappled with the fundamental issue of *how* and *why* such belts develop.

---

[1]M. Scott Wilkerson, lead author.

Figure 17.1 The Canadian Rocky Mountain Front Ranges, near Seebe, Alberta (Canada). The cliffs are Paleozoic strata of the McConnell Thrust sheet placed over forested Cretaceous foreland basin deposits. View to the north.

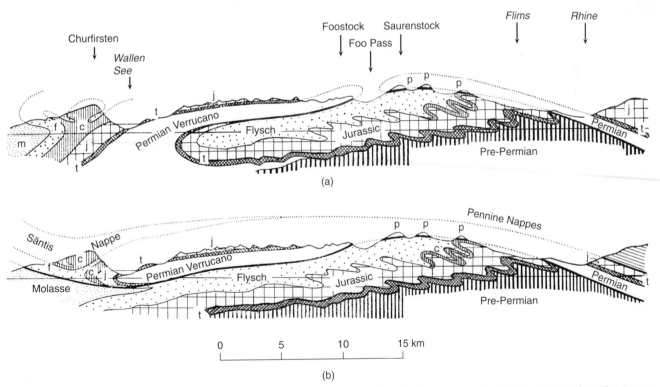

Figure 17.2 Early sketch of the "Glarner double fold" in the Swiss Alps as originally envisioned by Heim (a), and subsequently reinterpreted as a thrust nappe (b). m = Tertiary molasse; n, f = Tertiary and Cretaceous flysch; c = Cretaceous (mainly limestone); j = Jurassic (mainly limestone); t = Triassic; p = Permian.

Table 17.1   Fold-Thrust Belt Terminology

Allochthonous	"Out-of-place" rocks that have moved a large distance from their point of origin.
Backstop	A colliding mass in the hinterland.
Backthrust	A thrust on which the transport direction is opposite to the regional transport direction.
Blind thrust	A thrust that terminates in the subsurface.
Branch line	The line of intersection between two fault surfaces.
Break-forward sequence	A sequence of thrusts in which progressively younger thrusts form toward the foreland.
Break-thrust fold	A fold that initiates prior to thrusting, but later breaks so that a thrust cuts through its forelimb.
Cutoff	The line of intersection between a fault and a bedding plane.
Detachment	A subhorizontal fault.
Detachment fold	A fold that forms in response to movement above a subhorizontal fault, much like a rug that wrinkles above a slick floor.
Duplex	A type of thrust system where a series of thrusts branch from a lower to an upper detachment.
Fault-bend fold	A fold that forms in response to movement of strata over bends in a fault surface.
Fault-propagation fold	A fold that forms immediately in advance of a propagating fault tip (also called a *tip fold*).
Floor thrust	The lower detachment of a duplex.
Fold nappe	A thrust sheet that contains a recumbent fold.
Fold-thrust belt (FTB)	The geologic terrane in which upper-crustal shortening is accommodated by development of a system of thrust faults and related folds, generally above a regional detachment.
Footwall block	The body of rock beneath the fault.
Footwall cutoff	The intersection between footwall strata and the fault (truncations between BC in Figure 17.3).
Footwall flat	The portion of the footwall where bedding surfaces parallel the fault.
Footwall ramp	The portion of footwall where bedding surfaces truncate against the fault.
Foreland	The region closer to the undeformed continental interior.
Foreland basin	A sedimentary basin formed on the continent side of an FTB that forms because the weight of the FTB depresses the lithosphere.
Forethrust	A thrust on which the transport direction is the same as the regional transport direction for the FTB.
Frontal ramp	A ramp that strikes perpendicular to transport direction.
Hanging-wall block	The rock mass that has been transported above a fault surface.

Continued

In this chapter, we first describe the geometry of thrust faults and thrust systems, and the relationship of faults to folds. We then outline the key ideas that have been proposed to explain the mechanics by which such fold-thrust belts develop. We conclude by discussing techniques that are used to construct geometrically valid cross sections. The tectonic setting of fold-thrust belts was discussed throughout the preceding chapter, and in particular in Section 16.2.3. You'll notice that we present many new terms in this chapter (see Table 17.1). Perhaps fold-thrust belt geologists are the most "jargonistic" people in the business. Yet hopefully, you will find that the vocabulary ends up simplifying and clarifying the discussion.

17.2   GEOMETRY OF THRUSTS AND THRUST SYSTEMS

To commence our discussion of fold-thrust belt geometry, we first provide a cross section of a thrust fault as it cuts through a stratified sequence (Figure 17.3). Generally, thrusts cut upsection in the direction of displacement on the fault, and place older strata in the hanging wall over younger strata in the footwall. As depicted in Figure 17.3, the Pine Mountain Thrust of the southern Appalachians places older Paleozoic strata over Siluro-Devonian beds as the fault cuts upsection from the base of the Cambrian to flatten out into Siluro-Devonian strata. Displacement along thrust faults results in duplication of strata and raising the

Hanging-wall cutoff	The intersection between hanging-wall strata and the fault (truncations between DE in Figure 17.3).
Hanging-wall flat	The portion of the hanging wall where bedding surfaces parallel the fault.
Hanging-wall ramp	The portion of the hanging wall where bedding surfaces truncate against the fault.
Hinterland	The portion of an FTB closer to the high-grade metamorphic core of an orogen.
Horse	A body of rock in a duplex that is surrounded by faults.
Imbricate fan	A type of thrust system where a series of thrusts branch from a lower detachment without merging into an upper detachment horizon.
Imbricate thrust	A thrust that branches from a lower detachment without flattening into an upper detachment.
Klippe	An erosional outlier of a thrust sheet that is completely surrounded by exposure of the footwall.
Lateral ramp	A ramp that strikes parallel to transport direction.
Oblique ramp	A ramp that strikes oblique to transport direction.
Out-of-sequence thrust	A thrust that forms (or reactivates an older thrust) to the hinterland of preexisting thrusts.
Out-of-plane strain	The strain due to movement in or out of the plane of cross section.
Ramp-flat geometry	The geometry of a thrust that cuts upsection via a series of flats and ramps (giving the thrust a stairstep trajectory).
Regional transport direction	The dominant transport direction for several thrust sheets in a portion of an FTB with respect to geographic north.
Regional vergence direction	*See* Regional transport direction.
Roof thrust	The upper detachment of a duplex.
Tear fault	A nearly vertically-dipping fault that exhibits strike-slip motion.
Tectonic inversion	The reactivation of preexisting faults by a reversal of slip sense on the faults.
Thin-skinned thrust	A thrust that ultimately flattens into a subhorizontal detachment that separates a package of rock that deforms independently from the basement below.
Thrust fault (thrust)	A shallowly to moderately dipping ($< 30°$) contractional fault with primarily dip-slip reverse movement.
Thrust sheet	The hanging-wall block that is transported above the thrust.
Thrust system	An array of related thrusts that connect at depth to the same basal detachment.
Tip line	The line at which displacement on the thrust becomes zero.
Transport direction	The direction in which a thrust sheet moves with respect to geographic north.
Window (fenster)	An erosional hole through a thrust sheet that exposes the footwall.

layers above their pre-faulting elevation. For example, strata in the hanging wall of the Pine Mountain Thrust are approximately 2.5 km above their original pre-faulting position (Figure 17.3), whereas Paleozoic strata in the hanging wall of the McConnell Thrust have been elevated by about 5 km (Figure 17.1).

A typical thrust fault consists both of segments that lie approximately in the plane of bedding (called *flats*) and segments that cut across bedding (called *ramps*). These segments are linked, and therefore make the overall thrust look like a flight of stairs (Figure 17.3). Flats commonly exceed ramps in length and typically occur within incompetent strata (e.g., shale and salt). Ramps are more likely to form in stronger, more competent rocks such as sandstone, dolomite, and limestone. Before movement occurs on a thrust fault, each ramp and flat in the hanging wall lies adjacent to its respective footwall ramp and flat. Thus, the number of hanging-wall flats and ramps must exactly match the number of footwall flats and ramps. When displacement occurs, flats and ramps in the hanging wall and footwall may become juxtaposed across the fault. For example, as shown in Figure 17.3, AB is a hanging-wall flat on a footwall flat, BC is a hanging-wall flat on a footwall ramp, CD is a hanging-wall flat on a footwall flat, and DE is a hanging-wall ramp on a footwall flat.

As mentioned previously, thrusts typically cut upsection (forming ramps) in the direction of displacement on the fault. Over an entire fold-thrust belt, the direction of movement for most thrust sheets in a region is roughly the same, thereby defining a *regional transport direction*. For

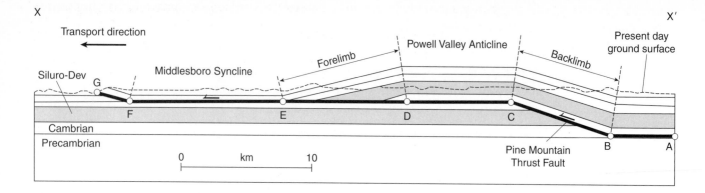

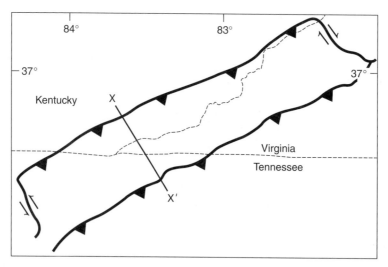

Figure 17.3 Cross section of Pine Mountain Thrust Sheet, southern Appalachians, that is simplified to illustrate terminology for describing hanging-wall and footwall structures. Location of cross section shown on map inset. Barbs point toward hanging wall of the thrust faults. Northeast and southwest ends of the Pine Mountain Thrust Sheet are bounded by tear faults.

Principles of Structural Geology by Suppe, John, © 1985. Adapted by permission of Prentice-Hall, Inc., Upper Saddle River, NJ.

example, the regional transport direction in the Appalachian fold-thrust belt is to the west or northwest, whereas the regional transport direction for the Canadian Rockies is to the east. This direction can be obtained using a variety of displacement-sense indicators, such as fold asymmetry, steps on slip-lineated surfaces, and inclination of cleavage. Although we often conceptualize that all thrusts in a fold-thrust belt are *forethrusts* on which movement is in the regional transport direction, there can be a minority of thrusts in a fold-thrust belt that verge in the opposite direction, forming *backthrusts*. A sketch of a quarry wall from southeastern New York illustrates the two most common occurrences of backthrusts, where they form to accommodate strain in the cores of synclines and where the front of a thrust sheet wedges between layers of strata (Figure 17.4). On the whole, however, the regional transport direction for fold-thrust belts is toward the continental interior (or craton; Figure 17.5). In accretionary wedges, which have the same structural geometry as continental fold-thrust belts, the general transport direction is toward the trench (Section 16.2.3).

Most thrust faults are not simple planes. Rather, they can be divided along strike into segments whose strike changes with respect to the transport direction. For most of the length of a thrust, the fault ramp strikes approximately perpendicular to the transport direction, forming a *frontal ramp*. Locally, however, especially near fault terminations, ramps may cut upsection laterally and strike at an acute angle to the transport direction, forming *lateral ramps* and *oblique ramps* (Figure 17.6). These segments can be distinguished, in that lateral ramps strike parallel to the transport direction, whereas oblique ramps strike at some other acute angle to the transport direction. If a lateral ramp dips vertically, we also call it a tear fault. *Tear faults* accommodate differential displacement between one part of a thrust sheet and another (see tear faults to the northeast and southwest of the Pine Mountain Thrust on Figure 17.3). Strike-slip movement dominates on lateral ramps and tear faults and may help to accommodate along-strike variations in displacement along the fault. On nonvertically-dipping oblique ramps, movement tends to be oblique slip, with the specific direction of movement being a function of the

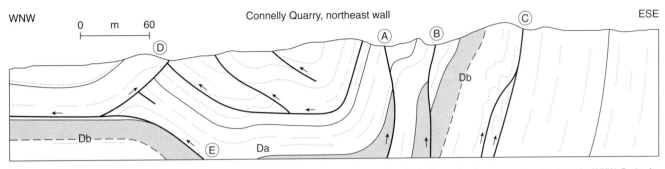

Figure 17.4 Cross-sectional sketch of the northeast wall of Connelly Quarry, southeastern New York. The regional vergence direction is to the WNW. Faults A, B, and C are out-of-the-syncline accommodation faults, and fault D accommodated the insertion of the hanging-wall anticline above fault E. Db is Devonian Becraft Limestone.

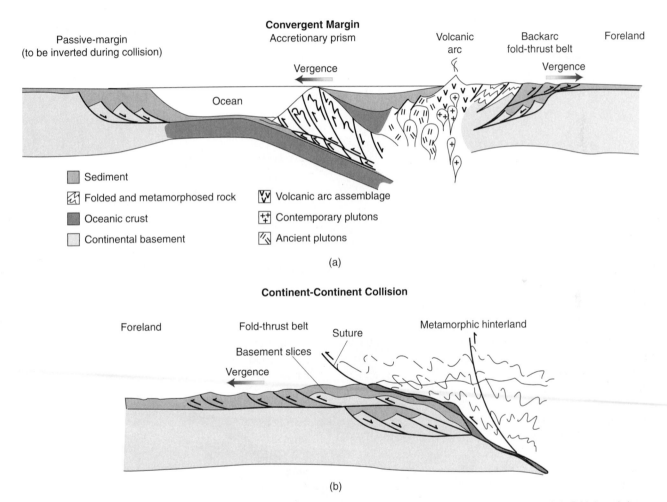

Figure 17.5 Tectonic settings of fold-thrust belts. (a) An accretionary prism, which verges toward the trench, is a type of fold-thrust belt. Fold-thrust belts also form in the backarc of an Andean-type convergent margin. (b) Fold-thrust belts in the foreland of a collisional orogen.

ramp angle and the degree of obliquity to the transport direction. Commonly, rotation of structures (e.g., bedding, fractures, fold axes) about a vertical axis occurs in the vicinity of an oblique ramp.

While active, many thrust faults lie entirely in the subsurface, whereas others (known as *emergent thrusts*) cut and displace the ground surface during movement. After

the thrusted region has been exhumed by erosion, it is not always possible to determine if a fault that presently intersects the ground surface was once emergent while it was active or not. Consequently, geologists distinguish *blind thrusts* that do not intersect the present-day ground surface (Figure 17.7) from those whose trace on the ground surface can be readily mapped. The 1994 Northridge earthquake in

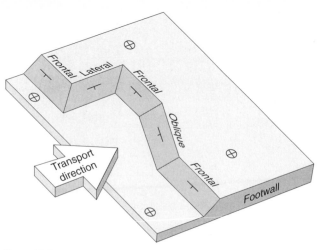

Figure 17.6 Three-dimensional block diagram illustrating different types of fault ramps (hanging wall removed). Tear faults are vertically-dipping lateral ramps.

California occurred on a blind thrust fault that had not previously been identified by any surface or subsurface expression. If erosion produces a hole in a thrust sheet, so that

at the ground surface we see an exposure of the footwall that is completely surrounded by hanging-wall rocks, then the exposure is called a *window* or *fenster* (which means window in German) (Figure 17.7). Alternatively, if erosion removes most of a thrust sheet so that you can map a remnant of the thrust sheet that is completely surrounded by footwall strata, then the 'island' of hanging wall is called a *klippe* (Figure 17.7).

Fold-thrust belts are not composed of a single thrust fault, but rather contain a family of related thrust faults, which tend to link to common *detachments* (Figure 17.8). These thrust systems of related faults develop because a limited number of horizons within a stratigraphic section (e.g., salt, shale) are apt to form regional detachments. It is common, therefore, for thrust faults in a given section to share the same basal detachment (or *floor thrust*), and to share a common upper detachment (or *roof thrust*), if one exists. Based on geometry, there are two endmember types of thrust systems, imbricate fans and duplexes. Faults that make up an *imbricate fan* branch upsection from a common floor thrust and terminate in a tip line without merging into an upper roof thrust (Figure 17.8a,b). Note that all faults

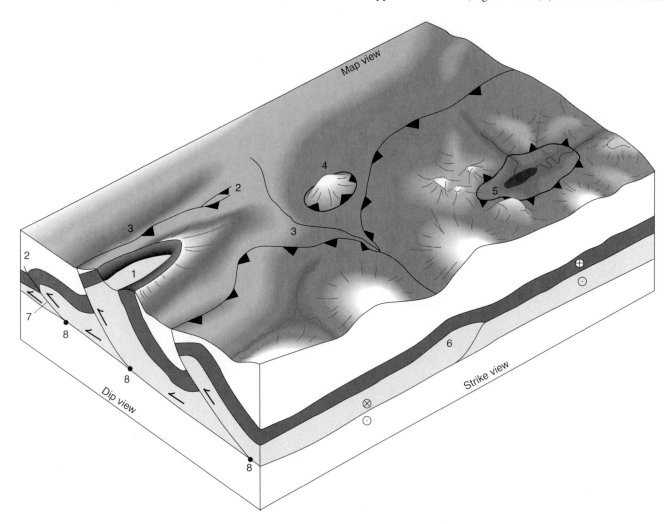

Figure 17.7 Block diagram of a series of thrusts. (1) Breached anticline where topography reflects the underlying structure, (2) tip lines in both map and cross-sectional view where fault displacement dies to zero, (3) thrust traces where faults intersect the ground surface (barbs point toward hanging wall), (4) klippe, (5) window, (6) lateral ramp, (7) blind thrust, and (8) branch lines where faults intersect underlying detachment.

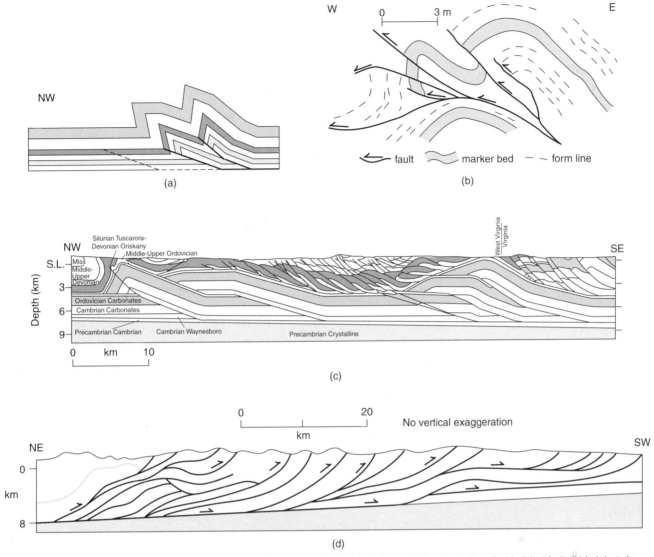

Figure 17.8 Different kinds of thrust systems. (a) Schematic cross-section of an imbricate fan containing two faults and an incipient fault. (b) Imbricate fan near Ozone, Tennessee. (c) Interpreted duplex through the Appalachian Valley and Ridge, Virginia and West Virginia. Observe that the roof thrust of the Cambrian-Ordovician duplex is the floor thrust for a higher level duplex. (d) Simplified cross section of the Canadian Rockies at the latitude of Calgary, Alberta. Thrusts involve only sedimentary strata above a detachment at the basement-cover contact.
(b) S. Mitra (1986).

must end in either a tip line or a branch line; that is, faults must either die out laterally along strike as well as up and down dip, or they must merge with another fault surface. In a *duplex,* by contrast, a series of ramps branch from a floor to a roof thrust (Figure 17.8c). Adjacent thrusts in a duplex completely surround deformed bodies of rock in three dimensions. These bodies are called *horses.* Depending on the spacing and relative displacement on thrusts in a duplex, the roof thrust of the duplex may be relatively planar or may be quite corrugated due to folding by the underlying horses of the duplex.

In both duplexes and imbricate fans, the faults composing the system do not initiate at the same time. Rather, new thrust faults tend to develop in a *break-forward sequence,* meaning that they become progressively younger toward the foreland (Figure 17.9). Although this break-forward pattern generally holds true, displacement at any

single time increment typically is partitioned between the latest thrust and thrusts immediately behind it. On occasion, a new, *out-of-sequence fault* may initiate to the hinterland of the preexisting faults, and crosscut earlier formed structures.

Before leaving the topic of thrust geometries, it is important to note that we have been limiting our discussion to structures that form above a subhorizontal basal detachment near the basement-cover contact (see Figure 17.8d). To emphasize that this subhorizontal detachment separates a package of rock that deforms independently from the basement below, we say that the thrusting is *thin skinned.* As we trace the detachment back into the hinterland (as has been done using seismic-reflection data in the Appalachians and in the Canadian Rockies), we see that eventually Precambrian crystalline rocks become involved in the thrust sheets (Figure 17.5). These rocks may be transported in a

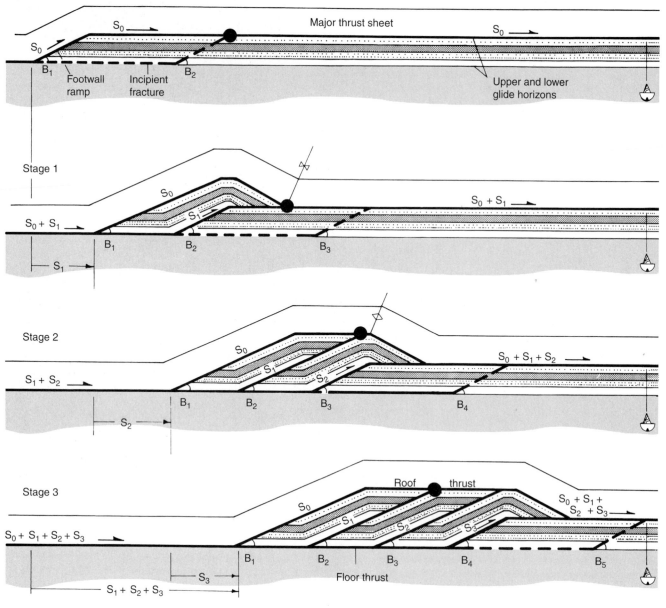

Figure 17.9 Progressive abandonment of footwall ramps during break-forward thrusting builds a duplex, consisting of a roof thrust, horses, and a floor thrust. The roof thrust undergoes a sequence of folding and unfolding, as shown by the change in geometry beneath the black dots. This geometric scenario assumes plane strain and kink folding.
S. E. Boyer and D. Elliott (1982).

'thin-skinned' manner on a subhorizontal fault over Paleozoic sedimentary rocks for perhaps hundreds of kilometers. Eventually, however, the thrust must root down into crystalline basement rocks, losing its thin-skinned character. Thin-skinned deformation contrasts with *basement-involved thrusting* in regions to the foreland of fold-thrust belts, such as the Wyoming Province of the Rocky Mountains (e.g., Wind River and Big Horn ranges), where steep thrust faults penetrate basement, but basement has not moved long distances on subhorizontal detachments over sedimentary strata. In such locations, sedimentary strata instead drape over the basement block. Some geologists refer to such structures as *thick-skinned.* When there is doubt as to whether or not a

structure is thin-skinned, it may be better to classify a fault as *basement involved* or *basement detached,* and then to describe the dip and/or geometry of the fault.

17.3 THRUST-RELATED FOLDS

So far, we have focused on the thrusts in fold-thrust belts. But what about the folds? "Fold" is in the name because part of the deformation in fold-thrust belts results from folding that occurs in association with faulting. We can categorize folds that occur in fold-thrust belts into three broad classes based on when they form: before, during, or after fault development.

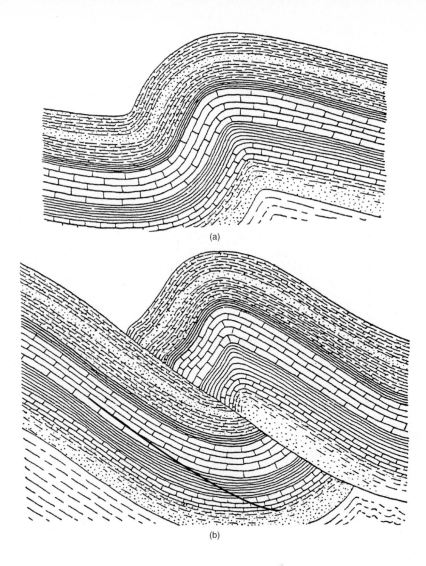

Figure 17.10 Break-thrust fold. Fold forms first (a) and then a fault breaks through its forelimb (b).

(a)

(b)

To understand folding that occurs before fault development, picture a fold forming as layers buckle in response to compressive stresses. Initially, no fault is present (Figure 17.10a). As the fold grows, it may tighten and develop an asymmetry. This tightening of the fold commonly results in a thrust breaking through the forelimb of the fold; hence, some geologists apply the name *break thrust* to such faults (Figure 17.10b). After the thrust has ruptured through the forelimb of an asymmetric anticline, it separates the anticline of the hanging wall from a relict syncline in the footwall. Strata in the hanging wall may then be translated along the fault.

What about the case of faults that develop before folding? As you may recall, thrust faults tend to form stairstep geometries as the fault cuts upsection through stratigraphy. *Fault-bend folds* form where hanging-wall strata move up and over a preexisting stairstep in a fault, deforming in order to conform to changes in dip (bends) of the fault surface

(Figure 17.11). If the strata did not fold, then a gap or overlap would develop between the fault surface and the thrust sheet. In its simplest form, the fault-bend fold model describes an anticline with kink-style hinges that forms and progressively grows above a footwall ramp as strata of the hanging wall begin to slide up and over the fault bends (Figure 17.12). As shown in Figure 17.12, the model predicts: (1) specific geometric relationships are maintained throughout the development of the anticline, (2) the anticline backlimb parallels the footwall ramp, (3) the anticline forelimb is shorter and steeper than the backlimb (although, in general, the fold is roughly symmetrical), and (4) the kinks that define the hanging-wall fold directly reflect bends in the fault surface (i.e., throughout the development of the fold, points X and Y delimit the footwall ramp and points X′ and Y′ delimit the corresponding hanging-wall ramp).

Another type of fold that develops after fault formation, often by modifying a preexisting fold, occurs along a

(a)

(b)

Figure 17.11 (a) Fault-bend fold related to bend in the McConnell Thrust, near Seebe, Alberta (Canada). The Paleozoic strata have been displaced over 5 km vertically and 40 km horizontally, and now lie above forested Cretaceous foreland basin deposits. (b) Fault-bend fold above a minor thrust in the Hudson Valley fold-thrust belt, Route 23 (New York).

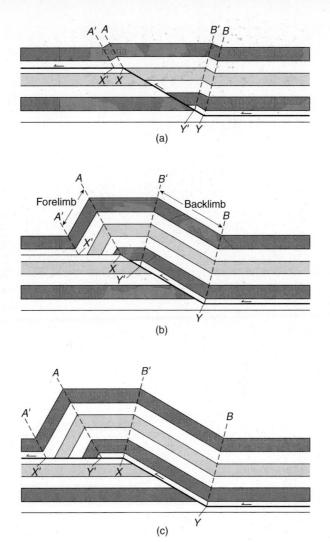

(a)

(b)

(c)

Figure 17.12 Model for progressive development of a simple fault-bend fold.

'locked' fault. Visualize a fault that develops and then stops growing, perhaps because a decrease in strain rate makes it easier for rock to deform by folding rather than by frictional sliding. Strain is then accommodated by fold development adjacent to the fault.

Folds also may develop concurrently with fault development. In these instances, a fold develops in advance of the tip of a propagating thrust fault. In outcrop-scale structures, folds may develop at both fault tips because the confining pressure is great enough that the upper free surface (i.e., the ground) does not influence fold development. In larger structures, however, a fold develops only in front of the upper tip as it propagates upsection toward the upper free surface, whereas the lower tip commonly links with a floor thrust rather than forming a second fold. Typically, the resultant folds are asymmetric in the direction of vergence, and they tighten with increased shortening to the point of developing a steep to overturned forelimb (Figure 17.13). Unless a fault breaks through the fold, all displacement along the fault is consumed in creating the fold and no displacement is transferred into the foreland. This model of *fault-propagation folds* predicts that the fault tip propagates upsection as the fold backlimb

and forelimb lengthen (Figure 17.14). Geometric constraints of this simple model necessitate that the fault tip and the merge point of two intersecting kink axes are at the same stratigraphic horizon, and that the interlimb angle of the fold remains the same through the evolution of the fold (Figure 17.14). Like the fault-bend fold model, the backlimb of a fault-propagation fold tends to parallel the dip of the fault, strata in the footwall remain flat lying, and strata move through the kink axes.

Not all folds in fold-thrust belts are underlain by ramps. Folding also occurs where strata are shortened above a subhorizontal detachment and wrinkle up like a rug that has been shoved across a slick wooden floor (Figure 17.15). Such *detachment folds* are particularly common in regions where substantial shale or salt occurs above the detachment, such as in the Jura Mountains of Switzerland. The weak rock (shale or salt) effectively flows into the core of the fold as the structure develops. In some cases, a thrust may develop at a late stage in the evolution of the fold and may cut across the forelimb of the already formed detachment fold.

Division of folds in fold-thrust belts into three classes clearly is a simplification. Today, most thrust-belt geologists

Figure 17.13 Asymmetric fold with a fault dying out upsection in the Lost River Range, (Idaho).

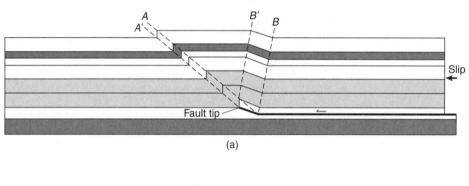

(a)

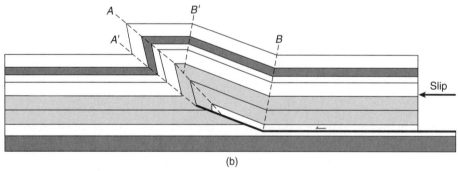

(b)

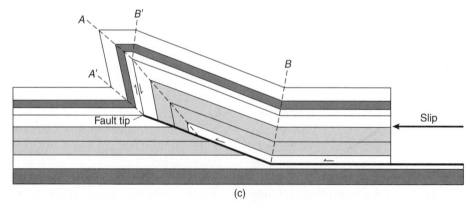

(c)

Figure 17.14 Model for progressive development of a simple fault-propagation fold.

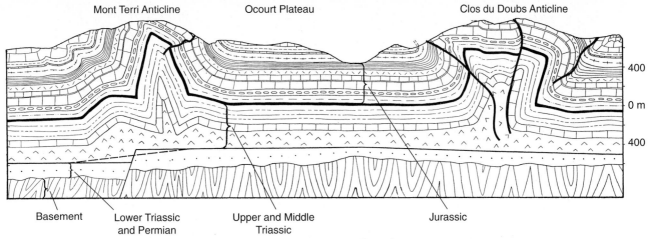

Mont Terri Anticline   Ocourt Plateau   Clos du Doubs Anticline

400

0 m

400

Basement   Lower Triassic   Upper and Middle   Jurassic
         and Permian   Triassic

Figure 17.15 Cross section of detachment folds in the foreland of the Jura Mountains (Switzerland).

feel that an evolutionary continuum exists between the three classes. That is, initial folding may set up the spacing for thrust-related folds. Further shortening may result in fold amplification and tightening, with the initiation and propagation of a fault further modifying the overall fold geometry. If displacement along the fault is sufficiently high, the fault may break through the fold forelimb, transporting the fold along the fault. Variations of this scenario undoubtedly occur as the rate of fold shortening and timing of fold initiation compete with the rate of thrust propagation and timing of thrust initiation to accommodate shortening under various conditions of strain rate and fluid pressure.

The folds that we have been describing so far form in portions of a fold-thrust belt that are closer to the continental interior (i.e., the foreland edge), where thrust sheets are detached from basement and deform under low-grade metamorphic conditions. We would be remiss in a discussion of fault-related folds if we didn't mention folds that form in the hinterland parts of fold-thrust belts as well. There, thrust sheets typically include basement and are deformed under medium- to high-grade metamorphic conditions. In fact, temperatures and pressures at depth in the hinterland are high enough for ductile deformation, and contraction along a thrust may be accompanied by development of large, regional, recumbent folds in the thrust sheet. These structures, called *fold nappes,* are particularly well exposed in the Helvetic Alps of Switzerland (Figure 17.2; see Section 19.2).

17.4 FOLD-THRUST BELTS IN MAP VIEW

Faults are not surfaces with infinite dimensions. That is, the map trace of a fault must terminate along its length, either because the fault merges with or is truncated by another fault, or because the magnitude of displacement across the fault decreases progressively along strike until, at the tip, displacement is zero (see Chapter 8). Beyond the termination of a fault, the shortening that is accommodated by slip on the fault may be accommodated instead by fold-

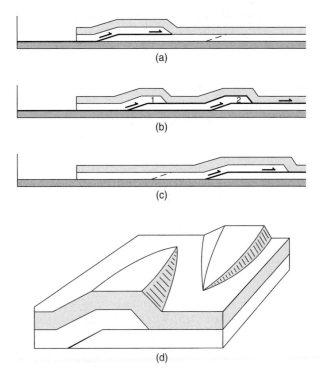

(a)

(b)

(c)

(d)

Figure 17.16 Concept of displacement transfer between thrusts in a relay zone. (a) All displacement on fault 1, none on fault 2. (b) Displacement equally partitioned between faults 1 and 2. (c) All displacement on fault 2, none on fault 1. (d) Block diagram illustrating the terminations of overlapping folds. Front face of the block is cross section in (a), and the rear face of the block is cross section in (c).

ing or by slip on neighboring faults. For example, as shown in Figure 17.16, a decrease in slip on fault 1 is matched by an increase in slip on fault 2. Effectively, slip transfers from one fault to another along a common floor thrust at a *transfer* or *relay zone,* much like a baton that is transferred from one relay racer to the next.

Commonly, the map trace of a major thrust or thrust system is convex toward the foreland (Figure 17.17a). As a rule of thumb, if you connect the termination points of the bowed fault trace with a straight reference line, the regional

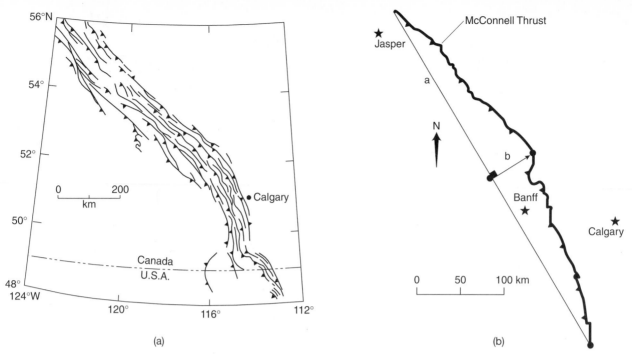

(a)   (b)

Figure 17.17   (a) Map pattern of a fold-thrust belt, showing typical curved traces of thrusts (Alberta Salient, Canadian Rockies). (b) The "bow-and-arrow rule," as applied to the McConnell Thrust in the Canadian Rockies. For most foreland thrust belts, the ratio b/a is roughly 0.1.

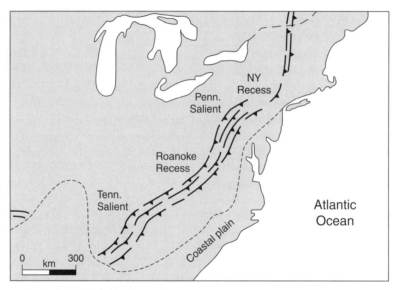

Figure 17.18   Map of the structural trends of the southern and central Appalachians, showing salients and recesses.

transport direction on the fault is in the direction perpendicular to this line, and the displacement magnitude is largest where the distance between the reference line and the fault trace is greatest. A vector drawn from the reference line to the thrust trace at this point is roughly 10% of the length of the reference line (the so-called bow-and-arrow rule; Figure 17.17b). Not all thrust traces, however, are simple bow-shaped arcs. In fact, if the along-strike change in displacement between adjacent regions along the trace is sufficiently large, then a lateral ramp may develop.

As illustrated by the Appalachians of eastern North America, regional map traces of fold-thrust belts typically are sinuous (Figure 17.18). Places where the belt bulges into the foreland are *salients,* and places where the belt has not propagated so far into the foreland are *recesses.* Map-view curves in a fold-thrust belt form for a variety of reasons, including interaction of the propagating fold-thrust belt with basement highs in the foreland, along-strike pinch-out of a favorable detachment horizon (e.g., a salt horizon), lateral variations in stratigraphic thicknesses, interaction of the belt with large strike-slip faults, and superposition of a second phase of shortening on the belt if the transport direction associated with the second phase is different from the first (Figure 17.19). If the curvature develops by actual bending of a previously straight belt, then the curve is an *orocline.*

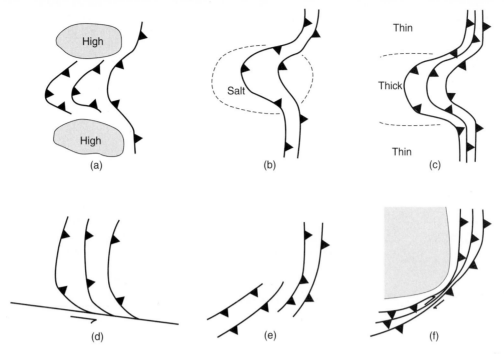

Figure 17.19 Possible processes leading to the formation of curving fold-thrust belts. (a) Interaction with basement highs in the foreland; (b) lateral pinch-out of a stratigraphic glide horizon; (c) lateral variation in stratigraphic thickness; (d) interaction with a strike-slip fault; (e) overprinting of two nonparallel thrust belts; (f) impingement against an irregular cratonic margin.

17.5 INTERNAL DEFORMATION OF THRUST SHEETS

The previous material should have given you a fairly good picture of what a fold-thrust belt looks like and what types of faults and folds you might observe in one. Most of the previously described structures are at the scale of entire thrust sheets (i.e., from meters to hundreds of kilometers). A significant proportion of the deformation in a fold-thrust belt, however, is accommodated by deformation within the thrust sheets themselves, at outcrop or grain scale. This *internal deformation* is manifested among several different types of fold and fault structures and strain features on a grain scale.

If a thrust sheet consists of alternating layers of contrasting competency, the layers may respond to regional shortening by deforming into a train of mesoscopic folds (Figure 17.20a). The wavelength of these folds is substantially shorter than the dimensions of the overall thrust sheet, and the style of folding depends on the nature of the interlayered strata (see Chapter 10). Under the pressure and temperature conditions typical for development of fold-thrust belts, sedimentary strata are susceptible to development of *cleavage* (Figure 17.20a). Cleavage formation, as we saw in Chapter 11, may result in a volume decrease of the thrust sheet if dissolved material enters the pore fluid and is flushed from the system.

Shortening within the thrust sheet may result in the development of numerous *wedge faults,* which are thrusts that ramp across a single bed (Figure 17.20b). Characteristically, wedge faults develop small fault-bend folds in both the hanging wall and footwall. In many thrust belts, a *conjugate*

system of strike-slip faults cuts thrust sheets. The acute bisectrix of the fault sets trends perpendicular to the traces of regional thrusts and folds, and thus slip on the faults accommodates shortening in the direction perpendicular to the trends of the thrusts, and stretching parallel to the trends of the thrusts. Locally, a fold-thrust belt may be diced up by a *mesoscopic fault array* with various orientations, but with slip magnitudes measured in only centimeters to meters. Movement on these faults effectively accommodates the overall deformation the thrust sheet is undergoing, with the relatively minor slip on individual faults combining to create a significant overall shortening perpendicular to regional thrust traces, and stretching parallel to regional thrust traces.

As a syncline becomes progressively tighter, the volume of rock being folded cannot fit in the core of the fold. To solve such space problems, local thrust faults (both forethrusts and backthrusts) develop that displace strata out of the core of the fold. Such *out-of-the-syncline faults* do not necessarily branch from the detachment beneath the fold, but rather die out in the core of the fold. Finally, in the foreland portion of a fold-thrust belt, where deformation occurs under relatively low P-T conditions, we have seen that pressure solution is a common deformation process, leading to formation of spaced cleavage. In units that contain relatively little clay, and thus are not amenable to pressure solution, penetrative strain may develop by intragranular deformation. For example, in limestone, calcite grains undergo twinning, and in quartz sandstone, the quartz grains develop deformation bands (see Chapter 9). In the hinterland portion of a fold-thrust belt, where deformation occurs under higher P-T conditions, rocks in thrust

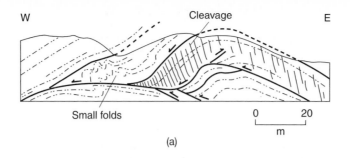

Figure 17.20 Mesoscopic structures within thrust sheets. (a) Sketch of Route 23 roadcut near Catskill (New York), illustrating mesoscopic-scale folds and cleavage. (b) Wedge thrust on New World Island (Newfoundland).

sheets deform internally, and large penetrative strains locally develop, as manifested, for example, by stretched grains and mineral lineations.

17.6 FOLD-THRUST BELT MECHANICS

Now that we have a feel for the geometry of structures that occur in a fold-thrust belt, we can explore the mechanisms of fold-thrust belt development. The movement of large thrust sheets that are tens of kilometers long (measured in the transport direction), but generally less than five kilometers thick, is seemingly a paradox. If you assume a reasonable value for frictional resistance to sliding on the underlying detachment (say, from experiments with dry rock), then the stress necessary to push the sheet over a horizontal surface or up a gentle incline greatly exceeds the failure strength of the intact rock composing the thrust sheet (see Chapter 8). Thus, you would expect the hinterland end of the thrust sheet to crush and buckle before the sheet as a

whole would move. Yet clearly, large thrust sheets do exist. So how do they move, and how do belts composed of many large thrust sheets develop? In this section, we'll briefly explain the development of key ideas that led to formulation of the current model for fold-thrust belt development.

The first issue to be addressed was whether friction between dry solid surfaces really resists thrust-sheet movement. In the late 1950s this assumption became contested.[2] It was pointed out that the force required to move a rectangular (in cross section) thrust sheet can be greatly diminished if fluid pressure in the detachment zone increases to values approaching lithostatic loads (Figure 17.21). Frictional resistance to sliding depends on the *effective normal stress* (σ_n^*) across the surface:

$$\sigma_n^* = \sigma_n - P_{H_2O} \qquad \text{Eq. 17.1}$$

---

[2]The pivotal ideas of M. K. Hubbert and W. W. Rubey were published during these years.

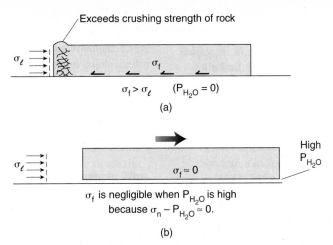

Exceeds crushing strength of rock

σ_ℓ σ_f

$\sigma_f > \sigma_\ell$ $(P_{H_2O} = 0)$

(a)

High P_{H_2O}

$\sigma_f \approx 0$

σ_ℓ

σ_f is negligible when P_{H_2O} is high
because $\sigma_n - P_{H_2O} \approx 0$.

(b)

Figure 17.21 When pushed from the rear (a), a rectangular thrust sheet sliding on dry rock would be crushed before overcoming frictional resistance. (b) High fluid pressure at the basal detachment decreases the effective stress and allows the thrust sheet to move under very small applied load. σ_ℓ is the stress resulting from horizontal loading, whereas σ_f is frictional resistance (other terms are defined in the text).

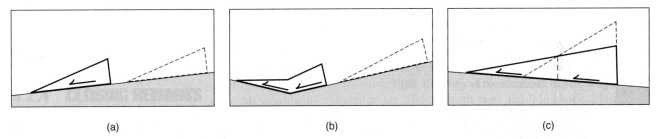

(a)   (b)   (c)

Figure 17.22 (a) The concept of gravity sliding. (b) Gravity sliding partly downslope and partly upslope. (c) The concept of gravity spreading.

where σ_n is the normal stress and P_{H_2O} is the fluid pressure. Substituting effective stress into the Mohr-Coulomb criterion (see Section 6.6.2, Equation 6.3) yields:

$$\sigma_s = C + \mu/\sigma_n^* \qquad \text{Eq. 17.2}$$

where σ_s is the shear stress necessary to initiate sliding, C is the cohesive strength of the fault surface, and μ is the coefficient of internal friction. As fluid pressure increases, the effective normal stress across the detachment decreases, because fluid in the detachment zone partially supports the weight of the thrust sheet. Therefore, the thrust sheet can be moved easily when fluid pressure is high even when the boundary load at the hinterland edge of the sheet is significantly lower than the failure strength of rock composing the sheet (Figure 17.21b).

But what initiates thrust motion? Initially it was assumed that a thrust sheet slides toward the foreland in response to gravity when the basal detachment dips slightly toward the foreland, just like slumps or landslides that slide downslope under the influence of gravity. This concept of *gravity sliding* became very popular, and in the 1960s, most structural geologists envisioned that development of fold-thrust belts occurred as thrust sheets glided down an incline created by uplift of the hinterland (Figure 17.22a and b). However, petroleum exploration of many fold-thrust belts provided seismic-reflection data showing that basal detach-

ments beneath almost all thrust systems dip toward the hinterland, not the foreland! To account for this contradiction, some structural geologists suggested that fold-thrust belt formation, instead, should be visualized as a process of *gravity spreading*, meaning that they form when a thick mass of rock collapses and spreads laterally under its own weight (Figure 17.22c). The main difference between this hypothesis and that of gravity sliding is that the gravity-spreading model suggests that the direction of dip of the topographic surface, not the dip of the basal detachment, is the controlling geometric feature responsible for thrust-fault movement. To visualize this, think of the movement of large continental ice sheets. As snow accumulates in the ice field, the topographic slope increases and the ice begins to spread laterally under its own weight. Similarly, it was suggested that orogenic belts move as the hinterland is uplifted due to magmatic processes. As a consequence, strata spread toward the foreland on hinterland-dipping thrust faults.

Thus, between the late 1950s and the mid-1970s, most thrust-belt geologists believed that gravity was the primary driving force for fold-thrust belt tectonics, and assumed that a *push-from-the-rear,* meaning a horizontal boundary load caused by relative foreland displacement of the backstop, was not a major factor in the development of fold-thrust belts. But in the late 1970s and early 1980s, models were

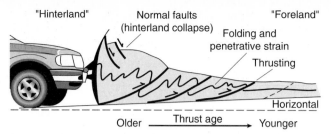

Figure 17.23 Snowplow analogy for fold-thrust belt development. The wedge of snow widens with continued shortening, younger thrusts generally initiating in a hinterland to foreland progression. While new thrusts are adding material at the toe of the wedge, the hinterland portions are developing penetrative strain, normal faults, and slump features.

developed that dusted off the push-from-the-rear concept and again applied it to thrust-belt mechanics. These models propose that thrust-belt mechanics is analogous to the static-force problem of pushing a wedge of snow (or sand) uphill in front of a moving bulldozer (Figure 17.23). The key modification in these models, which until then had not been incorporated into attempts at explaining thrust-belt mechanics, is that the wedge-shaped pile has a foreland-dipping topographic surface and a weak, hinterland-dipping basal detachment. Some model the deforming thrust wedge as a mass of perfectly plastic material, whereas others consider a thrust wedge as a granular aggregate in which component fragments frictionally slide past one another (materials that behave this way are called *Coulomb materials*). The latter view forms the contemporary concept of fold-thrust belt development, and is often referred to as the *critical-taper theory* (Figure 17.24).

In critical-taper theory, stresses in the wedge result both from a push-from-the-rear caused by displacement of a backstop and from *gravitational potential energy*. The gravitational potential energy, due to the elevation of the hinterland, creates both vertical and horizontal stresses (Figure 17.24a). As the backstop moves, the wedge deforms internally (by forming folds, faults, and penetrative strain), and as a consequence, its surface slope increases. When the wedge reaches a certain critical taper, ϕ_c (which is defined as the surface slope angle, α, plus the detachment dip, β), the wedge slides stably toward the foreland along the weak detachment. Movement occurs on the detachment because the coefficient of sliding ('internal') friction on the detachment is less than the coefficient of internal friction in the wedge. During foreland translation of the wedge, new material is added to the wedge at its toe by break-forward thrusting, causing the taper angle to decrease (Figure 17.24c and d). When this happens, stable sliding stops, and internal deformation occurs once again within the wedge. Over time, the wedge sustains internal deformation during periods of subcritical taper and renewed stable sliding as deformation builds taper to critical and supercritical taper levels. This cycle is maintained by a dynamic equilibrium among: (1) the addition of new material at the toe of the wedge (by break-forward thrusting), which decreases the surface slope, (2) the internal deformation within the wedge due to formation of duplexes at the base of the wedge, out-of-sequence thrusts in the wedge, folds, and penetrative strain,

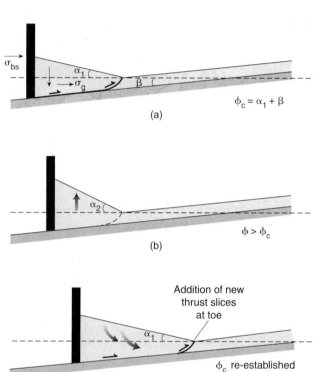

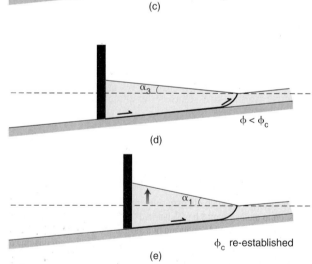

Figure 17.24 The critical-taper theory of fold-thrust belt mechanics. The critical taper (ϕ_c) is defined as the sum of the surface slope angle and the detachment slope angle. (a) Stress acting on wedge is partly a horizontal boundary load caused by the backstop (σ_{bs}), and is partly caused by gravity (σ_g). (b) If the backstop moves, the wedge thickens, so the surface slope increases, and the taper (ϕ) exceeds ϕ_c. (c) The wedge slides toward the foreland and new material is added to the toe; extension of the wedge occurs so that surface slope decreases. (d, e) If the surface slope becomes too small, thrusting at the toe stops, and the wedge thickens by penetrative strain or out-of-sequence thrusting.

which increases the surface slope, and (3) thinning of the wedge due to extensional deformation (i.e., hinterland collapse) and erosion of the wedge, which decreases the surface slope.

Returning to the snow analogy, imagine a snowplow standing at the end of a snow-covered driveway that slopes gently toward the street (Figure 17.23).[3] The plow lowers its blade and begins to move up the driveway toward the garage (the foreland in this analogy). As soon as the plow begins to move, a wedge-shaped pile of snow forms in front of the blade, and the surface of this pile slopes toward the foreland. A detachment forms beneath the pile, and an imbricate thrust develops at the front of the pile. A few meters in front of the plow, however, the snow remains totally unaffected. During the next increment of movement, the detachment propagates farther beneath the snow and above the driveways concrete ('basement') toward the foreland, and a new imbricate thrust develops. The thrust system now consists of two imbricate thrusts cutting upward from the basal detachment, with the imbricates forming in a break-forward sequence (i.e., the foreland imbricate is younger than the hinterland one). While the wedge adds new snow at its foreland edge, the hinterland portion of the wedge sometimes slides stably over the driveway for a while without changing (thereby decreasing the overall surface slope of the wedge), and sometimes thickens by a combination of folding, penetrative horizontal shortening, and out-of-sequence reactivation of thrusts (Figure 17.23). Because snow is not very strong, the snow wedge cannot thicken indefinitely. Eventually, the hinterland of the wedge collapses under its own weight by slumping or by formation of normal faults within it. This *hinterland collapse* results in a net thinning of the thicker part of the wedge, thereby maintaining a dynamic balance between wedge thickening and the proper surface slope for wedge translation.

In summary, fold-thrust belts can be modeled as wedges composed of a Coulomb material. Growth of a wedge occurs because stresses in the wedge exceed the strength of the wedge, and the stresses are a consequence both of a surface force in the hinterland and gravitational spreading. The wedge evolves such that a balance is established between foreland sliding of the wedge and addition of material at the toe. Addition of material at the toe, along with slumping, hinterland collapse, and erosion, tends to lower the topographic slope (reducing taper), whereas internal deformation of thrust sheets and out-of-sequence thrusting thicken the hinterland portion of the wedge and steepen the topographic slope (increasing taper). The observed critical taper reflects this balance, and depends on the material strength of the wedge, the resistance to sliding across the basal detachment, and the ratio of fluid pressure to overburden pressure both in the wedge and across the detachment. If the effective strength of the wedge is increased (either by increasing strength or by decreasing fluid pressure), then

the critical taper decreases. If the resistance to sliding on the basal detachment is increased (either by increasing the coefficient of sliding friction or by decreasing the fluid pressure), the critical-taper angle increases. Fold-thrust belts whose basal detachments lie in salt, a very weak lithology, have a critical taper, ϕ_c, as low as $1-2°$, whereas fold-thrust belts that have detachments in stronger rocks may have critical tapers as high as $8-10°$.

17.7 THE CONCEPT OF BALANCED CROSS SECTIONS

In this chapter, we have used a number of cross sections through fold-thrust belts. Cross sections tend to depict structures at depth, right down to the basal detachment, even though these regions are not exposed. Perhaps you've asked yourself the fundamental question, "How do people create such sections?" and, "How reliable are they?" Well, to begin with, it is important to remember that a cross section is an interpretation of the subsurface geology, and nothing more. We do not have access to a single cliff face that extends several kilometers into the crust to let us see exactly where formation contacts and faults are positioned. Cross-section interpretations are constrained by projecting surface geology into the subsurface, by interpreting seismic-reflection profiles, and by interpreting well data. Such data rarely provide a complete picture of subsurface geology, so we always must extrapolate when we make cross sections. We can determine, however, if the relations shown might be possible, which is one of the primary aims of balanced cross sections.

After decades of research, geologists have a fairly clear sense of what fold-thrust belts look like. With this picture in mind, we establish four fundamental criteria that determine whether a cross section at least has the possibility of being correct. The criteria are:

- Structures in the deformed-state cross section must look like structures that have been observed in outcrop or seismic profiles. We call cross sections that meet this criterion *admissible sections.*
- *Restored cross sections* (Figure 17.25), which represent the predeformation configuration of strata and location of faults in the region, must contain realistic looking structures. Restored sections are created by removing offsets on faults, by stretching out folds so that bedding returns to horizontal, and by removing the effects of penetrative deformation. For example, if a restored fault trace zigs and zags upsection, then there's something wrong with the section.
- The area of rock shown on the restored cross section must equal the area shown on the deformed-state cross section. Cross sections that meet this criteria are *area balanced.* If volume loss occurred during deformation (e.g., by pressure solution), this must be taken into account when restoring the section.

---

[3]The specific climate zone of the authors' homesteads is obvious.

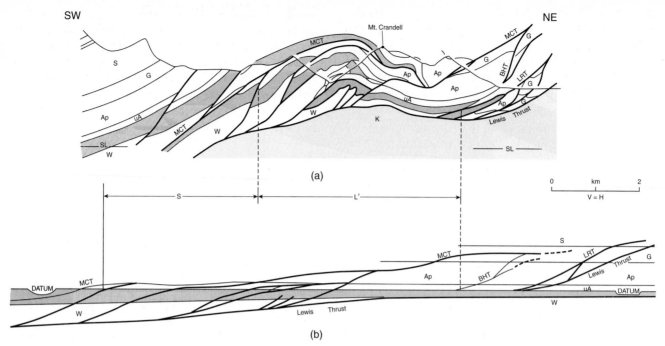

Figure 17.25 Restoration of a cross section through the Lewis Thrust Sheet. Comparison of the deformed state (a) and the restored cross section (b) allows you to determine how much minimum horizontal shortening has occurred in the fold-thrust belt. S. E. Boyer and D. Elliott (1982).

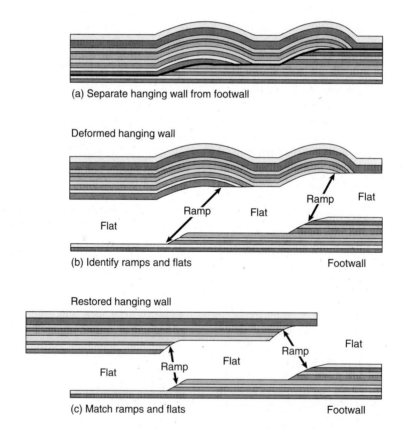

(a) Separate hanging wall from footwall

Deformed hanging wall

Ramp   Flat   Ramp   Flat

Flat

(b) Identify ramps and flats   Footwall

Restored hanging wall

Ramp   Flat   Ramp   Flat

Flat

(c) Match ramps and flats   Footwall

Figure 17.26 Diagram illustrating a quick-look technique for checking a cross section for potential problems. The key is to recognize ramps and flats in the deformed-state section and realize that hanging-wall flats and ramps must exactly match footwall ramps and flats in number and in stratigraphic composition.

- It must be possible to create the deformed-state cross section from the restored cross section in a *kinematically reasonable way.*

A cross section that meets all of the above criteria is a *balanced cross section.* The first two criteria in the list ensure that the section doesn't depict impossible structural geometries. The third criterion ensures that the configuration of structures shown on the deformed-state section does not imply that undocumented volume change occurred during deformation. The fourth criterion emphasizes that you do not really understand the geometry of a complex structure until you can figure out how the structure formed from undeformed rock. Lastly, keep in mind that although we are focusing on contractional terranes in this essay, these criteria also apply to balancing cross sections in extensional terranes.

Clearly, 'balancing' a cross section involves careful examination of the deformed-state section and construction of a restored section, which is then examined for probability and area balance. It is beyond the scope of this chapter to provide detailed guidelines for balancing cross sections; most structural geology laboratory manuals offer step-by-step instructions and exercises. We do point out that, in some cases, you can simply scan a deformed-state cross section to see if it has the potential to be balanced. To do this, first identify ramps and flats in each part of the cross section and count them (Figure 17.26). Are there the same number of ramps and flats in the hanging wall as in the footwall? These numbers must match, because in an admissible, restored cross section, the hanging wall fits over the footwall with no gaps or overlaps. Now, paying particularly close attention to the ramps, check to see if the same beds are truncated in the hanging wall as in the footwall. They must be, because the hanging-wall beds were originally adjacent to the footwall beds. Perhaps surprisingly, these two simple tests are able to highlight the majority of common cross-section errors that lead to construction of unbalanced cross sections.

Before getting too carried away with the value of cross-section balancing, you should keep in mind that not all cross sections have to balance. To be balanced, the section must be drawn parallel to the direction of transport in the fold-thrust belt. But two-dimensional balancing techniques cannot deal with cross sections over lateral or oblique ramps, or across strike-slip faults. In such settings, out-of-plane strain may occur and thus by definition, area balance in the cross-sectional plane is impossible. It may be necessary to draw cross sections of such regions, but don't expect them to balance. Similarly, in regions where much of the deformation has occurred by flow of rock, units may be transposed in the plane of the section and units may flow into or out of the section plane, so balance is again impossible. Moreover, even in low-grade rocks, pressure solution may cause significant volume change that may make area balance a challenge. Finally, keep in mind that checking a cross section for balance does not automatically ensure a "correct" interpretation or that the interpretation is unique (i.e., there may be other balanced interpretations that fit the data). Balancing procedures are simply meant to focus your attention on potentially problematic areas in the cross section that require geological explanation and/or reinterpretation.

17.8 CLOSING REMARKS

Fold-thrust belts are inherently fascinating geologic terranes. They contain all the components that make for a good scientific puzzle: intriguingly complex structures and potentially quantifiable relationships. Moreover, they yield beautiful mountain ranges that are fun places for field work and, not in the least, contain valuable resources. No wonder fold-thrust belts have been the focus of such intense research for so long, and will likely challenge the talents of geologists for years to come.

ADDITIONAL READING

Boyer, S., and Elliott, D., 1982, Thrust systems, *American Association of Petroleum Geologists Bulletin,* v. 66, p. 1196–1230.

Dahlstrom, C., 1969, Balanced cross sections, *Canadian Journal of Earth Science,* v. 6, p. 743–758.

Davis, D., Suppe, J., and Dahlen, F., 1983, Mechanics of fold-and-thrust belts and accretionary wedges, *Journal of Geophysical Research,* v. 88, p. 1153–1172.

Marshak, S., and Woodward, N., 1988, Introduction to cross-section balancing, *in* Marshak, S., and Mitra, G., eds., *Basic methods of structural geology:* Englewood Cliffs, NJ, Prentice Hall, p. 303–332.

McClay, K., 1992, *Thrust tectonics,* London, Chapman & Hall, 447 p.

Mitra, S., 1986, Duplex structures and imbricate thrust systems: Geometry, structural position, and hydrocarbon potential, *American Association of Petroleum Geologists Bulletin,* v. 70, p. 1087–1112.

Price, R., 1981, The Cordilleran foreland thrust and fold belt in the southern Canadian Rocky Mountains, *Thrust and nappe tectonics:* Geological Society of London, p. 427–448.

Rich, J., 1934, Mechanics of low-angle overthrust faulting as illustrated by Cumberland thrust block, Virginia, Kentucky, and Tennessee, *American Association of Petroleum Geologists Bulletin,* v. 18, p. 1584–1596.

Suppe, J., 1983, Geometry and kinematics of fault-bend folding, *American Journal of Science,* v. 283, p. 684–721.

Suppe, J., and Medwedeff, D., 1990, Geometry and kinematics of fault-propagation folding, *Eclogae geol. Helvetica,* v. 83, p. 409–454.

Tearpock, D., and Bischke, R., 1991, *Applied subsurface geological mapping:* Englewood Cliffs, NJ, Prentice Hall, 646 p.

Chapter 18
Strike-Slip Tectonics

18.1   INTRODUCTION

When people hear that you are studying to be a geologist, one of the likely questions you'll be asked in North America is, "Will California fall into the ocean?" Perhaps the legend of the vanishing state originated from the observed huge slumps forming along the Pacific coast during large earthquakes. Most of the tremors jolting California are due to sudden increments of movement along the San Andreas Fault (Figure 18.1), which is an example of a major *continental strike-slip fault zone*. Across the fault zone, the Pacific Plate moves north, relative to North America. Other impressive examples of continental strike-slip zones include the Alpine Fault in New Zealand, the Anatolian Faults in Turkey, the Chaman Fault in Pakistan, and the Red River and Altyn Tach Faults of China. Strike-slip faults also cut oceanic lithosphere, in particular along ocean ridges, but because *oceanic strike-slip fault zones* are hidden by the sea, they are not as well known as their continental kin.

Perhaps when reading about strike-slip faulting, you've come across some of these terms: wrench, tear, transfer, lateral ramp, transcurrent, and transform; for quick reference, they are defined in Table 18.1. Unfortunately, different authors use strike-slip fault terminology in different ways, leading to occasional confusion about their meaning.

In the last three chapters, we have already mentioned strike-slip faults on several occasions, because they occur as components of contractional and extensional tectonic regimes. In this chapter, we focus on these faults themselves. Specifically, we examine the structure and kinematics of various types of strike-slip faults (both oceanic and continental), the geometry of the subsidiary structures that form in association with these faults, and the tectonic settings in which they occur. We begin by clarifying the distinction between the two kinematic classes: transfer and transcurrent faults.

Figure 18.1   Trace of the San Andreas Fault, north of San Francisco (Tomales Bay California).

Table 18.1 Basic Names for Strike-Slip Faults

Lateral ramp	A surface connecting two non-coplanar parts of a thrust fault or normal fault (i.e., a ramp that is roughly parallel to transport direction).
Strike-slip	A general term for any fault on which displacement is parallel to the strike of the fault, in present-day surface coordinates. The use of this term is purely *geometric,* and has no genetic, tectonic, or size connotation. Strike-slip faults are subdivided into two *kinematic* classes (i.e., based on the nature of displacement): transcurrent and transfer faults.
Tear	A term used for strike-slip faults that accommodate differential displacement of a thrust sheet along strike (falls under the general category of lateral ramp).
Transcurrent	One of the two *kinematic* classes of strike-slip faults. Transcurrent faults have the following characteristics: they do not terminate at plate boundaries, but rather they die out along their length; the displacement across them must be less than the length of the fault; the length of the fault only increases with time and continued movement; displacement is greatest at the center of the fault trace and decreases toward its ends.
Transfer	One of the two *kinematic* classes of strike-slip faults. Transfer faults have the following characteristics: once formed, displacement across them can be constant along the length of the fault; displacement across them can be much greater than the length of the active fault; their length can be constant, increase, or decrease with time. Some authors restrict the name transfer fault to strike-slip faults linking rift segments. We do not.
Transform	Transform faults are transfer faults that serve as lithospheric plate boundaries. They have the following characteristics: once formed, displacement across them can be constant along the length of the fault; displacement across them can be much greater than the length of the active portion of the fault; they terminate at other plate boundaries; their length can be constant, increase, or decrease with time. Note also that a transform fault is a small circle around the Euler pole describing relative motion of the plates adjacent to the fault.
Wrench	A term used in reference to regional-scale, continental strike-slip faults. Its use implies that the strike-slip movement was accompanied by complex crustal deformation (generally transpression). We avoid the term, as it tends to be used inconsistently.

18.2 TRANSFER AND TRANSCURRENT FAULTS

The recognition of transfer faults as a class of strike-slip fault initially came from studies of the mid-ocean ridge system. In the 1950s, refined bathymetric maps of the ridge system showed that a ridge axis is segmented along strike, and that adjacent segments are not coplanar. At first, geologists believed that this pattern was due to strike-slip displacement on faults that crossed a once continuous ridge, as illustrated in Figure 18.2a. If this assumption is correct, then the fault shown in Figure 18.2a has a left-lateral sense of slip. But if the hypothesis of seafloor spreading is correct, then the story is different. To see this, redraw the ridge-fault-ridge system again (Figure 18.2b), but this time add arrows to indicate the direction of spreading on the ridge segments. If seafloor spreading is occurring, the motion shown on the fault in Figure 18.2b must actually be right-lateral! When the early advocates of plate tectonics theory[1] reached this conclusion, they realized that the strike-slip faults linking mid-ocean ridge segments are "a new class of faults" and gave them the name "transform faults." *Transform faults* do not displace segments of originally continu-

ous ridges relative to one another; rather, they initiate at the same time as the ridge itself and accommodate differential plate motion between the ridge segments.

Geologists most commonly use transform faults in the context of describing lithospheric plate boundaries; however, kinematically such faults are not restricted to this situation. The displacement characterizing transform faults also develops at smaller scales, connecting two veins, for example (Figure 18.3). Therefore, we use the general term *transfer fault* to describe this type of structure and reserve transform fault to its lithospheric equivalent. A fault is a transfer fault if it has the following geometric and kinematic characteristics:

- The active portion of a transfer fault terminates where it connects to other structures (generally faults) at its ends. For example, a transform fault linking two mid-ocean ridge segments terminates at the ridge axes. Transform faults may also link subduction zones and other transform faults. Some transfer faults offset continental rather than oceanic lithosphere. The San Andreas Fault and the Alpine Fault are examples of such continental transforms faults.
- The length of a transfer fault can be constant, increase, or decrease over time, depending on the relative movement of the endpoints of the fault with respect to one another. For example, in Figure 18.4a, the spreading rate on ridge segment A is the same as

[1]The paper in which transform faults were proposed was written by Canadian geologist J. Tuzo Wilson (in 1965). The sense of slip predicted for transform faults was confirmed a couple of years later by earthquake fault-plane solutions. Documentation of transform faults stand as one of the key proofs of plate tectonic theory.

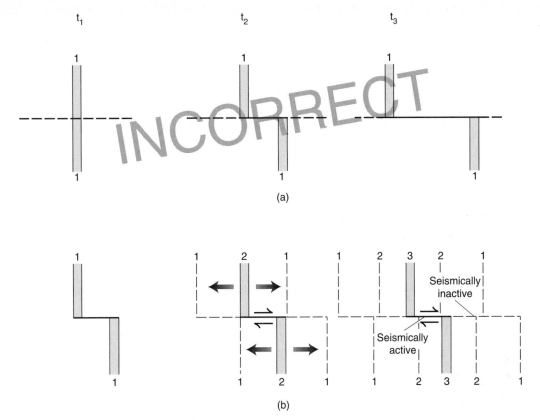

(a)

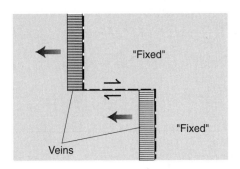

(b)

Figure 18.2 Diagrams showing (a) incorrect and (b) correct kinematic interpretation of oceanic transform (i.e., lithospheric transfer) faults. The history with time is indicated by three time steps. In (b) the locations of ocean floor anomalies are also shown.

Figure 18.3 Sketch of a mesoscopic transfer fault connecting two veins.

the spreading rate on ridge segment B. As a consequence, the length of the transform fault remains constant over time. In Figure 18.4b, the transform fault connects two triple junctions (T), each of which is the intersection between transform, ridge, and trench. As subduction of the ridge continues, the two triple junctions move apart, and the length of the transform grows. In Figure 18.4c, the transform fault links a ridge and a trench. In this case, the fault length decreases with time as it is subducted if the subduction rate is faster than the spreading rate.

• The amount of displacement along the length of a transfer fault, averaged over time, is constant, if the fault length is constant or decreasing. For example, the displacement at point X on the fault in Figure 18.5a is

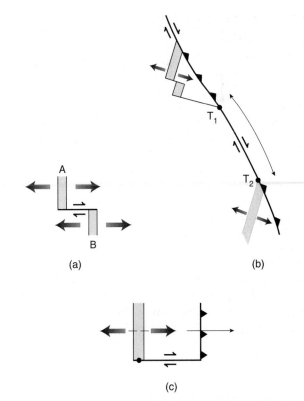

(a)

(b)

(c)

Figure 18.4 Examples showing how the length of a transform fault can change with time. (a) Length stays constant if spreading rates on both ridge segments are constant. (b) Length increases as the two triple junctions (heavy dots) move apart. (c) Length decreases if the spreading rate is less than the subduction rate.

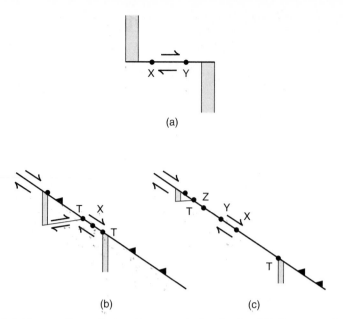

(a)

(b)   (c)

Figure 18.5   (a) If the transform initiated at the same time as the two ridge segments, then the slip at X is the same as the slip at Y. (b/c) If the transform length changes with time, then the amount of slip varies along the length. The slip at X is greater than the slip at Y and Z.

the same as the displacement at point Y on the fault. If the length of a transfer fault is increasing, however, the amount of displacement on the younger portions of the fault is less than the amount on the older portions, as illustrated in Figure 18.5b.

- Displacement across a transfer fault can be much greater than the length of the fault itself. For example, consider a 10 km long transform fault connecting two segments of the Mid-Atlantic Ridge. If there has been 1000 km of spreading on the ridge segments, there must be 1000 km of displacement on the 10 km long transform.
- Transform faults form small-circle plate boundaries around the pole of rotation (Euler pole) describing the relative motion between the two plates that are juxtaposed at the transform fault (Figure 14.13).

Transcurrent faults differ from transfer faults in a number of ways. To start with, transcurrent faults are not plate boundaries, and thus do not necessarily terminate at a plate boundary (or the mesoscopic equivalent of a plate boundary if you're looking at an outcrop-scale fault). In fact, transcurrent faults die out along their length. Transcurrent faults initiate at a point and grow along their length (Figure 18.6), so the displacement across a transcurrent fault is typically greatest near the center of the trace and reduces to zero at the ends of the fault. As a consequence, the displacement on a transcurrent fault is always less than the length of the fault.

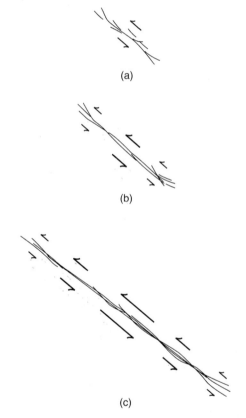

(a)

(b)

(c)

Figure 18.6   Growth of a transcurrent fault. As displacement increases, the length of the fault increases. The magnitude of slip at the ends of the fault is zero and the magnitude of slip near the center is maximum.

18.3   FEATURES OF CONTINENTAL STRIKE-SLIP FAULTS

Major strike-slip faults, meaning ones that have trace lengths in the range of tens to thousands of kilometers, are not simple planar surfaces, particularly when they occur in continents. Typically, such faults have locally curved traces, divide into anastamosing splays, and/or occur in association with subsidiary faults and folds. In this section,

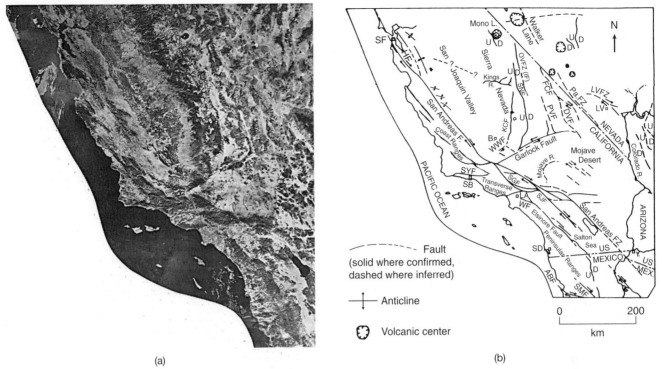

(a)

(b)

Figure 18.7 Fault of California, southwestern USA. Satellite photo (a) and labeled sketch of the area (b). Abbreviations for cities are: B = Bakersfield; LA = Los Angeles; LV = Las Vegas; SB = Santa Barbara; SD = San Diego; SF = San Francisco; Y = Yuma. Abbreviations for faults are: DVF = Death Valley Fault; FCF = Furnace Creek Fault; HF = Hayward Fault; KCF = Kern Canyon Fault; LVFZ = Las Vegas Fault Zone; OVFZ = Owens Valley Fault Zone; PVF = Panamint Valley Fault; SGF = San Gabriel Fault; SJF = San Jacinto Fault; SNF = Sierra Nevada Fault; SYF = Santa Ynez Fault; WF = Whittier Fault; WWF = White Wolf Fault.

we will describe the complexities of large continental strike-slip faults and suggest why these complexities occur.

18.3.1 Distributed Deformation

Commonly, a strike-slip zone acts as a region of distributed shear, which leads to formation of complex structures. During movement in strike-slip fault zones, slip is distributed among the many subparallel faults that compose the zone. For example, the San Andreas Fault "proper" is only one of about ten major faults and literally thousands of minor faults that slice up California (Figure 18.7). Together, this array of faults composes the San Andreas Fault Zone. Within a fault zone, faults may bifurcate and merge along strike (i.e., anastomose), thereby surrounding blocks of crust (Figure 18.7). The region between the endpoints of two parallel but non-coplanar faults is a *stepover* (see Section 18.3.2), so named because displacement in the fault zone "steps over" from one fault to the other. Individual faults in fault zones die out along strike.

Locally, large faults in the zone are bordered by arrays of smaller strike-slip, normal, or reverse faults. The most common subsidiary strike-slip faults define an *en echelon* array in which individual faults intersect the main fault at an angle of about 15°. These faults are called *Riedel shears* (R-shears), and possibly initiated prior to the formation of the main throughgoing fault (see Chapter 8). Other subsidiary strike-slip shears that initiate along large strike-slip faults include R'- and P-shears (Figure 18.8; see also 8.19b).

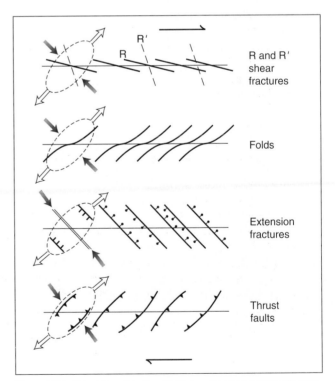

Figure 18.8 Orientation of fault and folds in a right-lateral strike-slip zone.

En echelon arrays of thrust faults, folds, and normal faults also cut the crust adjacent to major strike-slip faults. Typically, the thrust faults and folds trend at an angle of

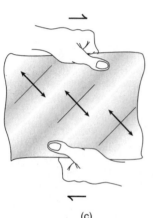

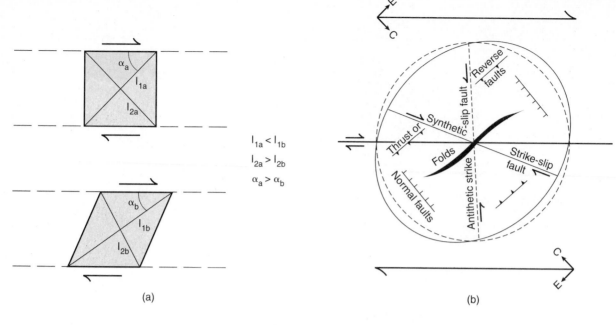

$$I_{1a} < I_{1b}$$
$$I_{2a} > I_{2b}$$
$$\alpha_a > \alpha_b$$

(a)

(b)

(c)

Figure 18.9 (a) Illustration of distributed strain in a strike-slip fault zone (map view), showing how the crust shortens in some directions and stretches in others. Note also that the reference lines rotate. (b) Composite diagram showing the relationship between subsidiary structure geometry, strain ellipse (map view), and shear sense. (c) Paper model of fold generation. (d) Side-looking radar image of *en echelon* anticlines in left-lateral shear (Darien Basin, eastern Panama). Width of view is ~50 km.
(b) Sylvester and Smith (1982).

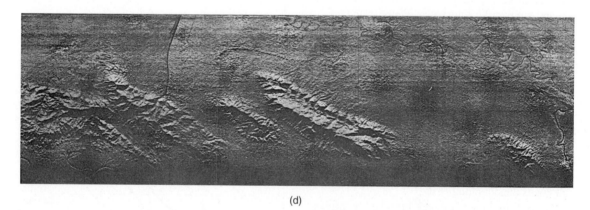

(d)

45° or less to the main fault, and the acute angle between subsidiary thrust faults and the main fault (or between the fold hinges and the main fault) opens in the direction of shear. Normal faults trend at an angle of 45° or more with respect to the main fault, but the acute angle between the subsidiary normal faults and the main fault opens opposite to the direction of shear (Figure 18.8).

To picture why *en echelon* arrays of thrusts, folds, and normal faults develop in a strike-slip zone, picture a block of crust that is undergoing shear in map view (Figure 18.9). The block may develop simple shear strain prior to the breakthrough of the main strike-slip fault, deforming a square into a rhomb while changing line length (Figure 18.9a). As we saw in Chapter 4, a reference circle drawn on

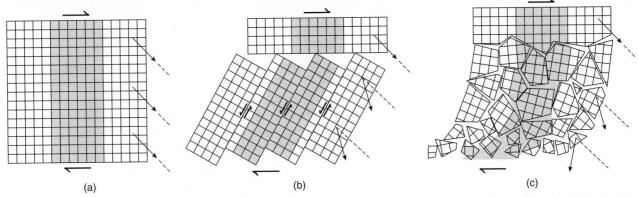

Figure 18.10 Mechanisms of block rotation in a right-lateral strike-slip zone. (a) An initially undeformed region. (b) Displacement is accommodated by slip on faults at a high angle to the zone (bookshelf model); rotation is the same for each block. (c) The zone is broken into smaller blocks that rotate different amounts; note large rotations that may occur locally. The amount of rotation is tracked by a hypothetical paleomagnetic declination in each block (solid arrows) relative to a reference direction (dashed line).

the block changes into an ellipse. In the direction parallel to the short axis of the ellipse, the crust shortens, and this shortening is accommodated by thrusting and folding (Figure 18.9b). You can simulate this process by shearing a piece of paper between your hands, as shown in Figure 18.9c. The ridges that rise in the center of the paper represent folds that develop because the top surface of the paper is unconfined and can buckle up into the air. In the direction parallel to the long axis of the ellipse, the crust stretches, and this stretching is accommodated by normal faulting. Eventually, the main strike-slip fault slices the block in two, and the two ends of the subsidiary faults and folds are displaced with respect to each other.

Progressive slip on strike-slip faults may result in rotation of crustal blocks about a vertical axis. To visualize this process, we return to the image of strike-slip faulting in Figure 18.9a. Notice that the cross in the middle of the original reference circle has rotated clockwise relative to its initial position. If the block itself is cut up by numerous parallel strike-slip faults, the overall rotation of the block can be accommodated by slip on the faults (Figure 18.10b), analogous to the bookshelf model for rotational planar normal faults (Chapter 8). Alternatively, small crustal blocks may rotate about a vertical axis like ball bearings (Figure 18.10c). Such rotations can be tracked by paleomagnetic analysis, which compares the declination in each crustal block with a stable reference direction (typically a cratonic direction) of the same age.

As we mentioned earlier, transcurrent faults do not continue forever, but die out along their length. Typically, transcurrent faults terminate by splaying into a fan of smaller faults (a *horsetail*) that curve away from the trend of the main fault. Depending on the direction of curvature with respect to the displacement sense, thrust or normal components of displacement occur in the horsetail, and these movements will be accompanied either by folding and uplift, or by normal faults and subsidence (Figure 18.11).

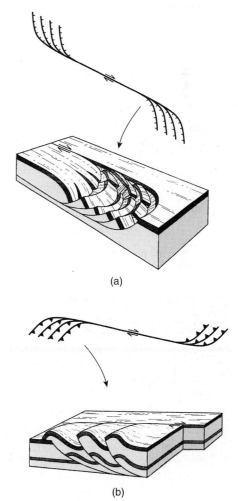

Figure 18.11 Terminations along strike-slip faults.

18.3.2 Transpression and Transtension

The trace of the San Andreas Fault for much of its length is not just a featureless line on the surface of the earth, nor is it a simple trough caused by preferential erosion of the

Figure 18.12   Folds developed in response to transpression along a segment of the San Andreas Fault (near Palmdale, California).

weak fault gouge and breccia that formed along the fault trace. The fault is commonly marked by a train of marshy depressions (called sag ponds), and by ridges rising as much as 100 m above the surrounding desert. Locally, intense folding accompanying the fault trace is visible (Figure 18.12). The presence of such topographic and structural features tells us that motion along the fault is not perfectly strike-slip. Ridges indicate that *transpression* is occurring at the fault, meaning that the strike-slip displacement is accompanied by shortening across the fault. This shortening results in thrusting and uplift within or adjacent to the fault zone. In places where the fault zone contains a thick band of weak gouge and breccia, these fault rocks are squeezed up into a fault-parallel ridge, much like a layer of sand would be squeezed up between two wood blocks that are pushed together (Figure 18.13a). In contrast, topographic depressions indicate that *transtension* is occurring in the fault zone, meaning that strike-slip is accompanied by extension across the zone. This extension results in normal faulting and subsidence (Figure 18.13b).

The dimensions of transpressive or transtensile structures forming along a strike-slip fault depend on the amount of cross-fault strain. Where relatively little transpressive or transtensile deformation has occurred, the cross-fault displacement results in relatively small ridges or sags with relief that is less than a couple of hundred meters. If, however, the amount of transpression or transtension on a fault has been active over a long time (e.g., millions of years), significant mountain ranges develop adjacent to the fault. For example, transpression along the Alpine Fault of New Zealand has resulted in uplift of the Southern Alps, a range of mountains that reaches an elevation of nearly

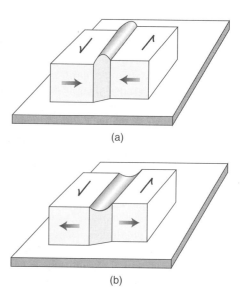

Figure 18.13   Simple wood block models illustrating the concept of transpression and transtension. (a) When blocks shear and squeeze together (transpression), weaker intervening material (e.g., sand) is pushed up. (b) When blocks shear and pull apart (transtension), intervening material sags.

4 km. Similarly, long-term transtension led to the development of the Gulf of California.

Seismic-reflection studies of large continental strike-slip faults indicate that faults associated with transpressional or transtensional zones within strike-slip systems are concave downwards, and merge at depth into a main vertical fault plane (Figure 18.14). Thus, in cross section, large continental strike-slip faults resemble flowers, in profile, with the petals splaying outward from the top of a

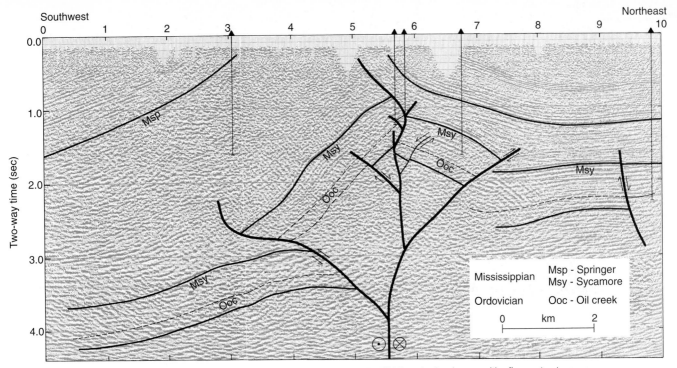

Figure 18.14 Reflection seismic profile across a strike-slip fault in the Ardmore Basin (Oklahoma), showing a positive flower structure. Harding and Lowell (1979).

stalk. This configuration of faults is, not surprisingly, referred to as *flower structure.*[2] On transpressive faults, apparent offset in cross section is in a thrust sense, and the result is a *positive flower structure;* whereas on transtensile faults, apparent offset in cross section is in a normal sense, and the result is *negative flower structure* (Figure 18.15).

In some cases, transpression or transtension occurs along the entire length of a fault zone because the zone is oblique to the vectors describing relative movement of blocks across the fault. Such a situation occurs when global patterns of plate motion change after the formation of the fault, but because a fault is a material plane in the Earth, it cannot change attitude once formed. Consequently, fault behavior rather than the geometry changes to accommodate the new motion. Alternatively, transpression and transtension may occur only at bends in a fault trace (see also Section 8.2.4). A bend along a strike-slip fault at which transpression occurs is a *restraining bend,* and a discrete bend at which transtension occurs is a *releasing bend* (Figure 18.16). At some bends, the region adjacent to the bend develops into a strike-slip duplex. A *strike-slip duplex* consists of an array of several faults that parallel the bend in the main fault and link segments of the main fault at either end. As viewed on a map, this array resembles a thrust-fault duplex (Figure 17.9) or a normal-fault duplex as viewed in

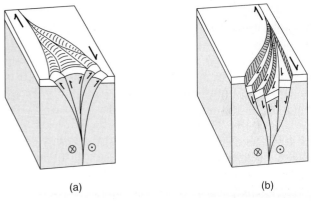

     (a)   (b)

Figure 18.15 Block diagrams illustrating (a) positive flower structure and (b) negative flower structure.

cross section, depending on the sense of bend and fault displacement. Strike-slip duplexes are contractional at restraining bends or extensional at releasing bends (Figure 18.16). The former yields a positive flower structure, whereas the latter forms a negative flower structure (Figure 18.15).

At restraining bends, pieces of crust on opposite sides of the fault push together, causing crustal shortening. If the shortening is substantial enough, a fold-thrust belt forms at the bend and a mountain range is uplifted. For example, the Transverse Ranges just north of Los Angeles in southern California developed because of shortening across a large restraining bend along the San Andreas Fault (Figure 18.7).

---

[2]Some authors prefer the term *palm structure* for the same arrangement of faults.

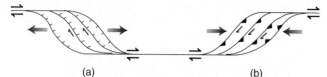

Figure 18.16 Releasing and restraining bends and strike-slip duplexes. (a) A transtensile duplex, (b) a transpressive duplex.

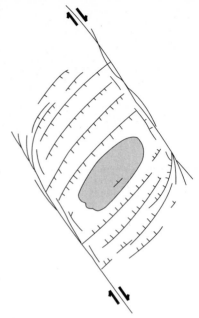

Figure 18.17 Schematic diagram of a pull-apart basin at a stepover.

At releasing bends, or stepovers, pieces of crust on opposite sides of the fault are pulling apart, normal faults develop, and a basin forms. Basins formed in this manner are called *pull-apart basins* (Figure 18.17). The amount of subsidence in a pull-apart basin depends on the size of the basin and on the amount of extension. Formation of small basins probably involves only brittle faulting in the upper crust. In contrast, formation of large basins probably involves thinning of the entire lithosphere, with the result that after extension has ceased, the floor of the basin thermally subsides (see Chapter 15), leading to development of a larger sedimentary basin. Examples of present-day pull-apart basins include the Dead Sea, at the border between Israel and Jordan, which subsided so far below sea level that it has the lowest elevation on Earth.

Because different segments of a fault may be subjected to transpression or transtension simultaneously, a region of crust moving along the fault may at one time be subjected to transtension, and then at a later time be subjected to transpression. During the transtension stage the crust is stretched and develops into a negative flower structure covered by a local sedimentary basin. When this crust is subsequently subjected to transpression, the negative flower structure is inverted and becomes a positive flower structure. As a consequence, the contents of the local basin are thrust up and over the margins of the former basin. Overall, blocks of crust moving along a major strike-slip fault rise and fall like a dolphin jumping and diving as it follows a ship.

18.3.3 Insights from Analog Models

In the shallow crust, continental strike-slip faults are complex deformation systems that include *en echelon* faults and folds, subparallel splays, Riedel shears and other subsidiary shears, contractional and extensional duplexes, restraining and releasing bends, and flower structures. Why are these zones so messy? A simple laboratory experiment, which we also used in Chapter 8, provides some insights.

Take a homogeneous clay slab and place it over two adjacent wooden blocks. The clay cake represents the weaker uppermost crust, and the wooden blocks represent stronger crust at depth. Now, push one of the blocks horizontally so that it shears past its neighbor. As the blocks move relative to one another, the clay cake begins to deform. Initially, deformation is manifested by arrays of small strike-slip faults (R-, R′-, and P-shears), as well as *en echelon* normal faults, thrust faults, and folds (Figure 18.18).

Throughout the zone, tiny extension fractures develop. Thus, when the main throughgoing strike-slip fault finally develops in the clay cake, it forms by the linking of smaller faults in an already intensely fractured zone. What this model illustrates is that the complexity of strike-slip fault zones in the uppermost crust is due to the weakness of the crust. In weak materials, deformation tends to be distributed over a broad zone. If, instead of using a homogeneous clay cake, we use a cake that is *inhomogeneous* (e.g., a cake containing embedded pebbles or preexisting fractures), then the deformation pattern would be even more complex, because perturbations to the local stress field caused by the inhomogeneities deflect the path of the fault, thereby causing bends and stepovers to develop. Such an inhomogeneous cake would be a closer analog, because continental crust contains many different rock types and preexisting structures. Nonetheless, our simple homogeneous clay-cake experiment is able to explain many first-order features that are observed in natural strike-slip zones.

18.3.4 A Note on Deep-Crustal, Strike-Slip Fault Geometry

In this chapter we primarily discuss the manifestation of major strike-slip faults in the near-surface regime of the Earth. But what do such faults look like at progressively greater depths in the crust? This question is difficult to answer conclusively, because we cannot directly examine active faults at depth. Based on exposed deep-crustal sections of ancient faults, the fault zone first becomes narrower as the crust becomes stronger. Then, below the brittle-plastic transition, the fault zone becomes wider again as the crust becomes weaker. This is the classic geometry of shear zones presented in Chapter 12 (Figure 12.2).

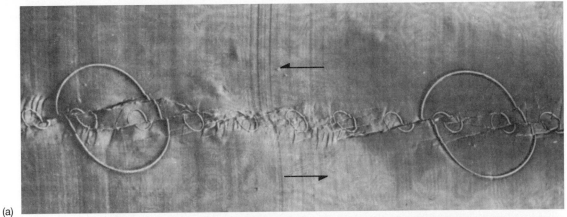

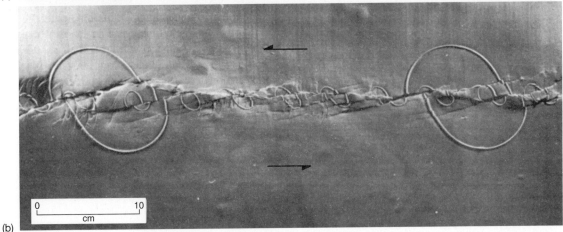

Figure 18.18 Map view of a clay layer that is progressively deformed by left-lateral strike-slip motion on underlying blocks. Note synthetic R shears. Width of view in each frame is ~55 cm.

Recent studies, however, show that the concept of a strike-slip fault zone cutting down through the entire crust as a continuous vertical zone of deformation may be inaccurate. There is growing evidence that some strike-slip faults are offset at *subhorizontal detachments* at depth, so that the strike-slip displacement in the upper crust does not lie directly over displacements in the lower crust and underlying mantle lithosphere. In the context of this model, dip-slip faults that bound zones of transtension and transpression ramp into this detachment, and the rotation of fault-bounded crustal blocks is confined to the crust above this detachment. Such major detachments, which are also hypothesized for extensional and contractional tectonics (compare with Chapters 15 and 16), present a new perspective of lithospheric structure in which the upper crust behaves radically different from the lower crust and upper mantle. These intriguing ideas and their implications await further study.

18.4 CONTINENTAL SETTINGS OF STRIKE-SLIP FAULTING

Until now, we have focused our discussion of strike-slip fault zones on the structural features that occur in these zones. In this and the next section, we turn our attention to the different plate tectonic settings where strike-slip fault zones occur (Figure 18.19).

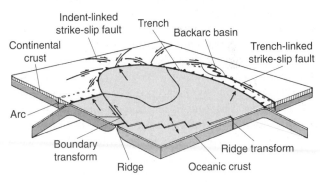

Figure 18.19 Plate tectonic settings of major strike-slip faults.

18.4.1 Oblique Convergence and Collision

Strike-slip faults occur along convergent margins where the vector describing the relative motion between the subducting and overriding plates is not perpendicular to the trend of the convergent margin (Figure 18.20a). At such *oblique convergent margins,* the relative motion between the two plates can be resolved into a component of dip-slip motion perpendicular to the margin, and a component of strike-slip motion parallel to the margin. Present-day examples of oblique convergent margins illustrate that the strike-slip component of offset affects a variety of locations across the margin, including the accretionary wedge, the volcanic arc, and the backarc region. Movement on strike-slip faults can dice up and displace these elements of convergent margins

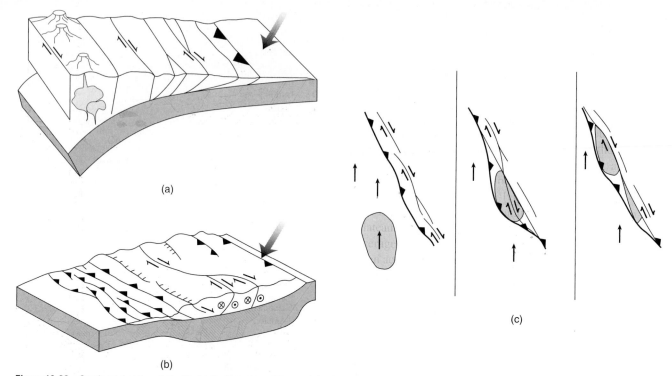

Figure 18.20 Continental strike-slip faults. (a) Faulting at an oblique subduction system; (b) strike-slip faulting at an oblique collisional margin; (c) stages of terrane accretion, showing oblique convergence, collision, and subsequent dismemberment of the terrane.

relative to one another, with the result that pieces may be missing or duplicated along a given cross section perpendicular to the margin.

Partitioning of movement into dip-slip and strike-slip components of motion also accompanies the oblique collision of two buoyant lithospheric masses (Figure 18.20b). The strike-slip component of motion displaces fragments of crust laterally along the orogen. For example, Wrangellia, an exotic crustal block that was incorporated into the western margin of North America during the Mesozoic, was sliced by strike-slip faults into fragments that were transported along the margin. As a result, pieces of Wrangellia are now found all the way from Idaho to Alaska (Figure 18.20c; see also Figure 16.27).

Strike-slip faulting also develops at collisional margins when one continent indents the other. For example, during the Cenozoic collision of India with Asia, India pushed northward into Asia, which resulted in a fan-shaped pattern of strike-slip faults in the east (Figure 16.24). The vise created when two continents converge may cause blocks of crust caught between the colliding masses to be squeezed laterally out of the zone of collision. The boundaries of these displaced blocks are strike-slip faults. At the western boundary of the Indian microcontinent, this relative motion is accommodated by sinistral strike-slip along the Chaman Fault Zone.

18.4.2 Fold-Thrust Belts

The traces of thrust faults that are perpendicular to the regional transport direction dominate the map pattern of fold-thrust belts. Locally, however, these belts contain strike-slip faults whose traces trend at a low angle to the regional transport direction (Chapter 17). Some of these *lateral ramps* cut detached thrust sheets into pieces that move relative to one another. Others accommodate along-strike changes in the position of a frontal ramp with respect to the foreland. For example, lateral ramps bound the two ends of the Pine Mountain Thrust Sheet in the southern Appalachians (Figure 18.21). Still other fold-thrust belts contain *conjugate strike-slip systems,* which simultaneously shortens thrust sheets across strike and stretches them along strike (e.g., the Makran fold-thrust belt in southern Pakistan and the Jura fold-thrust belt of Switzerland and France).

18.4.3 Rifts

Strike-slip faults in rifts accommodate offsets in the axis of rifting, as well as along-strike changes in the magnitude and vergence of extension (see Figure 15.15). For a particularly instructive example, we take a look at the Garlock Fault in southern California (Figure 18.22). The Garlock Fault, which is an ~250 km long left-lateral transfer fault, forms the northern border of the Mojave Desert and inter-

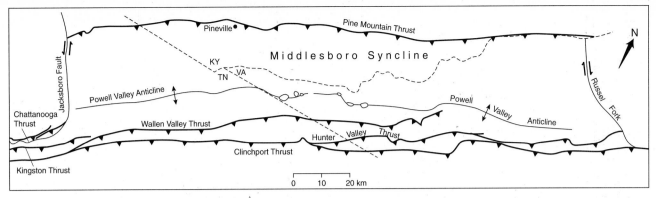

Figure 18.21 Map of the Pine Mountain Thrust System in the southern Appalachians, (eastern USA), showing lateral ramps (Jacksboro Fault and Russell Fork).

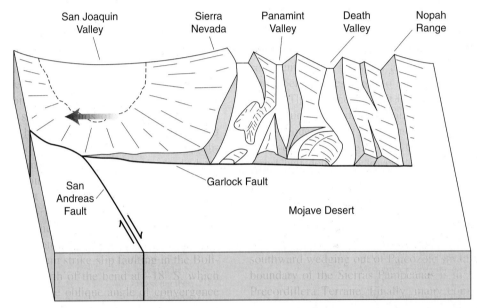

Figure 18.22 Block model of the Garlock Fault, southern California (USA), accommodating the offset from the extended Basin-and-Range Province to the north and the Mojave Desert to the south.

sects the San Andreas Fault at its western end. It defines the boundary between two portions of the Basin-and-Range Province that have not developed comparable extensional strains. Because of the greater magnitude of extension to the north of the Garlock Fault, the San Andreas Fault has been moved left-laterally north of its intersection with the Garlock Fault. This displacement created a major restraining bend in the right-lateral San Andreas Fault Zone, which is responsible for development of the Transverse Ranges just north of Los Angeles (Figure 18.7).

18.5 OCEANIC TRANSFORMS AND FRACTURE ZONES

All ocean ridges along which seafloor spreading occurs are segmented into pieces that range in length from about 10 to 100 km. Each segment is linked to its neighbor by an

oceanic transform fault, the length of which varies from 10 to 1000 km (Figure 14.10). Oceanic transform faults are relatively narrow (less than a few kilometers wide), but even considering this narrow width, the frequency with which they occur is so high that between 1% and 10% of the oceanic lithosphere is involved in transform-fault related deformation.

Why do oceanic transforms occur in the first place? To answer this question, first remember that formation of transform faults does not postdate the formation of the ridge axis itself. Transform faults do *not* offset originally continuous segments of the ridge, but rather they initiate in association with the ridge segments themselves (Figure 18.2). In the case of ocean ridges that initiated as continental rifts, transform segments are the end product of the evolution of accommodation zones connecting rift segments (Figure 15.15). In the case of ridges that initiate in oceanic

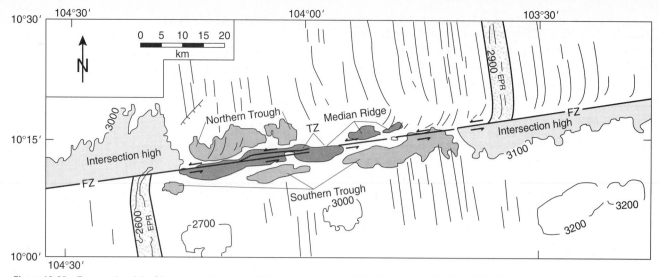

Figure 18.23 Topography of the Clipperton fracture zone (FZ) and transform zone (TZ) of the East Pacific Rise (EPR). Note intersection highs at ridge tips, and trough and ridges along the transform zone. Contours in meters below sea level.

crust due to the jumping of a spreading center from one locality to another, or due to the reorientation of a spreading center, transform faults are the end product of the interaction between two rift tips.

It is very important to realize that transform faults occur only between ridge segments; they do not extend beyond! There are, however, pronounced topographic lineaments, known as *fracture zones,* that extend beyond the tips of the ridge segments. Seismicity in the ocean floor shows that earthquakes occur along the ridge segments and along the transform faults between them, but that they are rare along fracture zones. Thus, fracture zones are not active fault zones.[3] Oceanic transforms have a second distinction. When you cross a transform boundary, you cross from one plate to another and there is differential movement (Figure 18.2). However, when you cross a fracture zone, you stay on the same plate, and, therefore, differential movement is not occurring. Nevertheless, because the age of the plate is different on opposite sides of a fracture zone, they are geological contacts that commonly have a topographic expression.

In modern oceans, transforms and fracture zones are linear features of topographic complexity, commonly including high escarpments, ridges, and narrow troughs (Figure 18.23). Along fracture zones, segments of the same plate, but with different ages, are in contact. Because depth of the ocean floor depends on its age, we expect to see a change in seafloor depth as we cross the fracture zone, with the older portion of the plate being deeper. Such a contrast in depth is observed away from the fracture zone, but is not so clear near the zone. Other factors come into play at the fracture

zone itself, including: (1) transpression and transtension across the fracture zone, leading to formation of flower structure, (2) serpentinization of olivine-rich rocks, which decreases the density of the crust and causes isostatic rise, and (3) heating of the plate opposite the tip of a ridge segment, causing the plate to become less dense and rise (creating the *intersection high;* Figure 18.23).

Direct insight into the structural character of transform faults and fracture zones comes from studies of exposed *ophiolites* (slices of oceanic lithosphere that have been emplaced on continental crust). Transform faults in ophiolites are zones of intense shear, manifested by brecciation and serpentinization. Serpentinite, a rock type formed by hydrothermal alteration of (ultra)mafic rocks, is relatively weak and, thus, tends to shear easily. Moreover, extensive slumping along the margins of fracture zones causes sediment and volcanic material to be mixed together, leading to the formation of sedimentary breccias and debris flows.

Examination of the global plate boundaries map in Figure 14.10 reveals that not all oceanic transform faults offset ridge segments. Some are plate boundaries that link non-coplanar subduction zones, or link subduction zones to ridges. Examples include the north and south border of the Scotia Plate and parts of the north and south border of the Caribbean Plate. The Alpine Fault of New Zealand connects along strike to oceanic transforms, resulting in a strike-slip system that links two oppositely-dipping subduction zones. At its north end the San Andreas Fault connects to the Mendocino transform fault that links the Juan de Fuca spreading ridge to the Cascadia subduction zone (the Mendocino triple junction). At its south end, the San Andreas Fault connects to a transform fault that terminates at a ridge segment in the Gulf of California.

Finally, lithospheric-scale transfer faults (or transforms) are not only a feature of ocean basins. Where they

---

[3]Some authors, therefore, prefer the term "inactive transform."

cross continental lithosphere, they are manifested as major strike-slip fault zones. You are already familiar with two prominent examples of such continental transforms, namely, the San Andreas Fault of California and the Alpine Fault of New Zealand, so we will not repeat the details here.

18.6 CLOSING REMARKS

With this chapter on strike-slip tectonics, we close our general discussion of tectonic settings. To this point, we have mainly focused on processes and terminology, with the goal of developing a vocabulary of concepts and words to describe tectonic settings. You will have noticed that we mentioned several features in more than one of the tectonics chapters. For example, strike-slip faults are not only described in this chapter, but are also mentioned in the context of extension and contraction, where appropriate. This necessary redundancy emphasizes that structures are not limited to one tectonic setting, as you will see in the remainder of this book. In Chapter 19 our structural and tectonic vocabulary will be applied to the examination of specific regions around the world.

ADDITIONAL READING

Aydin, A., and Nur, A., 1982, Evolution of pull-apart basins and their scale independence, *Tectonics,* v. 1, p. 91–105.

Freund, R., 1974, Kinematics of transform and transcurrent faults, *Tectonophysics,* v. 21, p. 93–134.

Naylor, M. A., Mandl, G., and Kaars-Sijpestein, C. H., 1986, Fault geometries in basement-induced wrench faulting under different initial stress states, *Journal of Structural Geology,* v. 7, p. 737–752.

Nelson, M. R., and Jones, C. H., 1987, Paleomagnetism and crustal rotations along a shear zone, Las Vegas Range, southern Nevada, *Tectonics,* v. 6, p. 13–33.

Sylvester, A. G., 1988, Strike-slip faults, *Geological Society of America Bulletin,* v. 100, p. 1666–1703.

Tchalenko, J. S., 1970, Similarities between shear zones of different magnitudes, *Geological Society of America Bulletin,* v. 81, p. 1625–1640.

Wilcox, R. E., Harding, T. P., and Seely, D. R., 1973, Basic wrench tectonics, *American Association of Petroleum Geologist Bulletin,* v. 57, p. 74–96.

Woodcock, N. H., and Fischer, M., 1986, Strike-slip duplexes, *Journal of Structural Geology,* v. 8, p. 725–735.

Chapter 19
Perspectives of Regional Geology—Essays

19.1 GENERAL INTRODUCTION

In Chapters 15 through 18, we introduced you to the great variety of tectonic settings in which deformation occurs. To put deformation in context, we have not limited the discussion to only structural features of these settings, but we have also presented basic information concerning petrologic and sedimentological features that accompany deformation. Most of our discussion focused on what happens at the three basic types of plate margins, in response to plate interaction. This ranges from what happens during the inception of a divergent margin (rifting) to the death of a convergent margin (collision ± strike-slip). Tectonism that occurs in these various settings is dramatic, and results in belts of deformation, regional metamorphism, and igneous activity. These belts are known as *orogens,* and the processes that create them are captured in the term *orogeny.*

Much of what is known about plate tectonic processes comes from study of Mesozoic and Cenozoic plate margins, rifts, and collision zones. Plate interactions, however, leave a permanent scar in the lithosphere, even long after the associated topography has eroded away, and thus through field study we are able to get a fairly thorough understanding of ancient plate tectonic events in our planet's history (Figure 19.1.1).

We feel that to understand the nature and consequences of plate tectonic processes, including orogeny, it is best to study natural examples. Thus, we devote the final chapter of this book to case studies of important deformation belts. Each case study is written by an expert, in his or her own style, so that you will get a flavor of how different geologists think and approach tectonics. When you talk with such seasoned geologists, chances are that they show great excitement about the area in which they work or have worked. This may stem from discovering key outcrops for understanding regional deformation or maybe new insights into an aspect of fundamental significance for crustal evolution. We have tried to preserve this excitement in the essays, which are not intended to be comprehensive review papers.

When you collect additional reading material on these and other examples, you will rapidly learn that views vary and, sometimes, contradict one another. The data, of course, stand, but interpretation always remains open to alternative views, which is especially the case for regional tectonics. You will find that the essays emphasize the

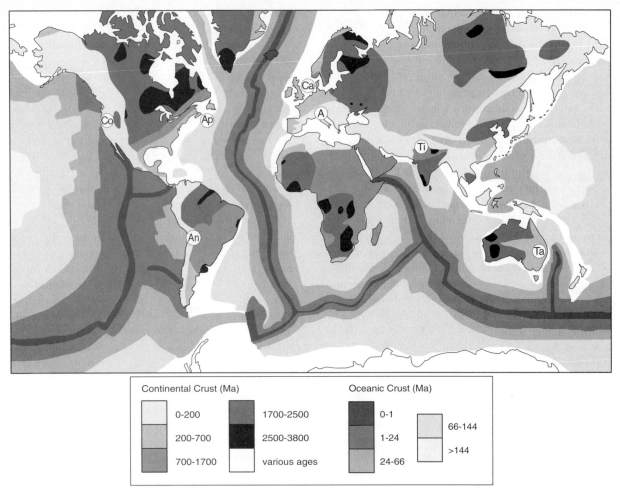

Figure 19.1.1 Areas of the world that were deformed during various geologic periods. In most continents the oldest deformed rocks (basement) are covered by younger sedimentary rocks (cover). Areas that are described in the essays are A: European Alps (Section 19.2), Ti: Himalaya/Tibet (Section 19.3), Co: North American Cordillera (Section 19.4), An: Andes (Section 19.5), Ca: Caledonides (Section 19.6), Ap: Appalachians (Section 19.7), Ta: Tasmanides (Section 19.8), Precambrian North America (Section 19.9), and North American cratonic cover sequence (Section 19.10).

large scale, yet the scenarios are generally based on smaller scale observations. This characteristic follows the approach we have taken throughout the book: observations on different scales are not separate and unrelated entities; rather, they form part of an integrated framework to study deformation.

19.1.1 Global Deformation Patterns

The most impressive deformation features that are exposed at the Earth's surface are concentrated along narrow belts at active plate margins. Today's active mountain belts, such as the European Alps (Section 19.2), the Tibetan region (Section 19.3), the North American Cordillera (Section 19.4), and the Andes of South America (Section 19.5) mark convergent plate boundaries. From a human perspective, the term 'active' mainly means earthquakes and volcanoes, but from a geologic perspective this term also implies continuing relative displacements at these margins over millions of years. Ancient mountain belts, such as the Caledonides (Section 19.6), the Appalachians (Section 19.7), and the Tasmanides (Section 19.8) were formed at plate

boundaries that have disappeared. We no longer see the mountainous physiography of orogens (Section 19.9) preserved (i.e., there is little or no related topography). Yet these flat, inactive regions of continental interiors (called *cratons*) preserve deeply eroded levels of once vast mountain belts that formed from tectonic processes similar to those active today. Not all deformation, however, takes place at plate margins, rifts, or collisional zones. For example, some of the great historical earthquakes occurred within continental interiors (e.g., the 1811/1812 New Madrid earthquakes in central North America). Tectonic activity is also seen by the occurrence of large intracratonic basins (some containing several km of sediment) and arches, and massive rift zones (such as the East African Rift and the Midcontinent Rift of North America). Thus, we conclude the essay series by discussing the fascinating deformational features of plate interiors (Section 19.10).

The continental regions of Earth preserve a long history, going back as far as 4 Ga, which is in marked contrast to today's ocean basins, where the oldest rocks are on the order of 200 Ma. It is perhaps ironic that, though many of

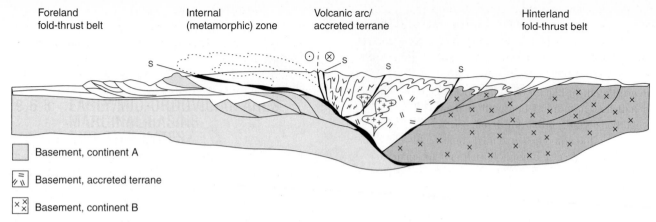

Foreland
fold-thrust belt

Internal
(metamorphic) zone

Volcanic arc/
accreted terrane

Hinterland
fold-thrust belt

☐ Basement, continent A

☐ Basement, accreted terrane

☐ Basement, continent B

Figure 19.1.2 Idealized section through a collisional orogen, showing foreland fold-thrust belt, metamorphic hinterland (with nappes), inverted passive margin, strike-slip plate boundary, accreted volcanic arc, accreted microcontinent, and sutures (S). Most if not all mountain belts contain several of the features shown in this ideal section, but none probably contains all of them. The diagram is based on observation in Phanerozoic mountain belts; most Precambrian belts preserve only the deeper crustal levels.

the fundamentals of plate tectonics were formulated from the study of today's ocean basins, if we wish to uncover the Earth's ancient plate tectonic history we have to focus our attention on the continents. When reading the essays you will find that we have a remarkably detailed understanding of this ancient history, especially for the Phanerozoic and increasingly so for the Proterozoic. The Archean, however, remains much less understood. This pattern merely reflects that rocks become sparser and have more complex histories with age, so our ability to study ancient tectonic history is limited. Yet, the pursuit of this understanding poses new challenges to field and laboratory geologists alike.

19.1.2 What Can We Learn from Regional Geology?

The remainder of the book contains regional geology perspectives. Each of these essays condenses a vast amount of information into relatively few pages. So the answer to the question posed in the header of this section could be: *a lot!* But it makes no sense to just read all the essays and memorize the respective history of these areas. Whereas the essays can be used as an introduction to continental tectonics of the world, several other purposes may also be served. We offer a few.

- The essays get you rapidly acquainted with fundamental geological aspects of some areas of the world. You can concentrate on one or two essays as a basis for a more in-depth study of a region.
- The essays show the various approaches that may be taken in the study of (ancient) mountain belts. When you look beyond the details of individual areas, you will find that stratigraphy, geochronology, geochemistry, geophysics, and other earth science disciplines are integrated to obtain an understanding of the region's tectonic history.
- The essays allow you to recognize fundamental features that are common to many areas. We will look at orogenic architecture as an example of one of these fundamental characteristics.

Orogenic architecture describes the broad geometry of a mountain belt (Figure 19.1.2). The details of each individual mountain belt differ, but many have several features in common. Generally you will find a thick wedge-shaped sedimentary sequence, containing marine carbonates if the area was located in the equatorial realm, which was deposited at the stable continental margin. Slivers of *ophiolite,* a rock assemblage containing ultramafics, gabbros, dikes, and pillow basalts, are remnants of ancient ocean floor[1] that are also preserved in an orogen. In fact, ophiolites are the pivotal evidence for the occurrence of modern-day plate tectonics in ancient mountain belts. Granites, associated with arc formation or melting of a thickened crust, are variably present. As the orogen evolves, marine clastics (sometimes called *flysch*) that are derived from the eroding mountain belt are deposited in foreland basins, and at the waning stages coarse continental clastics (sometimes called *molasse*) are laid down. In young orogenic belts we find that isolated slivers of basement rocks (*crystalline rocks*) and in some cases of mantle rocks have become exposed by brittle and ductile faulting. In ancient orogens this basement component significantly increases, and the sedimentary sequence is mainly preserved in metamorphosed and highly deformed rocks (*paragneiss*). The oldest mountain belts consist nearly entirely of deformed mid- to lower-crustal rocks of magmatic origin (*orthogneiss*). In a way, these ancient, deeply eroded orogens expose the roots of formerly active mountain belts, and in combination with currently active regions they provide a fairly complete section through orogenic crust.

Deformation is typically polyphase, and each phase in turn contains several fold generations. Within a single orogenic phase the deformation sequence may look something like this. Early contractional structures are thrusts that re-

---

[1]Geochemical evidence suggests that most ophiolites in mountain belts are, in fact, obducted backarc-basin oceanic lithosphere rather than main ocean basin.

peat stratigraphy, or large recumbent folds that repeat and locally invert stratigraphy. These thrusts often root in a detachment zone at depth. In metamorphic regions this early stage produced widespread *transposition*. These early structures are overprinted by upright folds that are associated with a regional axial plane foliation, whereas later generations are present as kink folds. These contractional features locally overprint evidence of the initial rifting stage (normal faulting) that formed the passive plate margin. In addition to early rifting, extensional structures may also form during the later stages of mountain building and subsequent unroofing (*synorogenic* and *postorogenic extension*, respectively; not shown in Figure 19.1.2).

Orogens are typically curved in map view (Figure 19.1.1), which may reflect the shape of the original plate margin or be a result of bending (*oroclinal bending*). Blocks with distinct lithologies and deformation history (tectono-stratigraphic blocks, nowadays called *terranes*) may be incorporated in the mountain belt, reflecting the accretion of oceanic plateaus, ocean islands, or fragments of disrupted continents to the active plate margin. The boundaries of these blocks are called *sutures*, and they may be marked by ophiolites. Other characteristics, such as progressive outboard-younging of deformation, may be present and you are encouraged to search for them in the essays that follow. While you may be interested in the geology of only one area, remember that knowledge of other areas quite often leads to understanding your own particular problem; that is why we need to study the literature and that is why we offer these regional perspectives.

19.1.3   Closing Remarks

You can imagine that a vast body of literature exists on regional deformation after some 150 years of mapping and associated laboratory work. The onset of the unifying plate tectonics concept in the 1960s also coincided with a publication explosion in the Earth sciences (all sciences, in fact). In the preface, the enormous volume of current literature was already mentioned, but mercifully each essay lists only some of the more informative references and makes no attempt to offer a comprehensive reading list. To these references we add two general textbooks that complement the information in the essays, and that include areas and topics not covered here. With all this information in hand, you should not find it too difficult to explore the literature on your particular area or topic of interest.

Of course, there are many other regions of tectonic interest in the world beyond those described in the essays that follow; our choices are merely a sampling of some reasonably well understood areas. Brand new discoveries continue to be made every day in these already well-studied regions, and many discoveries are waiting to be made in lesser known areas. As every scientist will tell you, our increasing knowledge (in our case, of deformation and tectonics) is usually accompanied by an increase in number of unanswered questions. This ensures the continued challenge for future generations of geologists. Happy reading!

ADDITIONAL READING

Condie, K. C., 1989, *Plate tectonics and crustal evolution* (third edition): Oxford, Pergamon Press, 476 p.

Windley, B. F., 1995, *The evolving continents* (third edition): Chichester, J. Wiley & Sons, 526 p.

19.2   The European Alps—John P. Platt[1]

19.2.1   INTRODUCTION

The European Alps (Figure 19.2.1) form part of a complicated system of arcuate mountain chains and young extensional basins created in the Mediterranean region during
convergence between the African and European Plates. The
Alpine chain is quite small in comparison to many mountain belts (it would fit easily into the state of California),
but it is probably the most intensively studied anywhere,
and is commonly regarded as the type example of a collisional mountain belt. The literature on the stratigraphy and
structure of the Alps extends back to 1840, and has been

published in French, German, Italian, and English, which
makes the study of the Alps both complicated and interesting. Alpine studies have contributed many basic concepts
to geology, as well as the words that express them (*nappe,
flysch,* and *molasse* are examples), though the meanings of
these words have changed in more general usage.

19.2.2   TECTONIC SETTING

In plate-tectonic terms, the history of the Alps began early in
Mesozoic time, when most of the continental masses of the
world were joined together in the supercontinent Pangaea.
During the Triassic, much of western Europe lay in the continental interior and experienced desert conditions. The associated deposition of continental clastics and evaporites was to
play an important role in later Alpine tectonism. Farther
south, in the shallow seas bordering the great oceanic gulf

---

[1]University College London, United Kingdom.

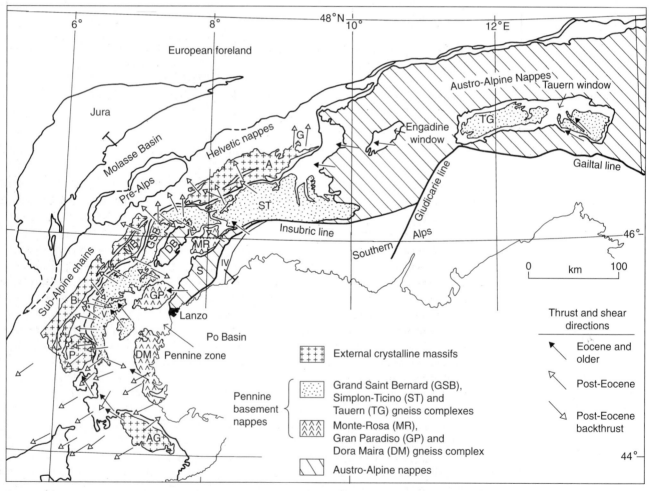

Figure 19.2.1   Sketch map of the Alps, showing the main tectonic units. Abbreviations: A = Aar Massif, AG = Argentera Massif, B = Belledonne Massif,
DB = Dent Blanche Nappe, DM = Dora Maira Massif, GP = Gran Paradiso Massif, GSB = Grand Saint Bernard Nappe, IV = Ivrea Zone, ST = Simplon Ticino gneiss
complex, TG = Tauern gneisses, V = Vanoise Massif. Thrust and shear directions are shown, as indicated by elongation lineations and sense of shear criteria.

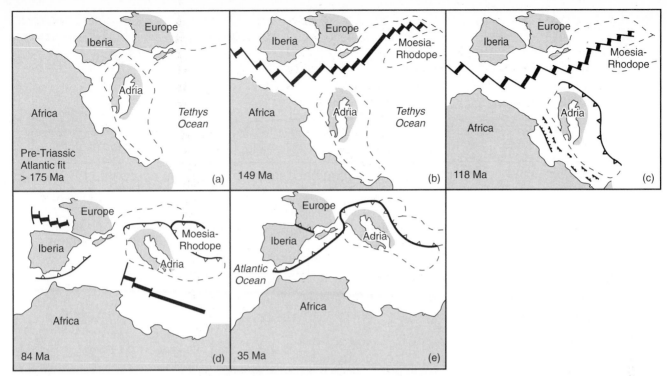

Figure 19.2.2 Plate-tectonic evolution of the Alpine region from Permian to Tertiary. Africa/Europe relative plate motions are from Dewey et al. (1989), based on analysis of Atlantic magnetic anomalies and transform faults. Each reconstruction corresponds to the time of a particular magnetic anomaly. (a) Pre-Triassic configuration: Africa, Europe, Iberia, and Adria all form parts of Pangaea, surrounding a gulf of the Tethys Ocean. (b) Late Jurassic. Opening of the southern and central Atlantic is transferred eastwards to open up the Neotethys Basin between Africa and Europe. Adria remains attached to Africa. (c) Mid-Cretaceous. Adria starts to rift away from Africa, and convergence begins on its northern margin. This causes high-P/low-T metamorphism of crustal rocks, now found in the Internal Zones of the Alps. (d) Late Cretaceous. Adria moves northwards away from Africa and towards Europe, forming an accretionary wedge on its northern margin. (e) Oligocene. Collision starts between Adria and Europe, to form the present collisional chain of the Alps.

From J. F. Dewey, et al., 1989, *Kinematics of the Western Mediterranean*, no. 45, pp. 265–284, Special Publications of the Geological Society, London. Used by permission.

known as Tethys, thick sequences of platform carbonates (now dolomitized) accumulated (Figure 19.2.2a).

Early in Jurassic time, Pangaea started to break up, with the opening of the central and southern Atlantic between Africa and the Americas. The rift extended into southern Europe between Iberia and Africa, opening a narrow strip of ocean that connected through to the Tethys (Figure 19.2.2b). It was the gradual closure of this ocean (sometimes known as Neotethys) that created the Alps, as well as other mountain chains around the Mediterranean.

The northern margin of the Neotethys was a classic passive continental margin that extended from southeastern France through Switzerland to Austria. It subsided steadily through the Mesozoic, accumulating several thousand meters of shallow-water limestone, passing oceanward into deeper water marl and shale (Figure 19.2.3). The southern margin of Neotethys in the region of the Alps was the rifted border of a continental block known as Adria. Adria is the continental crust that underlies the present shallow Adriatic Sea and adjacent areas, including northern Italy. The stratigraphy of the northern Adrian margin is dominated by thick sequences of Triassic platform carbonates (Figure 19.2.3) that are now spectacularly exposed in the southern Alps (the Dolomites).

The floor of the Neotethys Basin was at least in part oceanic, fragments of which are preserved as ophiolites

within the Alps. It also seems to have contained a considerable amount of stretched and thinned continental crust, which now forms large thrust slices of pre-Triassic crystalline rocks. Where this continental material lay within the Neotethys is uncertain. Some of it must have formed the stretched leading edges of the southern European and Adrian continental margins, but some may have been detached and isolated within the Neotethys (Figure 19.2.3).

During the Mesozoic, Africa (including Adria) moved southeast and then east relative to Europe, rotating slowly counterclockwise as it did so (Figure 19.2.2c). About the middle of the Cretaceous, Africa started to move northeast relative to Europe, and subduction of the Neotethys began on the northern margin of Adria (Figure 19.2.2d). From this time on, the Alps began to evolve by the progressive subduction and accretion of sediment and crustal slices from the floor of the Neotethys. At this stage the Alps could best be described as an accretionary wedge, similar to those that develop elsewhere above subduction zones. There is no evidence, however, for Cretaceous subduction-related magmatism. By Early Tertiary time, the Neotethys had closed, and the European continental margin entered the subduction zone (Figure 19.2.2e). This marks the start of continental collision, and shortening and deformation of both margins and of the Alps between continued until the Pliocene. Shortening now seems to have ceased.

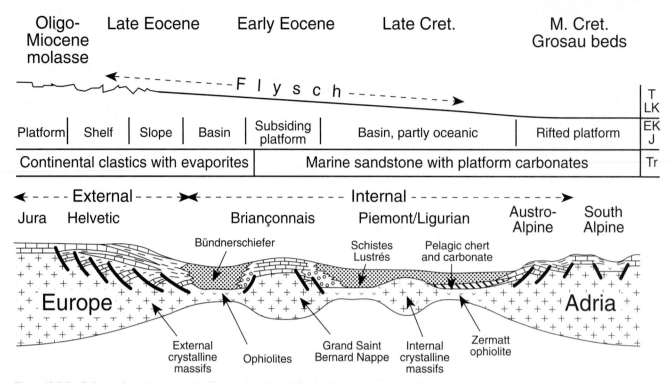

Oligo-Miocene molasse	Late Eocene			Early Eocene	Late Cret.	M. Cret. Grosau beds	
Platform	Shelf	Slope	Basin	Subsiding platform	Basin, partly oceanic	Rifted platform	T / LK / EK / J
Continental clastics with evaporites				Marine sandstone with platform carbonates			Tr

◄ - - - External - - - ►◄ - - - - - - - - - - Internal - - - - - - - - - - ►

| Jura | Helvetic | Briançonnais | Piemont/Ligurian | Austro-Alpine | South Alpine |

Bündnerschiefer · Schistes Lustrés · Pelagic chert and carbonate

Europe · Adria

External crystalline massifs · Ophiolites · Grand Saint Bernard Nappe · Internal crystalline massifs · Zermatt ophiolite

Figure 19.2.3 Palinspastic section across the Alps, to show the relative positions and tectonic settings of the various zones. The section is not drawn to scale, and the different parts did not lie along a straight line in the past (see Figure 19.2.2). The upper half of the diagram shows how depositional environments evolved with time in each zone. Tr = Triassic, J = Jurassic, EK = Early Cretaceous, LK = Late Cretaceous, T = Tertiary.

Convergence between Africa and Europe from mid-Cretaceous time onwards was dominantly northwards, and this is traditionally regarded as the direction of Alpine shortening. Structural data from thrust faults in the Alps, however, suggest that motion was everywhere roughly normal to the local trend of the chain right around the Alpine Arc (Figure 19.2.1). Taken at face value, this radial pattern of motion would require a large amount of extension along the trend of the Alps, for which there is little evidence. In fact, convergent motion in the Alps was probably separated into shortening normal to strike in the outer parts of the system, and shear parallel to strike (e.g., along east-west-trending segments of the Insubric Fault System) in the inner parts. Slip directions on thrust faults are therefore only indirectly related to the direction of relative plate motion. The only direction of plate motion that can easily explain the observed range of thrusting directions (Figure 19.2.1), however, is somewhere between northwest and west. This suggests that Adria may have moved independently of Africa during at least the later part of its history (Figure 19.2.2d and e).

19.2.3 STRUCTURE

From a structural point of view, the Alps can be divided into several regions with different structural styles and histories of deformation. These differences reflect differences in stratigraphy, tectonic setting, and the physical conditions and duration of deformation. The regions that lie on the outer part of the arcuate Alpine chain are known as the External Zones, and represent the former continental platform, shelf, and slope on the southern margin of Europe. The Internal Zones, on the other hand, include the now intensely deformed and metamorphosed contents of the Neotethys Ocean, as well as the leading (northern) margin of Adria. Because the direction of the Alpine subduction was southwards, beneath Adria, deformation started earlier in the Internal Zones and becomes younger towards the external side of the chain. The Internal Zones are also for the most part more deeply eroded, and hence show effects of metamorphism and a ductile style of deformation.

I will briefly discuss each of the main regions of the Alps in turn, starting with the youngest and simplest part of the system, on the external side of the chain.

(1) The Jura Mountains of northwestern Switzerland are morphologically quite distinct from the Alps, as the two are separated by the central Swiss plain. They are, however, the youngest expression of Alpine convergence, resulting from shortening of a thin platform-facies cover of mainly Jurassic limestones during the Late Miocene to Pliocene. This cover was detached from the underlying Paleozoic basement along the Triassic evaporite horizons, and the basement appears not to have been involved in shortening. The Jura is a classic example of thin-skinned tectonics, and

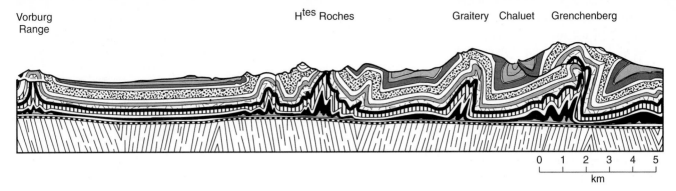

Vorburg
Range

H^tes Roches

Graitery　Chaluet　Grenchenberg

0　1　2　3　4　5
km

Figure 19.2.4　Section across part of the Jura Mountains, showing thin-skinned style of folding and thrusting above a basal décollement. Note the folded thrust at the southeast end of the section (to the right). The structure at depth is based on observations in the Grenchenburg railway tunnel. More recent seismic data confirm the lack of Alpine deformation in the underlying pre-Triassic basement.

shows a distinctive style that involved large-scale kink folds with both southeast and northwest vergence, disharmonic anticlines cored by Triassic evaporite and marl, and folded thrust faults (Figure 19.2.4). The northwestern margin of the Jura is firmly attached to European continental crust, so the shortening in the Jura requires the southern margin to have moved horizontally up to 40 km to the northwest. The décollement that carries this displacement passes beneath the Swiss plain and the main chain of the Alps, both of which therefore also moved to the northwest by the same amount during the formation of the Jura. This motion reflects the final stage of the convergence of Adria with Europe.

The shortening direction in the Jura, as indicated by stylolites, varies from north to west around the arc, mirroring the pattern in the Alps as a whole. In the case of the Jura, this appears to be accommodated by a component of arc-parallel

extension, achieved by slip-on conjugate sets of strike-slip faults that must be confined to the Mesozoic cover.

One of the puzzles of the Jura is why Alpine deformation jumped forward in Late Miocene time from the main Alpine front across the Swiss plain to the Jura. The Jura rises 1000 m or more above the Swiss plain, and has a fairly constant average elevation across its breadth. These features appear to violate the requirement that thrust wedges develop sequentially forward, maintaining a tapered geometry as they do so, but a section across the Swiss plain shows that a tapered geometry does in fact exist for the rocks above the Triassic detachment surface, which dips constantly southwards (Figure 19.2.5). The tapered wedge includes the Miocene foreland basin sediments in the Molasse Basin, which occupies the Swiss plain. Sedimentation has therefore contributed to the growth of the Alpine wedge, and has strongly influenced its mechanics.

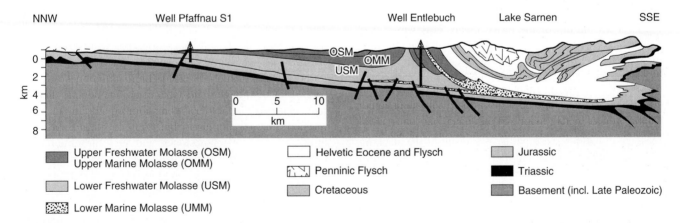

NNW　　　　Well Pfaffnau S1　　　　　　　　Well Entlebuch　Lake Sarnen　　　SSE

OSM
USM
OMM

km
0
2
4
6
8

0　5　10
km

Upper Freshwater Molasse (OSM)
Upper Marine Molasse (OMM)

Lower Freshwater Molasse (USM)

Lower Marine Molasse (UMM)

Helvetic Eocene and Flysch

Penninic Flysch

Cretaceous

Jurassic

Triassic

Basement (incl. Late Paleozoic)

Figure 19.2.5　Section through the Molasse Basin of Central Switzerland, based on an industrial section reproduced in Trümpy (1980). The apparently rootless thrust sheets in the southern part of the section are the Pre-alpine "klippen" (thrust outliers) of Penninic rocks. Thrust sheets belonging to the external margin of the Helvetic Zone appear at the extreme southern end of the section. Note the wedge-shaped geometry of the sedimentary sequence in the Molasse Basin, which creates a mechanical link between the Alps proper to the south and the Jura to the north.

The other factor that affects the mechanics of Alpine deformation in this area is the presence of Triassic evaporites at the base of the Mesozoic sequence. These are so much weaker than the overlying limestones that the taper required for stable sliding of the wedge is reduced to a degree or less.

(2) The Helvetic Zone forms the main frontal ranges of the Alps in Switzerland, and merges with the Jura west of Lake Geneva to form the Sub-Alpine chains in France. It is characterized by a stack of large thrust sheets of Mesozoic and Tertiary sediments, which were emplaced in mid-Tertiary time. These include both the former continental shelf and slope sequence, as well as the Helvetic Flysch: the marine clastic (mainly turbiditic) deposits of the Early Tertiary foreland basin. All these rocks have been folded on a variety of scales, the clay-rich rocks show slaty cleavage, and there is abundant evidence for both pressure-solution and crystal-plastic deformation in calcite. These ductile features reflect the fact that the present erosion level is up to 15 km below the original surface of the thrust wedge.

The Helvetic thrust sheets are commonly referred to as "nappes." *Nappe* is a word of French origin meaning sheet, and refers to the fact that these structures form a really extensive but relatively thin bodies of rock that have been translated relative to underlying rocks. Large-scale thrust tectonics was first discovered and described in the Helvetic Alps during the 1880s. Many of the Helvetic nappes include within them large-scale recumbent folds, including recumbent anticlines in the frontal parts, that were probably formed by tightening original ramp anticlines or fault-propagation folds. In the Morcles Nappe (Figure 19.2.6), the entire thrust sheet is occupied by a high recumbent anticline, with a strongly thinned overturned limb that lies in tectonic contact with the underlying rocks. This structure has been described as a fold-nappe, implying a close link between recumbent folding and thrusting. Note that nappes do not necessarily include recumbent folds, and recumbent folds do not necessarily form nappes.

Some of the Helvetic nappes (most notably the famous Glarus Nappe) dip northwards, so that the movement direction is down the dip of the thrust plane. This was sometimes interpreted to indicate that the nappes formed by sliding under gravity, but this is almost certainly incorrect: the thrusts clearly repeat the stratigraphic section, and when they were active dipped south. Their present orientation is a result of bending over major culminations (antiformal stacks) in the external part of the Alps (Figure 19.2.6). These culminations arose partly as a result of the stacking of the sedimentary cover, but more importantly, result from the imbrication and shortening of the underlying crystalline basement. This occurred after the cover nappes formed, so that the imbricated basement forms a large-scale duplex beneath them, bowing them up into a major anticlinorium. The basement slices crop out within the Helvetic Zone to form the External Crystalline Massifs.

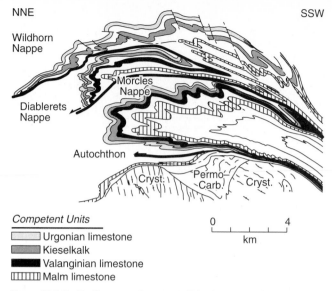

Figure 19.2.6 Profile across the western Helvetic nappes. Note that each of the three nappes shown is a separate thrust sheet, bounded by major faults, and together they form an antiformal stack. The nappes show considerable internal folding and ductile strain, and the Morcles Nappe is composed of a major recumbent fold with a highly stretched and thinned lower limb.

These form impressive ranges, including the Mont Blanc Massif, which reaches 4800 m.

The structural level currently exposed varies along the length of the Helvetic Zone, alternately exposing the imbricated basement and the overlying cover nappes (Figure 19.2.1). These along-strike culminations and depressions are not accompanied by significant gravity anomalies, which means that they reflect differing amounts of crustal thickening at depth. Some of the major basement slices, like the Mont Blanc Massif, which are up to 10 km thick, may in fact terminate laterally, and not continue beneath the depressions.

(3) The Pennine Zone is structurally the most complicated region of the Alps. It is made up of slices of both oceanic and continental rocks, together with sedimentary cover sequences that reflect a wide variety of depositional settings within the Neotethys (Figure 19.2.3). These include fragments of the pre-rifting Triassic carbonate platform, fault-scarp breccias related to Jurassic rifting, pelagic chert and limestone deposited on oceanic crust, thick sequences of deep-water marly limestones, and thick turbidite fans of Late Cretaceous age related to the active margin. Some of these cover sequences are found as thrust slices in the Pre-Alps, north of the Helvetic Zone (Figure 19.2.1). These formed the leading edge of the accretionary wedge, which was emplaced onto the European margin in the early stages of collision. They were subsequently isolated by the growth of the great Helvetic culminations.

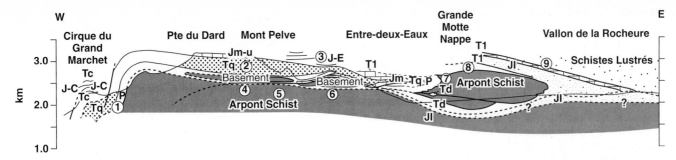

Figure 19.2.7 Structural section across the southern Vanoise Massif, in the Briançonnais domain of the French Penninic Alps, to illustrate the structural style of the Pennine Zone. The section is roughly normal to the trend of this part of the Alps, and shows thrust slices of Briançonnais basement (Arpont Schist) and cover overlain by a thrust sheet of internal Penninic calc-schists (Schistes Lustrés), cut and refolded by late east-directed backthrusts and backfolds. P = Permian mica schist, Tq = Triassic quartzite, Td = Triassic dolostone, Tc = Triassic carbonate rocks, Jl = Lower Jurassic siliceous calc-schist, Jm = Middle Jurassic calc-schist and quartzite, Ju = Upper Jurassic marble, C = Cretaceous to Paleocene chloritic calc-schist, E = Paleocene to Eocene graphitic phyllite. T1 = early northwest-directed thrusts. Notes on the section (circled numbers): (1) east-directed shear zone in the overturned limb of the major backfold with slivers of basement, P, Tc, and Schistes Lustrés at east end of section. (2) 200 m scale tight folding in Tq, in core of major backfold. (3) Early thrusts and isoclinal folds in right-way-up Jm to E, strongly overprinted by backfold-related deformation. (4) Arpont Schist overlain by 1 m of Jl, 10 m of Schistes Lustrés, and then by the inverted limb of the major backfold. (5) Mylonitic inverted limb of major backfold, with, from bottom up, Ju, Tc, Tq, P, basement. (6) Boudins of Td above Jl in sequence inverted by early thrusting, above Arpont Schist. (7) Permian and Arpont Schist in early thrust slice folded around the hinge of the major backfold. (8) Slivers and isoclinal folds of P, Tc, Ju, C, and E, at base of early thrust sheet. (9) Base of early thrust sheet of Schistes Lustrés has been strongly reactivated by late east-directed shear.

In contrast to the External Zones, basement and cover rocks in the main part of the Pennine Zone were imbricated together, and crystalline basement is found at all structural levels. The basement and cover slices were subsequently refolded, in some cases forming high recumbent folds, and they experienced very large ductile strains (Figure 19.2.7). They were also cut by late faults and shear zones: most notably by south- or east-directed backthrusts. On the small scale multiple generations of folds and foliations are common. This structural complexity, developed over a period of about 100 m.y., makes reconstruction of the pre-deformational geometry difficult and speculative.

Deformation in the Pennine Zone was accompanied by metamorphism, which falls broadly into two types. From about 100 Ma onwards, rock subducted on the southern margin of the Neotethys experienced metamorphism of high-P/low-T type, in the clancophane-schist or eclogite facies. Metamorphic pressures generally increase from the north or west side of the arc to the south and east, from about 700 to 2000 MPa (and exceptionally as much as 3000 MPa), indicating depths of burial between 25 and 70 km. The age of this metamorphism also increases from about 50 Ma in the north and west to about 100 Ma in the south and east, reflecting progressive subduction and accretion.

Between 40 and 25 Ma, much of the Pennine Zone was affected by regional metamorphism under moderate pressure (500–800 MPa) and at medium to locally high temperature (400–700°C). By this time the deeply subducted rocks had already returned partway towards the surface, and metamorphism was in the greenschist to amphibolite facies, partly or completely overprinting the earlier high-pressure history. This change in metamorphic environment may reflect the start of full-scale continental collision at about 40 Ma, accompanied by a decrease in convergence rate and a considerable increase in crustal thickness. It may also reflect a change in the underlying mantle structure, increasing heat flow from below.

A major tectonic problem in the Pennine Zone concerns the process by which the very deeply buried rocks were exhumed, which requires the material that originally lay above them to be removed. It is likely that 20 km or more of rock had been removed by erosion in Tertiary time from the Pennine Zone, but it is very unlikely that erosion can account for 70 km or more. Much of the exhumation history occurred before 40 Ma, at which time there is little evidence for the erosion of vast thicknesses of deep crustal and mantle rock. The alternative is that the original overburden was tectonically stretched and thinned. This may seem unlikely in an actively converging margin, but the mechanics of thrust wedges provide a possible explanation. Subduction and accretion of large amounts of crustal material at depth beneath the thrust wedge will increase the thickness, surface slope, and angle of taper so that it may need to extend horizontally to maintain a mechanically stable geometry (Figure 19.2.2c and d). There is increasing evidence from the Internal Zones of the Alps that some of the postsubduction deformation involved vertical shortening and horizontal extension.

(4) The Austro-Alpine nappes form a tectonic cover to the Pennine Zone, and are mainly exposed in the eastern Alps. They include both crystalline basement and Mesozoic cover derived from the leading edge of Adria. They were strongly deformed in mid-Cretaceous time, when the active margin was initiated, and the deeper parts of the complex experienced medium pressure metamorphism. There is also evidence for extensional deformation and strike-slip faulting in Late Cretaceous and Tertiary time.

Figure 19.2.8 Tectonic evolution of the central Alps. The sections are drawn to scale, and the cross-sectional area of continental crust is conserved. Note, however, that the earlier sections do not have the same orientation as the present-day section, nor do they necessarily lie along a straight line (see Figure 19.2.2). Lettered triangles indicate material points that can be followed through the deformation. (a) Mid-Cretaceous. Several hundred km of oceanic crust has been omitted from the three basins. (b) Late Cretaceous. Subduction along the margin of Adria is accompanied by high-P/low-T metamorphism in underplated slices. Structures are schematic. (c) Early Eocene. Continued underplating of continental and oceanic slices is accompanied by extension in the upper part of the thrust wedge, allowing exhumation of the high-P rocks. (d) Early Oligocene. The start of continental collision as the Helvetic margin of Europe enters the subduction zone. Medium P/T ratio metamorphism in the deeper parts of the thrust wedge. (e) Present day. Underthrusting and imbrication of the European continental margin has caused substantial uplift and erosion of the whole of the Alps, as well as backthrusting within the internal zones. Structures in the European basement are schematic; shortening amounts to 70 km. Shortening in the Helvetic cover is 120 km. The orogen as a whole has been shortened from around 800–1000 km to its present width of 150 km. A = Antrona ophiolite, AA = Austro-Alpine basement, C = Combin zone calc-schists, DLB = Dent Blanche Nappe, EM = External Crystalline Massifs, G = Gosau Beds, GSB = Grand Saint Bernard Nappe complex, H = Helvetic cover nappes, IV = Ivrea Zone, L = Lanzo and related peridotite massifs, M = Molasse, MR = Monte Rosa Nappe, NC = Northern Calcareous Alps, PA = Pre-alpine nappes (including Penninic Flysch), PF = Penninic Flysch, S = Sesia Zone, S-T = Simplon-Ticino gneiss complex, UG = Ultra-Helvetic cover and flysch, Z = Zermatt-Saas ophiolite and calc-schist, V = Valais calc-schist.

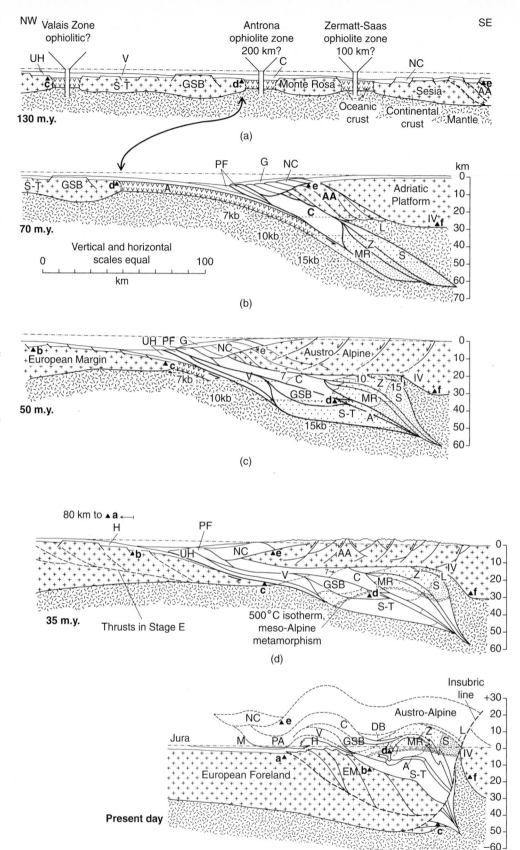

(5) The Internal Zones of the Alps are bounded to the south by a linked array of major faults (Figure 19.2.1), sometimes referred to collectively as the Insubric Line, which have complicated histories. Some show evidence for significant dextral displacement in Tertiary time, perhaps taking up some of the strike-parallel component of the west-northwest motion of Adria with respect to Europe. There is also evidence for northwest-directed underthrusting of Adria beneath the Alps along parts of this system during the later stages of convergence.

(6) South of the Insubric Line are the southern Alps, which form part of Adria proper that has been involved in Alpine deformation. Tertiary shortening of Mesozoic cover in the Dolomites has produced a south-directed thrust belt with several tens of kms of shortening. The great elevation of the Dolomites, however, probably reflects thrusting of Penninic rocks beneath them at depth. Farther west, the Ivrea Zone exposes what may be a complete section through the Southern Alpine crust, stretched and thinned during the Jurassic rifting event, and subsequently tilted during Alpine convergence.

In Figure 19.2.8 the tectonic evolution of the Alps is summarized in a series of schematic cross sections.

19.2.4 CLOSING REMARKS

In spite of its small size, the Alpine chain contains the deformed remains of two passive continental margins and an intervening ocean basin, and exhibits a remarkable range of deformational, metamorphic, and igneous phenomena. The key to Alpine structure, however, has always been the very clear and varied stratigraphy, which not only allows the elucidation of structures that would otherwise be obscure, but reveals the tectonic setting of the various regions of the Alps at past times.

ADDITIONAL READING

Bailey, E. B., 1935, Tectonic essays, mainly Alpine: Oxford, Oxford University Press.

Debelmas, J., and Lemoine, M., 1970, The Western Alps, palaeogeography and structure, *Earth Science Reviews,* v. 6, p. 221–256.

Dewey, J. F., Pitman, W. C., Ryan, W. B. F., and Bonnin, J., 1973, Plate tectonics and the evolution of the Alpine system, *Geological Society of America Bulletin,* v. 84, p. 3137–3180.

Ernst, W. G., 1973, Interpretative synthesis of metamorphism in the Alps, *Geological Society of America Bulletin,* v. 84, p. 2053–2078.

Milnes, A. G., 1978, Structural zones and continental collision, Central Alps, *Tectonophysics,* v. 47, p. 369–392.

Ramsay, J. G., 1981, Tectonics of the Helvetic nappes, *in* McClay, K. R., and Price, N. J., eds., Thrust and nappe tectonics, *Special Publication of the Geological Society of London,* v. 9, p. 293–310.

Trümpy, R., 1960, Palaeotectonic evolution of the central and western Alps, *Geological Society of America Bulletin,* v. 7l, p. 843–908.

———, 1980, Geology of Switzerland: Basel, Wepf & Co.

19.3 The Tibetan Plateau and Surrounding Regions—Leigh H. Royden and B. Clark Burchfiel[1]

19.3.1 INTRODUCTION

The collision of India with Asia about 50 m.y. ago and their subsequent convergence (Figure 19.3.1) has produced a spectacular example of active continent-continent collision. This immense region of continental deformation, which includes the Tibetan Plateau and flanking mountain ranges, contains the highest mountain peaks in the world, with many peaks rising about 8000 m. Indeed so many mountain peaks rise above 7000 m that many of these remain unnamed and uncounted. The Tibetan Plateau proper stands between about 400 and 500 m in elevation and covers a region about 1000 by 2000 km^2, and most of it has remarkably little internal relief. It is sobering to compare the size of this region to other mountain ranges; for example, the crustal mass of the Tibetan Plateau above sea level is about one hundred times greater than that of the western Alps.

The first synthesis of the tectonic evolution of Southeast Asia, including the Himalaya and the Tibetan Plateau, was constructed by the Swiss geologist Emile Argand in 1924. Many of our current ideas about this region are contained in his book *La tectonique de l'Asie,* which remains today one of the truly creative and imaginative works in the Earth sciences. Unfortunately, from about World War I until about 1980 much of this region (in China and the former Soviet Union) was closed to foreign scientists, and only since the 1980s has the area been open to international research. Thus, much of our knowledge of the geology of the Tibetan Plateau and surrounding regions must be considered as largely reconnaissance, particularly when compared to regions like the relatively small western Alps, where geologic research has been conducted by generations of Earth scientists for more than 150 years.

In spite of our relatively sketchy knowledge of the Tibetan region, this exotic, remote, and inaccessible area has long excited the interest of scientists and nonscientists alike. For Earth scientists, the rise of the Tibetan Plateau and the creation of the flanking mountain ranges and basins is a dramatic example of continental convergence and collision (Figure 19.3.1), and is exciting because of the youthful nature of the structures that accommodate deformation and the exceedingly rapid rates at which deformation is occurring. One of the great attractions of geological study in this

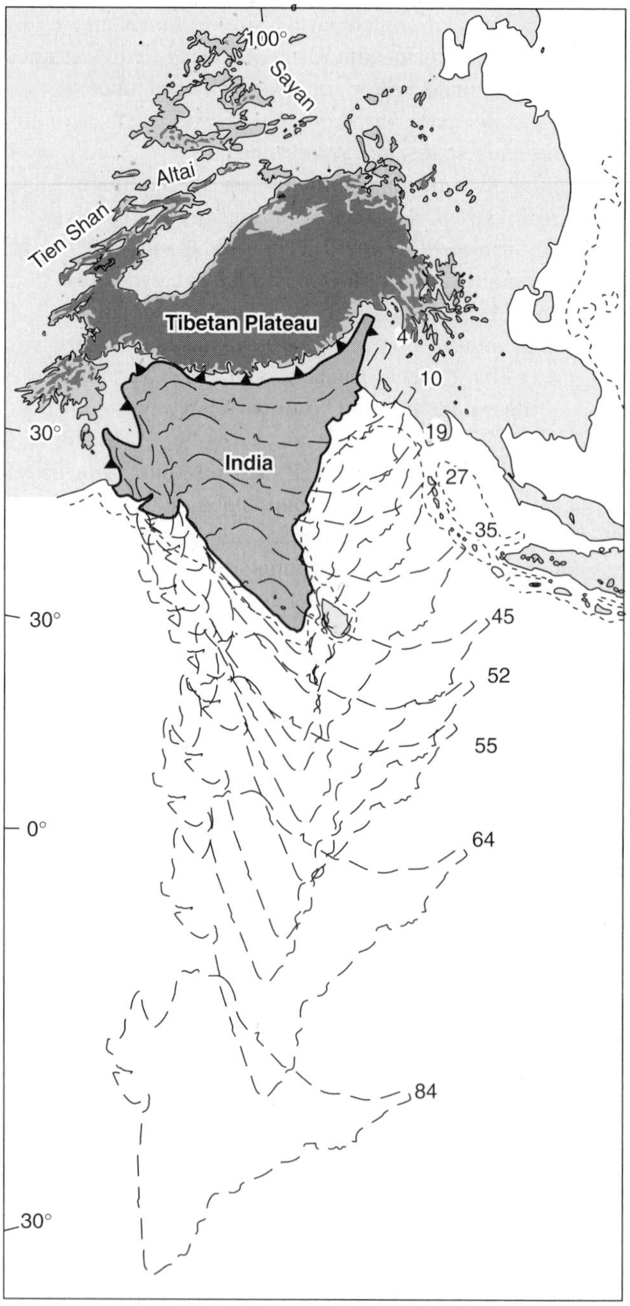

Figure 19.3.1 Generalized topographic map of Southeast Asia showing all the areas below 2 km (light gray), between 2 and 3 km (medium gray), and above 3 km (black) above sea level, and positions of India (no topography shown) with respect to Asia from Late Cretaceous time until the present. Numbers refer to millions of years before present. Note decrease in convergence rate from about 100 mm/yr to about 50 mm/yr at about 50 Ma.

region is the great promise that it holds for enhancing our knowledge of continental deformation processes, as exemplified by an enormous list of still unanswered questions (including crust-mantle interactions, driving mechanisms for local and regional deformation, and the interdependence

---

[1]Massachusetts Institute of Technology, Cambridge, Massachusetts.

of mountain building and global climate). In this essay we will try to summarize briefly what is known about the deformation history of the Tibetan Plateau and surrounding regions, outline some of the hypotheses and controversies that are unresolved today, and discuss some of the fundamental questions that will direct the future of geological and geophysical studies in this region.

19.3.2 PRECOLLISIONAL HISTORY

At the beginning of Mesozoic time, all of the continental land masses were assembled into a giant continent called Pangaea. At this time a Mesozoic ocean, called Tethys, formed a huge embayment into Pangaea from the east and separated the part of Pangaea that is now Eurasia (called Laurasia) from the part of Pangaea that is now the continents of Africa, India, Australia, and Antarctica (called Gondwana). Tethys was perhaps a few hundred kilometers wide in the westernmost Mediterranean region, but widened toward the east, so that at the longitude of Tibet Tethys was approximately 6000 km wide.

At the longitude of Tibet, several sizable continental fragments collided with Asia and became accreted to the Asian continental margin before the main collision between India and Asia (Figure 19.3.2). These continental fragments were rifted from Gondwana and moved northward. As each continental fragment was rifted from Gondwana, it opened a new ocean region between itself and Gondwana. At the same time, the old ocean north of each fragment was closed by subduction. For example, the first fragment to collide with Asia closed the Early Mesozoic Tethys (more properly called Paleotethys). The Indian subcontinent is the last fragment to have rifted off of Gondwana and moved northward, opening the Indian Ocean to the south of India and closing the Neotethyan ocean to the north. It is the closure of this Neotethyan ocean and the collision of India with Asia that has produced the 2500 km long Alpine-Himalayan mountain chain, along which continental collision and convergence are occurring today, and many parts of which are still incompletely closed (e.g., the Mediterranean region still contains Mesozoic seafloor of Tethys).

Prior to the main collision of India with Asia in Early Tertiary time, several thousand kilometers of the Neotethys Ocean floor was subducted beneath Asia along a north-dipping subduction boundary. Subduction was extremely rapid, resulting in convergence between India and Asia at about 100 mm/yr. Evidence for precollisional convergence and deformation is recorded within the Himalayan mountain belt and in the southernmost part of the Tibetan Plateau. For example, within the central part of southern Tibet, Jurassic ophiolites obducted onto the Asian continental margin attest to the existence of an ancient ocean region between India and Asia. A Late Cretaceous forearc sequence records subduction continuing at least into the

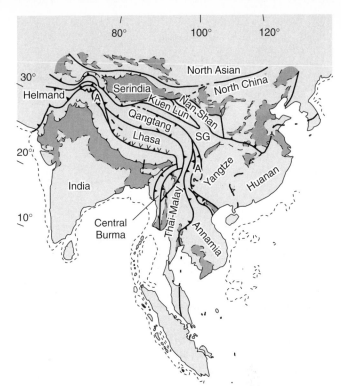

Figure 19.3.2 Major fragments in Southeast Asia during Late Paleozoic, Mesozoic and Early Cenozoic time. The sutures between fragments are shown by barbed lines, with the barbs indicating the upper plate of the subduction system. The suture between the Qangtang and Songpan Ganze (SG) fragments is of Mesozoic age and is thought to represent the closure of the Paleotethys Ocean. All sutures north of this are Paleozoic. The suture south of the Lhasa block is the Indus-Tsangpo Suture, and represents the closure of Neotethys. A stands for island-arc fragments, V shows the location of Cretaceous-Early Tertiary arc volcanism. Location of cross section in Figure 19.3.3 is also shown.

Cretaceous (Figure 19.3.3). North-dipping subduction of an oceanic region is also recorded by a Late Cretaceous to Early Cenozoic volcanic arc along the southern margin of Asia (e.g., the Gandese batholith near Lhasa in south-central Tibet; this lies just north of the Xigatze forearc basin in Figure 19.3.3). This arc volcanism shut off in Eocene time, presumably reflecting the time of collision and the cessation of subduction of oceanic lithosphere. In the western Himalaya, precollisional subduction of Tethys must have occurred partly offshore, south of the Asian continental margin, because a Jurassic-Cretaceous volcanic arc (the Kohistan Arc) was developed in a marine environment, and subsequently collided with the southern margin of Asia in latest Cretaceous time, somewhat before the time of the main India-Eurasia collision. It is likely that events before and around the time of collision were very complicated, but at present the geologic data to unravel these events has not been obtained, making the early evolution of the Himalayan system look deceptively simple.

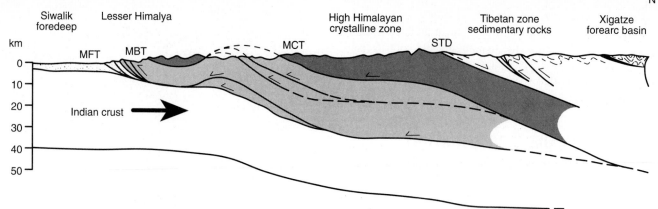

Figure 19.3.3 Schematic cross section through the central Himalaya (location in Figure 19.3.2). Heavy dots are Miocene-Quaternary molasse of the modern foredeep basin and contained within the outer thrust sheets near the Main Frontal Thrust (MFT). Light shading represents low-grade metamorphic rocks north of the Main Boundary Thrust (MBT) and south of the Main Central Thrust (MCT). Dark shading represents high-grade metamorphic and crystalline rocks north of the MCT. These are separated from sedimentary rocks of the Tibetan Zone (in white) by the normal South Tibetan Detachment (STD). Also shown are the Cretaceous Xigatze forearc deposits that rest on an ophiolitic basement (black sliver).

19.3.3 POSTCOLLISIONAL CONVERGENT DEFORMATION

At about 50 Ma, India collided with Asia, and at about the same time the convergence rate between India and Asia slowed from about 100 mm/yr to about 50 mm/yr. Since that time India has continued to move northward relative to stable Eurasia at about 50 mm/yr, giving a relative convergence of about 2500 km since the time of collision, and shortening and thickening the crust of Asia to produce the elevated Tibetan Plateau. Today, the continental crust beneath the high-standing plateau and flanking mountains is about 70 km, nearly double the more normal values of 30–40 km for continental crust. It is remarkable that in 1924, long before the advent of plate tectonics, Emile Argand already understood that an ocean had closed between India and Asia and that subsequent convergence between India and Eurasia has caused the intra-continental deformation of Southeast Asia. However, it is only in the last few decades that knowledge of the magnetic anomalies on the seafloor and paleomagnetism has allowed scientists to reconstruct the motions of India with respect to Asia in a quantitative manner, and to calculate the width of the subducted ocean.

The boundary where India and Asia first collided, often referred to as the Indus-Tsangpo Suture, is marked by a discontinuous belt of ophiolites, melange, and forearc sedimentary rocks of the Gandese magmatic arc. It has been impossible to reconstruct the early collisional history of this zone because postcollisional deformation involving north-directed backthrusting, strike-slip, and normal faulting has completely obscured the older collisional structures. There is also very little information about postcollisional deformation within the Himalayan Belt prior to about 250 Ma, although in the western Himalayan chain geologists have identified Early Cenozoic metamorphic events dated at around 35–45 Ma. This indicates that thrust faulting and crustal shortening were active in the western Himalaya at

about this time, and suggests that the creation of high topography within the western Himalaya may have already been underway. However, as we shall see, the time at which high topography developed within most of the Himalaya and Tibet is extremely controversial, not least because of its implications for global climate change.

The evolution of the Himalayan Orogen is better known from about 25 Ma to the present, although there is still much that we do not know about the orogen. A schematic north-south cross section through the Himalaya and into southern Tibet (Figure 19.3.3) shows that regional shortening is thought to have been taken up on thrust faults within sedimentary rocks north of the crest of the Himalaya (called the Tibetan zone sedimentary rocks) and on three major north-dipping thrust faults south of the crest of the Himalaya: the Main Central Thrust (MCT), the Main Boundary Thrust (MBT), and the Main Frontal Thrust (MFT). All of these structures appear to be roughly continuous for more than 200 km along the entire length of the Himalayan chain. Shortening across the Himalaya was further accommodated by intense ductile strain within high-grade and crystalline rocks as well as by folding and thrusting along less important faults.

The oldest of these thrust faults are probably within the Tibetan sedimentary sequence, but their age remains uncertain. The MCT, which may be younger, carries the high-grade crystalline and metamorphic rocks of the High Himalayan zone over the lower grade to unmetamorphosed sedimentary rocks of the Lesser Himalaya. So far, only one date has been obtained for thrusting on the MCT. This date was obtained on rocks just south of Qomolangma (Mt. Everest) in central Tibet, indicating that the MCT was active at 21 Ma. However, we do not yet know how long before or after this time the MCT was active, although in most places it is not active today. In addition, we do not know if this data can be extrapolated to the east or west along the strike of the MCT; it is possible that although the MCT appears continuous along the length of the Himalaya, it is of different ages and different character along different parts of the belt. Metamorphic pressure and temperature data from central Tibet show that at 20–25 Ma the MCT was at a depth of about 30 km and that temperatures of High Himalaya rocks near and just above the MCT were at least 500° to 650°C. Thus the rocks of the High Himalaya must have been brought to the surface from 30 km depth during the last 30 m.y.

South of the MCT, the MBT carries the low-grade to unmetamorphosed rocks of the Lesser Himalaya southward over mainly Cenozoic sedimentary rocks. In some places there is evidence that the MBT may be currently active, but the main shortening and convergence across the Himalaya today is probably absorbed by motion along the structurally lower MFT and by folding within its hanging wall. The MFT marks the southern limit of deformation along most of the Himalayan chain and carries rocks and sediments of the Himalaya southward over the Ganga foredeep basin. Sediments being deposited in the Ganga Basin today are similar to those now exposed by erosion within the Himalayan foothills, suggesting that the latter were also deposited in a foredeep position in front of the frontal thrust faults, and have been subsequently incorporated into the thrust belt.

Seismic studies and examination of the rate of advance of the Ganga Basin southward over the Indian foreland indicate that about 10 to 25 mm/yr of shortening is currently being taken up by thrust faulting and shortening within the Himalaya. Seismic studies also show that the main active thrust fault beneath the Himalaya dips gently northward by about 10°, until it becomes aseismic at a depth of about 18 km. At these depths we assume that this active, gently dipping thrust boundary is somewhat analogous to the Early Miocene MCT, but of course there are no direct observations to support this hypothesis.

Because the current rate of convergence between India and stable Eurasia averages about 50 mm/yr, this means that only about one-quarter to one-half of the convergence between India and Eurasia is accommodated by shortening within the Himalaya. The rest must be taken up to the north within and around the margins of the Tibetan Plateau. The way in which the remaining convergence is absorbed remains a highly controversial topic, for which there are perhaps more models than data. However, considerable amounts of convergence are clearly absorbed by thrusting and shortening within mountain belts that lie north of the plateau, particularly in the Pamir, Tien Shan, Qilian Shan, and southern Ningxia (e.g., Madong Shan) thrust systems (Figure 19.3.4). The rates of shortening across these belts are not very well known, but they are probably at least 10 mm/yr, perhaps considerably more. Although shortening across the Pamirs began in Early Cenozoic time, shortening across the other northern thrust belts probably began only in Late Miocene to Quaternary time. The onset of thrusting in these regions is primarily constrained by stratigraphic data because Cenozoic metamorphic rocks are not found in any of these young thrust belts.

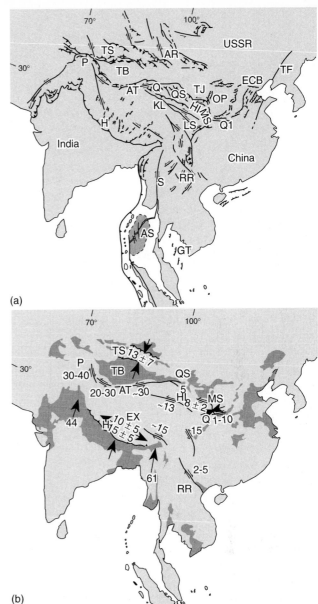

Figure 19.3.4 (a) Generalized Cenozoic tectonic map of Southeast Asia. Horizontal arrows indicate sense of slope on strike-slip faults, barbed lines indicate thrust faults in the continental areas and subduction zones in the oceanic areas with barbs on the upper plates, and ticked lines indicate normal faults. Lunate shapes are folds. AR = Altai Range; AS = Andaman Sea; AT = Altyn Tagh Fault; ECB = East China Basins; GT = Gulf of Thailand; H = Himalaya; HI = Haiyuan Fault; KL = Kun Lun Fault; LS = Longmen Shan; MS = Madong Shan; OP = Ordos Plateau; P = Pamirs; Q = Qaidam Basin; QS = Qilian Shan; QI = Qinling Mountains; RR = Red River Fault; S = Sagaing Fault Zone; TB = Tarim Basin; TF = Tan Lu Fault Zone; TJ = Tianjin Shan; TS = Tien Shan; X = Xianshuhe Fault; Y = western Yunnan. (b) Slip rates (in mm/yr) on active faults within Southeast Asia. Half arrows identify slip direction on strike-slip faults. Arrows that diverge indicate areas and directions of extension; arrows that converge indicate areas of shortening across thrust faults (and folds) with barbed lines. Large north-pointing arrows south of the Himalaya indicate direction and magnitude of present convergence between India and northern Asia.

The onset of uplift and shortening along the northern and northeastern margins of Tibet is probably best regarded as the northward growth of the Tibetan Plateau. We can

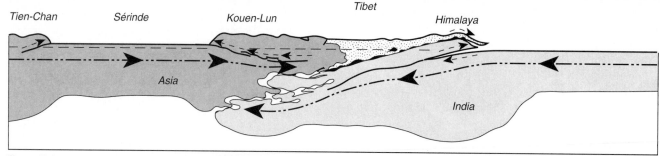

Figure 19.3.5 North-south cross section through Tibet and the Himalaya published by Emile Argand in 1924. Modern interpretations have not added much to our understanding of the deeper structure beneath Tibet. Light gray is Indian crust, dark gray is Asian crust, mantle material is white.

only surmise that the early history of rocks now contained within the central part of the plateau might have been similar to those now being incorporated into the plateau by shortening along its northern boundary. It is clear that the northward growth of the plateau has not been a smooth process. The map of present-day deformation north of the plateau shows a very irregular pattern of thrust belts, with a huge undeformed region, the Tarim Basin, sandwiched between the Tibetan Plateau proper and the very active region of shortening within the Tien Shan and overthrust from all sides. Active shortening, albeit at much slower rates, also occurs far to the north within the Altai ranges in north China and Mongolia. The northward growth of the plateau has not been a smooth process in time either. For example, shortening appears to have begun sometime in Late Pliocene to Early Quaternary time in much of the northeastern portion of the plateau. However, examination of individual ranges and thrust belts in this region does not reveal a steady northward progression of the deformation front, but rather a more haphazard onset of shortening in part caused by the reactivation of pre-Cenozoic structures. On a scale of hundreds of kilometers, the deformation along the northern margin of Tibet appears to be controlled largely by strength heterogeneities and preexisting structural trends within the Asian lithosphere.

The structure of the thick crust beneath the Tibetan Plateau is not known, although a number of different hypotheses have been presented, each linked to a different model for plateau growth. On the one hand, a number of authors, beginning with Argand (1924), believe that the crust beneath the Tibetan Plateau has been doubled by underthrusting of the Indian crust and lithosphere beneath the Asian crust (Figure 19.3.5). In some of these models it is suggested that essentially no deformation of the upper (Asian) crust has occurred, while others include moderate amounts of upper crustal deformation or deformation only within the northern part of the plateau. In contrast, others have suggested that the Indian lithosphere extends only beneath the Himalaya and southernmost Tibet, and that the doubling in thickness of the Tibetan crust was mainly the result of shortening and thickening of the Tibetan crust due to the India/Eurasia collision. To date there are little conclusive data that bear directly on this problem.

19.3.4 CRUSTAL SHORTENING AND STRIKE-SLIP FAULTING

A heated controversy surrounds two extreme, and basically incompatible, views of how the India/Eurasia collision and crustal shortening in Tibet are related to Cenozoic deformation throughout much of Southeast Asia. In the 1970s, a ground-breaking set of papers suggested that the postcollisional convergence of India and Asia has been responsible for most of the Cenozoic deformation of Southeast Asia. It was argued that although the Tibetan Plateau did indeed form by crustal shortening, the total crustal mass beneath the plateau is too little (by about 30%) to account for all of the convergence between India and Eurasia since the time of the collision. Using earthquake seismology and newly available satellite imagery, it was suggested that the remaining convergence was accommodated by eastward ejection of continental crust away from the plateau and out of the way of India as it moved northward toward Siberia (Figure 19.3.6a). Subsequent experiments on plasticine suggested that eastward extrusion of continental fragments was also responsible for the Early Cenozoic extension within the South China Sea and the Gulf of Thailand. This requires displacements of hundreds to a thousand kilometers on the large strike-slip faults within the Tibetan Plateau.

In a second very different model, 1980s workers used a numerical computer simulation to model Asia as a viscous sheet deformed by a rigid indenter (India). This numerical model predicts a zone of shortening and thickening crust that grows northward and slightly to the side of the northward moving indenter (Figure 19.3.6b). Because no deformation occurs beyond the boundaries of this zone of crustal shortening and thickening in front of the indenter, it was argued that only the Tibetan Plateau and flanking mountain belts have resulted from the collision of India and Asia. In this interpretation, the other regions of Cenozoic deformation in Southeast Asia are mainly unrelated to the India/Eurasia collision.

Both of these interpretations are based mainly on models, although the earlier work incorporated the geological and geophysical data available at that time. What do current data say about this debate? The study of slip rates on active faults shows that continental blocks within Tibet are indeed

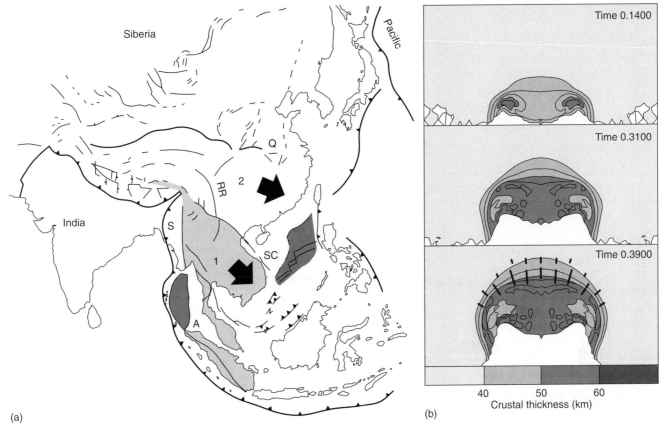

(a)

(b)

Figure 19.3.6 Two interpretations of the tectonic framework of Southeast Asia. (a) The interpretation of Tapponnier and coworkers, which emphasizes the eastward extrusion of two large crustal fragments bounded by major strike-slip faults as a result of the India-Eurasia convergence. In this interpretation, eastward movement of these fragments results in the extension on the Southeast Asian continental shelf and creation of oceanic crust in the South China Sea. The first crustal fragment to move (1) is shown in light gray and the second to move (2) is shown in coarse dots. RR = Red River Fault, S = Sagaing Fault, SC = South China Sea. (b) The interpretation of England and Houseman shows the progressive development of topography in Asia by computer modeling of a rigid indenter (India) deforming a viscous sheet (Asia). This model suggests that India-Eurasia may have little if any effect east or west of the Tibetan Plateau, and so is very different from the first interpretation. Contours are of increasing crustal thickness at different times (times are dimensionless, but the bottom panel would approximately correspond to the present topography of Tibet).

moving eastward at rates between 10 and 30 mm/yr, as predicted by the extrusion model. However, these motions record only a snapshot of present-day motions and do not answer the question of how much eastward ejection of material has occurred within Tibet. So far, the total offset on the large strike-slip faults has been determined only in a few places from geologic mapping. This indicates about 15 km of left slip on the Haiyuan Fault, 200 km of left slip on the Altyn Tagh Fault, 50 km of left slip on the Xianshuhe Fault (Figure 19.3.4 for locations). Thus, while all of these strike-slip faults are moving very fast, they are also very young, and the total offset on these faults, while large, is not of the magnitude predicted by the models. Whereas a definitive test of the extrusion model is not yet possible, the data do suggest that while extrusion does occur, it is of much smaller magnitude than that predicted by this model.

Strike-slip faults within the Tibetan Plateau also have a very interesting relationship to thrust faulting and crustal shortening occurring around the margins of the plateau. At the eastern end of some of the strike-slip faults, they merge with or end against the active thrust faults that rim the margins of the plateau. This is best illustrated for the Haiyuan

strike-slip fault in the northeastern corner of the plateau, where geologic mapping shows that 15 km of left slip on this east-west trending thrust fault is absorbed by 15 km of shortening on a north-south trending thrust belt in the Liupan Shan (Figures 19.3.4 and 19.3.7). In addition, the onset of strike-slip faulting on the Haiyuan Fault is of approximately the same age as the onset of shortening in the Madong Shan (Early Quaternary). Thus the left-slip motion on the strike-slip faults is absorbed by thrusting at the plateau margins, suggesting that lateral extrusion of crust within the plateau does not extend beyond the topographically high region.

Probably one of the reasons that so much emphasis has been placed on strike-slip faulting within the Tibetan Plateau and surrounding regions is that many of the studies of active faulting have used satellite photos to identify important faults. Faults that dip steeply and are straight show up very well on these photos, while gently dipping faults with a complicated outcrop pattern can be difficult to recognize. Therefore, the strike-slip faults within the plateau were recognized very quickly, while many of the important thrust faults have probably not yet been identified. Another

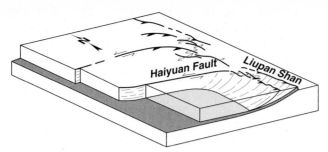

Figure 19.3.7 Schematic block diagram illustrating that strike-slip faults (such as the Haiyuan Fault shown here) end against thrust belts that bound the margins of the Tibetan Plateau (such as the Liupan Shan), and indicating how both thrust faults and strike-slip faults may be detached from the deeper crust by a zone of decoupling within the midcrust. See Figure 19.3.4 for locations.

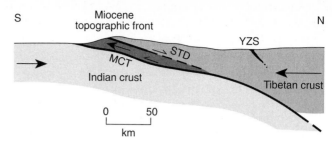

Figure 19.3.8 Diagrammatic cross section through the Himalaya and southern Tibet showing the southward ejection of a crustal wedge bounded below by the MCT and above by the South Tibetan Detachment (STD). Near Qomolangma (Mt. Everest) both faults were active at about 20 Ma, during the convergence of India. Faulting is thought to be due to gravitational collapse of the Miocene topographic front; the geometry shown at depth is speculative. YZS is the Indus-Tsangpo Suture Zone.

problem is that large amounts of shortening around the margins of the plateau occur not only by faulting but by folding, which is aseismic and not recorded by earthquakes. Thus it is easy to underestimate the rates of crustal shortening relative to the rates of strike-slip faulting from looking at first-motions from earthquakes on and around the plateau.

19.3.5 EXTENSION OF THE TIBETAN PLATEAU

Not all of the deformation of the Tibetan Plateau is compressional. A series of north-south striking grabens in southern Tibet is accommodating active east-west extension of the plateau at about 10 mm/yr (Figure 19.3.4). These grabens are very young, with extension beginning at about 2 Ma, giving a total amount of east-west extension of about 20 km. The significance of this extension is unclear and is also somewhat controversial. Some workers relate the east-west extension to the eastward extrusion of material from the Tibetan Plateau and argue that the presence of east-west extension supports their interpretation. In contrast, others have suggested that east-west extension of the plateau records the delamination (falling off) of the mantle lithosphere beneath the plateau sometime in the Late Miocene or Pliocene. They argue that the removal of the dense lithosphere from beneath the plateau caused uplift of the plateau surface (as cold, dense mantle lithosphere was replaced by hot, buoyant asthenosphere) and that the surface elevation of the plateau is now decreasing by east-west extension within the plateau. At present, the significance of these north-south trending grabens remains highly uncertain and will probably be a crucial point in developing new models for the deformation of the plateau. It is perhaps noteworthy that the grabens occur mainly in the western two-thirds of the plateau and that some of the grabens extend southward through the Himalaya and nearly to the Himalayan thrust front.

Within the High Himalaya there is also evidence for large-scale north-south extension that preceded the younger east-west extension and occurred on gently north-dipping

normal faults (the South Tibetan Detachment Zone) that parallel the MCT (Figure 19.3.3) and form a major structural break between the crystalline rocks of the High Himalaya and the Tibetan Sedimentary Zone rocks (Figure 19.3.3). Extension occurred simultaneously with thrusting on the Main Central Thrust Zone, as indicated by dates of 20 Ma on both fault systems near Qomolangma (Mt. Everest). The normal fault zone juxtaposes rocks of midcrustal levels in the footwall against shallow level sedimentary rocks in the hanging wall, and has a minimum displacement of 40 km (although the total displacement could be much greater). A north-south cross section through the fault shows that it bounds the top of a midcrustal wedge that was ejected southward at about 20 Ma (Figure 19.3.8). This wedge was bounded at its base by the Main Central Thrust Zone. Southward ejection of the wedge was from an area of high topography to an area of low topography and probably reflects gravitational collapse of the steep topographic slope along the southern margin of the Tibetan Plateau. Collapse may have occurred due to weakening of the crust by melting of granites at midcrustal depths, as suggested by the presence of ductilely deformed syn-kinematic granites along and just below the South Tibetan Detachment Fault. Several other areas within Tibet contain low-angle faults with normal displacement, but at present they are recognized only on short fault segments and their regional significance remains uncertain.

The occurrence of north-south extension of large magnitude along the southern margin of the plateau is somewhat surprising, since extension was not only contemporaneous with north-south shortening along the MCT but also occurred during continued north-south convergence of India and Eurasia at about 50 mm/yr. This indicates that the extensional processes that controlled deformation within the mid- to upper crust in this region were essentially decoupled from the convergent motions occurring deeper in the crust and within the mantle lithosphere. It is probable the mid- to upper-crustal deformation under many, if not all, parts of the plateau are decoupled from the lower crust and mantle. For example, the general style of deformation within the thrust belts that rim the northern and northeastern margins of the plateau, and in some cases balanced cross sections constructed across the thrust belts, show that the thrust faults sole out at depths of about 15 km. Moreover, the relationship of the large strike-slip faults (such as the Haiyuan Fault) within the plateau to adjacent thrust-fault systems also indicates that many of the strike-slip faults are probably also restricted to upper and mid-crustal depths (Figure 19.3.7). Thus the motions of crustal fragments within the plateau and along the margins of the plateau may not reflect the motion or deformation of the underlying lower crust and uppermost mantle.

19.3.6 CLOSING REMARKS

Many fundamental questions remain to be answered within the Himalayan-Tibetan region. Some of these appear straightforward, but will require many years of intensive field work—such as, How is active crustal deformation partitioned within the plateau? How can we unravel the active and young deformation of the plateau to learn about the pre-Pliocene history of the plateau and its growth through time? Finding the answers to other questions will require the use of sophisticated geophysical techniques in combination with field geological data—such as, How is the deformation partitioned vertically within the crust of the Tibetan Plateau? How is crustal deformation related to motions within the underlying mantle? On what length scale is crustal deformation related to motions within the mantle? On the scale of the plate boundary system as a whole, a region 1000 by 2000 km^2, it is likely that the averaged motions of the upper crust and the mantle lithosphere are reasonably similar. Is this also true at smaller scales, such as over regions a few hundred kilometers in length and width? How do we go about determining the degree of coupling between the crust and the mantle?

Lastly, one of the most controversial issues surrounding the geological evolution of the Tibetan Plateau is the history of uplift of the plateau surface to its present elevation of about 5 km. This is important not only to geologists, because it bears on the mechanisms by which the plateau has been deformed, but also to a wide range of scientists in fields from marine geology to climate modeling. Because the Tibetan Plateau is so large and stands so far above sea level, its uplift is thought to have influenced global circulation patterns within the atmosphere and to

have been responsible for the onset of the Indian monsoons, which did not exist prior to Late Miocene time. Uplift of the plateau can also be correlated with changes in faunal distributions within the oceans and with huge changes in the isotopic ratios of elements such as strontium within the worldwide oceans. Indeed, the record of plateau uplift and erosion, as preserved within the composition of marine sediments found worldwide, as well as the geologic record from the plateau itself, suggests that collisional events of the magnitude found in Tibet and the Himalaya are probably rare events, even on a geological time scale, and that collisional events of comparable magnitude may not have occurred within the last 600 m.y.

ADDITIONAL READING

Argand, E., 1977, Tectonics of Asia (translation by A. V. Carozzi): New York, Hafner Press, 218 p.

Burchfiel, B. C., and Royden, L. H., 1991, Tectonics of Asia 50 years after the death of Emile Argand, *Eclogae geol. Helv.,* v. 84, p. 599–629.

England, P. C., and Houseman, G. A., 1988, The mechanics of the Tibetan Plateau, *in* Shackleton, R. M., Dewey, J. F., and Windley, B. F., eds., Tectonic evolution of the Himalayas and Tibet, *Philosophical Transactions of the Royal Society of London* (A), v. 327, p. 379–320.

Lyon-Caen, H., and Molnar, P., 1985, Gravity anomalies, flexure of the Indian plate, and the structure, support and evolution of the Himalaya and Ganga Basin, *Tectonics,* v. 4, p. 513–538.

Molnar, P., and Tapponnier, P., 1975, Cenozoic tectonics of Asia; Effects of a continental collision, *Science,* v. 189, p. 419–426.

Molnar, P., 1988, A review of geophysical constraints on the deep structure of the Tibetan Plateau, the Himalaya and the Karakorum, and their tectonic implications, *in* Shackleton, R. M., Dewey, J. F., and Windley, B. F., eds., Tectonic evolution of the Himalayas and Tibet, *Philosophical Transactions of the Royal Society of London* (A), v. 326, p. 33–88.

Tapponnier, P., Peltzer, G., Le Bain, A. Y., Armijo, R., and Cobbold, P., 1982, Propagating extrusion tectonics in Asia, New insight from simple experiments with plasticine, *Geology,* v. 10, p. 611–616.

19.4   The North American Cordillera—Elizabeth L. Miller[1] and Phillip B. Gans[2]

19.4.1   INTRODUCTION

The broad mountainous region of western North America is known as the Cordillera (Spanish for mountain or mountain chain), an orogenic belt that extends from South America (the Andean Cordillera) through Canada (Canadian Cordillera) and into Alaska. The youthful topography of this impressive mountain belt is closely related to active crustal deformation as indicated by the current distribution of seismicity across the width of the orogenic belt (Figure 19.4.1). The current plate tectonic setting and the dominant style of deformation vary along the orogen from folding and faulting above an active subduction zone in the Pacific Northwest of the United States and the Aleutians in Alaska to strike-slip or transform motion along the Queen Charlotte (Canada) and San Andreas Fault (California), to extension and rifting in the Basin-and-Range Province of the western United States and Mexico's Gulf of California (Figure 19.4.1). Variations in structural style are also apparent across the orogen; for example, crustal shortening and strike-slip faulting in coastal California are concurrent with extensional faulting and basaltic volcanism in the adjacent, and inboard, Basin-and-Range Province. This great diversity in structural style along and across strike of the Cordillera is likely to have also characterized the past history of the orogenic belt, which was shaped by paleo-Pacific-North America plate interactions since the late Precambrian, making it the longest lived orogenic belt known on earth.

The Cordillera provides an excellent natural laboratory for studying the evolution of a long-lived active margin and the effects of subduction and plate boundary processes on the evolution of continental crust. However, the exact nature of the relationship between plate motions and continental deformation and mountain building remains elusive for a number of reasons. The theory of plate tectonics treats the Earth's lithosphere as a series of rigid plates that move with respect to one another along relatively discrete boundaries. This simplification applies well to oceanic lithosphere, which is dense and strong and thus capable of transmitting stresses across great distances without undergoing significant internal deformation. However, it does not apply well to continents, whose more quartzo-feldspathic composition and greater radiogenic heat flow make them inherently weaker. Displacements or strain within continental crust can accumulate at plate tectonic rates (1–15 cm/year) within narrow zones of deformation, or can take place more

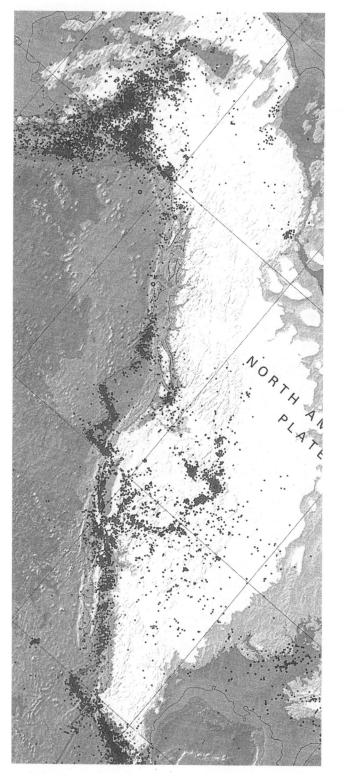

Figure 19.4.1   Digital topography, plate tectonic setting, and seismicity map of the North American Cordillera.

[1]Stanford University, Stanford, California.
[2]University of California, Santa Barbara, California.

slowly (mm to cm/year) across broad zones of distributed deformation. Thus, continents can accumulate large strains, thickening over broad distances during crustal shortening and thinning during extension. These strain histories are usually preserved in the geologic record because the inherent buoyancy of continental material prevents it from being subducted into the mantle. Subduction-related magmatism can lead to greater strain accumulation within continental crust by increasing temperatures and thus rheologically weakening the crust, allowing it to deform in a semicontinuous fashion. Because of these considerations regarding the thermal structure, composition, thickness, and rheology of continental crust, the response of the overriding continental plate to changes in subducting plate motion or to changes in the nature of plate interactions along a margin may be sluggish and vary with time and depth in a complex fashion. In other words, orogenic belts like the Cordillera may turn out to be at best imperfect recorders of past plate motions. Our understanding of these problems and the link between plate tectonics and continental deformation is evolving as more detailed geologic and geochronologic studies are carried out, providing quantitative information on the time-scale of events, the rates of geologic processes, and the ability to compare timing of events from one part of the belt to the next. Geophysical and petrologic studies remain key tools that help us understand how physical processes in the deeper crust and mantle are coupled to more easily studied deformation at shallower levels of the crust.

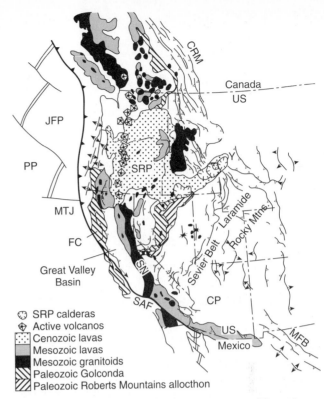

Figure 19.4.2 Simplified geologic/tectonic features map of the western United States and southern Canada. Abbreviations: JFP = Juan de Fuca Plate; PP = Pacific Plate; MTJ = Mendocino Triple Junction; SAF = San Andreas Fault; SRP = Snake River Plain; CP = Colorado Plateau; BR = Basin-and-Range Province; SN = Sierra Nevada.

19.4.2 GEOLOGIC AND TECTONIC HISTORY

Studies of modern active plate margins have played an important role in interpreting the more fragmentary evidence in the geologic record of the Cordillera. Based on these comparisons, it appears that all plate tectonic styles and structural regimes known to us, with the exception of continent-continent collision (like the Himalaya), played a part in the creation and evolution of the Cordillera. The initial formation of the Cordilleran margin goes back to the latest Precambrian. The Windemere Supergroup is a thick succession of shelf-facies clastic rocks deposited between about 730 Ma and 550 Ma whose facies and isopachs define the newly rifted margin of western North America after the breakup of the Rodinia supercontinent (see Section 19.9). The Windemere supergroup forms the lower part of a 15 km thick, dominantly carbonate shelf succession whose isopachs and facies boundaries closely parallel the trend of the Cordillera and are remarkably similar along the length of the Cordillera from Mexico to Alaska. This shelf sequence (including the older and more localized Belt-Purcell Supergroup) is now spectacularly exposed in the eastern Cordilleran fold-and-thrust belt, whose overall geometry and structure are controlled by the facies and thickness of this succession.

The Paleozoic history of the Cordillera has generally been described as one of continued passive-margin sedimentation and little active tectonism. However, there are a multitude of tectonic fragments of island arcs and backarc basin sequences, which range in age from Cambrian to Triassic, embedded in the western Cordillera (Figure 19.4.2). The common conception that these represent a collage of far-traveled terranes exotic to the Cordillera ('suspect terranes') is being reevaluated by numerous studies that indicate many of these sequences developed adjacent to, but offshore of, the western edge of the continental margin. Although these 'exotic' fragments are likely displaced from their site of origin by rifting, thrusting, and strike-slip faulting, their presence nonetheless argues convincingly for a long history of subduction of paleo-Pacific crust beneath the western edge of the North American Plate. Study of these accreted fragments suggests that during the Paleozoic, western North America looked much like the southwest Pacific today, with its fringing arcs separated from the main Australasian continental shelves by backarc basins (Mariana-type convergent margin). During the Paleozoic, the North American shelf itself experienced episodes of regional subsidence and uplift, but no significant deformation. Exceptions include deformation and intrusion of latest Proterozoic to Late Devonian granites along parts of the margin in Alaska and southern British Columbia, and the closure of deep-water, backarc basins by thrusting onto the shelf during the earliest Mississippian *Antler Orogeny* (Roberts

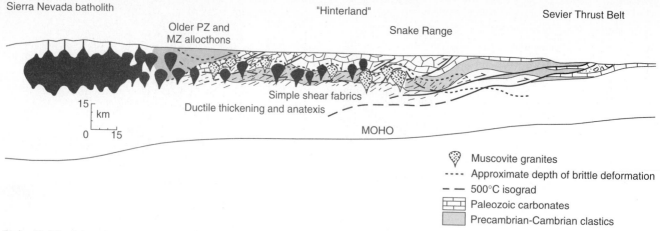

Older PZ and
MZ allocthons

Snake Range

15
km
0   15

Simple shear fabrics
Ductile thickening and anatexis

MOHO

Muscovite granites
- - - - Approximate depth of brittle deformation
— — 500°C isograd
Paleozoic carbonates
Precambrian-Cambrian clastics

Figure 19.4.3 Schematic crustal cross section of the western United States at the end of Mesozoic subduction-related arc magmatism and backarc crustal thickening.

Mountains Thrust) and during the Permo-Triassic *Sonoma Orogeny* (Golconda Thrust) in the western United States part of the Cordillera.

Magmatic belts related to eastward subduction beneath western North America are much better developed beginning in the Mesozoic. Arc magmatism of Triassic and Early Jurassic age (230–180 Ma) is recorded by thick sequences of mafic to intermediate volcanic rock erupted in an island arc (Alaska, Canada, and northern part of U.S.) or continental arc (southwestern U.S.) setting. Tectonism accompanying subduction during this time-span was generally extensional in nature, leading to rifting and subsidence of parts of the arc and continental margin. The Middle to Late Jurassic brought a dramatic change in the nature of active tectonism along the entire length of the Cordilleran margin. This time-span is characterized by increased plutonism during the interval 180–150 Ma and many hundreds of km of crustal shortening. This shortening closed intra-arc and backarc basins, accreted arc systems to the North American continent, and fundamentally changed the paleogeography of the Cordillera from a Marianas-type margin to an Andean-type margin, a tectonic framework that persisted throughout most of the latest Mesozoic and Cenozoic. In Canada, Middle Jurassic orogenesis has been attributed to the collision of the allochthonous Stikine Superterrane, (a terrane consisting of previously joined-together smaller terranes). However, in the United States portion of the Cordillera there is no evidence for collision, only eastward subduction of oceanic crust beneath the margin. The preferred explanation for this orogeny is that it is instead linked to rapid westward motion of North America with respect to a fixed hotspot reference frame. This westward motion occurred as the North Atlantic began to open and in effect caused the western margin of the continent to collide with its own arc(s) and subduction zone(s) and then to deform internally. Deformation associated with Middle Jurassic orogenesis (sometimes referred to as the Columbian Orogeny) began first in the region of elevated heat flow within and behind the arc and migrated eastward with time toward the continental interior.

In the Cretaceous, we have a better record of the nature of Pacific Plate motions with respect to North America (in large part from magnetic anomalies on the ocean floor), and it is possible to draw some inferences about the link between orogenic and magmatic events and the history of subduction beneath the active margin. There is a general lull in deformation and a lack of evidence for significant magmatism during the time interval 150–120 Ma, which possibly corresponds to a time of dominantly strike-slip motion along the Cordilleran margin. Particularly rapid rates of orthogonal subduction of the Farallon and Kula Plates occurred in the later part of the Cretaceous and Early Tertiary, resulting in major batholithic terranes of this age inboard of continental margin subduction zone complexes. Depending on the configuration of subducting oceanic plates, large components of margin-parallel, strike-slip faulting are also implied. This component of motion would have in general displaced parts of the margin and its accreted terranes northwards toward Alaska along right-lateral, strike-slip faults. How much, where, and along what faults are still controversial questions about this implied motion. In the western United States (Figure 19.4.2), increased rates of subduction resulted in the emplacement of the composite Cretaceous Sierra Nevada batholith. On the western (forearc) side of the batholith, depressed geotherms caused by the rapidly subducting slab led to high pressure-low temperature (blueschist) metamorphism within rocks now represented by the Franciscan Complex, interpreted as a subduction zone assemblage. The intervening Great Valley Basin (Figure 19.4.2) underwent a similar history of 'refrigeration' during rapid subduction. Sediments deposited in this basin and buried as deep as 10 km, and the section reached temperatures of only about 100°C, suggesting thermal gradients of 10°C/km or less. In contrast, heat flow in the arc and backarc region to the east was high and accompanied by major shortening, which migrated eastward with time and resulted in the well-known Sevier foreland fold-and-thrust belt (Figure 19.4.3). At any given latitude, there are typically a series of major thrusts that displace stratified Paleozoic-Mesozoic shelf sediments eastward, with minimum total displacement of 100–200 km. Along most of its length, the eastern front of the thrust belt closely follows the transition from thin cratonic to thicker shelf sequences, indicating important stratigraphic control over the structures produced. Deeper parts of the crust between the thrust belt and the magmatic arc were hot and mobile and underwent thickening by folding and ductile flow (Figure 19.4.3). Crustal thickening in turn precipitated crustal melting, now represented by a belt of unusual muscovite-bearing granites that lie just west of the main thrust belt (Figure 19.4.3). Because the orogen at this latitude was later reworked by Cenozoic extension and associated crustal thinning, the amount of crustal thickening during the Mesozoic is still controversial. Was the western United States like the Tibetan Plateau at the end of the Cretaceous, underlain by 70 km thick crust? Or was crustal thickening more modest, as evidenced by the ~50 km thick crustal root beneath the Canadian Cordillera today?

The post-latest Cretaceous history of the western United States segment of the Cordillera differs substantially from its neighboring segments to the north and south, where subduction-driven arc magmatism and crustal shortening continued uninterrupted into the Early to mid-Tertiary. Magmatism in the western United States portion of the belt shut off abruptly at about 80 Ma, although subduction continued, and in fact accelerated, achieving convergence rates of ~15 cm/yr. These north to south differences have been attributed to segmentation of the subducting slab, with an extremely shallow angle subduction beneath only the United States portion of the belt. This hypothesis is supported by evidence for rapid cooling of the Sierra Nevada batholith as it moved into a forearc position. As the crust of the arc and backarc was 'refrigerated,' it regained its rheologic strength and was thus able to transmit stresses for greater distances. During this time, deformation stepped far inboard to Utah, Colorado, and Wyoming, where crustal-penetrating reverse faults caused uplift of the Rocky Mountains during the latest Cretaceous to Eocene *Laramide Orogeny* (Figure 19.4.2). One classic example is the west-verging Wind River Thrust that exposed highly metamorphosed Precambrian basement rocks. Their uplift was contemporaneous with continued shortening in the foreland thrust belt in Arizona and Mexico to the south and in Montana and British Columbia to the north (Figure 19.4.2).

Plate motions between the oceanic Kula and Farallon Plates and North America changed again at the end of the Paleocene and the component of orthogonal convergence diminished rapidly. In the western United States, it is hypothesized that the shallowly-dipping slab either fell away into the mantle or gradually 'decomposed' due to conductive heating. Decompression melting of upwelling asthenospheric mantle into the previous region of the slab generated basalts that heated the base of the thickened continental crust, causing extensive assimilation and melting of crustal rocks. This magma mixing ultimately led to eruption of large volumes of intermediate to silicic volcanic

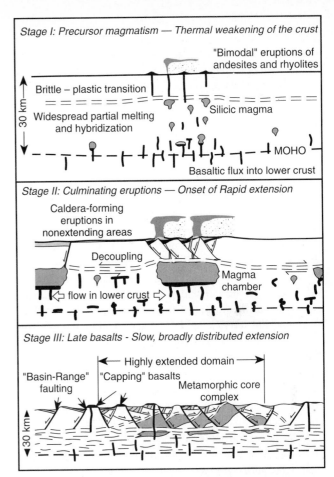

Figure 19.4.4 Schematic sequence of superimposed Cenozoic Basin-and-Range extension-related events leading to the present (mostly young) crustal structure of the orogenic belt.

rocks (Figure 19.4.4). Volcanism migrated progressively into the area of previous flat slab subduction, both southeastward from the Pacific Northwest and northward from Mexico. The large input of heat into the thick crust rheologically weakened it, triggering wholescale extensional collapse of much of the western United States, which resulted in the formation of the Basin-and-Range Province (Figure 19.4.2). This broad zone of continental extension wraps around the southern end of the unextended but (thermally) elevated Colorado Plateau and projects as a finger northward along the Rio Grande Rift on the eastern side of the plateau. To the west of the Basin and Range lies the unextended Sierra Nevada crustal block with its thicker crustal root and the virtually undeformed Great Valley sequence, underlain in part by oceanic crust refrigerated during the Mesozoic (Figure 19.4.2). Volcanism and seismicity are diffuse across this broad zone of continental extension and thermal springs abound. One of the most impressive physiographic features related to this young volcanism is the Snake River Plain depression, believed to represent the Miocene to Recent track of a mantle hotspot that now resides beneath Yellowstone National Park (Figure 19.4.2). The inception of extension in the western United States was diachronous and crudely followed and/or accompanied the inception of magmatism. Rapid, large-magnitude extension in restricted regions was commonly followed by more distributed, slower extension across larger regions (Figure 19.4.4). The present Basin-and-Range Province together with associated extension in the Rio Grande Rift and that north of the Snake River Plain reflect ~200–300 km of east-west extension that began in the Early to mid-Tertiary and continues today. The modern, regularly spaced basin-and-range physiogeography that lends the province its name is the surficial manifestation of the youngest system of major normal faults bounding large, tilted, upper crustal blocks (Figure 19.4.4).

Given the long history of ocean-continent plate interaction along the western margin of North America, it may seem surprising that the actual limits and present topography of parts of the Cordillera are largely dictated by the youngest events to affect the belt. For example, the Basin-and-Range Province includes all or parts of the Mesozoic magmatic arc, backarc, and thrust belt, older Paleozoic allochthons and sutures, and is also underlain by the Precambrian rifted western margin of North America. Despite the diversity of tectonic elements across the Basin and Range, the crust is uniformly 25–30 km thick and much of it stands >1 km above sea level, reflecting an anomalously thin and hot mantle lithosphere. The nearly flat Moho across this broad (600 km) extensional province, despite differences in the history of older shortening and younger extension, implies that the lower crust was capable of undergoing large-scale flow during extensional deformation (Figure 19.4.4). Thus, it seems clear that the present-day structure of most of the crust and perhaps the entire lithosphere across this region reflect only the youngest event to affect this long-lived orogenic belt. This would imply that if the upper 5–10 km of the crust were removed by erosion, we would probably see very little evidence for the previous 600 m.y. history of this orogenic belt. Convergence presently occurs beneath the Alaskan-Aleutian portion of the margin and beneath the Pacific Northwest, and transform boundaries separate the North America Plate and the Pacific Plate along most of the rest of the margin (Figure 19.4.1). In Alaska, shortening and associated diffuse seismicity occur in the overriding North American continent across a broad distance (1000 km). Large-magnitude subduction-zone earthquakes have occurred as recently as 1964 beneath Anchorage, and uplift by reverse faulting has generated some of the most spectacular and rapidly rising mountains of the Cordillera, including Denali (~6 km) in the Alaska Range. In the Pacific Northwest, folding and thrusting are active in the surficial part of the crust above the Cascadia subduction zone, but there have been no large subduction-zone earthquakes since the population of the region by Europeans. Detailed studies of contemporary deformation and paleoseismicity coupled with native oral tradition, however, suggest recurrence intervals of 300 years for such earthquakes, and raise the spectre of a very large (M > 8.0), devastating earthquake occurring in the near future beneath Portland, Seattle, and Vancouver.

In California, the relative motion between the Pacific and North American Plates is partitioned into strike-slip displacement along the San Andreas Fault, and into folding and thrusting related to shortening perpendicular to the San Andreas transform plate boundary (reflected by the recent Loma Prieta and Northridge earthquakes). The exact physical explanation for the observed strain partitioning and how deformation at the surface is coupled with strain at depth in the Earth's crust in such zones of continental deformation remain exciting and challenging problems for structural geologists.

19.4.3 CLOSING REMARKS

This brief discussion of the geologic and tectonic evolution of the Cordillera permits us to make several generalizations about the evolution of such orogenic belts. These generalizations are perhaps contrary to some widely held beliefs about the growth and evolution of continental crust but represent testable hypotheses as our data base grows.

Mountain building and orogeny (i.e., thickening of continental crust) are not necessarily the result of subduction and collision of allochthonous crustal fragments (terranes) along an active continental margin. Subduction occurred for long spans of time during the history of the Cordilleran margin, but, like the southwest Pacific, led mostly to rifting and backarc basin development. Although "subduction leads to orogeny" has a nice ring to it, true mountain building in the Cordillera appears to have occurred during finite time intervals of rapid convergence, magmatism, and absolute westward motion of North America.

The Cordillera has long been cited as a classic example of continental growth by the lateral accretion of allochthonous terranes. However, this mechanism is probably not the most fundamental or important process of crustal growth, unless it involves the addition of mature island arcs. Rifting, with formation of rift basins on existing continental shelves, along continental slopes, and within island arcs and the subsequent filling of these basins by thick prisms of sediment have contributed significantly to the formation of many terranes now incorporated in the Cordillera. Extensional thinning and reworking of continental crust or previously thickened orogenic crust, especially when accompanied by magmatic additions from the mantle, can serve to redistribute and remobilize crust across great portions of an orogen, often equaling or exceeding estimates of crustal shortening within the same belts. The best example of this is the reworking and changing of shape of the continent during Cenozoic extension in the western United States.

Magmatism is sometimes viewed as a process that is largely independent of deformation in mountain belts. The Cordillera provides excellent examples that magmatism is, instead, intricately tied to deformation of continental crust in that heating causes rheologic weakening. The rise of magmas advect heat to higher levels of the crust, permitting continents to undergo large-scale deformation, whether shortening or stretching. This is evidenced by the increasingly better documented eastward migration of magmatism and deformation in the Mesozoic of the Cordillera as well as the space-time relation between magmatism and extensional tectonism in the Cenozoic of the western United States.

ADDITIONAL READING

The information and ideas above are distilled from the author's own works and views and those of many others as represented in the various chapters of the books *The Cordilleran Orogen: Conterminous U.S.* and *The Geology of the Cordilleran Orogen in Canada,* two of the volumes of the Geological Society of America's Decade of North American Geology Series. A particularly helpful review of the evolution of the western United States is given in the first reference listed beneath:

Burchfiel, B. C., Cowan, D. S., and Davis, G. A., 1992, Tectonic overview of the Cordilleran orogen in the western United States, *in* Burchfiel, B. C., Lipman, P. W., and Zoback, M. L., eds., The Cordilleran Orogen: Conterminous U.S.: Boulder, Colorado, Geological Society of America, *The Geology of North America,* v. G-3, p. 407–480.

Coney, P. J., Jones, D. L., and Monger, J. W. H., 1980, Cordilleran suspect terranes, *Nature,* v. 288, p. 329–333.

Dumitru, T. A., Gans, P. B., Foster, Da. A., and Miller, E. L., 1991, Refrigeration of the western Cordilleran lithosphere during Laramide shallow-angle subduction, *Geology,* v. 19, p. 1145–1148.

Engebretson, D. C., Cox, A., and Gordon, R. G., 1985, Relative motions between oceanic and continental plates in the Pacific Basin, *Geological Society of America Special Paper* 206, 59 p.

Gans, P. B., Mahood, G. A., and Schermer, E., 1989, Synextensional magmatism in the Basin and Range Province, A case study from the eastern Great Basin, *Geological Society of America Special Paper* 233, 53 p.

Hoffman, P. F., 1991, Did the breakout of Laurentia turn Gondwanaland inside out? *Science,* v. 252, p. 1409–1419.

Miller, E. L., and Gans, P. B., 1989, Cretaceous crustal structure and metamorphism in the hinterland of the Sevier thrust belt, western U.S. Cordillera, *Geology,* v. 17, p. 59–62.

Monger, J. W. H., Price, R. A., and Tempelman-Kluit, D. J., 1982, Tectonic accretion and the origin of the two major metamorphic and plutonic welts in the Canadian Cordillera, *Geology,* v. 10, no. 2, p. 70–75.

Severinghaus, J., and Atwater, T., 1990, Cenozoic geometry and thermal state of the subducting slabs beneath western North America, in Basin and Range extensional tectonics near the latitude of Las Vegas, Nevada, ed. B. P. Wernicke, *Geological Society of America Memoir* 176, p. 1–22.

19.5   The Central Andes—Richard W. Allmendinger and Teresa E. Jordan[1]

19.5.1   INTRODUCTION

One of the great early advances of plate tectonics was the realization that mountain building is associated with activity at the margins of tectonic plates. The Andes represent the case of mountain building produced by the convergence of an oceanic and a continental plate, a relatively simple and elegant endmember in the spectrum of orogenesis; the other extreme is represented by continent-continent 'collision,' for example, the Himalayan-Tibetan system. Because of the association of the Andes with andesites and the chain of volcanoes that ring the Pacific, a widespread misconception among geologists outside of South America was that the Andes Mountains are primarily a volcanic edifice. In fact, it is now clear that the crustal thickening that produced the Andes is mostly structural in origin and that the volcanoes for which the mountain belt is best known sit on top of that structural welt.

During the 1980s, many investigators sought evidence of accretionary events in the Andes that might have been responsible for Andean mountain building. Underlying this search was the concept that a major mountain system produced by horizontal contraction must be produced by collision of two buoyant crustal masses. However, nearly a decade of concerted paleomagnetic study in Chile and Peru has turned up no evidence that accretion played even a minor role in building the modern Andes. Instead, as we describe below, the evidence suggests that material was removed from the continental margin during the Andean Orogeny.

Thus, the central Andes do represent mountain building by dominantly structural processes in a noncollisional setting, in which the oceanic Nazca Plate is being subducted beneath the continental South American Plate. The purpose of this essay is to describe those processes, as well as the general tectonic setting of the Andes. Several factors make the central Andes, located between 5° and 35° S latitude, a unique laboratory of orogenesis. Because deformation is active today, the governing boundary conditions can be identified and, in some cases, quantified. These include: (1) the plate convergence rate and obliquity, (2) the geometry of the subduction zone, and (3) the dynamic topography, which reflects the interaction of tectonic and climatic

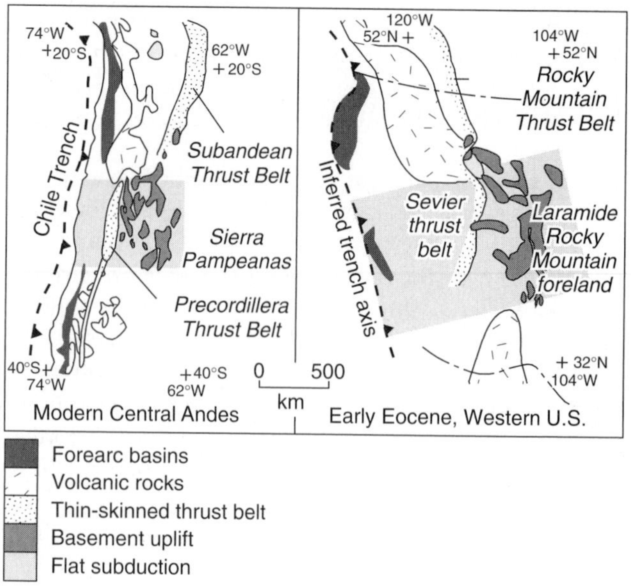

Figure 19.5.1   Comparison of the major tectonic elements of the modern central Andes and the Early Eocene of the western United States, both at the same scale.

processes. Furthermore, crustal seismicity gives us an idea of short-term strain rates as well as the distribution of modes of failure in the continental crust.

The modern Andes are commonly regarded as a modern analog for ancient mountain belts such as in the Mesozoic-Early Tertiary Cordillera of the western United States (Figure 19.5.1). Although the modern setting is quite simple, western South America has a complex pre-Andean history. Thus, the starting materials for Andean deformation were extremely heterogeneous and their responses to the stresses that produced the Andean deformation are equally varied. This factor becomes particularly important when one tries to decide whether a particular structural geometry owes its existence to the modern plate setting or to the ancient anisotropies of the continental crust.

19.5.2   THE ANDEAN OROGENY

Subduction of oceanic crust beneath western South America has occurred more or less continuously since the mid-Jurassic. The term "Andean Orogeny" refers collectively to all tectonism that occurred between the Jurassic and the Recent. Although subduction has been continuous, the obliquity and rate of convergence, as well as the dip of the subducted plate, have varied considerably. Thus, the style and distribution of mountain building has not been continuous or uniform. Here we focus mostly on structures developed during the last 30 m.y. of the Andean Orogeny, but we begin by briefly reviewing the older events.

---

[1]Cornell University, Ithaca, New York.

Arc-related igneous rocks of Jurassic and Cretaceous age occur along the present coastline in Chile and Peru, and, as one moves progressively farther inland (i.e., eastward), the arc becomes progressively younger. The Mesozoic magmatic arc is anomalously close to the present-day trench (within 75 km in some cases), leading to the conclusion that the Mesozoic forearc region has been stripped off the margin and has either been subducted or perhaps moved laterally.

In many areas, the first significant deformation of continental crust associated with the Andean subduction system was horizontal extension. Based primarily on interpretation of lithologic associations, the Jurassic and Cretaceous have long been suspected as a time of intra-arc or backarc rifting. Both in the northern Andes of Colombia and Ecuador and in the southernmost Andes of Tierra del Fuego, this rifting culminated in the production of new oceanic crust. In the central Andes, the clues are more subtle, but work in northern Chile at 27° S has yielded unique data on the geometry and kinematics of the extensional structures of this event. Those structures, low-angle normal faults, extensional chaos, domino blocks, and so on, are geometrically comparable to those seen in Cenozoic detachment terranes of the western United States.

Early Tertiary Andean orogenesis is commonly referred to as the "Incaic Orogeny." This phase of deformation is largely restricted to the present forearc in Chile and Peru. However, deep erosion on the eastern side of the central Andes northeast of La Paz, Bolivia, has revealed rocks that were metamorphosed (or cooled through about 300°C) at about 40 Ma, the same time as the mountain building farther west.

The morphologic edifice that we associate with today's Andes is a product of mountain building during just the last 30 m.y. Most of the surface-breaking structures associated with this phase of deformation are concentrated within the high topography and on the eastern side of the mountain range. Many workers refer to this young phase of deformation as the "Quechua Orogeny," although different workers use this term to refer to events of different ages. The topography of the Andes is the result of structural and thermal processes and is not simply due to piling up of volcanic rocks; in the central Andes, deformed Paleozoic to Cenozoic rocks are commonly found as high as 4 km or more, where they form the great bulk of the high plateau known as the Altiplano. Miocene to Recent volcanoes are perched on top of this plateau.

19.5.3 LATE CENOZOIC TECTONICS OF THE ANDES

The Nazca Plate is currently being subducted beneath western South America in a direction of 077 ± 12° at about 10 cm/y, a rate that varies little along strike of the plate boundary from the triple junction at 49° S to at least Ecuador at 0° of latitude (Figure 19.5.2). Nonetheless, the geometry of the subducting slab is highly variable for rea-

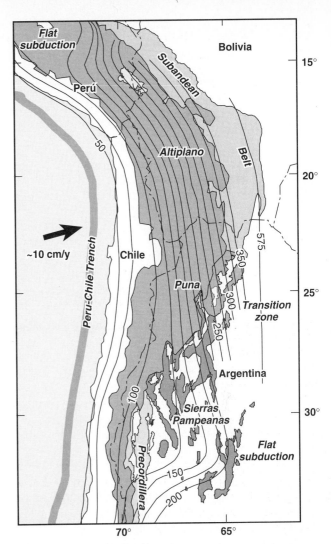

Figure 19.5.2 Tectonic provinces of the central Andes. Thin, smooth lines are contours on the Wadati-Benioff Zone (interval = 25 km). Medium gray region marks the area above 3 km elevation. Foreland thin-skinned thrust belts are shown in light gray; thick-skinned provinces in dark gray. Note that the thrust belt is not continuous but is intersected by the thick-skinned Sierras Pampeanas.

sons that are not totally understood. Between the triple junction and ~33° S, the slab dips ~30° E but can be tracked only to a position directly beneath the modern arc; subduction zone earthquakes beneath the backarc region are virtually unknown. From 33° to 28° S, the subducted plate dips at ~30° down to depths of ~100 km, but it is nearly flat farther east; nearly 600 km east of the plate boundary beneath the city of Córdoba, Argentina, the subducted plate is only about 200 km deep. This segment of the Andes has had no volcanism since the Late Miocene. To the north of 28° S, the plate gradually steepens and by ~24° S has resumed its uniform 30° E dip. This central steep-dipping segment correlates with the Central Volcanic Zone of the Andes as well as the high plateaus known as the Altiplano and Puna. The slab in this segment can be traced to a depth of nearly 600 km beneath the foreland of

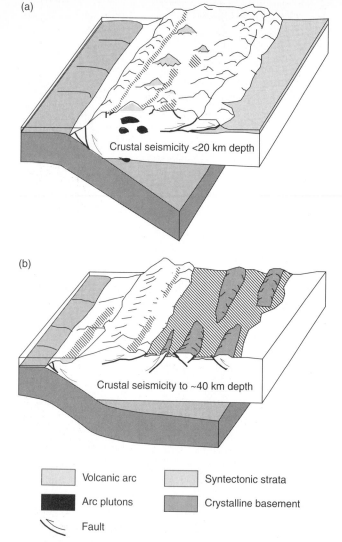

(a)

Crustal seismicity <20 km depth

(b)

Crustal seismicity to ~40 km depth

☐	Volcanic arc	☐	Syntectonic strata
■	Arc plutons	▨	Crystalline basement
⟋	Fault		

Figure 19.5.3 Schematic block diagrams showing the tectonic segmentation of the central Andes. (a) Steep subduction; (b) flat subduction.

the Andes, one over the nearly flat segment of the subducted Nazca Plate between 28° and 33° S and the other over the 30° east-dipping segment between 15° and 24° S, with brief references to other parts of the Andes.

The Andes from 15–24° S. This segment of the central Andes overlying a 30° east-dipping subducted slab most closely approximates most geologists' image of the 'type' Andean orogen (Figures 19.5.2 and 19.5.3a). A cross section from west to east across this segment would show the plate boundary, a forearc region dominated by the longitudinal valley of northern Chile, the active volcanic arc, a high continental plateau system bounded on the east by the eastern Cordillera, and the low-lying Subandean belt of thrusts and folds. The segment is dominated by the Altiplano-Puna Plateau, a region of more than 500,000 km² elevated to an average height of 3.7 km. Volcanism in this segment has been active for about the last 25 m.y. and, though focused in the western Cordillera, locally reaches the eastern edge of the plateau system. The distribution of volcanic rocks, modern morphology, ancient geomorphic surfaces, and structural geometry has led to the proposal of a two-stage uplift model for the plateau. During the Miocene, the crust beneath the current plateau was thermally weakened, resulting in horizontal shortening of the entire crust. At about 10 Ma, shortening mostly ceased in the plateau and in the eastern Cordillera and migrated eastward into the Subandean Belt, a thin-skinned foreland fold-thrust belt. At that point, the cold lithosphere of the Brazilian Shield began to be thrust beneath the mountain belt.

The Andes from 28–33° S. This segment overlies the southern flat subduction zone (Figures 19.5.2 and 19.5.3b). The High Andes (above 3 km elevation) are narrow but include Mt. Aconcagua, at 7 km the highest peak in the Western Hemisphere. Although they are largely composed of volcanic rock, they are not volcanic edifices but structural uplifts; magmatism has been lacking in this segment for the last 5–10 m.y. The magmatic history of the segment indicates that the subducted plate beneath Argentina shallowed between 16 and 6 Ma. Scanty evidence from the Peru flat subduction segment indicates shallowing of the slab beginning as recently as 2 Ma. The cessation of volcanism where subduction is flat is thought to be due to the virtual lack of asthenosphere between the subducted and overriding plates.

In the Argentine flat segment, the Sierras Pampeanas are thick-skinned basement uplifts with structural geometries very reminiscent of the Laramide Rocky Mountain foreland of the western United States (Figure 19.5.1). Like the Rocky Mountain foreland, many blocks of the Sierras Pampeanas display an exhumed Paleozoic-Early Mesozoic erosional surface that has been rotated, probably by movement on listric faults, during the last 5 m.y. Seismicity associated with the westernmost of these basement uplifts indicates brittle failure, at least on short time scales, to depths of 40 km, nearly the entire thickness of the continental crust. This is unusually deep for continental deformation and is well below the predicted depth for the brittle-plastic transition, even using the most conservative parameters for

the Andes. At 15° S beneath southern Peru, the subducted slab again flattens to a near horizontal altitude below 100 km depth. This northern 'flat slab' underlies most of Peru; like its southern counterpart, the High Andes here are quite narrow and recent volcanism is absent.

It has been proposed that subduction of the Nazca and Juan Fernandez oceanic ridges are responsible for flat subduction. The latter does coincide closely with the boundary between flat and steep subduction at ~33° S but the former is located 1–2° north of the flat-steep boundary beneath southern Peru. The plate kinematics clearly shows that the ridges have swept progressively southward along the plate boundary, but the continental geology shows no evidence of this effect.

Tectonic Segmentation of the Central Andes

The geology of the late Cenozoic Andes to a first order reflects the lateral segmentation of the subducting plate (Figure 19.5.3). Below, we concentrate on two swaths across the frictional and power law rheologies and heat flow on which those models are usually based. Because of extensive jungle cover, much less is known of the style of foreland deformation located over the Peruvian flat segment north of 15° S.

Although the thick-skinned basement uplifts are the most obvious features of the foreland overlying the Argentine flat slab, thin-skinned deformation also occurred between the Sierras Pampeanas and the High Andes in the Precordillera (Figure 19.5.2). Thrusting there began at about 20 Ma and has continued to the present where it has not been buttressed by the Sierras Pampeanas. Thus, as in the western United States, thin- and thick-skinned thrusting overlapped in time, and both were active during flat subduction and during a period without significant magmatism. Shortening is about 5% in the Sierras Pampeanas and >60% in the Precordillera.

Horizontal Extension and Strike-Slip Faulting

Horizontal extension within convergent mountain belts has been of considerable interest during the 1980s. Young extensional deformation in the central Andes is concentrated at the northern and southern ends of the Altiplano-Puna Plateau. In the Puna, horizontal extension is mostly related to strike-slip and oblique-slip faults, whereas at the northern end of the Altiplano normal faults dominate. In both areas, however, horizontal extension is oriented approximately north-south, that is, subparallel to the strike of the orogen. There is little firm evidence for significant extension perpendicular to the belt and the main part of the plateau system is neotectonically and seismically quiescent. There is probably significant strike-slip faulting in the Bolivian eastern Cordillera north of the bend at ~18° S, which may accommodate the more oblique angle of convergence in this part of the belt. Even though extension is most notable around the ends of the high plateau, it is not restricted to the high topography. In northwest Argentina, horizontal extension occurs at elevations as low as 900 m in the foreland, and in southern Peru normal faults are found on the Pacific coast.

South of 28° S, several long fault zones, including the Liquiñe-Ofqui Fault in southern Chile and the El Tigre Fault in western Argentina, had significant strike-slip movement during the Quaternary. These are thought to be due to the slight nonorthogonal convergence between the Nazca and South American Plates. The Atacama Fault System of northern Chile is probably best known for its strike-slip history, but recent work has shown that it has a much more complicated and protracted history, including dip-slip and strike-slip displacements during the Mesozoic to mid-Tertiary and mostly dip-slip motion since then.

Oroclinal Bending

The marked curvature of the Andes at about 18° S begs the question of whether it represents the initial shape of the continental margin or if it is a product of Andean deforma-

tion. The answer is probably "both." Paleomagnetic data derived from Mesozoic rocks along the west coast of South America show clockwise rotations south of the bend and counterclockwise rotations to the north. The main debate over these results centers on whether the rotations reflect *in-situ* block rotations or regional oroclinal bending. To date, several carefully mapped sites in the foreland provide evidence only for local, rather than regional, rotations. These preliminary results, however, do not preclude a model in which the curvature of the central Andes has been accentuated during the last 25 m.y. by laterally variable shortening, which is greatest within, and on the margins of, the Altiplano-Puna Plateau and decreases both to north and south.

Paleotectonic Control

Although structural style within the central Andes shows a broad correlation with the geometry of the subducted plate, as described above, preexisting heterogeneities within the continental crust also play an important role. Thin-skinned thrust belts are restricted to thick, wedge-shaped Paleozoic basins. The Subandean Belt is largely located within a previously undeformed, Paleozoic passive margin and foreland basin sequence, east of a zone of preexisting deformation now occupied by the eastern Cordillera. The Precordillera thrust belt deforms a lower Paleozoic passive-margin sequence of what is called the Precordillera Terrane, a narrow slice within western Argentina that bears marked similarities to the Lower Paleozoic of the Appalachians. The transition from the Subandean Belt to the thick-skinned deformation of the northern Sierras Pampeanas coincides with the southward wedging out of Paleozoic strata, and the western boundary of the Sierras Pampeanas is the boundary of the Precordillera Terrane. Finally, many complex local structures within the Bolivian Altiplano and in northwestern Argentina owe their geometries to reactivation of Late Cretaceous rift basins.

19.5.4 CRUSTAL THICKENING AND LITHOSPHERIC THINNING

Modern estimates suggest that magmatism contributes less than 10% to the total crustal thickening in the Andes during the last 25 m.y. Thus, the rest of the topography in the Andes must be accounted for by two mechanisms: thickening of the crust by deformation and thermally controlled thinning of the lithosphere giving rise to uplift. Most of the crustal shortening responsible for the present topography is manifest at the surface as the thin-skinned thrust belts that provide the interface between the Andes and the Brazilian Shield. The thrust belts of the central Andes, including the Subandean Belt and the Precordillera, are of considerable interest because they are among the few active examples of an antithetic (i.e., foreland-verging) foreland thrust belt in which the overall sense of shear is opposite to that in the associated subduction zone. The along-strike variations in

Table 19.5.1 Summary of Thrust Belt Characteristics, Bolivia and Argentina

Location	Width (km)	Shortening (km)	Topogr. Slope	Wedge Taper	Annual Precipitation (mm)
N. Bolivia 13–17° S	70	115 ± 20	3.5°	7° ± 1°	1000–2800
S. Bolivia 19–23° S	90–110	75 ± 10	0.5–1.0°	2.5° ± 1°	400–1000
Precordillera Argentina 29–33° S	40–60	105 ± 20	~2.5°	3.5° ± 1.5°	100–200

these thrust belts allow one to identify the key first-order associations of geometry, topography, shortening, and paleotectonic setting (Table 19.5.1). There is a general correlation among high critical wedge taper, width, and high shortening. In contrast, the order of magnitude variation in precipitation (and, presumably, erosion rate) shows no clear effect on the shortening.

Crustal thickening alone does not appear to be sufficient to explain the high plateau (the Altiplano) of the central Andes. An additional 1 to 1.5 kilometers is probably accounted for by thinning of the lithosphere beneath the plateau. Major unresolved problems include when this thinning occurred and how it relates to crustal shortening. As has been pointed out for the Alpine system, horizontal shortening should thicken not only the continental crust, but also the entire lithosphere. Yet beneath the plateau, the continental lithosphere must have thinned even as the crust was thickening. Furthermore, in the regions of flat subduction, the subducted plate is only 100 to 120 km beneath the surface, even though 'normal' continental lithosphere is thought to be on the order of 150 km thick. Because it takes many millions of years to thin the lithosphere by conduction alone, recent proposals have invoked delamination of the base of the lithosphere to produce the necessary thinning on the time scale implied by Late Cenozoic mountain building in the Andes.

19.5.5 CLOSING REMARKS

The Andes present a unique natural laboratory for studying mountain building that has occurred without the aid of a collision between two continental masses. The most important processes that have produced the modern topography of the Andes are structural shortening and lithospheric thinning. Volcanism, in contrast, is responsible mostly for the volumetrically minor topography above 4 to 4.5 km. Magmatism, nonetheless, probably plays a very significant role in determining the rheology of the crust, producing weak zones favored for faulting.

ADDITIONAL READING

Beck, M. E., Jr., 1988, Analysis of Late Jurassic-Recent paleomagnetic data from active plate margins of South America, *Journal of South American Earth Sciences,* v. 1, p. 39–52.

Dalziel, I. W. D., 1986, Collision and Cordilleran orogenesis: An Andean perspective, *in* Coward, M. P., and Ries, A. C., eds., Collision tectonics, *Geological Society of London Special Publication* 19, p. 389–404.

Isacks, B. L., 1988, Uplift of the central Andean plateau and bending of the Bolivian orocline, *Journal of Geophysical Research,* v. 93, p. 3211–3231.

Jordan, T. E., Isacks, B. L., Allmendinger, R. W., Brewer, J. A., Ramos, V. A., and Ando, C. J., 1983, Andean tectonics related to geometry of subducted Nazca plate, *Geological Society of America Bulletin,* v. 94, p. 341–361.

Mpodozis, C., and Ramos, V. A., 1990, The Andes of Chile and Argentina, *in* Ericksen, G. E., Ca—as Pinochet, M. T., and Reinemund, J. A., eds., Geology of the Andes and its relation to hydrocarbon and mineral resources: Houston, Texas, Circum-Pacific Council for Energy and Mineral Resources, *Earth Science Series* 11, p. 59–90.

Pardo-Casas, F., and Molnar, P., 1987, Relative motion of the Nazca (Farallon) and South American plates since Late Cretaceous time: *Tectonics,* v. 6, p. 233–248.

19.6 The Caledonides—Kevin T. Pickering[1] and Alan G. Smith[2]

19.6.1 INTRODUCTION

The Caledonides of Svalbard (Spitsbergen) and northwestern Europe mark the edges of ancient lithospheric plates, the subduction of Paleozoic oceans, and the collision of the adjacent continents. In the Cambrian to Silurian (ca. 570 to 408 Ma) the present-day Caledonides region consisted of essentially three large continental blocks separated by one or more oceans (Figure 19.6.1): Gondwana (South America and Africa being the most important continents in relation

[1]University College London, London, United Kingdom.
[2]University of Cambridge, Cambridge, United Kingdom.

Figure 19.6.1 Computer-generated plate reconstructions, based on a synthesis of paleomagnetic data, for the following time intervals: (a) Late Precambrian (600–570 Ma); (b) Cambrian (530 Ma); (c) Early Ordovician (480 Ma); (d) Middle Ordovician (460 Ma); (e) latest Ordovician (440 Ma); (f) Middle Silurian (420 Ma); and (g) Early Middle Devonian (390 Ma). European microplates are not shown on these reconstructions. Positions of major magmatic arcs are shown schematically.

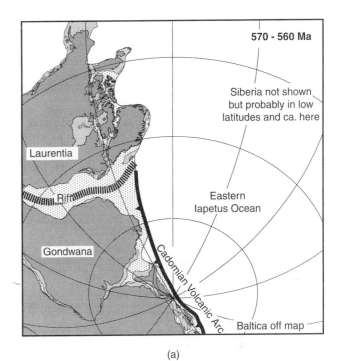
(a)

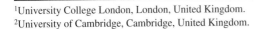
Legend to maps
- Outline of present coastlines
- Outline of normal continental crust
- Inferred thinned/anomalous thickness continental and arc crust
- Location of accreted volcanic arc terranes
- Subduction zone, showing polarity of subduction
- Foreland basin, showing polarity of underthrusting
- Active rift zone
- Transcurrent fault
- Oceanic spreading center

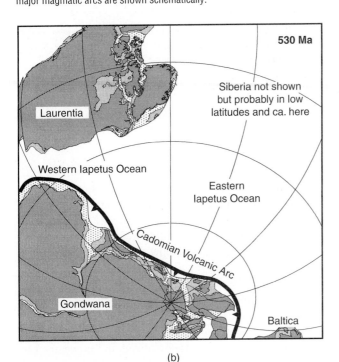
(b)

(c)   (continued)

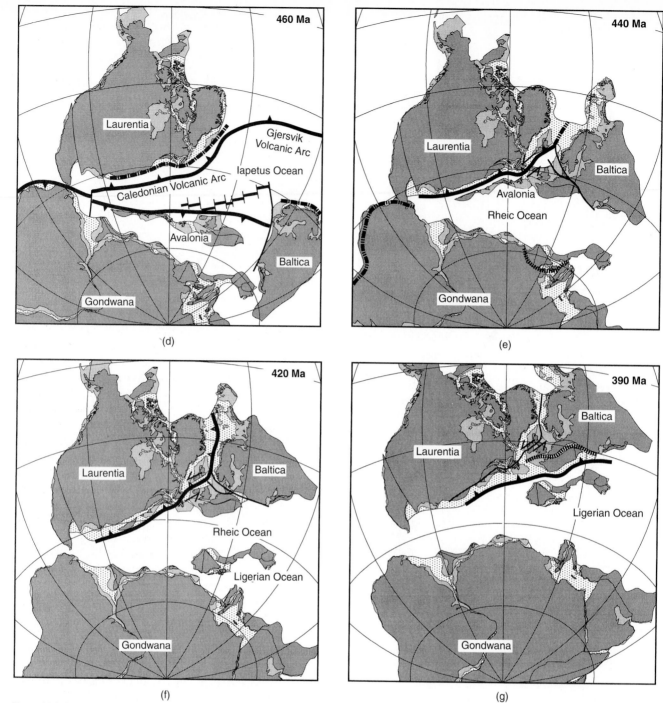

Figure 19.6.1 (continued)

to the Caledonides), Laurentia (North America, Greenland, and northwest Scotland), and Baltica (northwest Europe to the Ural Mountains in the east, and south to the region of the Tornquist-Teisseyre Lineament).

Three collisional belts formed between these three continents: the Caledonides of Norway, western Sweden and eastern Greenland lying between western Baltica and eastern Laurentia; a poorly exposed branch of the Caledonides under the North Sea, continuing into eastern Europe situated between northern Gondwana and southern Baltica; and the Caledonides of the British Isles between northwest

Gondwana and southern Baltica. Each collisional belt represents the site of a former ocean, which has been given the name the Iapetus Ocean. For convenience, the ocean between Baltica and Laurentia will be referred to as the Eastern Iapetus, that between Gondwana and Baltica as Tornquist's Sea, and the ocean between Gondwana and Laurentia as the Western Iapetus.

In detail, the histories of the collisions are complex. Continental slivers rifted away from the northern margin of Gondwana, and collided with the margins of Laurentia and Baltica before Gondwana itself collided. In Europe, the best defined of these slivers are eastern Avalonia (southern Britain and much of France); in North America they are western Avalonia (Avalon Peninsula and Gander Zone of Newfoundland, New Brunswick, and Nova Scotia), and the Piedmont terrane and the Carolina slate belt in the southern Appalachians. Thus, there was a stage in which short-lived new oceans existed, which can be regarded as parts of the Iapetus in the broad sense. One of them, the Rheic Ocean, lay between the northern margin of Gondwana and parts of central and western Europe to the south of eastern Avalonia, and will receive special attention here.

The orogenic belts formed by these collisions have long-established names related to their present-day geographic positions. The youngest phase of orogenesis (Silurian-Early Devonian) is known as Acadian in North America and Late Caledonian in eastern Greenland and the British Isles, and Ligerian in mainland Europe, except in Scandinavia where the name Scandian is used. In North America, the orogenic activity that peaked in the Middle Ordovician is known as Taconic, whereas it is the Early Caledonian in northwest Europe, and M'Clintock Orogeny in Svalbard (Spitsbergen). Late Precambrian orogeny is known as the Famantinian in South America, the Grampian in northwest Scotland, and Cadomian in Brittany (northwest France) and southern Britain. The processes that led to the eventual welding of the three supercontinents to form Pangaea in Permian time gave rise to a perfect match: there are no gaps in the reassembly, such as unfilled remnant ocean basins. The continents most unlikely matched perfectly before collision, and one must therefore expect processes to have taken place that allow imperfectly matching continents to fit together, such as thickening of continental crust, indentation by promontories, and lateral movement of slices.

Paleomagnetic poles from the stable parts of the three large continents provide the principal quantitative data for positioning the continents relative to one another in Early Paleozoic time. The Late Precambrian-Early Cambrian was a time when new continental margins formed along Laurentia, Baltica, and parts of Gondwana. By joining the original opposing margins together it is possible to make a plausible reassembly of the circum-Iapetus continents, which serves as a starting point for the evolution through the Paleozoic. Where paleomagnetic evidence is inconclusive, geological evidence permits an independent inference about the nature of each continental margin from its stratigraphic record.

19.6.2 LATE PRECAMBRIAN TO CAMBRIAN BREAKUPS AND ARCS

The Late Precambrian to Cambrian history of Laurentia, Baltica, and western South America reflects Late Precambrian breakup and the development of passive continental margins. In northwest Britain, the mainly Late Precambrian Dalradian Supergroup accumulated between Laurentia and South America (Gondwana). In West Africa, the earliest recorded tectonic event is the westward rifting of a continental fragment ~700 Ma ago, and the development of a rift-drift stratigraphy. Dike swarms, reflecting crustal extension, are known along the Laurentian margin in Labrador dated at ~615 Ma. Similar dike swarms are known from the Baltica margin in Scandinavia, dated at 665 Ma and approximately 640 Ma. In themselves, the dikes merely indicate stretching, but in all the areas studied they pass upward into or are closely associated with the development of extensive carbonate platforms (believed to have accumulated in low latitudes), and/or were intruded into very thick deep-water clastics that are interpreted as passive continental margin successions. The eastern Laurentian passive margin extended from Greenland through the Durness sequence of northwest Scotland to fringe most of the United States. Similarly, the Late Precambrian to Cambrian margin of Baltica is interpreted as a 200 km-wide passive margin sequence. A comparable passive margin of the same age is known in northwest Argentina and the western margin of the South American craton.

Margins formed by lithospheric extension have a characteristic subsidence curve that is determined both by the amount of stretching and the time it began. From such curves, the breakup and rifting ages of the new passive margins of Laurentia, Baltica, and northwest Argentina occurred between 625–555 Ma, in agreement with all the other evidence. Also, in western Newfoundland, a rift-drift transition has been proposed for western Newfoundland at about 570–550 Ma.

Eastern North America (Laurentia) and western South America (Gondwana) separated from each other in the Late Precambrian. This suggestion is supported by the otherwise puzzling distribution of Early Cambrian *olenellid* trilobites in northwest Argentina that are similar to those in eastern Laurentia and not known elsewhere. Rift events recorded in dike swarms from Baltica and Greenland suggest that it occurred about 100 million years earlier than for the Appalachians, that is, that the Eastern Iapetus Ocean was much older than the Western Iapetus Ocean.

By contrast with the passive margins of Gondwana, the Late Precambrian to Cambrian history of north and northwest Gondwana is one of arc formation and orogenesis. The orogeny, known as the Cadomian (approximately 650–500 Ma), is exposed on the northern edge of Gondwana, with the type area in Brittany, northwest France. It includes arc and arc-related rocks of approximately 650–500 Ma tectonothermal events, also exposed in southern Britain, Spain, southeast Ireland, and the 'Avalonian' of the northern Appalachians. To the east of Brittany, Cadomian deformation is recognized in Czechoslovakia and the south Carpathian-Balkan region; that is, this was a time of extensive crustal growth. The Cadomian events probably represent the vestiges of a major arc system formed by subduction under the northern and northwestern margin of Gondwana.

Many areas affected by the Cadomian Orogeny were later detached from the margin of Gondwana. They migrated across the intervening ocean to become 'exotic' terranes attached to Laurentia before Gondwana itself collided with them and Laurentia. Thus, during the Late

Precambrian, microcontinental blocks such as (eastern and western) Avalonia, the Piedmont terrane, and the Carolina slate belt, probably formed the outboard parts of Gondwana along a margin that bordered an existing ocean.

19.6.3 EARLY/MID-ORDOVICIAN ARCS, MARGINAL BASINS, AND OPHIOLITES

Throughout northwest Europe, regional chemical and isotopic signatures in the Ordovician-Devonian igneous suites north of the Iapetus Suture, or its inferred along-strike continuation, show a subduction-related affinity associated with both southeast- and northwest-directed subduction, active from Early Ordovician to Middle Silurian times. By Late Tremadoc to Early Arenig time (ca. 490 Ma) the Laurentian craton was fringed by marginal basins and arcs. Two subparallel arcs appear to have existed along much of the western/northwestern margin of the Iapetus Ocean: (1) An immature inboard arc that was developed mainly on continental crust northeastward from Ireland, but on oceanic crust off western Newfoundland and the Appalachians. (2) The second arc system was within the ocean, and developed above a north/northwest-dipping subduction zone associated with the destruction of the Iapetus Ocean sensu stricto. Destruction of the marginal basins took place throughout the Early Ordovician to Late Silurian interval along most of its length, but was restricted to Middle Ordovician time in Svalbard (M'Clintock Orogeny) and latest Ordovician-earliest Silurian time in central Newfoundland.

As in the Mesozoic Tethys, the Western Iapetus Ocean was associated with a relatively brief phase of arc development and oceanic spreading, rapidly followed by plate convergence leading to Ordovician, ophiolite obduction along the Laurentian margin immediately preceding arc-continent collision (Taconic to Early Caledonian). Early Ordovician ophiolites fall into three categories, from west to east: (1) Laurentian marginal basin oceanic crust, as in western Newfoundland and the Shetland Islands; (2) intra-arc, backarc marginal basin oceanic crust, as in the central Newfoundland, western Ireland, and Highland Border Group ophiolites and possible slivers along the Highland Boundary Fault Zone; and (3) forearc ophiolites, including accreted seamount material, as in eastern central Newfoundland and southwest Scotland, which may include seamount fragments. In the Scandinavian Caledonides, arc fragments appear to have been involved in collisional events in Early Ordovician times (Finnmarkian), but finally emplaced onto Baltica during the Silurian-Early Devonian.

The marginal basins that fringed Laurentia in the Early Ordovician were closed by the end of mid-Ordovician time. In North America their closure caused extensive east-west shortening (present coordinates), mainly accommodated by westward transport of thrust slices (Taconic Orogeny). The closure of the marginal basins was caused by arc-continent collision, driven by the proximal approach of the northern

margin of Gondwana (probably northern South America) to the Appalachian margin of Laurentia, and/or a major reorientation of relative plate vectors because of plate-tectonic processes associated with other ocean basins. The collision events along the Laurentian margin on the northern side of the Eastern Iapetus Ocean may be related to the narrowing of the ocean between Baltica and Laurentia such that by Late Llandeilo generic faunal links were established across a Galápagos-like island chain between both continents.

The maps of this period (Figure 19.6.2, 480–460 Ma) show western South America moving past eastern Laurentia. The motion appears to have been quite oblique, without actual collision but causing arc accretion in Middle Ordovician time in the absence of an intense Himalayan-style continent-continent collision between Laurentia and western Gondwana. South America or 'Occidentalia', which is bordered by a Cambrian carbonate platform similar to that of eastern North America, may have been an opposing continental margin to Laurentia during Late Ordovician time. In northwest Argentina, the early Paleozoic *olenellid* trilobite faunas in the Famantinian Orogen are similar to those found in eastern Laurentia, which may suggest geographic linkage between these areas.

As noted above, because of uncertainty about Baltica's position relative to other continents, it has been omitted from maps of 490 Ma and older periods. It is first shown on the 460 Ma map separated from Laurentia by a relatively narrow branch of the Eastern Iapetus Ocean. For compatibility with the events in Svalbard, the distance between the two continents is shown as decreasing between 460 and 440 Ma. This presumed narrowing of the ocean between Baltica and Laurentia is supported by the establishment of faunal links between the two continents by Late Llandeilo time (about 465 Ma) where none had existed previously.

Throughout northwest Scotland and northern Ireland, north of the Iapetus Suture, geological data, including regional chemical and isotopic signatures in the Ordovician-Devonian igneous suites, show a subduction-related affinity associated with northwest-directed subduction, active from Early Ordovician to Middle Silurian times. This subduction zone appears to have been active in accommodating the subduction of most of the Iapetus oceanic crust, although southeast-directed subduction may have been important locally, and even in the final stages of closure.

19.6.4 EARLY-MIDDLE ORDOVICIAN BREAKUP, RIDGE SUBDUCTION, CONTINENTAL FRAGMENTATION

In the Late Arenig, there was a second major episode of continental fragmentation in Gondwana (Phase II breakup, to distinguish it from the Late Precambrian Phase I breakup). The Avalonian, Piedmont, and Carolina slate belt terranes, all of which contain Cadomian-age arc basement, broke away from the northwestern edge of Gondwana. The partial separation of eastern Avalonia (including northwest

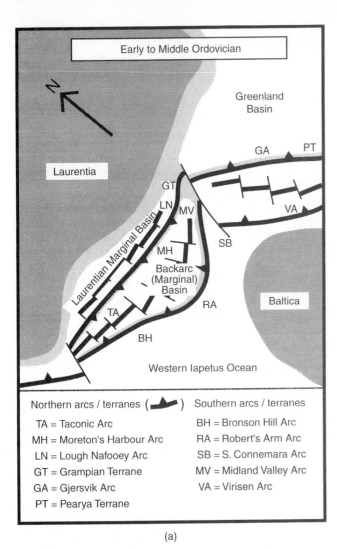

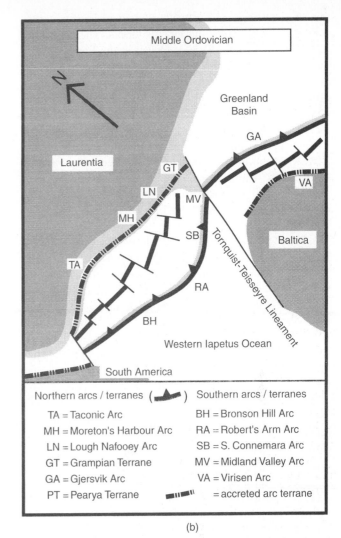

Figure 19.6.2 Schematic diagram showing (a) Early Ordovician arcs and marginal basins originally next to the Laurentian margin (eastern North America), and which were (b) subsequently accreted, in part, to Laurentia and South America (Taconic-Early Caledonian Orogeny), and Baltica (final emplacement in Scandian [= Late Caledonian] Orogeny).

France, also called Armorica) from Gondwana, during the Late Arenig, has been documented using sedimentological criteria.

In addition to the separation of these fragments, a new ocean basin may have been created on the eastern edge of Baltica in the present region of the western Urals. A more speculative view is that the Eastern Iapetus Ocean (i.e., Tornquist's Sea) was created at about the same time by the separation of southern Baltica from northern South America; that is, Phase II breakup could have resulted from Baltica rifting away from Gondwana along with Avalonia, and the Piedmont and Carolina slate terranes. After rifting, Baltica may have moved equatorward, from a more southerly position in Early Ordovician time. There is no clear evidence for the creation of new passive margins in these areas like those during the Late Precambrian to Cambrian age discussed above. Avalonia is not linked with Baltica at this time.

Much of the northern Gondwanan margin of the Iapetus Ocean was an Andean-type plate margin with a subduction polarity towards the continental interior. Remnants of this arc include the calc-alkaline igneous rocks of the English Lake District, southern Welsh Basin, and southern Ireland. The Welsh Basin was initiated as a marginal basin on the southern side of the Eastern Iapetus Ocean during the Arenig, at the same time as eastern Avalonia rifted off northwest Gondwana. The late Arenigian Phase II breakup of Gondwana immediately preceded the extensive ophiolite obduction initiated in the Llanvirn, which may be causally related.

During Cambrian to Early Ordovician time, northwest Britain was part of Laurentia and located at about 15–20° S. By contrast, the paleolatitude of eastern Avalonia (southern Britain) in Early Ordovician time was about 60° S. The paleolatitude of eastern Avalonia had changed to about 45° S in the Middle Ordovician, and approximately 15–25° S in the latest Ordovician-Early Silurian, suggesting a steady northward drift across the Iapetus Ocean. The latitudinal separation across the Iapetus Ocean in the Early Ordovician, between the part of Laurentia containing northwest Scotland (where eastern Avalonia eventually docked), and the Gondwanan margin with eastern Avalonia, changed from about 5000 km in the Late Tremadoc-Early Arenig to approximately 3300 km by the Llanvirn-Llandeilo. The underlying plate-tectonic causes for this northward motion do not have an immediately obvious explanation. The paleomagnetic data independently support the faunal arguments for the northward movement of eastern Avalonia across the Iapetus Ocean during the Late Ordovician.

Subduction-related igneous activity occurred in the British Isles, south of the Iapetus Suture, from the Tremadoc (earliest Ordovician) to earliest Caradoc (mid-Ordovician) with a rapid change to a more alkaline and per-alkaline signature and abrupt cessation in the Longvillian (mid-Caradoc). Ridge subduction, and the creation of a slab window below the northern margin of Gondwana, that is, eastern Avalonia, provide an elegant mechanism to explain: (a) the abrupt switch-off in subduction-related igneous activity in eastern Avalonia in the Early Caradoc; (b) the changed geochemical signature of the Caradoc compared to earlier igneous activity in eastern Avalonia; (c) the subduction of thermally warm ridge flanks millions of years prior to ridge subduction as a reason for the widespread Llandeilo hiatus, or thin stratigraphies, throughout much of eastern Avalonia; and (d) a fundamental cause for eastern Avalonia being transferred to a north-moving plate and its rifting away from Gondwana, as the ridge spreading center jumped southward of the microcontinent. An analogy can be found in the present-day Pacific where the small continental fragment of Baja California is now attached to the Pacific Plate and moving with it.

19.6.5 MIDDLE-LATE ORDOVICIAN COLLISIONS BETWEEN ARCS AND CONTINENTS

The protracted collision events contemporaneous with the Taconic Orogeny of North America culminated in high-grade metamorphism and major uplift at about 460–440 Ma in the Western Iapetus Ocean, along other parts of the Laurentian margin, and the associated marginal basins. Parts of the Scandinavian Caledonides record an approximately 450–435 Ma uplift history. Stable argon isotope studies of rocks from northern Sweden in the Upper Allochthon (Lower Koli Nappe, the Seve-Koli Shear Zone, the Seve Nappe) and the shear zones of the Middle Allochthon reveal high-grade metamorphism and associated deformation of the Seve units as a Late Cambrian-Early Ordovician event in which the rocks cooled below the respective closure temperatures for hornblende at approximately 490 Ma and muscovite at approximately 455 Ma. The structurally lower rocks

of the Middle Allochthon, inferred to have been more proximal to Baltica prior to emplacement, show only the approximately 430 Ma event(s), whereas the Upper Allochthon records the older Finnmarkian event(s). In the Seve Nappe, there is evidence for Middle to Late Ordovician 450–440 Ma shear zones, showing that a pre-Scandian deformation affected rocks outboard from or marginal to Baltica. During this time, Seve Nappe of different P-T-t histories were juxtaposed. Subsequently, during the Scandian Orogeny, the Seve and Koli Nappes were juxtaposed, and the Middle Allochthon mylonites formed as these nappes were emplaced over the Baltic Shield. All these tectonic units were assembled prior to regional cooling through the closure temperature of muscovite.

19.6.6 LATE ORDOVICIAN-SILURIAN CLOSURE OF THE EASTERN IAPETUS OCEAN

The Ordovician-Silurian history of the Midland Valley, Scotland, records the evolution of an arc and backarc basin. Throughout the Late Ordovician and Early Silurian, the Southern Uplands of Scotland and the along-strike Wexford-County Down area (Longford Down inlier) of Ireland were part of an active accretionary prism developed above a northward-dipping subduction zone on the northern margin of the Iapetus Ocean. For example, the Southern Uplands accretionary prism that developed over at least 50 million years, from the Llanvirn to Wenlock, and was associated with the Midland Valley forearc basin farther to the north. The arc massif of older metamorphic basement in the Grampian Highlands northwest of the forearc basin was capped by calc-alkaline arc volcanics and intrusive igneous suites, and supplied most of the sediments to the trench-forearc accretionary system to the south.

Closure of the Eastern Iapetus Ocean by oblique (overall sinistral) collision took place between the island arc(s) sandwiched between the converging continents of Laurentia and Baltica. The closure may have begun during the latest Llandeilo to Caradoc in the region of northern Norway-Svalbard, to incorporate M'Clintock-Finnmarkian orogenic crustal fragments. In places the setting probably resembled that between mainland Southeast Asia and northern Australia today. Elsewhere, continent-continent collision may have created a situation like that between the present-day Himalayas and the Bengal Fan of eastern India. Voluminous flysch sediments were shed away from the collision zone to form the axial-trench wedges of sandy turbidites preserved in the Ordovician tracts of the Southern Uplands accretionary prism, northwest Britain, akin to the present-day Bengal Fan.

The Silurian-Devonian Scandian Orogeny, caused by the collision of Laurentia and Baltica above a northwest-dipping subduction zone, resulted in the final emplacement of thrust sheets eastward onto the Scandinavian crystalline basement with its Cambrian-Ordovician shelf successions.

Collision of eastern Avalonia with Baltica and Laurentia occurred in the latest Ashgill-earliest Llandovery, with the microcontinent behaving as a rotating rigid indentor, probably in the region of present-day central Newfoundland and above a north-dipping subduction zone (Figure 19.6.3). Collision was oblique or 'soft', with a sinistral component along the margin. The major phase of bimodal Silurian magmatism in central Newfoundland, New Brunswick, and south-central Britain, and in western Ireland, implies a component of extension shortly after the initial collision. Extension is also suggested by the kinematic history of syn-deformation granites in northwest Britain. Even faunal evidence from Middle Silurian ostracodes suggests a phase of extension after which oblique convergence continued. The phase of mid-Silurian oblique extension was probably caused by the rotation of eastern Avalonia against Laurentia during the final stages of suturing. The main sinistral displacement of eastern Avalonia took place throughout the Silurian and up until the peak Acadian deformation in the Emsian (Early Devonian time).

Final welding of eastern Avalonia took place in the Wenlock, associated with a prolonged phase of deep-marine foreland-basin development. By Llandovery time, the Tornquist-Teisseyre Lineament formed a major, probably active, backarc strike-slip fault to the arc associated with closure of the Eastern Iapetus Ocean between Baltica and eastern Avalonia, and associated concealed continental fragments, as a submarine (subshelf) lineament. The southern continental plate boundary of Baltica remains poorly defined, but was south of the arc complex associated with northward subduction in the northwest European Caledonides.

Silurian-Devonian sinistral shear was associated with the amalgamation of the Western, Central, and Eastern provinces of Svalbard. These provinces have undergone pre-Devonian histories that involved complete separation of the terranes. Eastern Spitsbergen and Nordaustlandet, for example, may have originated along the Laurentian margin far to the south of their present position, to be juxtaposed against the Central Province along the Billesfjorden Fault Zone by the Late Devonian.

Within the Caledonian slate belt, south of the Iapetus Suture, there is a major arcuate trend in the orientation of the strike of cleavage, from an 'Appalachian' (northeast) trend in Ireland and Wales to a 'Tornquist' (east/southeast) trend, typical of northern Germany and Poland, and intermediate trends in northern England. Based on the clockwise cleavage transection of related folds, and associated sinistral displacement on strike-slip faults in northwest England, there was an episode of Late Caledonian ('Acadian') sinistral transpression. The microgranite dike swarm associated with the emplacement of the Shap granite intrudes folded and cleaved Silurian sediments, but the dikes themselves are weakly cleaved. In the English Lake District the formation of this cleavage was contemporaneous with the emplacement of various syn- and postdeformational igneous suites, dated at about 394–392 Ma or early Middle Devonian

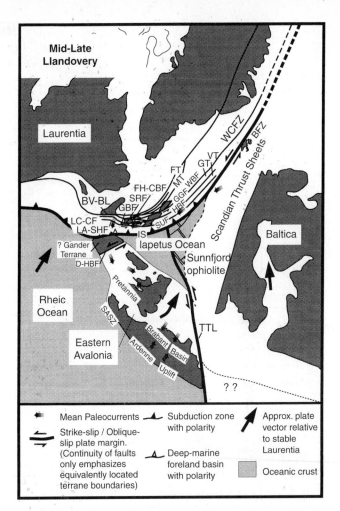

Figure 19.6.3 Mid-Late Llandovery reconstruction. Plate motion arrows shown for the Baltica and eastern Avalonia plates relative to a fixed Laurentian (North American) plate. Closely-spaced stipple outlines parts of Armorica, Britain, and the Gander Terrane, south of the Iapetus Suture, and fragments of western Newfoundland and northwest Britain with Laurentian crustal affinities prior to ocean closure. Major faults/lineaments are: BV-BL = Bay Verte-Brompton Lineament; LC-CF = Lobster Cove-Chanceport Fault; LA-SHF = Lukes Arm-Sops Head Fault; GBF = Galway Bay Fault; SRF = Skerd Rock Fault; FH-CBF = Fair Head-Clew Bay Fault; SUF = Southern Uplands Fault; HBF = Highland Boundary Fault; GGF = Great Glen Fault; WBF = Walls Boundary Fault; FT = Flannan Fault; MT = Moine Thrust; IS = Iapetus Suture (= Cape Ray-Reach Fault in Newfoundland); BFZ = Billefjorden Fault Zone (Svalbard); WCFZ = Western Central Fault Zone (Svalbard); TTL = Tornquist-Teisseyre Lineament; SASZ = South Armorican Shear Zone; D-HBF = Dover-Hermitage Bay Fault.

(Emsian). The cleavage arcuation (from north/northeast in western northern England to a more easterly trend further east) is explained as a consequence of the anticlockwise rotation of eastern Avalonia relative to Laurentia during collision and final suturing.

19.6.7 ORDOVICIAN-SILURIAN MAGMATIC ARCS ELSEWHERE IN EUROPE

In contrast to the Lower Paleozoic rocks of Norway and Sweden, rocks of this age elsewhere in continental Europe

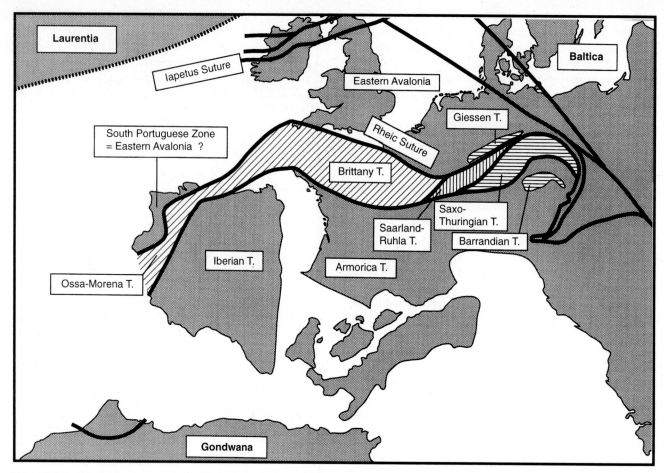

Figure 19.6.4 Principal European continental blocks during the Lower Paleozoic, and their sutures.

are largely concealed beneath younger strata or have been overprinted by intense Variscan and Alpine deformation. Exposure is poor and tectonic boundaries are difficult to define. Nevertheless, the Ordovician-Silurian outcrops of the European Caledonides can be assigned to three continental plates (Figure 19.6.4): (1) eastern Avalonia (considered above); (2) the southern parts of Baltica; and (3) parts of the northern margin of Gondwana (including, from west to east, the Brittany, Saarland-Ruhr, Tepla-Barrandian, Saxo-Thuringian, Bavarian, Gory Sowie, and Moravo-Silesian 'terranes', and, farther south, the Ibero-Armorica-Moldanubian 'terranes').

By Late Ordovician time, the Eastern Iapetus between eastern Avalonia and Baltica (also known as Tornquist's Sea) had probably closed by eastward/northeastward subduction. The southern margin of eastern Avalonia is marked by a major Early Paleozoic suture zone with ophiolites in Ibero-Armorica and Alpine Europe. The zone also includes arc-related igneous-volcanic suites and thick marine successions that can be traced through Iberia, into Armorica (along the South Armorican Shear Zone, or SASZ), and across to southern Austria. The SASZ includes thrust-bound slices of metasedimentary and igneous rocks metamorphosed to eclogite facies, 420–375 Ma blueschists, and a high temperature-low pressure migmatite belt, interpreted

as remnants of an accretionary complex formed above subduction zone, almost certainly active during the Silurian. There was also a major Late Ordovician to Silurian, 450–415 Ma, high-pressure event with little deformation in France, Iberia, and Morocco.

In the Ossa Morena Zone, central Iberia, the Middle Devonian emplacement of northeast-verging nappes was followed by major sinistral transpression. This created the central Iberian fold belt separating the Aquitaine-Cantabrian microcontinent to the east from the south Portuguese block to the west, which can be regarded as belonging to the southern part of eastern Avalonia.

Further east, from Early Silurian to Middle Devonian time, the mid-European and Tepla-Barremian 'terranes' appear to have been subject to considerable crustal extension and the extrusion of voluminous, within-plate, alkali basalts. In the eastern Alps of southern Austria, there are subduction-related volcanics and sediments formed in an island arc and active continental margin setting, probably vestiges of an arc on the edge of Gondwana, but their precise relation to Gondwana is unclear.

The Ligerian Orogeny involved the collision of these Gondwana-derived fragments with the southern margin of eastern Avalonia, itself a fragment broken off Gondwana at an earlier period. It involved the development and destruction of a volcanic arc complex, with a backarc, marginal basin to the north, prior to its incorporation and final destruction by continent-continent collision in the younger Variscan (or Hercynian) Orogeny. A Middle Devonian intermediate-pressure metamorphic event associated with major tectonism is well known not only in France but also in Morocco and more speculatively in Iberia. This event is interpreted as reflecting the amalgamation of other continental fragments such as Saxo-Thuringia and Moldanubia with eastern Avalonia in addition to those in Aquitaine and Cantabria.

The Rheic Ocean lay to the south of the Ligerian Orogenic Belt and north of Gondwana, that is, south of the Tepla-Barremian Plate, south of which the Ibero-Armorica-Moldanubian Plate was converging northward throughout the Devonian. Essentially, the Rheic Ocean is synonymous with the vestiges of the Eastern Iapetus Ocean. The ocean appears to have been closed by mid-Devonian (Givetian) time, though its closure may have been preceded by the creation of a small backarc basin to the north. All oceanic areas between Gondwana and Laurussia (Laurentia and Baltica) had been eliminated by Late Carboniferous (Namurian) time. Subduction of the Rheic Ocean crust probably was initiated in the Early Devonian, and ocean closure occurred mainly by the northward subduction of oceanic crust below the southern arc-related margin of the Baltica Plate (which includes cratonic Russia west of the Urals, a Permian collisional orogen).

The southern/southeastern margins of some of the microcontinental terranes that were accreted to Laurentia-Baltica during the Ordovician-Devonian are marked by Late Devonian ophiolites; for example, the approximately 400–375 Ma Lizard ophiolite obducted probably around 370 Ma. After unfolding the major Variscan flexure from Armorica to Iberia, and allowing for significant dextral offset along faults, it appears that the approximately 390 Ma Morais ophiolite, one of a series of crystalline complexes exposed in northern Iberia, represents the along-strike equivalent of the Lizard complex. These ophiolites were obducted northward as fragments of the Rheic or Ligurian Ocean crust.

19.6.8 POSTOROGENIC CONTINENTAL SEDIMENTATION AND IGNEOUS ACTIVITY

The collision of Baltica with Laurentia united them into a single continent known as Laurussia and created a high mountain chain. The collision of Gondwana and Laurussia to form Pangaea was not completed until Late Carboniferous time. Molasse accumulated in fault-controlled basins, preserved along the major tectonic lineaments and suture zones.

By the Early/Middle Devonian, most of the major strike-slip between terranes appears to have occurred along many lineaments; for example, the Great Glen Fault of northwest Scotland is a major strike-slip fault but does not

appear to have been significantly active during Old Red Sandstone (ORS) deposition. Furthermore, the ORS next to the fault shows a net offset today of 25–29 km in a dextral rather than the sinistral sense implied by the geometry of docking and collision. Post-ORS dextral offsets are also known in the Shetland Isles and are much larger—(of the order of 120 km). However, substantial movement along the Great Glen Fault must have occurred by the Late Silurian.

The high topography was supported by thickened continental crust. Temperatures in the lower part of the crust rose sufficiently to partially melt it and produce late-stage granites and granitoid bodies. These are the 'late-orogenic' and 'post-orogenic' intrusions that characterize the final stages of continent-continent collisions. By earliest Carboniferous time, the crust on the southern margin of Laurentia was extending and spreading laterally by 'gravitational collapse' and inducing faulting at the surface.

The tectonic vergence divide between major northward (Late Paleozoic coordinates) obduction and thrusting events associated with closure of the Western Iapetus Ocean, and southward tectonic transport onto the Baltic Shield due to closure of the Eastern Iapetus Ocean, was situated in the region of central Newfoundland to British Caledonides. This latter region probably was the site of a triple junction associated with transforms and spreading centers during the opening of the Western and Eastern Iapetus Oceans, but later, with ocean-basin destruction, a triple junction involving subduction zones and transforms. The obliquity of the collision events may have been a major contributing factor in the preservation of the low-grade slate belts, rather than a Himalayan-style collision (which occurred farther north) with high-grade metamorphic rocks being common at the surface.

19.6.9 CLOSING REMARKS

The Late Precambrian-Early Cambrian was marked by the opening of an essentially east-trending Western Iapetus Ocean between southeastern Laurentia and western South America. The Eastern Iapetus Ocean, separating the Greenland area of Laurentia from Baltica opened earlier than the western ocean, and was elongated north-south. Like the Mesozoic breakup of Pangaea, the approximately 620–570 Ma Phase I, and approximately 490–470 Ma Phase II, breakup of Gondwana may have been associated with plume activity, though the evidence is not clear. Continental breakup would probably have increased the length of the global ocean ridge system and reduced the mean age of the ocean floor. This would have caused a global rise in sea level, leading to widespread flooding of continents. This is consistent with the preponderance of wide Cambro-Ordovician shelf seas, commonly the sites for the accumulation of organic-rich muds (now pyrite-rich black shales). The lack of glaciogenic sediments or striated pavements, at least until the latest Ordovician (Ashgill), suggests that there were no substantial, if any, polar ice caps. The Cambro-Ordovician probably was a 'greenhouse' period induced by enhanced atmospheric CO_2 levels, in turn attributable to increased oceanic ridge and mantle plume activity.

Future research needs to better constrain the timing and nature of basin-forming events and their subsequent histories, which are possible through improved radio metric dating techniques, and structural and stratigraphic/sedimentologic studies. Re-evaluation and new measurements of paleomagnetic data using modern demagnetization techniques are giving an improved understanding of the movement history for continental fragments. Sophisticated geochemical and isotopic approaches are helping to define plate-tectonic settings of igneous rocks (suprasubduction zone, extensional intraplate, etc.), and to infer past global/regional climate from sediments. Undoubtedly, the Palaeozoic will remain an area of fruitful research.

ADDITIONAL READING

Bond, G. C., Nickeson, P. A., and Kominz, M. A., 1984, Breakup of a supercontinent between 625 Ma and 555 Ma: New evidence and implications for continental histories, *Earth and Planetary Science Letters*, v. 70, p. 325–345.

Franke, W., 1989, Tectonostratigraphic units in the Variscan belt of central Europe, *in* Dallmeyer, R. D., ed., Terranes in the Circum-Atlantic Paleozoic orogens, *Geological Society of America Special Paper 230*, p. 67–90.

Frisch, W., and Neubauer, F., 1989, Pre-Alpine terranes and tectonic zoning in the Eastern Alps, *in* Dallmeyer, R. D., ed., Terranes in the Circum-Atlantic Paleozoic orogens, *Geological Society of America Special Paper 230*, p. 91–100.

Harland, W. B., Armstrong, R. L., Cox, A. V., Craig, L. E., Smith, A. G., and Smith, D. G., 1989, A geologic time scale 1989: British Petroleum Company and Cambridge University Press.

Pickering, K. T., and Smith, A. G., 1995, Arcs and backarc basins in the Early Paleozoic Iapetus Ocean, *The Island Arc*, v. 4, p. 1–67.

Ryan, P. D., and Dewey, J. F., 1991, A geological and tectonic cross-section of the Caledonides of western Ireland, *Journal of the Geological Society,* London, v. 148, p. 173–180.

Salda, L. H. D., Cingolani, C., and Varela, R., 1992, Early Paleozoic orogenic belt of the Andes in southwestern South America: Result of Laurentia-Gondwana collision? *Geology,* v. 20, p. 617–620.

Soper, N. J., and Hutton, D. H. W., 1984, Late Caledonian sinistral displacements in Britain: Implications for a three-plate model, *Tectonics,* v. 3, p. 781–794.

Stephens, M. B., and Gee, D. G., 1985, A tectonic model for the evolution of the eugeoclinal terranes in the central Scandinavian Caledonides, *in* Gee, D. G., and Sturt, B. A., eds., The Caledonide Orogen— Scandinavia and Related Areas: Chichester, John Wiley and Sons, p. 953–978.

Sturt, B. A., and Roberts, D., 1991, Tectonostratigraphic relationships and obduction histories of Scandinavian ophiolitic terranes, *in* Peters, T. J. et al., eds., Ophiolite genesis and evolution of the oceanic lithosphere, *Sultanate of Oman, Ministry of Petroleum and Minerals,* p. 745–769.

Torsvik, T. H., Olesen, O., Ryan, P. D., and Trench, A., 1990, On the palaeogeography of Baltica during the Palaeozoic: New palaeomagnetic data from the Scandinavian Caledonides, *Geophysical Journal International,* v. 103, p. 261–279.

van der Pluijm, B. A., Johnson, R. J. E., and van der Voo, R., 1993, Paleogeography, accretionary history and tectonic scenario: A working hypothesis for the Ordovician and Silurian evolution of the northern Appalachians, *Geological Society of America Special Paper,* v. 75, p. 27–40.

Wilson, J. T., 1966, Did the Atlantic close and then re-open? *Nature,* v. 211, p. 676–681.

Zonenshain, L. P., Kuzmin, M. I., and Natapov, L. M., 1990, Geology of the USSR: A Plate-Tectonic Synthesis: Geodynamics Series, 21, Washington, D.C., American Geophysical Union.

19.7 The Northern Appalachians—Dwight C. Bradley[1]

on the fundamental contribution of stratigraphy in unravelling the tectonic history. To illustrate this approach I will focus on the rocks of New England and adjacent New York. In particular, I will show how information about the timing, location, and plate geometry of orogenic events can be gleaned from foreland basin sequences.

19.7.1   INTRODUCTION

The Appalachian Orogen of eastern North America extends more than 3000 km from Alabama to Newfoundland (Figure 19.7.1). The idea that the Appalachians represent the site of a long-vanished ocean basin was put forward in the classic paper by J. T. Wilson in the mid-1960s. The destruction of this ocean is recorded in three main phases of orogeny, known as the Taconic (Ordovician), Acadian (Silurian and Devonian), and Alleghanian (Carboniferous and Permian) orogenies. Many 'hard-rock' aspects of Appalachian geology (structure, metamorphism, plutonism, and geochronology) have been studied in great detail, and these disciplines have played a key role in formulating plate-tectonic interpretations. This essay, however, focuses

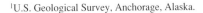

[1]U.S. Geological Survey, Anchorage, Alaska.

19.7.2   MAIN GEOLOGIC FEATURES

Along strike, the Appalachians are traditionally divided into the southern, central, and northern Appalachians (Figure 19.7.1). A simple four-fold scheme is often used for dividing the orogen across strike into subparallel zones (Figure 19.7.1): (1) Early Paleozoic North America and its deformed continental margin; (2) a complex belt containing various remnants of the Paleozoic Iapetus Ocean (deep-water sedimentary sequences, ophiolites, and arcs; sometimes called the *Central Mobile Belt*); (3) a Late Precambrian to Early Paleozoic microcontinent called *Avalonia;* and (4) a sliver of Early Paleozoic rocks formed along the African continental margin, called *Meguma.* Additional subdivisions of North America and the Central Mobile Belt in New England will be introduced below (see also Figure 19.7.2).

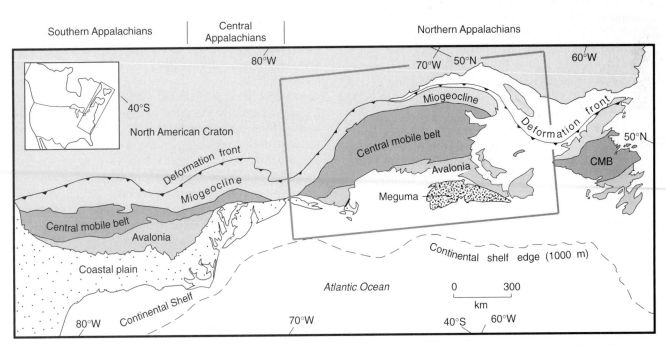

Figure 19.7.1   Map of the Appalachian Orogen showing four major lithotectonic zones and subdivision into northern, central, and southern Appalachians.

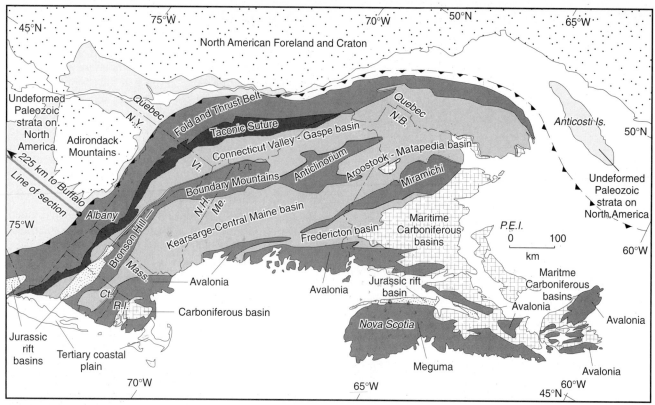

Figure 19.7.2 Map of the northern Appalachians showing major tectonic belts.

19.7.3 THE TACONIC OROGENY

Geology

The North American Margin. Cambrian and Ordovician events leading up to and including the Ordovician Taconic Orogeny are probably best understood in eastern New York state and western New England (Figure 19.7.2). This part of the northern Appalachian foreland is especially informative because Neogene doming of the Adirondack Mountains has exposed the Precambrian (~1.1 Ga) continental basement as well as the entire flat-lying Cambrian to Devonian cover sequence. Cambrian to Middle Ordovician shallow-marine quartzites and platformal carbonates of the foreland (in New York, these go by the names of Potsdam Sandstone and Beekmantown Group) form an eastward-thickening wedge or miogeocline. In Middle Ordovician time, the platform experienced gentle uplift, followed by rapid deepening, normal faulting, and inundation beneath a thick wedge of east-derived shale and graywacke turbidites (Utica Shale and Schenectady Formation in New York; together these compose what is informally known as the Taconic Flysch) (Figure 19.7.3). The flysch thickens and coarsens markedly toward the orogenic front in the east. The graywacke contains low-grade metamorphic and volcanic detrital grains that must have come from an easterly source—thought to be the Taconic Orogen—that had

formed on the site of the former miogeocline. The flysch basin is known as the Taconic foredeep.[2]

The western, frontal portion of the northern Appalachians is a fold-and-thrust belt (Figure 19.7.2) that formed primarily during the Taconic Orogeny. Thrust slices of Precambrian (Grenville) basement, Cambrian and Ordovician carbonates and quartzites, and Ordovician flysch—essentially the same succession as in the undeformed foreland—were displaced a relatively short distance with respect to the craton, and are described as parautochthonous. The thrust belt also includes a far-travelled, composite thrust sheet (Taconic Allochthon of eastern New York and western Vermont) containing deep-water Late Precambrian to Ordovician strata that structurally overlie platformal rocks of the same age range. The

---

[2]A foredeep (or foreland basin) is a sediment-filled depression that forms next to an orogenic belt, due to flexure of the foreland lithosphere beneath the orogenic load. During collision, a foredeep will migrate toward the craton, keeping ahead of the advancing orogenic load. In the process, older parts of the foredeep commonly get deformed as they are overridden by thrusts, and may become part of the thrust load itself. Typically, the older sedimentary fill of a foredeep consists of syntectonic, deep-marine turbidites (flysch), whereas the younger fill consists of shallow-marine or fluvial sandstone and conglomerate (molasse).

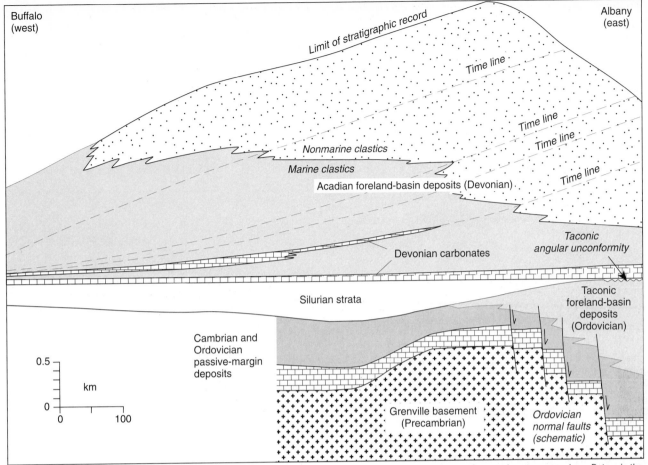

Figure 19.7.3 Cross section through Taconic and Acadian forelands of New York, showing two clastic wedges and two carbonate successions. Datum is the base of the Lower Devonian Helderberg Group.

Taconic Allochthon originally must have lain many tens of kilometers to the east because the Cambrian and Ordovician platformal sequence can be found on the east side of the Green Mountains, leaving no room for the allochthon except still farther east.

The Central Mobile Belt. Just east of the fold-and-thrust belt is a discontinuous zone of ophiolites that occur as fault slices ranging from outcrop-size to several kilometers thick. They are enclosed within complexly deformed and metamorphosed black shales and turbidites of deep-water origin (Rowe Schist). This belt is interpreted as the Taconic suture zone (Figure 19.7.2), demarking the boundary between rocks of North American origin to the west, and rocks of an island-arc complex to the east.

East of the suture lies the Connecticut Valley Basin (Figure 19.7.2). Deposition in it was essentially continuous from Ordovician to Devonian—without a break corresponding to Taconic deformation. Its position suggests that at the time of the Taconic Orogeny, it was a forearc basin. Ordovician volcanic and plutonic rocks (e.g., Ammonoosuc Volcanics and Highlandcroft Plutonic Suite in New Hampshire) along the Bronson Hill-Boundary Mountains Anticlinorium (Figure 19.7.2) are generally interpreted as marking the axis of a Taconic magmatic arc. In Maine, Ordovician

volcanics of the Bronson Hill-Boundary Mountains anticlinorial belt unconformably overlie two distinct terranes—an inboard one consisting of Precambrian gneiss (Chain Lakes Massif), and an outboard one consisting of a Cambrian ophiolite and volcanic sequence (Boil Mountain Complex and Jim Pond Formation), Cambrian melange (Hurricane Mountain Formation), and Cambrian and Early Ordovician flysch (Dead River Formation).

Plate Tectonics of the Taconic Orogeny

Plate-tectonic interpretations of events leading up to and including the Taconic Orogeny are fairly straightforward (Figure 19.7.4a–d). The Cambrian and Ordovician miogeocline is interpreted as a passive-margin platform that faced an ocean to the east (Iapetus); and the deep-water strata of the Taconic Allochthons are interpreted as rift, slope-rise, and early foredeep deposits. By Early Ordovician, the thermally subsiding passive margin was attached to some unknown width of oceanic crust. The Taconic Orogeny is interpreted as the result of closure of this tract of ocean crust at a relatively short-lived, east-dipping subduction zone, marked in New England by the ophiolite belt. This plate geometry led, inevitably, to collision between the passive margin and magmatic arc. The Taconic

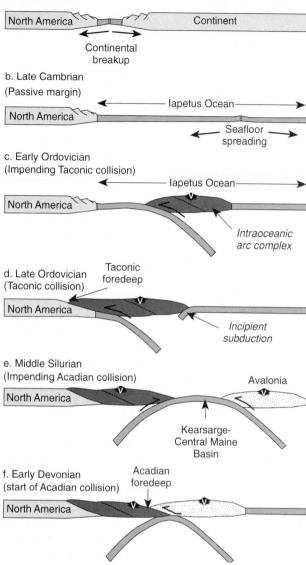

a. Early Cambrian

North America | Continent

Continental
breakup

b. Late Cambrian
(Passive margin)

North America — Iapetus Ocean —

Seafloor
spreading

c. Early Ordovician
(Impending Taconic collision)

North America — Iapetus Ocean —

Intraoceanic
arc complex

d. Late Ordovician
(Taconic collision)

Taconic
foredeep

North America

Incipient
subduction

e. Middle Silurian
(Impending Acadian collision)

Avalonia

North America

Kearsarge-
Central Maine
Basin

f. Early Devonian
(start of Acadian collision)

Acadian
foredeep

North America

Figure 19.7.4 Schematic diagram showing postulated Early and Middle Paleozoic tectonic evolution of the northern Appalachians. (a) Late Precambrian rifting. (b) Cambrian passive margin. (c) Early Ordovician impending collision with an island arc. (d) Middle and Late Ordovician arc-continent collision (Taconic Orogeny). (e) Silurian subduction leading to impending Acadian collision. (f) Early Devonian, during start of Acadian collision.

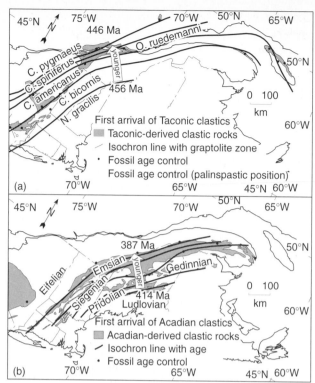

Figure 19.7.5 Maps of (a) the Taconic and (b) Acadian foredeeps showing age of inception of foredeep sedimentation with contour (isochron) lines. For the Taconic foredeep, the contour gradient is based entirely on autochthonous rocks, and therefore approximates the plate convergence rate (10–20 km/m.y.). In the case of the Acadian foredeep, the contours are based on deformed rocks that have been telescoped and displaced relatively toward the North American craton, and therefore yield only a minimum syncollisional plate convergence rate (about 5 mm/y).

formed Ordovician rocks with pronounced angular unconformity (the Taconic unconformity; Figure 19.7.3); (2) Ordovician graywacke contains detrital grains of already metamorphosed Taconic slate; and (3) beneath and immediately west of the Taconic Allochthon is an olistostromal melange with a matrix of chaotically deformed turbidites and black shales, and blocks derived from both the Taconic Allochthon and the miogeocline, which presumably were shed in front of the advancing thrust sheets. Graptolites within the matrix shales date the emplacement of the Taconic Allochthon to within a few million years (*C. spiniferus* graptolite zone; 450–446 Ma according to the DNAG time scale). Another important aspect of the Taconic Flysch is that its age systematically varies across strike, being older in the thrust belt and younger in the foreland. The area of flysch deposition (foredeep axis) presumably migrated toward North America because the thrust load also moved, and thus the rate of migration is approximately the plate convergence rate, which has been found to be 10–20 km/m.y. (Figure 19.7.5a). This is comparable to some of the slower rates of convergence in the present plate mosaic. By combining this information with the known duration of Ordovician arc magmatism (about 45 m.y. in Maine), the width of the ocean that closed during the Taconic Orogeny has been estimated at 500 to 900 km.

is commonly described as an "arc-passive margin collision," but in fact, the forearc that overrode and downflexed the passive margin was located perhaps 200 km west of the magmatic arc, which itself felt little if any Taconic deformation. The end result of the Taconic Orogeny was accretion of an arc onto the North American continental margin. The new margin was located somewhere farther east, probably just east of the Bronson Hill-Boundary Mountains Anticlinorium.

Much work on the Taconic Orogeny has focused on the relationships between flysch sedimentation and tectonics. Flysch along the frontal thrust zone preserves three lines of evidence that the Taconic Orogeny was an Ordovician event: (1) Silurian and Devonian strata overlie de-

This is considerably less than the full width of Iapetus, which has been estimated at 4000 km on the basis of paleomagnetic and faunal data.

19.7.4 THE ACADIAN OROGENY

Geology of the Acadian Orogeny

The North American Margin. By the Early Devonian, a slowly subsiding carbonate platform (Helderberg Group of New York) had been reestablished across the earlier Taconic foredeep and the eroded front of the Taconic Mountains. This platform shows that the Taconic foredeep had rebounded to near sea level, from estimated water depths of 1–3 km. In Middle Devonian, the rate of subsidence increased dramatically, as a second east-derived clastic wedge (the Catskill delta of New York) prograded across the foreland (Figure 19.7.3). West-directed paleocurrents indicate that the source of the Catskill clastics was a new mountain belt to the east—the Acadian Orogen. Conglomerate clasts in the eastern Catskills (a few kilometers west of the thrust front) include Catskill-like sandstones that must have been deposited in a more easterly foredeep basin, one that was uplifted and eroded away during the Devonian.

The Central Mobile Belt. Within the mountain belt, stratigraphic studies reveal the presence of two Silurian magmatic arcs (one on the Bronson Hill-Boundary Mountains Anticlinorium, one on Avalonia) and two deep-water basins (Connecticut Valley and Kearsarge-Central Maine Basins). During the Devonian, each of these belts except Avalonia was inundated by outboard-derived flysch deposited in a foredeep that eventually migrated to its final position in the Catskills.

Stratigraphic relations along the Bronson Hill-Boundary Mountains Anticlinorium and Kearsarge-Central Maine Basin are an important key to unravelling the tectonic history of the Acadian Orogeny in New England. Along the anticlinorial belt, Silurian sedimentary rocks are mainly quartzites and impure limestones (e.g., Clough Quartzite and Fitch Formation in New Hampshire; Ripogenus Formation in Maine) that contain shallow-water fossils such as stromatoporoids, corals, and brachiopods. These units record an interval of shallow-marine deposition at slow to moderate subsidence rates along the Taconic-modified margin of North America, in sharp contrast with what came next. Another key feature of the anticlinorial belt is the presence of volcanic rocks representing all four stages of the Silurian and the first three of the Devonian. These volcanic rocks compose the Piscataquis magmatic belt, an arc that was built on the older, Ordovician arc. Deeper-water flysch sedimentation commenced in Devonian time (Littleton Formation in New Hampshire, Seboomook Group in Maine), followed, at least in Maine, by progradation of a deltaic complex (Matagamon Sandstone). The Devonian clastics are partly younger than, partly coeval with, and partly older than, magmatism along the Piscataquis Belt.

The Kearsarge-Central Maine Basin—the main source of controversy regarding Acadian tectonics in New England—is underlain deep-marine strata that were folded, metamorphosed, and intruded by plutons during the Devonian. The stratigraphy is broadly divisible into: (1) a lower sequence derived from the Taconic-modified margin of North America that lay to the northwest (e.g., Silurian Rangeley, Perry Mountain, and Smalls Falls Formations in Maine), and (2) an upper sequence derived from outboard sources (Upper Silurian Madrid Formation and Lower Devonian Carrabassett Formation in Maine).

Avalonia. The next major belt to the southeast is a complex belt of rocks that have been collectively assigned to the Avalon zone or "Avalonia." Recent findings reveal that Avalonia is not a distinct entity but rather a composite of several terranes, some of which have Precambrian basement. From the standpoint of Acadian tectonics, three points are salient. (1) In Maine, a thick succession of Silurian through Lower Devonian volcanics, and related plutonic rocks, compose a long-lived magmatic arc that shut off during the Early Devonian, at virtually the same time as Acadian deformation was happening in the Kearsarge-Central Maine Basin. (2) There is no evidence of a Devonian clastic wedge in the Avalon zone. (3) Acadian deformation and metamorphism in Avalonia were mild, in stark contrast with the intense Acadian tectonism experienced by the Kearsarge-Central Maine Basin.

Plate Tectonics of the Acadian Orogeny

The Acadian Orogeny is much more controversial than the Taconic Orogeny. A generally (but still not universally) accepted starting point is the interpretation that the Taconic Orogeny resulted in the accretion of an arc to the edge of North America. The main problem is whether or not the Kearsarge-Central Maine Basin is the site of a second ocean that also closed by subduction, leading to an Acadian collision. The mode of closure of the Kearsarge-Central Maine Basin is also controversial; options include subduction beneath its northwestern margin, subduction beneath its southeastern margin, subduction beneath both margins, or deformation of the basin fill without true subduction of the basement.

The distribution and age of Acadian foredeep deposits provide several clues to Acadian plate tectonics. An influx of Devonian clastics, derived from outboard sources, has been recognized in the Kearsarge-Central Maine Basin, Piscataquis magmatic belt, Connecticut Valley-Gaspe Basin, locally within the parautochthonous thrust belt, and in the undeformed foreland (Catskill delta). These clastics were deposited in a migrating foredeep basin that advanced in concert with an advancing Acadian orogenic load. The load must have originally been outboard of the oldest part of the clastic wedge, that is, in about the present location of Avalonia. The clastics occur on both sides of the Piscataquis magmatic axis, as well as along its axis, but were not derived from the arc. This extremely unusual stratigraphic pattern implies that the foredeep was superimposed on an active arc, and that subsidence was the result of flexure of the arc beneath a thrust load that lay farther outboard (southeast). The obvious candidate, once again, is Avalonia. The pattern of foredeep sedimentation and the presence of two magmatic arcs is therefore best explained by a plate geometry like that shown in Figure 19.7.4e and f.

The Devonian clastics were clearly diachronous, both across strike, younger toward the craton, and along strike, younger toward the south. By analogy with the Taconic foredeep, the rate of migration should approximate the plate convergence rate during Acadian collision (Figure 19.7.5b). This rate, unfortunately, is not so readily determined, because Acadian shortening has partly closed the gaps between what originally were widely spaced sections. Nonetheless, we can say with reasonable confidence that the plate convergence rate in northern New England during Acadian collision must have been at least 5 km/m.y., and likely was much faster.

19.7.5 POST-ACADIAN TECTONICS

The stratigraphic record in the northern Appalachian foreland ends with the Catskill clastics, but in the central and southern Appalachians, there is yet a third east-derived clastic wedge, of Carboniferous age. These clastics were derived from, and deformed during the Alleghanian Orogeny, which resulted from the final collision between Africa and North America. In the northern Appalachians, the Carboniferous was for a time dominated, instead, by strike-slip tectonics. Most of the faults strike northeast-southwest or east-west. Avalonia and Meguma are juxtaposed across such a fault system in Nova Scotia. Episodic motion along numerous anastomosing fault strands caused subsidence of transtensional basins, and uplift of areas of transpressional deformation. The overall setting of this strike-slip regime is unresolved, but probably relates to oblique convergence between Africa and North America, prior to the final Alleghanian collision that completed the assembly of Pangaea around the end of the Carboniferous. Mesozoic breakup of Pangaea produced a series of rift basins, three of which are shown in Figure 19.7.2. When Africa drifted away, a piece was left behind—the Meguma terrane.

19.7.6 CLOSING REMARKS

Ordovician and Devonian flysch sequences in the northern Appalachians have long been snubbed as monotonous, but they're quite interesting when you get to know them. As discussed here, they contain important clues to quantifying ancient plate motions. Ultimately, however, this sort of analysis rests on those no longer glamorous geologic specialties that are threatened with extinction: regional mapping, stratigraphy, and paleontology.

ADDITIONAL READING

Bradley, D. C., 1983, Tectonics of the Acadian Orogeny in New England and adjacent Canada, *Journal of Geology,* v. 91, p. 381–400.

Bradley, D. C., 1989, Taconic plate kinematics as revealed by foredeep stratigraphy, Appalachian orogen, *Tectonics,* v. 8, p. 1037–1049.

Osberg, P. H., Tull, J. F., Robinson, P., Hon, R., and Butler, J. R., 1989, The Acadian orogen, *in* Hatcher, R. D., Jr., Thomas, W. A., and Viele, G. W., eds., The Appalachian-Ouachita Orogen in the United States: Boulder, Colorado, Geological Society of America, *The Geology of North America,* v. F-2, p. 179–232.

Rankin, D. W., Drake, A. A., Jr., Glover, L., III, Goldsmith, R., Hall, L. M., Murray, D. P., Ratcliffe, N. M., Read, J. F., Secor, D. T., Jr., and Stanley, R. S., 1989, Pre-orogenic terranes, *in* Hatcher, R. D., Jr., Thomas, W. A., and Viele, G. W., eds., The Appalachian-Ouachita Orogen in the United States: Boulder, Colorado, Geological Society of America, *The Geology of North America,* v. F-2, p. 7–100.

Rast, N., 1989, The evolution of the Appalachian chain, *in* Bally, A. W., and Palmer, A. R., eds., The Geology of North America—An Overview: Geological Society of America, *The Geology of North America,* v. A, p. 323–348.

Rowley, D. B., and Kidd, W. S. F., 1981, Stratigraphic relationships and detrital composition of the medial Ordovician flysch of western New England: Implications for the tectonic evolution of the Taconic Orogeny: Journal of Geology, v. 89, p. 199–218.

Stanley, R., and Ratcliffe, N., 1985, Tectonic synthesis of the Taconian Orogeny in western New England, *Geological Society of America Bulletin,* v. 96, p. 1227–1250.

van der Voo, R., 1988, Paleozoic paleogeography of North America, Gondwana, and intervening displaced terranes: Comparisons of paleomagnetism with paleoclimatology and biogeographical patterns, *Geological Society of America Bulletin,* v. 100, p. 311–324.

Wilson, J. T., 1966, Did the Atlantic close and then re-open? *Nature,* v. 211, p. 676–681.

19.8   The Tasman Belt—David R. Gray[1]

19.8.1   INTRODUCTION

The Tasman Orogenic Belt, a north-south trending composite Paleozoic orogenic belt along the eastern margin of Australia (Figure 19.8.1), evolved over 400 m.y. from the Cambrian to the Triassic period (Figure 19.8.2). During that time eastern Australia underwent an important period of continental accretion, which added approximately 30% to the size of the ancient Australian cratonic core. The Tasman Orogenic Belt consists of deformed tracts of deep-marine sedimentary and volcanic rocks that occur to the east of the exposed Precambrian cratonic crystalline basement marked by the Tasman Line (Figure 19.8.1). Part of the Gondwana margin, the Tasman Orogenic Belt belongs to a Paleozoic orogenic system that extended some 20,000 km from the northern Andes, through the Pacific margin of Antarctica to eastern Australia. In eastern Australia there are three north-south trending deformed belts (Kanmantoo, Lachlan/Thomson, and New England fold belts; Figure 19.8.1), which are distinguished by their lithofacies, tectonic settings, timing of orogenesis, and eventual consolidation to the Australian craton (Figure 19.8.2 and Table 19.8.1). Boundaries between the three belts are not exposed and generally they are covered by younger sequences. The inner belt, the Kanmantoo and the western part of the Lachlan, shows pronounced curvature and structural conformity with the promontories and recesses in the old cratonic margin (the Tasman Line) (Figure 19.8.1). Outboard of this, the central and eastern belts of the Lachlan, and the New England Belt, have more continuous trends that truncate the inner belt trends and thus show no relationship to the old cratonic margin.

The Tasman Orogenic Belt of eastern Australia formed in part by massive telescoping and strike-slip translation largely within a passive margin sequence but without the rotations and large-scale translations that are typical of the North American Cordillera. Transpressive convergence with far-field transmission of stress accompanied by lithospheric (A-type) subduction is related to collision and amalgamation of an array of variably moving crustal segments and microplates along a complex transform plate boundary involving south or southeast translation of the Australian craton. The present distribution of fold belts and the observed structural patterns within them (Figure 19.8.3) are a response to the changing character of the Gondwana margin/plate

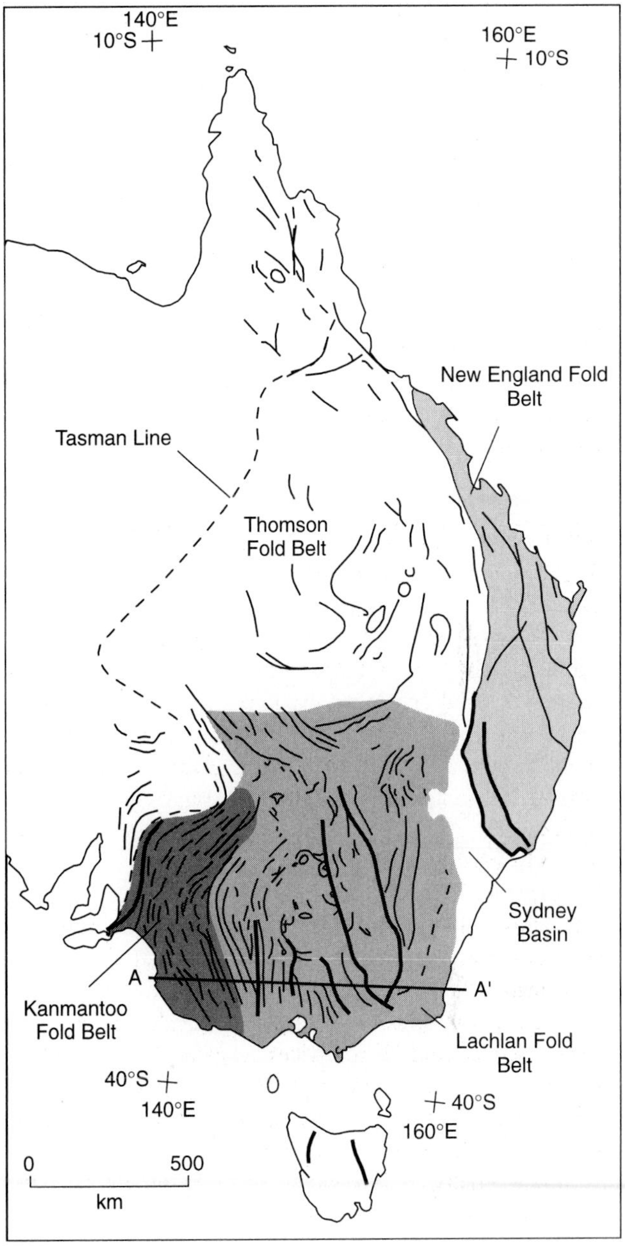

Figure 19.8.1   Map of eastern Australia showing the major elements of the Tasman Orogenic Belt (the Kanmantoo, Lachlan/Thomson, and New England fold belts), the Tasman Line, which defines the western limit of Precambrian cratonic crystalline basement, and structural and aeromagnetic trend lines (heavy lines are major fault traces). The Sydney Basin is a foreland basin to the New England Fold Belt. A-A′ is section in Figure 19.8.3.

boundary from the Cambrian to the Triassic. Differences in crustal architecture relative to other orogens may be due to the size, nature, age, and lithospheric-crustal densities of the colliding and accreting masses. Structural features of the Tasman Orogenic Belt include the superposition of different age thrust belts sharing a common midcrustal detachment, linked contractional and strike-slip faults, and a marked

[1]Monash University, Melbourne, Victoria, Australia.

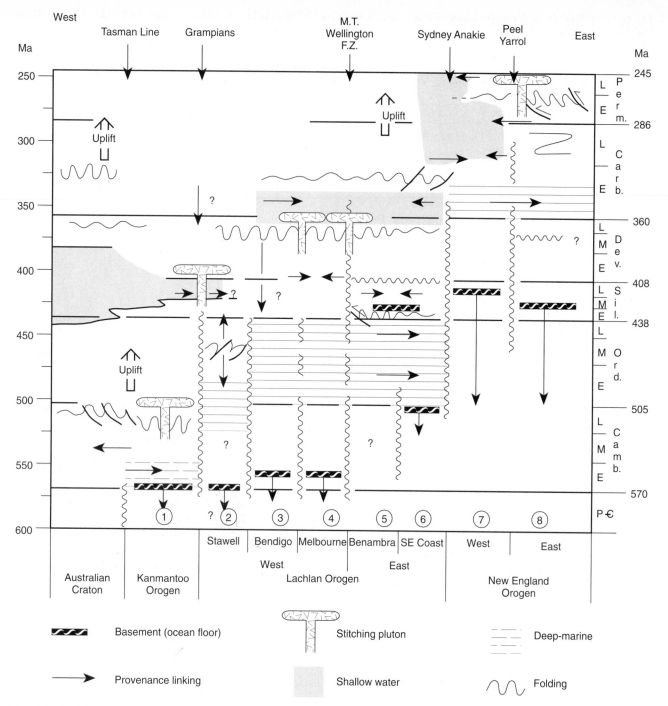

Figure 19.8.2 Schematic time-space chart for the southern part of the Tasman Orogenic Belt showing the relationships between sedimentation, deformation, and plutonism over an approximate 400 m.y. history and the progressive west to east cratonisation.

pattern of alternating extensional (Late Silurian, Late Devonian, and Early Carboniferous) and compressional (Late Ordovician-Early Silurian, Early Devonian, Middle Devonian, and middle Early Carboniferous) deformation over the history of the belt. Extensional periods are marked by localized development of rift basins (half grabens) accompanied by extensive granitic magmatism and silicic volcanism. Subsequent basin collapse occurs by reactivation/inversion of the extensional faults and folding of the basin sequence.

19.8.2 CRUSTAL STRUCTURE AND MAIN TECTONIC ELEMENTS

Crustal thicknesses in the Tasman Orogenic Belt range from 40 to 52 km under the eastern highlands of Australia, with upper crustal thickness varying from 35–36 km in central and northeast Victoria to 26 km in central New South Wales based on seismic refraction and reflection studies. In southeastern New South Wales the velocity-depth structure is characterized by an upper crust with P-wave velocities of

Table 19.8.1 Fold Belts and Subprovinces of the Tasman Orogenic Belt (see Figures 19.8.1 and 19.8.2).

| | Kanmantoo | Lachlan | | | New England |
		Western	Central	Eastern	
Main plutonism	Late Ordovician	Late Devonian	Late Silurian	Late Carboniferous	Late Permian-Early Triassic
Tectonic vergence	West-directed thrusting	East-directed thrusting	Overall strike-slip with southeast-directed thrusting	East-directed thrusting	West-directed thrusting
Terminal folding	Late Cambrian-Early Ordovician	Middle Devonian	Middle Silurian	Early Carboniferous	Mid-Permian
Main facies	Platform to deep water passive-margin sequence	Quartz-rich turbidite passive-margin sequence	Quartz-rich turbidite passive-margin sequence	Platform carbonates and clastics with rhyolites and dacitic tufts	Volcanogenic clastics
Initial record	Basic volcanics	Cambrian basic volcanics	Tremadocian chert	Ordovician andesitic volcanics	Ordovician basic volcanics

5.6–6.3 km/sec, a lower crust with velocities between 6.7–7.4 km/sec, and a low velocity zone at a depth of about 15–20 km. In central Victoria upper-crustal velocities range 5.3–5.9 km/sec, with a change to 6.3 km/sec at a depth of 17 km.

The three preserved north-south trending deformed tracts (Figure 19.8.1) represent three distinct tectonic settings: (1) deformed intracratonic rift (Adelaide/Kanmantoo Fold Belt), (2) deformed passive-margin sediment prism (Lachlan Fold Belt), (3) deformed arc-subduction complex belt (New England Fold Belt). These three major tectonic elements show a progressive eastward younging of accretion along the evolving Australian continental margin,

which is defined by their respective peak deformations of Early Ordovician, mid-Devonian, and Permian-Triassic age (Figure 19.8.2).

The Kanmantoo Fold Belt is an arcuate, craton-verging thrust belt (Figure 19.8.4, section A-A′) with foreland-style folds and detachment-style thrusts (external zone) to the west and a metamorphic hinterland (internal zone) to the east characterized by polyphase deformation, amphibolite grade metamorphism with local development of kyanite-sillimanite assemblages, and intrusion of syn- and post-tectonic granites. During the Late Preambrian/Early Ordovician (the Delamerian Orogeny), allochthonous sheets consisting of northwest-verging duplexes (deformed

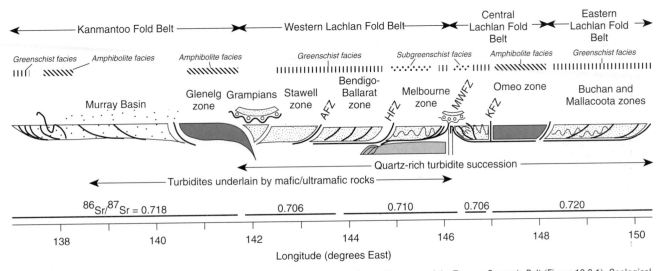

Figure 19.8.3 Schematic structural profile at approximately 37.5° S latitude across the southern part of the Tasman Orogenic Belt (Figure 19.8.1). Geological zones, cover basins, directions of tectonic transport, and variations in metamorphic grade and $^{86}Sr/^{87}Sr$ ratios for granitic rocks are shown. The Sr $^{86/87}$ ratios of the granites reflect compositional variations in the lower crust. Chevron folding and steep faults define the structural style at the present level of exposure (see Figure 19.8.6, section A-A′).

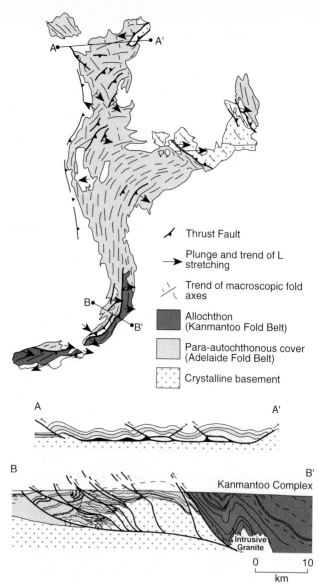

Figure 19.8.4 Adelaide and Kanmantoo Fold Belts. Form-surface map showing trends of macroscopic fold axes, thrust fault traces, trend and plunge of stretching lineations. Section A-A' shows detachment-related folds in the external zone (Adelaide Fold Belt), whereas section B-B' shows thrust relations between metamorphic hinterland (Kanmantoo Complex of Kanmantoo Fold Belt) and the Adelaidean shelf sequence of the external zone (Adelaide Fold Belt).

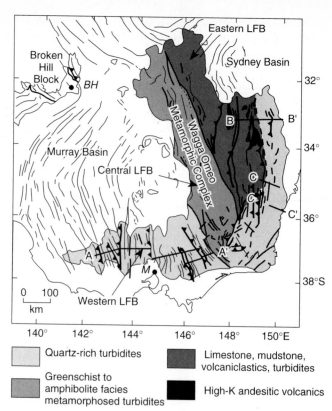

Figure 19.8.5 Structure map of the Lachlan Fold Belt showing the three major geological divisions (western, central, and eastern belts), the cover sequences of the Murray and Sydney Basins (white regions), structural and aeromagnetic trend lines (medium lines), major faults (heavy lines with barbs: contraction faults; heavy lines without barbs: strike-slip faults), and the distribution of the Ordovician volcanic rocks (dark areas). M = Melbourne; C = Canberra; BH = Broken Hill.

Cambrian Kanmantoo Group) were emplaced over the less deformed and metamorphosed shelf sequence (Adelaidean) of the external zone (Figure 19.8.4, section B-B'). High-grade assemblages that formed at around 3–5 kbar (~10–17 km depth) are spatially and temporally confined to aureoles of syn-kinematic granites that are conformably aligned with the structural grain.

Rapid unroofing (~10 km in approximately 20 m.y.) of the belt is suggested by juxtaposition of these high-grade rocks and their syntectonic granites (520–490 m.y.) with undeformed, high-level silicic granites and volcanics intruded at 486 m.y. This provides a source for the extensive Ordovician turbidite sequences of the Lachlan Fold Belt to the east.

The Kanmantoo Belt was tectonically active from the Late Precambrian (~650–600 m.y.) to the Early Ordovician (~500 m.y.). It consists of a deformed Upper Proterozoic Adelaidean intracratonic rift sequence of marine to deltaic sandstones and shales, lagoonal evaporites, dolomites, and limestone, transgressed by Lower Cambrian shelf sediments transitional into deep-water sandstones and mudstones of the Cambrian Kanmantoo Group (Figure 19.8.2).

The Lachlan Fold Belt (Figure 19.8.1) is a Middle Palaeozoic fold belt with a 200 m.y. history. Perhaps the most enigmatic of the three belts, it has undergone a complex amalgamational and deformational history with interplay of compressional (Late Ordovician-Early Silurian, Early Devonian, Middle Devonian, and middle Early Carboniferous) and extensional (Late Silurian, Late Devonian, and Early Carboniferous) events (Figure 19.8.2). Long-lived subduction in the mid-Paleozoic is envisaged along the Gondwana margin, but there is no evidence for major collision. Surface structures have been used to infer collisional processes, and Ordovician volcanics in New South Wales (Figure 19.8.5) have been used to define the plate margin setting. There is no craton-verging thrust belt, no preserved platform (miogeoclinal) successions, and no metamorphic hinterland, apart from the wedge-shaped Wagga-Omeo metamorphic belt, which is located in the central-eastern part of the fold belt (Figure 19.8.5). Thrust

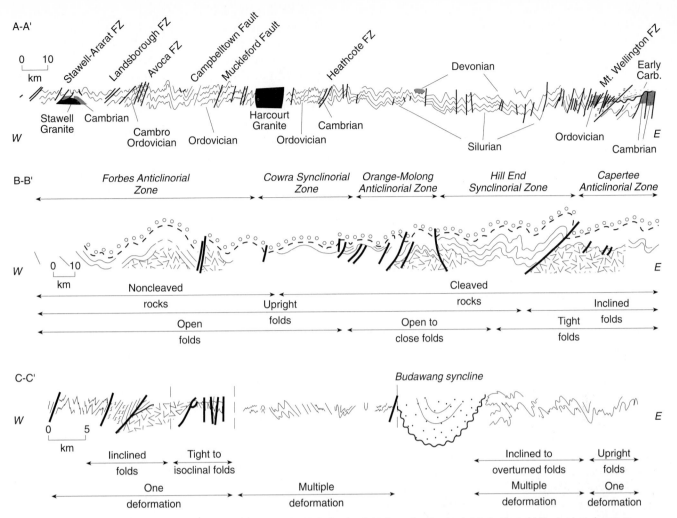

Figure 19.8.6 Structural profiles for various parts of the Lachlan Fold Belt. Section A-A′ shows the structural style for the turbidite-dominated western Lachlan Fold Belt, whereas B-B′ shows the eastern Lachlan Fold Belt with anticlinoria (dominated by Ordovician andesites and shallow water deposits) and intervening synclinoria (dominated by turbidites). Section B-B′ requires inversion of preexisting extension faults to explain the facies variations across faults. Section C-C′ is from the turbidite-dominated eastern part of the eastern belt. Polydeformed tracts relate to contact strain effects within the Ordovician 'basement' due to folding of the Late Devonian molasse sequences (e.g., Budawang syncline).

slices or windows exposing Proterozoic basement within the Lachlan Fold Belt have not been recognized, apart from those exposed in Tasmania. Furthermore, the distribution and ages of granites do not fit simple orogenic models involving either A or B type subduction.

The major feature of the Lachlan Fold Belt is the similarity of sedimentary facies and overall structural style across this segment of the Tasman Orogenic Zone (Figure 19.8.5). The region is dominated by a monotonous succession of Ordovician through Devonian quartz-rich turbidites. These are laterally extensive over 800 km present-width and have a current thickness upwards of 10 km. The western Lachlan Fold Belt consists of accreted fragments of a very large submarine sediment dispersal system associated with the Gondwana margin during the Early Paleozoic, a system dimensionally comparable to the present-day Bengal Fan in the Bay of Bengal. Linear north-south trending, fault-bounded Cambrian meta-volcanic belts, composed of

boninite (high-Mg andesites), low-Ti andesite, and tholeiite of oceanic affinities define the boundaries of the most important accreted fragments and are considered to underlie the quartz-rich turbidite succession. The eastern belt consists of shoshonitic (high-K) volcanics, mafic volcaniclastic rocks and limestone, as well as quartz-rich turbidites and extensive black shale in the easternmost part (Figure 19.8.5).

Tight to open chevron folds (accommodating between 50–70% shortening) cut by predominantly west-dipping, high-angle reverse faults are part of different thrust systems within the Lachlan Fold Belt (Figure 19.8.6). Chevron folds are upright and gently plunging but become inclined and polydeformed approaching major faults. Faults in the western part of the Lachlan Fold Belt are brittle faults, but have high strain zones of varying widths showing intense development of crenulation cleavages associated with variably but generally steeply plunging mesofolds and microfolds.

Overprinting cleavages within these zones indicate complex fault movements with early thrusting followed by wrench movements.

Midcrustal level detachments occur at the base of the Ordovician and within the Cambrian successions in the western belt, with deep crustal seismic profiling indicating a depth to detachment of approximately 15 km. The allochthonous nature of the imbricated upper-crustal sequence is supported by the lack of coincidence between lower-crustal basement terranes and the upper-crustal structural zones.

Metamorphism is greenschist facies or lower across the Lachlan Fold Belt, except in the fault-bounded Wagga-Omeo and several smaller (Cooma, Cambalong, Jerangle, and Kuark) metamorphic complexes (Figure 19.8.5) where high-temperature low-pressure metamorphism is characterized by andalusite-sillimanite assemblages. Such assemblages are typical of thermal metamorphism, but here they occur on a regional scale. Peak metamorphic conditions in the Wagga-Omeo zone are T~ 700°C and P~ 3–4 kbar. Erosional unroofing of the metamorphic complex in the Middle-Late Silurian necessitates shallow overburden and high geothermal gradients on the order of 65°C/km. It is apparent that regional metamorphism and felsic magmatism throughout the fold belt took place under very little cover, suggesting a shallow to midcrustal heat input for melting.

Granites cover up to 36% of the exposed Lachlan Fold Belt. Regional aureole, contact aureole, and subvolcanic field associations as well as S and I types based on geochemistry/mineralogy have been recognized. Most granites are post-tectonic and are undeformed and have narrow (1–2 km wide) contact aureoles. Some of these are subvolcanic granites associated with rhyolites and ash flows of similar composition. The regional aureole types are less common and are associated with the high-T/low-P metamorphism, migmatites, and Kspar-cordierite-andalusite-sillimanite gneisses (e.g., Cooma, Cambalong, and Kuark Belts; Figure 19.8.5).

The shape distribution of the granites suggests three major granite provinces, which presumably reflect the mode and timing of emplacement and state of stress in the mid- to lower crust. Elongate north-northwest to north-trending granites define the Wagga-Omeo metamorphic belt and the major part of the eastern Lachlan Fold Belt in New South Wales. Many of these granites are syntectonic, showing internal deformation and emplacement associated with deep crustal shear zones. In the western Lachlan Fold Belt post-tectonic granites of the Central Victorian magmatic province, the largest granitic bodies are east-west trending and have elongated form. The remaining granites are smaller and are more equant in shape. Spacing of the granitic bodies in Victoria fits a diapiric emplacement model with a source depth at 12–24 km and possibly deeper in a crust that was at least 35 km thick.

The New England Fold Belt is the youngest and most easterly part of the Tasman Belt (Figure 19.8.1). A collage of deformed and imbricated terranes (Figure 19.8.7), it con-

sists of largely Middle to Upper Paleozoic and Lower Mesozoic marine to terrestrial sedimentary and volcanic rocks, as well as strongly deformed flysch, argillite, chert, pillow basalts, ultramafics, and serpentinites. The New England Fold Belt was tectonically active from the Late Devonian to the mid-Cretaceous (~95 m.y.) (Figure 19.8.2), and activity of this convergent margin involves arc, forearc, and accretionary complexes.

Widespread climactic Permian-Triassic deformation involving west-directed thrusting, interleaving, and imbrication of the arc magmatic belt (Connors-Auburn Belt), fore-arc (Yarrol-Tamworth Belt), and oceanic assemblages including subduction complexes (Wandilla-Gwydir Belt) and ophiolite (Gympie Belt), consolidated the terranes into Australia and caused the development of a Permo-Triassic

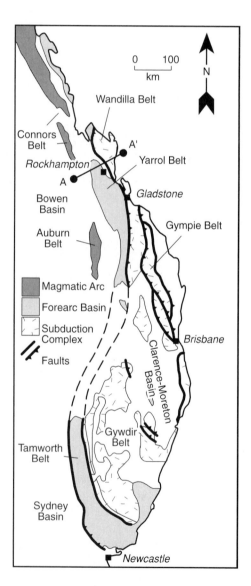

Figure 19.8.7 Generalized map of the New England Fold Belt showing the main tectonic units, including the magmatic arc belt (Connors-Auburn Belt), forearc (Yarrol-Tamworth Belt), oceanic assemblages including subduction complexes (Wandilla-Gwydir Belt) and ophiolite (Gympie Belt).

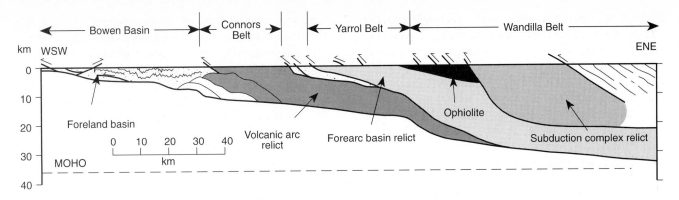

Figure 19.8.8 Section through the New England Fold Belt showing west-directed thrusting and structural interleaving of tectonic units. See Figure 19.8.7 for location of Section A-A' and legend.

foreland basin (Sydney-Bowen Basin) (Figure 19.8.8, Section A-A' in Figure 19.8.7). Outboard subduction complex assemblages show a strong thrust-related fabric, polyphase deformation, and greenschist to amphibolite facies metamorphism.

19.8.3 TECTONIC SETTING

The progressive west to east younging of the exposed crustal segments in the Tasman Orogenic Belt (see Table 19.8.1) has led to concepts of large-scale continental accretion for eastern Australia. Most tectonic evolutionary models require arc-continent collisional interaction along the Gondwanan continental margin in the Paleozoic, although there is some dispute about the existence of an "arc" in the Ordovician based on the shoshonitic character of the Ordovician andesitic rocks. Such high-K rocks commonly form during postorogenic extensional collapse of tectonically thickened crust. Amalgamation was either by backarc accretion in a marginal sea setting or in a passive margin setting, but must have involved some form of plate interaction.

As in most orogenic belts, problems relate to the recognition of paleogeographic elements in the deformed rock record and to their assembly into a coherent framework defining a plate tectonic setting. In the Tasmanides, tectonic settings for the Kanmantoo and New England Fold Belts are more definite compared to the Lachlan Fold Belt where major disputes relate to: (1) the tectonic setting in the Ordovician and Cambrian periods: continental rift or oceanic setting? (2) the basement to the voluminous quartz-rich turbidite succession; (3) the hinterland-directed (away from the craton) thrusts that dominate much of the belt.

Varying effects inboard of the arc have been related to changes in the rate and dip of subduction, and to plate margin setting (subduction to intracontinental transform setting), and even to movement of a triple junction, as has been proposed for western North America. For the Ordovician, most workers have argued for an island-arc setting with the western Lachlan Fold Belt part of a backarc basin,

the eastern Lachlan being part of a disrupted, segmented, and imbricated island-arc sequence represented by the Ordovician volcanics. Also, observations from modern fan systems indicate that large parts of such fans accumulate on oceanic rather than continental crust (e.g., Bengal Fan). The geochemical affinities of the metamorphosed mafic volcanics also indicate an oceanic setting associated with one or more island arcs. Thrusting over and against a small plate (former oceanic plateau) has been argued to explain the noncraton-directed thrusts in an ocean-continent collisional setting and local obduction associated with deeper continental underthrusting of the microplates. The interaction of hot and cold crust in the collisional process may also be important. Disrupted blocks of rigid, thick, cold continental crust surrounded by less rigid and thinner crust can lead to thrust belts of limited extent and varying position. Clearly, details on the positions, transport direction, and magnitude of thin-skinned fold-thrust belts relative to the 'stable' craton, in this case Gondwanaland, must be an important consideration in any proposed tectonic models.

My present research aims to establish the sequence and kinematics of deformation by examination of polydeformation within the major fault zones and major structural zones/blocks within the Lachlan Fold Belt. Interactions between structural blocks is reflected by the fault zone structure and deformation at the margins of or within blocks, whereas within-block structural vergence is shown by regional fold asymmetry and major fault dip directions.

ADDITIONAL READING

Cas, R. A. F., 1983, Palaeogeographic and tectonic development of the Lachlan Fold Belt of southeastern Australia, *Geological Society of Australia Special Publication* 10, 104 p.

Coney, P. J., Edwards, A., Hine, R., Morrison, F., and Windrum, D., 1990, The regional tectonics of the Tasman orogenic system, eastern Australia, *Journal of Structural Geology,* v. 12, p. 519–543.

Gray, C. M., 1990, A strontium isotopic traverse across the granitic rocks of southeastern Australia: Petrogenetic and tectonic implications: Australia, *Journal of Earth Sciences,* v. 37, p. 331–349.

Gray, D. R., and Willman, C. E., 1991, Thrust-related strain gradients and thrusting mechanisms in a chevron-folded sequence, Southeastern Australia, *Journal of Structural Geology,* v. 13, p. 691–710.

Fergusson, C. L., and Glen, R. A., 1992, eds., The Palaeozoic eastern margin of Gondwanaland: Tectonics of the Lachlan Fold Belt, Southeastern Australia and related orogens, *Tectonophysics,* v. 214, 461 p.

Finlayson, D. M., Collins, C. D. N., and Denham, D., 1980, Crustal structure under the Lachlan fold belt, Southeastern Australia, *Physics of the Earth and Planetary Interiors,* v. 21, p. 321–342.

Powell, C. McA., Li, Z. X., Thrupp, G. A., and Schmidt, P. W., 1990, Australian Paleozoic paleomagnetism and tectonics—I. Tectonostratigraphic terrane constraints from the Tasman Fold Belt, *Journal of Structural Geology,* v. 12, p. 519–543.

Veevers, J. J., 1986, Phanerozoic Earth history of Australia, *Oxford Monographs on Geology and Geophysics* No. 2, Oxford, Oxford Science Publications.

19.9   Tectonic Genealogy of North America—Paul F. Hoffman[1]

19.9.1   INTRODUCTION

Continents are complex tectonic aggregates, evolved over billions of years. Today's continents are drifted fragments of the mid-Phanerozoic supercontinent Pangaea, more or less reshaped by accretion and ablation at subduction zones. The creation of Pangaea involved the fusion of many older continents, each of which was itself a drifted fragment of some still older continental assembly. Orogenic belts, marking the sites of ocean opening and closing, are the basic elements used to unravel the tectonic genealogy of continents.

Today's giant continent, Eurasia, was assembled in the Phanerozoic eon (545–0 Ma) through the piecemeal convergence of many pre-Phanerozoic continental fragments, roped together by Phanerozoic subduction complexes. The assembly of Eurasia is ongoing. Other continents are fragments of former giant continents assembled at various times in the pre-Phanerozoic. The southern continents, for example, are derived from Gondwanaland, which was assembled in the Neoproterozoic era (1000–545 Ma). North America (Figure 19.9.1) is the largest fragment of a continent assembled in the Paleoproterozoic (2500–1600 Ma). Other fragments exist in Eurasia and probably elsewhere. Since the Paleoproterozoic, North America (Laurentia, exclusive of fragments lost to Europe when the Atlantic opened) has twice collided to form supercontinents (all continents gathered together). The older collision is represented by the Mesoproterozoic (1600–1000 Ma) Grenville Orogen, and the resulting supercontinent, named Rodinia, had an approximate age span of 1050–750 Ma. The younger supercontinent is Wegener's Pangaea, which had an age span (liberal interpretation) of 300–150 Ma. It was conjoined with North America along the Appalachian Orogen and its connections around the Gulf of Mexico (Ouachitas), East Greenland (Caledonides), and Arctic Canada (Franklin). Ancestral North America participated in most of the salient tectonic events of the past three billion years.

The Phanerozoic evolution of the continents is reasonably well understood and key Phanerozoic orogenic belts are

[1]Harvard University, Cambridge, Massachusetts.

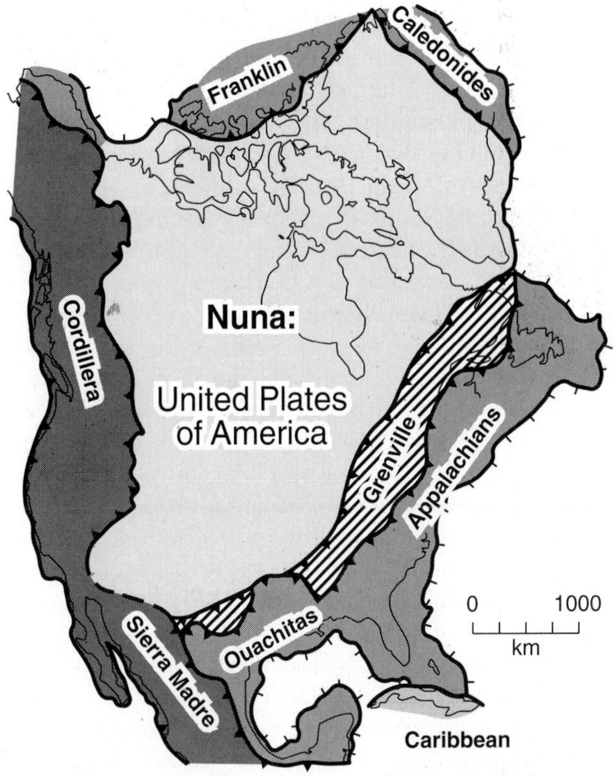

Figure 19.9.1   Simplified orogenic structure of North America: a Paleoproterozoic nucleus, Nuna, is discontinuously bordered by the Mesoproterozoic Grenville and Racklan (far northwest) orogens; the Paleozoic Ouachita, Appalachian, Caledonide, and Franklin orogens, and the Mesozoic-Cenozoic Cordilleran and Caribbean orogens. Greenland is restored to its pre-rift (>90 Ma) position.

the subjects of other chapters in this book. Pre-Phanerozoic Earth history is less well known, despite having produced over 80% of existing continental crust, but it is a subject in healthy ferment. Before discussing the role of ancestral North America in pre-Phanerozoic continental evolution, I should clarify some terminology. A collisional orogen implies a fusion of mature (>200 million years old) continental blocks. An accretionary orogen implies the addition of juvenile (<200 million years old) oceanic material—accretionary prisms, magmatic arcs, and volcanic plateaus for the most part. For brevity, I will use the numerical geon time scale, where geon 0 equals 0–99 Ma, geon 1 equals 100–199 Ma, geon 10 equals 1000–1099 Ma, and so on. The divisions of the Proterozoic eon, defined above, are those recognized by the International Union of Geological Sciences and differ slightly from the parallel divisions in the Geological Society of America DNAG time scale.

19.9.2 PHANEROZOIC (545–0 MA) OROGENS AND PANGAEA

The stable interior of North America is framed by two great Phanerozoic orogenic systems. The Cordilleran system (Figure 19.9.1) borders the Pacific Ocean basin and is still active. The Pacific continental margin first opened in geon 7, but the main phase of tectonic accretion occurred in geon 1, coeval with rapid northwesterly drift of the continent (relative to hotspots) following the breakup of Pangaea and opening of the North Atlantic basin. Cordilleran crust that was thickened during Mesozoic accretion collapsed in extension in the Paleogene, when convergence between North America and the Pacific basin slowed. Dextral strike-slip deformation became increasingly important in the Neogene, when North America began to override the East Pacific spreading ridge. The Cordillera will not die until the Pacific basin closes. If the Atlantic continues to open, North and South America will eventually collide with eastern Asia, which by then will have incorporated Australasia. The result will be a new supercontinent dubbed "Amasia."

The other Phanerozoic orogenic system formed during the Paleozoic assembly of Pangaea. It includes the Appalachian, Ouachita, Caledonide, and Franklin Orogenic Belts (Figure 19.9.1). They evolved from a continuous continental margin that opened diachronously in geons 6 and 5. Parts of the margin collided with island arcs and became active margins in geon 4, and by the end of geon 3 the northern Appalachian and Caledonide sectors had collided with Baltica and the southern Appalachian and Ouachita sectors had done the same with northwest Gondwanaland. The resulting orogenic system was dismembered when the North Atlantic basin opened.

19.9.3 NEOPROTEROZOIC (1000–545 MA) OROGENS AND GONDWANALAND

Gondwanaland (Figure 19.9.2), the former giant continent that broke up in the Mesozoic to form Africa, South America, Antarctica, Australasia, and southern Eurasia, was assembled in the Neoproterozoic era. Gondwanaland was an aggregate containing at least five older continents—West Africa, Amazonia, Congo, Kalahari, and East Gondwanaland (Australia, East Antarctica, India). They were welded together by a network of Neoproterozoic collisional orogens and bordered by Neoproterozoic-Paleozoic accretionary orogens. Pre-Phanerozoic North America lacks Neoproterozoic orogens, but displaced continental slivers that originated on the northwest margin of Gondwanaland were incorporated into the Appalachians in Middle and Late Paleozoic time. A two-way land trade apparently occurred in the Middle Paleozoic, suggesting a glancing encounter between eastern North America and western South America. Part of the southern Appalachians ended up in northwest Argentina, and a strip of northern South America (the Avalon terrane) was added to the coast of New England and eastern Canada. Later, a piece of northwest Africa (the Florida Peninsula and panhandle)

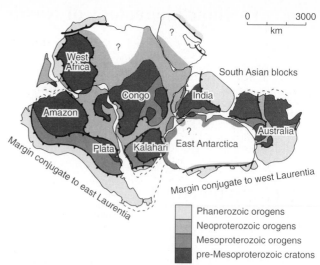

Figure 19.9.2 Aggregate structure of Gondwanaland, cemented by Neoproterozoic collisional and accretionary orogens. The reconstruction is well constrained by dated Mesozoic-Cenozoic seafloor magnetic anomalies. Dashed lines show edges of extensive modern continental shelves. Margins conjugate to Laurentia based on the Rodinia restoration in Figure 19.9.3. Note the disaggregated nature of Mesoproterozoic orogenic segments and the Neoproterozoic orogenic belt entering East Antarctica opposite Sri Lanka.

was transferred to North America during the climactic Late Paleozoic collision with Gondwanaland.

19.9.4 MESOPROTEROZOIC (1600–1000 MA) OROGENS AND RODINIA

The Mesoproterozoic Grenville Orogen lies inboard of the Appalachians and extends for 5000 km from Mexico to Labrador (Figure 19.9.1). It truncates Archean and Paleoproterozoic structural fabrics and tectonic boundaries to the northwest, implying that the orogen evolved as a rifted or sheared continental margin. The orogen comprises an outer (northwestern) zone consisting of reactivated Archean and Paleoproterozoic basement, and an inner (southeastern) zone of juvenile Mesoproterozoic crust. The outer zone is characterized by northwest-directed crustal-scale thrust shears and exposes metamorphic rocks that underwent >30 km of post-1050 Ma exhumation. Thrusting occurred in geons 11–10 and is presumably related to accretion of the inner zone and terminal collision of ancestral North America with an outboard continent(s). The hypothetical Grenvillian hinterland must have broken away when the Iapetus paleocean basin opened, initiating the Paleozoic Appalachian orogenic cycle. Likely candidates for the Grenvillian hinterland are the Amazonia-Plata craton of South America (Figure 19.9.2) and the Baltica craton of northern Europe. Both are flanked by Grenville-age orogenic and plutonic belts, consistent with those cratons belonging to the overriding plate of the terminal Grenvillian collision.

Rifted segments of orogenic belts of Grenville age occur throughout Gondwanaland, except for the West African craton (Figure 19.9.2). In addition to the belts in

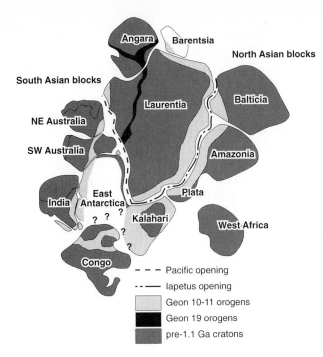

Figure 19.9.3 Hypothetical reconstruction of Rodinia, showing lines of Neoproterozoic opening of the Pacific and Iapetus ocean basins. As shown, Rodinia is an aggregate of cratons cemented by geon 10–11 (Grenvillian) orogens. Present-day north arrows for Laurentia (North America plus Rockall Bank and northwest Britain) and Kalahari cratons point to top and bottom of page, respectively.

South America, mentioned above, deeply eroded, geon 11–10 orogenic belts occur in central and southern Africa, including Madagascar, in Sri Lanka and the Eastern Ghats of India, and in East Antarctica and Australia (Figure 19.9.2). Under current investigation is the hypothesis that all or most of these segments originally belonged to a continuous, 10,000 km long system analogous to the Neogene Alpine-Himalayan system. The product of these collisions was the supercontinent Rodinia (Figure 19.9.3), the exact configuration of which is still conjectural but which should make a structurally compatible restoration of the Grenvillian orogenic segments. The Rodinia reconstruction (Figure 19.9.3) implies that Gondwanaland was turned inside-out following the breakup of Rodinia at about 750 Ma. The requisite anticlockwise rotation of East Gondwanaland, relative to Laurentia, and convergence with West Gondwanaland before about 500 Ma is consistent with paleomagnetic data.

19.9.5 PALEOPROTEROZOIC (2500–1600 MA) OROGENS AND NUNA

The vast region between the Grenville and Cordilleran Orogens (Figure 19.9.1), including the marginal parts of the orogens, was assembled in Paleoproterozoic time. It incorporates at least four Archean microcontinents—Churchill, Superior, Nain, and Slave (Figure 19.9.4). Paleomagnetic data indicate >4000 km of Late Paleoproterozoic convergence between the Churchill and Superior cratons, broadly

contemporaneous with significant relative motion between the Churchill and Slave cratons. The Wyoming craton may be an extension of the Churchill or a fifth independent microcontinent: the mutual boundary is buried by thick Phanerozoic platform cover. The Churchill has two major divisions—Rae and Heane (Figure 19.9.4)—previously thought to have fused in the Paleoproterozoic. However, recent studies indicate an earlier, Late Archean time of assembly. The overall Paleoproterozoic assembly has been called the United Plates of America (UPA). It is commonly believed to be continuous with Baltica (Figure 19.9.4) and an appropriate name for the entire continent assembled by the end of the Paleoproterozoic is Nuna, an Eskimo name for the lands bordering the northern oceans and seas.

The subduction zones that accommodated the converging Archean microcontinents dipped prevailingly beneath the Churchill continent. This had far-reaching structural and magmatic consequences. The Churchill margins have well-developed Paleoproterozoic plutonic belts, representing the eroded roots of continental magmatic arcs. These are lacking on the Churchill-facing margins of the Superior, Nain, and Slave cratons. The Churchill continent was far more severely and extensively deformed by the collisions than were the cratons. Large-scale strike-slip and oblique-slip shear zone systems developed as weak Churchill crust was extruded laterally in response to indentation by the three more rigid cratons (Figure 19.9.4). The Churchill also experienced unique intraplate magmatic events during and after the collisions. Ultrapotassic alkaline volcanism occurred in a 100,000 km^2 area west of Hudson Bay close to the time of the Churchill-Superior collision. Almost 80 million years later, after most of the collision-related deformation and metamorphic unroofing had occurred, the same region underwent high-silica rhyolite volcanism and associated rapakivi-type granite emplacement over an area of 250,000 km^2. Analogies have been drawn between the Churchill hinterland and the Neogene Tibetan Plateau, hinterland of the Himalayan collisional orogen.

The margins of the cratons facing the Churchill are characterized by large-scale thrust and nappe structures, kinematically directed away from the Churchill hinterland. Paleoproterozoic sedimentary and volcanic rocks deposited on the rifted margins of the cratons are discontinuously preserved and exposed. Volcanism and dike swarms related to initial continental breakup occurred mainly in geons 21–20. Subsequent passive-margin sediments (platformal carbonates and mature fine clastics) are overlain disconformably by foredeep sediments (ironstones, black shales, graywacke turbidites, and redbeds in complete ascending sequence). The change occurred as the leading edge of each craton entered a peri-Churchill subduction zone. The stratigraphic transition and hence the onset of collision can be precisely dated if suitable material (e.g., airborne volcanic ash layers) is present.

The structurally higher levels of the overthrust belts bordering the Superior craton are composed of quasi-oceanic material. At the Ungava syntaxis (Figure 19.9.4) in

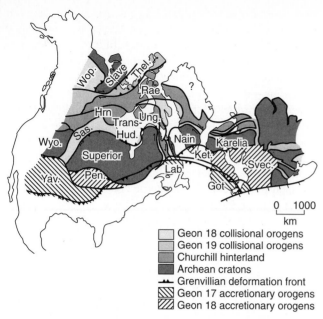

Figure 19.9.4 Existing aggregate structure of Nuna: the United Plates of America and its extension in Baltica. Nuna was cemented by geon-18 orogens, which are truncated at the present margins of Nuna and by the peripheral geon-17 accretionary orogens. Other extensions of Nuna exist on other continents or have been destroyed. Got. = Gothian Orogen; Hrn = Hearne craton; Ket. = Ketilidian Orogen; Lab. = Labrador Orogen; Pen. = Penokean Orogen; Sas. = Saskatoba syntaxis; Svec. = Svecofennian Orogen; Thel. = Thelon Orogen; Trans-Hud. = Trans-Hudson Orogen; Ung. = Ungava syntaxis; Wop. = Wopmay Orogen; Wyo. = Wyoming craton; Yav. = Yavapai Orogen.

Legend:
- Geon 18 collisional orogens
- Geon 19 collisional orogens
- Churchill hinterland
- Archean cratons
- Grenvillian deformation front
- Geon 17 accretionary orogens
- Geon 18 accretionary orogens

0 1000 km

Phanerozoic collisional orogens, but lithologically they are richer in volcanic rocks, consistent with higher mantle temperatures in the Paleoproterozoic.

The timing of the Paleoproterozoic collisions is best constrained by U-Pb geochronology of (1) the passive-margin to foredeep stratigraphic transition, described earlier, (2) the cessation of arc magmatism or change from arc-type to collision-type magmatism on the Churchill margins, and (3) the exhumation of metamorphic rocks having "clockwise" pressure-temperature-time trajectories of collisional origin. The geochronological data show that the Slave-Churchill collision occurred first, beginning about 1970 Ma. The Nain-Churchill collision occurred about 100 million years later and the Superior craton joined the assembly at about 1840 Ma.

Around the time the Superior, Nain, and Slave cratons collided, nucleating Nuna, juvenile Paleoproterozoic crust began to be accreted onto their trailing margins (i.e., those facing away from the Churchill hinterland). The respective accretionary orogens are the Penokean of the Great Lakes region, the Ketilidian of South Greenland and adjacent Labrador (where it is called Makkovik), and the Wopmay around Great Bear Lake, Northwest Territories (Figure 19.9.4). All three orogens evolved from passive margins that collided with island arcs and were converted to Andean-type margins as a result of subduction-polarity reversal (i.e., subduction zones first dipped away from the cratons and later beneath them). Arc-continent collision in the Wopmay Orogen occurred at 1.88 Ga, almost 90 million years after the Slave-Churchill collision. In the Ketilidian Orogen, less certain chronometrically, the arc-continent collision is placed at about 1.84 Ga, about 30 million years after the Nain-Churchill collision. Arc-continent collision in the Penokean Orogen occurred about 1.85 Ga, close to the time of the Superior-Churchill collision. The confluence of cratons and accretion at their trailing margins suggests a large-scale pattern of lithospheric flow converging on the Churchill hinterland. A possible explanation is that the hinterland was situated over a vigorous downwelling region of the sublithospheric mantle self-sustained by the descent of cold oceanic slabs.

Accretion on the (present) southern and southeastern margins of Nuna was renewed in geon 17, following apparent truncation of geon 18 structures on those margins. The accretionary orogens of geon 17 are principally exposed in Labrador and the southwestern United States, but, based on studies of drill cores, they also make up the buried basement across most of the southern Midcontinent (Figure 19.9.4). Juvenile crust, at least 1300 km wide, was accreted in geon 17 between the Wyoming craton and the Grenville Orogen of West Texas. The accreted material proved to be a fertile source for Mesoproterozoic crustal-melt granites and rhyolites, which are extensively encountered in the Midcontinent subsurface.

---

northern Quebec, parts of an oceanic plateau (1.92 Ga), an imbricated ophiolite (2.00 Ga), and an immature island arc (1.87–1.83 Ga) were thrust southwards across the rifted margin onto the Superior craton. The ophiolite, one of the world's oldest, includes a sheeted dike complex, ultramafic cumulates, and volcanic suites chemically and isotopically correlated with mid-ocean ridge and ocean-island basalts. At the Saskatoba syntaxis (Figure 19.9.4) in northern Saskatchewan and Manitoba, the craton is tectonically juxtaposed by juvenile island-arc-type volcanic and plutonic rocks (1.93–1.85 Ga) and derived metasediments. The craton at first formed the structural footwall but was later thrust toward the hinterland over Paleoproterozoic juvenile rocks of the intervening Trans-Hudson Orogen (Figure 19.9.4). The northeast-facing lateral margin of the craton and the arcuate embayment between the Ungava and Saskatoba syntaxes (Figure 19.9.4) contain allochthonous mafic sill-sediment complexes formed in syncollisional pull-apart basins (rhombochasms). The thrust-fold belts on the lateral margins of all three indented cratons are flanked by crustal-scale strike-shear zones bordering the hinterland (Figure 19.9.4). Structurally, the belts have much in common with

The widely held belief that Nuna originally included Baltica (Baltic Shield and East European Platform) is based on proposed continuity between the Archean Nain and Karelia cratons, the Ketilidian and Svecofennian Orogens of geon 18, and the Labrador and Gothian Orogens of geon 17 (Figure 19.9.4). A more tentative connection is that proposed between Nuna and the Angara craton of Siberia, based on extensions of the Slave-Churchill collision zone (Thelon Orogen) across the Arctic. Even more tenuous links exist between Nuna and major geon-18 orogenic belts in northern and western Australia. In addition, there are tantalizing similarities in Late Paleoproterozoic to Early Mesoproterozoic platform cover sequences worldwide that have contributed to the notion, as yet undemonstrated, that the UPA was part of a giant continent predating Rodinia.

19.9.6 ARCHEAN CRATONS AND KENORLAND

Archean cratons have dimensions that make their plate tectonic settings often difficult to determine. The largest, one continuously exposed, is the Superior craton (Figure 19.9.5) of the Canadian Shield, measuring 2500 km east-west and 1500 km north-south. It was constructed in the latter half of geon 27. It exposes, in zonally varying proportions, deformed plutonic and volcanic rocks of island-arc and oceanic-plateau affinities and derived sediments. The regional tectonic strike swings from east-west, in the western and southeastern parts of the craton, to north-south in the northeastern part (Figure 19.9.5). In the northwest of the craton, a long-lived composite protoarc had evolved since geon 30. Bilateral accretion of juvenile material onto the protoarc began about 2.75 Ga. A progressive southward docking sequence of south-facing volcanic arc and accretionary prism couplets occurred until 2.69 Ga. Accretion was terminated by the collision of the entire assemblage with an old (~3.6 Ga) continent, the Minnesota River Valley (MRV) terrane (Figure 19.9.5). The southward docking sequence and the structural evidence of persistent dextral strike-slip displacement accompanying the docking events implies overall oblique northwest-directed subduction. In the northeast of the craton, where the tectonic strike is north-south, a relatively high proportion of plutonic rocks, deep level of exhumation, and absence of strike-slip displacements indicate a less oblique subduction regime. The tectonic zonation and associated structural grain of the craton is clearly truncated at its western, southeastern, and northeastern margins (Figure 19.9.5), showing that the entire craton is but a rifted fragment of a Late Archean continent, referred to as Kenorland.

There is a good possibility that the Wyoming craton (Figure 19.9.4) was originally connected to the south margin of the Superior craton in Kenorland. The Wyoming craton is old, like the MRV, and it preserves erosional rem-

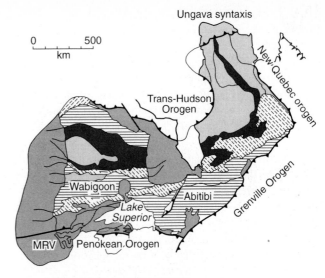

2.75 - 2.68 Ga juvenile terranes:
deeper crust exhumed in Paleoproterozoic
sediment-dominated accretionary prisms
volcanic-dominated island arcs and plateaus

2.73 - 2.69 Ga magmatic arc within older proto-arc
3.0 - 2.8 Ga volcanic-rich composite proto-arc
ca 3.6 Ga continental foreland

post-Archean cover on Superior Craton

Figure 19.9.5 Accretionary structure of the Superior craton, the largest Archean craton in Nuna (Figure 19.9.4). Note change in tectonic strike from east-west to north-south and the truncation of tectonic boundaries at the edge of the craton. MRV = Minnesota River Valley terrane.

nants of a highly distinctive early Paleoproterozoic sequence of uraniferous conglomerates, tropically weathered quartz arenites, and glacial diamictites remarkably similar to the Huronian sequence on the north shore of Lake Huron in the southern Superior craton. Separation of the Superior and Wyoming cratons occurred in the latter half of geon 21, about 300 million years before the two cratons were reunited in Nuna.

A remarkable number of cratons worldwide underwent major accretion or collision in geon 27. It was a time of oblique plate convergence and rapid continental accretion in the Yilgarn craton, the largest in Australia, and the time of collision between the Kaapvaal and Zimbabwe cratons of southern Africa, to name just two of the more important overseas examples. In North America, the Hearne Province of the Churchill hinterland was accreted in geon 27, but the Slave and Nain cratons were assembled in geon 26. Precise U-Pb dating of large-scale mafic dike swarms, emplaced during breakup events associated with mantle plumes, holds promise as a means of identifying formerly contiguous cratons. This should yield important insights into the extent and configuration of Kenorland, possibly the first giant continent.

19.9.7 CLOSING REMARKS

North America consists of an aggregate nucleus, assembled in the Paleoproterozoic, discontinuously bordered by Meso-proterozoic and Phanerozoic orogens. Parts of North America participated in perhaps five giant continents: Kenorland, assembled in geons 27–26; Nuna, assembled in geons 18–17; Rodinia, assembled in geons 11–10; Gondwana-land, assembled in geons 6–5; and Pangaea, assembled in geons 3–2.

The recurrence interval for giant assemblies seems to have become shorter with time—900, 700, 500, and 300 million years. This may merely reflect our inferior knowledge of the earlier part of the record. If real, on the other hand, how can it be reconciled with secular cooling of the Earth, resulting in mantle convection of diminishing vigor. It seems unlikely that continental collisions have become more frequent. However, collision zones tend to re-open, particularly when young. This tendency could have been even stronger in the distant past. To make a giant continent piecemeal, the disassembly processes must be checked. Therefore, the growing incidence of giant continents may reflect increased stability, rather than increased frequency, of continental collisions.

ADDITIONAL READING

Card, K. D., 1990, A review of the Superior province of the Canadian shield, a product of Archean accretion, *Precambrian Research,* v. 48, p. 99–156.

Condie, K. C., and Rosen, O. M., 1994, Laurentia-Siberia connection revisited, *Geology,* v. 22, p. 168–170.

Dalziel, I. W. D., 1992, On the organization of American plates in the Neoproterozoic and the breakout of Laurentia, *GSA Today,* v. 2, p. 237–241.

Dalziel, I. W. D., Dalla Salda, L. H., and Gahagan, L. M., 1994, Paleozoic Laurentia-Gondwana interaction and the origin of the Appalachian-Andean mountain system, *Geological Society of America Bulletin,* v. 106, p. 243–252.

Gorbatschev, R., and Bogdanova, S., 1993, Frontiers in the Baltic shield, *Precambrian Research,* v. 64, p. 3–21.

Gower, C. F., Rivers, T., and Ryan, A. B., eds., 1990, Mid-Proterozoic Laurentia-Baltica: Special Paper 38, *Geological Association of Canada,* St. John's, 581 p.

Hoffman, P. F., 1988, United plates of America, the birth of a craton: Early Proterozoic assembly and growth of Laurentia, *Annual Reviews of Earth and Planetary Sciences,* v. 16, p. 543–603.

———, 1989, Speculations on Laurentia's first gigayear (2.0–1.0 Ga), *Geology,* v. 17, p. 135–138.

———, 1991, Did the breakout of Laurentia turn Gondwanaland inside-out? *Science,* v. 252, p. 1409–1412.

Lewry, J. F., and Stauffer, M. R., eds., 1990, The Early Proterozoic Trans-Hudson orogen of North America, Special Paper 37, *Geological Association of Canada,* St. John's, 505 p.

Lucas, S. B., Green, A., Hajnal, Z., White, D., Lewry, J., Ashton, K., Weber, W., and Clowes, R., 1993, Deep seismic profile across a Proterozoic collision zone: Surprises at depth, *Nature,* v. 363, p. 339–342.

Moores, E. M., 1991, Southwest U.S.—East Antarctic (SWEAT) connection: A hypothesis, *Geology,* v. 19, p. 425–428.

Powell, C. McA., Li, Z. X., McElhinny, M. W., Meert, J. G., and Park, J. K., 1993, Paleomagnetic constraints on timing of the Neoproterozoic breakup of Rodinia and the Cambrian formation of Gondwana, *Geology,* v. 21, p. 889–892.

Reed, J. C., Jr., Bickford, M. E., Houston, R. S., Link, P. K., Rankin, D. W., Sims, P. K., and Van Schmus, W. R., eds., 1993, Precambrian: Conterminous U.S.: Boulder, Colorado, Geological Society of America, *The Geology of North America,* v. C-2, 657 p.

Rivers, T., Martignole, J., Gower, C. F., and Davidson, A., 1989, New tectonic divisions of the Grenville province, southeastern Canadian shield, *Tectonics,* v. 8, p. 63–84.

Sengör, A. M. C., Natal'in, B. A., and Burtman, V. S., 1993, Evolution of the Altaid tectonic collage and Paleozoic crustal growth in Eurasia, *Nature,* v. 364, p. 299–307.

Thurston, P. C., Williams, H. R., Sutcliffe, R. H., and Stott, G. M., eds., 1991, Geology of Ontario: Parts 1 and 2, *Ontario Geological Survey Special Volume 4,* 709 p. (Part 1), 1525 p. (Part 2).

van der Voo, R., 1988, Paleozoic paleogeography of North America, Gondwana, and intervening displaced terranes: Comparisons of paleomagnetism with paleoclimatology and biogeographical patterns, *Geological Society of America Bulletin,* v. 100, p. 311–324.

Williams, H., Hoffman, P. F., Lewry, J. F., Monger, J. W. H., and Rivers, T., 1991, Anatomy of North America: Thematic geologic portrayals of the continent, *Tectonophysics,* v. 187, p. 117–134.

19.10   The U.S. Continental Interior—Stephen Marshak and Ben A. van der Pluijm

19.10.1   INTRODUCTION

If you've ever flown from coast to coast in the United States on a clear day, you can't help but notice the contrast in topographic relief that distinguishes mountain belts of the east and west from the Great Plains of the interior. The former contain dramatic slopes and deeply incised canyons, whereas the latter contains a vast checkerboard of flat farmland in which roads run without a curve for tens of kilometers. Topographic contrasts between plains and mountains emphasize fundamental differences in the geology of these regions. We can subdivide North America into two types of geologic provinces: (1) Phanerozoic orogens, which are linear belts that have been uplifted to elevations in excess of a couple of kilometers, and have been affected by relatively intense deformation (folding, faulting, and fabric development), and locally by metamorphism and igneous activity during the Phanerozoic; and (2) the craton, which is the portion of the continent that has acted as a relatively stable and rigid mass, at least since the end of the Precambrian. In North America, the Phanerozoic orogenic belts lie between the craton and the ocean, so we also refer to this cratonic region as the continental interior or the Midcontinent. Note that our categories lump together many types of continental crust (Table 19.10.1).

Geologists divide the North American craton into two components: the shield, in which Precambrian basement crops out at the Earth's surface, and the continental-interior platform, in which a relatively thin veneer of Phanerozoic strata covers the basement. The region that is now the craton was itself the locus of intense tectonism in the Precambrian, as demonstrated by basement rocks containing tectonite fabrics, shear zones, and folds. In this essay we do not look at the basement's history (see Section 19.9), but focus our attention on Phanerozoic tectonism in the craton. Stratigraphy of the continental-interior platform provides the primary record of this tectonism, so we base our discussion on features of Midcontinent USA, where the stratigraphic record has been preserved.

To structural geologists raised on a diet of spectacular folds, faults, and fabrics exposed in Phanerozoic orogens, structural features of Midcontinent United States may seem

Table 19.10.1   Categories of Continental Crust

Active Orogens	Those portions of the continental crust in which tectonism (faulting ± volcanism ± uplift) currently takes place or has taken place in the recent past (Cenozoic). Such orogens tend to be linear belts, in that they are substantially longer than they are wide. Examples are the North American Cordillera and the European Alps.
Active Rift	A region where the crust is currently undergoing extension, so that crustal thicknesses are less than the average crustal thickness; for example, the Basin-and-Range Province.
Continental Shelf	A belt fringing continents in which a portion of the continent has been submerged by the sea. Water depths over shelves are generally less than a few hundred meters. Continental shelves are underlain by passive-margin basins, which form when continental crust first gets stretched during rifting, and then subsides. Sediment washed off the adjacent land buried the sinking crust. Examples are the east coast and Gulf coast of the United States.
Craton	The portion of continents that acts as relatively stable and rigid crustal blocks at least since the end of the Precambrian; in other words, geologists do not consider continental interior regions to be parts of orogens. However, cratons include Precambrian mountain belts that are now fully eroded. Cratons may be divided into two types: shields, where Precambrian (crystalline) rocks are exposed, and platforms, where Precambrian rocks are covered by unmetamorphosed Phanerozoic strata.
Inactive Phanerozoic Orogens	Orogenic belts that were active in the Phanerozoic, but are not active today. Some inactive orogens, however, have been uplifted during the Cenozoic, so they are topographically high regions. Examples include the Appalachians and the Tasmanides.

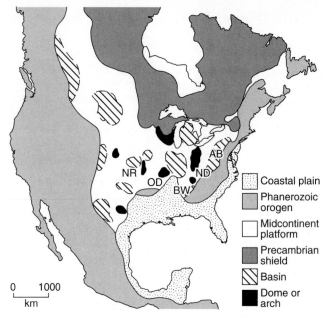

Figure 19.10.1 Regional map of North America showing the distribution of Phanerozoic orogenic belts, shield, continental-interior (Midcontinent) platform, and basins and domes.

Legend:
- Coastal plain
- Phanerozoic orogen
- Midcontinent platform
- Precambrian shield
- Basin
- Dome or arch

tame and uninteresting. But in reality, the notion of a "stable" Midcontinent is far from true. Certainly, the region is stable relative to the surrounding orogens, but it is not barren of tectonism. The Midcontinent contains four classes of tectonic structures: (1) epeirogenic structures, (2) midcontinent fault and fold zones, (3) regional joint systems, and (4) penetrative strain. *Epeirogeny* refers to gradual vertical movements affecting broad regions of continents or ocean basins. Epeirogeny affects regional patterns of stratigraphic thickness and facies, but does not tilt strata by more than a few degrees. In the Midcontinent, epeirogeny has yielded regional-scale basins, domes, and arches with areas greater than ~10,000 km^2 (Figure 19.10.1). We call these features epeirogenic structures. In addition to epeirogenic structures, the Midcontinent contains belts of relatively localized deformation (less than ~100 km wide) in which faulting displaced stratigraphic contacts and/or folding warped bedding, creating dips that typically exceed 30°. In fact, bedding within these zones is locally overturned. Joints cut all strata of the Midcontinent. Some of the joints compose systematic sets with consistent orientation over broad regions. Cleavage in Midcontinent strata is rare to nonexistent, but the rocks have been internally strained very slightly. This strain has been recorded by fossils and calcite-twinning.

19.10.2 EPEIROGENIC STRUCTURES

Stratal thickness varies significantly with position in the Midcontinent region, thereby defining discrete basins, in which strata reach a maximum thickness of about 5 km, and arches or domes along which thickness decreases to zero. Thinning of strata along arches and domes reflects periods of nondeposition and/or erosion that occurred when the arches or domes were high relative to adjacent basins. In effect, the lateral variations in sediment thickness that we observe in the Midcontinent indicates that there has been differential epeirogenic uplift and subsidence of regions of the Midcontinent with respect to one another. The general positions of basins and domes in the Midcontinent have remained fixed through the Phanerozoic. For example, the Illinois Basin has been a basin since its initial formation in the Late Proterozoic or Early Cambrian. Thus, these structures represent permanent features of the North American continental-interior lithosphere.

Lack of stratigraphic evidence, either due to nondeposition or erosion, makes it impossible to constrain the rate of uplift of arches and domes, but we can constrain rates of epeirogenic movement in basins by looking at sedimentary thickness as a function of time. Studies of many basins demonstrate that basins have continued to subside through most of the Phanerozoic, perhaps due to slow cooling of lithosphere beneath the basins, but the rate of subsidence for a basin varies with time. For example, pulses of rapid subsidence occurred in the Michigan and Illinois basins during Ordovician, Late Devonian-Mississippian, and Pennsylvanian-Permian time. In fact, during these times, rates of subsidence exceed by a factor of 2 the rate of subsidence that could be accomplished solely in response to eustatic (i.e., global) sea-level rise. Notably, the timing of some, but not all, epeirogenic movements in the Midcontinent roughly corresponds with the timing of major orogenic events along the continental margin.

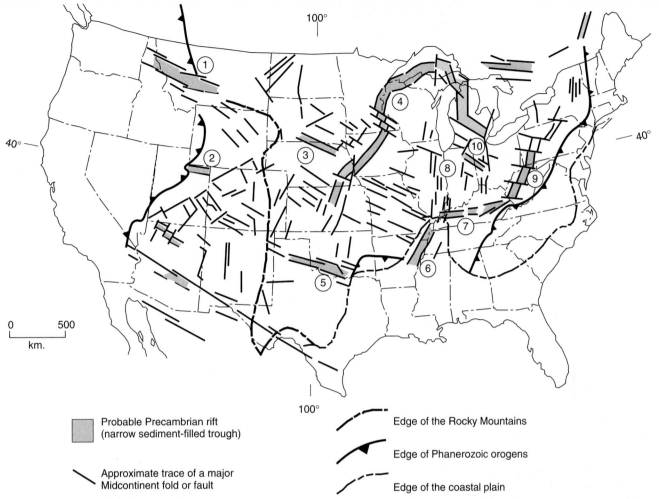

Figure 19.10.2 Map of the United States showing the distribution of Midcontinent fault-and-fold zones and documented intracratonic rifts. 1 = Beltian embayment, 2 = Uinta trough, 3 = Nebraska sag, 4 = Midcontinent rift, 5 = Southern Oklahoma aulocogen, 6 = Reelfoot rift, 7 = Rough Creek graben, 8 = La Salle deformation belt, 9 = Rome trough, 10 = Michigan arm of the Midcontinent rift.

19.10.3 FAULTING AND FOLDING IN THE MIDCONTINENT

We cannot see Midcontinent folds and faults directly because younger sediments cover most of them. Thus, interpretations of Midcontinent structures rely substantially on drill-hole data (compiled on isopach and structure-contour maps), on potential-field studies (gravity- and magnetic-anomaly maps), and on rare seismic-reflection data. Geologists can infer the position and trace of a Midcontinent fault-and-fold zone from steps and ridges on structure-contour maps and/or from distinct linear anomalies, or offsets of anomalies, on potential-field maps. Though the results of such studies are sometimes sketchy and incomplete, overall they do provide a consistent image of what Midcontinent structures look like at depth. It is clear that Midcontinent structures share a number of characteristics that distinguish them from the fault-and-fold systems of orogenic belts.

Individual Midcontinent fault-and-fold zones affect regions up to 500 km in length and 100 km in width (Figures 19.10.2 and 19.10.3). Larger zones typically include numerous noncoplanar faults that range in length

from <5 km to as much as 50 km. At their tips, these faults overlap with one another in a relay fashion (see Chapter 8). Locally, a band of *en echelon* subsidiary fault segments borders principal fault traces. In the upper few kilometers of the crust, Midcontinent faults dip steeply, and typically numerous splays diverge from a larger fault, thereby creating fault arrays that resemble the flower structures that form along major strike-slip fault systems or inverted rifts. At depth, major faults decrease in dip (i.e., some faults appear to be listric) and may penetrate basement. There is no evidence demonstrating that Midcontinent faults merge at depth with a regional detachment in the sedimentary sequence. Some but not all faults border rift basins that contain an anomalously thick sequence of sediments and volcanics. The largest of these intracratonic rifts, the Midcontinent Rift, consists of two principal arms, one running from Lake Superior into Kansas and the other running diagonally across Michigan. Faults along these rifts initiated as normal faults, but later reactivated as thrust or strike-slip faults.

Major faults in the continental interior locally have throws of as much as 2–3 km, but more typically throws are

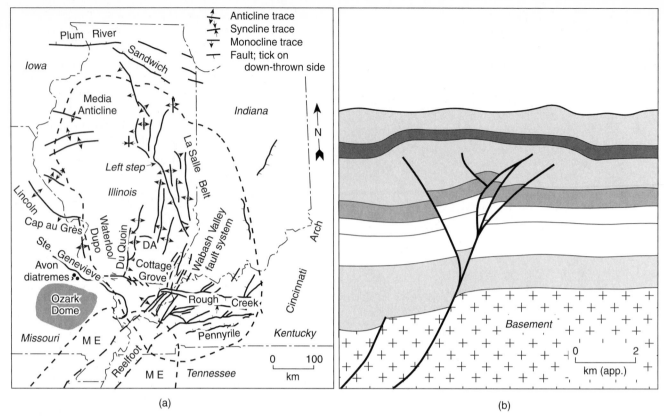

(a)　　　　　　　　　　　　　　　　　　　　(b)

Figure 19.10.3 Structural styles of Midcontinent fault-and-fold zones. (a) Map of the Illinois basin region showing the map traces of fault-and-fold zones. (b) Schematic cross section showing typical flower structure and monoclinal folds formed over basement faults.

less than ~1 km. Strike-slip displacements across continental-interior faults are difficult to document because of lack of shear-sense indicators on fault surfaces and lack of recognizable offset markers. But geologists have found strike-slip lineations on faults exposed in coal mines, the *en echelon* map pattern of subsidiary faults adjacent to larger faults resembles the *en echelon* faulting adjacent to continental strike-slip faults, and steeply plunging folds border some faults. All these features suggest that strike-slip has probably occurred on some Midcontinent faults.

In general, folds of the continental interior are monoclinal in profile, meaning that they have one steeply dipping limb and one very shallowly dipping to subhorizontal limb. In this regard, Midcontinent folds resemble the Laramide drape folds of the Rocky Mountain region and Colorado Plateau, though at a smaller scale. Locally, oppositely facing monoclinal folds form back to back, creating boxlike anticlines. Though geologists have not yet obtained clear images of many Midcontinent folds at depth, several studies do document that folds lie above steeply-dipping faults and that structural relief on folds increases with depth. Fault-related folds are not the only folds to develop in the Midcontinent. Tilted strata also form as a consequence of differential compaction over buried basement highs, and because of development of initial dips along buttress unconformities. Compaction folds also develop around reefs and sand lenses higher in the stratigraphic section.

Detailed study of spatial variations in the thickness and facies of a stratigraphic unit relative to a structure, documentation of the timing of unconformity formation, as well as documentation of local slump-related deformation adjacent to a structure, permits determination of whether a fold and fault zone was tectonically active during a specified time interval. Unfortunately, timing constraints for movement on Midcontinent structures are not tight, but they do demonstrate that structures, in general, became active during more than one event in the Phanerozoic. Activity appears to have been particularly intense during times of orogenic activity along the continental margin, but occurred at other times as well.

Faults and associated folds in the continental interior do not have random orientations, but rather display dominant trends over broad regions of the craton. As indicated by the map in Figure 19.10.3b, most fault-and-fold zones in the continental interior either trend north-south to northeast-southwest, or east-west to northwest-southeast. As we noted previously, faults tend to cluster in distinctive belts, some of which are hundreds of kilometers long. Along-strike linkage of fault-and-fold zones seems to define transcratonic belts of tectonic reactivation, which localize seismicity today.

19.10.4 PENETRATIVE STRAIN AND REGIONAL JOINTING

Despite the lack of notable regional cleavage within strata of the Midcontinent, the occurrence of deformation features such as calcite twinning in limestone units indicates that layer-parallel shortening did develop in these rocks. As we noted in Chapter 9, twinning accumulates strain in calcite under the relatively low pressure and temperature conditions that are characteristic of the upper crust. Regional studies of calcite-twinning in Midcontinent limestones indicate that the maximum shortening direction remains fairly constant over broad regions, and trends roughly perpendicular to orogenic fronts, though more complex shortening patterns occur in the vicinity of intracratonic fault-and-fold zones. For example, in the eastern United States Midcontinent, strain magnitudes due to twinning rarely exceed 6% and typically range between 0.5–3%, and appear to decrease progressively from the Southern Appalachians-Ouachitas front toward the cratonic interior. This geometry indicates that layer-parallel shortening in strata is a response to collisional orogeny along the continental margin; that is, the Late Paleozoic collision of Africa with North America that created the fold-thrust belt of the Appalachian-Ouachita system.

No one has yet compiled joint data for the entire Midcontinent, but the literature does provide data from numerous local studies. Joint frequency diagrams suggest that there are two dominant joint sets (one trending generally northwest and one trending generally northeast) and two less prominent sets (one trending east-west and one trending north-south) in the Devonian strata of northern Michigan. Similar, but not exactly identical, trends have been documented in Ohio, Indiana, Illinois, and Wisconsin. Taken together, regional studies suggest that systematic vertical joint sets do occur in platform strata of the Midcontinent, and that in general there are east-west sets, northwest sets, north-south sets, and northeast sets, but that orientations change across regions, and that different sets dominate in different locations.

19.10.5 CAUSES OF EPEIROGENY

Over the years, geologists and geophysicists have proposed many mechanisms for continental interior epeirogeny, particularly that leading to basin subsidence. The candidates that may explain epeirogeny are briefly discussed below and illustrated in Figure 19.10.4.

Intracratonic basins may form because of thermal contraction due to cooling of an unsuccessful rift that opened during the Late Proterozoic. When active extension ceased, the rift cooled and subsided, much like the rifts that underlie passive margins. Such epeirogeny will continue through the Phanerozoic at ever decreasing rates. Convective movement in the mantle may cause upward flow of hot material in the asthenosphere that heats the overlying lithosphere, thereby decreasing its density and causing it to rise. Considering the thermal heterogeneity of the asthenosphere, drift of plates might cause the plate to move over relatively hotter areas (e.g., a hotspot) with the effect that these areas would go up. Similarly, cooling of a plate by juxtaposition with cooler asthenosphere below might cause it to sink.

Changes in stress state in the lithosphere may cause epeirogenic movement in many ways. For example, as the differential stress increases, plastic deformation occurs more rapidly, so that the viscosity of the lithosphere decreases. Thus, an increase in differential stress effectively weakens the lithosphere. Denser masses in the crust, such as mafic volcanics in a rift, which were previously supported by the flexural strength of the lithosphere, would sink, whereas less dense masses, such as a granite pluton, would rise. Thus, differential epeirogenic movements reflect the heterogeneities of the crust. Changes in horizontal stress magnitude may also cause epeirogeny by effectively buckling the lithosphere or by amplifying existing depressions (basins) or rises (arches). Changes in the stress state in the continental interior may also cause tilting of regional-scale, fault-bounded blocks of continental crust relative to one another, creating regional triangular-shaped depressions or uplifts.

Creation of a load, such as from a volcano or a stack of thrust sheets, results in flexural loading on the surface of the continent. This causes a broad depression, much like pushing down on the end of a (thin) wooden plank causes a broad curve to develop in the plank. A surface load does not simply sink into the lithosphere, as a brick sinks into water. This effect is due to the flexural strength of the lithosphere, which causes the load to be distributed over a broader region. The levering effect of the portion of the lithosphere that is depressed creates an outer swell on the cratonic-interior side of the depression. Recent studies demonstrate that the effects of load emplacement along the margin of a continent due to formation of a fold-thrust belt may cause such epeirogenic uplift and subsidence well into the interior of the continent.

Subduction of oceanic lithosphere can cause epeirogenic movement because it is equivalent to placing a positive mass anomaly under the continent. Thus, subduction over time pulls the continent down.

Changes in the ratio of crustal thickness to lithosphere mantle thickness changes the surface elevation of the continent, because continental elevation is controlled by isostasy. Since crust and mantle do not have the same density, changes in their relative proportions in the lithosphere will cause a rise or fall of the ground surface as the region moves to attain regional isostatic equilibrium. Thickening of the crust, perhaps due to strain in response to tectonic compression, may cause the surface to rise. Similarly, delamination and sinking of the mantle part of the lithosphere could also lead to a rise in surface elevation.

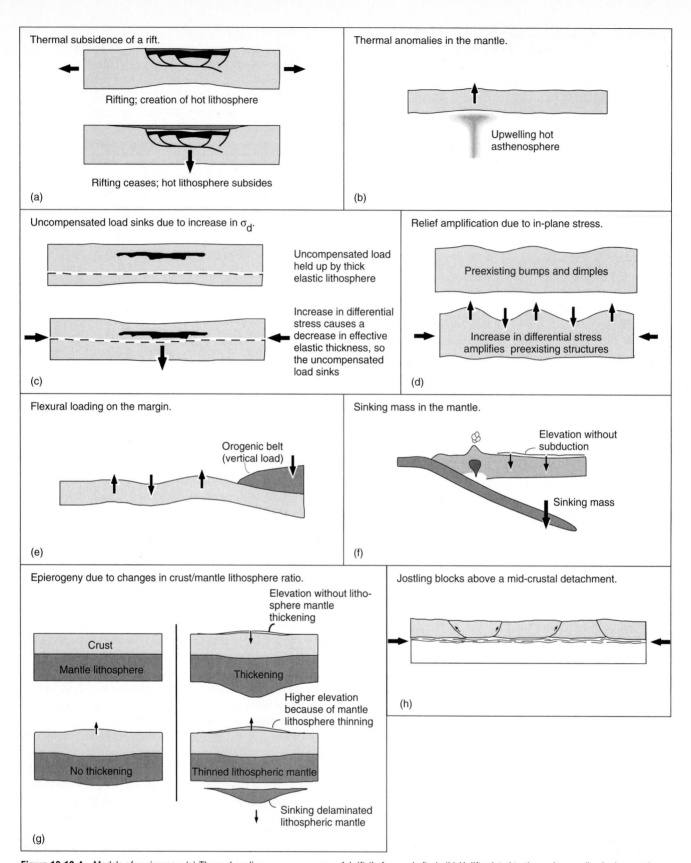

Figure 19.10.4 Models of epeirogeny. (a) Thermal cooling over an unsuccessful rift (before and after). (b) Uplift related to thermal anomalies in the mantle. (c) Vertical movement of an uncompensated load due to changes in the elastic thickness of the lithosphere. (d) Amplification of preexisting bumps and dimples due to in-plane stress. (e) Flexural loading of a lithospheric margin. (f) Epeirogeny related to subduction. (g) Epeirogeny due to changes in the ratio of crustal thickness to lithospheric mantle thickness. (h) Epeirogeny due to tilting of regional fault-bounded blocks.

19.10.6 SPECULATIONS ON MIDCONTINENT FAULT-AND-FOLD ZONES

As we mentioned previously, geologists have identified numerous narrow rifts in the Midcontinent, based on the occurrence of narrow anomalously thick packages of sediments and volcanics. These rifts appear to have initiated in Precambrian time, and subsequently controlled patterns of sedimentation during the Paleozoic. Most other Midcontinent fault-and-fold zones do not appear to border distinct riftlike basins but nevertheless share a number of distinctive structural characteristics with known rifts. These enigmatic Midcontinent fault-and-fold zones, like rift border faults, have been reactivated repeatedly during the Phanerozoic, and, regionally, the enigmatic fault-and-fold zones display trends that parallel those of rifts. Such similarity suggests that many other Midcontinent fault-and-fold zones may also have initiated during the same tectonic events that caused rift basin formation, that is, in response to crustal extension. Perhaps the geometry of the structures was controlled, in part, by preexisting basement fabrics and sutures.

Regardless of their origin, Midcontinent fault-and-fold zones represent reactivation of long-lived basement weak zones. Because the two dominant trends of these structures are nearly at right angles to each other, they outline roughly rectangular blocks of less faulted crust (Figure 19.10.5). Thus, perhaps a way to envision Midcontinent faulting and folding is to think of the basement as a mosaic of fault-bounded blocks that jostle relative to one another in response to changes in the stress state in the continental interior. Presumably, the long-lived faults bounding these blocks are weak enough that even the relatively low differential stresses characteristic of time periods when major collisional orogenies have not affected the continental margins can trigger movement. Contemporary seismicity in the Midcontinent may represent just such movement. Note in Figure 19.10.5 that the majority of intracratonic seismic zones occur along a linkage of northwest-trending fault zones that extend diagonally across the interior of the country from Idaho to Carolina, suggesting that this belt of structures has been particularly susceptible to reactivation. When collision does occur, differential stress in the interior increases, and the magnitude of displacement on block boundaries may increase. Also, epeirogenic movements effectively bend the crust, and thus may trigger additional fault movement as the crust accommodates the changing radius of curvature. Depending on the geometry of the stress field during a given time period, blocks may move apart slightly (are reactivated as rifts), move together slightly (are reactivated as contractional deformation zones), or move laterally relative to one another (are reactivated as zones of strike slip). Most likely, movements tend to be transpressional or transtensional, for the belts need not be oriented appropriately for simple thrust, reverse, or strike-slip faulting to occur.

Midcontinent strata have not been metamorphosed, but they have undergone regional diagenesis in response to continental-scale brine migration. Limited observations

Figure 19.10.5 Block model of intracratonic tectonism in North America. Stars indicate seismically active regions. The same concept applies to the Laramide Province of the Cordillera.

suggest that some diagenetic phenomena (e.g., resetting of paleomagnetic poles) and ore deposits appear to be localized along Midcontinent fault-and-fold zones. This would imply that the fault-and-fold zones are regions of enhanced permeability through which mineral-laden, crustal fluids are flushed up into shallower crustal levels where they react with cooler strata.

19.10.7 CLOSING REMARKS

The Midcontinent of the United States and, by analogy, other continental-interior platform regions of cratons, have been tectonically active throughout the Phanerozoic. Deformation is manifested primarily by differential epeirogeny (resulting in formation of basins, arches, and domes) and by reactivation of local fault-and-fold zones. Epeirogenic structures develop in response to stress and/or thermal changes in the lithosphere that decrease the effective elastic thickness of the lithosphere and allow uncompensated loads to move, and/or to flexural loading of the continental margin. The pattern of fold-and-fault belts in the U.S. Interior suggests that three structural trends dominate, east-west, northwest-southeast, and north-south to northeast-southwest. Therefore, these zones outline roughly rectangular blocks of crust that move with respect to one another when the stress field in the continental interior changes. The zones themselves may have initiated as rifts during the Proterozoic, and were reactivated repeatedly in the Phanerozoic, both during times of collisional orogeny at the margins and times when collision was not occurring. They are preserved because the area has not been subjected to regional metamorphism. Movement, which is most active at the corners of blocks where two fold-and-fault zones intersect, may cause intracontinental seismic activity. Because the zones have re-ruptured frequently throughout the Phanerozoic, they are zones of high permeability that could have acted as fluid conduits and thereby localized ore deposits.

The tentative tone of this essay emphasizes that geologists need to obtain much more data on structures in cratonic interiors before we can confidently explain them and assess their significance. However, it is becoming increasingly clear that plate interiors are not tectonically 'dead' regions. Rather, they are sensitive recorders of large-scale, plate-tectonic activity and stress patterns, without the disturbance of this record that characterizes collisional orogens at plate margins.

ADDITIONAL READING

Bally, A. W., 1989, Phanerozoic basins of North America, *in* Bally, A. W., and Palmer, A. R., eds., The Geology of North America—An Overview: Boulder, Colorado, Geological Society of America, *The Geology of North America,* v. A, p. 397–446.

Bond, G. C., 1979, Evidence for some uplifts of large magnitude in continental platforms, *Tectonophysics,* v. 61, p. 285–305.

Cathles, L. M., and Hallam, A., 1991, Stress-induced changes in plate density, Vail sequences, epeirogeny, and short-lived global sea level fluctuations, *Tectonics,* v. 10, p. 659–671.

Craddock, J., Jackson, M., van der Pluijm, B. A., and Versical, R. T., 1993, Regional shortening fabrics in eastern North America: Far-field stress transmission from the Appalachian-Ouachita orogenic belt, *Tectonics,* v. 12, p. 257–264.

Gurnis, M., 1992, Rapid continental subsidence following the initiation and evolution of subduction, *Science,* v. 255, p. 1556–1558.

Howell, P. D., and van der Pluijm, B. A., 1990, Early history of the Michigan basin: Subsidence and Appalachian tectonics, *Geology,* v. 18, p. 1195–1198.

Karner, G. D., 1986, Effects of lithospheric in-plane stress on sedimentary basin stratigraphy, *Tectonics,* v. 5, p. 573–588.

Lambeck, K., 1983, The role of compressive forces in intracratonic basin formation and mid-plate orogenies, *Geophysical Research Letters,* v. 10, p. 845–848.

Marshak, S., and Paulsen, T., 1996, Midcontinent fault and fold zones, USA: A legacy of Proterozoic intracratonic extensional tectonism? *Geology,* v. 24, p. 151–154.

Park, R. G., and Jaroszewski, W., 1994, Craton tectonics, stress and seismicity, *in* Hancock, P. L., ed., *Continental deformation:* Oxford, Pergamon Press, p. 200–222.

Paulsen, T., and Marshak, S., 1995, Cratonic weak zone in the U.S. continental interior: The Dakota-Carolina corridor, *Geology,* v. 22, p. 15–18.

Quinlan, G. M., and Beaumont, C., 1984, Appalachian thrusting, lithospheric flexure, and the Paleozoic stratigraphy of the eastern interior of North America, *Canadian Journal of Earth Sciences,* v. 21, p. 973–996.

Sloss, L. L., 1963, Sequences in the cratonic interior of North America, *Geological Society of America Bulletin,* v. 74, p. 93–114.

Appendix A
Spherical Projections

Spherical projections are primarily used in structural geology to present orientation data for structural elements (such as bedding planes and hinge lines) and crystallographic orientation data (such as c-axis orientations) in two-dimensional space. Generally, we use only the lower hemisphere, which can be imagined as slicing the Earth in half along a plane containing the poles (i.e., a meridian). The Earth's lines of latitude and longitude are projected in this sectional plane, which produces a gridded net.

On a spherical projection, a plane appears as an arc. To picture this, imagine that the half-sphere represented by the spherical projection is a bowl. Now pass your hand through a point in space that represents the center of the sphere, so that it intersects the surface of the bowl. The trace of the intersection between the plane and the bowl is a curved line. A line is represented as a point in spherical projection. To picture this, imagine passing your finger through a point in space in the center of the sphere to where it intersects the surface of the bowl.

In the equal-area net (or Schmidt net; A/a), the projection is such that the lines of latitude and longitude become elliptical arcs. The main advantage of this type of spherical projection is that area of a $1° \times 1°$ grid segment does not change with position on the net; a $1° \times 1°$ grid segment occupies the same area at the center of the net as it does at the edge (hence the name equal-area net). Thus, the equal-area net is particularly useful to analyze the distribution of spatial data. In contrast, the equal-angle net (or Wulff net; A/b) projects lines of latitude and longitude as segments of circular arcs. As a consequence, the area size varies with position on the net. The main advantage of the equal-angle net lies in the fact that angular relationships are preserved (hence its name), which also means that circular elements are not distorted in this projection (as opposed to the equal-are projection where circles become ellipses).

The procedures for plotting planar and linear elements are essentially the same for both the equal-area and the equal-angle nets, and you should consult a laboratory manual for step-by-step instructions. In addition, a spate of projection programs is available for personal computers, which also allow sophisticated data analysis methods (such as data contouring and clustering analysis).

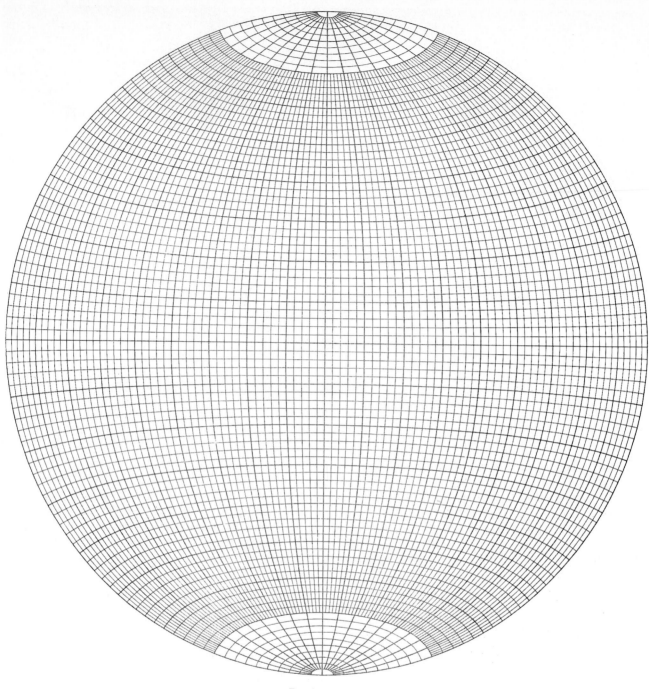

Equal-area net
(a)

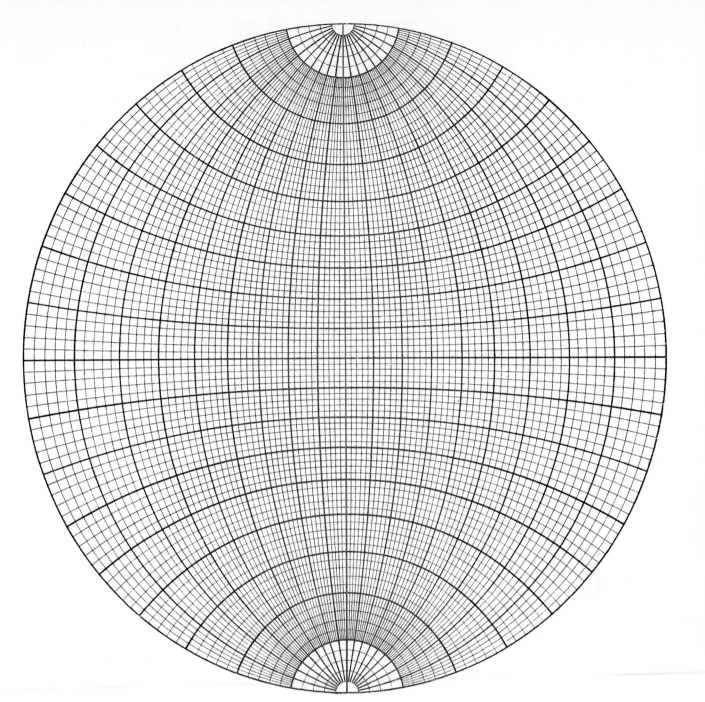

Equal-angle net
(b)

Appendix B
Geologic Time Scale

Eon	Era	Period	Age at boundary
Phanerozoic	Cenozoic	Quaternary	2 Ma
		Tertiary	66 Ma
	Mesozoic	Cretaceous	144 Ma
		Jurassic	208 Ma
		Triassic	245 Ma
	Paleozoic	Permian	286 Ma
		Carboniferous (Pennsylvanian and Mississippian)	360 Ma (320 Ma and 360 Ma)
		Devonian	408 Ma
		Silurian	438 Ma
		Ordovician	505 Ma
		Cambrian	545 Ma
Proterozoic	Neoproterozoic		1000 Ma
	Mesoproterozoic		1600 Ma
	Paleoproterozoic		2500 Ma
Archean	Late		3000 Ma
	Middle		3400 Ma
	Early		4000 Ma (~ age of oldest rock)

Ages from a 1983 compilation for the Decade of North American Geology (DNAG) project of the Geological Society of America, with slight modification in older ages.

Credits

PHOTOS

Chapter 1
Figure 1.1: The Royal Collection © Her Majesty Queen Elizabeth II; **Figure 1.9:** Steve Marshak

Chapter 2
Figure 2.1: John G. Dennis; **Figure 2.3b:** Steve Marshak; **Figure 2.4:** © The McGraw-Hill Companies, Inc./Doug Sherman, photographer; **Figure 2.5:** Steve Marshak; **Figures 2.6, 2.7:** Ben van der Pluijm; **Figure 2.8:** Steve Marshak; **Figure 2.10:** Ben van der Pluijm; **Figure 2.11:** Steve Marshak; **Figure 2.13:** Ben van der Pluijm; **Figure 2.14:** Steve Marshak; **Figure 2.15:** Courtesy Henry McQuillan; **Figure 2.20:** Steve Marshak; **Figure 2.21:** Ramsay & Herber, Fig. 3.15, Academic Press, 1983; **Figure 2.22:** Steve Marshak; **Figure 2.23:** J. C. Crowell; **Figure 2.24a:** Photo by N. H. Darton, U.S. Geological Survey; **Figure 2.25:** Steve Marshak

Chapter 3
Figure 3.5a: from Daubree, G. A. 1879. *Etudes Synthetiques de Geologie Experimentale.* Paris: Dunod; **Figure 3.11 a–b:** Courtesy of Mary Lou Zoback

Chapter 4
Figure 4.1: Reproduced by permission of the Director, British Geological Survey © NERC; **Figures 4.2, 4.21:** Ben van der Pluijm; **Figure 4.25 a–b:** Heim, A. 1978 *Untersuchungen liber den Mechanismus der Gebirgsbildung.* Basle: Schwabe; **Figure 4.27:** Ben van der Pluijm

Chapter 6
Figure 6.1a: Steve Marshak; **Figure 6.1b:** Ronadh Cox; **Figure 6.17:** Steve Marshak

Chapter 7
Figure 7.1a: U.S. Geological Survey; **Figure 7.1b:** John G. Dennis; **Figures 7.1c, 7.2 a–b, 7.5a:** Steve Marshak; **Figure 7.7:** © David McGeary; **Figure 7.22a:** Steve Marshak; **Figure 7.26:** Ramsay & Herber, Fig. 13.18, Academic Press, 1983; **Figure 7.28:** Courtesy of the Royal Canadian Air Force

Chapter 8
Figure 8.2a: Carol Simpson; **Figure 8.2b:** A. Keith, U.S. Geological Survey; **Figure 8.2c:** © C. C. Plummer; **Figure 8.15a:** Steve Marshak; **Figure 8.15b:** Ben van der Pluijm; **Figure 8.16:** Jerry Magloughlin; **Figures 8.17, 8.18a, 8.22, 8.23:** Steve Marshak

Chapter 9
Figure 9.1: U.S. Coast Guard; **Figure 9.2:** from Griggs, D. T., and Handin, J. Observations of fracture, a hypothesis of earthquakes. *Rock Deformation, a Symposium. Geol. Soc. Am. Mem.* '79:347–364, 1960; **Figure 9.7:** Ben van der Pluijm; **Figure 9.11:** Steve Marshak; **Figures 9.15, 9.21 a–d:** Ben van der Pluijm; **Figure 9.22:** from Phakey, P., Dollinger, G., and Christie, J. M. *Am. Geophys. Union, Geophys. Monogr.* 16:117–138, 1972. Photomicrograph courtesy J. M. Christie; **Figures 9.25, 9.26, 9.27:** Ben van der Pluijm; **Figure 9.37:** W. T. Lee, U.S. Geological Survey

Chapter 10
Figure 10.1: From Haller, J. Die Strukturelemente Ostgroenlands Zwischen 74 und 78 N. 1956. *Medd Groenland* 154:(3):153; **Figure 10.5:** © R. Y. Anderson; **Figure 10.7:** © John S. Shelton; **Figure 10.8:** Mary Hill; **Figures 10.18, 10.19:** Ben van der Pluijm; **Figure 10.20:** Ronadh Cox; **Figure 10.21:** Ben van der Pluijm; **Figure 10.22:** John G. Dennis; **Figure 10.26:** Steve Marshak; **Figure 10.37:** Sorby, H. C. On the origin of slaty cleavage. *Edinburgh New Philosophical Journal* 55:137–148, 1853; **Figure 10.40:** Topographical Survey of Switzerland

Chapter 11
Figure 11.4: J. B. Woodworth, U.S. Geological Survey; **Figure 11.5 a–b:** John G. Dennis; **Figure 11.8:** Steve Marshak; **Figure 11.10:** A. Keith, U.S. Geological Survey; **Figure 11.11a:** Ben van der Pluijm; **Figure 11.11b:** Reprinted from *Tectonophysics,* 78 Weber, K. in Lister et al, eds. "The Effect of Deformation on Rocks Tectono Physics" 291–306 © 1981 with kind permission of Elsevier Science—NL; **Figure 11.12:** John G. Dennis; **Figure 11.14:** R. L. Harris, Jr.; **Figure 11.16:** Steve Marshak; **Figure 11.18:** Ben van der Pluijm; **Figure 11.22:** Reproduced by permission of the Director, British Geological Survey. © NERC; **Figure 11.26:** Steve Marshak; **Figure 11.28:** John G. Dennis; **Figure 11.29:** Nei-Che Ho.

Chapter 12
Figures 12.1, 12.3, 12.5a: Ben van der Pluijm; **Figure 12.5b:** Carol Simpson; **Figure 12.9:** Ben van der Pluijm; **Figure 12.12:** Courtesy of Mikrotektoniek Collection, University of Utrecht; **Figures 12.15, 12.20, 12.25, 12.27d:** Ben van der Pluijm; **Figure 12.28:** Jerry Magloughlin

Chapter 13
Figure 13.1a: Ben van der Pluijm; **Figure 13.4a:** Mary Ellen Tuccillo; **Figure 13.6 a–c:** Cees Passchier

Chapter 14
Figure 14.1: Photo by NASA

Chapter 15
Figure 15.2: Courtesy of A. J. Sutcliffe, Geological Museum, London; **Figure 15.12:** Steve Marshak; **Figure 15.28:** Photo by NASA

Chapter 16
Figures 16.1, 16.10, 16.16: Ben van der Pluijm

Chapter 17
Figures 17.1, 17.11a: M. Scott Wilkerson; **Figure 17.11b:** Steve Marshak; **Figure 17.13:** M. Scott Wilkerson; **Figure 17.20b:** Ben van der Pluijm

Chapter 18
Figure 18.1: Robert E. Wallace and Parke D. Shavely, Jr., U.S. Geological Survey; **Figure 18.7a:** Goddard Space Flight Center, January 1976; **Figure 18.9d:** From R. E. Wilcox, T. P. Harding and D. R. Seely, 1973 reprinted by permission of the American Association of Petroleum Geologists, Imagery acquired by Westinghouse Electric Corporation, under contract from U.S. Army TOPOCOM, Ft. Belvoir, Virginia; **Figure 18.12:** John G. Dennis; **Figure 18.18 a–b:** From R. E. Wilcox, T. P. Harding and D. R. Seely, 1973 reprinted by permission of the American Association of Petroleum Geologists, Imagery acquired by Westinghouse Electric Corporation, under contract from U.S. Army TOPOCOM, Ft. Belvoir, Virginia

Chapter 19

Figure 19.4.1: The Dynamic Planet, U.S. Geological Survey (map)

LINE ART
Chapter 1

Figure 1.2: Source: J. J. Scheuchzer, 1716, *Nature-Historie des Schwetzerlandes.* Part 1: Beschreibung der Elementen, Grenzen und Bergen des Schwetzerlandes: Bodmer, Zurich. **Figure 1.3:** Source: G. P. Scrope, 1825, *Considerations on Volcanoes,* W. Philips, p. 270. **Figure 1.4:** After B. Isacks, et al., 1968, "Seismology and the New Global Tectonics" in *Journal of Geophysical Research,* 73, pp. 5855–5899, American Geophysical Union.

Chapter 2

Figure 2.2: From A. H. Bouma, 1962, *Sedimentology of Some Flysch Deposits,* Elsevier Science Publishers B.V., Amsterdam. Used by permission of the author. **Figure 2.17:** Redrawn from W. E. Galloway, et al., *Atlas of Major Texas Reservoirs,* Bureau of Economic Geology, University of Texas at Austin, 1983. Used by permission. **Figure 2.19:** H. Closs, 1936, *Einführung in die Geologie,* Gebr. Borntraeger Publishers Berlin · Stuttgart, Germany. **Figure 2.24b:** Adapted from E. M. Shoemaker, 1960, "Penetration mechanics of high velocity meteorites," *21st Int. Geol. Congress,* Norden, v. 18.

Chapter 3

Figure 3.2: Source: Modified from Hobbs, Means and Williams, *An Outline of Structural Geology,* p. 7, copyright © John Wiley & Sons, Inc., New York. **Figure 3.10:** Source: Modified from Hafner, in *Geological Society of America Bulletin,* vol. 62, pp. 373–398, 1951. **Figure 3.11:** M. L. Zoback, "First and second order patterns of stress in the lithosphere: the world stress map project" in *Journal of Geophysical Research,* pp. 11703–11728, copyright 1992 by the American Geophysical Union. Courtesy of M. L. Zoback. **Figure 3.12:** Source: G. Ranalli, *Rheology of the Earth,* copyright 1986 Allen and Unwin, Boston, figs. 12.1 and 12.2.

Chapter 4

Figure 4.14: From J. G. Ramsay and M. I. Huber, "The techniques of modern structural geology, vol. 1, *Strain Analysis,* figs. 11.9 and 11.0, copyright © 1983 Academic Press, London. Reprinted by permission. **Figure 4.16:** C. Richter and B. A. van der Pluijm. **Figure 4.18:** Source: E. Cloos, "Oolite Deformation in the South Maintain Fold, Maryland," in *Geological Society of America Bulletin,* 58 pp. 843–918, 1947. **Figure 4.24:** Modified with permission from J. G. Ramsay, *Folding and Fracturing of Rocks,* copyright 1967 McGraw-Hill, Inc. All Rights Reserved. **Figure 4.26:** From J. G. Ramsay and M. I. Huber, 1983, *The*

Techniques of Modern Structural Geology, Vol. 1: Strain Analysis, Academic Press, London. Reprinted by permission. **Figure 4.28:** Reprinted from *Journal of Structural Geology,* Vol. 15, C. Richter, B. A. van der Pluijm, and B. A. Housen, "The quantification of crystallographic preferred orientation using magnetic anisotrophy," Pages 113–116, Copyright 1993, with permission from Elsevier Science Ltd., The Boulevard, Langford Lane, Kidlington OX5 1 GB, UK. **Figure 4.29:** J. P. Craddock, et al., *Tectonics,* 12:257–264, no. 1, 1993, copyright by the American Geophysical Union. **Figure 4.30:** O. A. Pfiffner and J. G. Ramsay, *Journal of Geophysical Research,* 87:311–321, 1982, copyright by the American Geophysical Union.

Chapter 5

Table 5.3: Source: R. D. Hatcher, Jr., *Structural Geology—Principles, Concepts, and Problems,* 2nd Edition, 1995, Prentice-Hall. **Table 5.5:** Sources: Several sources, including D. L. Turcotte and G. Schubert, 1982, *Geodynamics—Applications of Continuum Physics to Geological Problems,* John Wiley & Sons, NY. **Figure 5.5:** Source: G. Ranalli, *Rheology of the Earth,* copyright 1987 Allen and Unwin, Boston. **Figure 5.6:** From Handin, John W., "Strength and Ductility" in *Geological Society of America Memoir 97,* pp. 223–290, copyright 1966 John W. Handin. **Figure 5.7:** Source: C. T. Walker and John G. Dennis, "Explosive phase transitions in the mantle," *Nature,* vol. 209, pp. 182–183, 1966 Macmillan Magazines, Ltd. **Figure 5.8:** After H. C. Heard, "Transition from brittle fracture to ductile flow in Solenhogen limestone as a function of temperature, confining pressure and interstitial fluid pressure," *Geological Society of America Memoir 79,* pp. 193–226, 1960. **Figure 5.9:** From F. A. Donath, "Some information squeezed out of rock," *American Scientist* vol. 58, pp. 54–72, 1970. Used by permission. **Figure 5.10:** After H. C. Heard, "Transition from brittle fracture to ductile flow in Solenhogen limestone as a function of temperature, confining pressure and interstitial fluid pressure," *Geological Society of America Memoir 79,* pp. 193–226, 1960. **Figure 5.11:** After D. T. Griggs and J. Hardin, in *Rock deformation—A Symposium,* GSA Memoir 79, 282 pp., 1960. **Figure 5.12:** Source: H. C. Heard, "Effect of large changes in strain rate in the experimental deformation of Yule marble: in *Journal of Geology,* 71:162–195, University of Chicago, 1963. **Figure 5.13:** Source: H. C. Heard and C. B. Raleigh, "Steady-state flow in marble at 500 to 800 C," *GSA Bulletin,* 83:935–956, 1972. **Figure 5.14:** Sources: (a) L. H. Robinson, "The effect of pore and confining pressure on the failure process in sedimentary rock," *Colorado School Mines Quarterly* 50:177–199, 1959; (b) F. Donath, "Some information squeezed out of rock," *American Scientist* 58:54–72, 1970. **Figure 5.15:** Source: D. T. Griggs, 1967, "Hydrolitic weakening of quartz and other silicates,"

Geophysical Journal 14:19–31, Royal Astronomical Society, Oxford, England. **Figure 5.16:** After John W. Handin, "Strength and Ductility" in *Geological Society of America Memoir 97,* pp. 223–290, 1966. **Figure 5.18:** Source: Rutter, *Geology Today,* 9:61–65, 1993. **Figure 5.19:** Reprinted from *Journal of Structural Geology,* Vol. 2, J. G. Ramsay, "Shear zone geometry: a review," Pages 83–99, Copyright 1980, with permission from Elsevier Science Ltd, The Boulevard, Langford Lane, Kidlington OX5 1GB, UK.

Chapter 6

Figure 6.4d–g: Source: T. Engelder, *Stress Regimes in the Lithosphere,* Princeton University Press, 1993. **Figure 6.16:** From C. H. Scholz, 1990, *The Mechanics of Earthquakes and Faulting.* Copyright © 1990 Cambridge University Press. Reprinted with the permission of Cambridge University Press.

Chapter 7

Figure 7.4: Adapted from *Tectonophysics,* vol. 104, D. Bahat and T. Engelder, "Surface morphology on cross-fold joints of the Appalachian Plateau, New York, and Pennsylvania," pp. 299–313, Copyright 1984, with kind permission of Elsevier Science—NL, Sara Burgerhartstraat 25, 1055 KV Amsterdam, The Netherlands. **Figure 7.6c:** After Pollard and Aydin, "Progress in understanding jointing over the last century," *Geological Society of America Bulletin,* vol. 100, pages 1181–1204, 1988. **Figure 7.13a:** T. Engelder and P. Geiser, *Journal of Geophysical Research,* vol. 85, pp. 6319–6341, 1980, copyright by the American Geophysical Union. **Figure 7.13c** From John G. Dennis, *Structural Geology: An Introduction.* Copyright © 1987 The McGraw-Hill Companies, Inc. All Rights Reserved. Reprinted by permission.

Chapter 8

Figure 8.9c: Based on J. E. Gill, "Fault Nomenclature," *Royal Society of Canada Transactions,* 35:71–85, 1941. **Figure 8.13:** Source: J. G. Ramsay and M. I. Huber, *The Techniques of Modern Structural Geology,* Vol. 2: Folds and Fractures, copyright 1987 Academic Press Ltd., London, p. 507. **Figure 8.14c:** Source: P. A. Cowie and C. H. Scholz, "Displacement—length scaling relationships for faults: data synthesis and conclusion," *Journal of Structural Geology* 14:1149–1156, 1992 Pergamon Press. **Figure 8.18c:** F. Arthaud and M. Mattauer, 1969, *Bulletin Société Geologiqué de France,* Series 7, Vol. 11, pp. 738–744. Reprinted by permission of Société Geologiqué de France, 77 rue Claude Bernard, Paris. **Figure 8.25:** Source: Modified from C. H. Scholz, 1990, *The Mechanics of Earthquakes and Faulting,* p. 29, Cambridge University Press, Cambridge, and other sources. **Figure 8.27:** After

M. K. Hubbert, "Mechanical Basis for Certain Familiar Geologic Structures" in *Geological Society of America Bulletin* pp. 355–372, 1951. **Figure 8.29:** Source: C. H. Scholz, *The Mechanics of Earthquakes and Faulting,* 1990 Cambridge University Press, fig. 3.18, p. 126. **Figure 8.31:** C. H. Scholz, "Microfracturing and the inelastic deformation of rock in compression," *Journal of Geophysical Research* 73:1417–1432, 1968 American Geophysical Union.

Chapter 9

Figure 9.10a: From H. J. Frost and M. F. Ashby, 1982, *Deformation-mechanism maps.* The plasticity and creep of metals and ceramics, Pergamon Press, 1982. Used by permission of the author. **Table 9.1:** Source: Data from H. R. Wenk (ed.), 1985, *Preferred Orientation in Deformed Metals and Rocks: An Introduction to Modern Texture Analysis,* 1985 Academic Press, Orlando. **Figure 9.16:** Source: Schedl and van der Pluijm, *Journal of Geological Education* 36:111–121, copyright 1988 National Association of Geology Teachers, Inc. **Figure 9.18:** After R. H. Groshong, Jr., in 1972, "Strain calculated from twinning in calle," *GSA Bulletin,* v. 83, pp. 2025–2038. Used by permission. **Figure 9.28:** Sources: (a) J. P. Poirier, *Creep in Crystals: High-temperature Deformation Processes in Metals, Ceramics and Minerals,* copyright 1985 Cambridge University Press, Cambridge; (b) J.L. Urai, et al., 1986, "Dynamic recrystallization of minerals" in *Mineral and Rock Deformation: Laboratory Studies* (The Paterson Volume), ed. by Hobbs and Heard, Geophysical Monograph 36, pp. 161–200. **Figure 9.29:** After S. White, "The effects of strain on the microstructures, fabrics, and deformation mechanisms in quartzites," *Philosophical Transactions of the Royal Society of London,* series A., v. 283, pp. 69–86. **Figure 9.31:** After M. F. Ashby and R. A. Verrall, "Micromechanisms of flow and fracture, and their relevance to the rheology of the upper mantle," *Philosophical Transactions of the Royal Society of London,* series A, v. 288, pp. 59–95. **Figure 9.32:** Source: T. G. Langdon, pp. 219–232 in H. R. Wenk, ed., *Preferred Orientation in deformed metals and rocks: An introduction to modern texture analysis,* Academic Press, Orlando, 1985. **Figure 9.33:** Source: E. H. Rutter, "The kinetics of rock deformation by pressure solution, *Philosophical Transactions of the Royal Society of London,* v. 283, pp. 203–219. **Figure 9.34:** Source: E. H. Rutter, "The kinetics of rock deformation by pressure solution, *Philosophical Transactions of the Royal Society of London,* v. 283, pp. 203–219. **Figure 9.35:** Source: M. F. Ashby and R. A. Verrall, "Micromechanisms of flow and fracture, and their relevance to the rheology of the upper mantle," *Philosophical Transactions of the Royal Society of London,* Series A, v. 288, pp. 59–85, 1978. **Table 9.5:** Sources: Rutter (1974); Schmid, et al. (1977); and S. M. Schmid, 1982, "Laboratory experiments on rheology and deformation

mechanisms in calcite and their application to studies in the field," *Mitt. Geol. Inst. ETH Zurich* 241, pp. 1–105.

Chapter 10

Figure 10.3: From F. J. Turner and L. E. Weiss, *Structural Analysis of Metamorphic Tectonites.* Copyright © 1963 The McGraw-Hill Companies, Inc., All Rights Reserved. Reprinted by permission. **Figure 10.6e–f:** Source: G. J. Borradaile, 1976, " 'Structural facing' (Shackleton's rule) and the Palaeozoic rocks of the Malagmide complex near Velez Rubio, SE Spain," *Proc. Kon. Nederl. Akademie Wetenschappen,* v. 7–9, pp. 330–336. **Figure 10.10:** From Richard, M. J., 1971, "A classification diagram for fold orientations," *Geological Magazine,* vol. 108, pp. 23–26. Copyright 1971 Cambridge University Press. Reprinted by permission of Cambridge University Press. **Figure 10.11:** From J. G. Ramsay, "The Geometry and Mechanics of Formation of Similar Type Folds," in *Journal of Geology,* vol. 70, pp. 309–327. Copyright © 1962 by The University of Chicago. Used by permission. **Figure 10.12:** Modified with permission from J. G. Ramsay, *Folding and Fracturing of Rocks.* Copyright © 1967 The McGraw-Hill Companies, Inc. All Rights Reserved. **Figure 10.24:** From J. G. Huber and M. I. Huber, *The Techniques of Modern Structural Geology,* Vol. 2: Folds and Fractures, Academic Press, London. Reprinted by permission. **Figure 10.25:** Reprinted from *Journal of Structural Geology,* Vol. 2, R. L. Thiessen and W. D. Means, "Classification of fold interference patterns: a reexamination," Pages 311–316, Copyright 1980, with permission from Elsevier Science Ltd, The Boulevard, Langford Lane, Kidlington OX5 1GB, UK. **Figure 10.30:** Source: J. B. Curie, et al., "Development of folds in sedimentary strata," *Geological Society of America Bulletin,* v. 73, pp. 655–674. **Figure 10.31:** From J. H. Dietrich, *Canadian Journal of Earth Sciences,* v. 7, figure 5–7; copyright © 1970 National Research Council of Canada. Reprinted by permission of NRC Research Press. **Figure 10.32:** From John G. Ramsay, *Folding and Fracturing of Rocks.* Copyright © 1967 The McGraw-Hill Companies, Inc. All Rights Reserved. Reprinted by permission. **Figure 10.33:** From John G. Ramsay, *Folding and Fracturing of Rocks.* Copyright © 1967 The McGraw-Hill Companies, Inc. All Rights Reserved. Reprinted by permission. **Figure 10.36:** From John G. Ramsay, *Folding and Fracturing of Rocks,* Copyright © 1967 The McGraw-Hill Companies, Inc. All Rights Reserved. Reprinted by permission. **Figure 10.37:** Source: H. C. Sorby, "On the Origin of Slaty Cleavage," in *Edinburgh New Philosophical Journal,* vol. 55, pp. 137–148, 1853. **Figure 10.38:** Reprinted from *Tectonophysics,* vol. 106, P. Y. Hudleston and T. B. Holst, "Strain analysis and fold shape in a limestone layer and implications for later rheology," pp. 321–347, 1984, with kind

permission of Elsevier Science—NL, Sara Burgerhartstraat 25, 1055 KV Amsterdam, The Netherlands.

Chapter 11

Figure 11.20: Source: Etheridge, Wall, and Vernon, *Journal of Metamorphic Geology,* vol. 1, pp. 205–226. **Figure 11.23:** Adapted from G. J. Borradaile, "Transected folds; a study illustrated with examples from Canada and Scotland," *Geological Society of America Bulletin,* vol. 89, pp. 481–493, 1978. **Figure 11.24:** Adapted with permission from J. G. Ramsay and M. I. Huber, *The Techniques of Modern Structural Geology,* Vol. 1: Strain Analysis, pp. 236–280. Copyright © 1987 Academic Press.

Chapter 12

Figure 12.2: Sources: Based on various sources, including R. H. Sibson, "Fault rocks and fault mechanisms," *Journal of the Geological Society of London,* 133:190–213; and C. H. Scholz, 1990, *The Mechanics of Earthquakes and Faulting,* Cambridge University Press, Cambridge. **Figure 12.4a:** Natural Resources Canada, Courtesy of the Geological Survey of Canada (Hanmer, S. and Passchier, C., 1991, "Shear-sense indicators: a review," Paper 90–17. Reproduced with the permission of the Minister of Public Works and Government Services Canada, 1996. **Figure 12.4b:** Source: S. Marshak and G. Mitra, 1988, *Basic Methods of Structural Geology,* Prentice-Hall, Upper Saddle River, NJ. **Figure 12.8:** Reprinted from *Journal of Structural Geology,* Vol. 6, G. S. Lister and A. W. Snoke, "S-C mylonites," Pages 617–638, Copyright 1984, with permission from Elsevier Science Ltd, The Boulevard, Langford Lane, Kidlington OX5 1GB, UK. **Figure 12.16:** Source: C. Simpson and D. G. De Paor, 1993, "Strain and kinematic analysis in general shear zones," *Journal of Structural Geology,* v. 15, pp. 1–20. **Figures 12.22 and 12.23:** From R. D. Law, in *Deformation Mechanisms, Rheology and Tectonics,* ed. by R. J. Knipe and E. H. Rutter, Geological Society Special Publication, v. 54, pp. 335–352, 1990. Used by permission. **Figure 12.24:** Source: Williams, 1982, *Geologische Rundschau,* v. 72, p. 602. **Figure 12.26:** Source: Hudleston and Lan, 1993, *Journal of Structural Geology,* v. 15, figs 7b, 7c. **Figure 12.29:** Modified with permission from Hudleston, 1986, *Journal of Geological Education,* v. 34, p. 24. Copyright © 1986 National Association of Geology Teachers.

Chapter 13

Figure 13.4b–d: From M. E. Tuccillo, et al., 1992, "Thermobarometry, geosynchronology, and the interpretation of P-T-t data. . ." *Journal of Petrology,* 33:1225–1259, Oxford University Press, Oxford, UK. By permission of Oxford University Press. **Figure 13.5:** From E. C. Zwart, 1962, "On the determination of polymetamorphic mineral associations. . . ," *Geologische Rundschau,* vol. 52, pp. 38–65.

Used by permission of Springer-Verlag GmbH & Co. KG, Germany. **Figure 13.7:** After Spry, 1963, *Journal of Petrology,* v. 4, pp. 211–222. Used by permission of Oxford University Press, Oxford. **Figure 13.9:** Source: Gunter Faure, *Principles of Isotope Geology,* 2d ed., fig. 8.2. Copyright © 1986 John Wiley & Sons, New York. **Figure 13.10:** Source: Gunter Faure, *Principles of Isotope Geology,* 2d ed., fig. 8.5. Copyright © 1986 John Wiley & Sons, New York. **Figure 13.11:** Courtesy of M. A. Cosca.

Chapter 14

Figure 14.3: R. S. Hart, et al., *Journal of Geophysical Research,* vol. 82, pp. 1647–1654, 1977, copyright by the American Geophysical Union. **Figure 14.5b:** Modified from *Journal of Volcanology and Geothermal Research,* vol. 4, M. Y. Dewitt and C. Stern, "Pillow Talk," pp. 55–80, Copyright 1978 with kind permission of Elsevier Science—NL, Sara Burgerhartstraat 25, 1055 KV Amsterdam, The Netherlands. **Figure 14.7:** Source: Mooney and Meissner, *EOS Transactions,* 72, pp. 537–541, 1991. **Figure 14.8:** Dziewonski, *Journal of Geophysical Research,* 89, pp. 5929–5952, 1984, copyright by the American Geophysical Union. **Figure 14.9:** From P. Keary and F. J. Vine, *Global Tectonics.* Copyright © 1990 Blackwell Scientific Publications Ltd., Oxford, UK. Reprinted by permission. **Figure 14.10:** From C. M. R. Fowler, 1990, *The Solid Earth: An Introduction to Global Geophysics,* © 1990 Cambridge University Press. Reprinted by permission of Cambridge University Press. **Figure 14.12:** Source: P. Keary and F. J. Vine, 1990, *Global Tectonics,* Blackwell Scientific Publications, Oxford, fig. 5.5. **Figure 14.13:** Modified from C. M. R. Fowler, 1990, *The Solid Earth: An Introduction to Global Geophysics,* © 1990 Cambridge University Press. Reprinted by permission of Cambridge University Press. **Figure 14.14:** Modified from C. M. R. Fowler, 1990, *The Solid Earth: An Introduction to Global Geophysics,* © 1990 Cambridge University Press. Reprinted by permission of Cambridge University Press. **Figure 14.15:** After A. Cox and R. B. Hunt, *Plate Tectonics: How it Works,* copyright © 1986 Blackwell Scientific Publications, Oxford. Used by permission. **Figure 14.16:** Courtesy of Henry N. Pollard from H. N. Pollack, et al., *Reviews of Geophysics,* 31, pp. 267–280, 1993, copyright by the American Geophysical Union.

Chapter 15

Figure 15.1: Modified from B. U. Haq and W. B. Van Eysinga, *Geological Time Table,* 4th ed., 1987 Elsevier Science Publishers, Amsterdam, The Netherlands. Used by permission of the author. **Figure 15.4c:** From P. A. Ziegler, "Faulting and graben formation in western and central Europe" in *Royal Society of London Philosophical Transactions,* pp. 13–143, fig. 13, copyright 1982. Used by

permission. **Figure 15.7:** Source: Based on figures in Gibbs, 1984, "Structural evolution of extensional basin margins," *Journal of the Geological Society of London,* v. 141, pp. 609–620, copyright 1964. **Figure 15.9:** From P. A. Ziegler, "Graben Formation in Europe," *Royal Society of London Philosophical Transactions* A305, pp. 113–143, 1982. Used by permission. **Figure 15.10:** Modified from W. Bosworth, "A model for three-dimensional evolution of continental rift basins, north-east Africa" in *Geologische Rundschau,* v. 83, no. 4, pp. 671–688. Copyright © 1994 Springer-Verlag GmbH & Co., KG, Germany. Used by permission. **Figure 15.13:** Redrawn from *Journal of Structural Geology,* Vol. 11, G. S. Lister and G. A. Davis, "The origin of metamorphic core complexes and detachment faults during Tertiary continental extension in the northern Colorado River Basin," Copyright 1989, with permission from Elsevier Science Ltd, The Boulevard, Langford Lane, Kidlington OX5 1GB, UK. **Figure 15.14:** Source: Based from Rosendahl, "Architecture of continental rifts with special reference to East Africa," *Annual Review of Earth and Planetary Sciences,* v. 15, pp. 445–503, fig. 3. **Figure 15.16:** After W. Bosworth, 1994, "A model for the three-dimensional evolution of continental rift basins, north-east Africa," *Geologische Rundschau,* v. 83, pp. 671–688, Springer-Verlag. **Figure 15.20a:** Source: Data from W. W. Atwood, *The Physiographic Provinces of North America,* 1940 Ginn and Company. **Figure 15.20b:** Source: Data from T. H. Dixon, et al., 1989, "Topographic and volcanic asymmetry around the Red Sea: Constraints on rift models," *Tectonics,* 8:1193–1216. **Figure 15.22:** K. C. MacDonald, "Mid-ocean ridges: fine scale tectonic, volcanic and hydrothermal processes with the plate boundary zone." With permission from the *Annual Review of Earth and Planetary Sciences,* Vol. 10, 1982, by Annual Reviews Inc. **Figure 15.24:** Modified from A. W. Erxleben and G. Carnahan, 1983, AAPG Studies in Geology Series, No. 15, Vol. II, 1983, *Detached Sediments in Extensional Provinces/Growth Faults: Slide Ranch Area, Starr County, Texas.* Reprinted by permission of the American Association of Petroleum Geologists, Courtesy Tenneco Oil Co.

Chapter 16

Figure 16.3: Modified from W. R. Dickinson and D. R. Seely, 1979, AAPG Bulletin, Vol. 63, No. 1, pp. 2–31, *Stratigraphy and Structure of Forearc Regions.* Reprinted by permission of the American Association of Petroleum Geologists, after Ernst, 1970; Grow, 1973; Marlow et al., 1973. **Figure 16.5a:** Source: Based on Turcotte et al., 1978, "An elastic-perfectly plastic analysis of the bending of the lithosphere at a trench," *Tectonophysics,* v. 47, p. 202, fig. 5. **Figure 16.7:** After E. R. Oxburgh and D. L. Turcotte, "Thermastructure of Island Arcs," *Geological Society of America Bulletin* 81:1665–1688, 1970. Used by permission.

Figure 16.8: From Thornburg and Kulm, *Journal of Sedimentary Petrology,* v. 57, pp. 55–74, fig. 1. Reprinted by permission of SEPM, Society for Sedimentary Geology. **Figure 16.9:** Moore, G. F., Karig, D. E., Shipley, T. H., Taira, A., Stoffa, P. L., and Wood, W. T., 1991. Structural framework of the ODP leg 131 area, Nankai Trough. *In* Taira, A., Hill, I., Firth, J., et al., *Proc. ODP, Initial Reports,* 131: College Station, TX (Ocean Drilling Program), 15. **Figure 16.24:** From P. Tapponnier, et al., "Propagating extrusion tectonics in Asia," *Geology* 10:611–616, Geological Society of America, 1982. Used by permission of the author. **Figure 16.25:** Reprinted with permission from *Nature,* Paul Tapponnier and Peter Molnar, v. 264. Copyright 1976 Macmillan Magazines Limited. **Figure 16.27:** Reprinted with permission from *Nature,* Peter J. Coney, et al., "Cordilleran suspect terranes," v. 288. Copyright 1980 Macmillan Magazines Limited.

Chapter 17

Figure 17.2: After E. B. Bailey, *Tectonic Essays, Mainly Alpine,* copyright © 1935 Oxford University Press, Oxford. Reprinted by permission of Oxford University Press. **Figure 17.4:** From S. Marshak and J. Tabor, "Structure of the Kingston orocline, in the Appalachian foldthrust belt, New York," *GSA Bulletin,* v. 101, pp. 683–701. Used by permission. **Figure 17.8b:** S. Mitra, 1986, AAPG Bulletin, Vol. 63, No. 1, *Duplex Structures and Imbricate Thrust Systems: Geometry, Structural Position, and Hydrocarbon Potential.* Reprinted by permission of the American Association of Petroleum Geologists. **Figure 17.9:** S. E. Boyer and D. Elliott, 1982, AAPG Bulletin, Vol. 66, No. 9, 1982, *Thrust Systems.* Reprinted by permission of the American Association of Petroleum Geologists, modified from Boyer, 1978. **Figure 17.10:** From B. Willis, *Geologic Structures.* Copyright © 1923 The McGraw-Hill Companies. All Rights Reserved. Reprinted by permission. **Figure 17.12:** From J. Suppe, 1983, "Geometry and kinematics of fault-bend folding," *American Journal of Science,* v. 283, pp. 684–721. Reprinted by permission of *American Journal of Science.* **Figure 17.14:** J. Suppe and D. Medwedeff, 1990, "Geometry and kinematics of fault-propagation folding," *Eclogae Geologicae Helvetiae,* v. 83, pp. 409–454. Used by permission. **Figure 17.15:** H. P. Laubscher, "Die Fernschubleypothese der Jurafalting" in *Eclogae Geologiae Helvetiae,* 1961 Swiss Geological Society. Used by permission. **Figure 17.17:** From D. Elliott, "The energy balance and deformation mechanisms of thrust sheets," *Philosophical Transactions of the Royal Society,* v. A283, pp. 289–312, 1976. Used with permission. **Figure 17.25:** S. E. Boyer and D. Elliott, 1982, AAPG Bulletin, Vol. 66, No. 9, 1982, *Thrust Systems.* Reprinted by permission of the American Association of Petroleum Geologists, modified from Douglas, 1952.

Chapter 18

Figure 18.8: Modified from A. G. Sylvester, 1988, "Slip-strike faults," *Geological Society of America Bulletin,* v. 100, pp. 1666–1703. Used by permission of the author. **Figure 18.9b** A. G. Sylvester and R. R. Smith, 1982, AAPG Bulletin, Vol. 60, No. 12, *Tectonic Transpression and Basement-Controlled Deformation in San Andreas Fault Zone, Salton Trough, California.* Reprinted by permission of the American Association of Petroleum Geologists, adapted from Harding, 1974, p. 1291. **Figure 18.10:** Source: M. R. Nelson and C.H. Jones, "Paleomagnetism and crustal rotations along a shear zone, Las Vegas Range," *Tectonics,* v. 6, pp. 13–33, 1987, copyright by the American Geophysical Union.

Figure 18.11: After *Tectonophysics,* vol. 21, R. Freund, "Kinematics of transform and transcurrent faults," pp. 93–134, Copyright 1974 Elsevier Science—NL, Sara Burgerhartstraat 25, 1055 KV Amsterdam, The Netherlands. **Figure 18.14:** T. P. Harding and J. D. Lowell, 1979, AAPG Bulletin, Vol. 63, No. 7, *Structural Styles, Their Plate-Tectonic Habitats, and Hydrocarbon Traps in Petroleum Provinces.* Reprinted by permission of the American Association of Petroleum Geologists, adapted from unmigrated interpretation by R. F. Gregory and E. C. Lookabaugh, 1973. **Figure 18.19:** From N. H. Woodcock, 1986, *Philosophical Transactions of the Royal Society,* v. A317, 1986. Used with permission. **Figure 18.21:** Modified with permission from S. Mitra, 1988, *Geological Society of America Bulletin,* v. 100, fig. 1. **Figure 18.22:** From G. A. Davis and B. C. Burchfiel, "The Garlock Fault, an Intracontinental Transform," *GSA Bulletin,* v. 84, pp. 1407–1422, 1973. Used by permission of the author. **Figure 18.23:** I. Barany and J. A. Karson, "Basaltic bracchias of the Clipperton fracture zone (east Pacific): Sedimentation and tectonics in a fast-slipping oceanic transform," *Geological Society of America Bulletin,* v. 101, pp. 204–220, 1989. Used by permission.

Chapter 19

Chapter 19 Essays 19.1 and 19.10 are original and copyrighted to this book; Essay 19.7, "A Stratigraphic Approach to Appalachian Tectonics," by Dwight C. Bradley, is in the public domain. All other essays in the chapter are copyrighted to the individual contributors and used by permission. **Figure 19.1.2:** Source: R. Hatcher, Jr. and R. T. Williams, *Geological Society of America Bulletin,* v. 97, 1986, pp. 975–985. **Figure 19.2.1:** From Platt, in *Geological Society of America Bulletin,* vol. 97, pp. 1037–1053. Used by permission of the

author. **Figure 19.2.4:** Source: A. Buxtorf, 1916, *Die Geologie des Juragebirges,* v. 27, p. 184, Verhandl. Naturforsch. Gesell. Basel. **Figure 19.2.5:** Source: BEB Gewerkschaften Brigitta und Elwerath, Hannover, 1979. **Figure 19.2.6:** From J. G. Ramsay, 1981, "Tectonics of the Helvetic nappes" in K. R. McClay and N. J. Price, eds., *Thrust and Nappe Tectonics,* Special Publication of the Geological Society of London, no. 45, pp. 135–152, fig. 4. Used with permission. **Figure 19.2.7:** After John P. Platt, et al., 1989, *Thrusting and backthrusting in the Brian onnais domain of the Western Alps,* Special Publication of the Geological Society of London, no. 45, pp. 135–152, fig. 4. Used with permission. **Figure 19.2.8:** From Platt, in *Geological Society of America Bulletin,* vol. 97, pp. 1037–1053, 1986. Used by permission of the author. **Figure 19.3.1:** J. F. Dewey, et al., 1989, "Tectonic evolution of the India/Eurasia collision zone," *Eclogae Geologicae Helvetiae,* v. 82, pp. 683–734. Used by permission. **Figure 19.3.2:** B. C. Burchfiel and L. H. Royden, 1991, "Tectonics of Asia 50 years after the death of Emile Argand," *Eclogae Geologicae Helvetiae,* v. 84, pp. 599–629. Used by permission. **Figure 19.3.3:** L. H. Royden, "The tectonic expression of slab pull at continental convergent boundaries," *Tectonics,* v. 12, pp. 303–325, 1993, copyright by the American Geophysical Union. **Figure 19.3.4ab:** B. C. Burchfiel and L. H. Royden, 1991, "Tectonics of Asia 50 years after the death of Emile Argand," *Eclogae Geologicae Helvetiae,* v. 84, pp. 599–629. Used by permission. **Figure 19.3.5:** Source: E. Argand, 1924, *La tectonique de l'Asie,* Proc. 13th Int. Geol. Congr. Brussels 1924, pp. 171–372. **Figure 19.3.6a:** After P. Tapponnier, et al., "Propagating extrusion tectonics in Asia: New insight from simple experiments with plasticine," *Geology,* v. 10, pp. 611–616, 1982. Used by permission of the author. **Figure 19.3.6b:** From P. C. England and G. A. Houseman, "The mechanics of the Tibetan Plateau, in: Tectonic Evolution of the Himalayas and Tibet, ed. by R. M. Shackleton, J. F. Dewey, and B. F. Windley, *Philosophical Transactions of the Royal Society of London* (A), v. 327, pp. 379–420. Used by permission. **Figure 19.3.7:** Modified from P. Zhang, et al., in *Geological Society of America Bulletin,* vol. 102, pp. 1484–1498, 1990. Used by permission. **Figure 19.3.8:** B. C. Burchfiel and L. H. Royden, 1991, "Tectonics of Asia 50 years after the death of Emile Argand," *Eclogae Geologicae Helvetiae,* v. 84, pp. 599–629. Used by permission. **Figure 19.4.2:** Compiled from B. C. Burchfiel, et al., "Tectonic overview of the Cordilleran orogen in the western United

States," in B. C. Burchfiel, et al., eds., *The Cordilleran Orogen: Conterminous U.S.,* GSA, The Geology of North America, v. G–3, pp. 407–480. Used by permission of the author. **Figure 19.4.3:** After E. L. Miller and P. B. Gans, 1989, "Cretaceous crustal structure and metamorphism in the hinterland of the Sevier thrust belt, western U.S. Cordillera," *Geology,* v. 17, pp. 59–62. **Figure 19.4.4:** After P. B. Gans, et al., 1989, *Synextensional magmatism in the Basin and Range Province: A case study from the eastern Great Basin,* GSA Special Paper 233. **Figure 19.5.1:** Source: Jordan, et al., 1983, *Episodes,* vol. 1983, no. 3, pp. 20–26, International Union of Geological Sciences. **Figure 19.5.3:** Modified from Jordan, et al., 1983, *GSA Bulletin,* vol. 94, pp. 341–361. **Figure 19.6.2:** Source: K. T. Pickering and A. G. Smith (1995). **Figure 19.6.3:** Adapted by permission of the Royal Society of Edinburgh from *Transactions of the Royal Society of Edinburgh: Earth Sciences,* volume 79, part 4 (1988), pp. 361–382. **Figure 19.6.4:** Source: Pickering and Smith (1995). **Figure 19.8.2:** Redrawn from *Journal of Structural Geology,* Vol. 12, C. Powell, et al., eds., "Australasian Tectonics," Pages 553–565, Copyright 1990, with permission from Elsevier Science Ltd, The Boulevard, Langford Lane, Kidlington OX5 1GB, UK. **Figure 19.8.3:** Adapted from *Tectonophysics,* vol. 158, G. L. Clarke and R. Powell, "Basement-cover interaction in the Adelaide Foldbelt, South Australia," pp. 209–222, Copyright 1989; and from *Tectonophysics,* vol. 214, R. J. F. Jenkins and M. Sandiford, "Observations on the tectonic evolution of the southern Adelaide Fold Belt," pp. 27–36, Copyright 1992 with kind permission of Elsevier Science -NL, Sara Burgerhartstraat 25, 1055 KV Amsterdam, The Netherlands. **Figure 19.8.4:** Modified from *Tectonophysics,* vol. 158, G. L. Clarke and R. Powell, pp. 209–222, Copyright 1989; and *Tectonophysics,* vol. 214, R. J. F. Jenkins and M. Sandiford, pp. 27–36, Copyright 1992, with kind permission of Elsevier Science Publishers—NL, Sara Burgerhartstraat 25, 1055 KV Amsterdam, The Netherlands. Amsterdam. **Figure 19.8.6:** Modified from C. L. Fergusson, "Lithology and structure of the Wandilla Terrane," *Australian Journal of Earth Sciences,* vol. 40, pp. 403–414, © 1993, with permission of Blackwell Science Pty. Ltd., Australia. **Figure 19.8.7b:** Modified from *Tectonophysics,* vol. 214, C. L. Fergusson and R. A. Glen, "The Palaeozoic eastern margin of Gondwanaland. . . ," Copyright 1992 with kind permission of Elsevier Science—NL, Sara Burgerhartstraat 25, 1055 KV Amsterdam, The Netherlands. **Figure 19.10.2:** S. Marshak and T. Paulsen, *Geology,* in press, 1996, copyright by the American Geophysical Union.

Index

flow, 78–79, 180. *See also* flow of rocks; rheologic relationships
flower structure, 174, 397
flow foliation, 30
flow laws, 192–93
flow of rocks
  confining pressure, effect of, 87–88, 92–93
  deformation apparatus, 86–87
  pore-fluid pressure, effect of, 90–91, 92–93
  scaled experiments, 86
  strain rate, effect of, 89–90, 92–93
  temperature, effect of, 88–89, 92–93
  value of experiments on natural rocks, 86
  work hardening-work softening, effect of, 91–92
flow stresses, 85
fluid-assisted diffusion, 185
fluid-pressure driven structures, 5
fluid-rock ratio, 253
fluids
  and faulting, 169–70
  tensile crack growth, effect of fluids on, 118–19
flute clasts, 16
flysch, 15, 406, 410, 446, 448, 449
foam microstructure, 198
foam structure, 196
focus, earthquake, 176
fold axis, 208, 209, 211
fold enveloping surface, 278
fold generation, 221, 225–26
fold-hinge lineations, 256–57
fold interference types, 222, 223
fold nappes, 368, 379
fold profile plane, 209, 211
folds and folding
  anatomy of folded surface, 208–13
  buckle folds, 227–30, 235–36
  classification, 213–16
  defined, 207
  enveloping surface, 216–17
  fault related, 162–64
  in fault zones, 163–64
  flexural slip/flow, 231, 233, 235
  fold facing, 211–13
  fold shape
    modification, 232–34, 235
    in profile, 215–16
  fold size, 216
  fold style, 225
  fold symmetry and fold vergence, 217–18
  fold systems, 216–18
  foliations in, 253–56
  interference patterns, 222–25, 226
  kinematic models, 230–35
  mechanics, 226–30
  multilayers, 230
  neutral-surface, 231–32, 233, 234–35
  overview, 207–8, 236–37
  passive versus active, 226–27
  in selected regions. (*see* regional geology)
  sequence of events, 235–36
  shear, 232
  special fold geometries, 218–20
  superposed, 221–26
  terminology, 211
  thrust related folds, 374–79

fold shape
  modification, 232–34, 235
  in profile, 215–16
fold symmetry, 217–18
fold-thrust belts (FTB), 173
  in Appalachian Mountains, 446–47
  balanced cross sections, concept of, 385–87
  in Central Andes, 433–34
  defined, 368
  internal deformation of thrust sheets, 381–82
  in map view, 379–81
  mechanisms of development, 382–85
  overview, 366–68
  strike-slip faults in, 400, 401
  in Tasman Orogenic Belt, 453–55
  terminology, 368–69
  thrust related folds, 374–79
  thrusts and thrust systems, geometry of, 368, 369–74
fold transposition, 277–80
fold vergence, 217, 218
foliations
  classification, 241, 242
  cleavage, defined, 240, 242
  crenulation cleavage, 249–50
  C-S and C-C′ structures, 269–70
  defined, 9, 240, 241
  deflected, 272, 273
  disjunctive cleavage, 243–45
  in faults and fault zones, 253–56
  gneissic layering and migmatization, 250–52
  mylonitic foliation, 240, 252, 263, 269–70
  pencil cleavage, 245–46
  phyllitic cleavage and schistosity, 247–48, 249
  slaty cleavage, 246–47
footwall blocks, 146, 368
footwall cutoffs, 153, 368
footwall flats, 368
footwall ramps, 368
force
  defined, 36, 37–38
  relationship between stress and, 38–39
  terminology, 36
forced folds, 164
forearc basins, 345, 346, 352–53
forearc regions, 345, 346
foreland, 358, 368
foreland basins, 368
foreset beds, 15
forethrusts, 368, 370
form lineations, 256–57
fossils, 72–73
fracture, 98, 99
fracture front, 101
fracture tip, 101
fracture toughness, 113
fracture trace, 101
fracture zones, 99, 402
fracturing, 268–69
  classification based on, 5
  defined, 79, 94
frame of reference, 54
Franklin Orogenic Belt, 459, 460
Frank-Read sources, 192
friction, 110
frictional-plastic transition, 264

frictional regime, 263, 264
frictional sliding, 102, 103, 110
  classification based on, 5
  criteria, 117
  pore pressure and, 119–20
frontal ramp, 368, 370, 372

G

gabbro, 306, 337, 338
Galápagos Ridge, 337
Garlock Fault (California), 400–401, 401
general linear behavior in rheologic relationships, 84
general noncoaxial strain accumulation, 58
general shear, 58
general strain, 63
generation, fold, 221, 225–26
geobarometry, 283, 286–87
geochemistry and elemental abundances, 305
geochronology, 283, 291–95
geologic structures
  classification, 4–6
  defined, 4, 12
  terminology, 9
Geologic Time Scale, Appendix B
geology
  historical survey, 2–4
  structural, 8
  terminology, 9
geosynclinal theory, 4, 313
geotectonic theory, 313
geothermal gradient, 28, 283, 296, 298
geothermobarometry, 283, 286–88
geothermochronology, 283
geothermometry, 283, 286–87
Glarner thrust (Switzerland), 364, 367
glide planes, 186–87
gneiss, 250–52
gneissic layering, 242
gneissosity, 240
Gondwana
  in Caledonides development, 435–40, 442, 443
  and North American tectonic genealogy, 459, 460–61, 464
  in Tasman Belt development, 454, 457
  in Tibetan Plateau tectonics, 417
gouge, 265
gouge zone, 167
grabens, 173, 422
graded beds, 14–15
grain, 266
grain-boundary diffusion, 184–85
grain-boundary sliding superplasticity, 200–201
grain-size sensitive creep, 201
grain-tail complexes, 266–68
Grand Unified Theory, 37
granodiorite, 306
gravitational collapse, 340
gravitational force, 37
gravitational potential energy, 384
gravity anomalies, 305
gravity sliding, 383
gravity spreading, 23, 383
Great Valley Basin (North America Cordillera), 426
Gregory Rift (Kenya), 324